Organic Electronics

Edited by
Christof Wöll

Related Titles

Schwoerer, M., Wolf, H. C.

Organic Molecular Solids

2007
ISBN: 978-3-527-40540-4

Brütting, W. (ed.)

Physics of Organic Semiconductors

2005
ISBN: 978-3-527-40550-3

Böer, K. W.

Survey of Semiconductor Physics
2-Volume Set

2002
ISBN: 978-0-471-35572-4

Hadziioannou, G., van Hutten, P. F. (eds.)

Semiconducting Polymers
Chemistry, Physics and Engineering

1999
ISBN: 978-3-527-29507-4

Müllen, K., Wegner, G. (eds.)

Electronic Materials: The Oligomer Approach

1998
ISBN: 978-3-527-29438-1

Organic Electronics

Structural and Electronic Properties of OFETs

Edited by
Christof Wöll

WILEY-VCH Verlag GmbH & Co. KGaA

The Editor

Prof. Dr. Christof Wöll
Lehrstuhl für Physikalische Chemie I
der Ruhr-Universität Bochum
Universitätsstrasse 150
44780 Bochum
Germany

Library of Congress Card No.:
applied for

British Library Cataloguing-in-Publication Data
A catalogue record for this book is available from the British Library.

Bibliographic information published by the Deutsche Nationalbibliothek
The Deutsche Nationalbibliothek lists this publication in the Deutsche Nationalbibliografie; detailed bibliographic data are available on the Internet at <http://dnb.d-nb.de>.

Composition Druckhaus Thomas Müntzer, Bad Langensalza

Printing Strauss GmbH, Mörlenbach

Bookbinding Litges & Dopf GmbH, Darmstadt

Printed in the Federal Republic of Germany
Printed on acid-free paper

ISBN: 978-3-527-40810-8

Contents

Organic Electronics. Structural and Electronic Properties of OFETs. Edited by Christof Wöll

ISBN: 978-3-527-40810-8

Foreword

Introduction

After the ground breaking discovery of electrical charge carrier transport in polymers in the late 1980s by Alan J. Heeger, Alan G. MacDiarmid and Hideki Shirakawa [1–3], who were awarded the Nobel Prize in chemistry in 2000, the question arose as to whether organic materials would also find applications as organic semiconductors. This field really started to attract major attention after the demonstration of the first organic light emitting device (OLED) in 1987 by Tang and Van Slyke [4]. Since then, the field of organic electronics has stimulated a tremendous research interest into organic semiconductors.

Today, owing to the constant improvements of the particular properties of molecular materials – including the synthesis of new compounds – organic semiconductors have made their way from a rather exotic and academic topic studied by a few specialists in molecular physics to a mature research field [5–8]. There are already numerous applications of organic semiconductors and a number of products have reached the market. Probably the first commercial products employing charge transport in organic materials and thus utilising the semiconducting properties of organic materials were laser printers. In this application the photoactive organic material is deposited on the imaging drum. After the charging of this photoactive organic layer by means of a corona discharge, selected areas are illuminated using the intense light of a laser beam. The drum is then brought into contact with toner particles that adhere to the charged parts of the drum but not to the illuminated regions. In the last step the toner particles are transferred to a sheet of paper and fixed there using a temperature treatment, thus yielding high quality prints.

In the future, there are several potential properties of organic electronics that may become very important. The most often quoted prospect for organic electronics is related to their cost. Many companies invest a substantial amount of money and effort into developing functioning organic electronic devices, motivated by the prospect of "cheap" electronics. The vision of being able to simply print electronic circuits on a substrate, using existing print technology, is very attractive. Organic materials can be processed at low temperatures (below 200 °C), thus allowing the combination of organic electronics into the flexible

Organic Electronics. Structural and Electronic Properties of OFETs. Edited by Christof Wöll

ISBN: 978-3-527-40810-8

plastic substrate. This particular property carries a huge potential with regard to the fabrication of e-paper (electronic paper) and the manufacture of functional clothing. The prospect of integrating electronic circuitry (e.g. organic solar cells with the corresponding electronics) into dresses and coats thus allowing mobile electronic equipment to be charged is certainly very attractive [9].

Organic Electronics: Using Molecules as Semiconductors

Today, applications related to organic electronics can be divided into essentially three major fields, namely organic light emitting devices (such as OLEDs), organic field effect transistors (OFETs) and organic solar cells. The last is still in its infancy but is receiving a substantial and strongly increasing amount of attention. Despite their lower efficiency compared with silicon-based devices, organic solar cells offer the advantage of low cost large area production, which might lead to a cost-effective use. Organic light emitting devices have so far had the largest impact in the field of organic electronics. After intense research on OLEDs throughout the last decade, these devices have already entered the market and are presently being used mainly for display applications, e.g. in car radios (Pioneer) or in mobile phone displays. Recently, also large-scale displays based on OLEDs have reached the market, e.g. in TV screens, computer laptop displays or mobile DVD-player monitors (see [10] for the first commercial OLED TV).

In contrast to such electro-luminescent devices, in many instances the realisation of organic field effect transistors (OFETs) for applications requires that these OFETs can be operated at a minimum switching speed, which in turn requires fairly large charge carrier mobilities of the organic materials used in the device. Unfortunately, these charge carrier mobilities are found to depend very sensitively on a number of parameters, including the degree of ordering, contaminations due to charge carrier traps, and other structural imperfections such as domain boundaries. A real breakthrough with regard to understanding and optimising charge carrier mobilities in organic materials has thus not been achieved as yet and electronic devices using organic semiconductors as an active material are still in the laboratory rather than on the market. Charge carrier mobilities are, however, not the key issue for all applications. For the electronic circuitry required to control a mobile OLED display, parameters such as low operation voltage and stability are, for example, more important.

Organic Field Effect Transistors – Prototype Devices in Organic Electronics

This book demonstrates fundamental physical and chemical aspects in organic electronics by mainly concentrating on one fundamental device, the organic field effect transistor. This important device is prototypical for organic electronics and was originally designed in analogy to field effect transistors using conventional semiconducting materials, Si, Ge, GaAs. A typical design of such a device is shown in Figure 1. Whereas the general concept of operation is similar to inorganic thin film FET devices, there are some distinct differences: one is fairly strong charge localisation, limiting the carrier transport to the first few layers above the gate electrode [11, 12], and the particular nature of the organic/inorganic interfaces.

The analysis of OFET device characteristics allows many of the key problems in organic electronics to be stressed, ranging from fundamentals (nature of the charge transport) to applied issues (long-term stability of organic molecules). All articles collected in this book are in one way or the other related to organic field effect transistors.

OFETs are the basis of all logic devices that are required to control, for example, the intensity of a display pixel (the so called *all organic display*) or the realisation of a radio frequency identification (RFID) tag. Since for the latter applications only limited frequency bands are available (essentially 13.56 MHz or 900 MHz) this places rather strong constraints on the required switching properties of the OFETs.

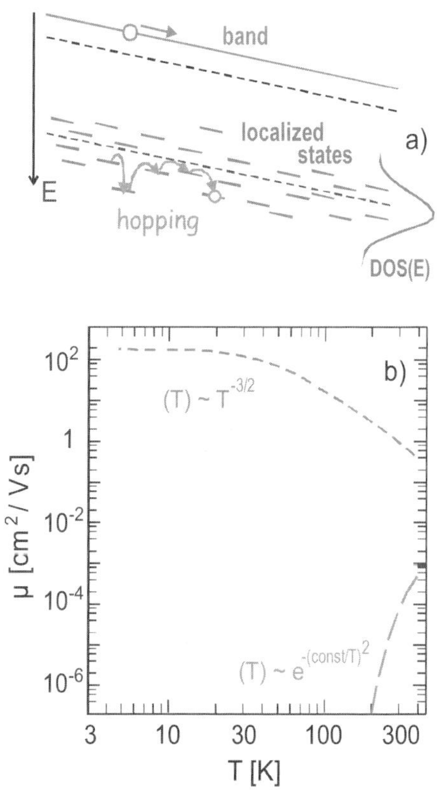

Figure 1 Schematic presentation of the two different mechanisms governing charge transport in organic semiconductors. The so-called hopping transport assumes a thermally activated hopping between charge-traps and is always present, the resulting mobilities are very small. For certain materials a mechanism yielding much higher mobilities has been observed, which is commonly referred to as "band-like", thus implying a similarity to band-transport observed in conventional semiconductors.

To reach this frequency is in no way trivial, and in particular requires a minimum mobility of charge carriers within the organic field effect transistors. This important issue will be discussed below. It has to be noted, however, that in view of the technological applications, a reliable and constant device performance may be even more important than a peak performance of the individual circuits.

Before we start to address particular properties of organic field effect transistors, or present the limitations and the prospects of improving the performance of these devices, we would like to point out that the interest in OFETs goes far beyond solving technological problems. With regard to determining charge carrier mobilities in organic materials and, more importantly, understanding the physics behind, in particular, the band-like transport observed in organic semiconductors, OFETs play a key role in experimentally determining these parameters. While there are other ways to determine conductivities, charge carrier mobilities, and temperature dependences in organic materials, for most of the molecules investigated so far these parameters have been determined using organic field effect transistors.

The Three Key Aspects of Organic Electronic Devices

With regard to applications as organic semiconductors, there are several properties of molecular materials that have been improved in the past and which still need to be optimised further. The most important characteristic of an organic semiconductor is the charge carrier mobility. The processability of organic compounds comes second; the question of whether high-performance thin organic films can be prepared in a straightforward fashion is very important not only for technological applications but also for fundamental studies. Third, the formation of electrical contacts where either electrons or holes are injected into the organic materials is also critical when it comes to fabricating a functioning device – for this reason the interaction between organic molecules and metals is a topic of key interest with regard to the development of functioning devices. In the following these three main aspects – which are also the key topic of the articles collected in this book – will by briefly discussed.

(i) Charge Carrier Mobilities in Organic Semiconductors

Charge carrier mobilities are, as already noted above, a key parameter describing the performance of a semiconductor. This quantity describes the mobility of charges, electrons or holes in the presence of an electric field. The most straightforward way to measure these charge carrier mobilities is to literally measure the speed of the charge carriers in a semiconductor in the presence of an electric field. A classical method, which can also be used for organic semiconductors, is to excite charge carriers at a defined point in space, e.g. by a laser pulse, and then measure the time needed to travel a certain distance, using a time-of-flight technique. This method is well established for conventional semiconductors and has also successfully been used for organic semiconductors, see the chapter by Pflaum et al. (Chapter 25) for a more detailed description of this technique. A major disadvantage of the time-of-flight method to determine mobilities in organic materials is the fact that rather large millimetre-sized single crystals are required in order to apply this technique. Unfortunately, in the case of organic semiconductors, the fairly small amounts of an organic material available and the difficulties encountered in obtaining single crystals of millimetre-sizes frequently prohibit the application of the method. It should also be noted that in the presence of defects (domain boundaries, impurities, vacancies), the applicability of the technique is severely hampered or even made impossible because the signals then become so broadened that a distinct time-of-flight can no longer be determined.

Because of these fundamental difficulties in applying the time-of-flight method for arbitrary molecular materials, today most mobilities known for molecular materials have in fact been extracted from the electrical characteristics of organic field effect transistors. There are several methods to extract the electron and hole mobilities from the electrical device characteristics of an OFET. For a more thorough description of the different procedures, the reader is referred to the paper by Scheinert and Paasch [13].

Quite often the mobilities determined for a given molecule, e.g. rubrene or pentacene, differ for different devices in different laboratories, reflecting the problems related to the extraction of the mobility data from the electrical characteristics. Typically, these characteristics are not only defined by the mobilities but also by the contact resistance and of course the presence of domain boundaries defects and impurities in the organic semiconductor. The determination of the true intrinsic mobility of an organic semiconductor is still a challenge, which has only been overcome in a very few cases.

Over the last two decades the charge carrier mobilities of organic materials have shown a rather rapid increase; today organic field effect transistors can be built with performances [14, 15] that can compete with FETs based on polycrystalline silicon [16].

Today, the search for new organic materials, both molecules and polymers, largely proceeds on a trial and error basis; see the review by Anthony [17]. There are very few theoretical guidelines that allow to predict charge carrier mobilities in a molecule or a polymer. Very often one finds the suggestion that charge mobilities are large if there is a substantial overlap between the systems of flat aromatic molecules. It should be noted, however, that rubrene, one of the molecules with the highest charge carrier mobility known today, is in fact not a flat aromatic molecule, but is distinctly non-planar; in the bulk there is basically no overlap between the core tetracene units of adjacent rubrene molecules. In addition to the lack of theoretical guidelines, there is also the severe problem of the measurement of the intrinsic mobilities, for a new molecule that has been synthesised it can be quite difficult to build OFETs that yield the same mobilities. Since the performance of an OFET device is also severely affected by phenomena occurring at the electrodes (charge injection), the presence of defects and morphological parameters, it is of utmost importance to determine these mobilities reliably. Whereas in most polymeric materials and many molecular systems there is clear evidence that charge transport at room temperature is brought about by a hopping-type transport, there is also clear evidence for at least some molecular materials that a band-like transport is present. The main difference from the hopping transport is that the mobilities either increase or at least stay constant when lowering the temperature, whereas the hopping of electrons or holes from one molecule to the next is strongly activated by temperature, thus leading to an increase of mobilities with temperature. While there are many models for a hopping transport of charge carriers in organic materials [18], only recently have *ab initio* calculations describing a band-like transport in organic materials become available [19, 20]. The main result of these calculations is that the relative orientation of the molecules and also the morphology of organic semiconductor films are very important since these electronic properties are highly anisotropic and depend strongly on the relative orientation of the molecules.

(ii) Processability of Organic Compounds for Applications in Organic Electronics

In general, for the realisation of OFETs two classes of materials are available: polymers and small molecules (so-called oligomers). Polymers exhibit the advantage that the deposition of the molecular material on a substrate is much more straightforward. Polymers at that stage are typically liquid and have – of course – a very low vapour pressure. Therefore the coating of a substrate can simply be performed by spin-coating a suitable substrate possibly followed by a drying process. In addition, polymers are suitable for printing techniques and therefore existing technology can be used to print structures with semiconducting polymers on substrates. Owing to their low vapour pressure it is impossible or at least very difficult to prepare polymers using organic molecular beam deposition, OMBD. As a result it is very difficult to obtain thin polymer layers that exhibit a high degree of crystalline order. Generally, probably as a result of the pure ordering, polymeric organic semiconductors exhibit fairly small carrier mobilities, typically less than 0.1 cm^2/Vs.

Oligomers on the other hand can be synthesised in a more pure form. More importantly, because of their high vapour pressure, these smaller molecules are compatible with OMBD and in many cases allow the preparation of highly ordered crystalline films with a high degree of crystallinity. Such well-defined, highly ordered thin layers of organic molecules make it possible to carry out detailed studies on structure/property interrelations, which can be used to identify the microscopic physical properties of such devices.

Note that recently soluble precursors of some oligomers have also been synthesised, which, in addition, allow the fabrication of highly ordered polycrystalline films, thus combining the advantage of flexible processing (i.e. printing) and their superior electronic properties [17].

Organic Molecular Beam Deposition (OMBD)
The fabrication of organic thin layers from small organic molecules using OMBD is in principle fairly straightforward, but considerable problems are frequently observed. A particularly interesting case is rubrene, a molecule that belongs to the class of organic materials for which mobilities have been reported [21, 22]. However, when using rubrene to fabricate ultrathin organic layers substantial problems with regard to nucleation and obtaining homogeneous films are observed [23, 24]. These problems could be traced back to the fact that the rubrene molecules adopt a different conformation in the bulk and in the gas phase. Whereas in the bulk the tetracene backbone of the molecule is essentially planar, for the free molecule the steric repulsion between the four phenyl units attached to the tetracene backbone cause a substantial tilt when depositing on the surface. In the initial stage of the deposition, the molecules maintain the conformation of the free molecule so that nucleation is significantly delayed and thin, well ordered, or even epitaxial films cannot be achieved.

(iii) Metal/Organic and Oxide/Organic Interfaces

Another important aspect in connection with the design and fabrication of OFETs is the role of interfaces, either between the OSC (organic semiconductor) and the dielectric or OSC and electrode. The structural properties of the dielectric/OSC interface have important consequences for the charge transport along the active layer just above the gate and at the metal/OSC interface. The injection of charge carriers, electrons or holes takes place, which significantly contributes to the overall performance of the OFET device. In particular, the relative position of the electronic levels of the metal and the molecules at the interface are fairly important for the injection properties. Originally it had been thought that in simple cases this electron level alignment could be understood by just aligning the respective ionisation potentials and then simply using the electronic structure determined for the systems separately. Unexpectedly, even for supposedly simple systems, such as a saturated hydrocarbon, for the prototype of an unreactive molecule deposited on a noble metal like gold there are significant deviations from this simple model, which goes back to Mott. The simplest way to demonstrate these problems is to consider the work function. Using this simple Mott-model mentioned above, the adsorption of a saturated hydrocarbon on gold should not lead to a change in the work function. It has been discovered earlier, however, by Seki and coworkers [25] and Kahn and coworkers [26] that there can be rather substantial changes in the work function even in this simple case, which, of course, have important consequences for the electronic level alignment at the metal/organic interface. By using precise *ab initio* electronic structure calculations, recently it was possible to un-

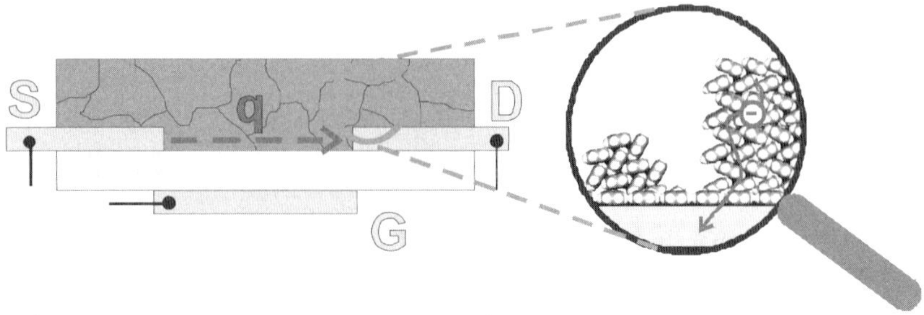

Figure 2 Scheme of a bottom-contact OFET illustration of the current research topics including the structure and morphology of organic semiconductor films as well as charge carrier transport and injection mechanism, which are the subject of this book.

ravel the origin of these unexpected phenomena: there are considerable exchange phenomena (or Pauli repulsions [27]), which have been dubbed the "cushion" effect [28]. Therefore, the prediction of the relative positions of electronic levels at metal/molecule interfaces is not straightforward and requires more detailed investigations [29, 30].

OFETs and Organic Electronics

Organic electronics and the preparation of organic thin films for applications in organic electronics have recently not only been the topic of a number of review articles [31–33], but in addition a number of journals have recently devoted special issues to this subject (*J. Mat. Res.* [34], *Phys. Stat. Sol.* [35, 36], *Chem. Mater.* [37], *Chem. Rev.* [38]). Also, several books [7, 8, 39, 40] have been published in the last few years on this topic. These are all strong indicators that that this field constitutes a "hot" topic in current research.

To provide a more fundamental understanding of the basic mechanism and properties of such OFETs, a national research initiative (supported by the DFG within the framework of the focus program 1121 OFET) was established in 2001. In this interdisciplinary research network various aspects of OFETs, including the synthesis of new materials, preparation and characterisation of organic thin films, characterisation of device properties and the development of new device concepts, have been addressed The results obtained by the various groups collaborating in this effort provide the basis for the present book.

Organization of the Book

As already mentioned above, in the present book the results obtained during six years of research within the framework of a national focus program entitled "Organic field effect transistors: Structural and dynamic characteristics" are presented. We have augmented the book with two contributions from companies who are either about to place products on the market or have definitive plans to do so.

Section I: Industrial Applications

The first section of the book is devoted to industrial applications. In two articles written by two of the major companies active in this field, PolyIC and Evonik, the applications that presently attract the most interest from a commercial point of view are described. At the same time, the key problems related to the manufacturing of "cheap electronics" through a printing process are addressed. These two chapters provide an excellent introduction to the more applied aspects of the field and also define the framework for the following chapters in the book, which all address problems that in one way or the other are related to producing organic field effect transistors and to improving their performance and stability.

Section II: Molecular Compounds

In the second section of the book entitled "Molecular compounds" there are four papers describing molecular aspects of the materials needed to fabricate organic field effect transistors. From these contributions it becomes clear that there are quite a few different routes to producing the basic material for an organic semiconductor. Today the search for molecules that allow the fabrication of organic field effect transistors with superior properties is an important topic. In particular, in the contribution from the Müllen-group (Chapter 4, by Tsao et al.) it becomes clear that it is not only the desire to reach high-charge carrier mobilities, but also to improve the processing properties of the compounds as well as their stability against environmental influences (i.e. degradation or oxidation), but which class of molecules will be incorporated in future organic semiconductors is not yet clear. There are a number of different molecular classes that are presently being investigated. One of the most of important classes of materials, oligothiophenes are discussed in the chapters from the Scheinert and Jaehne group (Chapter 5, by Haubner et al.) and in the contribution from the Scherf and Neher groups (Chapter 6, by Zen et al.). Phthalocyanines have also been of interest for such applications, mostly because of the rather large flexibility of the molecular properties, which can be changed by slightly modifying the composition, and are discussed in the contribution from the Wöhrle and Schlettwein groups (Chapter 3, by Brinkmann et al.).

Section III: Structural and Morphological Aspects

The third part of the book concentrates on structural and morphological aspects of thin organic layers deposited on solid substrates, either metals or oxides. Here the emphasis is not – as in the previous part – on the particular compounds but on the ways to fabricate well defined organic thin layers from such molecules suitable for applications in organic electronics. The contributions in this section of the book are all based on molecules that are fairly well understood, in some cases even for decades. The experimental work described in the chapters contained in this section make it clear that it is in no way a trivial process to produce well defined organic thin layers suitable for applications in organic electronics from a given molecule. For virtually all molecule/substrate combination examples, a careful optimisation of processing parameters is required and in many cases the detailed characterisation of organic films made from materials that are considered to be well known carries a number of surprises. Some of these surprises result from the phenomena that occur when molecules generally considered as "soft" matter are deposited on a solid substrate that typically is considered to be "hard".

A particularly important problem is the fabrication of electrical contacts between metal and organic materials. While the deposition of organic molecules on a metallic substrate is fairly straightforward, the opposite deposition of metals on an organic material is significantly non-trivial. A very interesting approach to this problem that uses a rather sophisticated procedure is presented in the contribution from the Terfort group (Chapter 7, by Thomas et al.).

For the characterisation of the organic/metal interphase in the onset of organic growth on solid substrates the technique of X-ray scattering is fairly significant. The most important aspects of such studies are described in the paper by the Schreiber group (Chapter 9, by Gerlach et al.). In the article by Joshi et al. (Chapter 10) particular aspects of the structure and morphology of thiophenes are discussed. A more general overview of organic molecular beam deposition (OMBD) is presented in the chapter by Witte and Wöll (Chapter 11) where results of other techniques are also discussed.

Of course, a proper understanding of the interphase between a metal and an organic adlayer requires a careful experimental investigation of the first layer of molecules in direct contact with the metal, because in this case the different environment of the molecule may cause distortion and corresponding electronic structure changes. This is the topic of the article from the Tautz group (Chapter 12, by Soubatch et al.) where a careful characterisation of molecules in the first monolayer, in particular with scanning tunnelling microscopy, is presented.

Surface enhanced Raman spectroscopy is a particularly useful spectroscopic method used to characterise the interface between metal and organics. In the chapter from the Scholz and Zahn groups (Chapter 13, by Salvan et al.) a fairly comprehensive review of work carried out in this field is provided.

The active organic layer in OFET devices is frequently deposited on thin silicon oxide layers on top of a silicon substrate. These SiO_2-layers are rarely structurally well characterised. For this reason the investigation of organic adlayers grown on well defined model substrates for oxides is mandatory. Hence the results described in the contribution from the Umbach group (Chapter 14, by Sachs et al.) for the growth of organic molecules on ordered, well defined sapphire substrates are very interesting.

In the article by the Pflaum and Sokolowski groups (Chapter 8, by Voigt et al.) a related dielectric, Al_2O_3 deposited on a transparent conductor, indium tin oxide, ITO, is used to fabricate pentacene-based OFETs. Such devices are particularly interesting in connection with the manufacture of drivers for OLED displays.

Section IV: Device Performance and Characterisation

The fourth section of the book is entitled "Device performance and characterisation" and describes the next level of complication, namely producing organic thin layers with electrodes attached to them that can be used to determine the device characteristics of at least prototype devices. Presently, there are a number of different designs for organic field effect transistors and there are also a number of parameters with regard to which of them need to be optimised. This section of the book is the largest and reflects, in part, the fact that there are quite a few different, and sometimes conflicting, requirements for devices. It will also become clear in this section that there is a pronounced difference between organic electronic devices based on monomeric materials (i.e. single molecules with a molecular mass below about 400 amu) and polymeric materials. Basically, monomeric compounds can be deposited on a given substrate using molecular beam deposition techniques, allowing for the preparation of fairly well defined systems. This approach, on the other hand, has the drawback of being rather complicated and typically requiring ultrahigh vacuum equipment. Using polymeric materials, on the other hand, has the advantage of a fairly straightforward sample preparation; typical samples can be prepared by spin coating the respective polymer on a pre-patterned substrate. This approach, however, carries the disadvantage that the polymeric thin films are typically not very well defined structures and that the interface in particular at the electrodes can be less well defined. With the exception of gold, which does not form an oxide stable under ambient conditions, all other metals are covered by such an oxide layer, which makes predictions about the precise structural arrangement at the molecule substrate interface impossible or at least very difficult.

In the contribution by Pflaum et al. (Chapter 25) the charge carrier mobilities in organic compounds are addressed, as pointed out in the Introduction, mobilities are one of the key parameters describing the suitability of a given molecular compound for organic electronics. The focus in this article is on oligomers. Also in the contribution from the Nickel group (Chapter 15) oligomers are used to fabricate devices, in this case pentacene is used. Also in this article, the use of a novel technique, recording the photo response, is used to characterise device performance. In the chapter by Scheinert and Paasch et al. (Chapter 16) fairly thorough general considerations about the general aspects in polymer based field effect transistors are presented. The contribution from the Brütting group (Chapter 17, by Opitz et al.) focuses on the question of whether charge carrier transport can be obtained for both polarities, a stringing requirement which has to be fulfilled, for example for building logic devices within organic electronics. In the paper from the Hilleringmann group (Chapter 18, Diekmann et al.) an interesting approach is pursued where organic materials are also used for the dielectric. As pointed out earlier, the metallisation of molecular compounds are a very interesting topic with regard to device manufacturing. Here, the contribution by Scharnberg, Faupel and Adelung (Chapter 19) describes a rather interesting approach where radioactive labelling is used to study the interdiffusion of metal into the films. For the characterisation of real devices, potentiometry based on scanning probe instruments is a very promising method. In the chapter by the Scholz and Zahn group (Chapter 20, by Scholz et al.) results from applying this approach on carrier mobilities and injection barriers within real devices are described. Also in the contribution from the Schmeißer group (Chapter 21, by Müller et al.) real devices are at the focus of interest. Here a combination of microscopic and spectroscopic methods is used to characterise the corresponding interphases. In the contribution from the Wagner group (Chapter 22, by Hoppe et al.) the focus is again on thiophenes, one of the most promising materials within organic electronics, where in particular the implications of minimising the size of the operating devices are discussed. In a working organic field effect transistor one of the most important parts, aside from the organic semiconductors, is the gate dielectrics. In the contribution from Dresden (Chapter 23, by Plötner et al.) the fabrication and performance of aluminium oxide gate films are described. Also in the chapter from the Schmechel and von Seggern group (Chapter 24, by Benson et al.) the dielectric/organic semiconductor interphase is the topic.

Section V: Novel Devices

The book ends with a section called "Novel devices" where either non-standard organic materials are used to produce organic field effect transistors (e.g. carbon nanotubes), or organic materials are used to fabricate devices other than organic field effect transistors. One of the examples described in the article from the Kowalsky group (Chapter 27, by Hefczyc et al.) are non-volatile memory devices fabricated from organic materials. The contribution from the Molenkamp and Geurts group (Chapter 28, by Michelfeit et al.) demonstrates a particularly interesting application at present, namely building organic field effect transistors employing spin-polarised transport.

Another particularly interesting approach for novel devices is described in the contribution from the Kern group (Chapter 26, by Burghard et al.) where instead of single molecules or polymers, carbon nanotubes are used as the active semiconducting material. The particular properties in contacting and addressing these nanotubes to obtain a functioning device are described in detail in this contribution.

Acknowledgement

We would like to thank the German Science Foundation, DFG, for funding of the research projects described here within the framework of the National Focus Programme (Schwerpunktprogramm SPP 1121: "Organic field effect transistors"). In particular we would like to acknowledge the support of Dr. Klaus Wefelmeier, the programme officer at the DFG in charge of this programme, for the support and advice provided throughout the years. We would also like to thank the referees who have evaluated the project three times and have provided valuable hints and suggestions for the work of the different groups in the field. Mrs. Knödlseder-Mutschler has provided very competent technical support through the years in managing the OFET-project. We thank her also for helping to compile this book.

Bochum, March 2009
Gregor Witte and Christof Wöll

References

1. C. K. Chiang, M. A. Druy, S. C. Gau, A. J. Heeger, E. J. Louis, A. G. MacDiarmid, Y. W. Park, and H. Shirakawa, J. Am. Chem. Soc. **100**, 1013 (1978).
2. C. K., Chiang, C. R. Fischer, Y. W. Park, A. J. Heeger, H. Shirakawa, E. J. Louis, S. C. Gau, and A. G. MacDiarmid, Phys. Rev. Lett. **39**, 1098 (1977).
3. H. Shirakawa, E. J. Louis, A. G. MacDiarmid, C. K. Chiang, and A. J. Heeger, J. Chem. Soc. Chem. Commun. 579 (1977).
4. C. W. Tang, and S. A. Van Slyke, Appl. Phys. Lett. **51**, 913 (1987).
5. W. Brütting (ed.), Physics of Organic Semiconductors (Wiley-VCH, Berlin, 2005).
6. I. H. Campbell, and D. L. Smith, Physics of Organic Electronic Devices, in Solid State Physics, vol. 55 (Academic Press, San Diego, 2001).
7. R. Farchioni, and G. E. Grosso, Organic Electronic Materials: Conjugated Polymers and Low Molecular Weight Organic Solids (Springer, Berlin, 2001).
8. M. Pope, and C. E. Swenberg, Electronic Processes in Organic Crystals and Polymers (Oxford University Press, New York, 1999).
9. M. Hamedi, R. Forchheimer, and O. Inganas, Nature Mater. **6**, 357 (2007).
10. Sony, www.oled.at/sony-xel-1-oled-tv.
11. G. Horowitz, Adv. Mater. **10**(5), 365 (1998).
12. T. Muck, V. Wagner, U. Bass, M. Leufgen, J. Geurts, and L. Molenkamp, Synth. Met. **146**(3), 317 (2004).
13. S. Scheinert, and G. Paasch, Phys. Stat. Sol. a-Appl. Res. **201**(6), 1263 (2004).
14. D. J. Gundlach, Y. Y. Lin, T. N. Jackson, S. F. Nelson, and D. G. Schlom, IEEE Electron. Dev. Lett. **18**, 87 (1997).
15. H. Klauk, M. Halik, G. Zschieschang, G. Schmid, W. Radik, and W. Weber, J. Appl. Phys. **92**, 5259 (2002).
16. C. D. Dimitrakopolous, and P. R. L. Malenfant, Adv. Mater. **14**, 99 (2002).
17. J. E. Anthony, Chem. Rev. **106**, 5028 (2006).
18. H. Bässler, Phys. Stat. Sol. b-Basic Res. **175**(1), 15 (1993).

19. K. Hannewald, and P. A. Bobbert, Appl. Phys. Lett. **85**(9), 1535 (2004).
20. V. Coropceanu, J. Cornil, D. A. da Silva, Y. Olivier, R. Silbey, and J. L. Bredas, Chem. Rev. **107**(5), 2165 (2007).
21. R. W. I. de Boer, M. E. Gershenson, A. F. Morpurgo, and V. Podzorov, Phys. Stat. Sol. a-Appl. Res. **201**(6), 1302 (2004).
22. V. C. Sundar, J. Zaumseil, V. Podzorov, E. Menard, R. L. Willett, T. Someya, M. E. Gershenson, and J. A. Rogers, Science **303**(5664), 1644 (2004).
23. D. Käfer, L. Ruppel, G. Witte, and C. Wöll, Phys. Rev. Lett. **95,** 166602-1 (2005).
24. D. Käfer, and G. Witte, PCCP **7,** 2850 (2005).
25. H. Ishii, K. Sugiyama, E. Ito, and K. Seki, Adv. Mater. **11**(8), 605 (1999).
26. I. G. Hill, A. Rajagopal, A. Kahn, and Y. Hu, Appl. Phys. Lett.. **73**(5), 662 (1998).
27. P. S. Bagus, V. Staemmler, and C. Wöll, Phys. Rev. Lett. **89**(9), 0961041-3 (2002).
28. G. Witte, S. Lukas, P. S. Bagus, and C. Wöll, Appl. Phys. Lett. **87**, 263502 (2005).
29. R. Caputo, B. Prascher, V. Staemmler, P. S. Bagus, and C. Wöll, J. Phys. Chem. A **111**, 12778 (2007).
30. N. Koch, J. Phys.: Condens. Matt. **20**, 184008 (2008).
31. S. R. Forrest, Chem. Rev. **97,** 1793 (1997).
32. F. Schreiber, Phys. Stat. Sol. a-Appl. Res. **201**(6), 1037 (2004).
33. G. Witte, and C. Wöll, J. Mater. Res. **19**(7), 1889 (2004).
34. F. Faupel, C. Dimitrakopoulos, A. Kahn, and C. Wöll (eds.), Organic Electronics, J. Mater. Res. **19** (2004).
35. W. Brütting (ed.), Physics of Organic Semiconductors, Phys. Stat. Sol. (a) **201** (2004).
36. C. Wöll (ed.), Organic Electronics – Structural and Electronic Properties of OFETs, Phys. Stat. Sol. (a) **205** (2008).
37. Chem. Mater. **16** (2004).
38. Chem. Rev. **107**(4) (2007).
39. M. Schwoerer, and H. C. Wolf, Organische Molekulare Festkörper (Wiley-VCH, Weinheim, 2005).
40. Z. Bao, and J. Locklin (eds.), Organic Field-Effect Transistors (CRC Press, New York, 2007).

List of Contributors

R. Adelung
Functional Nanomaterials
Institute for Materials Sciences
University of Kiel
24118 Kiel
Germany

H.-J. P. Adler
Institute of Macromolecular
Chemistry and Textile Chemistry
Technical University of Dresden
01062 Dresden
Germany

Ralf Anselmann
Evonik Industries
Creavis Technology and Innovation
Paul-Baumann-Straße 1
45764 Marl
Germany

Kannan Balasubramanian
Max-Planck-Institute for
Solid State Research
Heisenbergstraße 1
70569 Stuttgart
Germany

T. Balster
School of Engineering and Science
Jacobs University Bremen
Campus Ring 1
28759 Bremen
Germany

Eike Becker
Technical University of
Braunschweig
Schleinitzstraße 22
38106 Braunschweig
Germany

Lars Beckmann
Technical University of
Braunschweig
Schleinitzstraße 22
38106 Braunschweig
Germany

Niels Benson
Institute of Materials Science
Darmstadt University of Technology
Petersenstraße 23
64287 Darmstadt
Germany

Robert Blache
PolyIC GmbH & Co. KG
Tucherstraße 2
90763 Fuerth
Germany

Organic Electronics. Structural and Electronic Properties of OFETs. Edited by Christof Wöll

ISBN: 978-3-527-40810-8

H. Brinkmann
Institute of Applied Physics
Justus-Liebig-University Gießen
Heinrich-Buff-Ring 16
35392 Gießen
Germany

Markus Bronner
Institute of Physics
University of Augsburg
86135 Augsburg
Germany

Wolfgang Brütting
Institute of Physics
University of Augsburg
86135 Augsburg
Germany

Marko Burghard
Max-Planck-Institute for
Solid State Research
Heisenbergstraße 1
70569 Stuttgart
Germany

Y. Burkov
Brandenburg University of
Technology at Cottbus
Department of Applied Physics
and Sensors
P.O. Box 101344
03013 Cottbus
Germany

M. Casu
University of Würzburg
Institute for Physical and
Theoretical Chemistry
Auf der Morgenstelle 8
72076 Tübingen
Germany

X.-D. Dang
Technichal University of Dresden
Institute of Physical Chemistry and
Electrochemistry
01062 Dresden
Germany
and
Department of Chemistry
University of California
Santa Barbara, CA 93106
USA

T. Diekmann
Department EIM-E,
Sensor Technology
University of Paderborn
Warburger Straße 100
33098 Paderborn
Germany

L. M. Eng
Institute of Applied Photophysics
Technical University of Dresden
01062 Dresden
Germany

F. Faupel
Chair for Multicomponent Materials
Institute for Materials Sciences
University of Kiel
24118 Kiel
Germany

W.-J. Fischer
Technical University of Dresden
Institute of Semiconductors and
Microsystems
01062 Dresden
Germany

Walter Fix
PolylC GmbH & Co. KG
Tucherstraße 2
90763 Fuerth
Germany

Alexander Gerlach
Institute of Applied Physics
University of Tübingen
Auf der Morgenstelle 10
72076 Tübingen
Germany

J. Geurts
University of Würzburg
Institute of Physics
Am Hubland
97074 Würzburg
Germany

L. Gisslen
Walter Schottky Institute
Technical University of Munich
Am Coulombwall 3
85748 Garching
Germany

A. Goryachko
Brandenburg University of
Technology at Cottbus
Department of Applied Physics
and Sensors
P.O. Box 101344
03013 Cottbus
Germany

J. Grenzer
Institute of Ion Beam Physics
and Materials Research
Forschungszentrum Dresden-
Rossendorf
01328 Dresden
Germany

S. Grigorian
FB7 – Physics Department
University of Siegen
57068 Siegen
Germany

K. Haubner
Institute of Macromolecular
Chemistry and Textile Chemistry
Technical University of Dresden
01062 Dresden
Germany

Artur Hefczyc
Technical University of
Braunschweig
Schleinitzstraße 22
38106 Braunschweig
Germany

K. Henkel
Brandenburg University of
Technology at Cottbus
Department of Applied Physics
and Sensors
P.O. Box 101344
03013 Cottbus
Germany

A. Herasimovich
Solid State Electronics and Center
of Micro- and Nano-Technologies
Technical University of Ilmenau
P.O. Box 100565
98684 Ilmenau
Germany

U. Hilleringmann
Department EIM-E,
Sensor Technology
University of Paderborn
Warburgerstraße 100
33098 Paderborn
Germany

F. Holch
University of Würzburg
Experimental Physics II
Am Hubland
97074 Würzburg
Germany

A. Hoppe
School of Engineering and Science
Jacobs University Bremen
Campus Ring 1
28759 Bremen
Germany

I. Hörselmann
Institute of Solid State Electronics
Technical University of Ilmenau
P.O. Box 100565
98684 Ilmenau
Germany

E. Jaehne
Institute of Macromolecular
Chemistry and Textile Chemistry
Technical University of Dresden
01062 Dresden
Germany

Hans-Hermann Johannes
Technical University of
Braunschweig
Schleinitzstraße 22
38106 Braunschweig
Germany

S. Joshi
FB7 – Physics Department
University of Siegen
57068 Siegen
Germany

C. Kelting
Institute of Applied Physics
Justus-Liebig-University Gießen
Heinrich-Buff-Ring 16
35392 Gießen
Germany

Klaus Kern
Max-Planck-Institute for
Solid State Research
Heisenbergstraße 1
70569 Stuttgart
Germany
and
Institute of Physics of Nanostructures
École Polytechnique Fédérale de
Lausanne
1015 Lausanne
Switzerland

D. Koehler
Institute of Applied Photophysics
Technical University of Dresden
01062 Dresden
Germany

Wolfgang Kowalsky
Technical University of
Braunschweig
Schleinitzstr. 22
38106 Braunschweig
Germany

Stefan Kowarik
Institute of Applied Physics
University of Tübingen
Auf der Morgenstelle 10
72076 Tübingen
Germany

Eduardo J. H. Lee
Max-Planck-Institute for
Solid State Research
Heisenbergstraße 1
70569 Stuttgart
Germany

Mark A. M. Leenen
Evonik Industries
Creavis Technology and Innovation
Paul-Baumann-Straße 1
45764 Marl
Germany

D. Lehmann
Institute of Physics
Technical University of Chemnitz
09107 Chemnitz
Germany

Th. Lindner
Leibniz Institute for Solid State and
Materials Research IFW Dresden
P.O. Box 270016
01171 Dresden
Germany

C. Loppacher
Institute of Applied Photophysics
Technical University of Dresden
01062 Dresden
Germany

S. Makarov
Institute of Organic and
Macromolecular Chemistry
University of Bremen
Leobenerstraße NW 2
28334 Bremen
Germany

D. Mandal
Brandenburg University of
Technology at Cottbus
Department of Applied Physics
and Sensors
P.O. Box 101344
03013 Cottbus
Germany

Christian Melzer
Institute of Materials Science
Darmstadt University of Technology
Petersenstraße 23
64287 Darmstadt
Germany

S. Meyer
3rd Department of Physics
University of Stuttgart
70550 Stuttgart
Germany

M. Michelfeit
University of Würzburg
Institute of Physics
Am Hubland
97074 Würzburg
Germany

L. W. Molenkamp
University of Würzburg
Institute of Physics
Am Hubland
97074 Würzburg
Germany

T. Muck
School of Engineering and Science
Jacobs University Bremen
Campus Ring 1
28759 Bremen
Germany

K. Müllen
Max-Planck Institute for
Polymer Research
Ackermannweg 10
55128 Mainz
Germany

A.-D. Müller
Anfatec Instruments AG
Melanchthonstraße 28
08606 Oelsnitz (V)
Germany

F. Müller
Anfatec Instruments AG
Melanchthonstraße 28
08606 Oelsnitz (V)
Germany

Jan Müller
Faculty of Chemistry
Philipps-University Marburg
Hans-Meerwein- Straße
35032 Marburg
Germany

K. Müller
Brandenburg University of Technology at Cottbus
Department of Applied Physics and Sensors
P.O. Box 101344
03013 Cottbus
Germany

Dieter Neher
Institute of Physics
University of Potsdam
Karl-Liebknechtstraße 24–25
14476 Potsdam
Germany

Bert Nickel
Ludwig-Maximilians-University
Department of Physics and CeNS
Geschwister-Scholl-Platz 1
80539 Munich
Germany

J. Niemax
Qimonda Memory Products Development Center
133 Changyang Street
Suzhou Industrial Park
215126 Suzhou
P.R. China

Andreas Opitz
Institute of Physics
University of Augsburg
86135 Augsburg
Germany

G. Paasch
Leibniz Institute for Solid State and Materials Research IFW Dresden
P.O. Box 270016
01171 Dresden
Germany

B. A. Paez
Institute of Physics
Technical University of Chemnitz
09107 Chemnitz
Germany

I. Paloumpa
Brandenburg University of Technology at Cottbus
Department of Applied Physics and Sensors
P.O. Box 101344
03013 Cottbus
Germany

M. Paul
University of Würzburg
Experimental Physics II
Am Hubland
97074 Würzburg
Germany

J. Pernpeintner
University of Würzburg
Experimental Physics II
Am Hubland
97074 Würzburg
Germany

J. Pflaum
Institute of Physics
University of Stuttgart
Pfaffenwaldring 57
70550 Stuttgart
Germany

J. Pflaum
University of Würzburg
Department of
Experimental Physics VI
Am Hubland
97074 Würzburg
Germany

U. Pietsch
FB7 – Physics Department
University of Siegen
57068 Siegen
Germany

Patrick Pingel
Institute of Physics
University of Potsdam
Karl-Liebknechtstraße 24–25
14476 Potsdam
Germany

W. Pisula
Max-Planck Institute for
Polymer Research
Ackermannweg 10
55128 Mainz
Germany
and
Degussa GmbH
Process Technology & Engineering,
Process Technology New Processes
Rodenbacher Chaussee 4
63457 Hanau-Wolfgang
Germany

W. Plieth
Technichal University of Dresden
Institute of Physical Chemistry and
Electrochemistry
01062 Dresden
Germany

M. Plötner
Technical University of Dresden
Institute of Semiconductors and
Microsystems
01062 Dresden
Germany

H. J. Räder
Max-Planck Institute for
Polymer Research
Ackermannweg 10
55128 Mainz
Germany

S. Richter
Technical University of Dresden
Institute of Semiconductors and
Microsystems
01062 Dresden
Germany

A. Rouhanipour
Max-Planck Institute for
Polymer Research
Ackermannweg 10
55128 Mainz
Germany

S. Sachs
University of Würzburg
Experimental Physics II
Am Hubland
97074 Würzburg
Germany

G. Salvan
Institute of Physics
Technical University of Chemnitz
09107 Chemnitz
Germany

M. Scharnberg
University of Kiel
Chair for Multicomponent Materials
Institute for Materials Sciences
24118 Kiel
Germany
and
University of Kiel
Functional Nanomaterials
Institute for Materials Sciences
24118 Kiel
Germany

S. Scheinert
Solid State Electronics and
Center of Micro- and
Nano-Technologies
Technical University of Ilmenau
P.O. Box 100565
98684 Ilmenau
Germany

Ullrich Scherf
Macromolecular Chemistry
University of Wuppertal
Gauss-Straße 20
42097 Wuppertal
Germany

D. Schlettwein
Institute of Applied Physics
Justus-Liebig-University Gießen
Heinrich-Buff-Ring 16
35392 Gießen
Germany

Roland Schmechel
Nanostrukturtechnik
University of Duisburg-Essen
Bismarckstraße 81
47057 Duisburg
Germany

D. Schmeißer
Brandenburg University of
Technology at Cottbus
Department of Applied Physics
and Sensors
P.O. Box 101344
03013 Cottbus
Germany

G. Schmidt
University of Würzburg
Institute of Physics
Am Hubland
97074 Würzburg
Germany

K. Schmidt
PolyIC GmbH & Co. KG
Tucherstraße 2
90763 Fuerth
Germany

G. Schnurpfeil
Institute of Organic and
Macromolecular Chemistry
University of Bremen
Leobenerstraße NW 2
28334 Bremen
Germany

A. Schöll
University of Würzburg
Experimental Physics II
Am Hubland
97074 Würzburg
Germany

R. Scholz
Walter Schottky Institute
Technical University of Munich
Am Coulombwall 3
85748 Garching
Germany

Frank Schreiber
Institute of Applied Physics
University of Tübingen
Auf der Morgenstelle 10
72076 Tübingen
Germany

Stefan Sellner
Institute of Applied Physics
University of Tübingen
Auf der Morgenstelle 10
72076 Tübingen
Germany
and
Max-Planck-Institute for
Metal Research
Heisenbergstraße 3
70569 Stuttgart
Germany
and
School of Engineering and
Applied Sciences
Harvard University
29 Oxford Street
Cambridge, MA 02138
USA

M. Sokolowski
Institute for Physical and
Theoretical Chemistry
University of Bonn
Wegelerstraße 12
53115 Bonn
Germany

S. Soubatch
School of Engineering and Science
Jacobs University
28759 Bremen
Germany
and
Institute for Bio- and Nanosystems 3
Forschungszentrum Jülich
53425 Jülich
Germany

Jürgen Steiger
Evonik Industries
Creavis Technology and Innovation
Paul-Baumannstraße 1
45764 Marl
Germany

F. S. Tautz
School of Engineering and Science
Jacobs University
28759 Bremen
Germany
and
Institute for Bio- and Nanosystems 3
Forschungszentrum Jülich
53425 Jülich
Germany

R. Temirov
School of Engineering and Science
Jacobs University
28759 Bremen
Germany
and
Institute for Bio- and Nanosystems 3
Forschungszentrum Jülich
53425 Jülich
Germany

A. Terfort
Faculty of Chemistry
Philipps-University Marburg
Hans-Meerweinstraße
35032 Marburg
Germany

Heiko Thiem
Evonik Industries
Creavis Technology and Innovation
Paul-Baumannstraße 1
45764 Marl
Germany

Heidi Thomas
Faculty of Chemistry
Philipps-University Marburg
Hans-Meerweinstraße
35032 Marburg
Germany

A. K. Tripathi
Holst Centre/TNO
High Tech Campus 31
5605 KN Eindhoven
The Netherlands

H. N. Tsao
Max-Planck Institute for
Polymer Research
Ackermannweg 10
55128 Mainz
Germany

O. Tsaryova
Institute of Organic and
Macromolecular Chemistry
University of Bremen
Leobenerstraße NW 2
28334 Bremen
Germany

Andreas Ullmann
PolyIC GmbH & Co. KG
Tucherstraße 2
90763 Fuerth
Germany

E. Umbach
University of Würzburg
Experimental Physics II
Am Hubland
97074 Würzburg
Germany
and
Forschungzentrum Karlsruhe
76133 Karlsruhe
Germany

M. Voigt
Institute for Physical and
Theoretical Chemistry
University of Bonn
Wegelerstraße 12
53115 Bonn
Germany

Heinz von Seggern
Institute of Materials Science
Darmstadt University of Technology
Petersenstraße 23
64287 Darmstadt
Germany

P. Vrdoljak
University of Würzburg
Experimental Physics II
Am Hubland
97074 Würzburg
Germany

V. Wagner
School of Engineering and Science
Jacobs University Bremen
Campus Ring 1
28759 Bremen
Germany

Ralf Thomas Weitz
Max-Planck-Institute for
Solid State Research
Heisenbergstraße 1
70569 Stuttgart
Germany

G. Witte
Physical Chemistry I
Ruhr-University Bochum
44780 Bochum
Germany

D. Wöhrle
Institute of Organic and
Macromolecular Chemistry
University of Bremen
Leobenerstraße NW 2
28334 Bremen
Germany

Ch. Wöll
Physical Chemistry I
Ruhr-University Bochum
44780 Bochum
Germany

D. R. T. Zahn
Institute of Physics
Technical University of Chemnitz
09107 Chemnitz
Germany

Achmad Zen
Institute of Materials Research
and Engineering
Agency for Science,
Technology and Research (A*STAR)
3 Research Link
117602 Singapore

Color Plates

Figure 1.1 RFID tag completely printed roll-to-roll, RFID chip: approx. 2 cm × 3 cm.

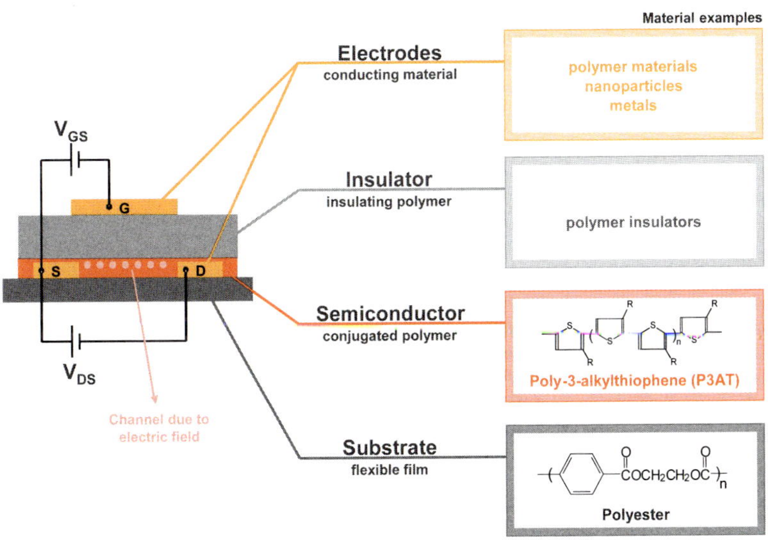

Figure 1.2 Transistor design.

Organic Electronics. Structural and Electronic Properties of OFETs. Edited by Christof Wöll

ISBN: 978-3-527-40810-8

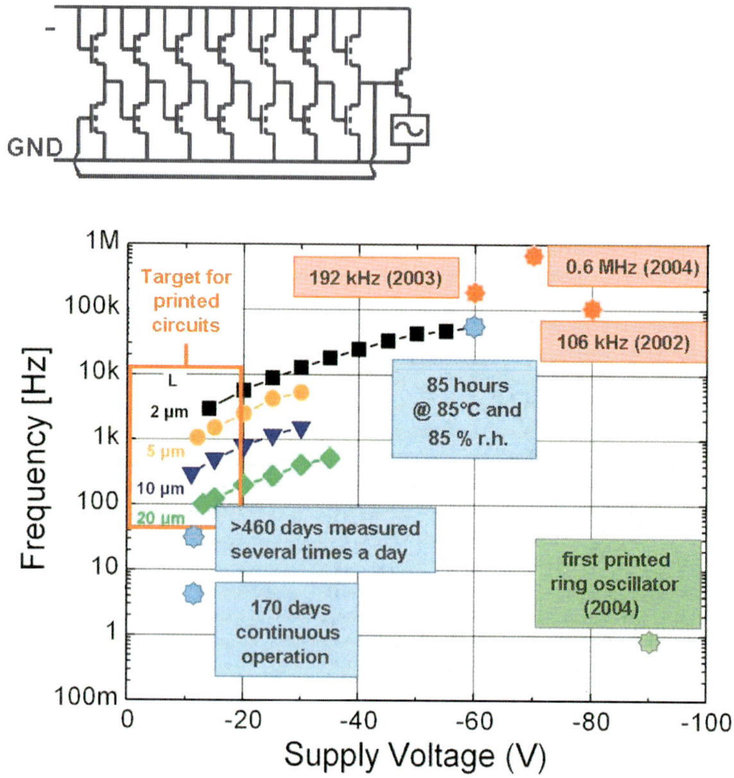

Figure 1.4 Principal electronic circuit of a seven stage ring oscillator (left) and an overview of fundamental investigations on our ring oscillators (right). *L* represents the channel length of the FETs.

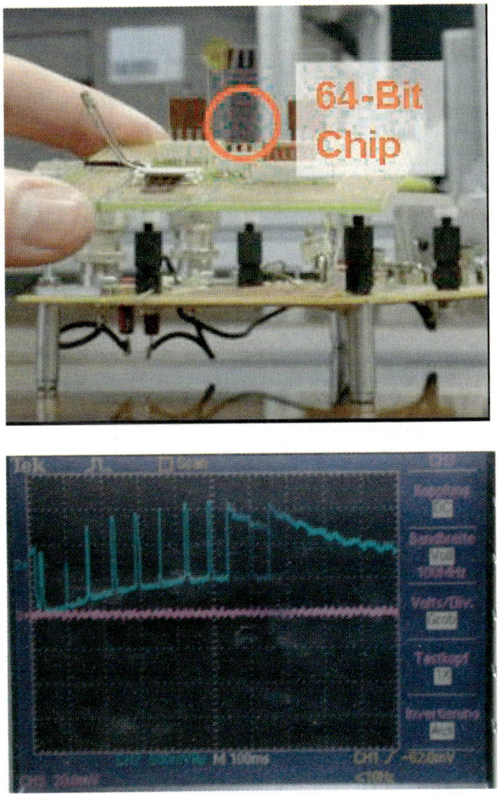

Figure 1.5 Hybrid setup of the 64-bit chip (left) and the output signal of the chip (right).

Figure 1.7 Roll-printed electronics.

Figure 1.8 RFID tags as an example of use in brand protection.

Figure 2.1 Printed silver patterns on foil.

Figure 2.2 Printing rolls for printing electronic structures on paper (source: printed systems).

Figure 2.3 Functional printed electronic cards for "CROSSLINK" technology (source: printed systems).

Figure 2.5 Item-level tagging with fully printed RFID-tags.

Figure 2.6 Silver 30 SN screen printing ink.

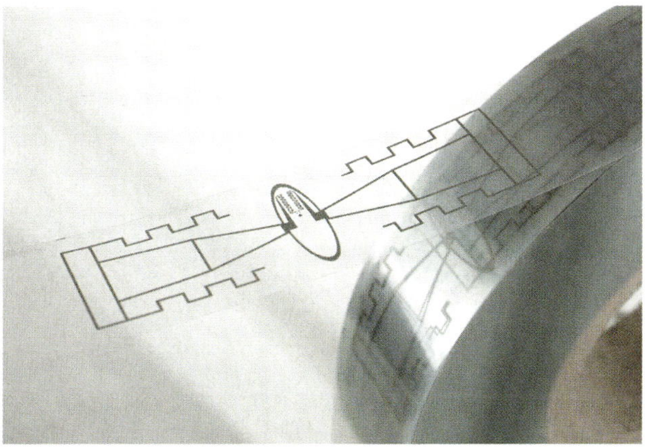

Figure 2.7 Printed RFID-antenna.

Figure 2.8 ITO nanopowder and dispersion (http://www.advancednanomaterials.com).

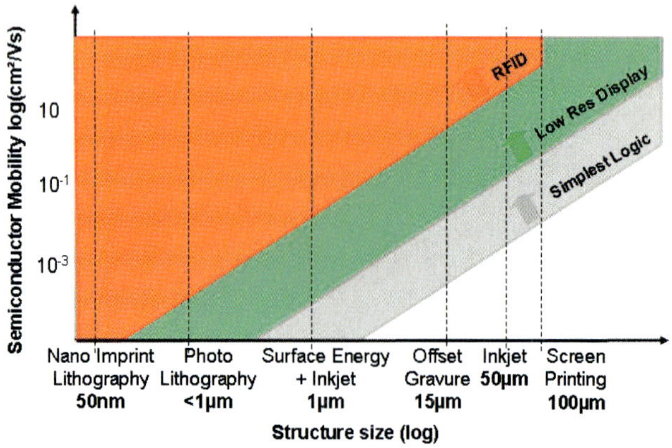

Figure 2.9 Semiconductor charge carrier mobility and printing technique resolution requirements for printed electronics.

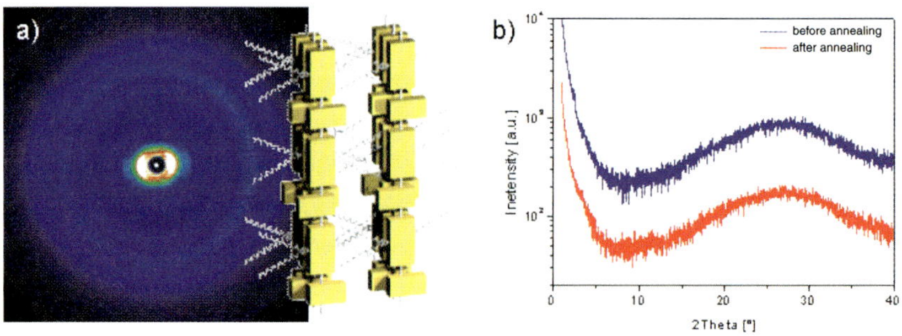

Figure 4.9 2DWAXS and the deduced schematic ordering of extruded BTZ-CDT copolymer (a), Bragg scattering of drop-cast BTZ-CDT copolymer film (b).

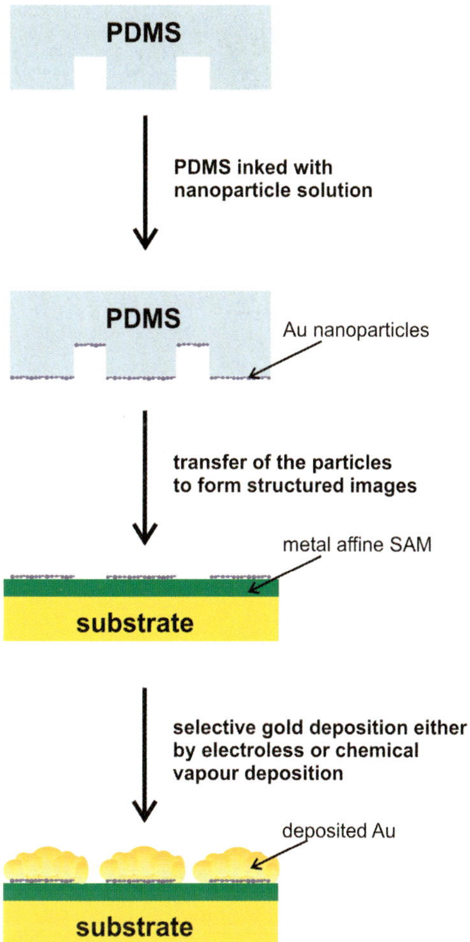

Figure 7.1 Schematic procedure for μCP of nanoparticles followed by selective gold deposition either by electroless deposition (ELD) or by chemical vapour deposition (CVD).

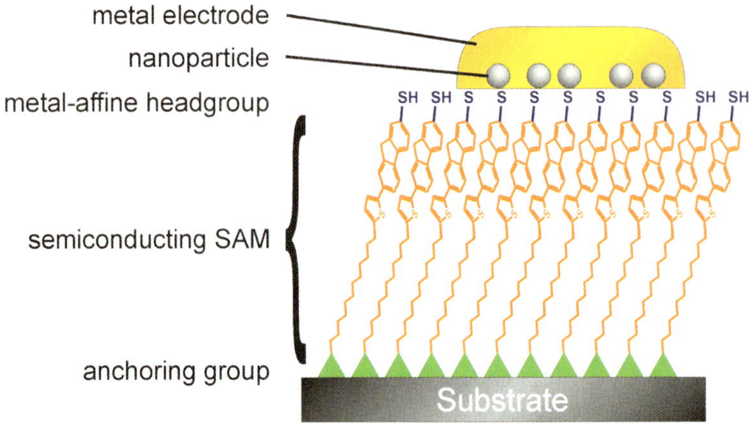

Figure 7.3 Schematic of the situation at the electrode/SAM interface: The use of palladium nanoparticles for the deposition of gold electrodes results in hetero-metallic interfaces at which band distortion in the organic material might build up due to field gradients.

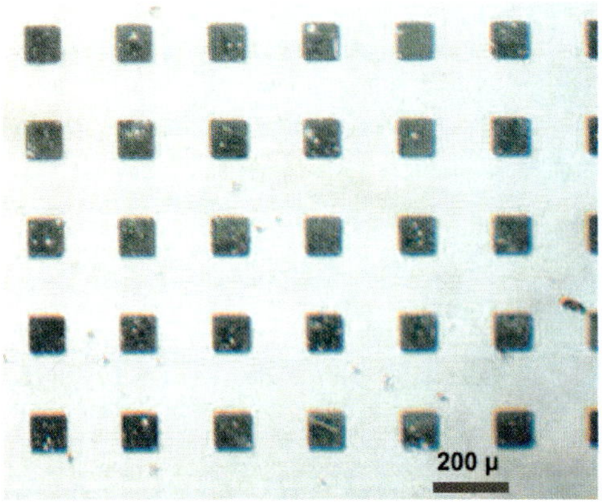

Figure 7.6 Optical micrograph of an array of gold squares deposited onto MPS-modified glass using micro-contact printing of Pd nanoparticles followed by CVD with $((CH_3)_3PAuCH_3)$ at 75 °C.

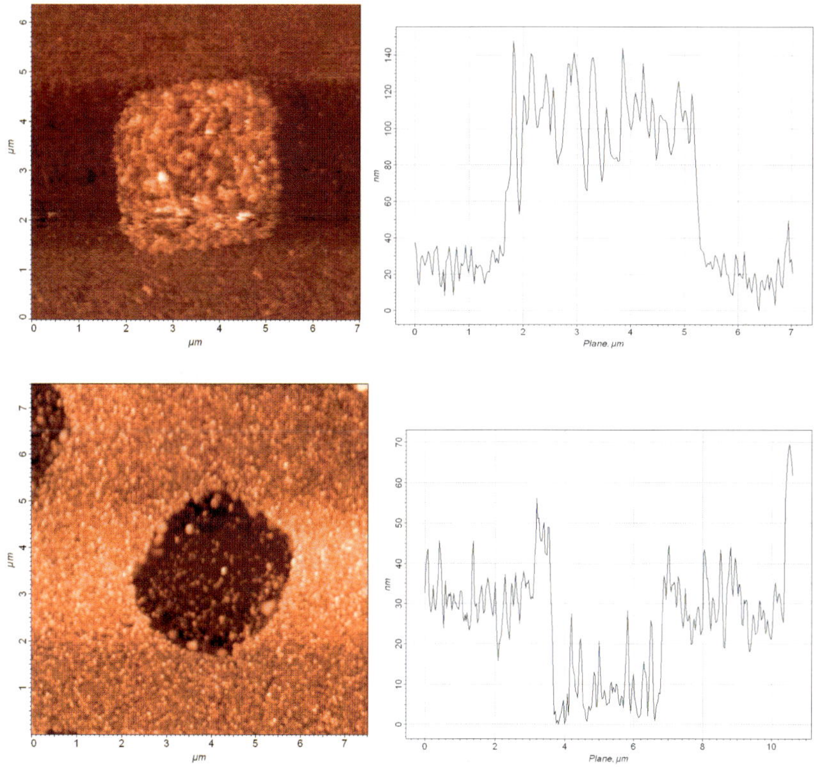

Figure 7.7 AFM micrographs and cross-sections of 3 μm × 3 μm squares obtained after the CVD of $((CH_3)_3P)AuCH_3$ onto printed nanoparticles. Upper: SAM of 10, deposited from EtOH, printed with Pd nanoparticles; lower: SAM of 11, deposited from EtOH, printed with GSH-NP (Au).

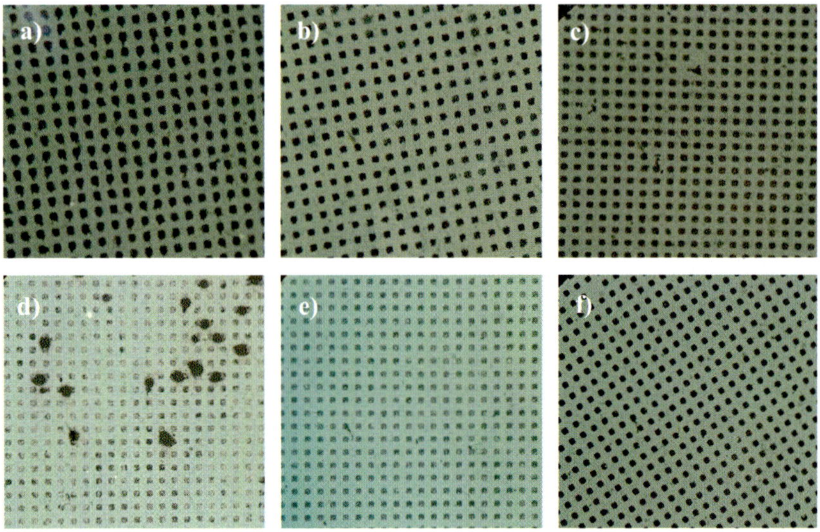

Figure 7.8 Optical micrographs of an array of 3 μm squares on different SAMs patterened by the μCP/ELD process. (a) SAM **2** (THF, MEE-NP), (b) SAM **2** (THF, Pd-NP), (c) SAM **10** (EtOH, MEE-NP), (d) SAM **3** (EtOH, GSH-NP), (e) SAM **3** (EtOH, citrate-NP) and (f) SAM **5** (THF, citrate-NP).

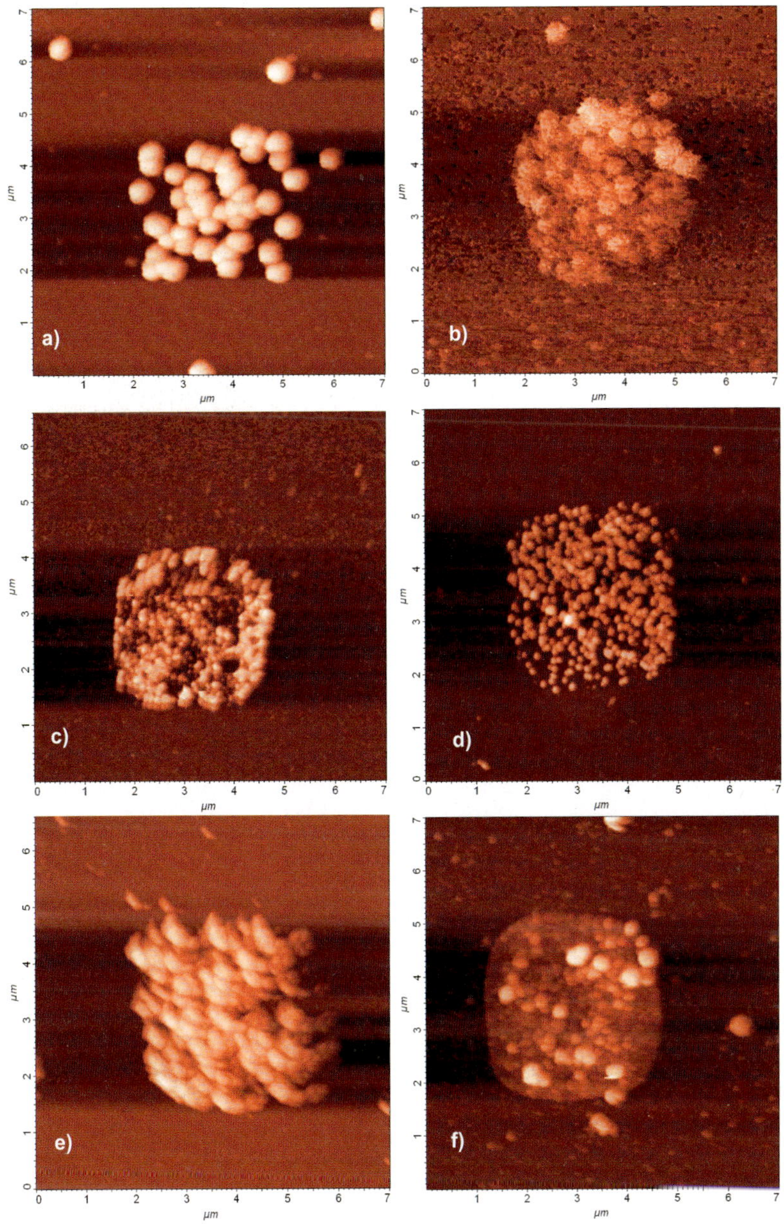

Figure 7.9 AFM micrographs of 3 μm squares deposited by the combined μCP/ELD process:(a) SAM **2** (THF, GSH-NP), (b) SAM **2** (EtOH, Pd-NP), (c) SAM **3** (EtOH, citrate-NP), (d) SAM **4** (EtOH, GSH-NP), (e) SAM **5** (THF, Pd-NP) and (f) SAM **10** (EtOH, MEE-NP).

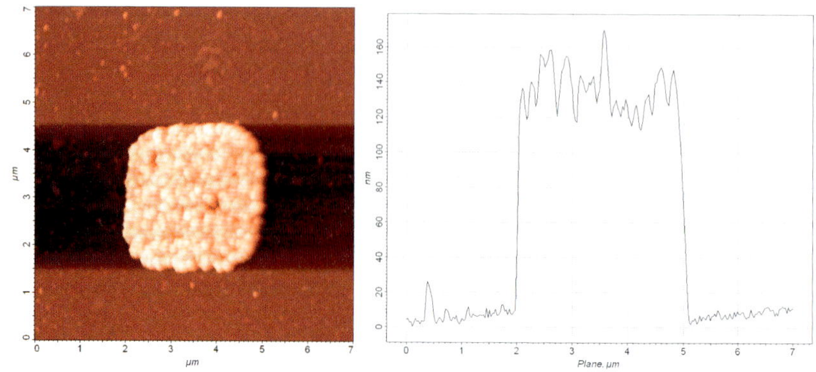

Figure 7.10 AFM images of an electrolessly deposited gold island on SAM 5 (THF) using citrate-NP as seeds.

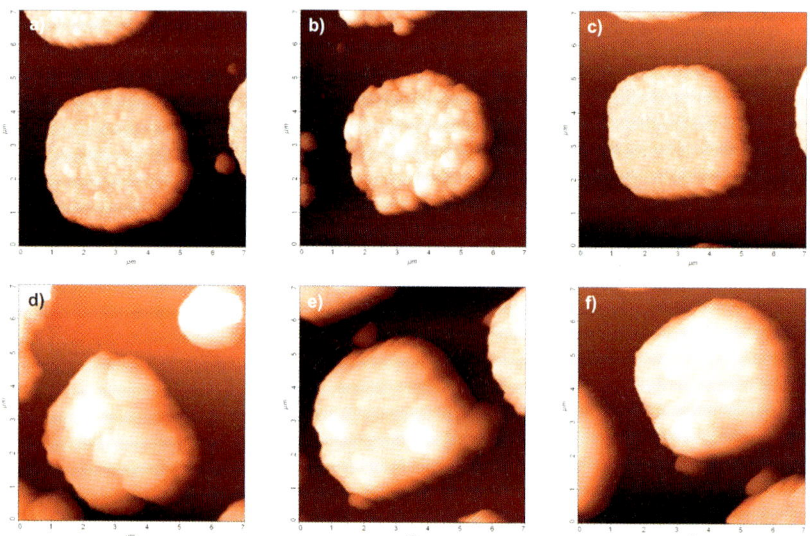

Figure 7.11 AFM micrographs of ELD substrates obtained by printing MEE-NPs onto: (a) SAM **7** (EtOH), (b) SAM **8** (EtOH), (c) SAM **10** (EtOH), (d) SAM **2** (THF), (e) SAM **9** (THF)and (f) SAM **6** (THF).

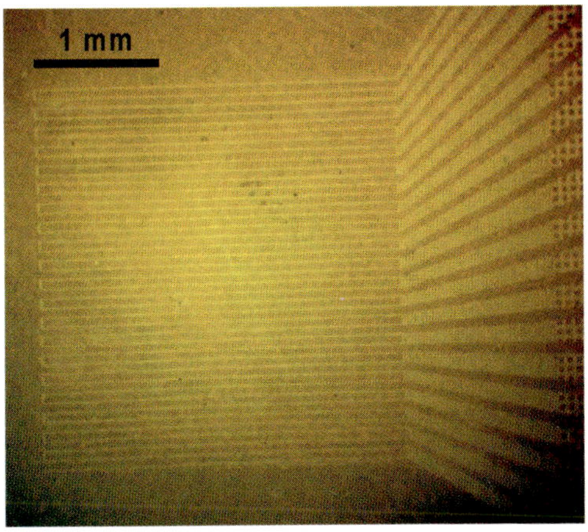

Figure 7.12 Pattern obtained using the MEE-NP and the sodium gold sulfite/hydroxylamine bath.

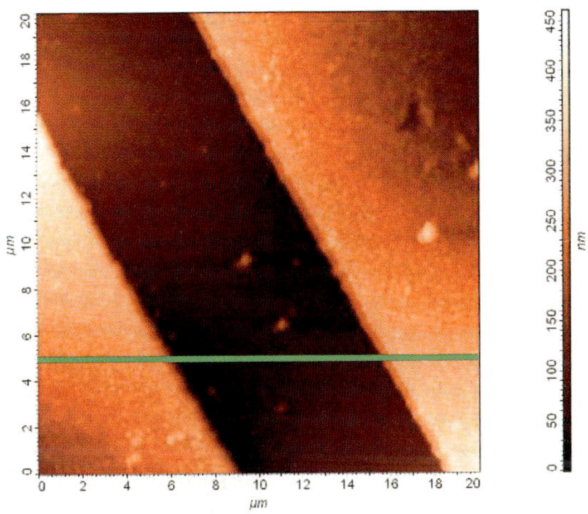

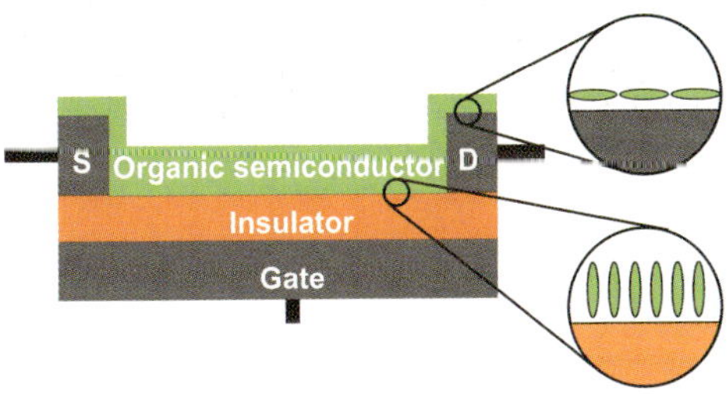

Figure 12.1 Schematic of an organic field effect transistor with its relevant interfaces. The molecules are shown in green.

Figure 12.3 Submolecular STM contrast of two PTCDA molecules adsorbed on the Ag(111) surface.

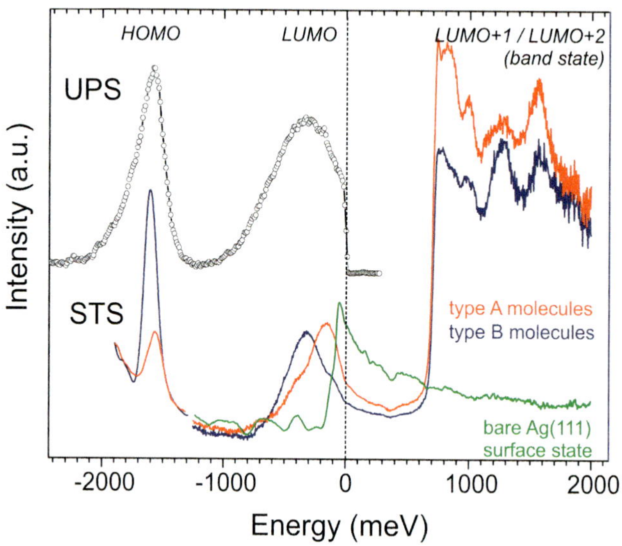

Figure 12.4 UPS and STS spectra of PTCDA/Ag(111) [27, 30, 35].

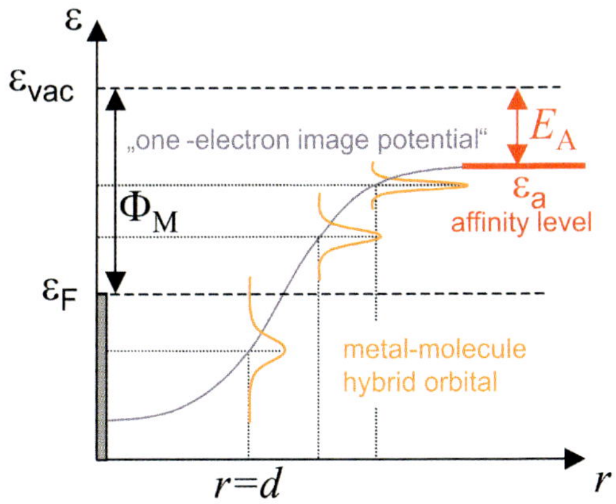

Figure 12.5 Schematic representation of the chemisorption process of an electron-accepting molecule. r is the distance coordinate between molecule and surface, d the equilibrium, after [42, 43].

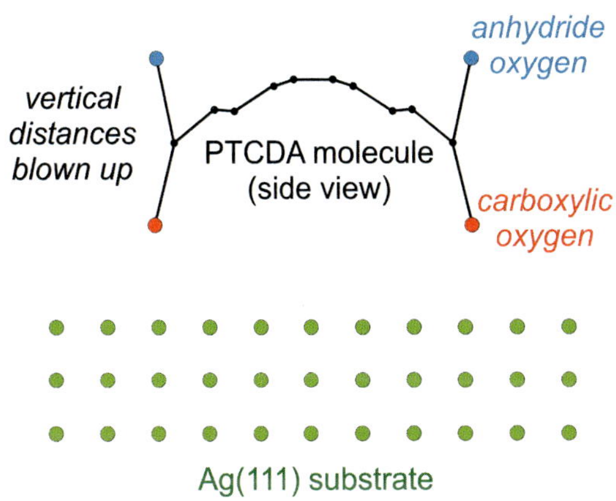

Figure 12.6 Schematic representation of the distortion of PTCDA on Ag(111). The structure has been calculated by density functional theory in the local density approximation [33].

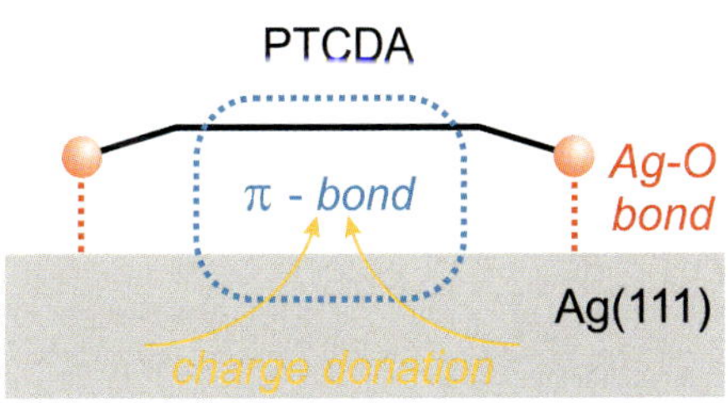

Figure 12.7 Schematic representation of the bonding interaction of PTCDA with Ag(111) [27, 33].

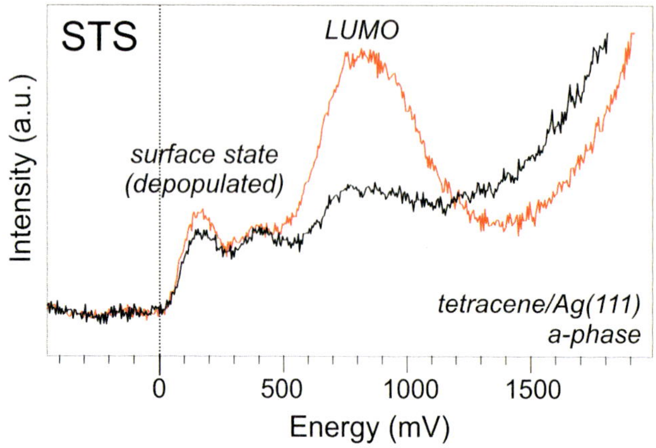

Figure 12.8 STS spectra of tetracene/Ag(111), α-phase, recorded at two different positions inside the molecule.

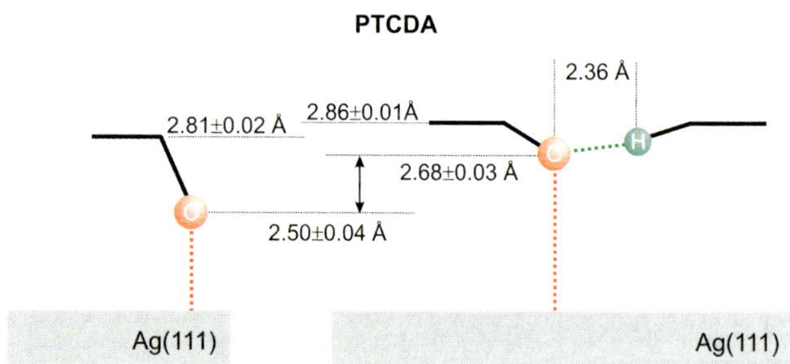

Figure 12.9 Schematic representation of structural changes of PTCDA on ordering in the herringbone phase [35]. Left: disordered low temperature phase. Right: herringbone phase.

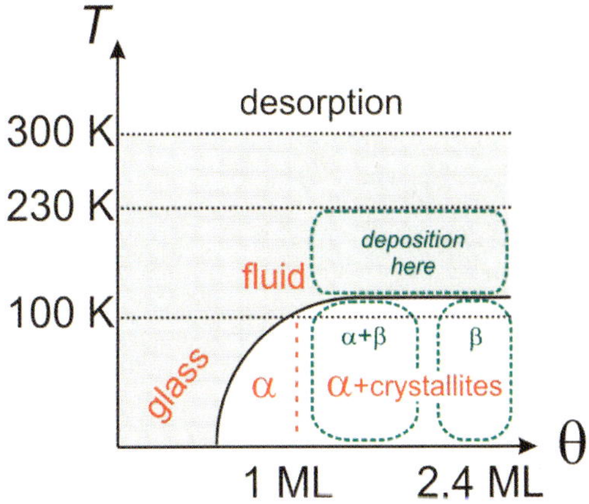

Figure 12.10 Schematic phase diagram of tetracene on Ag(111) (after Langner et al. [69]). Stable phases are shown in red, metastable phases and their preparation parameters in green.

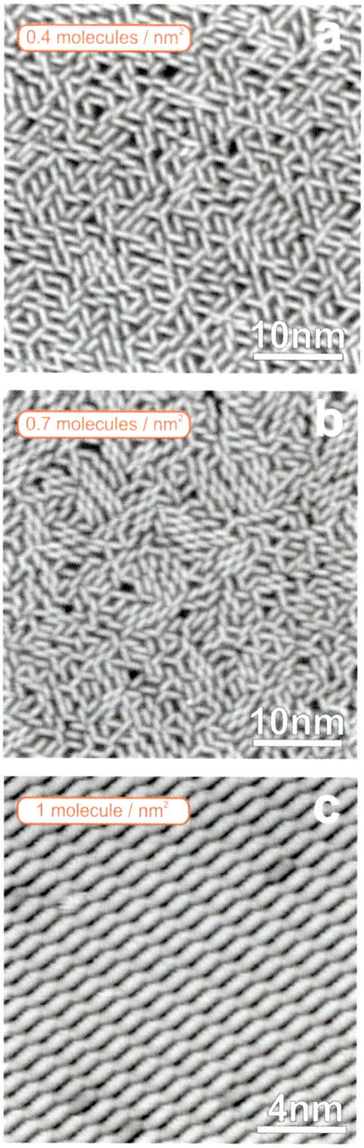

Figure 12.11 STM images of tetracene on Ag(111) at various coverages as indicated. Parameters: (a) 40 × 40 nm^2, 16 pA, 1.3 V. (b) 40 × 40 nm^2, 22 pA, 0.8 V. (c) 16 × 16 nm^2, 0.1 nA, 1.5 V.

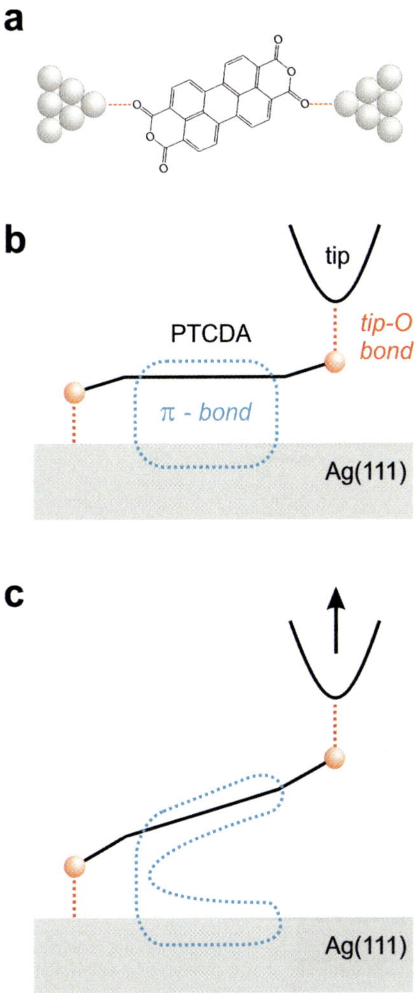

Figure 12.12 (a) Idealised single-molecule transport experiment. (b) Single-molecule transport experiment with optimal structural control based on a (1) epitaxial molecular monolayer with well-characterised adsorbate-substrate bond and (2) a specific, chemically well-defined tip-molecule contact. (c) Mechanically gated single-molecule wire based on experiment in b and STM tip retraction.

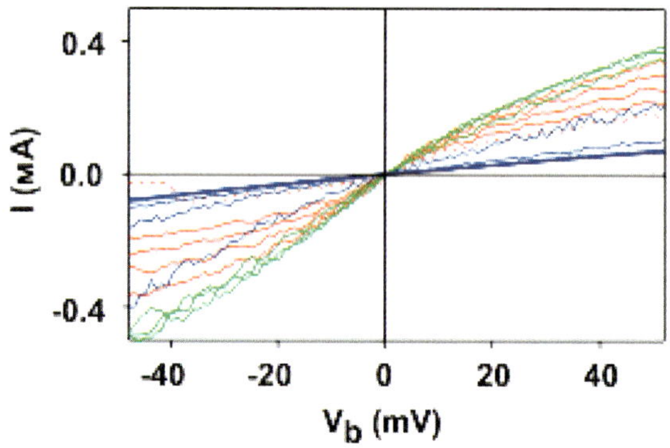

Figure 12.13 Current voltage characteristic of the device in Figure 12.12c [31]. Curves of different colours refer to different vertical tip positions and hence "streching states" of the junction.

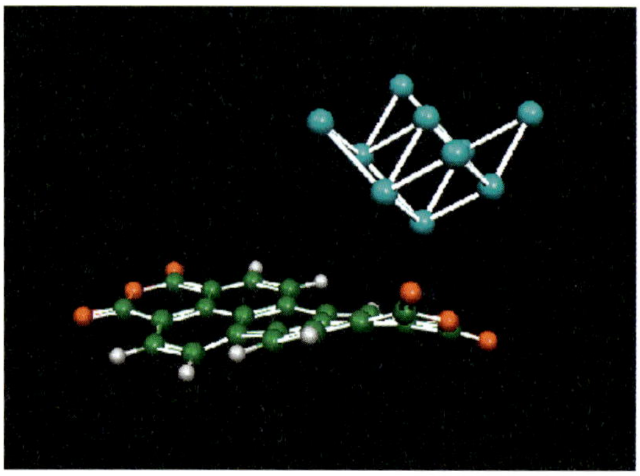

Figure 12.14 Simulation of the PTCDA/Ag(111) bond cleavage by tip retraction (calculated by local density approximation of density functional theory) [41].

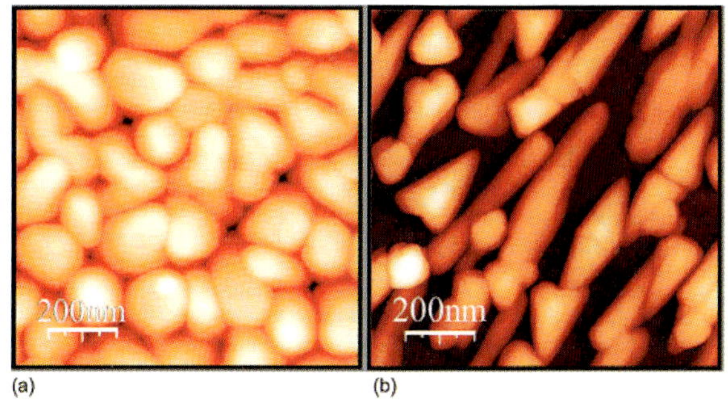

Figure 13.1 AFM images of PTCDA (a) and DiMePTCDI (b) films with a nominal thickness of 20 nm grown on S-GaAs(100).

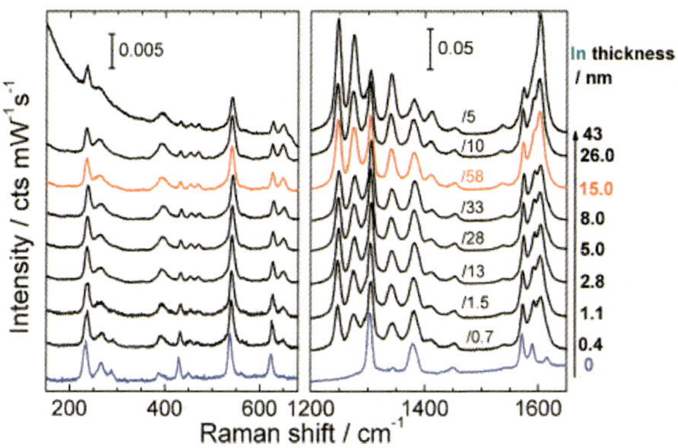

Figure 13.2 Raman spectra of PTCDA acquired upon successive deposition of In onto a 15 nm PTCDA film. The spectra normalisation is done with respect to the intensity of the C–C stretch modes (1572 cm^{-1}).

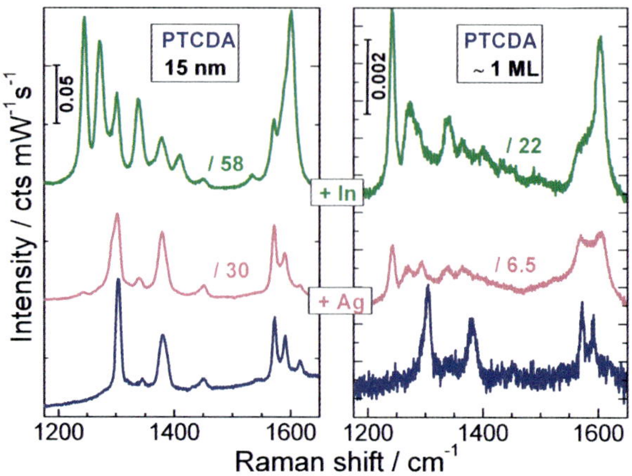

Figure 13.3 Raman spectra of a 15 nm (left) and of a ML (right) PTCDA film clean (bottom) and covered with nominally 15 nm of silver (middle) and indium (top) each.

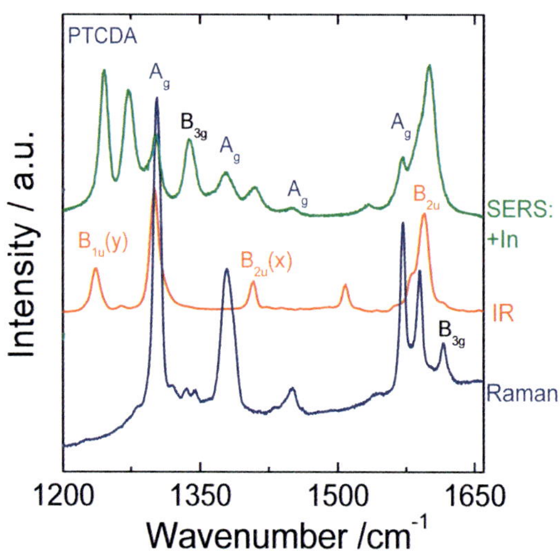

Figure 13.4 Example of the activation of infrared active modes in the Raman spectrum for the case of indium deposition onto PTCDA.

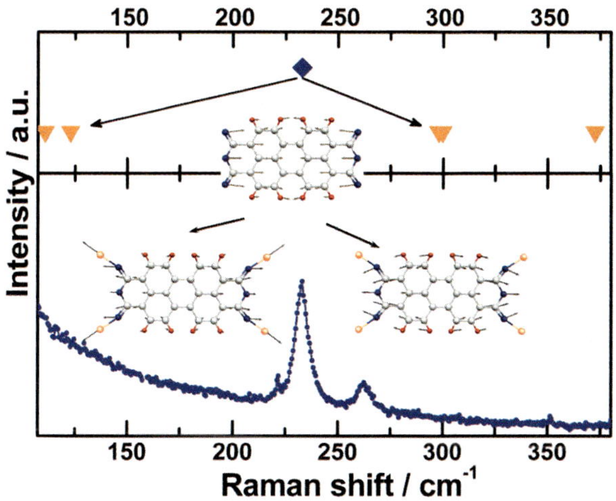

Figure 13.5 Raman spectrum of a PTCDA crystal along with calculated frequencies for a single molecule (rhombus) and for a In_4PTCDA complex (triangles).

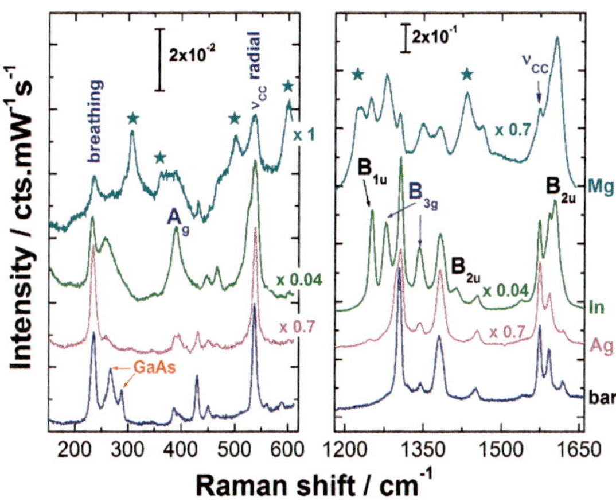

Figure 13.6 Raman spectra for In (5 nm), Ag (4,5 nm) and Mg (5 nm) coverages on 15 nm thick PTCDA films, compared with the spectrum of the bare PTCDA film in the spectral region of the internal breathing mode (left) and in the spectral region of C–H deformation and C=C stretching modes (right).

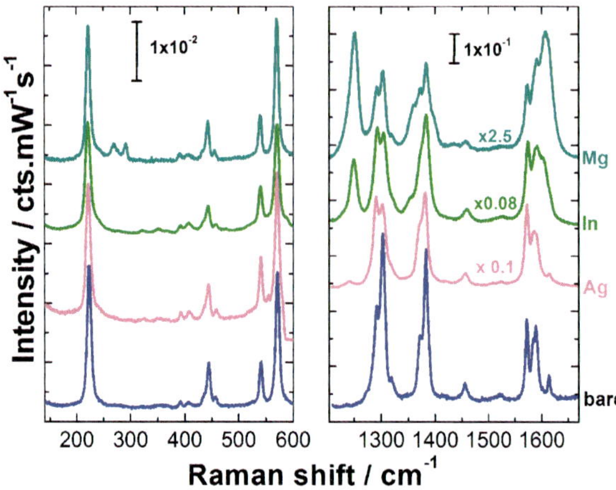

Figure 13.7 Raman spectra for In (5 nm), Ag (4,5 nm) and Mg (6 nm) coverages on 15 nm thick DiMe-PTCDI films, compared with the spectrum of the bare DiMe-PTCDI film.

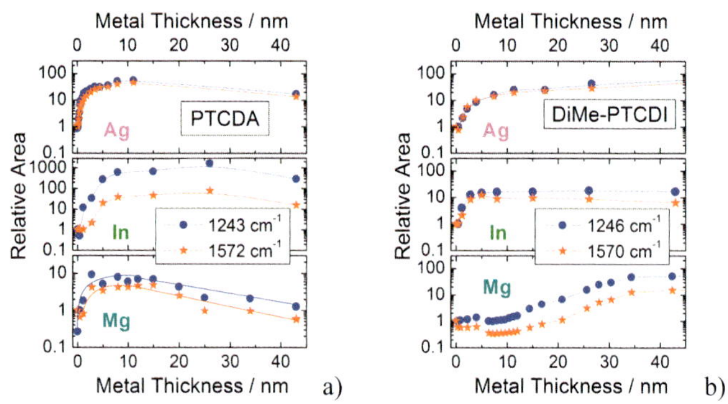

Figure 13.8 Enhancement factors of the B_u mode (1243 cm^{-1} in PTCDA and 1246 cm^{-1} in DiMe-PTCDI) and of the C-C stretching A_g mode (1572 cm^{-1} in PTCDA and 1570 cm^{-1} in DiMe-PTCDI) for PTCDA (a) and DiMe-PTCDI (b) as a function of the metal coverage.

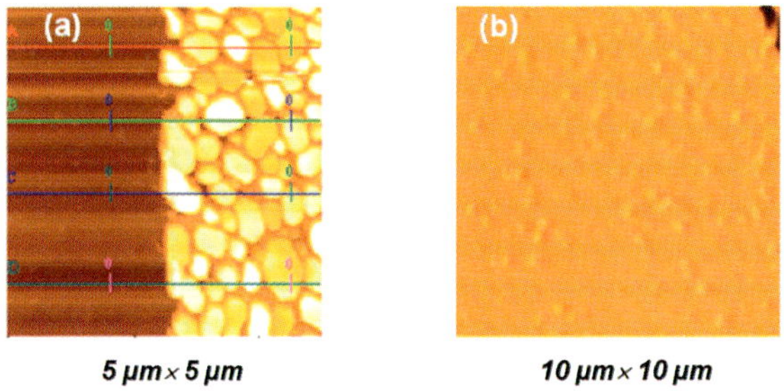

Figure 13.9 AFM topographic images of a 30 nm thick In film on PTCDA (a) (the left part of the image corresponds to the substrate and the right part of the image corresponds to the In film grown on PTCDA) and of a 113 nm thick Mg film on PTCDA (b).

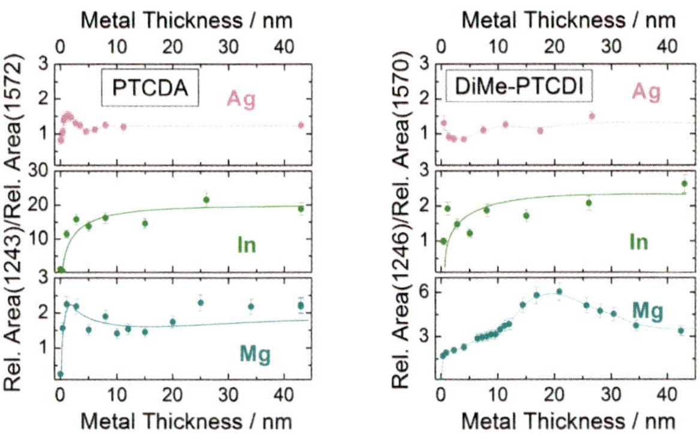

Figure 13.10 Ratio between enhancement factor of the B_u mode (1243 cm^{-1} in PTCDA and 1246 cm^{-1} in DiMe-PTCDI) and of the C–C stretching A_g mode (1572 cm^{-1} in PTCDA and 1570 cm^{-1} in DiMe-PTCDI) as a function of the metal coverage.

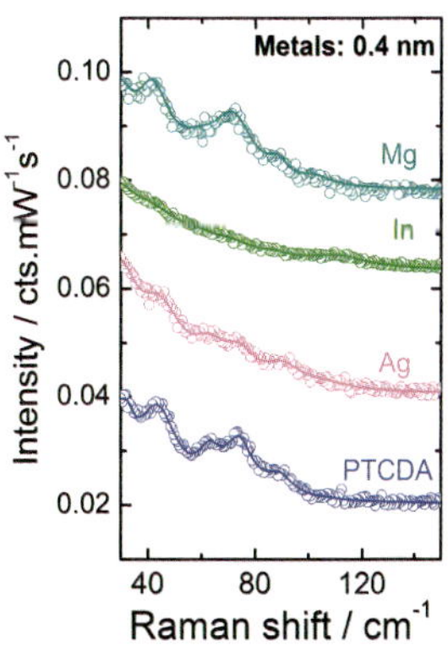

Figure 13.11 Raman spectra of 15 nm thick PTCDA films covered with Ag, In and Mg in the region of the external modes.

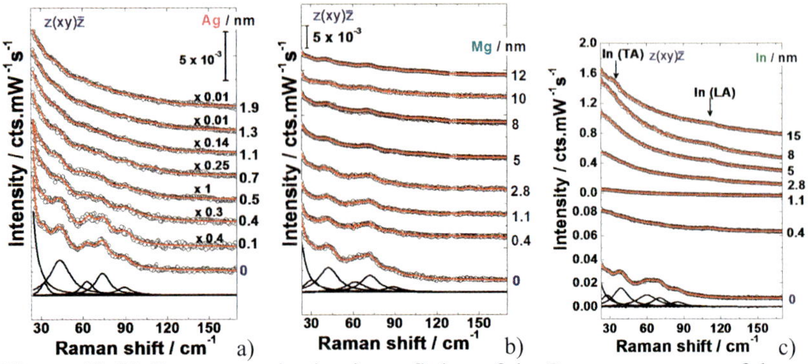

Figure 13.12 Raman monitoring in the external mode region upon metal deposition: Ag (a), Mg (b), and In (c). The experimental spectra are shown by open symbols and the fitted spectra by red lines. The Lorentzian functions used for curve fitting of the Raman spectrum of the pure PTCDA film are shown by lines in the lower part of the Figs. The spectra are normalised for Ag/PTCDA for a better observation of the phonons.

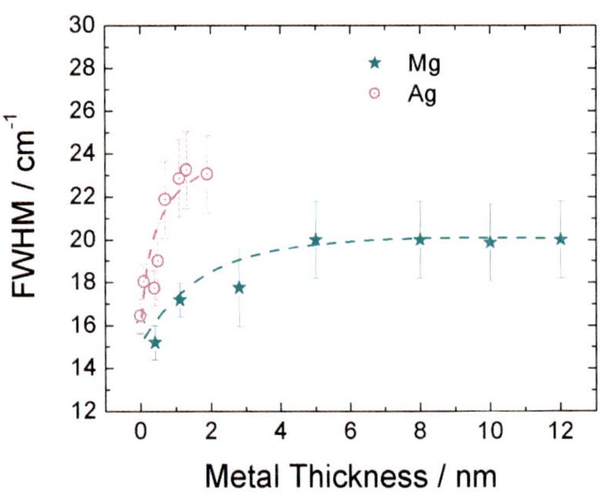

Figure 13.13 Evolution of the FWHM of the external mode at 41 cm^{-1} as a function of the metal coverage relative to the initial values before the metal deposition: for Ag (a) and Mg (b). The dashed lines are guidelines for the eyes.

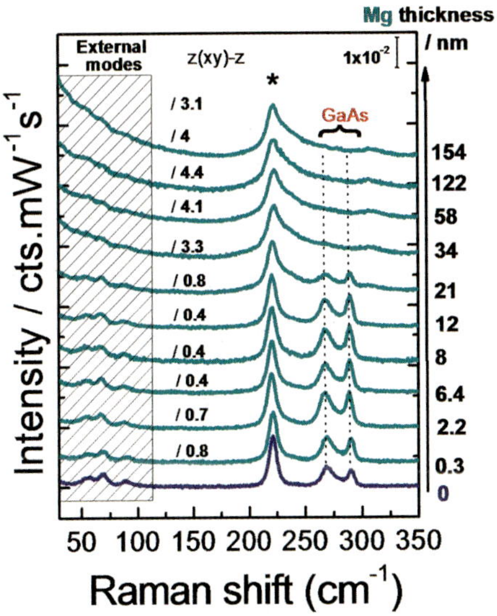

Figure 13.14 Raman spectra of DiMe-PTCDI upon Mg deposition in the region of external modes and GaAs phonons. The spectra are normalised with respect to the intensity of the breathing mode at 221 cm^{-1}.

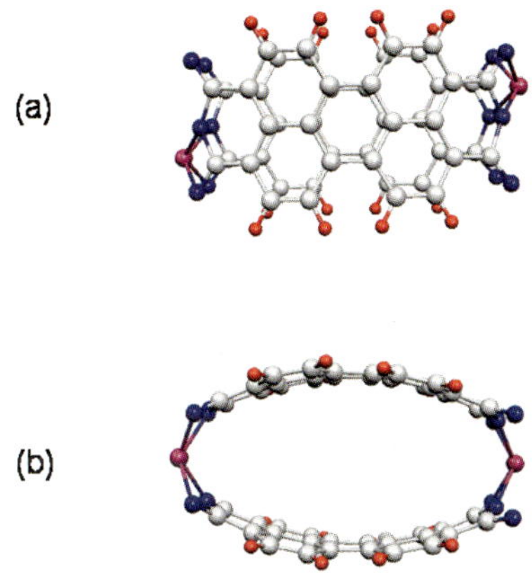

Figure 13.16 Model geometry of intercalated Mg ions, optimised at the B3LYP/DZ level, (a) perspective view along the molecular normal, (b) view along a crystal plane. This geometry was used for the calculation of the Huang-Rhys factors visualised in Figure 13.15.

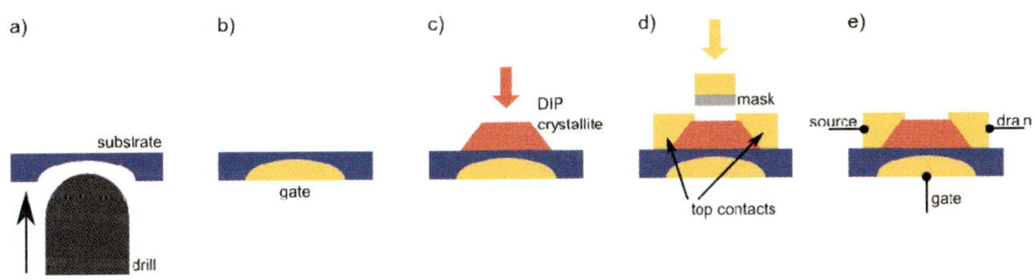

Figure 14.2 Illustration of our micro-OFET concept: (a) Thinning of the sapphire substrate from the backside to achieve a thin gate dielectric; (b) mounting of the metal gate electrode by thermal evaporation; (c) deposition of the active DIP layer by OMBD; (d) application of the source and drain top contacts by Au deposition using a lithographic mask; e) electrical characterization.

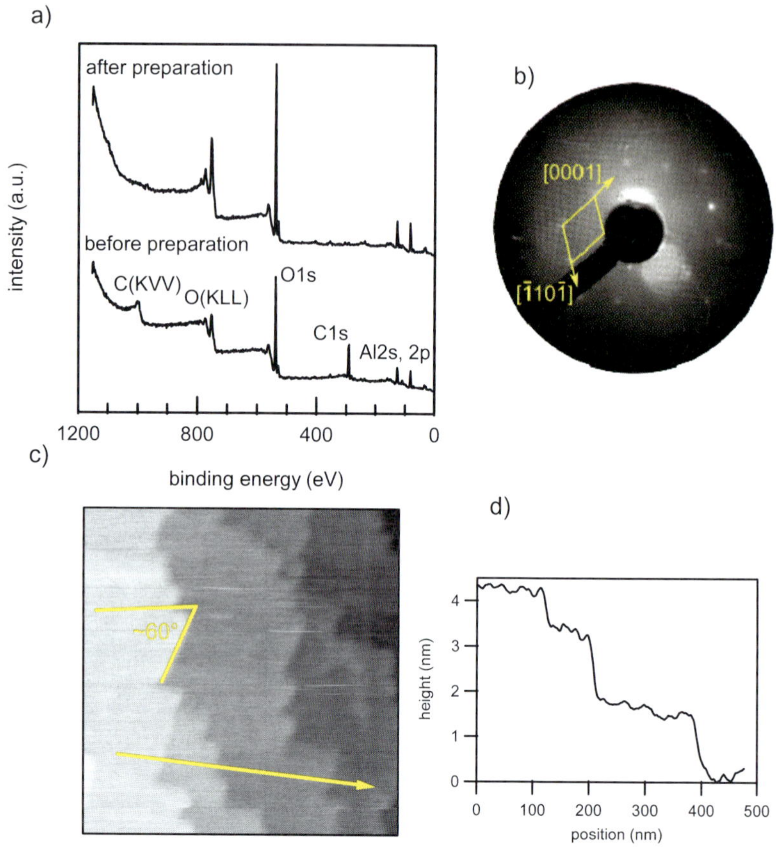

Figure 14.3

a) XPS survey scans of the sapphire substrate before (bottom) and after (top) UHV preparation by sputtering and annealing. The most prominent signals are denominated.

b) LEED picture of the sapphire substrate after UHV preparation. The unit cell and the lattice directions of the sapphire surface are indicated by arrows.

c) Contact mode AFM picture of a sapphire substrate after UHV preparation (window size: 500 nm × 500 nm).

d) Height profile along the line indicated by the arrow in Figure 14.3c.

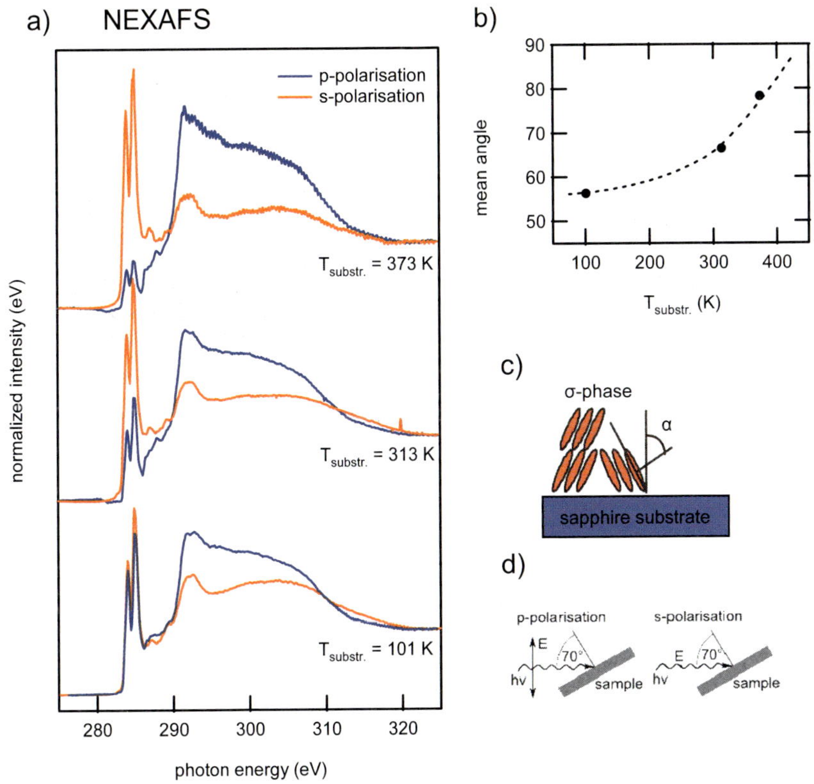

Figure 14.4

a) C K-NEXAFS spectra of DIP samples prepared at different substrate temperatures (373, 313, and 101 K), recorded in the partial electron yield (PEY) mode with p- (blue) and s-polarisation (red) of the incident X-rays.

b) Plot of the average molecular tilt angle versus substrate temperature deduced by evaluation of the NEXAFS dichroism from Figure 14.4a. The red curve is a guideline to the eye.

c) Illustration of the molecular tilt angle for growth in the σ-phase (from [24]).

d) Experimental geometry of the NEXAFS experiment.

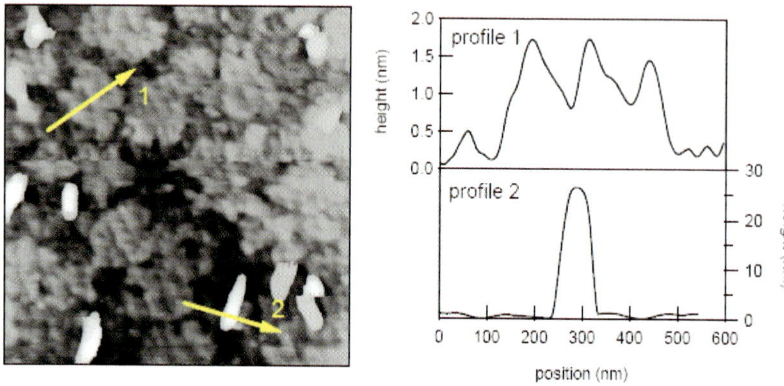

Figure 14.5 Tapping mode AFM picture (window size: 2 μm × 2 μm) of a nominally 2 nm thick DIP film deposited on a sapphire surface at a substrate temperature of 330 K. The profiles were extracted along the lines indicated by arrows in the AFM picture.

a)

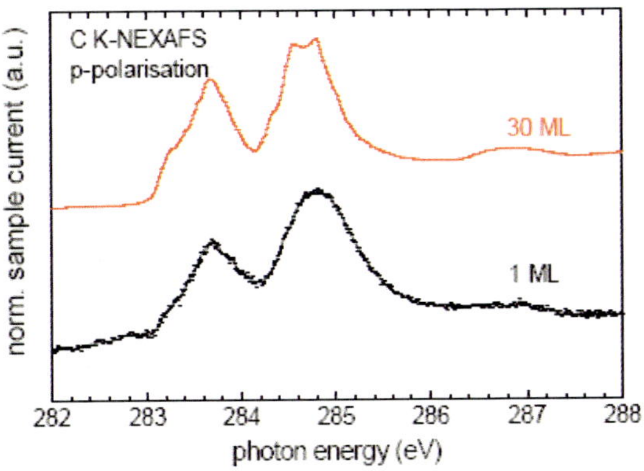

b)

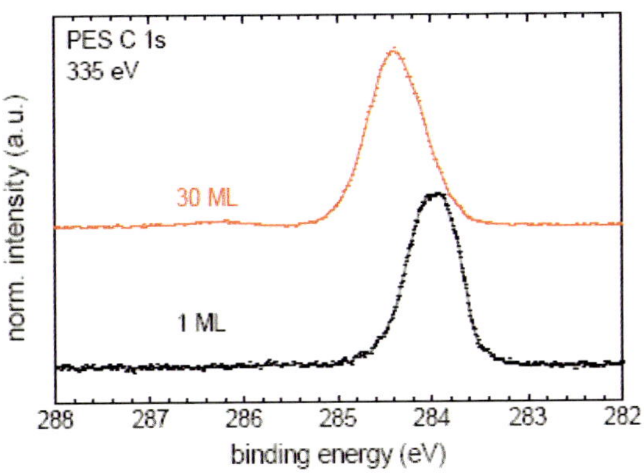

Figure 14.6

a) Pre-edge regime of high-resolution C K-NEXAFS spectra of a 30 ML (top, red) and a 1 ML thick (bottom, black) DIP film on Au(111). The spectra were recorded by measuring the sample current with p-polarisation of the incident X-rays.

b) High-resolution C 1s XPS spectra of a 30 ML (top, red) and a 1 ML thick (bottom, black) DIP film on Au(111). The spectra were recorded using a photon energy of 335 eV in normal emission geometry.

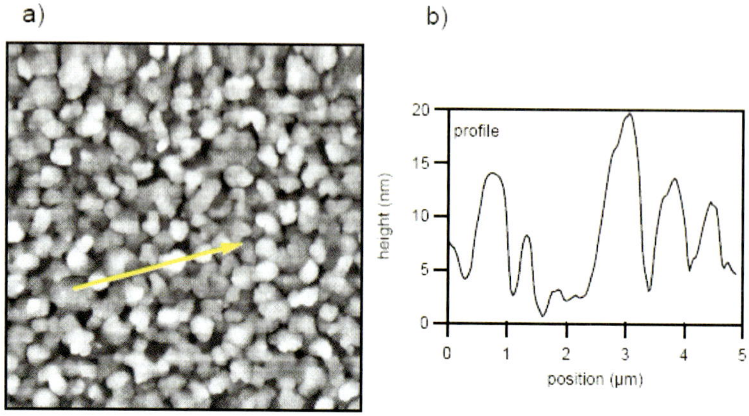

Figure 14.7
a) Contact mode AFM picture (window size: 10 μm × 10 μm) of an Au contact with 10 nm thickness evaporated on a DIP film.
b) Height profile along the direction indicated by the arrow in Figure 14.7a.

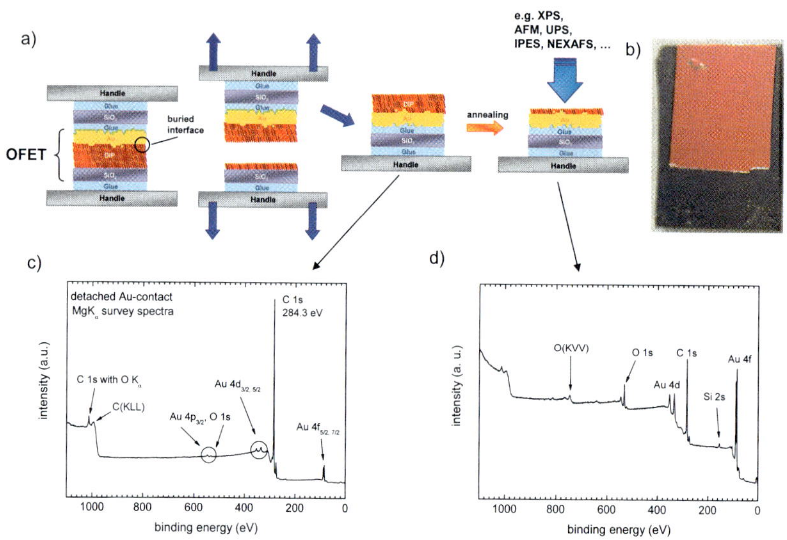

Figure 14.8
a) Illustration of the lift-off technique applied to investigate the buried Au/DIP interface; for details see text.
b) Picture of a separated Au top contact mounted on a sample holder for spectroscopic analysis.
c) XPS survey scan of the separated Au top contact recorded with a MgKα laboratory source. The most prominent signals are indicated.
d) XPS survey scan of the separated Au top contact after annealing at 470 K which leads to a partial desorption of the DIP layer.

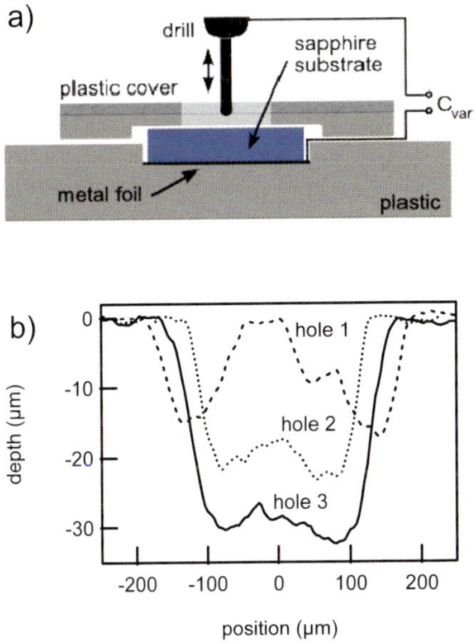

Figure 14.9
a) Illustration of the set-up used for high-precision drilling and thickness monitoring of the blind hole for the gate electrode; for details see text.
b) Profilometer scans of 3 different blind holes established in sapphire with the set-up sketched in Figure 14.9a.

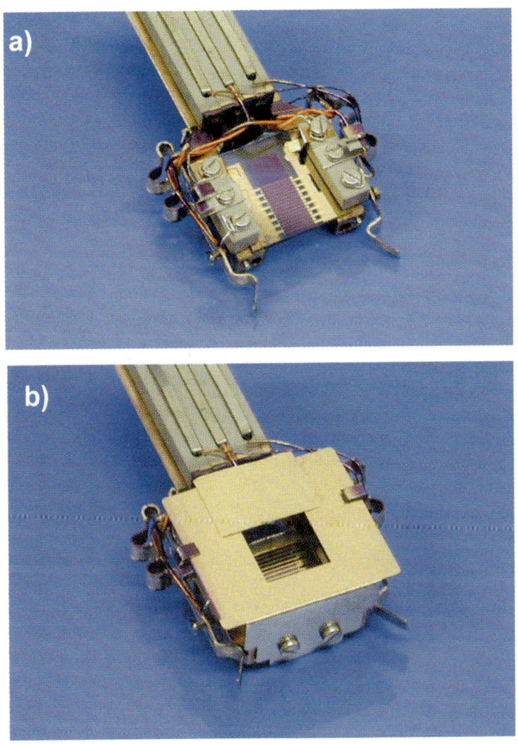

Figure 14.10 Picture of the transferable sample holder dedicated for *in situ* device fabrication, IV-measurements, and surface analysis in UHV without (Figure 14.10a) and with (Figure 14.10b) a mask to evaporate a metal finger structure. In Figure 14.10a a CuPc film contacted by a Au finger structure is mounted on the holder.

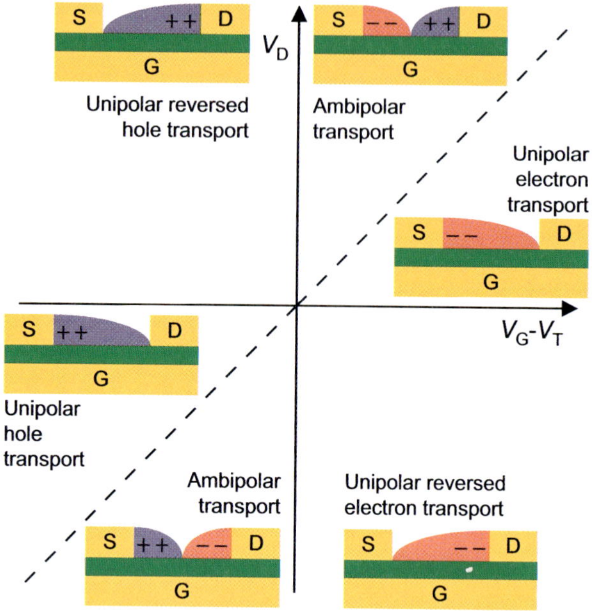

Figure 17.1 Sketch for the unipolar and ambipolar operation regimes of an organic field-effect transistor. V_D denotes the drain voltage, V_G and V_T are gate and threshold voltage, respectively.

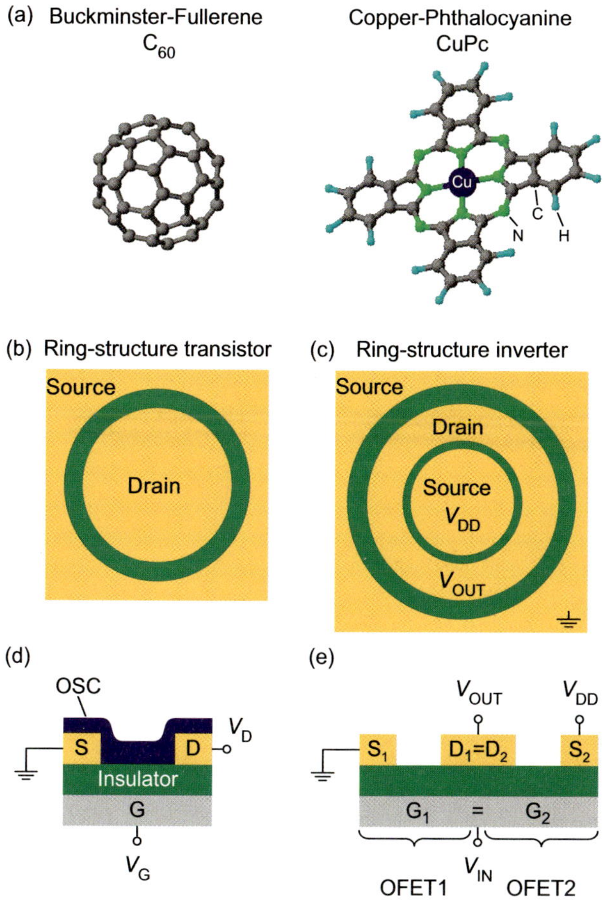

Figure 17.2 (a) Chemical structures of the used materials: fullerene (C_{60}) and copper-phthalocyanine (CuPc). Top view (b) and cross section (d) of the ring-type transistor in bottom-gate and bottom-contact geometry. Top view (c) and cross section (d) of the ring-type inverter. The silicon substrate acts as the gate electrode for transistors and inverters.

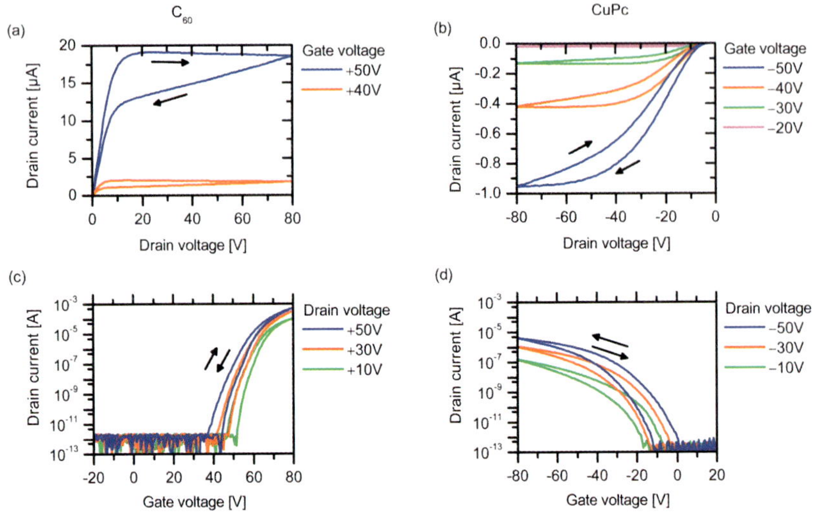

Figure 17.3 Output and transfer characteristics of unipolar field-effect transistors with neat C_{60} (a, c) and neat CuPc (b, d) films. The substrates were treated with O_2-plasma and the films evaporated at 375 K substrate temperature. The direction of the hysteresis is indicated by arrows. (Figure adopted from Ref. [27].)

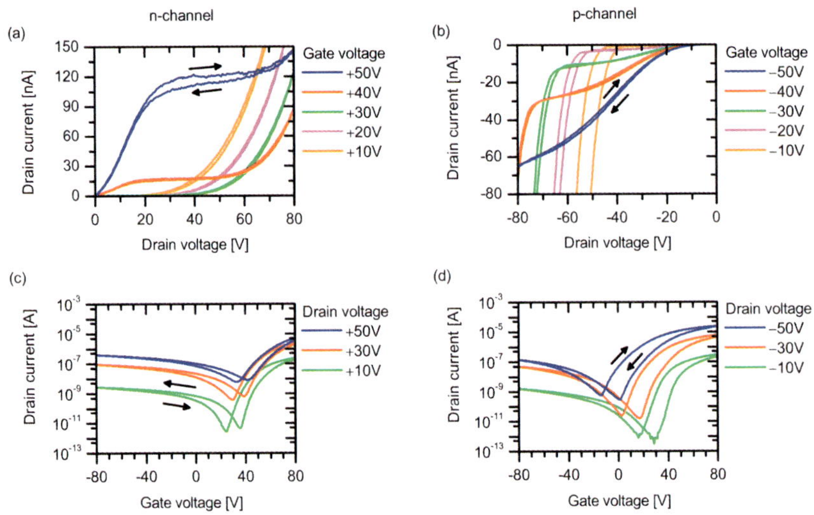

Figure 17.4 Output and transfer characteristics of ambipolar field-effect transistors for a mixing ratio between C_{60} and CuPc of 1 : 1 measured in the n-channel regime (a, c) as well as the p-channel regime (b, d). The substrate was O_2-plasma treated and the film evaporated at 375 K substrate temperature. The direction of the hysteresis is indicated by arrows. (Figure adopted from Ref. [27].)

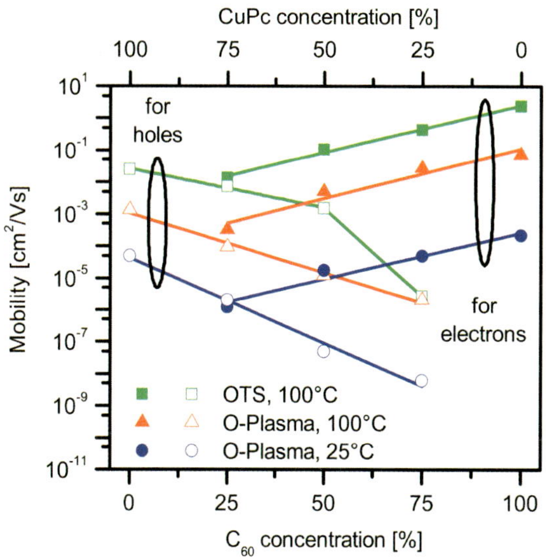

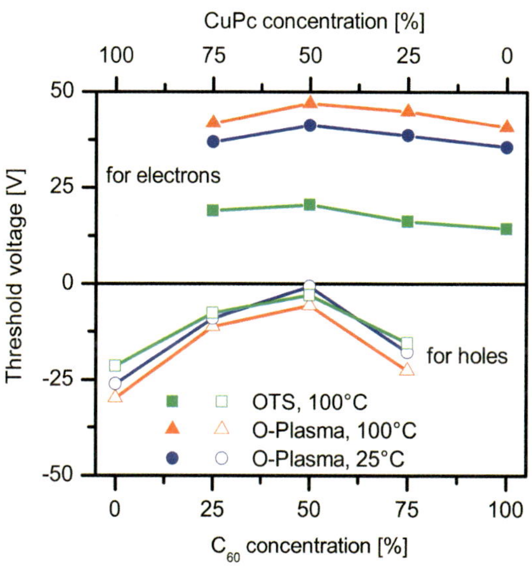

Figure 17.5 Mobility (left) and threshold voltage (right) as determined from transfer characteristics in the saturation regime of OFETs with different composition, substrate temperature and substrate treatment. The filled symbols are related to the electron transport, the open symbols to the hole transport.

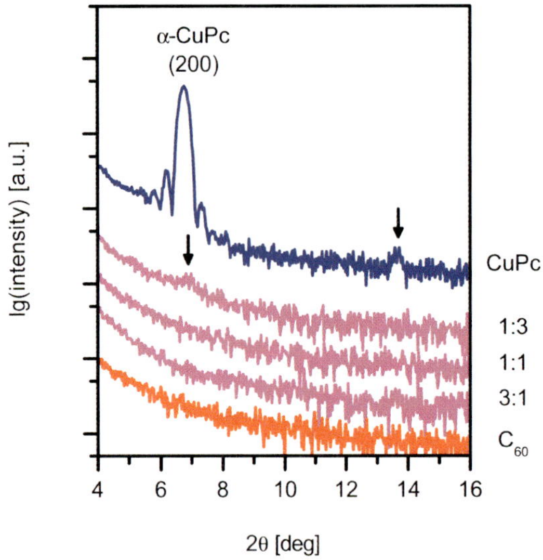

Figure 17.6 X-ray diffraction patterns for neat CuPc and C_{60} films as well as for three mixed films grown at 375 K. The arrows indicate the weak (200) peak in the sample with a mixing ratio of 1 : 3 and the second order diffraction peak of the neat CuPc sample.

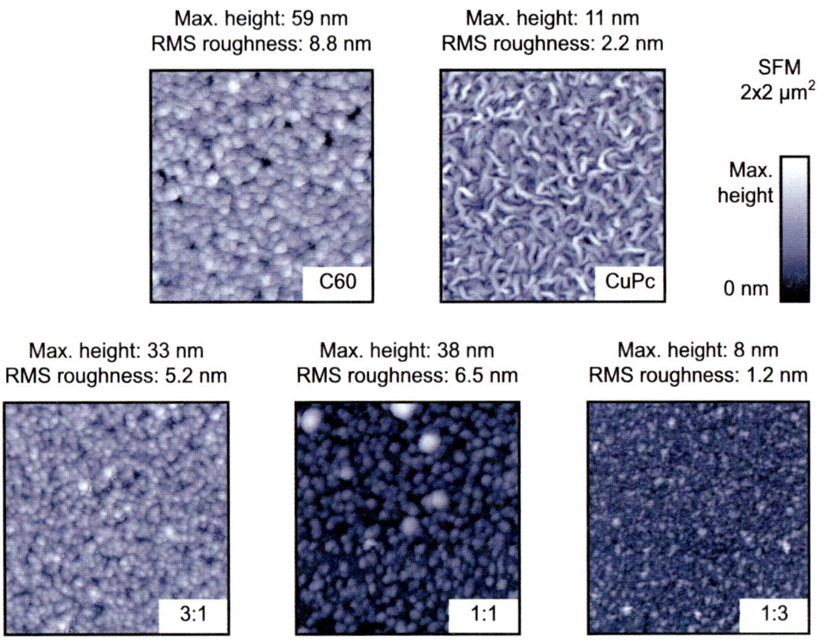

Figure 17.7 Scanning force microscopy images taken in non-contact mode for neat C_{60} and CuPc films as well as for three mixed films grown at 375 K. The total image size is $2 \times 2\ \mu m^2$. The max. height is given as the difference between the lowest value (dark blue) and the highest value (white) in each of the images.

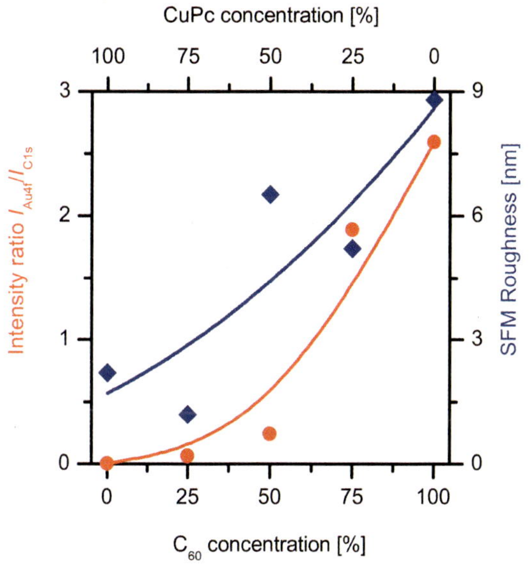

Figure 17.8 Comparison of the RMS roughness measured by SFM and the ratio of the substrate to film signal measured by XPS. The solid lines are to guide the eyes.

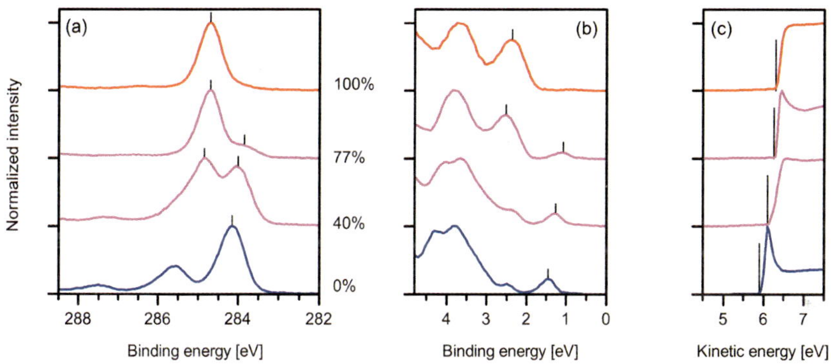

Figure 17.9 XPS and UPS measurements of neat and different mixed films of C_{60} and CuPc: (a) C1s spectra (excitation: monochromated Al K), (b) HOMO levels (excitation: He II) and (c) secondary electron cut-off (excitation: He I and $V_{Bias} = -2$ V). The numbers on the right side of the diagram (a) give the C_{60} concentration as determined from the measurements shown in part (a). (Figure adopted from Ref. [53].)

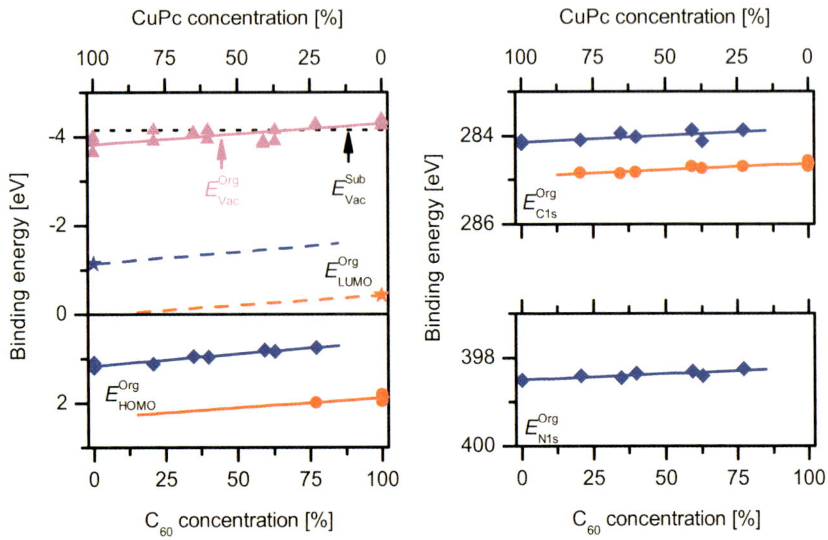

Figure 17.10 Position of the vacuum level, the high energy edge of the HOMO level, and the C1s and N1s core levels as a function of the C_{60}/CuPc mixing ratio. The ◆ and the ● symbols are representing the measured values for C_{60} and CuPc related levels, respectively, and the ▲ symbols the measured work functions. The solid lines are linear fits of the corresponding measured values. The constant dotted line is the vacuum energy of the gold substrate (covered by a conductive hydrocarbon layer). The LUMO levels of the neat materials are marked with the ★ symbols. The dashed lines are extrapolated LUMO levels for the blends. (Figure adopted from Ref. [53].)

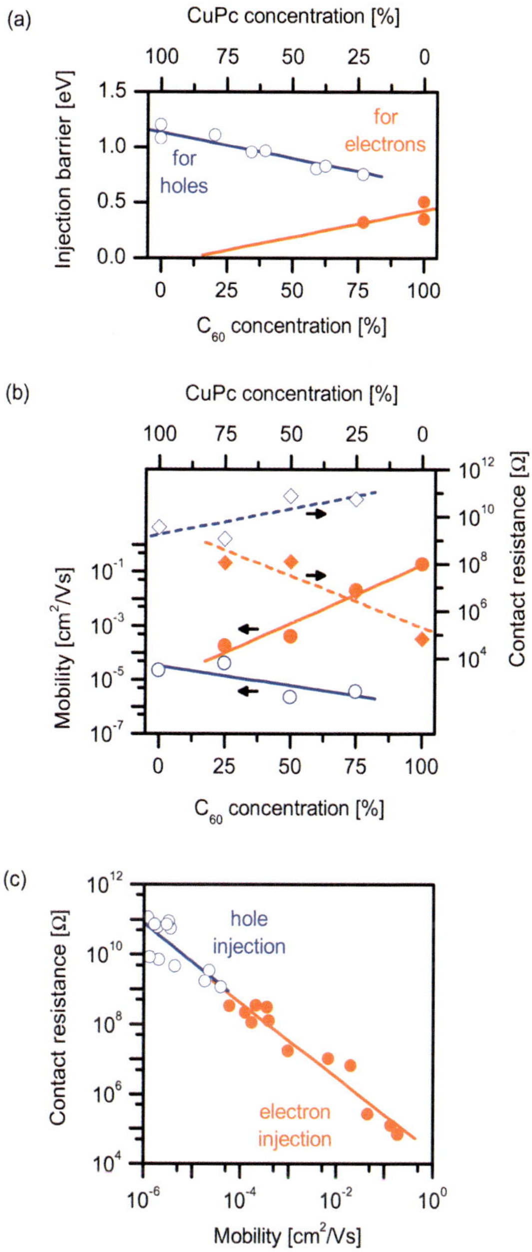

Figure 17.11 (a) Injection barrier determined from UPS (compare with Figure 17.10). (b) Mobility (solid line) and contact resistance (dashed line) as determined from transfer characteristics in the linear regime for $|V_G - V_T| = 33$ V. (c) Relation between mobility and contact resistance for different $|V_G - V_T| > |V_D|$ and different mixing ratios (0, 25, 50, 75 and 100% C_{60}). The straight lines are linear fits related to the vacuum level shift in part (a) and to guide the eyes in the parts (b) and (c).

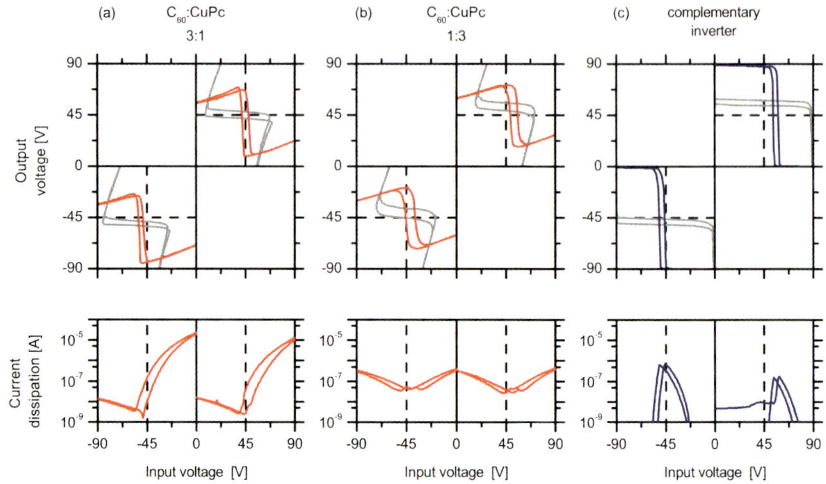

Figure 17.13 Characteristics of ambipolar inverters with mixing ratios of 3 : 1 (a) and 1 : 3 (b) and a complementary inverter (c) consisting of a neat C_{60} and a neat CuPc transistor. The driving voltage is $V_{DD} = \pm 90$ V. Transfer characteristics (top) and current dissipation (bottom) are shown. The substrates were O_2-plasma treated and the films evaporated at 375 K substrate temperature. The grey lines are shown to explain the noise margin (see text). (Figure adopted from Ref. [27].)

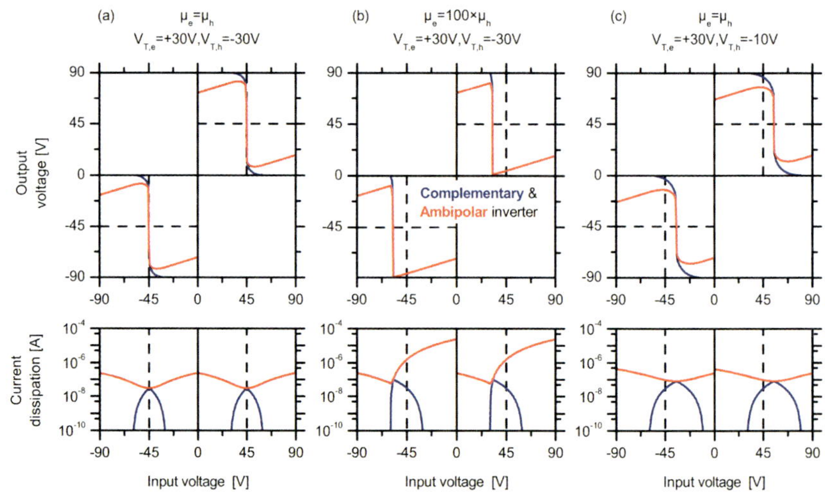

Figure 17.14 Simulations of ambipolar and complementary inverter transfer characteristics and current dissipation with (a) symmetric mobility and threshold voltage for p- and n-channel in comparison with (b) asymmetric mobilities and (c) asymmetric threshold voltages.

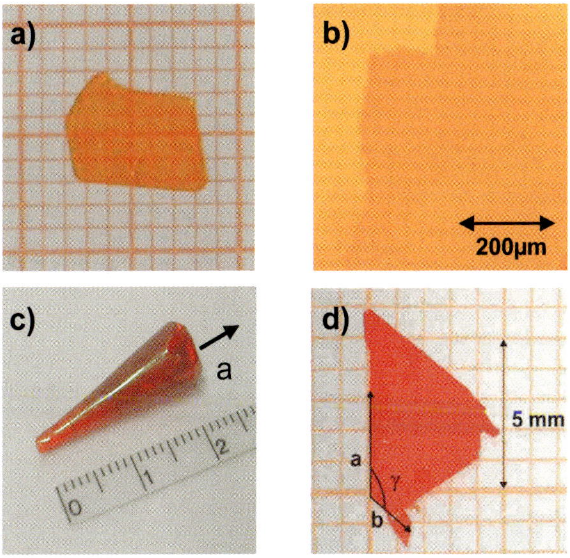

Figure 25.2 Photographs of a sublimation grown tetracene crystal with millimetre extensions along the a- and b-directions (a). Interference contrast imaging of the corresponding (001) surface reveals flat terraces of several hundred micrometers in width (b). Tetracene bulk crystals can be prepared from saturated vapour in a standard Bridgman-setup (c). In this case, the conical volume of 1 cm^3 is aligned with the cone axis along the a-direction. Similar to the oligoacenes, diindenoperylene crystals prepared by sublimation under streaming hydrogen show platelet-like morphologies with an [001] surface normal (d). The smooth terraces of 100 μm are a precondition for reliable FET measurements.

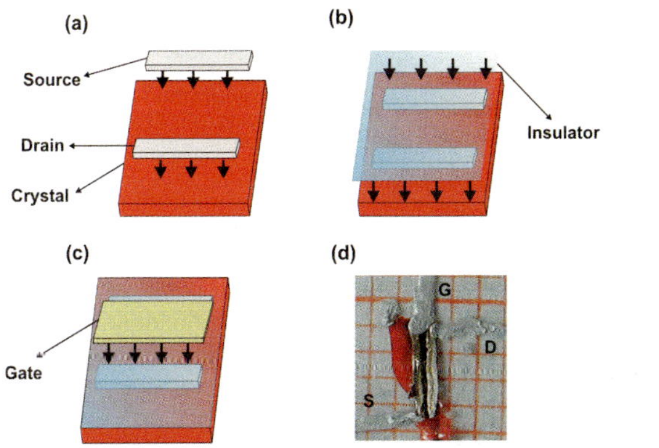

Figure 25.3 Scheme of the three major steps of FET fabrication. After preparing source drain contacts by Ag paste (a) the gate insulator is fabricated by a two-step temperature process of para-cyclophane (b). Finally, by thermal evaporation of Au the top-gate contact of 20 nm thickness is deposited (c). An image of the resulting structure is shown in (d) for a DIP crystal FET.

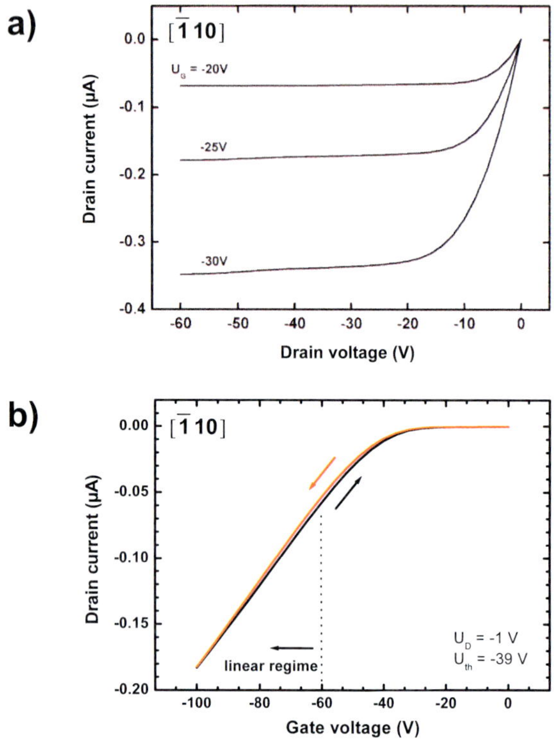

Figure 25.5 Input curve (a) and transfer-curve (b) measured on a tetracene single crystal FET along the $[\bar{1}10]$. According to Eqs. (4) and (5) the hole mobility can be deduced from different parts of the curve and is consistently estimated to $\mu = 0.78$ cm^2/Vs. The small hysteresis of the transfer curve indicates a minor influence of trapping and charging effects above the threshold voltage.

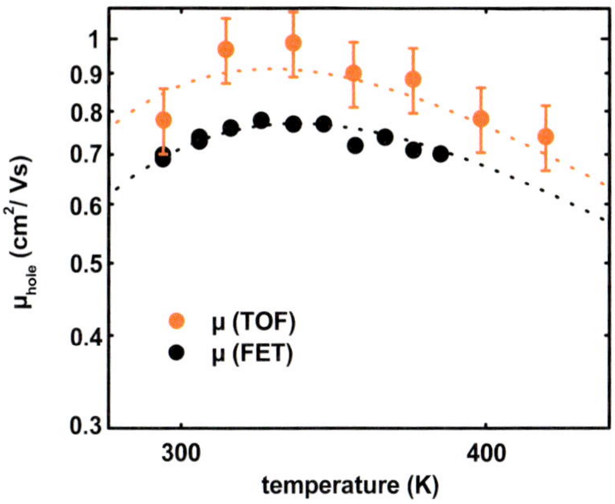

Figure 25.6 Hole mobility of tetracene single crystals measured at the surface (FET) and in the bulk (TOF). Both curves resemble a temperature behaviour that can be ascribed to the multiple-shallow-trapping and release of charge carriers. Parameter of the HL fits for TOF (FET) are $E_T = 200$ meV (190 meV), $n_T = 3 \times 10^{-4}$ (7×10^{-4}) and $\mu_{int} = 1.5$ cm^2/Vs (1.4 cm^2/Vs). Remarkably, the surface mobility in the plane of highest π-overlap, is lower than that along the [001] direction of smallest π-orbital overlap.

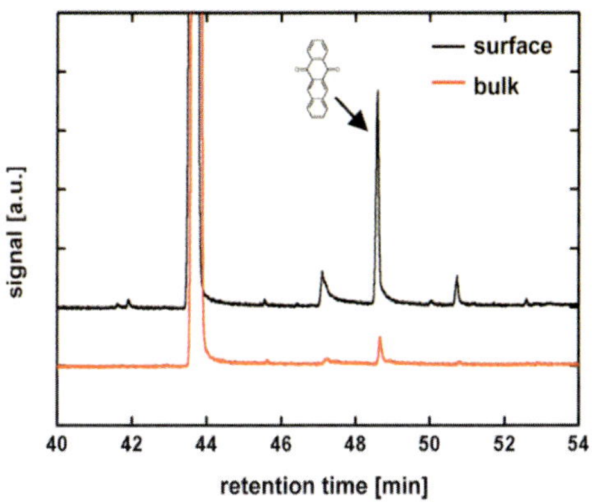

Figure 25.7 Depth dependent profile of the chemical composition of tetracene sublimation crystal. Specifically, the stable oxidation product 5,12-tetracene-quinone, indicated by the arrow, is concentrated to a much larger amount at the surface than in the bulk.

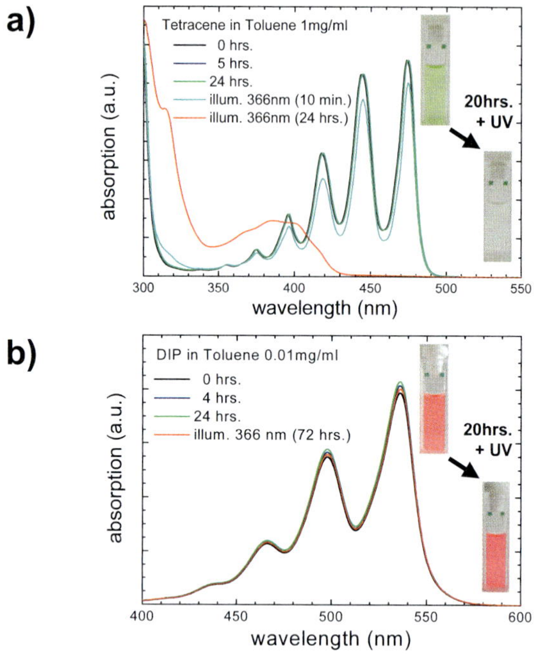

Figure 25.8 UV-Vis absorption spectra of tetracene (a), and diindenoperylene (b), both solved in toluene at saturation concentration. As indicated by the curves as well as by the cuvettes in the insets, there is a strong chemical degradation of tetracene whereas DIP remains intact, even after UV illumination at 366 nm for 3 days. Furthermore, the oxidation product generated upon UV illumination of tetracene can be identified as 5,12-tetracenequinone by its spectral signature (absorption maximum at 390 nm).

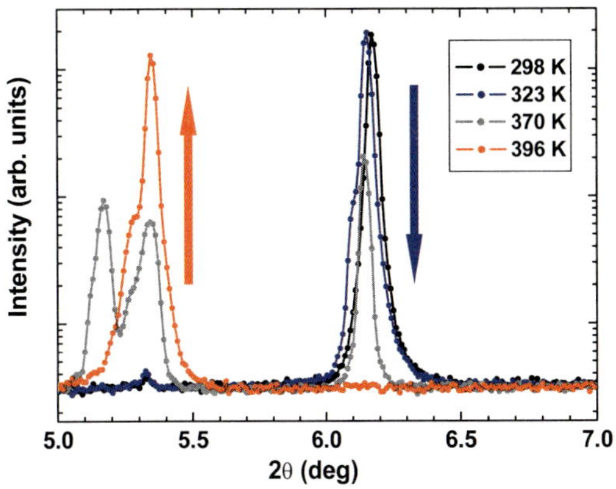

Figure 25.9 X-ray structural characterisation of DIP single crystal along the c′-direction at different temperatures. As indicated by the measurements performed in specular θ-2θ geometry, the structural phase transition is completed at around 400 K. The [001] lattice spacing of 1.68 nm (2θ = 5.31°) in the high temperature phase coincides with that of DIP thin films grown on weakly interacting substrates such as SiO_2 at room temperature.

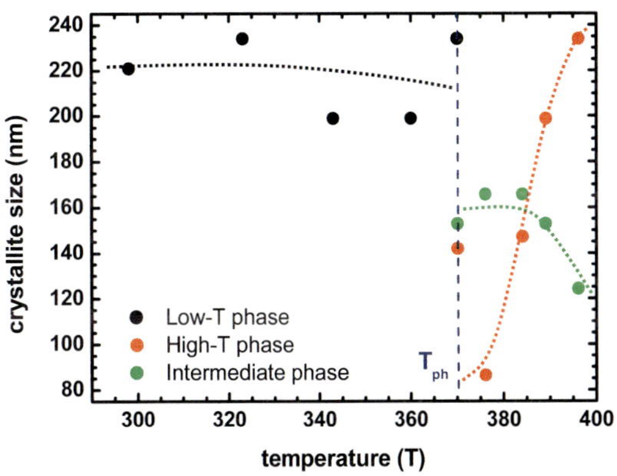

Figure 25.10 Temperature dependent evolution of the crystallite sizes in the respective structural phases of DIP. Above the transition point the previous crystallite size of about 220 nm is almost restored. In the vicinity of the phase transition, strong effects on the carrier transport might be expected from structural inhomogeneities that appear.

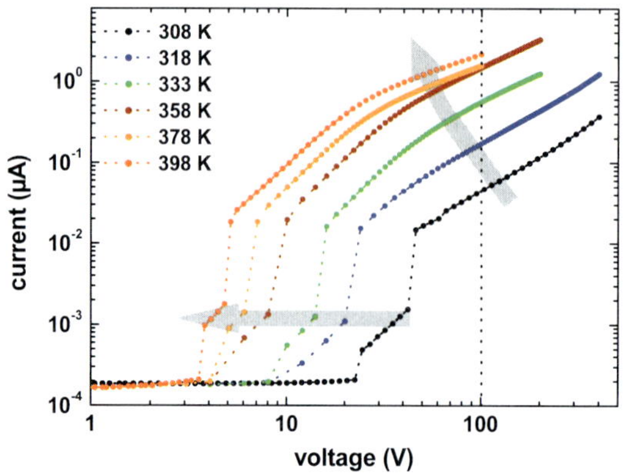

Figure 25.11 SCLC measurements along the c′-direction of a diindenoperylene single crystal at various temperatures. The mobility of the holes has been adjusted according to Eq. (5) at around 100 V where slope of the $I(V)$-curve approximates 2 at room temperature. The estimated values present a lower limit with respect to the intrinsic hole mobility in DIP. A shift of the trap-related current jump towards lower voltages and a change in the $I(V)$-slopes can be detected at elevated temperature (indicated by the arrows).

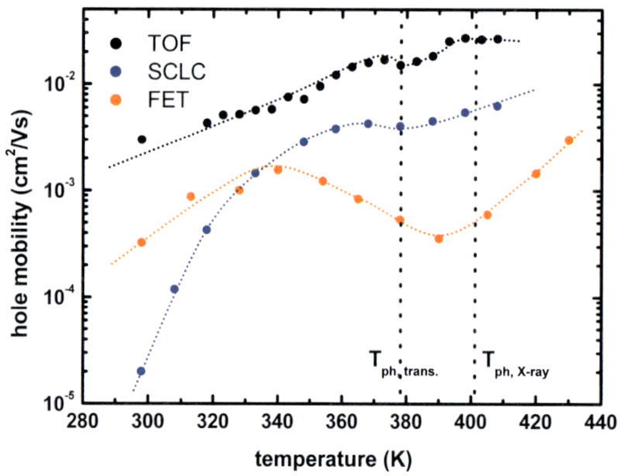

Figure 25.12 TOF, SCLC and FET measurements on DIP single crystals as a function of temperature. The dotted curves present a guide-to-the-eyes. The electronic characterisation by TOF and SCLC studies were performed along the c′-direction whereas hole transport in FETs occurs along the (ab)-plane. In addition, the phase transition temperatures deduced from transport data, $T_{ph,\,trans}$, and from X-ray structural analysis, $T_{ph,\,X\text{-}ray}$, are indicated by the dotted vertical lines.

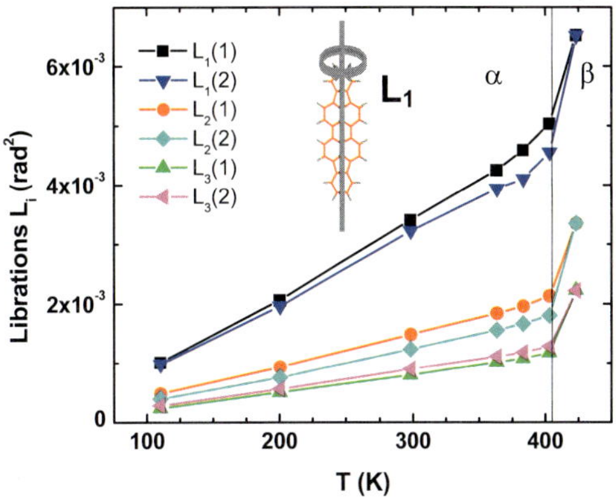

Figure 25.13 Temperature dependence of the fundamental libration modes determined by X-ray diffraction on DIP single crystals. The libration mode L_1 around the long molecular axis (inset) experiences the largest variation in amplitude with increasing temperature and, therefore, might significantly distort the charge carrier movement in the DIP (001) plane. The two molecules in the unit cell of the α-phase are indicated by (1) and (2); the basis of the β-phase is generated by just one molecule [56].

Section I
Industrial Applications

Organic Electronics. Structural and Electronic Properties of OFETs. Edited by Christof Wöll

ISBN: 978-3-527-40810-8

1
Organic Transistors as a Basis for Printed Electronics

W. Fix, A. Ullmann, R. Blache, and K. Schmidt

1.1
Introduction

The idea of manufacturing integrated circuits by means of roll-to-roll printing technologies has fascinated researchers and developers since the late 1990s [1–4]. Electronics being printed like newspaper, in large quantities, on a daily basis, opens up entirely new areas of application for micro-electronics: from radio-operated labels for replacing the optical barcode to intelligent packaging and objects capable of processing and displaying information, as well as for anti-counterfeiting, such as in the case of pharmaceuticals. The trend towards highly cost-efficient electronics which are simultaneously available in great numbers ultimately leads to printed electronics, since there is no structuring or layering process that is faster than printing.

The great challenge for printed electronics is to develop electronic inks that can be used for printing purposes, while having suitable electric semiconducting or conducting functions. Printing inks are typically enriched with numerous additives in order to enhance printability. Although such additives do not change the visual impression, they generally affect the electronic properties of the material.

So, how does one obtain a semiconducting ink to begin with? It requires a semiconducting material that can be processed into ink, i.e. brought into solution or dispersion. In this regard, organic semiconductors, first and foremost semiconducting polymers, can contribute one of their strengths, as they can easily be brought into solution. Organic transistors are thus the basic element for printed electronics.

The objective of printed electronics is to create new mass markets for cost-efficient electronics, without attempting to compete with silicon electronics, which would be a hopeless endeavour anyway.

In recent years, major progress has been made in the development of printed electronics [5–10], the highlight so far being the introduction in September 2007 of a 13.56 MHz RFID transponder [11] (Figure 1.1) that is completely printed roll-to-roll (except for the antenna). Despite these achievements, basic

Organic Electronics. Structural and Electronic Properties of OFETs. Edited by Christof Wöll

ISBN: 978-3-527-40810-8

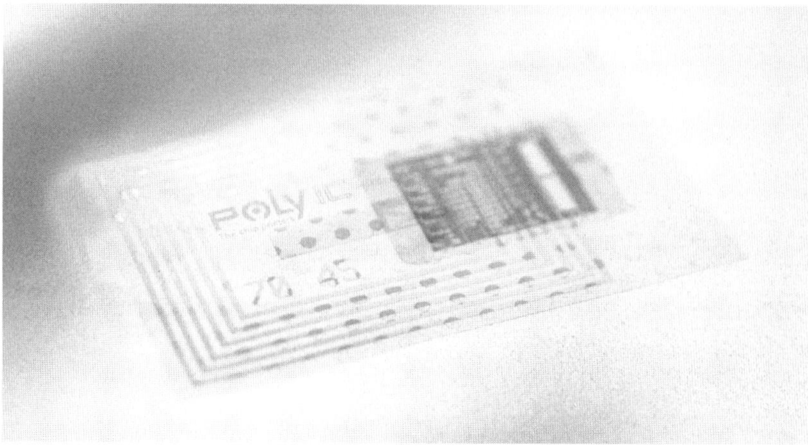

Figure 1.1 RFID tag completely printed roll-to-roll, RFID chip: approx. 2 cm × 3 cm. (see Colour plates p. XLV)

knowledge and understanding concerning the functionality of polymer transistors are still lacking.

1.2 What is an Organic Transistor?

A field-effect transistor (FET) basically consists of four different components: an electrically conducting material for the so called source/drain and gate contacts, an insulating material as the gate dielectric and a semiconducting material, as well as a substrate functioning as the carrier. An FET is already referred to as being "organic", when the semiconducting layer is merely composed of organic molecules or polymers, although all components may be replaced by organic materials.

Hence, the term "organic transistor" is not specifically defined, but rather serves as a generic term for a great variety of transistor concepts: from a transistor with only one organic semiconductor to a fully organic transistor [1, 12]. These various concepts can be categorised according to semiconductor material (small molecules or polymers), the share of inorganic components (substrate, insulator, electrodes) and the design (top-gate or bottom-gate electrode).

Figure 1.2 illustrates the typical design of a printed transistor: the source and drain electrodes are mounted on a polyester foil, followed by the semiconducting layer of polymer (i.e. polythiophenes), the insulating layer of polymer insulators is on top and, as the final layer, the gate electrode.

A great challenge in the production of organic transistors is to find a suitable combination of materials and solvents, which do not attack or dissolve each other.

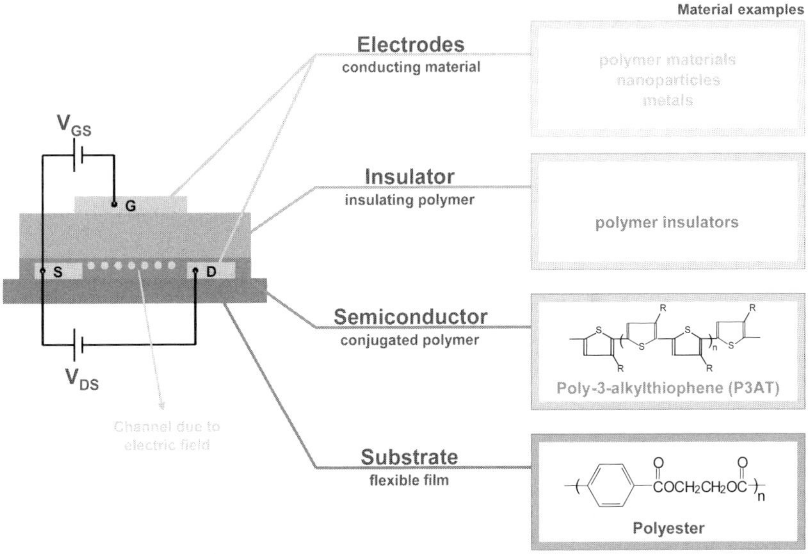

Figure 1.2 Transistor design. (see Colour plates p. XLV)

1.3 How Does an Organic Transistor Work and How Does it Distinguish Itself from a Conventional One?

The basic functionality of organic transistors is very simple and comparable to thin-film transistors (TFT) (Figure 1.2). Without a gate voltage being applied, no current flows between the source and drain electrode, since the semiconductor layer is intrinsic and, consequently, non-conducting. Once a gate voltage is applied, a very narrow conductive channel forms at the semiconductor/insulator interface due to the accumulation of charge carriers, so that current can flow from the source to drain. The current level depends on the gate voltage, which determines the number of charge carriers, as well as on the charge carrier mobility (both material properties of the semiconductor), which in turn determines the charge carrier velocity.

Thus far, only organic transistors exist, which are based on the principle of charge carrier accumulation. In contrast, inorganic transistors are based almost exclusively on the principle of charge carrier inversion, i.e. a p-doped layer is embedded between two n-doped areas (source and drain electrodes). The gate electrode creates an n-channel (= inversion) in this layer, causing current to flow from source to drain. Although charge carrier accumulation is also possible in inorganic transistors, it has the disadvantages of greater leakage or off-currents, and limits the possibilities of adjusting transistor characteristics through well controlled doping. The process of charge carrier inversion, which is more advantageous for applications in general, has not yet been observed or proved in the case of organic transistors.

The type of charge carrier is another distinctive feature. While for inorganic semiconductors the charge carrier type is adjusted by doping, in the case of organic semiconductors it is a material property. For Si or GaAs-FETs, n-type semiconductors have more favourable properties compared with p-type ones, while the opposite applies to organic semiconductors. In general, organic p-type semiconductors are capable of conducting current more quickly, and are also considerably more stable than n-type semiconductors. For this reason, most organic FETs are p-type transistors.

For all FETs the switching speed of a transistor is limited by the transit time of the charge carrier from source to drain (= channel length L). The unit of measurement for the charge carrier velocity per electric field is the mobility μ of the charge carriers. The shorter the channel length L and the greater the charge carrier mobility, the faster the transistors switch, with the channel length even being squared: maximum switching frequency $f \sim \mu * U_{ds}/L^2$ (where U_{ds} is the drain–source voltage) [13].

The channel length is limited by the process technology, while the charge carrier mobility is basically a material property, but also depends on the degree of order in the organic semiconductor. The greatest hole mobilities to date, which amount to several cm^2/Vs, were observed in vapour-deposited small organic molecules [1, 4, 14], in polymers they are typically between 10^{-4} and 0.1 cm^2/Vs (for comparison: crystalline silicon can have a hole mobility of up to 500 cm^2/Vs) [15–23].

Electrical measurements on PFETs confirm the expected drain current increase with gate voltage, the adequate saturation of the drain–source current and a high on/off ratio (Figure 1.3a). Both factors, but low off-currents in particular, are important for digital circuits. The transfer curve, i.e. the drain–source current versus the gate voltage (Figure 1.3b) results in a threshold voltage of more than +3 V. For gate voltages $V_{GS} > +3$ V, the transistor is turned off, whereas at negative voltage, it starts to become conductive. The crucial aspect is the change in current per gate voltage. This so called transconductance is determined by the charge carrier mobility and the transistor geometry. Thus, at a given geometry the charge carrier mobility can be extracted from the transconductance signal (Figure 1.3b).

1.4
Basic Logical Integrated Circuits: Ring Oscillators

The electrical performance of individual transistors is fairly good, however the goal for commercial applications is to have lots of these transistors working together in an integrated plastic circuit and executing well defined logic operations. Since inverters are simple basic logic elements based on only two transistors, they are well suited to prove the logic capability of organic FETs. In

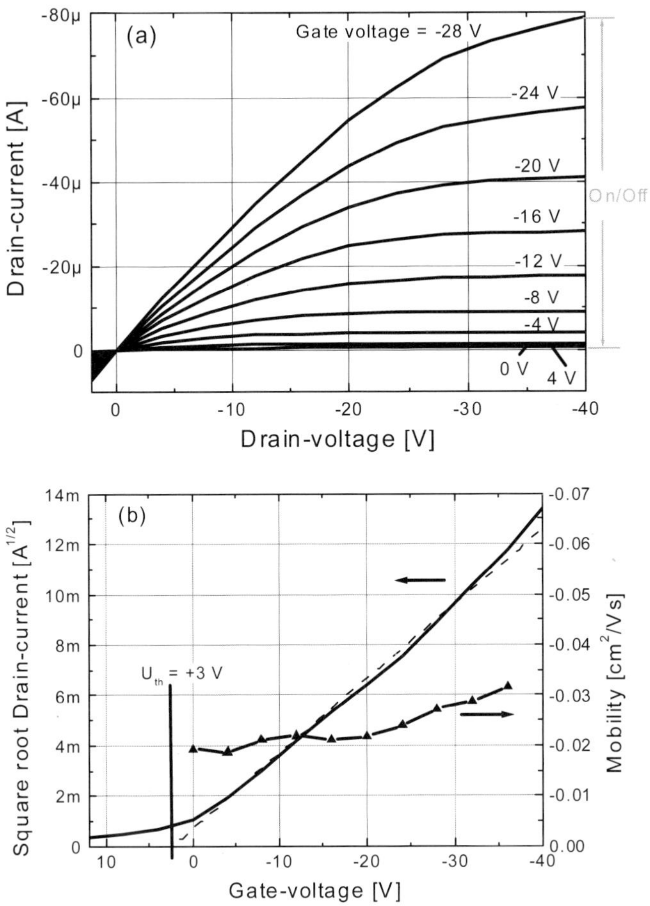

Figure 1.3 Starting and transfer curve, mobility vs. gate voltage [24].

order to build up more complex logic circuits these inverters have to show signal amplification and, most importantly, their output signal has to be able to drive a subsequent inverter stage. In particular, the last requirement can be tested with a ring oscillator. The device consists of an odd number of inverters connected with each other in series. Figure 1.4 shows the principal electronic circuit of a seven stage ring oscillator.

If a low voltage (logical "0") is applied to the input of the first inverter stage the signal is transformed into a high voltage signal (logical "1"), which is applied to the input of the second inverter. In this way the logical information passes through the complete inverter chain until a logical "1" is generated at the output of the last inverter of the chain. This signal is now coupled back to the input of the first inverter stage. If the output signal of each inverter in the oscillator circuit is able to drive the following stage the ring oscillator starts

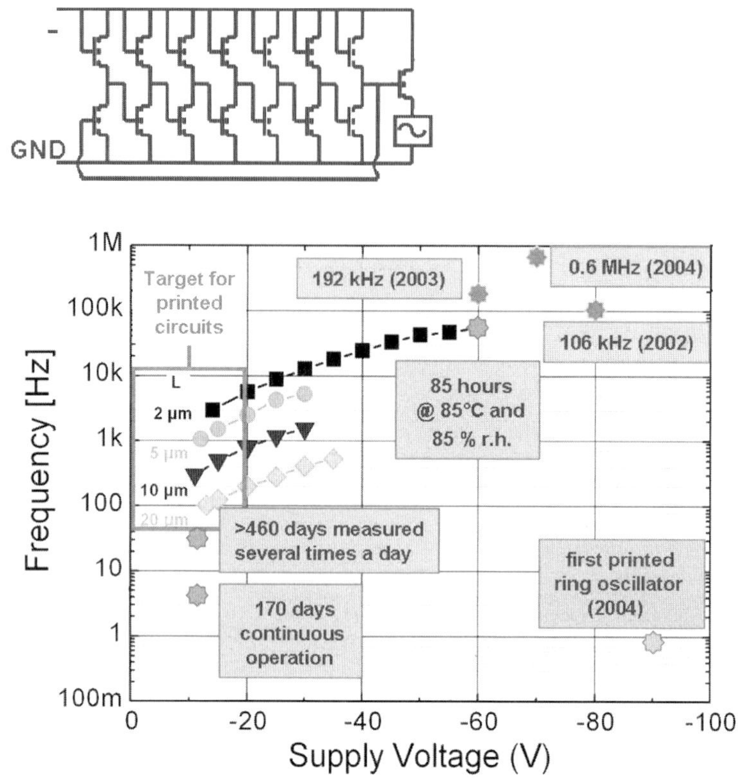

Figure 1.4 Principal electronic circuit of a seven stage ring oscillator (left) and an overview of fundamental investigations on our ring oscillators (right). *L* represents the channel length of the FETs. (see Colour plates p. XLVI)

oscillating at a certain frequency, which is directly correlated with the switching speed of the individual inverter stages. The oscillating behaviour can be visualised with an output FET, whose gate is connected to the feedback loop of the inverter chain, in combination with a fast current amplifier and an oscilloscope. A ring oscillator sets decisive standards in many aspects for a logic circuit: (i) oscillation shows the logical capability of the circuit, (ii) the onset voltage at which the ring oscillator starts oscillating denotes the minimum supply voltage required for logical elements, (iii) the oscillation frequency reflects the stage delay of the inverters, and last but not least (iv) the on/off ratio as well as the shape of the signal characterises the quality of the signal in terms of symmetry and noise, etc.

Figure 1.4 gives a brief overview of our work on integrated polymer ring oscillators based on polythiophene. As expected, a strong dependence of the oscillation frequency on the transistor channel length (*L*) and the supply voltage is observed [24]. No significant degradation could be detected even after passing a triple 85 test. Here, the device under test is stressed for a minimum of 85 h at 85 °C and 85% relative humidity. Then the circuit is dried at 60 °C for 3 h before it is measured again. Although the 2 µm ring oscillator was stressed in this way for 92 h only a slight decrease in oscillation frequency (from 56 kHz down to 47 kHz) and in the amplitude by 10% was observed [25]. In addition, the ring oscillators were still working even after more than 460 days when measured several times a day and after 170 days under continuous operation, respectively. In order to study limiting factors for device operation we further increased the supply voltage. Thus a maximum frequency of 106 kHz could be measured for the 2 µm device, which was by far the fastest polymer ring oscillator at that time. Due to improved materials and circuit design the ring oscillator frequency could be gradually increased over 192 kHz in 2003 [24, 26] and up to 0.6 MHz in the year 2004, which is still the world record for organic ring oscillators [26]. Since the signal has to pass through the seven stage inverter chain twice within one oscillation period, an oscillator frequency of 0.6 MHz results in an inverter stage delay of only 120 ns.

Also in 2004, the first completely printed ring oscillator circuit was published by our group [27]. Due to its large transistor geometries the device was working at 0.8 Hz and –90 V supply voltage. However, further improvements in printing techniques and materials make the target value for commercial applications (the marked square area in Figure 1.4) quite feasible for completely printed devices based on our standard production process.

1.5 Complex Organic Circuits: the 64-Bit RFID Tag

For complex circuits such as a 64-bit RFID Tag, a combination of different sub-circuits is necessary (Figure 1.5) [28]. The analogue part of a RFID Tag consists of an antenna with a capacitor forming a HF (13.56 MHz) resonant circuit where a rectifier is connected to transform the HF reader signal into a DC voltage. This voltage is needed as the voltage supply for the digital part of the RFID Tag. The digital circuit for 64-bit is designed in an 8-bit architecture, including a ring oscillator as the clock, a 128-bit counter and a multiplexer with an attached 64-bit WORM. As a start sequence of the 64-bit RFID Tag the first memory block output is "100...", followed by specific sequences for the remaining memory blocks. After sending 64 data-bits, the signal is "0" for the following 64-bits (sync-bits) due to the 128-bit counter, which blocks each second transponder chip signal. The minimum required supply voltage for this 64-bit RFID chip is 14 V. In the upper picture a hybrid setup of the 64-bit RFID can be seen. The lower picture shows the demodulated reader signal, receiving the 64-bit signal, followed by 64 "0" bits. This signal was transmitted at a reading distance of 3 cm (inductive coupling between reader and transponder). This proves the feasibility of polymer RFID tags.

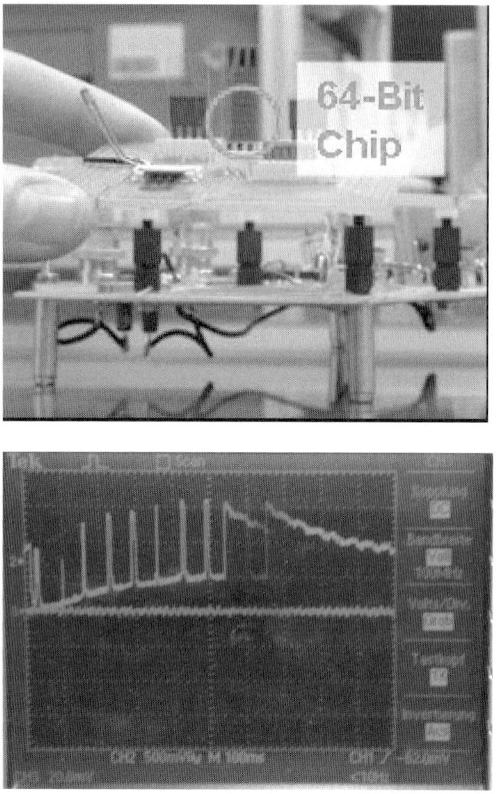

Figure 1.5 Hybrid setup of the 64-bit chip (left) and the output signal of the chip (right).
(see Colour plates p. XLVII)

1.6
Organic CMOS Circuits

Although most of the results on organic circuits published so far are based solely on p-type organic transistors [29], in recent years also organic CMOS circuits have been investigated [30]. As in silicon technology, CMOS means complementary logic, consisting of both n-type and p-type transistors. The main challenges for realising organic CMOS circuits are finding a suitable n-type transistor setup that offers stable performance under ambient conditions comparable to that of the p-type transistors and establishing a fabrication process for integrating both transistor types on the same substrate.

If both needs are satisfied, organic CMOS circuits like a seven stage ring oscillator shown in Figure 1.6 can be realised. The output signal of such a circuit depicted on the lower proves the feasibility of organic CMOS circuits. The signal shows an amplitude corresponding quite well to GND and V_{DD}, a clear sign of the robustness of organic CMOS circuits. This robustness could be the most important feature of organic CMOS circuits fabricated by printing methods, because it could overcome fluctuations within the printing process that lead to

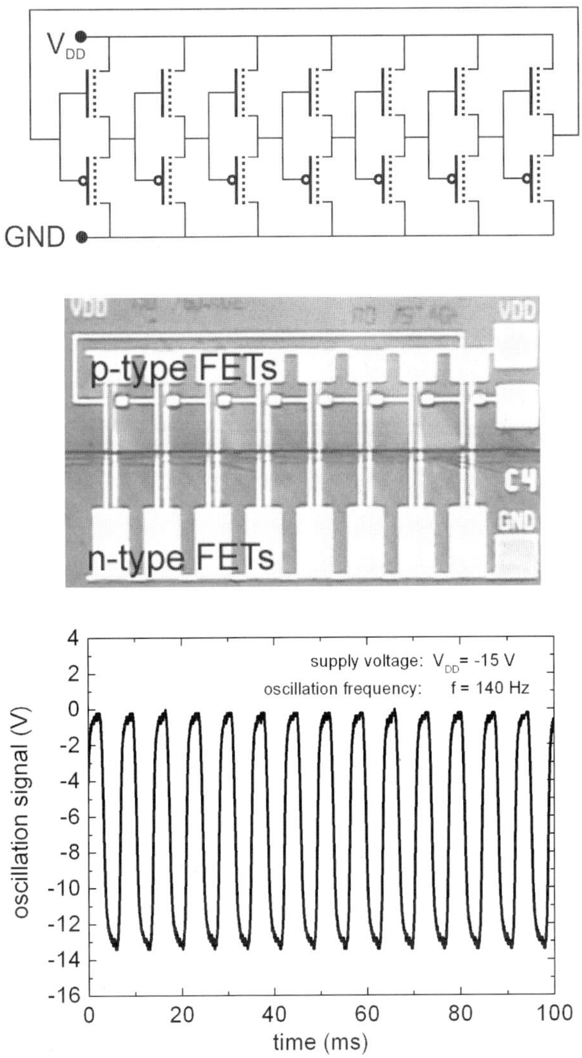

Figure 1.6 Schematic, photograph and output signal of a seven stage organic CMOS ring oscillator.

inexact feature sizes resulting in performance variations between the different transistors.

1.7 Printing Electronics

Modern printing machines are high-tech devices that have little in common with historic printing presses. Today, resolutions and register precision of 20 μm and less can be attained. In terms of printing speed, more than 500 m per minute are achieved. This would suffice to print a surface that equals the annual chip production volume of an entire silicon factory within less than an hour. This equation is simplified, of course, but the use of continuous printing processes nevertheless opens up new worlds of manufacturing electronic circuits. In principle, these printing processes can always be applied to plastic chip production, provided that soluble materials such as polymers are used. In this case, the dissolved polymer can be regarded as "electronic ink". Instead of the usual colours, conducting, semiconducting and insulating polymer ink is then printed.

Unfortunately, this process is not quite as easy as it seems at first glance. The production ("formulation") of electronic inks is difficult, since the electronic properties are easily lost due to the use of additives. These additives are necessary in order to adjust the formulations to the individual printing processes. Furthermore, conventional printing of images is optimised for visual inspection with the human eye. Thus, resolutions around 100 μm are sufficient on the one hand and, on the other, the images are composed of individual adjacent or slightly overlapping pixels, which fuse in the eye of the viewer. When printing plastic chips, completely different prerequisites need to be taken into consideration. First of all, contiguous lines are required for the drain/source electrodes with resolutions in the μm area. The semiconductor and the insulator need to be on top of each other in the form of very thin, homogenous and defect-free layers, while the gate structure needs to be aligned as precisely as possible with respect to the drain/source structures. These are tremendous requirements for both machines and materials.

Figure 1.7 Roll-printed electronics. (see Colour plates p. XLVIII)

1.8
Application and Future Prospects

RFID tags are employed for various applications and fields of use: Depending on the customer's needs, the focus is on anti-theft systems, proof of authenticity, logistics tracking or indicator functions, for which combined multifunctional tags are well within the bounds of possibility. The first products are already in use.

RFID tags also prove their worth in presence detection: be it in production, distribution and warehouse logistics or in the sales and services sector. Even when they are solely required for logistic control functions, RFID labels already include the trustworthy "certificate of authenticity". In wireless identification, seemingly perfect forgeries no longer have the chance of going undetected (Figure 1.8).

With printed polymer electronics [31] another vision could soon become a reality: Automatically appearing information symbols, which display colourful effects when dipped into the activating field, are likely to find a plethora of applications in the foreseeable future. The tremendous appeal of such visualisation aids will create entirely unprecedented and highly original applications (e.g. interactive packaging)!

One development is already a certainty for the future: item level tagging, i.e. the marking of individual goods, all the way to single yoghurt cups, will be accomplished in future development stages of this technology. For this reason, the 96-bit electronic product code™ is being developed as the replacement for the optical bar code.

Figure 1.8 RFID tags as an example of use in brand protection.
(see Colour plates p. XLVIII)

The prospect of being able to print electronics directly onto products or their packaging is even more visionary. The technical challenges that still exist are, however, also related to manufacturing aspects, given that the tagging of low-value mass products should not notably increase their price. Moreover, polymer circuits still have considerable potential in terms of material and technology.

1.9 Summary and Prospects

By employing conducting and semiconducting polymers, the technology of printed electronics opens up a vast field of novel electronic products. If the expectations regarding price and performance are met, the vision of electronics that are available everywhere could become true. Polymer electronics will not bring forth new supercomputers, but they will establish themselves in the form of products with intelligent packaging and electronic paper, all the way to plastic chips in shirts and on yoghurt cups.

However, there are still several obstacles to be overcome in order for this electronic revolution to take place. A particularly important aspect is the physical understanding of polymer transistors, especially as regards charge transport in polymer layers and the influence of interfaces on the transistor characteristics. Building on this understanding, simulation models need to be developed, which are indispensable as the basis for complex circuits. Additionally, extensive interdisciplinarity between physics, chemistry and printing technology is required, in order to push the boundaries of current printing technology, with the aim of producing the high-performance circuits from the laboratory using continuous roll-to-roll printing processes.

This goal can only be achieved as a collective effort. A new type of electronics will then be at our disposal to simplify our lives in many areas.

Acknowledgements

Special thanks go to all who contributed to the results described, as well as to the BMBF (Federal Ministry of Research and Education) and the EU, which promotes this development with funded projects.

References

1. C. D. Dimitrakopoulos, et al., BM J. Res. Dev **45**, 11 (2001).
2. H. Sirringhaus, et al., Science **280**, 1741 (1998).
3. C. M. Hart, et al., Proc. ESSCIRC, pp. 30–34 (1998).
4. H. Sirringhaus, Adv. Mater. **17**, 2411 (2005).

5. G. Cho, "All Printed RFID Tags", Proceedings of Printed Electronics Europe, Cambridge, UK (2007).
6. A. Hübler, "Mass-printed Electronics onto Markets", Proceeding of Organic Electronic Conference, Frankfurt (2007).
7. S. Burns, "Paper-like Displays Enable by Flexible AM OTFT Backplane", Proceeding of Plastic Electronics, Frankfurt (2006).
8. K. Dimmler, "Developments in printed RFID Tags", Proceedings of Organic Electronic Conference, Frankfurt (2006).
9. K. Dimmler, "Printed RFID", Proceedings of Organic RFID, San Diego (2006).
10. F. Padinger, "A New Generation of Semiconductor Foundries for Organic Electronic Devices", Proceedings of Organic Electronic Conference, Frankfurt (2006).
11. W. Fix, "Polymer Based 13 MHz RFID Transponders", Organic Electronics Conference, Frankfurt (2007).
12. W. Clemens, et al., Phys. J. **2**, 31 (2003).
13. S. M. Sze, "Physics of Semiconductor Devices" (J. Wiley & Sons, New York, 1981).
14. T. W. Kelley et al., in "Organic and Polymeric Materials and Devices", edited by P. W. M. Blom, N. C. Greenham, C. D. Dimitrakopoulos, and C. D. Frisbie (Mater. Res. Soc. Symp. Proc. 771, Warrendale, PA, 2003), p. 169, L6.5.
15. H. Sirringhaus, et al., Nature **685**, 857 (1999).
16. G. Dicker, et al., J. Chem. B **108**, 17818 (2004).
17. Z. Bao, et al., Appl. Phys. Lett. **69**, 4108 (1996).
18. A. Zen, et al., Adv. Funct. Mater. **14**, 757 (2004).
19. R. J. Kline, et al., Adv. Mater. **15**, 1519 (2003).
20. R. J. Kline, et al., Macromolecules **38**, 3312 (2005).
21. A. Ullman, et al., "High Performance Organic Field-Effect Transistors and Integrated Inverters", in Electronic, Optical and Optoelectronic Polymers and Oligomers, edited by G. E. Jabbour and N. S. Sariciftci (Mat. Res. Soc. Symp. Proc. **665**, Warrendale, PA, 2002), p. 265, C7.5.
22. H. Sirringhaus, et al., Nature **401**, 685 (1999).
23. C. Goh, et al., Apl. Phys. Lett. **86**, 233220 (2005).
24. W. Clemens, et al., J. Mat. Res. **19**, 1963 (2004).
25. J. Ficker, et al., J. Appl. Phys. **94**, 2638 (2003).
26. W. Fix, et al., "Fast and Stable Polymer Electronic Circuits", edited by J. DeMaria, Symposium on Optical Science and Technology, No. 5217-01 (SPIE, San Diego, CA, 2003).
27. A. Knobloch, et al., J. Appl. Phys. **96**, 2286 (2004).
28. A. Ullmann, et al., Proceedings of International Conference on Organic Electronics, Eindhoven (2007).
29. W. Fix, et al., Appl. Phys. Lett. **81**, 1735 (2002).
30. B. Yoo, et al., Appl. Phys. Lett. **88**, 082104 (2006).
31. M. Boehm, et al., "Printable Electronics for Polymer RFID Applications", IEEE International Solid-State Circuits Conference, Digest of Technical Paper, Session 15.1 (2006).

2
Printable Electronics: Flexibility for the Future

Mark A.M. Leenen, Volker Arning, Heiko Thiem, Jürgen Steiger, and Ralf Anselmann

2.1
Introduction

When people think about electronics, what first comes to mind are chips and circuits fabricated out of silicon. Concerning high-end applications like computer processors and memories, this will still be the case in the near future. But more and more, a new kind of electronics will enter daily life: electronics that have been mass-printed, on flexible surfaces. Small electronic games on cornflakes packages, intelligent tags on all products you buy on the internet or at the supermarket, newspapers and magazines displayed on compact, bendable E-book readers etc. are just the first examples coming to mind. Printed electronics technology has outgrown the laboratory and is emerging at this very moment. And Evonik Industries is involved in making it happen.

This chapter will start off with a view on printed electronics market forecasts and exciting new arising products. After that, several aspects of printing are addressed. Printed electronics is a field where many disciplines meet, and large-scale printing of electronic devices is not trivial. Subsequently, material requirements for printed electronics are explored extensively. Chemists are able to tune a wide variety of material properties, essential because printed electronics requires high quality, high purity materials with very specific physical properties. Throughout the chapter, special attention is given to field-effect transistors (FETs). Finally, before concluding, a view of Evonik's unique approach towards printed electronics will be presented.

2.2
Printed Electronics Market Forecasts

Several market studies, like those of IDTechEx [1, 2], clearly visualize the bright near-future for printed electronics. Currently, printed electronics is more and more finding its way from laboratories into commercial products. Over the

Organic Electronics. Structural and Electronic Properties of OFETs. Edited by Christof Wöll

ISBN: 978-3-527-40810-8

Figure 2.1 Printed silver patterns on foil. (see Colour plates p. XLIX)

last 3 years, the printed electronics market size approached 1 billion US dollars. With a forecasted market size of almost 10 billion US dollars in 2011, over 30 billion US dollars in 2015 and 96 billion US dollars in 2020, printed electronics will be a booming market.

However, as is the case with any new technology, printed electronics market entrance and growth were delayed, because of fundamental issues in this complex technology area that are taking more time and effort to solve than anticipated. OLED lighting is a good example of a technology in which fundamental problems were underestimated and market entrance and growth were expected earlier. The same trend is apparent, but less pronounced, for sensors and also for OLED display, a market that is clearly growing at this moment. Current fundamental issues are expected to be overcome shortly, and the latest printed electronics market size estimations even surpass earlier estimations, underlining the great expectations for printed electronics' near future.

Responsible for this increase in market size estimation are two market developments: first, the market size of printed photovoltaics is expected to grow faster than previously anticipated. And second, a number of new printed electronics submarkets were introduced in the latest market forecasts, which were not taken into account in earlier forecasts. The emerging of these new submarkets is a logical development of a technology that is as versatile as printed electronics, continually finding its way into more market areas.

2.3 New Products

Printed electronics will not replace silicon electronics. Instead, it will be applied in new fields inaccessible to (poly)crystalline silicon. Two main qualities of printed electronics are responsible for its ability to venture into these new application fields: flexibility, and low-cost mass electronics production by utilization of printing.

2.3.1 Advantages of Printed Electronics

(Poly)crystalline silicon is a rigid material, and the electronic properties of a silicon film strongly depend on the crystal grain size. The smaller the grains are, the more grain boundaries a charge carrier will encounter. Grain boundaries are charge carrier trap sites and they greatly limit device performance. Processes to increase grain size require high temperatures and are incompatible with flexible substrates [3]. Another disadvantage of (poly)crystalline silicon lies in the rigidity of the crystal lattice. If such a silicon film, applied on a flexible substrate, is bent, the crystalline structure is damaged and more grain boundaries will be created. This is detrimental to the electronic film properties and results in an unstable device performance.

In the form of amorphous silicon, thin flexible silicon films can be fabricated as well. Because of the absence of a crystal lattice, electronic performance of amorphous silicon is multiple orders of magnitude worse than the performance of (poly)crystalline silicon, but still high enough for applications like TFT backplanes. However, processing temperatures required to obtain a-Si films of this high performance are not compatible with flexible substrates either.

An advantage of organic materials, most notably polymers but also crystalline materials like rubrene [4], is that films of the material can be stressed without breaking. Upon bending, a layer of such a material will retain its electronic properties, because there is no covalent crystal lattice to be disturbed. Processing temperatures are compatible with flexible substrates, for the practical reason that the thermal stability of the organic materials is comparable to that of the plastic substrates. This thermal compatibility with flexible substrates gives organic electronics the opportunity to access new markets inaccessible to silicon. Devices on flexible substrates can be directly applied on for instance clothing or tent canvas.

Concerning costs, silicon electronics is and will remain an expensive technology [5]. Production of electronic-grade silicon is an expensive process, as are the subsequent vacuum evaporation and lithography steps needed to make chips out of the material. Due to the high costs, application of silicon in large-scale low-end electronics is not likely.

Printed electronics reduces the costs of produced modules, compared to silicon, because expensive production technologies are replaced by printing. Large areas can be covered in a small amount of time. Electronic performance is not as good as the performance of (poly)crystalline silicon but good enough for a whole range of applications. Printed electronics will be applied in new fields, fields for which silicon electronics has always been too expensive.

An important opportunity for printed electronics lies in the inexpensive customization of circuit design. Key to the feasibility of silicon electronics are the large volumes in which the electronics are produced. Producing small volumes of specifically designed circuits increases their price even further, since the manufacturing costs are then divided over a smaller amount of produced devices. In case of printed electronics, this price increase for smaller volumes of product is less significant, since manufacturing costs are much lower than for silicon electronics.

In the end, combining low-cost mass printing with the use of flexible substrates, electronics can be mass-produced in a roll-to-roll manner, much like newspapers are produced daily. The greatly reduced electronics production costs can bring electronics to areas where it has not been able to go before.

2.3.2
Passive Elements

A first application field for printed electronics is passive elements. An innovative company active in this field is printed systems (http://www.printed-systems.de).

Printed systems aims at printing electronically functional polymers on paper (Figure 2.2). Data stored in this polymer layer can be read with a card-reader, and the paper cover, which also protects the polymer layer, can be coloured and designed according to customer wishes (Figure 2.3). Such paper cards with electronically stored data on them can be applied in for instance marketing and entertainment.

Menippos was the first company to use the technology of printed systems in a new product, the electronic card game Hurra Fussball (http://www.hurra-fussball.de), which won the 2006 IDTechEx printed electronics award in the category of best commercialization of printed electronics.

On each card, 16 bit of information is stored, which is used in a computer game by reading the card with the card reader. More than 700,000 cards were

Figure 2.2 Printing rolls for printing electronic structures on paper (source: printed systems). (see Colour plates p. XLIX)

Figure 2.3 Functional printed electronic cards for "CROSSLINK" technology (source: printed systems). (see Colour plates p. L)

brought on the German market in a few months time. With Hurra Fussball, the printed electronics community has shown that even with existing technology, printed electronics can already be marketed successfully.

2.3.3 TFT-Backplanes

Another important application field for printable electronics are backplanes for active-matrix displays. These backplanes, which contain up to several millions of transistors, ideally will be fully printed. When flexible plastic substrates are used, the displays can be bendable and rollable while retaining their original performance. The stacked layers of materials are extremely thin, in the order of hundreds of nanometers, and therefore light-weight: ideal properties for an E-book reader.

An example of an innovative E-reader is Polymer Vision's Readius® (http://www.polymervision.com), which will enter the market by fall 2008 [6]. It features a rollable E-ink display with an organic active layer, incorporated in a mobile phone (Figure 2.4). In this way, the screen can be larger than the mobile phone itself, resulting in the mobile phone with the largest screen on the market today, while the phone itself is only standard size.

2.3.4 RFID Tags

A third important application for printed electronics is the Radio Frequency Identification tag, or RFID-tag in short. This tag, the follow-up of the barcode, is a transponder, consisting of a small chip and an antenna. The antenna is

Figure 2.4 Polymer Vision's Readius® (source: Polymer Vision).

mostly printed from metal particle inks. On the chip, small amounts of information can be written, which can be read contactlessly with a signal in the radio frequency range (13.56 MHz is a commonly applied frequency). Applying these tags to products would supply each product with its own unique code.

Each product can be tracked and identified anywhere it goes, applied for instance in luggage labelling on airports. The Dutch Schiphol airport, in cooperation with KLM, has tested the use of RFID-tags to minimize luggage loss, and decided to extend the RFID-tag use because of satisfying results [7].

Counterfeits, either without tag or with the wrong code, can be identified easily as well. In this way, for instance VIP concert tickets can be protected. Many tags can be read in a few seconds, for example at the check-out of a supermarket, where waiting in line will be history. You just move your filled shopping cart past a detector and every item in your shopping cart will have been identified at the moment you have passed through.

However, in order to make the tagging of each single product feasible, costs should be below 1 cent per tag. It will be very hard for silicon electronics to reach this price domain, but large-area mass printing is envisioned to open up this possibility.

Printed RFID-tags have already been demonstrated by companies active in the printed electronics industry. At the Organic Electronics Conference (OEC) 2007 in Frankfurt, in each conference-ticket an RFID-tag fabricated by PolyIC (http://www.polyic.com) was incorporated [8].

Recently, the Holst centre (http://www.holstcentre.com) demonstrated a 64 bit passive RFID-tag approaching technological requirements for item-level tagging [9]. The data could be read from 10 cm distance with a 780 bit/s data readout, fivefold higher as current state of the art bit rate in plastic electronics.

Summarizing, the first printed electronics products are already available and many more are at the doorstep of the world market. In the future, printed electronics will be everywhere.

Figure 2.5 Item-level tagging with fully printed RFID-tags. (see Colour plates p. L)

2.4 Printing Considerations

Printed electronics is a research area where many different disciplines meet. For example, formulation of printable inks of synthesized semiconductors, conductors or insulators requires knowledge of chemistry and physics, while printing of these materials into defined and homogeneous thin films requires printer engineering, whereas design of functional devices is done by electronics engineers, etc.

Furthermore, several different printing techniques are being adapted for printing of electronic devices, like flexography, inkjet, offset, screen and gravure printing. Printing techniques are chosen on the basis of their lateral resolution, printing speed (throughput) and material and layer requirements. Also, for each printing technique a separate ink has to be formulated, since most notably ink viscosity has to be tuned.

Earlier in this chapter, printing of electronics has been compared to roll-to-roll printing of newspapers. Indeed, this is the envisioned fabrication method for printed electronics, but there is a large difference in requirements between newspapers and electronics.

For printed electronics, large areas have to be printed with electronic functionality. This requires homogeneous layers of electronically active materials, of a defined thickness and with defined surface and interfaces. These layers have to be printed and patterned with a high resolution, to obtain as much electronic functionality out of the smallest possible area. For newspapers, the only requirement is that people can read the printed text or can see what's on a printed picture. A low resolution still suffices for such a purpose, and defects on the few micrometer scale are acceptable. Contrary to that, a defect on such a scale in an electronic device would completely erase any device functionality. If only one transistor out of the hundreds of transistors present on the chip of an RFID-tag does not work, no information can be read from the tag.

In printed electronics, multiple functional layers are printed on top of each other. Each layer needs to be positioned in certain defined structures on top of the other. For example, a gate electrode in a FET needs to be positioned between the source and drain electrodes. If the gate electrode is mispositioned, the field-effect on the channel is not optimal anymore and device performance is reduced. This so-called "registration" requirement to the roll-to-roll printing process is much more stringent for printed electronics. Thermal post-treatment of a printed device may cause the substrate and all layers on it to shrink and may influence the registration as well, further complicating production.

Considering the printable ink, choice of solvent is essential to obtain a homogeneous printed film. Also, the concentration of the material in the ink should be just right, to be sure that the required film thickness is achieved after printing. A too high concentration could lead to clogging material inside the print head (inkjet printing) or material build-up on the printing cylinder (flexography, gravure, offset) or mesh (screen printing) too.

A full treatment of all technological problems associated with printing electronics (compared to printing newspapers) would be out of the scope of this text, but it is clear that printed electronics requires high standards and resolution in printing techniques, and carefully engineered inks. With such stringent requirements, the delay in technological development and subsequent market entrance of printed electronics was unavoidable.

2.5 Materials

Next to the printing process, also materials for printed electronics have to be carefully engineered. Materials for printed electronics can be subdivided into conductors, semiconductors and dielectrics. However, it is important not to look at the single materials, but to combinations of materials with matching properties. An optimized transistor for instance has electrodes of a conductor plus a dielectric with properties that match those of the semiconductor, resulting in optimal device performance. Evonik aims to supply such system solutions for the printed electronics market.

Important to mention is that materials that show a good performance in the laboratory, subsequently have to be adapted for a larger scale production. Instead of fabricating one small device in a glovebox, large areas of functional film have to be printed under ambient conditions. Device performance of lar-

Figure 2.6 Silver 30 SN screen printing ink. (see Colour plates p. LI)

ge-scale produced devices rarely matches small-scale laboratory performance. Indeed, upscaling might be the biggest challenge of the printed electronics community.

2.5.1 Conductors

Conductors have a number of requirements to be met. If we take electrodes on a semiconductor in a FET device as an example, the work function of the electrode should be suitable for either hole (p-type) or electron (n-type) injection into the semiconductor layer. Just as important for charge carrier injection is the interface between the conductor and the semiconductor, which should be as defined as possible. Next to that, printing the conductor rather than vacuum-depositing it should still give the electrode a sufficient conductivity and stability.

Solution processable conductors come in three main classes: metals, metal oxides or organics. Printed metals, most notably as electrodes or RFID antennae, are mainly obtained by using particle-based inks. Silver or gold particle inks are commercially available. Evonik has developed silver printing pastes, such as Silver 30 SN, a formulation of silver flakes in organic solvents, optimized for screen printing (Figure 2.6).

Curing time after printing is in the order of seconds in a hot air drier. Sheet resistance of a 25 μm thick film is lower than 14 mΩ, giving a conductivity sufficient for RFID antennae (Figure 2.7).

Inks of metal precursors, containing metal atoms with organic ligands, are also used for solution deposition of conductors. Like with metal particles or flakes, a thermal post-treatment is necessary to obtain highly conductive metal films. Metal particles need to be sintered together, while in case of metal precursors the insulating organic ligands have to be removed. For flexible substrate compatibility, processing temperatures should not exceed 150 °C.

Figure 2.7 Printed RFID-antenna. (see Colour plates p. LI)

The most well-known example of conducting metal oxides is indium tin oxide (ITO). Substrates with pre-patterned ITO bottom electrodes are heavily used in device research, for instance for the fabrication of solar cells or light-emitting diodes. These pre-patterned substrates are fabricated through sputtering, which requires additional lithography and etching steps. The ITO layers have a surface roughness in the nanometer range, but have spikes too [10]. These spikes are often higher than the thickness of the layers which have been deposited on top of the ITO, increasing leakage current and the risk of short-circuits. Polishing pre-patterned substrates removes the ITO spikes, but at the cost of a further time-consuming processing step.

Evonik has developed an ITO nanopowder, VP AdNano® ITO (Figure 2.8), and has formulated this nanopowder into dispersions for antistatic coatings. Both the nanopowder and the dispersion are commercially available. Current-

Figure 2.8 ITO nanopowder and dispersion (http://www.advancednanomaterials.com). (see Colour plates p. LI)

ly, research is in progress towards printable inks based on this nanopowder, for printing structured transparent ITO electrodes. After depositing the ink, a post-processing step is still required to improve film resistance down to 100 Ω/sq, but it involves no costly lithography, etching or polishing steps. Surface roughness is around 10–15 nm for an 800 nm thick film, but contrary to sputtered ITO, no spikes are present at the surface.

Another advantage of solution-processed ITO over sputtered ITO is its low waste of indium. Indium is known to be an expensive metal, and recycling of non-deposited indium is an important contributor to indium supply. In case of solution-processed ITO, there is no significant loss of material, contrary to the sputtering process. Nevertheless, Evonik is also developing alternative, cheaper transparent conductive oxides for the printed electronics market.

Conducting polymers like commercially available PEDOT:PSS are the third class of printable conductors. However, their conductivity (maximum 500 S/cm for PEDOT:PSS [11]) is several orders of magnitude lower than the conductivity of metals. The advantages of PEDOT:PSS are its transparency, flexibility and low-temperature post-processing; the thermal treatment is only necessary to remove residual solvent, no sintering is required.

2.5.2
Dielectrics

In general, a dielectric should have a high dielectric strength. For TFT devices, a high capacitance C_i is also of importance. This means for a FET device that, at lower gate voltages, higher charge density can be induced and the threshold voltage of the device will be reduced. A high capacitance can be achieved by using a high-k material, a material with a large dielectric constant. Capacitance is also increased if the dielectric film thickness is reduced, however care must be taken that the film does not become too thin and shorts may arise. All in all, development of dielectrics for printed electronics forms an important research field for materials science and device physics.

Thermally grown SiO_2 on a silicon wafer is often used as a convenient dielectric, but this is not possible for printed electronics on flexible substrates. Instead, solution-processable insulating polymers are most suited, like PVP, PET, PP or PMMA. Evonik actively develops polyacrylates for a whole range of applications. This includes the most well-known polyacrylate, PMMA (polymethylmethacrylate, plexiglas®). PMMA is a commonly used dielectric in printed electronics. Its solubility in many common solvents and accompanying excellent processability have made PMMA into a standard in the field. Moreover, because of the large range of solvents which can be used, PMMA is very suitable for multilayer devices, in which multiple layers are solution-deposited on top of each other. It is of utmost importance that solution-depositing a layer on top of another layer does not dissolve the latter; in case of PMMA, an orthogonal solvent that does not dissolve the underlying layer can be picked from a wide collection of solvents.

PMMA's high specific resistance of 10^{15} Ω/cm secures a low leakage current through the dielectric. The dielectric constant of PMMA is around 3.0 at 100 kHz at room temperature, in range with other polymers. PMMA is not very hygroscopic: 0.3% w water is taken up by a PMMA film [12, 13]. The dielectric constant does not change over time due to water uptake, securing a constant device performance. Because PMMA films are amorphous, crystalline regions are absent, securing a homogeneous film density and a homogeneous film refractive index. Its transparency is also an advantage for display backplane applications.

Dielectrics should not be taken for granted. Using a different dielectric in a FET device will influence device performance [14]. For each semiconductor material, a certain insulator will perform better than others. Dielectrics and semiconductors are not engineered separately; it is the combination of two materials that will determine the device performance. Evonik does not only develop dielectrics and semiconductors on their own but combines them into high-performance material systems for the printed electronics market.

For dielectrics, the choice of solvent is also essential for a satisfying film morphology. The film should be homogeneous, without pinholes, and should have a smooth and defined surface. The interface between the dielectric and the semiconductor is very important, since that is where charge transport takes place in a FET. Defects should be kept at a minimum, since defects are trap sites and traps induce hysteresis, a difference in device current between a forward and a backward voltage sweep. This hysteresis can be taken advantage of in a memory device, but for FETs it should be as low as possible.

2.5.3 Semiconductors

Semiconductors come in two classes: organic (carbon based) molecules on the one hand, and inorganic compounds on the other. Both have their advantages and disadvantages; indeed, many research groups aim to combine both material classes into hybrid devices, bringing the best of both worlds together [15, 16]. Inorganic materials in general have a superior environmental stability and performance (charge carrier mobility). Organic materials on the other hand have a better processability and physical properties can be more easily tailored chemically.

For all semiconductor materials, a high purity is essential. Impurities can trap charge carriers or disturb film ordering, thereby limiting charge transport through the active layer. Furthermore, the material should be environmentally stable, meaning no degradation upon heating or irradiation, and no susceptibility towards oxidation by wet oxygen. For organic semiconductors, environmental stability requirements have been reported in an excellent article by Philips research [17].

Detailed charge carrier mobility requirements versus the resolution of relevant printing and patterning technologies are schematically shown in Figure 2.9.

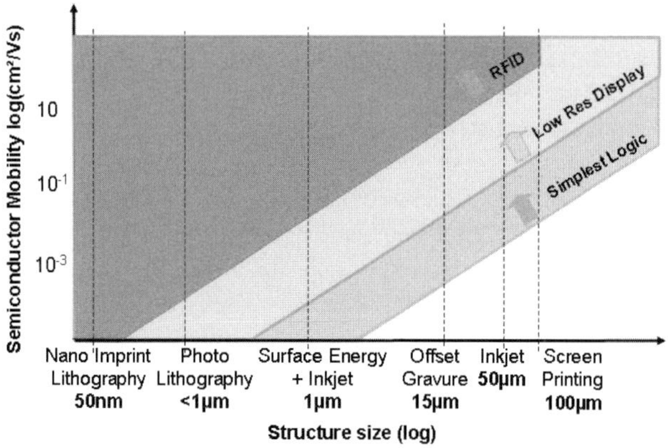

Figure 2.9 Semiconductor charge carrier mobility and printing technique resolution requirements for printed electronics. (see Colour plates p. LII)

The higher the charge carrier mobility of the semiconductor, the lower are the resolution requirements of the printing methods that can be used to print products with increasing circuit complexity. Requirements for the simplest logic, an oscillator frequency of 10 Hz, are already met by even currently available organic semiconductors (charge carrier mobility of 0.1–1 cm^2/Vs) and low-resolution printing techniques. But for a low resolution display (100 rows of pixels, 10 Hz), even the resolution of inkjet printing barely suffices when currently available organic semiconductors are applied. Requirements for RFID-tags (13.56 MHz, 128 kHz modulation frequency) are even more stringent: both high resolution patterning technologies as well as semiconductors with a high charge carrier mobility are needed.

2.5.3.1 Organic Semiconductors

Polymers are mainly known for their use as electrical insulators, but since the discovery of charge transport in organic molecules in the 1970's by Heeger, MacDiarmid and Shirakawa [18] (2000 Nobel Prize in chemistry), a tremendous amount of research into these conjugated molecules has taken place. The bulk of organic semiconductors are most suitable for hole-transport (p-type); n-type organic semiconductors tend to have environmental stability problems. Nevertheless, lately stability of n-type organic semiconductors is improved by chemical modification of most notably perylene-based molecules [19]. Research activity in n-type and ambipolar organic semiconductors has been increasing over the past years [20].

Organic p-type materials for printed electronics can be divided in two main classes: polymers and small molecules. These two material classes have different physical properties, leading to a difference in performance and processability. Small molecules in general have a better FET performance than polymers, because they are crystalline, resulting in highly ordered semiconductor films.

Charge carrier mobility is mainly limited by grain boundaries (omitting injection barriers at the electrodes). Polymers on the other hand are semicrystalline at most, and charge carrier mobility is limited by the amorphous domains. Several excellent reviews on state of the art organic p-type materials have been written [21, 22].

Currently, conjugated organic molecules are mainly applied in organic light-emitting diodes (OLEDs), organic solar cells and organic field-effect transistors (OFETs). Each device requires materials with different physical properties. For field-effect transistors, a high charge carrier mobility is key. To compete with amorphous silicon, a charge carrier mobility of at least 0.5 cm^2/Vs is desired. For some applications, like electrophoretic displays, materials with lower mobilities in the order of 10^{-2} cm^2/Vs would suffice.

2.5.3.2 **Inorganic Semiconductors**

Recently in thin film electronics research, inorganic (particulate) films have received a lot of attention [3, 23]. High charge carrier mobilities (mainly n-type) and good environmental stability make inorganic materials highly favourable for electronics applications. Their generally problematic processability, however, limits current application in printed electronics.

Inorganic particles are not soluble, but they can be dispersed in different solvents by using the right additives. These dispersions can be printed, to obtain a film of particles. Charge carriers have to be transported from particle to particle, and therefore have to cross a lot of particle interfaces, which greatly limits conductivity. Therefore, particles are thermally sintered together after solution deposition, resulting in a continuous inorganic film. In such a way, inorganic films can be solution deposited.

One example of inorganic (particulate) materials is zinc oxide (ZnO). ZnO is a non-toxic and transparent material and a good n-type semiconductor. Charge carrier mobilities obtained in FETs exceed the amorphous silicon benchmark of 1 cm^2/Vs [24, 25]. However, to obtain these high field-effect mobilities, the ZnO films have to be annealed at temperatures incompatible with flexible substrates.

The challenge in this field at the moment is to develop a process to form a film of an inorganic material with good electronic properties and a high homogeneity of the achieved performance over large areas. Of course the process should be compatible with flexible substrates, i.e. with processing temperatures that do not exceed 150 °C.

Experiments in the laboratories of Evonik towards this goal are in progress. With new material systems, charge carrier mobilities of 0.5 cm^2/Vs have been achieved in devices that were processed at ambient conditions. A typical current-voltage characteristic is shown in Figure 2.10.

On/Off ratios are in the range of 10^5–10^6, and the threshold voltage is around 15 V. In this example case, the dielectric layer thickness was 230 nm, and the device has been annealed at a temperature above 150 °C. Currently, promising efforts are underway to reduce processing temperatures below 150 °C.

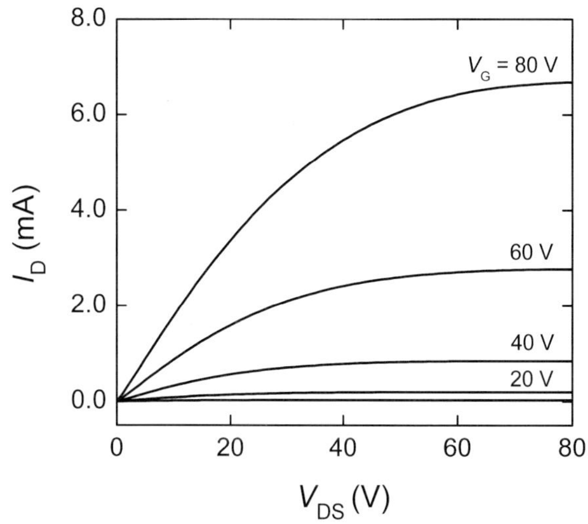

Figure 2.10 Typical current-voltage characteristic of new inorganic material system. Channel length 20 μm, channel width 10 mm.

Summarizing, both organic and inorganic semiconductors suitable for mass-printed electronics are in development. Several materials with excellent physical properties have already been synthesized and applied in printed electronic devices, but nevertheless semiconductor design and synthesis remain important research topics. There is still room for improvement of not only material performance and environmental stability (organics) or processability (inorganics); also costs of semiconductors and processing thereof still can be further optimized.

2.6 Creavis Science-to-Business Approach

Printed electronics is a good example of a young and complex market. Before being able to successfully enter this market, necessary fundamental knowledge and technological experience has to be established. Therefore, a large amount of valuable time has to be invested first before one can actively participate in a new market.

Creavis, part of Evonik, has its own unique approach towards the fast and successful entrance of markets that are relatively new to Evonik. The time between an invention and the corresponding market-ready end-product is reduced tremendously, by housing closely collaborating fundamental and applied research, as well as marketing and sales, under one roof in the science-to-business centre. Creavis' close collaboration with universities and industry along the entire value chain is also an essential contribution to the fast development of new products.

Starting from Evonik's base competencies in synthesis, formulation and production of fine chemicals, closely related new markets are explored. Additional fundamental and process know-how necessary for the new market is established, and process upscaling is considered in an early stadium as well. Evonik has profound knowledge of both producing high-quality chemicals, and formulating these chemicals into functional dispersions. Entry into the printed electronics market by development of functional materials and formulation of high-quality inks is a logical expansion of its applied research.

Research is conducted in-house, as well as by a multitude of scientific cooperation partners, like universities and research institutes. Industrial partners include innovative companies like printed systems. Creavis participates actively in multiple European and BMBF-financed projects. The Nanomatch project (http://www.nanomatch.eu), in which new hybrid devices are developed, and the MaDriX project (http://www.madrix-projekt.de) that focuses on the development of printed RFID-tags, are just two examples. Next to such projects, Creavis is also a member of international research and development networks, like the Organic Electronics Association (OEA) and the Dutch Polymer Institute (DPI). Membership of these networks enables Creavis to keep a close eye on the market, in order to respond as fast as possible to new trends and demands.

As a result, its science-to-business approach and its large scientific and industrial network have enabled Creavis to grow into an active participant in the printed electronics market today.

2.7 Conclusion

Currently the printed electronics field is in an exciting time in which first products, like Hurra Fussball, have been introduced on the market and many more, like the Readius®, are following. Estimations of the printed electronics market continue to predict a multi billion dollar market size within less than 5 years.

Printing electronics is a multidisciplinary field that requires high standards in material quality, ink, printer and device engineering. Many fundamental problems have already been solved, but still many remain as well, underlining the need of future fundamental research in all facets of printing electronics.

Key to the success of printed electronics will be high-performance materials. Printable conductors with high conductivities, environmentally stable, high-performance semiconductors and matching dielectrics have to be further developed to increase the potential of printed electronics. These materials have to be formulated into inks, suitable for high-resolution and fast printing techniques. Device architectures have to be optimized to obtain as much functionality out of the least circuit complexity. In this way, printed electronics will find its way in all facets of daily life.

Based in an excellent scientific and industrial network and utilizing the advantageous concentration of all R&D activities and resources along the entire value chain under one roof, Evonik aims be an important player in the upcoming printed electronics market, as a supplier of high performance, high quality materials systems and basic electronic components.

Acknowledgements

The authors thank the European Commission for financing through the Human Potential Programme (Marie-Curie RTN NANOMATCH, Grant No. MRTN-CT-2006-035884) and the BMBF for financing through the MaDriX project. We also greatly acknowledge the European Union and the state of Northrhinewestphalia for financial support.

References

1. IDTechEx, Organic & Printed Electronics Forecasts, Players & Opportunities 2007–2027 (2007).
2. IDTechEx, Organic Electronics Forecasts, Players, Opportunities 2005–2025 (2005).
3. Y. Sun, and J. A. Rogers, Adv. Mater. **19**, 1897 (2007).
4. A. L. Briseno, R. J. Tseng, M.-M. Ling, E. H. L. Falcao, Y. Yang, F. Wudl, and Z. Bao, Adv. Mater. **18**, 2320 (2006).
5. IDTechEx, Printed electronics vs. Silicon, Electronicsweekly.com, 21 March (2007).
6. Polymer Vision press release, 22.01.08.
7. http://www.nu.nl/news/769929/32/rss/KLM_rust_bagage_uit_met_chips.html.
8. PolyIC Press release, 25.09.07.
9. Holst Centre press release, 05.02.08.
10. K.-B. Kim, Y.-H. Tak, H.-G. Park, K.-H. Lee, and J.-R. Lee, Mater. Res. Soc. Symp. Proc. **708**, 89 (2002); Thin Solid Films **411**, 12 (2002).
11. http://www.clevios.com.
12. for an overview, see S. Gross, D. Camozzo, V. di Noto, L. Armelao, and E. Tondello, European Polymer Journal **43**, 673 (2007).
13. http://www.azom.com/details.asp?ArticleID=788.
14. J. Veres, S. Ogier, G. Lloyd, and D. M. de Leeuw, Chem. Mater. **16**, 4543 (2004).
15. D. B. Mitzi, C. D. Dimitrakopoulos, J. Rosner, D. R. Medeiros, Z. Xu, and C. Noyan, Adv. Mater. **14**, 1772 (2002).
16. H. Nakanotani, M. Yahiro, C. Adachi, and K. Yano, Appl. Phys. Lett. **90**, 262104 (2007).
17. D. M. de Leeuw, M. M. J. Simenon, A. R. Brown, and R. E. F. Einerhand, Synth. Met. **87**, 53 (1997).
18. www.nobel.se/chemistry/laureates/2000/public.html
    H. Shirakawa, Angew. Chem. Int. Ed. **40**, 2574 (2001).
    A.G. MacDiarmid, Angew. Chem. Int. Ed. **40**, 2581 (2001).
    A. J. Heeger, Angew. Chem. Int. Ed. **40**, 2591 (2001).
19. B. A. Jones, A. Facchetti, M. R. Wasielewski, and T. J. Marks, J. Am. Chem. Soc. **129**, 15259 (2007).
20. J. Zaumseil, and H. Sirringhaus, Chem. Rev. **107**, 1296 (2007).
21. J. E. Anthony, Chem. Rev. **106**, 5028 (2006).

22. S. Allard, M. Forster, B. Souharce, H. Thiem, and U. Scherf, Angew. Chem. **47**, published online March 20th (2008).

23. J. J. Schneider, and J. Engstler, Nachrichten aus der Chemie **55**, 634 (2007).

24. D. Redinger, and V. Subramanian, IEEE Transactions on electronic devices **54**, 1301 (2007).

25. B. S. Ong, C. Li, Y. Li, Y. Wu, and R. Loutfy, J. Am. Chem. Soc. **129**, 2750 (2007).

Section II
Molecular Compounds

Organic Electronics. Structural and Electronic Properties of OFETs. Edited by Christof Wöll

ISBN: 978-3-527-40810-8

3
Fluorinated Phthalocyanines as Molecular Semiconductor Thin Films

H. Brinkmann, C. Kelting, S. Makarov, O. Tsaryova, G. Schnurpfeil, D. Wöhrle, and D. Schlettwein

3.1
Introduction

Phthalocyanines (Pc) show a long standing record of interest in both basic research and applications referring to their electrical and photoelectrical properties [1]. The main technical application in this context is their use as photoconducting pigment particles in electrophotographic printing applied in most of the present photocopiers and laser printers. Because of the high extent of planarity in the molecules and because of additional binding interactions of the hetero(N)atoms and the central metal ions, phthalocyanines readily form aggregates [2], crystalline particles [3] and thin films [4, 5].

Although the charge carrier mobility in phthalocyanine thin films is typically low, crystals could be grown that showed a field-effect mobility of up to 1 $cm^2 V^{-1} s^{-1}$ [6]. Phthalocyanines are of considerable interest because of the numerous possibilities of chemical modifications both of the central group and different substituents at the ligand [7]. Chemical modification leads to characteristic changes in the molecular structure opening the possibility of systematic tuning of the microstructure in the solid state [2–5]. The preferred crystal structure, e.g. is most directly modified by choice of the central metal ion, in particular by an additional axial ligand, like the O-atom in phthalocyanines with VO, TiO, etc. or the halogen atom in phthalocyanines with InCl, AlF, etc. as central groups, but also by the presence of chemical substituents of H-atoms in the ligand of phthalocyanines. F-atoms have turned out to be most efficient in this context, leading to characteristically different microstructures in films of incompletely ($F_{14.5}Pc$) [8, 9] or completely ("per") fluorinated phthalocyanines ($F_{16}Pc$) [10] when compared to unsubstituted Pc. For thin evaporated films the choice of substrate (metal, inorganic oxide insulator, organic polymer insulator) and the evaporation conditions of the films are decisive parameters for the film microstructure [8–14]. Not only the crystal structure but also the growth mode of thin films of $F_{16}Pc$ is altered significantly by the presence of

Organic Electronics. Structural and Electronic Properties of OFETs. Edited by Christof Wöll

ISBN: 978-3-527-40810-8

the F-atoms [14–16], leading to the formation of conductive layers in the monolayer thickness range. A clear influence of the interface with the substrate on the observed current was also reported for other organic thin films [17]. Thickness-dependent conduction measurements are a valuable tool to analyse such effects [14–18] and distinguish between the formation of conductive layers with a characteristic increase in the monolayer regime [15, 18] or the filling of interface traps characterised by a current independent on film thickness [17, 19].

Aside from a pronounced influence of chemical modifications on the microstructure and intermolecular coupling in thin films, also the position of energy levels in phthalocyanines can be shifted over a wide range of about 1.5 eV when electron-donating or electron-withdrawing substituents or heteroatoms are present in the ligand. This was shown for a number of phthalocyanines in molecular orbital calculations of individual molecules [20, 21], by experimental shifts of the electrochemical potential of dissolved molecules in solution [22] or by shifts of the molecular ionisation energy obtained by photoelectron spectroscopy for molecules in the gas phase [23]. Such changes for isolated molecules are also leading to corresponding shifts in the electrochemical redox potential of phthalocyanine thin films in contact to an inert electrolyte [24, 25], or in the ionisation potential obtained by photoelectron spectroscopy [23, 26–29] or the electron affinity obtained by inverse photoemission spectroscopy [30–33] at thin phthalocyanine films. Although solid-state effects are superimposed on the molecular changes in these experiments (intercalation of counterions in the electrochemical studies; polarisation effects in the photoelectron and inverse photoemission spectroscopy studies) leading to additional spectral broadening and shifts, the trends of individual molecules are clearly preserved enabling also quantitative correlations for thin films. In these studies, $F_{16}Pc$ are characterised by a strong stabilisation of more than 1 eV for occupied and unoccupied electronic levels relative to the unsubstituted Pc.

The combination of well-defined tunable crystal structures and strongly stabilised electronic energy levels in $F_{16}Pc$ explains their rather good success as n-type molecular semiconductor thin films with electron mobilities for the copper complex $F_{16}PcCu$ of 0.03 $cm^2\ V^{-1}\ s^{-1}$ or 0.08 $cm^2\ V^{-1}\ s^{-1}$ in optimised thin films in OFET structures, even stable under air [34, 35]. $F_{16}Pc$ are also interesting because of moderate intermolecular coupling energies large enough to allow a high electron mobility, but small enough to allow tuning of the crystal structure in solid films by choice of substrate and preparation conditions [5, 10–16, 34–36]. A rather small lattice energy was also seen in the strong dependence of the optical absorption spectrum on the temperature of films [36].

In this contribution we go a step further and include recent work on partly fluorinated phthalocyanines F_4Pc and F_8Pc (Figure 3.1) in the discussion to explore further the possibilities that fluorination of phthalocyanines offers in the structural and energetic tuning possibilities in n-conducting phthalocyanine films and to fill the gap both in structural and energetic respect between traditional unsubstituted Pc and perfluorinated $F_{16}Pc$ thin films for applications as semiconducting channels in OFET. The Zn complexes were chosen as a typical representative of divalent phthalocyanines and for comparison purposes with earlier electrochemical and spectroscopic work [23–27].

3.2 Experimental

3.2.1 Chemical Synthesis

Phthalonitrile (Aldrich), 4,5-dichlorophthalonitrile (Aldrich), 4-fluorophthalonitrile (Lancaster) and tetrafluorophthalonitrile (Aldrich) were sublimed for purification. Zinc(II) acetate (Aldrich; water free, 99.99%), potassium fluoride (Aldrich), 18-crown-6 (Fluka) and 1,8-diazabicyclo[5.4.0]undec-7-ene (DBU, Aldrich) purchased in the highest available purity were used without further purification. The solvents used for the preparations (reagent grade) were dried, distilled and stored under dry conditions. All syntheses were carried out under high purity and dry nitrogen.

3.2.1.1 **Phthalocyaninato Zinc(II) (PcZn)**

PcZn was prepared after a slightly modified known procedure [37]: DBU (0.76 g, 5 mmol) was added to a mixture of sublimed phthalonitrile (0.64 g, 5 mmol) and dry zinc(II) acetate (0.23 g, 1.25 mmol) in dry n-pentanol (20 mL). The mixture was heated for 36 h under reflux. Methanol (50 mL) was added. The isolated product was washed with water and methanol, and then treated with methanol in a Soxhlet apparatus overnight. Yield 0.54 g (75%). The dried PcZn was then purified by zone sublimation at 10^{-7}–10^{-6} mbar and 370 °C. UV/Vis (thf): $\lambda_{max} = 666$ nm. MS (DCI, negative, NH_3): *m*/*z* 576 (M^-).

3.2.1.2 **2,29,20,2-Tetrafluorophthalocyaninato Zinc(II) (F_4PcZn)**

4-Fluorophthalonitrile (0.5 g, 3.4 mmol) and dry zinc(II) acetate (0.17 g, 0.94 mmol) in 1-chloronaphthalene (10 mL) were heated under reflux for 20 h. The cold reaction mixture was diluted with methanol (40 mL) and stirred for 1 h. The blue coloured product was isolated, washed at first with methanol and then with DMF followed by precipitation of the partially soluble phthalocyanine by addition of methanol. Following filtration the product was again washed with methanol and dried at 150 °C i. vac. Yield 0.25 g (45%). The blue-reddish F_4PcZn was purified by zone sublimation at 10^{-7}–10^{-6} mbar (350 °C) and obtained as a mixture of regioisomers. UV/Vis (thf): $\lambda_{max} = 660$ nm. MS (DCI, negative, NH_3): *m*/*z* 648 (M^-).

3.2.1.3 **4,5-Difluorophthalonitrile**

4,5-Dichlorophthalonitrile (Lancaster; 1 g, 5 mmol), potassium fluoride (3 g, 50 mmol; dried i. vac. over P_2O_5 at 100 °C over night) and 18-crown-6 (1.4 g, 5 mmol) in DMF (6 mL) were stirred at 100 °C for 3 h. The conversion was controlled by TLC. After cooling the reaction mixture was poured into water. The isolated filtrate was washed with water, dried and sublimed i. vac. at 100 °C. Yield 0.45 g (54%) of white crystals. TLC ($CHCl_3/SiO_2$) showed traces of 4-chloro-5-fluoro-phthalonitrile which were removed by repeating the reaction of the obtained phthalonitrile (0.42 g, 2.5 mmol) with KF (0.3 g, 5 mmol) and 18-crown-6 (0.14 g, 0.55 mmol) in DMF (1.2 mL) in the same manner as before. ^1H NMR (200 MHz, $CDCl_3$, TMS): 7.70 (*t*, 7.9 Hz) ppm.

3.2.1.4 **2,29,20,2-,3,39,30,3-Octafluorophthalo-cyaninato Zinc(II) (F_8PcZn)**

4,5-Difluorophthalonitrile (0.66 g, 4 mmol) and dry zinc(II) acetate (0.2 g, 1.1 mmol) in 1-chloronaphthalene (10 mL) were heated under reflux for 20 h. The cold reaction mixture was diluted with methanol (40 mL) and stirred for 1 h. The blue coloured product was isolated, washed at first with methanol and then with DMF followed by precipitation of the partially soluble phthalocyanine by adding methanol. After filtration the product was again washed with methanol and dried at 150 °C i. vac. Yield 0.45 g (62%). The blue-reddish F_8PcZn was purified by zone sublimation at 10^{-7}–10^{-6} mbar and 370 °C. UV/Vis (thf): λ_{max} = 652 nm. MS (DCI, negative, NH_3): *m/z* 720 (M^-).

3.2.1.5 **1,19,10,1-,2,29,20,2-,3,39,30,3-,4,49,40,4--Hexadecafluoro-phthalocyaninato Zinc(II)**

($F_{16}PcZn$) $F_{16}PcZn$ was obtained after a slightly modified known procedure [38]. Sublimed tetrafluorophthalonitrile (0.48 g, 2.4 mmol) and dry zinc(II) acetate (0.110 g, 0.6 mmol) was intensively mixed in a mortar. The mixture was filled in a glass vessel, three times flushed with nitrogen and vacuum, and finally the glass ampoule was sealed under vacuum. After heating for 1 h at 180–190 °C the blue product was isolated and washed with acetone and petrol ether (bp. 140 °C; 800 mL). The product was treated in a Soxhlet apparatus at first with water and then with petrol ether to remove impurities. Yield 0.45 g (85%). The dried $F_{16}PcZn$ was purified by zone sublimation at 10^{-7}–10^{-6} mbar and 350 °C. UV/Vis (thf): λ_{max} = 668 nm. MS (DCI, negative, NH_3): *m/z* 864 (M^-).

3.2.2 **Calculation of Energy Levels**

Semiempirical molecular orbital (MO) calculations of individual molecules of zinc(II) phthalocyanines PcZn, F_4PcZn, F_8PcZn and $F_{16}PcZn$ (metal with closed shell electron configuration) in vacuum were carried out with the commercially available Hyperchem program, Release 7.0 in the RHF (Restricted Hartree-Fock) approximation using the PM 3 (Parameterised Method 3; a reparameterisation of AM1 = Austin Model 1) parameter set. For all compounds a full geometry optimisation was carried out using PM3 until a gradient of less than 0.001 kcal/(Å mol) was reached. Based on this structure single point calculations were performed by using the PM3 method (convergence limit less than 0.001; RHF; next lowest state) to yield the appropriate positions of the highest occupied molecular orbitals (HOMO) and lowest unoccupied molecular orbitals (LUMO).

3.2.3
Thin Film Preparation and Measurements

Glass substrates (Menzel Gläser, Germany) were purified by washing with acetone (Rotipuran, 99.98%). Polyimide (PI) films for the growth experiments were prepared by spin-coating two layers of a precursor solution of poly(3,3′,4,4′-biphenyltetracarboxylic dianhydride-co-1,4-phenylenediamine) in 1-methylpyrrolidone (Aldrich 43118-4) with a thickness of 1.1 μm each. After deposition the polyimide received baking at 300 °C in a nitrogen atmosphere to complete cross-linking and evaporate residual solvent. For the electrical conduction measurements 100 nm thick silver electrodes were Ar^+-RF-sputtered (Leybold Z 400, 10^{-3} mbar, 1500 V) onto these substrates using a shadow mask to leave a gap of 40 μm × 10 mm. For the field-effect – measurements, thicker PI films were prepared on ITO-coated glass (Flachglas, Germany, 250 Ω/□) to utilise ITO as gate electrode with evaporated Au electrodes also leaving a gap of 40 μm × 10 mm. The electrodes were contacted by low temperature soldering and mounted in the deposition chamber. For the optical transmission experiments the substrates were used without metal contacts.

PcZn, F_4PcZn, F_8PcZn and F_{16}PcZn purified by zone sublimation as described above were evaporated from BN crucibles (Kurt J. Lesker) resistively heated by a Ta wire on the substrates held at 313 K under high vacuum conditions (base pressure 10^{-8} mbar). The deposited amount was monitored by changes in the resonance frequency of a quartz crystal microbalance (Sunny 30, 03-41(A)) mounted next to the substrate and the thickness equivalent calculated based on a density of 1.63 g cm^{-3} for PcZn calculated from the crystallographic data [3]. The evaporation rates were controlled by adjusting the heating current and were adjusted to 0.1–0.4 nm min^{-1}.

Conductance measurements were performed during film growth at 10 V applied voltage corresponding to a field of 2500 V cm^{-1} with a Keithley 487 picoammeter for films grown on glass and a Keithley 617 electrometer for films grown on PI (higher sensitivity needed). Following deposition of the molecular semiconductor the development of current along the thin film was measured at 298 K with elapsing time without breaking the vacuum. Subsequent gas dosing was performed with oxygen (Air Liquide, 99.995%) through a precision leak valve.

Field-effect measurements were performed with a Keithley 487 picoammeter which also acted as the source-drain voltage source and a Grundig PN 300 voltage source to apply the gate voltage. The domains of positive or negative drain voltage were measured independently. A leak current less than 10^{-10} A was measured through the polyimide layer between source and gate contact or between drain and gate contact independently. In the presented FET data an experimental shift in the drain voltage of about 1.3 V was observed and corrected in Figure 3.16.

UV/Vis absorption spectra were measured in transmission by use of a Tec5 Evaluation Line spectrometer during deposition. For the measurements the films were briefly rotated out of the molecular deposition beam and perpendicular to the light path of an optical fibre mounted to the vacuum system.

3.3 Results and Discussion

3.3.1 Synthesis and Molecular Characterisation

Zinc(II) phthalocyanines either unsubstituted or with different numbers of fluorine atoms as substituents were employed for the investigations: PcZn, F_4PcZn, F_8PcZn, F_{16}PcZn (Figure 3.1). The synthesis of PcZn and F_{16}PcZn was described before [37, 38]. In order to obtain phthalocyanines of high purity the preparation procedures had to be slightly modified. All phthalocyanines were obtained by cyclotetramerisation of the corresponding fluoro-substituted phthalonitriles with carefully dried zinc(II) acetate. The commonly used method for the preparation of PcZn from the unsubstituted phthalonitrile

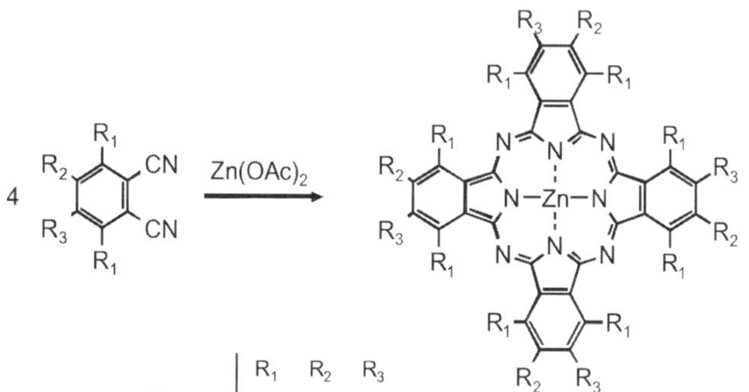

	R_1	R_2	R_3
PcZn	-H	-H	-H
F_4PcZn	-H	-F	-H
F_8PcZn	-H	-F	-F
F_{16}PcZn	-F	-F	-F

Figure 3.1 Synthesis and structure of the employed zinc(II) phthalocyanines.

(with DBU in pentanol) cannot be used for the fluorinated phthalonitriles. Under the basic conditions pentanolate will substitute the fluoro groups. Therefore the cyclotetramerisations of the fluoro-substituted phthalonitriles were either carried out in mass or in an inert solvent at higher temperatures. All phthalocyanines were purified by zone sublimation at 350–370 °C under vacuum. Purity was controlled by DCI mass spectra and electronic spectra in solution. The high purity was estimated from measurements of the total ion current and single ion traces in the mass spectra for all samples. No hints for impurities were observed. Such a high purity is essential for the determination of relevant electric properties.

The positions of HOMO/LUMO energy levels of single molecules of the zinc(II) phthalocyanines PcZn, F_4PcZn, F_8PcZn and $F_{16}PcZn$ were determined by the commercially available Hyperchem Program, Release 7.0. As expected, the energy calculated for the HOMO and LUMO shifts to lower energy (more negative values) with increasing electron-withdrawing properties [39, 40] which means with increasing number of fluoro substituents (Figure 3.2). The order of increasing negative values of HOMO/LUMO (lower energy) is as follows: PcZn > F_4PcZn > F_8PcZn > $F_{16}PcZn$. We note that the frontier orbital gap remains almost constant at 1.8 eV according to these calculations.

Ultraviolet photoelectron spectroscopy (UPS) and inverse photoemission spectroscopy (IPES) confirmed the energetic staircase that was calculated for the individual molecules to be also valid in thin films (Table 3.1). Although small differences in the absolute positions of energy levels are obvious when the different sets of experiments are compared the general trend as calculated in Figure 3.1 is clearly confirmed in the different experiments on the different substrates. In particular it is interesting to note that the energy gap between the highest occupied and lowest unoccupied electronic levels is quite constant for

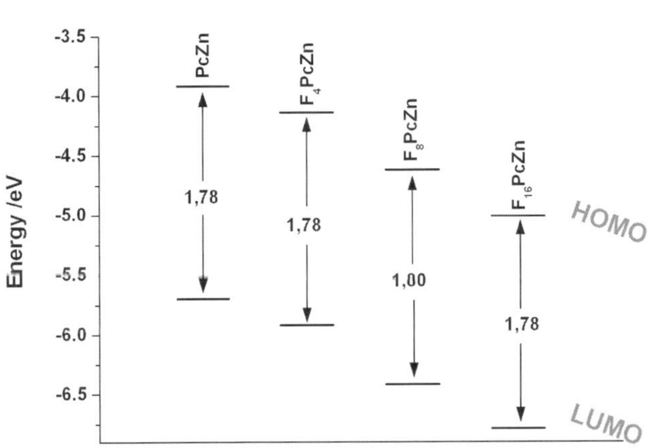

Figure 3.2 Influence of successive fluorination on the frontier energy levels of individual molecules of zinc(II) phthalocyanines calculated on the semiempirical level [26, 27].

Table 3.1 Experimental positions of highest occupied electronic levels (IP, determined by UPS) and lowest unoccupied electronic levels (EA, determined by IPES) in thin films of partly fluorinated phthalocyanines.

material	IP (UPS, onset)	EA (IPES, onset)	References
PcZn[1]	4.8_8		[23]
F_{16}PcZn[1]	6.2_8		[23]
PcZn[2]	4.8_5		[29]
F_4PcZn[2]	5.0_4		[29]
F_8PcZn[2]	5.0_3		[29]
F_{16}PcZn[2]	6.0_1		[29]
PcCu[1]	5.0_5		[28]
F_4PcCu[1]	5.7_0		[28]
F_{16}PcCu[1]	6.1_1		[28]
F_{16}PcCu[1]	6.3	4.8	[30]
PcCu[2]	4.8_2	2.6_5	[33]
F_4PcCu[2]	5.5_5	3.6_0	[33]
F_{16}PcCu[2]	6.3_2	4.5_2	[33]
PcCu[1]	5.2_0	3.1_6	[31]
F_8PcCu[1]	6.0_6	3.9_1	[31]
F_{16}PcCu[1]	6.3_9	4.4_6	[31]

[1]on Au; [2]on Si.

the differently fluorinated phthalocyanines and was well predicted even quantitatively by the semiempirical calculations of the first excited electronic state of the molecules. Since both orbitals are predominantly C2p in character, the influence of the central metal is negligible, which is also confirmed by the constant UPS values when the Cu and Zn complexes are compared. It is therefore a fair assumption that the LUMO level also in thin films of the Zn complexes follows the sequence PcZn > F_4PcZn > F_8PcZn > F_{16}PcZn as calculated for the individual molecules.

3.3.2
Thin Evaporated Films of Zinc(II) Phthalocyanines with a Different Degree of Fluorination

Thin films of PcZn, F_4PcZn, F_8PcZn, and F_{16}PcZn were evaporated in high vacuum on either glass or PI on glass with evaporated Ag electrodes. The current across the insulating 40 μm gap was measured during deposition at a small electric field of 2500 V cm^{-1} to detect the formation of electrical pathways and hence to distinguish between layer formation and island growth. Since changes of the conduction were also observed subsequent to film growth in some of the present experiments, the current was measured continuously for an extended period of time while the films were still kept in high vacuum at 298 K.

In the present series of experiments, i.e. on the smooth glass or PI surfaces we found both, island growth beyond a percolation limit or the formation of an initial conduction layer in the monolayer thickness range, or a combination of these two characteristics. For PcZn on both glass or PI such a combination was found, with an initial peak of the current in the monolayer range (specific conductivity $\sigma = 7 \times 10^{-6}$ S cm^{-1} at 5 nm on glass and $\sigma = 2 \times 10^{-5}$ S cm^{-1} at 3 nm on PI) speaking for a conductive thin film formed initially, which then stabilised to a linear increase with increasing amount of deposited PcZn, leading to $\sigma = 1 \times 10^{-6}$ S cm^{-1} at 50 nm on both glass and PI [41]. Subsequent to deposition, the conductance of the PcZn films on glass increased only slightly by 8% within 14 h on glass and almost insignificantly on PI indicating only little mobility of the PcZn molecules in these films at room temperature [41].

Figure 3.3 shows typical results for the growth of F_4PcZn on glass and PI. The measured current was clearly smaller when compared to the growth of PcZn. The conduction resulted from percolating islands ($\sigma = 2 \times 10^{-9}$ S cm^{-1} at 50 nm on glass) and an initial conductive layer was not detected. Similar characteristics were observed during growth on PI. A percolation limit at larger av-

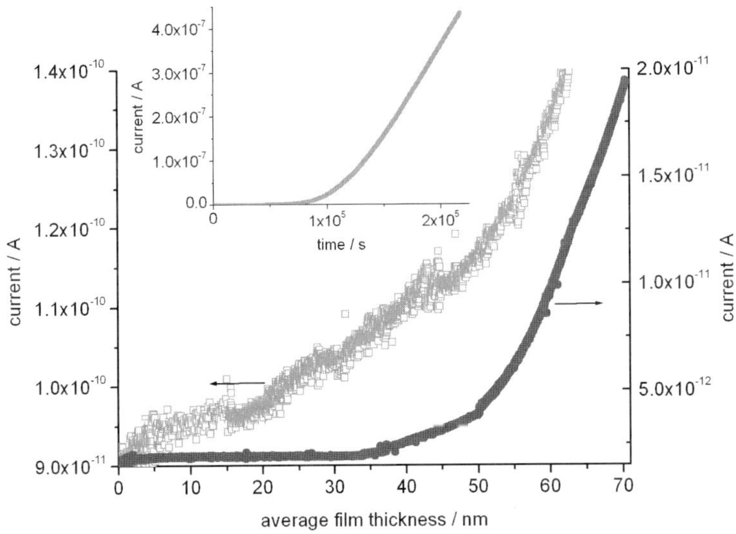

Figure 3.3 Current measured during deposition of F_4PcZn on glass (□, left axis) and on PI (●, right axis) at 313 K. The inset shows the change of current across a 53 nm thin F_4PcZn film on glass at 298 K following the deposition and kept at 9×10^{-7} mbar.

erage film thickness and a further decreased current ($\sigma = 2 \times 10^{-10}$ S cm^{-1} at 50 nm) was determined. Following the deposition we observed an increase of the current and the current was therefore monitored during storage at 298 K in high vacuum. The inset of Figure 3.3 shows the results at F_4PcZn on glass. The increase occurred slowly in the beginning but at an increased rate after about one day (86400 s) reaching a $\sigma = 3.3 \times 10^{-5}$ S cm^{-1} at 50 nm, which is 4 orders of magnitude larger when compared to the freshly deposited film. We assign this change to a ripening of the films since interaction with contaminations from the rest gas (see below) would lead to an opposite change.

The F_4PcZn films consisted of well-defined particular domains of about 70 × 20 nm size, considerably smaller than those of PcZn (Figure 3.4) but well connected to allow electrical conduction along the film.

F_8PcZn showed similar growth characteristics and comparable percolation limits as F_4PcZn at, however, slightly higher currents pointing at a higher specific conductivity of $\sigma = 3 \times 10^{-7}$ S cm^{-1} at 50 nm on glass and $\sigma = 8 \times 10^{-9}$ S cm^{-1} at 50 nm on PI (Figure 3.5). Subsequent to the deposition on glass the current was further monitored (see inset) and showed an initial decrease followed by a late and slow increase, which compared with F_4PcZn was quite insignificant.

For films of F_8PcZn the transition from a non-conducting film up to 15 nm and a rather well-conducting film beyond can also be seen in the comparison

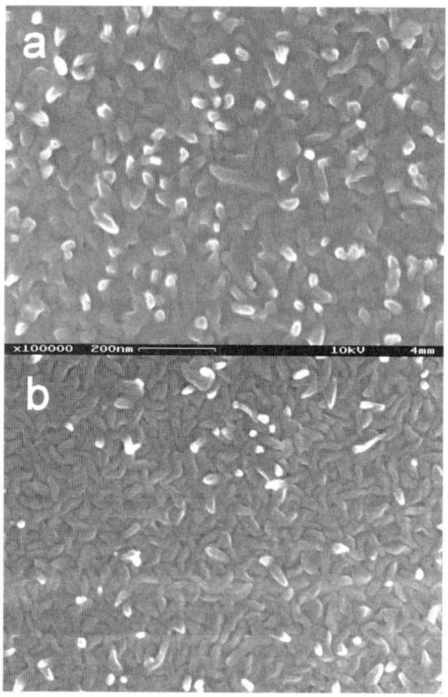

Figure 3.4 Scanning electron microscopy images of 50 nm thin films of PcZn (a) and F_4PcZn (b) on glass.

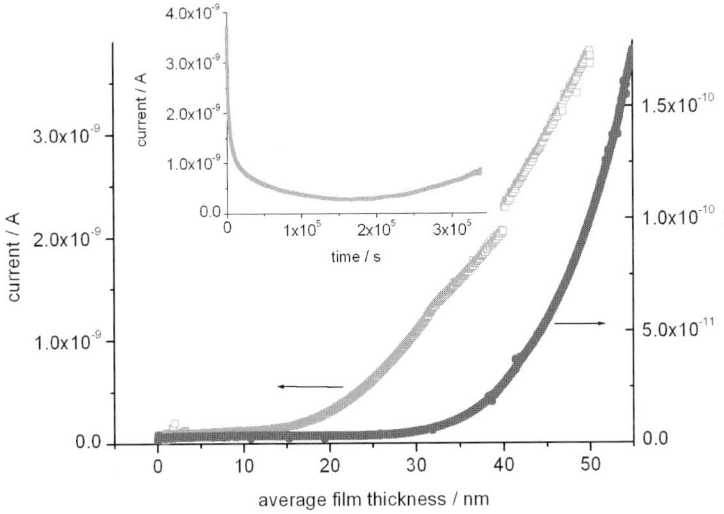

Figure 3.5 Current measured during deposition of F_8PcZn on glass (□, left axis) and on PI (●, right axis) at 313 K. The inset shows the change of current across a 50 nm thin F_8PcZn film on glass at 298 K following the deposition and kept at 9×10^{-7} mbar.

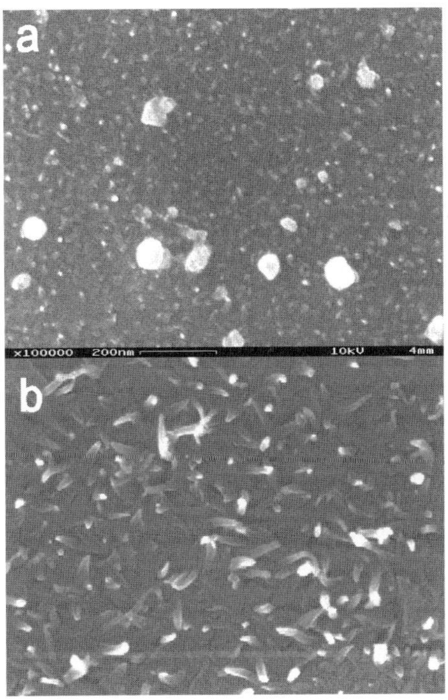

Figure 3.6 Scanning electron microscopy images of 15 nm (a) and 50 nm (b) F_8PcZn on glass.

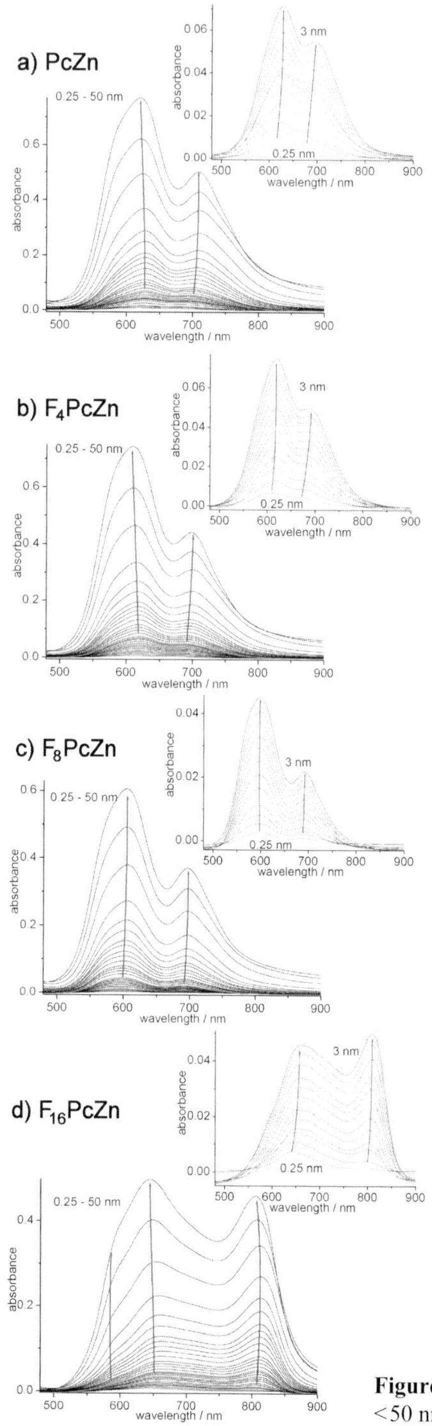

Figure 3.7 Optical absorbance during growth <50 nm on glass.

of SEM images of these two conditions (Figure 3.6). While a film of 15 nm average film thickness consisted of small particles (or upright oriented needles) of around 20 nm diameter, the thicker film showed better interconnected needles of at least 100 nm length, explaining the higher specific conductivity. Also seen is the high mobility of the molecules since the large block-like particles in the thinner film were transformed to the needles in the thicker film. A quite dynamic situation seems to exist even at temperatures of 298–313 K.

In parallel experiments the film growth on glass or PI was analyzed by optical transmission spectroscopy to probe the intermolecular coupling of transition dipoles. This allows an analysis of the film microstructure from the monolayer range up to dense films. In Figures 3.7 and 3.8 spectra in the range of the characteristic Q-band of phthalocyanines are depicted for the growth of PcZn, F_4PcZn, F_8PcZn, and $F_{16}PcZn$ on glass or PI. Although clear differences were observed in the splitting pattern especially for thin films in the monolayer range and up to 3 nm, these differences do not account for the large differences observed in σ since the thicker films, e.g. average film thickness 50 nm of PcZn, F_4PcZn, and F_8PcZn show identical splitting patterns and hence almost identical band shape of a typical Pc α crystal structure [4]. The characteristically different spectra in the monolayer range, however, can be used to discuss the different growth characteristics observed on PI when compared with glass.

On glass (Figure 3.7) the spectra in the sub-monolayer range already closely resemble those of thicker films and the main contributions to the α-band are already present. At 3 nm average film thickness the spectra of PcZn, F_4PcZn, and F_8PcZn were composed of the two bands contributing to the α-structure but at different intensity ratios. Beyond 3 nm only subtle shifts were observed but the intensity ratio approached that of the α-structure. At 50 nm average film thickness the spectra of PcZn, F_4PcZn, and F_8PcZn were almost undistinguishable. Only the shoulder on the high energy side of the band (570 nm) was developed to a different extent.

On PI on the other hand (Figure 3.8), the optical absorbance of PcZn, F_4PcZn, and F_8PcZn in the sub-monolayer range showed a contribution from uncoupled monomers characterised by a narrow absorption band at 670–680 nm. Clearly different spectra were observed at 3 nm average film thickness. Films of F_4PcZn and F_8PcZn on PI showed a band at 800 nm which was not detected for PcZn on PI nor for any of the three on glass (see, however spectra of $F_{16}PcZn$ discussed below). Such different coupling of transition dipoles points at a new molecular arrangement of fluorinated phthalocyanines on PI. In the thickness range between 3 nm and 50 nm significantly larger shifts were observed on PI than on glass. At 50 nm a spectrum almost identical to that on glass was obtained for PcZn, but not for F_4PcZn and F_8PcZn for which the band at 800 nm developed as a persistent contribution to optical absorbance additional to the bands of the α-structure.

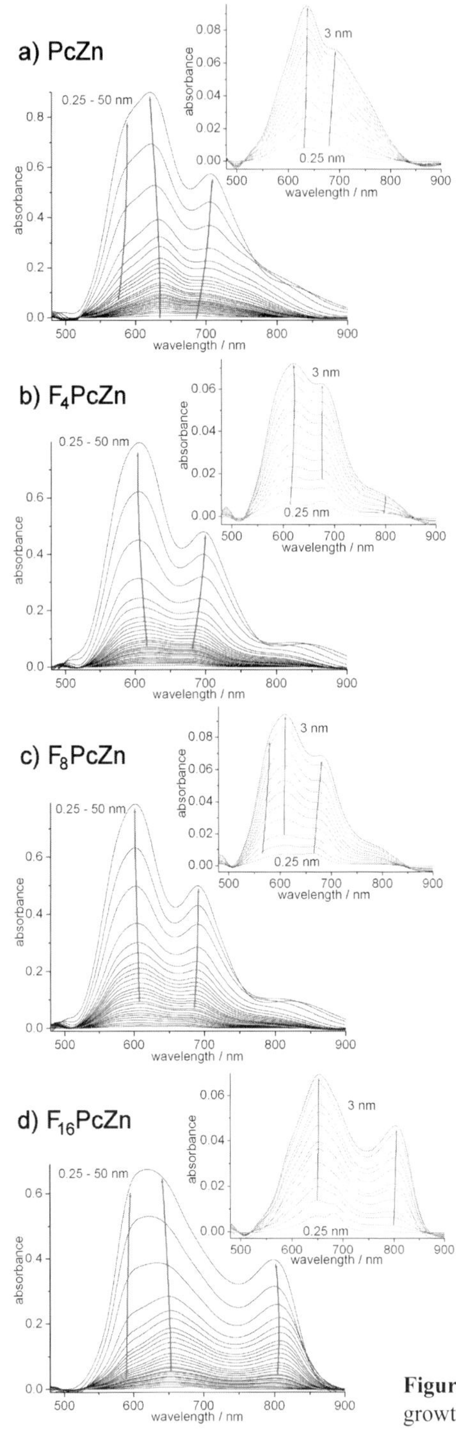

Figure 3.8 Optical absorbance during growth < 50 nm on PI.

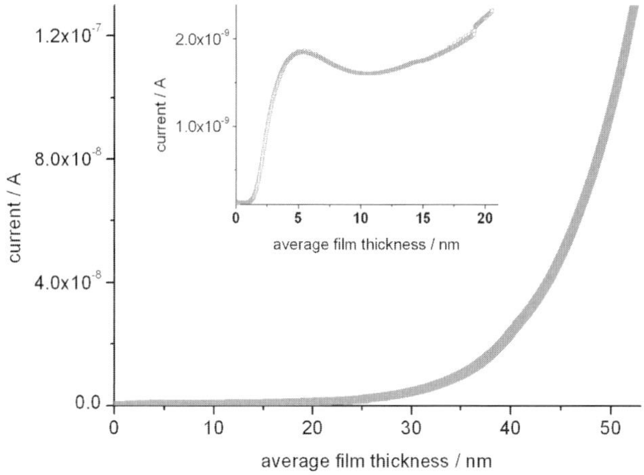

Figure 3.9 Current measured during deposition of $F_{16}PcZn$ on glass. The inset shows details in the early stage of deposition.

3.3.3 Growth of $F_{16}PcZn$ Thin Films

The current along growing films of $F_{16}PcZn$ on glass is dominated by the percolation of islands starting at about 20 nm average film thickness (Figure 3.9) leading to $\sigma = 8 \times 10^{-6}$ S cm^{-1} at 50 nm. A small contribution from a conductive layer intermediately formed around 5 nm average film thickness can be detected at high sensitivity (inset of Figure 3.9) with $\sigma = 1 \times 10^{-6}$ S cm^{-1} at 5 nm. Such ultra thin conductive layers were reported also earlier for the deposition of $F_{16}Pc$ [14–16] and have also directly been monitored for the growth of PTCDA (perylene tetracarboxilic acid dianhydride) on Ag(111) by means of photoelectron emission microscopy (PEEM) [42].

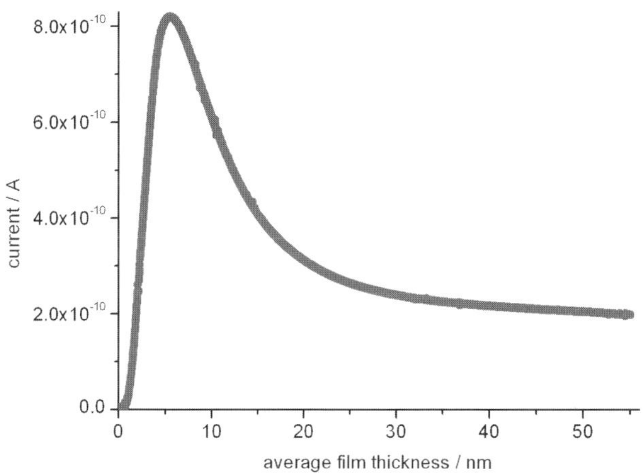

Figure 3.10 Current measured during deposition of $F_{16}PcZn$ on PI.

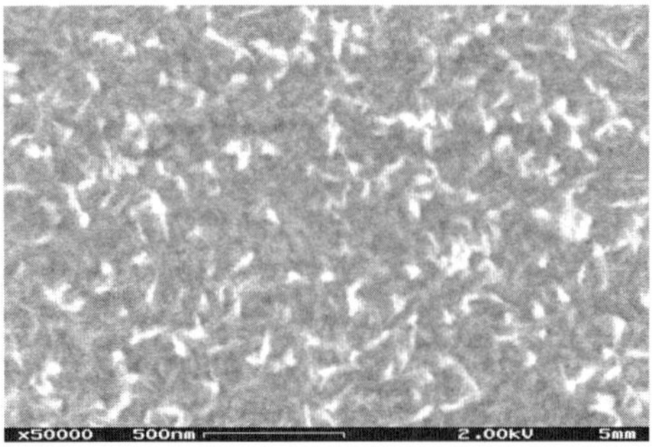

Figure 3.11 Scanning electron microscopy image of 50 nm $F_{16}PcZn$ on glass.

Ultra thin conductive layers clearly dominate the current across thin films of $F_{16}PcZn$ when deposited on PI (Figure 3.10) with a similar conductivity of $\sigma = 6 \times 10^{-7}$ S cm^{-1} at 5 nm. A lack of percolating islands, however, leads to a drop in $\sigma = 2 \times 10^{-8}$ S cm^{-1} at 50 nm.

The higher currents and hence specific conductivity found for $F_{16}PcZn$ on glass compared to F_4PcZn or F_8PcZn are consistent with a well connected array of crystalline needles as depicted in Figure 3.11.

The different growth and conduction characteristics of $F_{16}PcZn$ when compared to the other fluorinated Pc are accompanied by a characteristically different molecular order in the $F_{16}PcZn$ films and hence different intermolecular coupling. These differences and also the difference between the growth of $F_{16}PcZn$ on glass or PI can be followed in the optical absorbance data (Figures 3.7d and 3.8d). The spectra of the perfluorinated Pc thin films show a characteristically different splitting pattern of the Q-band speaking in favour of a different crystal structure [15].

Spectra of $F_{16}PcZn$ on glass differ from those on PI, most significantly in the lower thickness regime up to 3 nm, but also beyond and up to 50 nm. The band at 800 nm which had turned out earlier to be characteristic for a coplanar head-to-tail crystalline arrangement of $F_{16}Pc$ molecules [10, 15] is found more pronounced on glass than on PI. For submonolayer coverage, however, a band shifted to higher transition energy (640 nm) characteristic for a cofacial arrangement of $F_{16}Pc$ molecules [10] was detected before the band structure of solid $F_{16}PcZn$ developed. On PI, on the other hand a monomer absorption is clearly observed at submonolayer coverage and is clearly detected at least up to 3 nm average film thickness.

A common feature among $F_{16}PcZn$ and the partially fluorinated F_4PcZn and F_8PcZn can only be seen in the contribution of a band at 800 nm for films grown on PI.

3.3.4
Response to Oxygen from Air

The response of the electrical conduction in organic semiconductor thin films to an exposure to oxygen gas is of interest since interaction with oxygen can only be avoided under optimum preparation conditions and complete sealing of devices. To decide if these costly measures are needed the response of materials to oxygen should be known. Further, the response to oxygen is of general interest to discuss the conduction type of a material. Oxygen has proven a small and easily diffusing electron acceptor molecule which is interacting with a number of organic semiconductor thin films, increasing the hole concentration in p-conductors by dopant interactions or decreasing the electron concentration in n-conductors by compensation reactions.

In Figure 3.12 the characteristics are shown during oxygen exposure of a PcZn thin film. Before the expected increase of current was detected caused by p-type doping of the film according to $PcZn + O_2 \rightarrow PcZn^+ + O_2^-$, a significant decrease of the current was observed [41]. This decrease was assigned to a compensation reaction with ionised donator sites present in the film since deposition. Such n-type doping of presumably clean phthalocyanine films under the chemically reducing conditions of high vacuum had been reported earlier for thin films of other unsubstituted phthalocyanines (TiOPc and PcMn [43, 44]) or single crystals of FePc or CuPc [45] and was also accompanied by a change from n- to p-conductance caused by oxygen. For simple unsubstituted phthalocyanine thin films like PcZn, however, such characteristics were reported in [41] for the first time.

Quite contrasting characteristics were found for the fluorinated phthalocyanines F_4PcZn, F_8PcZn and $F_{16}PcZn$, for which oxygen consistently induced a decrease of conductivity at, however, different extents depending on the degree of fluorination.

The introduction of 4 fluorine atoms (F_4PcZn, Figure 3.13) was sufficient to suppress the clear increase in the current that was observed for PcZn under extensive exposure to oxygen (Figure 3.12).

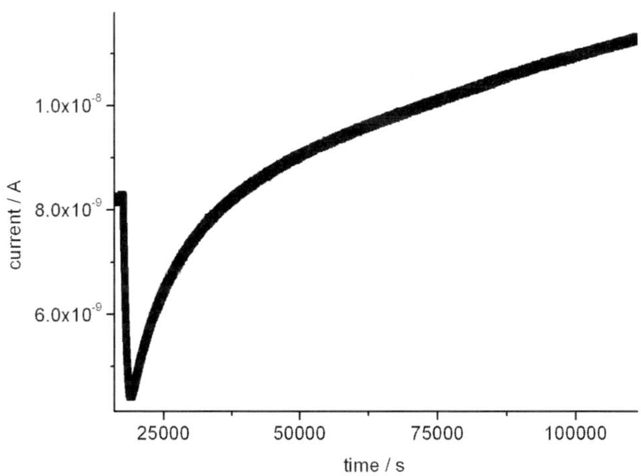

Figure 3.12 Current during exposure of a 50 nm thin PcZn film on glass to a partial pressure of 0.1 mbar oxygen (reproduced from [41] with permission).

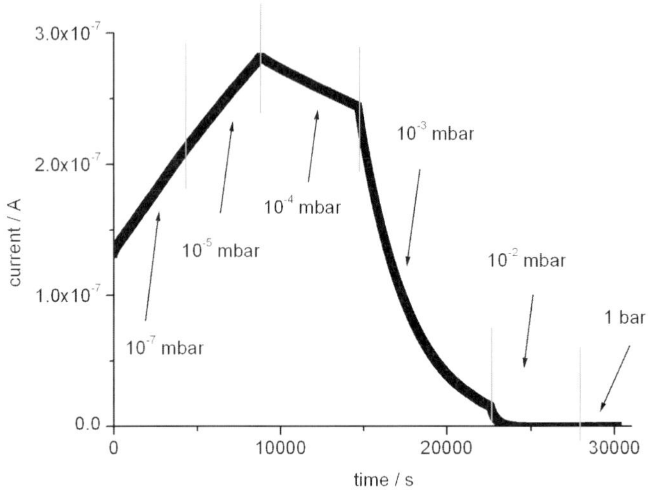

Figure 3.13 Current of a 50 nm thin F_4PcZn film on glass during exposure to oxygen.

The increase of conductivity for pressures smaller than 10^{-4} mbar is explained by the ongoing ripening process already depicted in the inset of Figure 3.3. At 10^{-4} mbar the current decreased and a partial pressure of 10^{-3} mbar led to almost complete compensation of excess electrons in the films and hence almost complete suppression of the conductivity. At normal pressure any conducting electrons are trapped by O_2 and therefore no current could be measured.

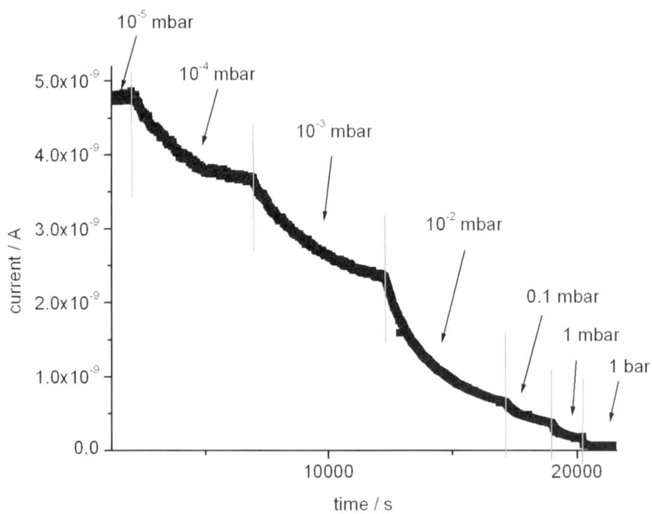

Figure 3.14 Current of a ripened 50 nm thin F_8PcZn film on glass during exposure to oxygen.

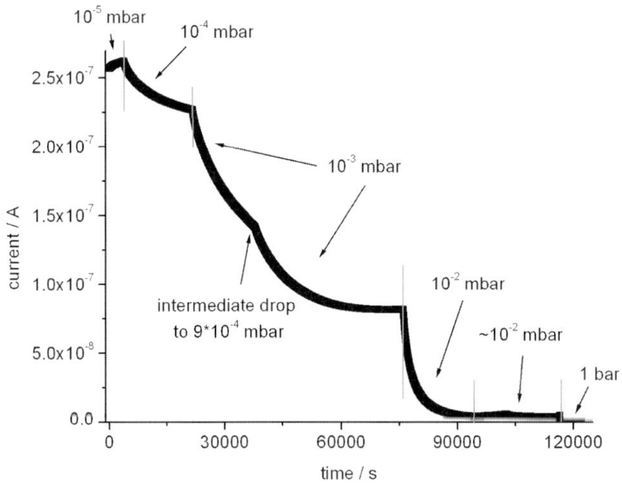

Figure 3.15 Current of a 50 nm thin $F_{16}PcZn$ film on glass during exposure to oxygen.

A stabilised n-conduction was observed for F_8PcZn (Figure 3.14) and $F_{16}PcZn$ (Figure 3.15) following each step of ex-posure. Whereas F_4PcZn thin films showed almost no saturation at different oxygen pressures, F_8PcZn and $F_{16}PcZn$ thin films showed clear saturation at different applied oxygen pressures. This leads to adjustable conductivities at intermediate partial pressures. Considerably larger partial pressures close to 1 bar were needed to completely suppress the conductivity of $F_{16}PcZn$ and F_8PcZn.

At oxygen pressures close to atmospheric pressure, the fluorinated phthalocyanines F_4PcZn, F_8PcZn and $F_{16}PcZn$ had all lost their conductivity to values below the detection limit of the present experiments. If the materials turn out to be of practical value as semiconductors, complete sealing of the devices appears essential.

3.3.5 Measurements of the Field Effect

Films of $F_{16}PcZn$ on PI turned out to be most useful as candidates for OFET measurements among the compounds in the present series of experiments. $F_{16}PcZn$ showed one of the highest specific conductivities. Superior growth characteristics led to conductive layers in the monolayer range (Figure 3.10). The latter finding is of particular relevance in OFET studies since the interface with the dielectric layer represents the active part of the semiconductor film. Compared to earlier studies [34], however, films of $F_{16}PcZn$ proved to be less stable against contaminations by air than films of $F_{16}PcCu$. Therefore, the following field effect measurements on PI as a dielectric material were performed with $F_{16}PcCu$ as semiconductor thin film. Further, films of $F_{16}PcCu$ already in vacuum showed significantly larger specific conductivities in our present studies when compared with films of F_{16}PcZn. On glass we measured $\sigma = 1 \times 10^{-3}$ S cm^{-1} at 5 nm and $\sigma = 1 \times 10^{-4}$ S cm^{-1} at 40 nm (factor of 50 and 1000 relative to F_{16}PcZn). On PI $\sigma = 1 \times 10^{-6}$ S cm^{-1} was observed at 5 nm and $\sigma = 6 \times 10^{-8}$ S cm^{-1} at 50 nm (factor of 2 and 3 relative to F_{16}PcZn).

Figure 3.16 shows the results of current-voltage measurements across the channel (40 μm) of a 40 nm thin evaporated F_{16}PcCu film on PI as a gate dielectric with an ITO layer beneath PI as gate electrode.

Positive applied gate voltages led to increased drain currents by an accumulation of excess electrons and negative gate voltages led to decreased drain currents by a depletion of excess electrons, clearly confirming the n-type conduction in F_{16}PcCu films even at air. Leak currents through the PI gate insulator of less than 10^{-10} A (Section 3.2.3) proved insignificant on the current scale of Figure 3.16 at applied drain voltage, and quite consistent with the observed values at zero drain voltage. Saturation of the drain current with the drain voltage was not reached. The increasing slope of the drain current at higher drain voltages points at space charge limitation. The slopes of the plots in the corresponding transfer characteristics (see inset for $V_{ds} = 9.3$ V) were used as a measure of the transconductance and to determine the charge carrier mobility as proposed for this case [46]. An effective electron mobility of $\mu = 2 \times 10^{-2}$ cm^2 V^{-1} s^{-1} was thereby determined from the present experiments. This value of μ compares well with values up to $\mu = 8 \times 10^{-2}$ cm^2 V^{-1} s^{-1} in earlier

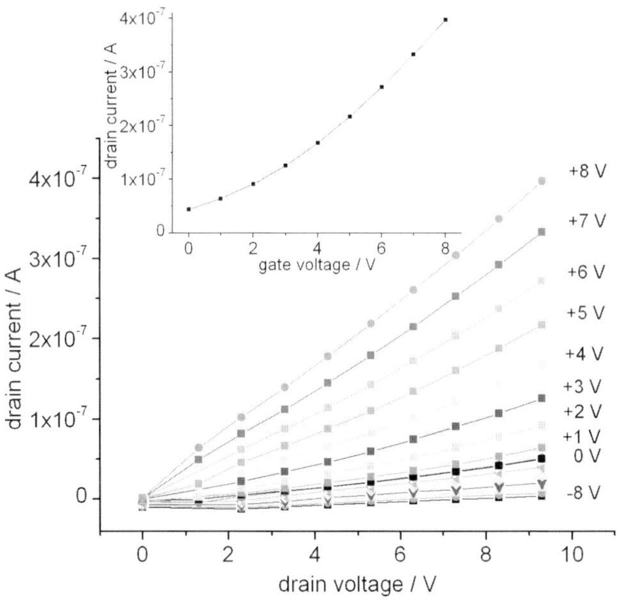

Figure 3.16 Output characteristics of a 40 nm evaporated F_{16}PcCu film between Au electrodes on PI as gate dielectric measured at air. The gate voltage is given as a label on the plots. The inset shows the transfer characteristics at a drain voltage $V_{ds} = 9.3$ V.

experiments at $F_{16}PcCu$ evaporated on Si-based devices with optimised contact formation [34, 35]. Now that we have reached similar values on polymer insulators efficient air-stable all-organic n-type OFET can be developed based on a phthalocyanine semiconductor.

3.4 Conclusions

Unsubstituted, partly fluorinated and perfluorinated zinc(II) phthalocyanines were prepared and carefully purified by zone sublimation. Quantum chemical calculations of single molecules and a comparison to UPS/IPES measurements of thin films clearly showed that HOMO and LUMO energy levels shift to lower energy with increasing degree of fluorination: PcZn > F_4PcZn > F_8PcZn > $F_{16}PcZn$.

Thin films of PcZn, F_4PcZn, F_8PcZn and $F_{16}PcZn$ showed a systematic influence of the degree of fluorination on the electrical and optical characteristics. The optical spectra revealed quite constant intermolecular coupling patterns for films of PcZn, F_4PcZn and F_8PcZn, only the perfluorinated $F_{16}PcZn$ showed a characteristically different pattern. When studied in different thickness regimes films on glass showed constant spectral characteristics of crystalline films whereas the spectral characteristics for films on PI changed with increasing film thickness. Monomeric species were detected for films in the monolayer range speaking for embedding of phthalocyanine molecules in the polymer film followed by crystal formation for thicker films.

Film growth for PcZn and $F_{16}PcZn$ showed a clear contribution of a conductive layer in the monolayer thickness regime reaching $\sigma \cong 10^{-5}$ S cm^{-1} for PcZn and $\sigma \cong 10^{-6}$ S cm^{-1} for $F_{16}PcZn$. The growth of F_4PcZn and F_8PcZn was dominated by island formation and significantly lower values of σ were reached. Subsequent to deposition the electrical conductance increased significantly although the samples were kept in high vacuum and at ambient temperature. Such changes were observed to the largest extent at films of F_4PcZn, which had shown by far the smallest σ as prepared. These observed changes following deposition indicate ripening reactions of the films even at room temperature. Since these changes have not been observed earlier when films were studied on thermal SiO_x following a photolithographic preparation we conclude that in the present case on the smoother glass and PI surfaces metastable films were formed and that, also due to the lower lattice energy of the fluorinated phthalocyanines, molecules were mobile enough to form stable microstructures subsequently, even at room temperature.

The change from p- to n-conduction was already observed following the first step of fluorination, from PcZn to F_4PcZn. If ambipolar transport in phthalocyanines is aimed at, unsubstituted phthalocyanines seem to be the best choice [47] since hole conduction could not be achieved for any of the presently studied fluorinated phthalocyanines F_4PcZn, F_8PcZn or $F_{16}PcZn$ even upon exposure to the strong and readily reacting acceptor oxygen. The rather small stabilisation of energy levels already in F_4PcZn by about 0.2–0.3 eV relative to PcZn seems sufficient to hinder oxidation of F_4PcZn molecules by oxygen molecules.

Dopant interactions that lead to n-type conduction of F_4PcZn, F_8PcZn and $F_{16}PcZn$ in vacuum, however, are efficiently compensated by interaction with oxygen leading to significant decrease of the conductivity in air. Films of F_4PcZn proved to be most sensitive and the conductivity was suppressed completely, whereas films of F_8PcZn and $F_{16}PcZn$ were reduced by donors stable enough not to be oxidised by oxygen. A residual conductivity was kept up to significantly higher partial pressures of oxygen. Films of $F_{16}PcCu$ provided an even higher conductivity and superior stability of the dopant interaction against oxygen and therefore seem the best choice as candidate for phthalocyanine-based n-type OFET. In a test device with PI as gate dielectric a mobility was reached that open the door for the development of all-organic air-stable OFET with n-type channels.

Acknowledgements

The authors are grateful to the groups of D. R. T. Zahn (TU Chemnitz) and W. Jaegermann (TU Darmstadt) for their cooperation in photoelectron and inverse photoemission spectroscopy as well as to DFG for continuous financial support of this joint work (DFG Schl340/4 and Wo237/32).

References

1. P. Gregory, J. Porphyrins Phthalocyanines **4**, 432 (2000).

2. A. W. Snow, in: The Porphyrin Handbook, Vol. 17, edited by K. M. Kadish, K. M. Smith, and R. Guilard (Academic Press, San Diego, 2003), pp. 129–176.

3. M. K. Engel, in: The Porphyrin Handbook, Vol. 20, edited by K. M. Kadish, K. M. Smith, and R. Guilard (Academic Press, San Diego, 2003), pp. 1–242.

4. M. J. Cook and I. Chambrier, in: The Porphyrin Handbook, Vol. 17, edited by K. M. Kadish, K. M. Smith, and R. Guilard (Academic Press, San Diego, 2003), pp. 37–127.

5. N. R. Armstrong, D. Alloway, D. Schlettwein, A. Back, K. W. Nebesny, and P. A. Lee, in: Encyclopedia of Surface and Colloid Science, edited by A. Hubbard (Marcel Dekker, New York, 2006), pp. 4349–4363.

6. Y. Shirota and H. Kageyama, Chem. Rev. **107**, 953 (2007).

7. N. B. McKeown, in: The Porphyrin Handbook, Vol. 15, edited by K. M. Kadish, K. M. Smith, and R. Guilard (Academic Press, San Diego, 2003), pp. 61–124.

8. S. Isoda, S. Hashimoto, T. Ogawa, H. Kurata, S. Moriguchi, and T. Kobayashi, Mol. Cryst. Liq. Cryst. **247**, 191 (1994).

9. S. Hashimoto, S. Isoda, H. Kurata, G. Lieser, and T. Kobayashi, J. Porphyrins Phthalocyanines **3**, 585 (1999).
10. D. Schlettwein, H. Tada, and S. Mashiko, Langmuir **16**, 2872 (2000).
11. J. O. Osso, F. Schreiber, V. Kruppa, H. Dosch, M. Garriga, M. I. Alonso, and F. Cerdeira, Adv. Funct. Mater. **12**, 455 (2002).
12. M. I. Alonso, M. Garriga, J. O. Osso, F. Schreiber, E. Barrena, and H. Dosch, J. Chem. Phys. **119**, 6335 (2003).
13. O. D. Gordan, M. Friedrich, W. Michaelis, R. Kröger, T. Kampen, D. Schlettwein, and D. R. T. Zahn, J. Mater. Res. **19**, 2008 (2004).
14. W. Michaelis, D. Wöhrle, and D. Schlettwein, J. Mater. Res. **19**, 2040 (2004).
15. D. Schlettwein, K. Hesse, H. Tada, S. Mashiko, U. Storm, and J. Binder, Chem. Mater. **12**, 989 (2000).
16. C. Kelting, W. Michaelis, A. Hirth, D. Wöhrle, and D. Schlettwein, J. Porphyrins Phthalocyanines **10**, 1179 (2006).
17. C. Tanase, P. W. M. Blom, M. Mulder, and D. M. de Leeuw, J. Appl. Phys. **99**, 103702 (2006).
18. T. Muck, J. Fritz, and V. Wagner, Appl. Phys. Lett. **86**, 232101 (2005).
19. L.-L. Chua, J. Zaumseil, J.-F. Chang, E. C.-W. Ou, P. K.-H. Ho, H. Sirringhaus, and R. H. Friend, Nature **434**, 194 (2005).
20. N. Kobayashi and H. Konami, in: Phthalocyanines Properties and Applications, Vol. 4, edited by C. C. Leznoff and A. B. P. Lever (Wiley-VCH, New York, 1996), pp. 343–404.
21. A. Ghosh, P. G. Gassman, and J. Almlöf, J. Am. Chem. Soc. **116**, 1932 (1994).
22. M. L'Her and A. Pondaven, in: The Porphyrin Handbook, Vol. 16, edited by K. M. Kadish, K. M. Smith, and R. Guilard (Academic Press, San Diego, 2003), pp. 117–169.
23. D. Schlettwein, K. Hesse, N. E. Gruhn, P. Lee, K. W. Nebesny, and N. R. Armstrong, J. Phys. Chem. B **105**, 4791 (2001).
24. D. Schlettwein, N. I. Jaeger, and T. Oekermann, in: The Porphyrin Handbook, Vol. 16, edited by K. M. Kadish, K. M. Smith, and R. Guilard (Academic Press, San Diego, 2003), pp. 247–283.
25. K. Hesse and D. Schlettwein, J. Electroanal. Chem. **476**, 148 (1999).
26. U. Weiler, T. Mayer, W. Jaegermann, C. Kelting, D. Schlettwein, S. Makarov, and D. Wöhrle, J. Phys. Chem. B **108**, 19398 (2004).
27. T. Mayer, U. Weiler, C. Kelting, D. Schlettwein, S. Makarov, D. Wöhrle, O. Abdallah, M. Kunst, and W. Jaegermann, Sol. Energy Mater. Sol. Cells **91**, 1873 (2007).
28. H. Peisert, M. Knupfer, and J. Fink, Appl. Phys. Lett. **81**, 2400 (2002).
29. U. Weiler, Dissertation, TU Darmstadt (2005), Chap. 6.
30. C. Shen and A. Kahn, J. Appl. Phys. **90**, 4549 (2001).
31. R. Murdey, N. Sato, and M. Bouvet, Mol. Cryst. Liq. Cryst. **455**, 211 (2006).
32. M. Gorgoi, W. Michaelis, T. U. Kampen, D. Schlettwein, and D. R. T. Zahn, Appl. Surf. Sci. **234**, 138 (2004).
33. D. R. T. Zahn, G. N. Gavrila, and M. Gorgoi, Chem. Phys. **325**, 99 (2006).
34. Z. Bao, A. J. Lovinger, and J. Brown, J. Am. Chem. Soc. **120**, 207 (1998).
35. M. Ling and Z. Bao, Org. Electron. **7**, 568 (2006).
36. D. Schlettwein, H. Graaf, J.-P. Meyer, T. Oekermann, and N. I. Jaeger, J. Phys. Chem. B **103**, 3078 (1999).
37. D. Wöhrle, G. Schnurpfeil, and G. Knothe, Dyes Pigm. **18**, 91 (1992).
38. S. Hiller, D. Schlettwein, N. R. Armstrong, and D. Wöhrle, J. Mater. Chem. **8**, 945 (1998).
39. J. H. Zagal, M. A. Gulppi and G. Cardenas-Jiron, Polyhedron **18**, 2255 (2000).

40. G. Schnurpfeil, A. K. Sobbi, W. Spiller, H. Kliesch and D. Wöhrle, J. Porphyrins Phthalocyanines **1**, 159 (1997).

41. H. Brinkmann and D. Schlettwein, in: Organic Electronics – Materials, Devices and Applications, Mater. Res. Soc. Symp. Proc., Vol. 965E, edited by F. So, G. B. Blanchet, and Y. Ohmori (MRS, Warrendale, PA, 2007), 0965-S06-25.

42. H. Marchetto, U. Groh, Th. Schmidt, R. Fink, H.-J. Freund, and E. Umbach, Chem. Phys. **325**, 178 (2006).

43. H. Tada, H. Touda, M. Takada, and K. Matsushige, Appl. Phys. Lett. **76**, 873 (2000).

44. J.-P. Meyer and D. Schlettwein, Adv. Mater. Opt. Electron. **6**, 239 (1996).

45. R. W. I. deBoer, A. F. Stassen, M. F. Craciun, C. L. Mulder, A. Molinari, S. Rogge, and A. F. Morpurgo, Appl. Phys. Lett. **86**, 262109 (2005).

46. G. Horowitz, Adv. Mater. **10**, 365 (1998).

47. J. Zaumseil and H. Sirringhaus, Chem. Rev. **107**, 1296 (2007).

4
Novel Organic Semiconductors and Processing Techniques for Organic Field-Effect Transistors

H. N. Tsao, H. J. Räder, W. Pisula, A. Rouhanipour, and K. Müllen

4.1
Introduction

Organic electronics have gained considerable interest due to their potential applications in cheap, flexible, and large area electronic devices, e.g. bendable displays or radio frequency identification (RFID) tags. For the realisation of such devices, field-effect transistors (FETs) are essential. Major advances have been made over the past years in terms of transistor performance, reaching charge carrier mobilities close to or even exceeding that of amorphous silicon, currently the workhorse of large area electronics [1]. Organic single crystals grown from rubrene [2] and polycrystalline pentacene [3] films are the most prominent examples. Such highly ordered molecular systems provide charge carrier mobilities well above that of amorphous silicon, namely $\mu_{sat} > 1$ cm^2/Vs. However, the fabrication of such crystalline structures requires complicated machinery. Furthermore, large area production is rather difficult and complicated and would necessitate large and expensive vacuum systems.

A more attractive way leading towards cheap, simple, and large area electronics is processing from solution, for example by drop-casting, spin-coating, or inkjet printing [4, 5]. Such procedures call for soluble organic semiconductors that at the same time show good intermolecular order to minimise charge carrier scattering and in this way to guarantee high transistor performance [6]. However, working with soluble compounds considerably adds complexity in terms of self-assembly and morphology. In order to make a molecule soluble, alkyl chains or other bulky substituents have to be introduced. These substituents have a high steric hindrance, leading to an increased occupation of space and thus preventing close intermolecular packing [7]. This in turn triggers poor charge carrier transport which is even more imparted by the isolating nature of the substituents [8]. Hence, it is important to understand the interplay between chemical structure, solubility, self-assembly, and the resulting device performance in order to gain powerful tools for the realisation of high performance organic electronic components.

Organic Electronics. Structural and Electronic Properties of OFETs. Edited by Christof Wöll

ISBN: 978-3-527-40810-8

With this goal in mind, we mainly focused on three types of chemical compounds, namely small molecules, polymers, and giant graphene-like polycyclic aromatic hydrocarbons (PAHs). Unsubstituted small molecules like rubrene, pentacene or oligothiophenes have a strong tendency to pack into crystalline structures, in this way providing highly ordered single crystals or polycrystalline films leading to high transistor performance. However, those materials can only be processed by vapor phase deposition in vacuum. Making such small molecules soluble by adding alkyl substituents and at the same time maintaining molecular order is thus a promising way towards cheap, large area and easily produced state of the art organic electronics [9–11]. Therefore, we studied the self-assembly of soluble discotic HBC-C_{12} and devised a new processing technique to control the molecular alignment from solution with the aim to induce highly ordered active layers for high performance field-effect transistors. Polymers on the other hand cannot be grown into highly ordered single crystals necessary for enhanced charge carrier mobilities. However, polymers have the advantage of fast charge carrier transport along their conjugated backbones, in this way enabling them to reach high transistor performance [12]. Together with the fact that polymers can in general be made soluble, this kind of material serves as a good candidate for solution processed organic electronics. Taking advantage of these points, we employed a donor-acceptor copolymer in a FET via solution processing, resulting in high charge carrier mobility. The polymer film surprisingly exhibited no macroscopic order. Finally, we introduce a novel method that allows the deposition, alignment and purification of large, insoluble and non-volatile molecules without destroying them along the process, which would likely happen during sublimation of such systems. Having a way to handle large molecules is extremely important since it paves the way for testing a class of molecules with theoretically promising electronic properties which could not be applied in devices so far.

4.2 Molecular Alignment from Solution Through the Zone-Casting Technique

A prerequisite for high device performance is a trap free charge carrier transport. Trapping sites can occur at many places in the transistor. Especially those situated at the semiconductor/insulator interface have been studied to degrade transistor behavior [13, 14]. More importantly, poor molecular order significantly induces trapping sites and has a deep impact on the functionality of the device [6]. Therefore, the ideal organic semiconductor layer should consist of molecules that self-assemble into highly ordered, long-range superstructures with a close intermolecular packing distance.

Polycyclic aromatic hydrocarbons (PAHs), e.g. hexa-*peri*-hexabenzocoronene HBC (Figure 4.1a), are disk-shaped molecules, the so called discotics, that contain an aromatic core which provides an extended delocalised π-orbital necessary for charge carrier transport [15]. Hence, this type of material is generally particularly attractive for high performance transistors. Substitution around the aromatic core by flexible aliphatic chains gives a tool for controlling solubility, thermal behavior and self-organisation of such a system [16]. Typically, the discotics stack on top of each other due to the π-interaction of adjacent molecules, in this way forming columnar superstructures that arrange into two-dimensional arrays (Figure 4.1b) [17]. Introduction of the proper aliphatic chains influences molecular ordering on the surface. The discotic molecule either lies with its aromatic core on the surface, the so called face-on arrangement (Figure 4.2a) or stands with the edge of the aromatic core on the substrate, the so called edge-on structure (Figure 4.2b). In fact, charge carrier mobility of up to 1.1 cm^2/Vs between neighbouring molecules have been measured by the PR-TRMC method [18]. The face-on arrangement is particularly useful in organic light emitting diodes or solar cells [19–21] (Figure 4.2a), whereas the edge-on arrangement is beneficial for field-effect transistors (Figure 4.2b). Such a versatility of HBCs together with their tendency to self-assemble into highly ordered large π-orbital systems hold considerable promise for the application in organic electronics. Following this idea, we synthesised and employed a soluble six fold dodecyl substituted HBC-C_{12} (Figure 4.1a) in a field-effect transistor.

HBC-C_{12} can be dissolved in most common organic solvents and is thus suitable for solution deposition. As a very simple first processing, this compound was drop-cast on bottom gate, bottom contact field-effect transistor substrates which consist of highly n++ doped silicon wafers covered with a thermally grown 150 nm SiO_2 dielectric layer. This layer was treated with HMDS in order to prevent charge carrier trapping at the semiconductor/dielectric interface. Source and drain gold contacts were patterned via conventional photolithography, with channel lengths of 10 μm and widths of 5 mm. Drop-casting was performed from a 1 mg/mL HBC-C_{12} solution dissolved in toluene.

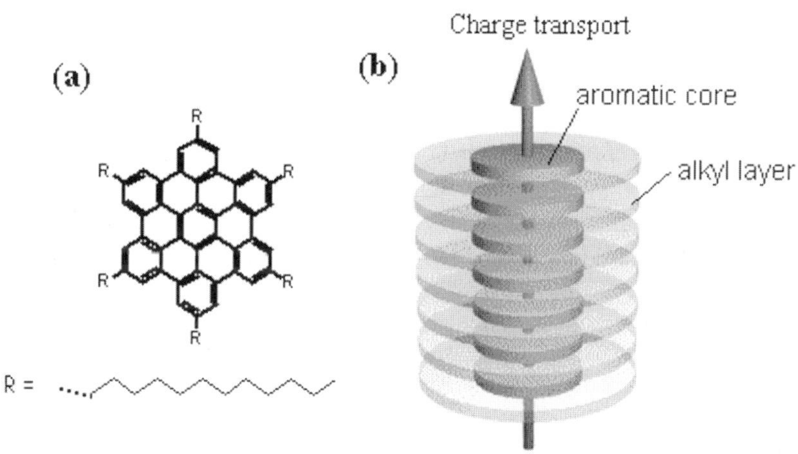

Figure 4.1 Chemical structure of HBC-C_{12} (a) and charge transport along the columnar HBC-C_{12} stack (b).

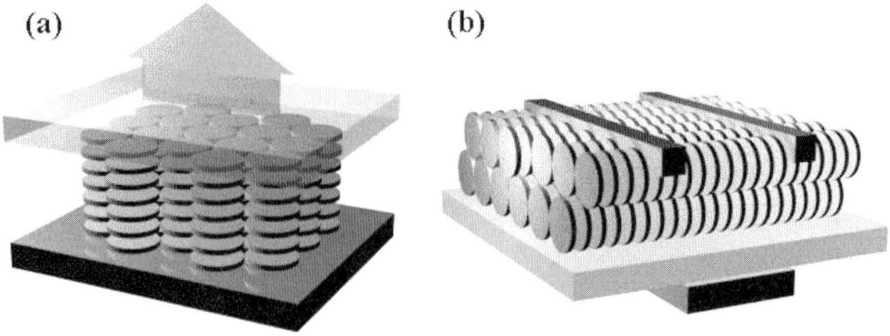

Figure 4.2 Face-on (a) and edge-on (b) arrangement of HBC-C_{12} columnar stacks.

The resulting drop-cast film consists of fibre like structures. Figure 4.3a illustrates an AFM image of such fibres. Figure 4.3b depicts fibres randomly distributed over the whole transistor substrate, with the yellow parallel stripes being the source and drain electrodes. Each microfibre consists of the typical columnar stacks that form due to the π-stacking interaction of the HBC-C_{12} molecules. Charge carrier transport takes place along the axis of the columns (Figure 4.1b). The transistor behaviour is depicted in Figure 4.4. From the transfer curve taken at a source drain voltage $V_{SD} = -40$ V a saturated hole mobility $\mu_{sat} = 3 \times 10^{-4}$ cm²/Vs was extracted, together with a source drain current on/off ratio of $I_{on}/I_{off} = 2 \times 10^5$. This moderate charge carrier mobility can be related to discotics macroscopically such that the columns connecting the source and drain electrodes well should lead to a higher charge carrier mobility and thus to a better device performance.

In order to align the molecules in one direction from the solution, we devised a zone-casting technique whose principle is based on observations performed during the solvent evaporation of a drop-cast solution [22]. One can observe that the deposited material is mainly oriented in the direction of the solvent evaporation. Theoretically, at adequate conditions it should be possible to obtain from simple drop-casting a homogeneous film with a radial alignment of the superstructure. However, this is quite difficult, since the concentration

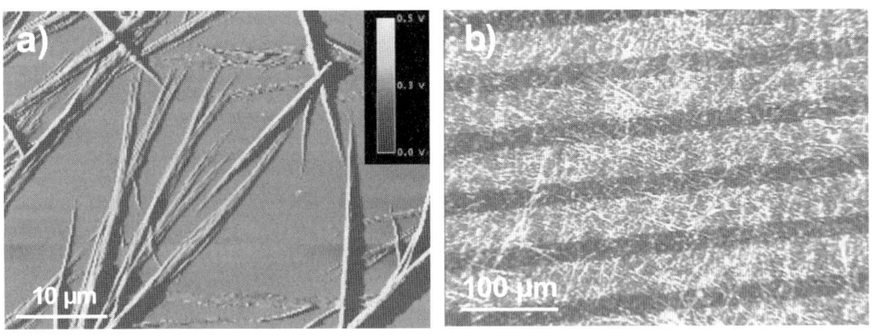

Figure 4.3 Image from atomic force microscopy of drop-cast HBC-C_{12} (a). Drop-cast HBC-C_{12} resulting in randomly oriented microfibres (b) as evidenced by an optical microscope.

within the solution drop changes with ongoing evaporation of the solvent and material precipitation. Therefore, it is not possible to achieve stationary conditions and a homogenous film morphology by drop-casting. At close inspection of the drop-cast layer, one can notice that most of the material is deposited at the rim of the precipitated drop, whereas the film interior is poor of material. This behaviour can be explained by the thermodynamic factors and kinetic aspects in solution.

The orientation of large ordered areas without the application of additional pre-oriented layers necessitate a processing technique that enables the alignment of the material directly during the deposition on a desired substrate at stationary conditions. As a consequence, the zone-casting technique has been developed which is based on simple solution deposition at steady conditions. The principle of this technique is presented in Figure 4.5a. A solution is spread out by means of a nozzle onto a moving support such as glass or silicon. Thereby, a meniscus is formed between the nozzle and the support. During the solvent evaporation, a concentration gradient adjusts within the meniscus. When the critical concentration is attained, the material begins to precipitate or to nucleate from the solution and then crystallises directionally onto the moving support, in this way forming the aligned thin layer. The film morphology can be controlled by the processing parameters which were e.g. the evaporation temperature and polarity of the solvent, concentration, temperatures of the heating blocks, solvent flow, substrate velocity etc. The optimised zone-casting conditions strongly depend on the solution behaviour of the material. Therefore, prior to the processing it is important to obtain a fundamental understanding about the processes such as self-aggregation taking place in solution and during the self-organisation in drop-cast films on surfaces. These findings are essential for the application of the appropriate processing parameters. Generally, the size or length of the surface layer prepared by the zone-casting processing is only restricted to the limited volume of the applied syringe. Theoretically, it is possible to zone-cast films of unlimited length, as long as the steady state conditions are not changed during the process.

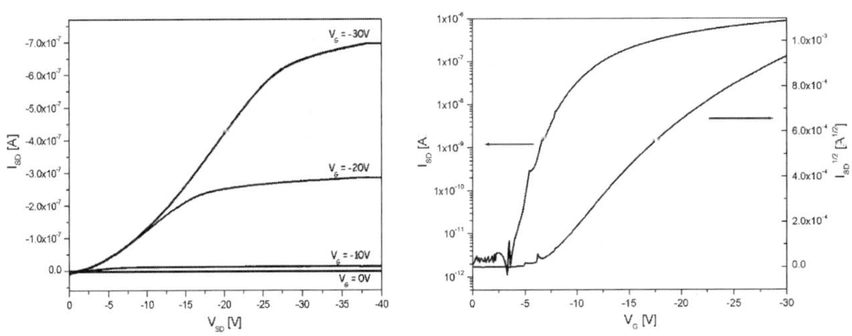

Figure 4.4 Output curves (a) and transfer curve (b) taken at $V_{SD} = -40$ V of drop-cast HBC-C_{12} film.

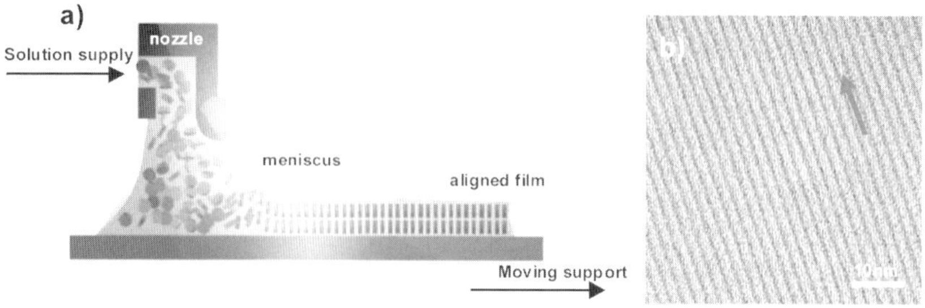

Figure 4.5 Schematic illustration of the zone-casting of HBC-C_{12} (a), HR-TEM image displays uniaxially oriented columns after zone-casting (b).

In this aspect, zone-casting was performed for HBC-C_{12} for the fabrication of FETs. Hereby, HBC-C_{12} was dissolved in tetrahydrofuran at a concentration of 0.25 mg/mL and zone-cast on HMDS treated 200 nm thick SiO_2. The resulting films were 25 nm thin and showed macroscopically ordered columnar structures as indicated by HR-TEM measurements in Figure 4.5b, which has been also reported for other discotic systems consisting of larger aromatic cores [23, 24]. The structural analysis revealed that the discotics arrange edge-on on the substrate into columnar superstructures, with the columns exhibiting a herringbone packing [25–28]. These devices were completed by evaporating source and drain gold contacts with a channel length and width of 25 µm and 1.6 mm respectively on top of the films. A charge carrier mobility of up to $\mu_{sat} = 0.01\ cm^2/Vs$ and an $I_{on}/I_{off} = 10^4$ was determined, indicating a mobility increase of almost two orders of magnitude [29] (Figure 4.6). This method clearly implies how important control of macroscopic molecular order is for the realisation of high performance field-effect transistors. The reason why the charge carrier mobility is limited to $\mu_{sat} = 0.01\ cm^2/Vs$ despite the high long range molecular order lies in the columnar structure of the HBC-C_{12}. This discotic molecule self-assembles into one dimensional columnar stacks, limiting

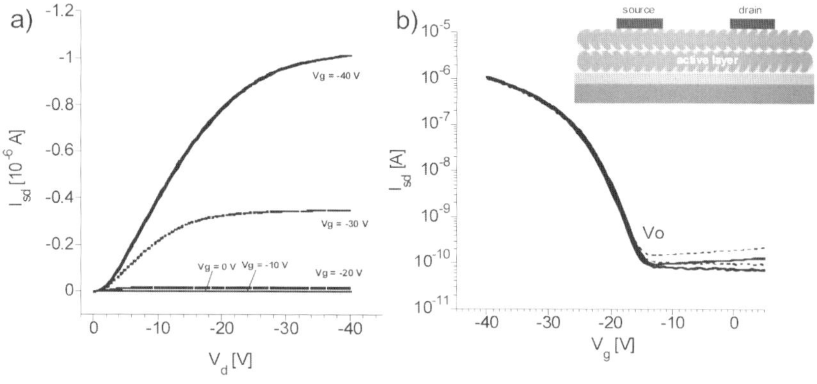

Figure 4.6 Output curves (a) and transfer curves (b) of zone-cast HBC-C_{12} FETs. Inset is a schematic representation of the zone-cast HBC-C_{12} FET.

charge carrier transport in one dimension and in this way preventing a more pronounced device performance. Hence, employing this zone-casting method to align rod-like molecules should enable highly packed two dimensional charge carrier transport, leading to improved device performance [30].

4.3
Solution Processed Donor–Acceptor Copolymer Field-Effect Transistors

As discussed previously, discotic materials are attractive in the sense that they contain a large π-system core, providing favourable intermolecular π-stacking and thereby allowing enhanced charge carrier transport along the columnar axis. However, the downturn is the strictly one dimensional transport, limiting the possible pathways the charge carriers can take. This issue makes the current flow very prone to structural defects within the columns. One single disorder within a column can already act as a significant scattering site for the charge carriers, making this whole system very sensitive. Hence, for discotics, processing techniques like zone-casting have to be applied in order to achieve macroscopic order and thus high device performance.

Semiconducting polymers on the other hand have become attractive candidates for organic electronics. They can be also easily solution processed and show charge carrier mobilities larger than 0.1 cm^2/Vs without the need for uniaxial alignment. The first polymer that achieved such a performance was regioregular poly(3-hexyl-thiophene) (P3HT) [31]. Spin-coated P3HT films consist of lamella-like ordered semicrystalline microstructures made of crystallites separated by amorphous material [32]. The well-packed semicrystals are responsible for the high mobility. Other semicrystalline and polycrystalline polymers have evolved over the past years reaching charge carrier mobilities up to 0.6 cm^2/Vs fabricated by simple spin-coating [33–35].

Motivated by these findings, we synthesised a benzothiadiazole-cyclopentadithiophene (BTZ-CDT) donor–acceptor copolymer shown in Figure 4.7 as an active component for high performance FETs without the need for molecular alignment techniques like zone-casting. Such a system allows easy, cheap, and fast solution processing generally needed for the commercial production of organic electronics. This copolymer was simply drop-cast from a 10 mg/mL 1,2,4-trichlorobenzene solution on bottom gate, bottom contact field-effect transistor substrates with channel lengths and widths of 10 μm and 5 mm respectively. The resulting films were 150 nm thick and yielded a mobility of $\mu_{sat} = 4 \times 10^{-4}$ cm^2/Vs in the saturation regime and an $\mu_{lin} = 7 \times 10^{-5}$ cm^2/Vs

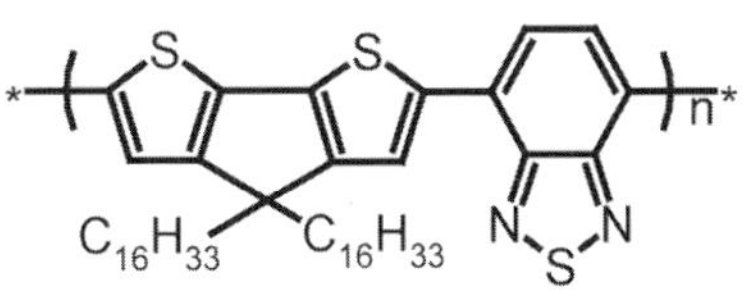

Figure 4.7 Chemical structure of BTZ-CDT donor–acceptor copolymer with $M_n = 1 \times 10^4$ g/mol.

in the linear regime. After thermally annealing the devices at 200 °C for 2 h, the transistors showed a saturated mobility of up to $\mu_{sat} = 0.17$ cm²/Vs and a linear mobility of $\mu_{lin} = 0.001$ cm²/Vs together with an on/off ratio of $I_{on}/I_{off} = 10^5$ as extracted from the FET curves illustrated in Figure 4.8 [36], making this polymer applicable for cheap, easily processed and large area electronics.

In order to understand the reason for this high performance, we performed structural analysis to gain information about molecular packing of this polymer. In this aspect, Bragg scattering was applied on the drop-cast film. Surprisingly, no macroscopic order could be found, with the film being totally amorphous (Figure 4.9b), both before and after thermal annealing. This finding lead us to conclude that the thermal treatment did not trigger an increased order in the film but instead caused the extraction of residual solvents that acted as trapping sites. That is why the charge carrier mobility increased after thermal annealing. Very apparently, there must be some other more important factor than long range order that promotes high charge carrier mobility. To clarify this issue, a closer look was taken at the supramolecular arrangement. This was done by 2-dimensional wide angle X-ray scattering (2D-WAXS) on an extruded filament of the BTZ-CDT copolymer. Reflections appeared in the pattern, which were characteristic for molecular organisation (Figure 4.9a). However, relatively diffuse reflections and the lack of higher order ones implied pronounced disorder. The equatorial scattering intensities at small angles described the orientation of the polymer chains along the shearing direction. The position of the reflections was related to the lateral distance of 2.66 nm between the polymer chains. They are arranged in a lamella structure consisting of the aromatic rigid backbone separated by disordered aliphatic side chains filling the periphery, as schematically presented in Figure 4.9a. More crucially, the wide-angle located also in the equatorial of the pattern was attributed to the exceptionally small π-stacking distance of 0.37 nm between macromolecules in comparison to other conjugated polymers. This distance coincided with similar values reported for all conjugated polymers revealing also a high performance in a device, especially with charge carrier mobilities exceeding

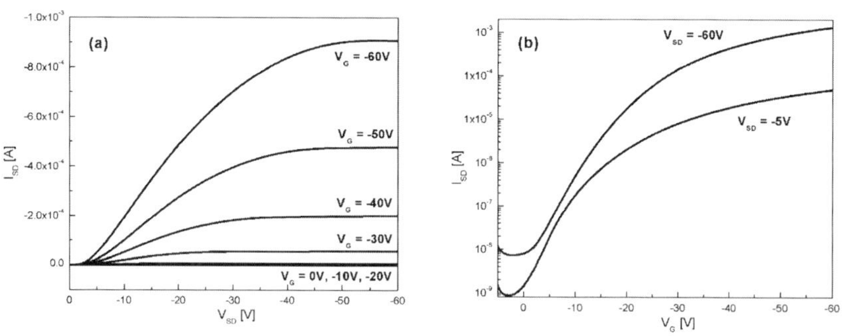

Figure 4.8 Output curves (a) and transfer curve (b) of BTZ-CDT donor–acceptor polymer FET.

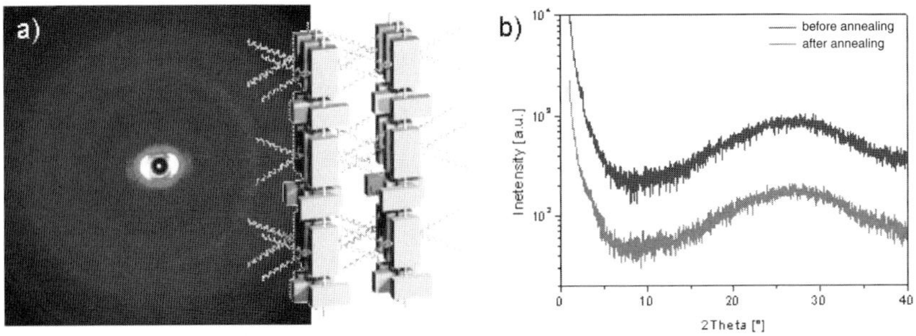

Figure 4.9 2DWAXS and the deduced schematic ordering of extruded BTZ-CDT copolymer (a), Bragg scattering of drop-cast BTZ-CDT copolymer film (b). (see Colour plates p. LII)

$\mu_{sat} = 0.1\ cm^2/Vs$ [33–35]. This fact urges us to believe that a close π-stacking distance plays a crucial role in designing high performance organic semiconductors, more important than achieving a long range order. Surely, controlling both close π-stacking and a macroscopic order will lead to very good device characteristics.

4.4 Processing of Giant Graphene Molecules by Soft-Landing Mass Spectrometry

The processability of (macro)molecules into ultra-pure and highly ordered structures at surfaces is of fundamental importance for the realisation of high performance FETs. As mentioned before, large conjugated molecules contain large delocalised π orbitals leading to strong π-stacking and thus improved charge carrier transport and increased charge carrier density.

Unsubstituted nanographenes are of particular interest for such an application, also allowing three dimensional charge carrier transport due to the lack of insulating alkyl substituents for example. Unfortunately, larger molecules commonly imply lower processability due to either their low solubility in liquid media or the occurrence of thermal cracking during vacuum sublimation. Thus, the nanographenes have to be solubilised by long alkyl chains to be processable, however, with the drawback of "diluting" the electroactive material by insulating components, surrounding the semiconductive nanographene columns by an insulating mantle (Figure 4.10). The search for novel strategies to process and characterise giant building blocks is therefore a crucial goal in building organic field-effect transistors with excellent performance. A possible solution for this dilemma is our recently developed method of MALDI soft-landing for molecules, so far considered as unprocessable. Soft-landing has the potential to be a new general route to process extraordinarily large molecules, i.e. synthetic nanographenes [37], into ultra-pure crystalline architectures at surfaces. Our method relies on soft-landing of ions [38] generated by solvent-

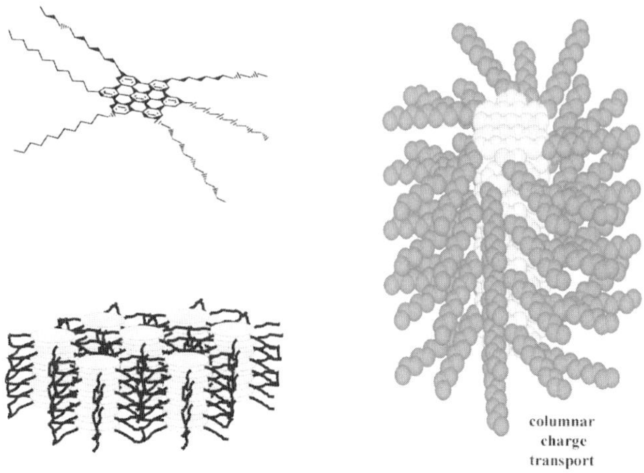

Figure 4.10 Insulating mantles caused by solubilising alkyl chains decreasing the dimensionality of charge transport mechanism in columnar structures of nanographenes.

free matrix assisted laser desorption/ionisation (MALDI) mass spectrometry. First, the unsubstituted nanographenes are transferred to the gas phase, accelerated by strong electric fields, purified according to their mass-to-charge ratio and finally decelerated and softly deposited at surfaces (Figure 4.11).
Then, the thin films are characterised by matrix assisted laser desorption mass spectrometry and scanning probe microscopic techniques. In the case of unsubstituted HBC scanning tunnelling microscopy revealed the formation of ordered nanoscale semiconducting supramolecular architectures at the surface of HOPG [39] (Figure 4.12a). The arrangement of HBC at the surface is charac-

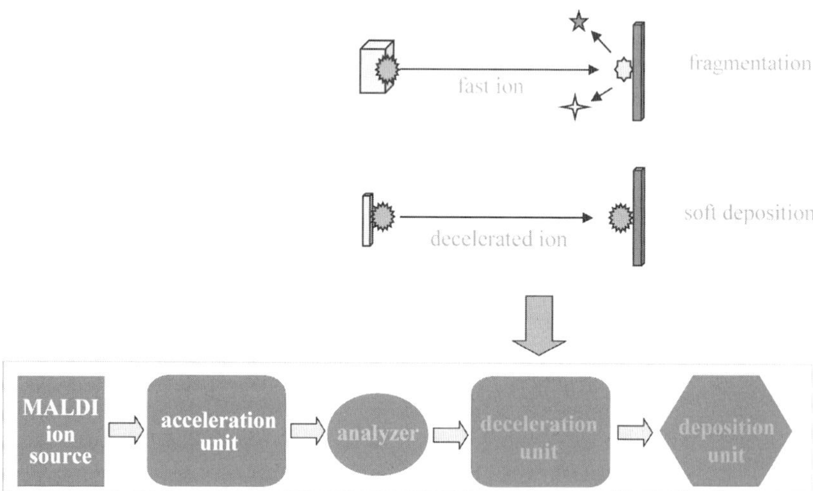

Figure 4.11 Principle of the soft-landing method.

terised by molecules standing nearly perpendicular on the substrate and reveals that our soft-landing deposition causes an unusual packing of HBC in comparison to conventional deposition techniques, such as drop-casting and vacuum sublimation. We attribute this "edge-on" alignment to an induced dipole moment in the planar π-system of the HBC disc [40], because of the high substrate surface potential of 8 kV during the soft landing experiment which causes an alignment of the molecular planes normal to the surface (Figure 4.12b). Such "edge-on" packing at surfaces obtained by soft-landing, which is similar to that produced by the recently re-visited zone-casting [29], is ideal for applications in organic electronics, such as Field Effect Transistors (FETs) [30]. This is particularly important since our deposition approach is viable also with conductive substrates covered by a dielectric thin layer, i.e. the typical support for a FET.

Our soft-landing approach has two advantages over all other processing procedures, including zone casting: (i) it enables a high degree of purity of molecules deposited at surfaces; (ii) it does not suffer from the limitation of the size of the molecules which can be processed into highly ordered architectures at surfaces [41].

The validity of our soft-landing method was tested by extending the investigation to the larger $C_{96}H_{30}$ nanographene. A layered structure, analogous to that shown in Figure 4.12a for HBC ($C_{42}H_{18}$) was found by SFM investigations. The thickness of the layer suggests an "edge-on" packing of the larger molecule on graphite analogous to that observed with the smaller HBC molecule.

Since the soft-landing bottom-up growth of polycrystalline layered ultrathin films is not limited by the solubility or volatility of the molecule, it is ideal for producing ultra-pure ordered architectures of giant molecular systems with enhanced and novel functionalities. This method can be extended to the controlled layer-by-layer deposition of different components up to 3D assemblies.

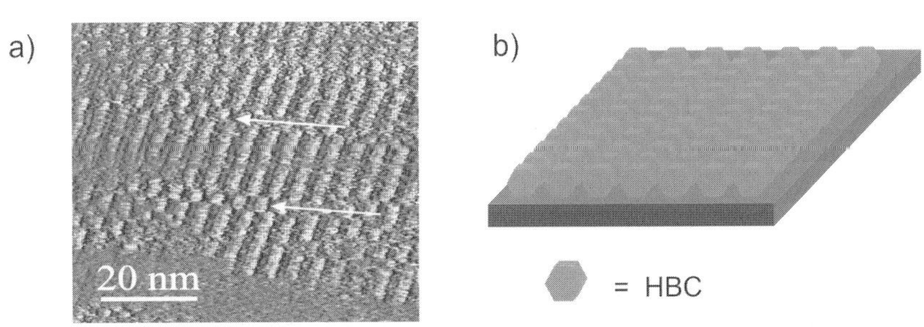

Figure 4.12 (a) STM image of HBC on HOPG surface after soft-landing. The arrows indicate a dislocation in the periodic lattice. (b) Schematic representation of the molecular packing with molecules "edge-on" the surface.

4.5 Conclusion

We have successfully synthesised novel soluble organic semiconductors and employed them in field-effect transistors. Hereby, we devised a new zone-casting technique that allows the macroscopic orientation of molecules, particularly of discotic HBC-C_{12} from solution, leading to increased charge carrier mobility due to improved molecular arrangement. Furthermore, a BTZ-CDT copolymer was synthesised and applied by simple drop-casting in a transistor, leading to a charge carrier mobility of 0.17 cm^2/Vs and an on/off ratio of 10^5. Surprisingly, no macroscopic order could be observed in this copolymer film, prompting us to believe that close π-stacking is a very crucial factor governing high transistor performance. Finally, with the goal to deposit giant insoluble molecules which, due to their large π-systems, are very promising candidates for enhanced charge transport in electronic devices, a soft-landing method was introduced. By this method, highly oriented and pure polycyclic aromatic hydrocarbons were processed, in this way paving a new path for high performance organic electronics.

Acknowledgements

This work was supported by the Deutsche Forschungsgemeinschaft, DFG-Schwerpunktprogramm Organische Feldeffekt-Transistoren: strukturelle und dynamische Eigenschaften (SPP1121).

References

1. C. D. Dimitrakopoulos and P. R. L. Malenfant, Adv. Mater. **14**, 99 (2002).
2. V. C. Sundar, J. Zaumseil, V. Podzorov, E. Menard, R. L. Willett, T. Someya, M. E. Gershenson, and J. A. Rogers, Science **303**, 1644 (2004).
3. H. Klauk, M. Halik, U. Zschieschang, G. Schmid, W. Radlik, and W. Weber, J. Appl. Phys. **92**, 5259 (2002).
4. H. Sirringhaus, T. Kawase, R. H. Friend, T. Shimoda, M. Inbasekaran, W. Wu, and E. P. Woo, Science **290**, 2123–2126 (2000).
5. H. Sirringhaus, Adv. Mater. **17**, 2411–2425 (2005).
6. V. Lemaur, D. A. da Silva Filho, V. Coropceanu, M. Lehmann, Y. Geerts, J. Piris, M. G. Debije, A. M. van de Craats, K. Senthilkumar, L. D. A. Siebbeles, J. M. Warman, J.-L. Bredas, and J. Cornil, J. Am. Chem. Soc. **126**, 3271–3279 (2004).
7. M. Kastler, W. Pisula, D. Wasserfallen, T. Pakula, and K. Müllen, J. Am. Chem. Soc. **127**, 4268 (2005).
8. J. M. Warman, J. Piris, W. Pisula, M. Kastler, D. Wasserfallen, and K. Müllen, J. Am. Chem. Soc. **127**, 14257 (2005).
9. M. M. Payne, S. R. Parkin, J. E. Anthony, C.-C. Kuo, and T. N. Jackson, J. Am. Chem. Soc. **127**, 4986 (2005).
10. P. T. Herwig, and K. Müllen, Adv. Mater. **11**, 480 (1999).

11. A. Afzali, C. D. Dimitrakopoulos, and T. L. Breen, J. Am. Chem. Soc. **124**, 8812 (2002).

12. P. Prins, F. C. Grozema, F. Galbrecht, U. Scherf, and L. D. A. Siebbeles, Phys. Chem. C **111**, 11104 (2007).

13. M. Yoon, C. Kim, A. Facchetti, and T. J. Marks, J. Am. Chem. Soc. **128**, 12851 (2006).

14. L.-L. Chua, J. Zaumseil, J. Chang, E. C.-W. Ou, P. K.-H. Ho, H. Sirringhaus, and R. H. Friend, Nature **434**, 194 (2005).

15. M. D. Watson, A. Fechtenkötter, and K. Müllen, Chem. Rev. **101**, 1267–1300 (2001).

16. W. Pisula, M. Kastler, D. Wasserfallen, M. Mondeshki, J. Piris, I. Schnell, and K. Müllen, Chem. Mater. **18**, 3634 (2006).

17. J. Wu, W. Pisula, and K. Müllen, Chem. Rev. **107**, 718–747 (2007).

18. A. M. van de Craats, J. M. Warman, K. Müllen, Y. Geerts, and J. Brand, Adv. Mater. **10**, 36 (1998).

19. L. Schmidt-Mende, A. Fechtenkötter, K. Müllen, E. Moons, R. H. Friend, and J. D. MacKenzie, Science **293**, 1119 (2001).

20. W. Pisula, Ž. Tomović, B. El Hamaoui, M. D. Watson, T. Pakula, and K. Müllen, Adv. Funct. Mater. **15**, 893 (2005).

21. W. Pisula, M. Kastler, B. El Hamaoui, M.-C. García-Gutiérrez, R. J. Davies, C. Riekel, and K. Müllen, Chem. Phys. Chem. **8**, 1025 (2007).

22. A. Tracz, T. Pakula, and J. K. Jeszka, Mater. Sci. (Poland) **22**, 415 (2004).

23. A. Tracz, J. K. Jeszka, M. D. Watson, W. Pisula, K. Müllen, and T. Pakula, J. Am. Chem. Soc. **125**, 1682 (2003).

24. W. Pisula, Z. Tomović, M. Stepputat, U. Kolb, T. Pakula, and K. Müllen, Chem. Mater. **17**, 2641 (2005).

25. D. W. Breiby, O. Bunk, W. Pisula, T. L. Sølling, A. Tracz, T. Pakula, K. Müllen, and M. M. Nielsen, J. Am. Chem. Soc. **127**, 11288 (2005).

26. D. W. Breiby, F. Hansteen, W. Pisula, O. Bunk, U. Kolb, J. W. Andreasen, K. Müllen, and M. M. Nielsen, J. Phys. Chem. B **109**, 22319 (2005).

27. J. Piris, M. G. Debije, N. Stutzmann, B. W. Laursen, W. Pisula, M. D. Watson, T. Bjornholm, K. Müllen, and J. M. Warman, Adv. Funct. Mater. **14**, 1053 (2004).

28. J. Piris, W. Pisula, A. Tracz, T. Pakula, K. Müllen, and J. Warman, Liq. Cryst. **31**, 993 (2004).

29. W. Pisula, A. Menon, M. Stepputat, I. Lieberwirth, U. Kolb, A. Tracz, H. Sirringhaus, T. Pakula, and K. Müllen, Adv. Mater. **17**, 684 (2005).

30. P. Miskiewicz, M. Mas-Torrent, J. Jung, S. Kotarba, I. Glowacki, E. Gomar-Nadal, D. B. Amabilino, J. Veciana, B. Krause, D. Carbone, C. Rovira, and J. Ulanski, Chem. Mater. **18**, 4724–4729 (2006).

31. H. Sirringhaus, P. J. Brown, R. H. Friend, M. M. Nielsen, K. Bechgaard, B. M. W. Langeveld-Voss, A. J. H. Spiering, R. A. J. Janssen, E. W. Meijer, P. Herwig, and D. M. de Leeuw, Nature **401**, 685 (1999).

32. S. Hugger, R. Thomann, T. Heinzel, and T. Thurn-Albrecht, Colloid Polym. Sci. **282**, 932 (2004).

33. B. S. Ong, Y. Wu, P. Liu, and S. J. Gardner, J. Am. Chem. Soc. **126**, 3378 (2004).

34. I. McCulloch, M. Heeney, C. Bailey, K. Genevicius, I. MacDonald, M. Shkunov, D. Sparrowe, S. Tierney, R. Wagner, W. Zhang, M. L. Chabinyc, R. J. Kline, M. D. McGehee, and M. F. Toney, Nature Mater. **5**, 328 (2006).

35. Y. Li, Y. Wu, P. Liu, M. Birau, H. Pan, and B. S. Ong, Adv. Mater. **18**, 3029 (2006).

36. M. Zhang, H. N. Tsao, W. Pisula, C. Yang, A. K. Mishra, and K. Müllen, J. Am. Chem. Soc. **129**, 3472 (2007).

37. M. D. Watson, A. Fechtenkötter, and K. Müllen, Chem. Rev. **101**, 1267–1300 (2001).

38. V. Franchetti, B. H. Solka, W. E. Baitinger, J. W. Amy, and R. G. Cooks, Int. J. Mass Spectrom. Ion Phys. **23**, 29–35 (1977).

39. H. J. Räder, A. Rouhanipour, A. M. Talarico, V. Palermo, P. Samorì, and K. Müllen, Nature Mater. **5**, 276–280 (2006).

40. A. Cristadoro, G. Lieser, H. J. Rader, and K. Müllen, Chem. Phys. Chem. **8**, 586–591 (2007).

41. Hitherto, the limit of the deposition is determined by the mass range of the high resolution mass spectrometer employed. For our case it is 4500 Da.

5
Assembly, Structure, and Performance of an Ultra-Thin Film Organic Field-Effect Transistor (OFET) Based on Substituted Oligothiophenes

K. Haubner, E. Jaehne, H.-J. P. Adler, D. Koehler, C. Loppacher, L. M. Eng, J. Grenzer, A. Herasimovich, and S. Scheiner

5.1
Introduction

Oligo- and polythiophenes represent attractive materials for thin-film organic field effect transistors (OFETs) due to their wide range of chemical modification possibilities [1–6]. Much work in this area was initially devoted to investigate the properties of different types of α,ω- and/or β,β′-substituted thiophene derivatives, as well as to the fabrication of thin film transistors (TFT) [1, 6–10]. The electrical, optical, and mechanical properties of such materials dramatically depend on their molecular conformation, packing, and orientation in the thin film. Many studies have investigated OFETs prepared from different oligothiophenes by vacuum deposition [1, 6, 7, 9]. Despite obvious advantages of this method to form well-ordered films with a good "run to run" reproducibility, there is one drawback – the high initial cost of the process. Here, we investigated OFETs fabricated with the novel oligothophene DCNDBQT from solution as well as by vacuum deposition, and compare these results.

Early endeavours to obtain soluble, well-ordered thin films with good electrical conductivity were not successful. The reason was the wrong orientation of the molecules with respect to the substrate as well as the reduced ability to form well-ordered structures in the solid film. Generally, the maximum charge transport in TFT is realised in the direction perpendicular to the plane of the thiophene rings due to the good π–π-overlap favouring electron transport [1, 3].

In a recent work the properties of a spin–coated β,β′-substituted sexithiophene derivative – β,β′-dihexylsexi-thiophene (DHST) – was investigated [11]. Although these films showed a fine crystalline structure only a very low field effect mobility (FEM) was observed. This could be explained by the herringbone packing, where the distance between two molecules was found to be 7.6 Å, which does not allow a good π–π-overlap, and therefore, caused the low FEM [12].

Organic Electronics. Structural and Electronic Properties of OFETs. Edited by Christof Wöll

ISBN: 978-3-527-40810-8

Additionally, DHST molecules were oriented parallel to the substrate, whereas, α,ω-substituted oligothiophenes (e.g., α,ω-DHST) were vertically aligned leading to a higher mobility [11].

Kim et al. used a self-assembly monolayer on an insulator substrate as the template for a targeted deposition of a soluble region-regular poly(3-hexylthiophene) (P3HT) [13]. The orientation of the upper P3HT layer was determined by its interaction with the terminal amino group of the SAM.

Our approach was to design molecules with appropriate properties (perpendicular orientation, solubility, and chemical stability). Thus, we synthesised α,ω-dicyano-β,β′-dibutylquaterthiophene (DCNDBQT) as the molecular semiconducting layer. While the short alkyl chain should improve the solubility, the α,ω-substituted cyano groups were introduced to enable lamellar stacking (see Scheme 5.1).

To improve the perpendicular orientation of the film, 5-cyano-2-(butyl-4-phosphonic acid)-3-butylthiophene (CNBTPA) (Scheme 5.1) was also designed to work as a template layer in order to promote a defined orientation of the oligothiophene layer. Compared to other concepts using a template layer [13] the designed molecules exhibit a similar structure (alkyl chain length, position of the cyano group), and therefore, the same required space. The combination of α,ω- and β,β′-substitution in the oligothiophene is expected to result in a material with high stacking order and a high FEM.

A schematic representation of the thin film formation from DCNDBQT with CNBTPA as the template layer is shown in Scheme 5.3. First, the template layer was deposited by a self-assembly process from solution on the oxide substrate. The phosphonic acid group was chosen as the anchor to the support, whereas the terminal cyano group is able to form hydrogen bonds to the α-H of the oligothiophene ring [2, 14]. The butyl chain in β-position of the thiophene ring also promoted the solubility and, hence, initiated good molecular packing.

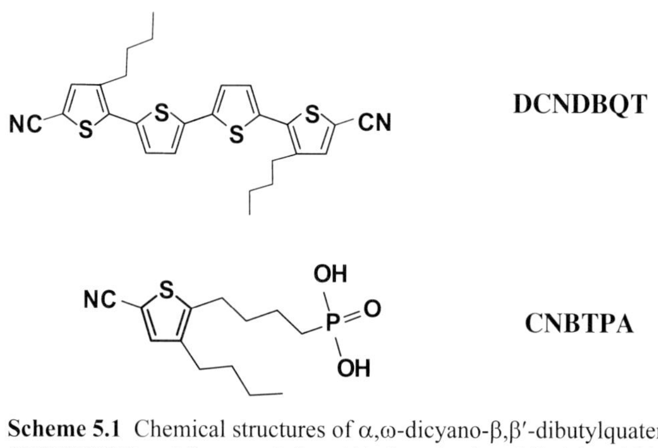

Scheme 5.1 Chemical structures of α,ω-dicyano-β,β′-dibutylquaterthiophene (DCNDBQT) and 5-cyano-2-(butyl(4-phosphonic acid))-3-butylthiophene (CNBTPA).

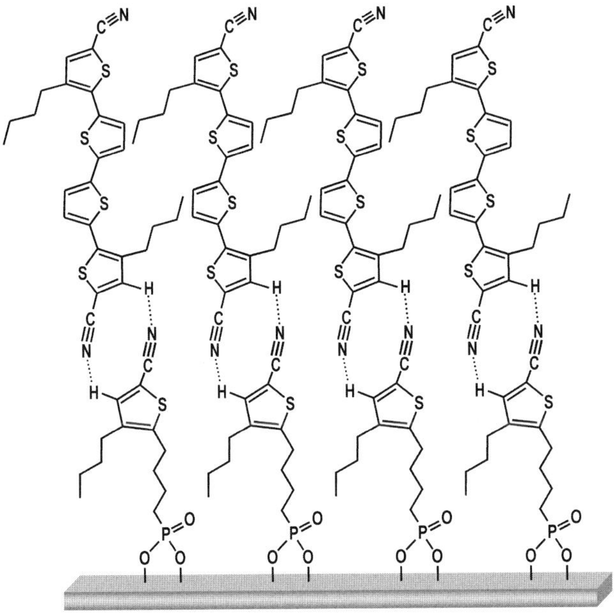

Scheme 5.2 Schematic representation of DCNDBQT with CNBTPA SAM on a TiO_2 oxide substrate.

The adsorbed CNBTPA layer now works as the template to support the vertical alignment of further DCNDBQT molecules, which are applied by vacuum sublimation.

Figure 5.1 illustrates the calculated structure of the DCNDBQT molecule. The thiophene units as well as the alkyl chains are mostly aligned in the plane, which would allow good lamellar packing.

In this chapter we investigate and discuss the thermal, optical, electrical properties of the oligothiophene derivatives by means of differential scanning calorimetry (DSC), thermal gravimetric analysis (TGA), and UV–Vis spectroscopy. The thin films of these compounds produced by solution cast and vacuum deposition methods are characterised by AFM measurements in contact and non-contact mode, and by X-ray diffraction. Finally, an ultra-thin OFET is built, and the transistor characteristics are determined.

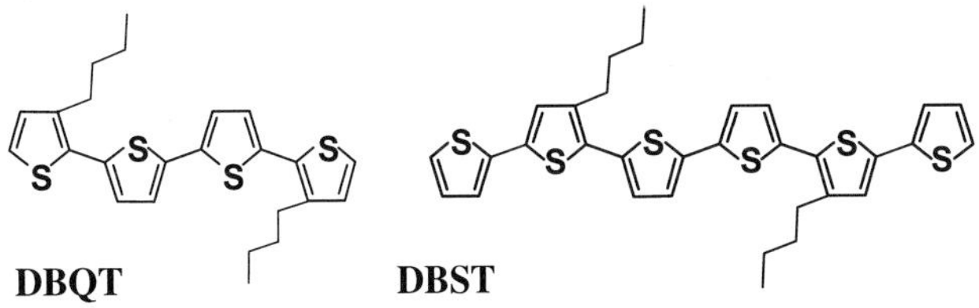

DBQT **DBST**

Scheme 5.3 Chemical structures of β,β′-dibutylquaterthiophene (DBQT) and β,β′-dibutylsexithiophene (DBST).

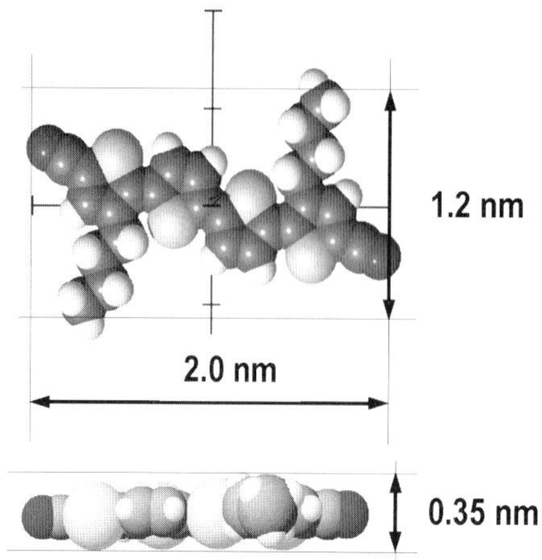

Figure 5.1 Calculated molecular dimension of the DCNDBQT molecule.

5.2 Experimental

The synthesis and characterisation of the oligothiophene derivatives will be published in a separate paper [15].

5.2.1 General Procedures

UV–Vis spectra were taken on a Perkin-Elmer Lambda 35 in different organic solvents in 1 cm cuvettes. Optical band gap energies E_g^{opt} were calculated from absorption edges and are given in eV.

Mass spectra were recorded with the Esquire-LC 00084 and with MALDI-TOF MS Bruker Biflex IV. DSC and TGA measurements were carried out using a TA instruments DSC 2920. The DSC as well as the TGA measurements were performed at a scan rate of 5 K/min under nitrogen atmosphere.

Cyclic voltammetry experiments were carried out on an EG & G Princeton Applied Research Model potentio-stat/galvanostat at 20 ± 1 °C under argon atmosphere. Electrochemical experiments were carried out in a three-compartment three-electrode glass cell with an aqueous saturated calomel electrode as the reference electrode. Platinum sheets were used both as the working electrode ($A = 1$ cm^2) and as the counter electrode ($A = 2.83$ cm^2). The potentials are reported versus SCE (saturated calomel electrode). Determination of the HOMO and LUMO energy levels was done using the following equations [16]:

$$\text{HOMO} = -e((E_{\text{ox1}}^{1/2}) + 4.4)\,, \tag{1a}$$

$$\text{LUMO} = -e((E_{\text{red1}}^{1/2}) + 4.4)\,, \tag{1b}$$

where $E^{1/2}_{ox1}$ and $E^{1/2}_{red1}$ are the formal oxidation and reduction potentials versus SCE, respectively.

Non-contact atomic force microscopy (nc-AFM) measurements of evaporated DCNDBQT were performed under ultra high vacuum conditions (base pressure 10^{-10} mbar) with a commercial cryogenic AFM/STM (atomic force microscope/scanning tunnelling microscope) [17]. The microscope is additionally equipped with a homebuilt vibration isolation stage composed of six copper beryllium springs, which was essential to obtain molecular resolution. nc-AFM measurements were carried out at liquid nitrogen temperature (sample temperature 82 K) over a maximum scan range of 8 μm. Spin-coated DCNDBQT and other molecules were also investigated in air using a commercial AFM in tapping mode as well as being measured with a home-built nc-AFM. Both AFMs were working in air at room temperature.

nc-AFM was performed with homebuilt electronics based on the principle of phase locked loop (PLL) techniques [18]. In our setup, a micro fabricated cantilever carrying a tip at its front end is excited at its eigenfrequency and a constant phase between the excitation signal and the oscillation signal. The cantilever oscillation amplitude is kept constant within a few Angstroms by a digital proportional/integral feedback controller. A second feedback controller was used to keep the distance between tip and sample surface constant via keeping the frequency shift of the cantilever oscillation at a preset value.

For all nc-AFM measurements, a Kelvin probe force microscopy (KPFM) feedback controller was additionally activated for simultaneous topographic imaging [19]. In order to compensate for electrically or electronically induced artefacts, an ac voltage was applied between tip and sample and used in combination with lock-in techniques and a feedback controller to compensate for the contact potential difference (CPD) between tip and sample. With this method, nc-AFM is assured to image the sample topography without any artefacts originating from different local surface potentials [20].

For measurements of thin films of DCNDBQT on TiO_2/Si substrates X-ray scattering methods (reflectometry and diffuse scattering) to study the film parameters like thickness, interface structure, and ordering were used, as well as X-ray diffraction in grazing incidence mode was applied to find a possible crystalline phase. These measurements were done using D5005 (θ–2θ-geometry) and D5000 (θ–θ-geometry) diffractometers, respectively. Both diffractometers use a Göbel-mirror to enhance both the collimation (2 mrad), and the intensity of the incoming X-ray beam.

5.2.2
Sample Preparation

All substrates were ultrasonically cleaned with toluene and ethanol for 5 min. Organic residues were removed by treatment with an UV-ozone lamp for 20 min.

DCNDBQT was deposited on the substrate (20 nm TiO_2 on silicon, Paul-Scherrer Institute, Switzerland) by spin-coating (5 s at 200 rot/min followed by 20 s at 2000 rot/min) from xylene solution at 50 °C on a pretreated substrate (50 °C). The used concentration was 30 mg/mL. Then the substrates were annealed at 80 °C for 2 h.

DCNDBQT was physically vapour deposited (PVD) under ultra high vacuum conditions (UHV) with a pressure of $\sim 5 \times 10^{-8}$ mbar. The substrates (TiO_2/Si) were annealed to 180 °C for 20 minutes in order to remove any water layer from the surface. The molecules were evaporated from homebuilt boron nitride crucibles at an evaporation rate of 1.0–1.5 ml min^{-1} with the substrates kept at room temperature. The deposition rate was monitored with a quartz microbalance and a frequency counter.

5.2.3 OFET Device Fabrication

DCNDBQT organic field effect transistors (OFETs) were fabricated on a highly doped n-Si wafer with 30 nm silicon dioxide. Firstly, the silicon surface was rinsed with DI-water, acetone and iso-propanol in order to remove small particles and organic impurities. Secondly, the substrate was treated with oxygen plasma and silanised for 26 hours at 60 °C by hexamethyldisilazane (HMDS) in order to improve the OFET performance [21]. As source–drain contacts of the bottom contact transistors (BOC) gold was used, which was evaporated through a shadow mask on the silicon dioxide (see Figure 5.2).

Then the organic material was deposited under vacuum. For the deposition process the device was shortly exposed to air during transportation from the vacuum box into and out of the evaporation chamber. The DCNDBQT layer thickness was measured using a Quartz Microbalance and found to be approximately 10 monolayers for a complete and dense integral film. As a last step, source–drain contacts were evaporated through the shadow mask for the top contact transistor configuration (TOC, see Figure 5.2), which was done under

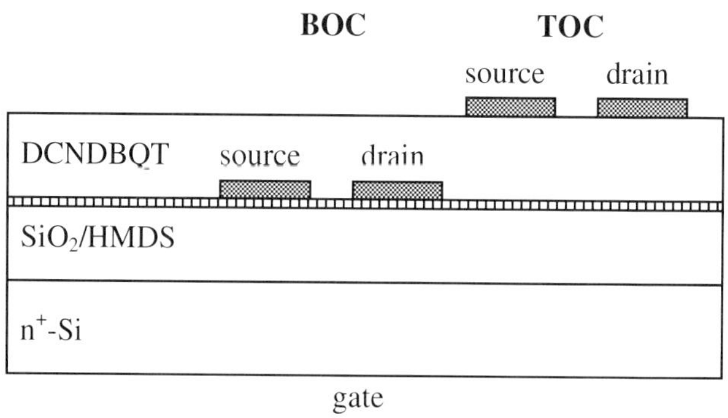

Figure 5.2 A cross-sectional diagram of the organic thin film transistor structure.

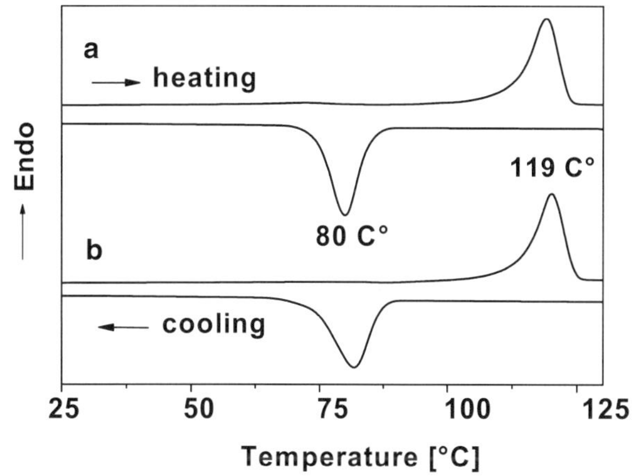

Figure 5.3 Differential scanning calorimeter thermograms of DCNDBQT, displaying (a) the first heating to 150 °C and cooling from 150 °C, and (b) the fifth heating and cooling.

vacuum and nitrogen atmosphere. As the source–drain contact materials we used gold, aluminium, and calcium. Transistor geometrical parameters were the channel length of $L = 25$ μm and width of $W = 1000$ μm. As the gate contact, we used the back of the n-Si substrate. The measurements were performed in-situ immediately after evaporation of the TOC contacts.

5.3 Results and Discussion

5.3.1 Bulk Characterisation

The thermal stability of the novel compounds DCNDBQT and CNBTPA were compared to two other thiophene molecules widely used in literature: β,β′-dibutylquaterthiophene (DBQT) and β,β′-dibutylsexithiophene (DBST). As a first characterisation we used DSC and TGA. DCNDBQT has a higher melting point compared to DBQT and a lower one than DBST. This can be explained by different steric influences of the α,β-substituted groups and the number of thiophene units to molecular symmetry and steric structure of the thiophene derivatives.

Figure 5.3a and b show the first and fifth DSC thermograms recorded for DCNDBQT.

These molecules exhibit one endothermic peak during heating at 119 °C which is due to melting of the material, and one recrystallisation peak at about 80 °C. The same behaviour is seen in the fifth cycle indicating that the material is thermally stable, and therefore annealing, spin-casting, and vacuum deposi-

Table 5.1 Thermal characterisation of thiophene compounds.

compound	T_D (°C)	T_M (°C)
DBQT	230	63.8 [22], 64.5
DBST	270	146–148 [22]
DCNDBQT	236	119.0
CNBTPA	–	114.9

T_D = decomposition temperature, T_M = melting temperature

tion processes do not alter or degrade the substance. The close melting points of DCNDBQT at 119 °C and CNBTPA at 114 °C provide a good opportunity for annealing the solution-cast as well as evaporated thin films of DCNDBQT on the adsorbed template layer.

The data are summarised in Table 5.1.

The absorption spectra of DCNDBQT dissolved in several solvents of different polarities are displayed in Table 5.2.

By comparison of the absorption maxima in the solvents used no significant shifts of the values could be observed. Repeated measurements even when waiting for some days revealed that the compounds were stable in solution and did not alter with time. These results indicated that our compound is neutral in the ground state.

Additionally, we used the known oligothiophene derivatives – β,β-dibutylquaterthiophen (DBQT) [22] and β,β′-dibutylsexithiophene (DBST) [22] to compare their properties with DCNDBQT (see Scheme 5.3).

Table 5.2 shows the optical properties of all three thiophene oligomers in dependence of the used solvents.

DBQT and DBST were insoluble in acetonitrile and, therefore, optical data of the oligomers measured in dichloromethane were used for comparison. The data show that the kind of solvent has no big influence on the value of absorption maxima.

In Figure 5.4 the absorption spectra of the three oligomers DBST, DBQT, and DCNDBQT dissolved in dichloromethane are compared to the spin-coated film of DCNDBQT are presented.

Table 5.2 Optical properties (maximum absorption (nm)) of thiophene derivatives.

compound	solvents			
	CH_3CN	CH_2Cl_2	$CHCl_3$	toluene
DBQT	–	378, 373 [22]	377	378
DBST	–	419, 412 [22]	418	419
DCNDBQT	396	397	398	397

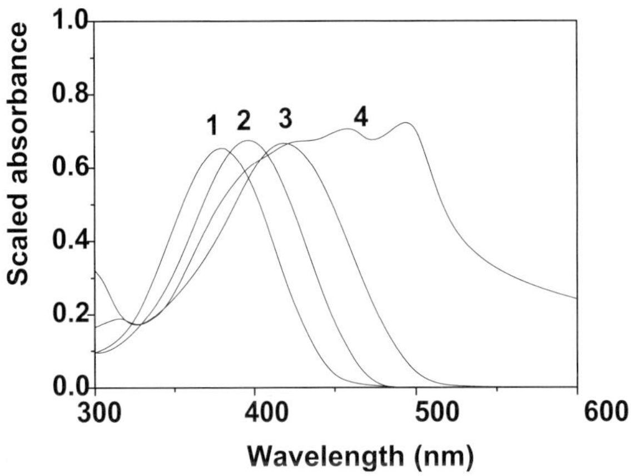

Figure 5.4 Absorption spectra of DBQT (1), DCNDBQT (2), DSQT (3) in CH_2Cl_2, and a solid DCNDBQT thin film (4).

The maximum absorption increased with increasing numbers of thiophene units. Surprisingly, the attachment of the electron-withdrawing cyano end groups produced a red shift of 20 nm compared to the unsubstituted β,β-dibutylquaterthiophene. This shift can be explained with the predominantly mesomeric effect of the cyano-groups, and therefore, the effect of electronegativity is not so strong.

DFT calculations of the DCNDBQT molecule revealed that the degree of π-orbital delocalisation of the cyano groups is very low compared to the thiophene ring system. Depending on the used calculation method, approx. 4.6% to 7.1% of the π-orbitals are situated at the CN-groups [23]. Thus, the π-system of the DCNDBQT is not weakened by the cyano groups.

The absorption spectrum of the film is very broad exhibiting three maxima at 434 nm, 465 nm and 490 nm resulting in absorption edges of 445 nm, 470 nm, and 499 nm, respectively. This bathochromic shift and the fine structure for the solid film indicate the appearance of a close-packed film structure. The absorption spectra recorded in solvents and film did not change during time, revealing that these materials exhibited a good environmental stability.

Optical band gap energies E_g^{opt} could be calculated from the absorption spectra (see Table 5.3). The values of 2.48 eV to 2.78 eV were in the range of semi-

Table 5.3 UV–Vis absorption and band gap energy (from solution).

compound	absorption edge (nm)	E_g^{opt} (eV)
DBQT	445	2.78
DCNDBQT	470	2.64
DBST	499	2.48

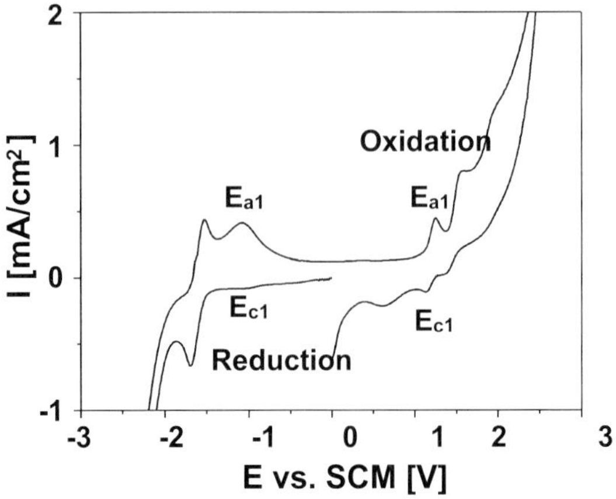

Figure 5.5 Cyclic voltammogram of DCNDBQT in AcCN.

conductor materials ($\Delta E_g < 3.0$ eV). These values also prove the minor influence of the cyano groups, because the optical band gap of DCNDBQT is lower than that for DBQT.

Cyclic voltammetry measurements revealed that in the case of DCNDBQT, different CV curves were obtained depending on the solvent used. In dichloromethane the CV exhibited two reversible oxidation and one irreversible reduction processes (not shown). The irreversibility of the reduction process can be explained by the conditions of recording the electrochemical measurements (scan rate, electrolyte, electrode material), which play an important role on the reversibility of the processes [24].

When the processes are (quasi)reversible, it is possible to calculate the formal potentials ($E^{1/2}$) [5, 24]. Measurements under the same conditions in acetonitrile revealed one reversible reduction potential, one quasireversible reduction and two quasireversible oxidation potentials (see Figure 5.5).

Therefore, we used for calculation of the HOMO and LUMO levels the CV data measured in acetonitrile. All electrochemical data are summarised in Table 5.4.

Table 5.4 Electrochemical properties of DCNDBQT.

compound	DCNDBQT
$E^{1/2}_{ox1}$ (V)	–1.18
$E^{1/2}_{ox2}$ (V)	–1.50
$E^{1/2}_{red1}$ (V)	–1.07
$E^{1/2}_{red2}$ (V)	–1.61
E^{el}_{g} (eV)	–2.25
HOMO (eV)	–5.58
LUMO (eV)	–3.33

The HOMO and LUMO energies were calculated with the use of Eq. (1a) and (1b) [16]. The electrochemical band gap values E_g^{el} were determined from the difference between HOMO and LUMO levels and found to be 2.25 eV, which is in the range of semi-conducting materials.

Optical and electrochemical band gaps are not equal. Optical measurements provide an estimation of the energy difference of the frontier orbitals, which correlate to the (optical) band gap E_g^{opt}, the absolute energetic positions of the HOMO and LUMO levels can be derived from redox potentials.

Electrochemical stability of DCNDBQT was examined by measuring repeated CV cycles and absorption spectra after electrochemical measurements (not shown). No changes were observed. These data indicated the high stability of DCNDBQT.

5.3.2
Film Characterisation

The oligothiophene films were prepared by solution cast methods and by vacuum deposition. Depending on the cast procedure, different layer thicknesses could be achieved. Preparation from solution either by drop casting or spin-coating resulted in films ranging from 50 nm to 200 nm in thickness. Vacuum sublimation yielded very thin layers in the range of 10 nm to 15 nm, which nevertheless allowed to build-up and operate an OFET structure of molecular thickness as demonstrated in this chapter.

The investigation of the degree of molecular alignment in the thin films was performed by AFM measurements in contact as well as in non-contact mode. Figure 5.6a and b presents contact AFM images of the spin-coated DCNDBQT film on TiO_2.

The film quality (e.g., uniformity, smoothness, defined thickness) strongly depended on the cast conditions. By varying the concentration, temperature, solvent, substrate, spin-coater regime, and annealing temperature, different film morphologies could be achieved. Best results were obtained with xylene as the solvent on a preheated substrate at 50 °C. High resolution images of these films did not exhibit any nano structures compared to the evaporated films (see Figure 5.7b). Spin coated films with and without the template layer have a donut-like structure showing big differences in height of about 39 nm, which were visible from the error-image of topography measurements (Figure 5.6b).

The thin film consisted of large particle agglomerates, and most of the agglomerates were disconnected. XRD measurements revealed (image not shown) an amorphous structure without distinct X-ray reflection peaks. In conclusion the molecules deposited by solution cast methods are randomly oriented on the Si/TiO_2 substrate.

In contrast, both vacuum-deposited films on TiO_2 with and without template layer showed a well-ordered and terraced structure with a step height of 1.5 ± 0.1 nm (see Figures 5.7 and 5.8). The data were determined with height distri-

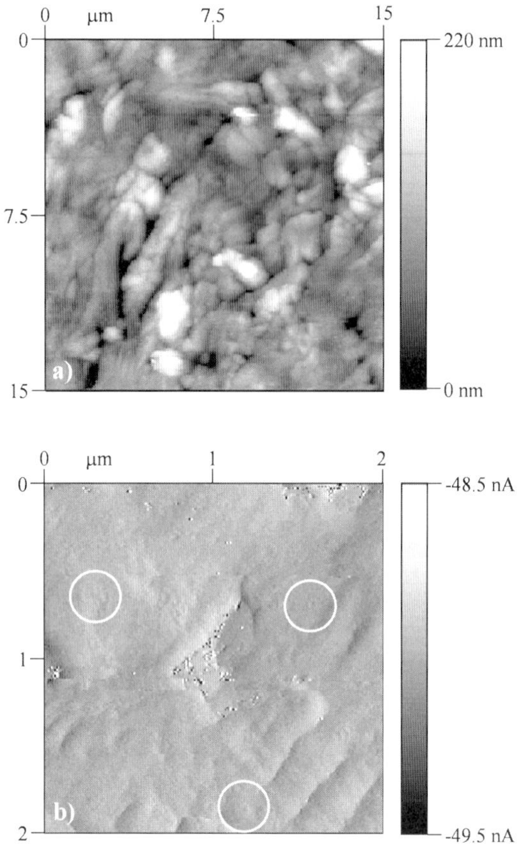

Figure 5.6 (a) Contact AFM image of DCNDBQT film deposited by spin-coating (2000 rot/min from xylene at 50 °C, 2 h annealing at 80 °C) without template layer on TiO_2; (b) error-image of topography.

bution in the framed areas, which corresponds very well with the long axis of the molecule and the X-ray data, confirming that the molecules are arranged on the substrate in the desired upright orientation.

When DCNDBQT was deposited onto the bare substrate (usually no template layer) we observe a double row structure as shown in Figure 5.7. In these rows the imaged structures are about 0.7 nm separated from each other, while the periodicity of the rows measures 2.0 nm. We determined a repeat unit (a = 2.0 nm, b = 0.7 nm, c = 1.5 nm), which could contain two molecules when assuming a densely packed structure. This is due to the molecular structure of DCNDBQT, taking into account that the butyl groups are tilted out of the molecule plane. The unit cell is not orthogonal, but shows an angle of ~78° between the surface vectors $\boldsymbol{a}$ and $\boldsymbol{b}$. For better visualisation the $\boldsymbol{a}$ and $\boldsymbol{b}$ parameters are multiplied by factor 4.

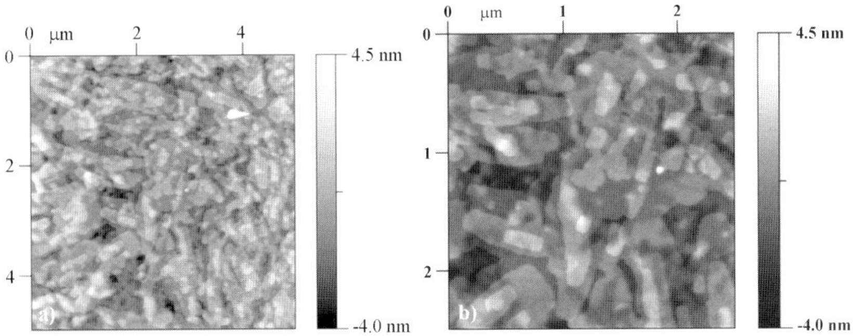

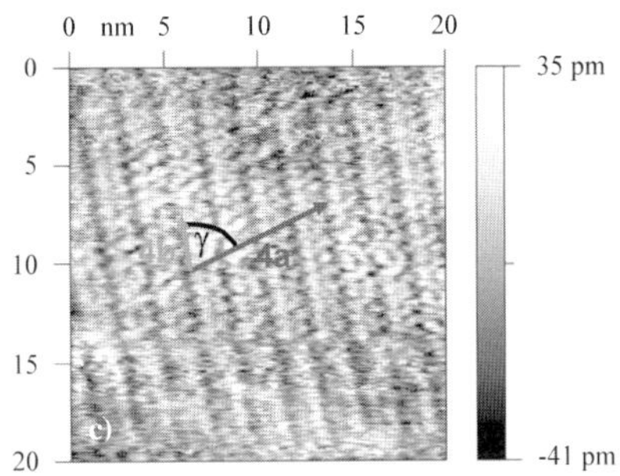

Figure 5.7 Non-contact AFM images of DCNDBQT without template layer evaporated on TiO_2: (a) 5 µm image of the prepared sample, (b) zoomed image showing the terrace structures, (c) high resolution image with a unit cell of $a = 2.0$ nm, $b = 0.7$ nm, $c = 1.5$ nm.

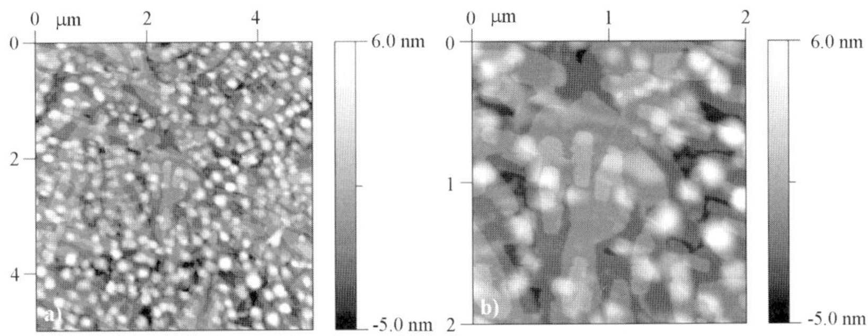

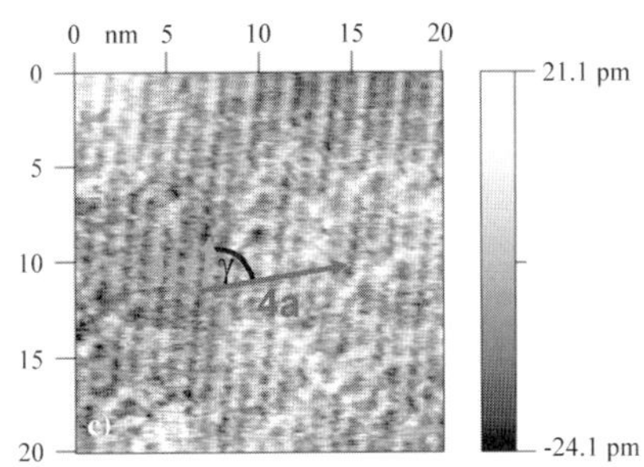

Figure 5.8 Non-contact AFM images of DCNDBQT evaporated on TiO_2 with a CNBTPA template layer: (a) 5 µm overview of the prepared sample, (b) zoomed image (terraces and the hill like structures), (c) high resolution image with a DCNDBQT unit cell of $a = 2.0$ nm, $b = 0.7$ nm, $c = 1.5$ nm.

Measurements of evaporated DCNDBQT on SiO_2 substrates revealed that similar structures were formed (images not shown). The step height was determined to be approx. 2.0 ± 0.4 nm indicating that again the molecules are standing upright on this substrate as well. On the contrary, when we evaporated DCNDBQT onto a nonreactive substrate like graphite (not shown) we detected structures, where the molecules are laying flat on the support with a step height of 0.16 ± 0.02 nm. Therefore, it can be concluded that evaporated DCNDBQT molecules stand upright on oxide substrates like TiO_2 and SiO_2, whereas they lay flat on hydrophobic samples like graphite.

We repeated the DCNDBQT deposition onto TiO_2 substrate that was first covered with a self-assembly monolayer of CNBTPA. Figure 5.8 displays some of these results. The data for DCNDBQT packing are the same as already reported for Figure 5.7 having no DCNDBQT film promoting SAM layer.

Again, we assert that the imaged structures are forming double rows, and also the distances were the same, indicating that under a nanoscopic view, there is no difference between the molecular growth with and without a template layer. However, looking at the micrometer length scale the sample without template layer had a roughness of 1.3 nm, which resulted from the well-ordered terrace structure, while the sample with template layer showed a slightly higher roughness of 2.1 nm, which might result from the large number of hill-like structures in addition to the terraces.

From the high resolution images of Figures 5.7 and 5.8 as well as from the XRD data we propose a model of the unit cell with $a = 2.0$ nm, $b = 0.7$ nm, and $c = 1.5$ nm. It contains two almost upright molecules, which are slightly tilted towards the surface normale as shown in Figure 5.9.

Figure 5.10 displays the X-ray pattern of the DCNDBQT thin film prepared by vacuum deposition. For this investigation the same sample was used, which was measured before by nc-AFM under UHV conditions.

The diffractograms contain in the low angle region a number of Kiessing fringes demonstrating the flatness of the film. From the angular distance of these fringes, an integral film thickness of 13 nm was calculated. Close to the

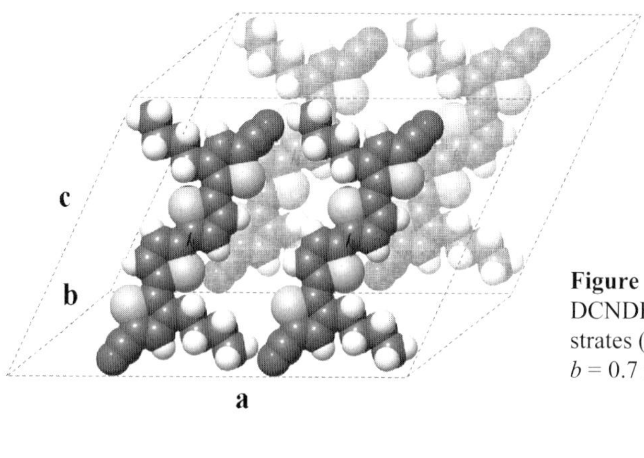

Figure 5.9 Unit cell of DCNDBQT on oxidic substrates ($a = 2.0$ nm, $b = 0.7$ nm, $c = 1.5$ nm).

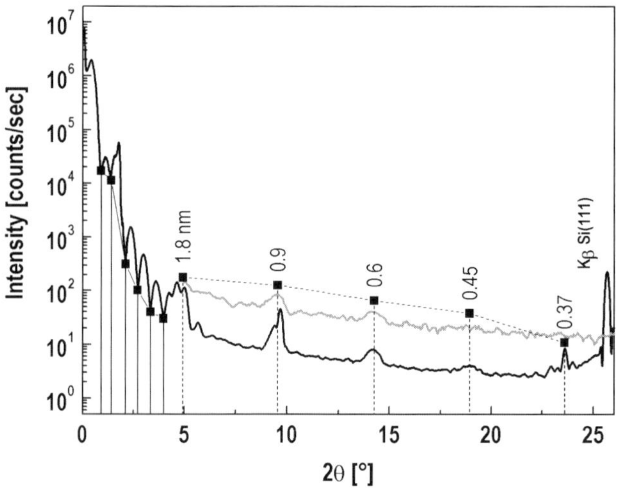

Figure 5.10 X-Ray diffractometry of a DCNDBQT thin film.

2θ-angles at 5 degrees, 10 degrees, 14 degrees, and 19 degrees, a series of reflection peaks was found corresponding to a super structure. Following the Bragg formula:

$$D = \nu\lambda / \sigma\iota\nu(2\theta/2)/2\,, \tag{2}$$

a thickness of 1.8 nm was calculated (see black curve in Figure 5.10).

It is shown that DCNDBQT is almost vertically aligned to the substrate indicating a highly ordered film. In comparison with the AFM results, which were performed in vacuum, the film properties seem to change when the layers were exposed to air. This could be confirmed, when the measurements were repeated after 50 days (see grey curve in Figure 5.10). Whereas the film thickness and roughness (not shown) did not change, we found an intensity decrease of the super structure peaks and an increase of the background scattering pointing out a reduction of the intermolecular ordering.

5.3.3 OFET Performance Characteristics

Electrical measurements were performed using a Keithley 4200 SMU-analyser under dry nitrogen.

Figure 5.11 shows the output characteristics of the top contacted OFET structure using gold source–drain contacts. The drain current I_D is plotted as function of the drain voltage V_{DS} with reversibly sweeping the applied gate voltage V_G from −12 V to −4 V.

Our OFET tested here is made up of pure DCNDBQT molecules as the molecular semi-conducting layer, and shows good saturation for all values V_{GS}. For small V_{DS}, I_D behaves quasi linear, while for large V_{DS} a low non-linear be-

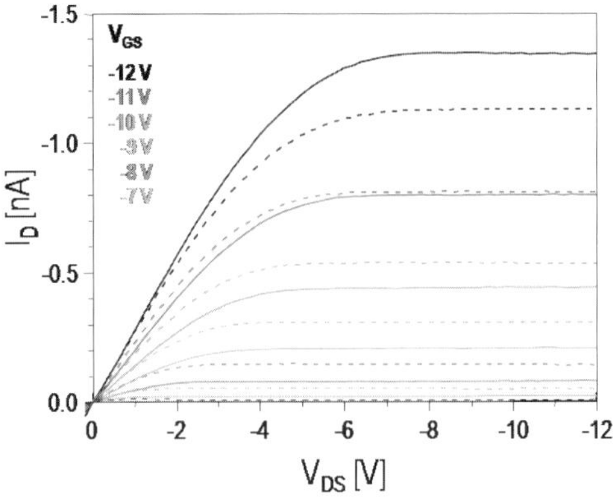

Figure 5.11 Output characteristics of the top contact OFET with gold source–drain contacts. Good saturation properties are observed for all gate voltages. Also note the forward/backward hysteresis.

haviour of the output characteristic is observed: Both electronic traps within the bulk organic thin film and/or mobility anisotropy are possible reasons for the reported non-linearity [25].

Also, we observed hysteresis effects (see forward and backward runs in Figure 5.11) most probably due to impurities in the oligomer layer, as already indicated by AFM and X-ray data.

The hole mobility μ_h was calculated using the Shockley equation for linear (Eq. (3a)) and saturation (Eq. (3b)) regimes [26–28] of the OFET correspondingly:

$$I_D = \mu \frac{W}{L} C_{ox} \left((V_{GS} - V_T) V_{DS} - \left(\frac{V_{DS}^2}{2} \right) \right), \tag{3a}$$

$$I_D = \mu \frac{W}{L} C_{ox} \frac{(V_{GS} - V_T)^2}{2}, \tag{3b}$$

where C_{ox} is the oxide capacitance of the substrate, L the channel length, and W the channel width.

The mobilities were determined from the slope of the linear portion of $|I_{D,sat}|^{1/2}$ as a function of V_G above the threshold voltage V_T and found to measure approximately 1.6×10^{-5} cm²/Vs and 2.3×10^{-5} cm²/Vs.

The transfer characteristics of the OFET structure with gold source–drain contacts for the linear regime at $V_{DS} = -3$ V and for saturation at $V_{DS} = -12$ V are presented in Figure 5.12a and b, respectively.

The transistor with DCNDBQT as an active layer has good performance parameters such as (i) a threshold voltage $V_T \approx -5.5$ V, (ii) an inverse sub-threshold slope of 0.65 V/dec, and (iii) an on/off ratio $>10^3$. Also a hysteresis effect was detected in the transfer characteristics $\Delta V_{GS} \approx 0.7$ V.

As already mentioned above, the transfer characteristics of an OFET structure can be tested both in the BOC and TOC configuration. However, when performing our measurements in the BOC design using gold contacts resulted in no drain current at all. A probable reason for this is the channel for-

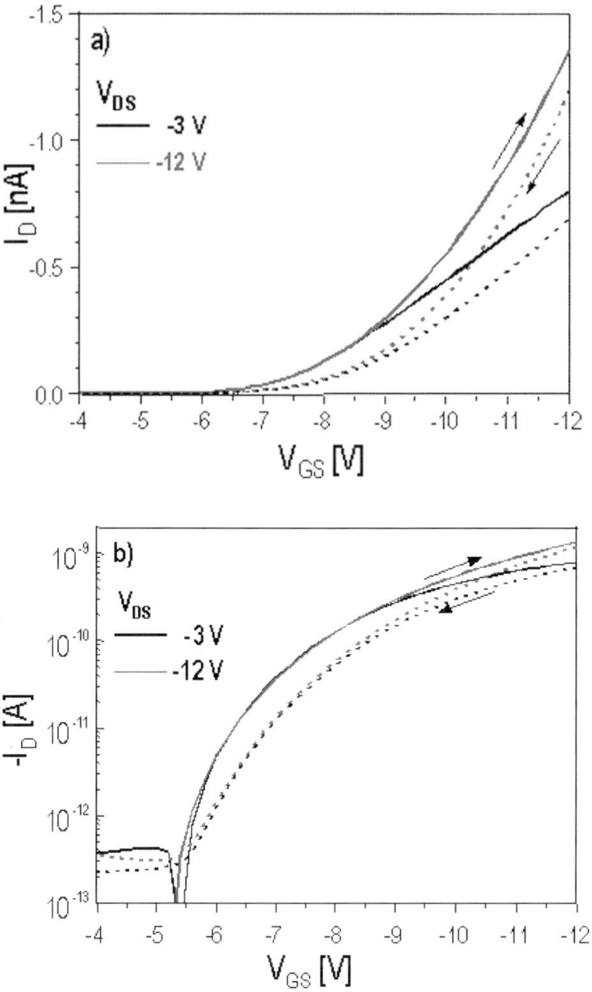

Figure 5.12 Transfer characteristics of the top contact OFET with DCNDBQT oligomer and gold source–drain contacts: (a) Enlarged section of transfer characteristic; (b) complete characteristic (note the different scales).

mation impossibility at the DCNDBQT/SiO_2/HMDS interface (see Figure 5.2) in consequence of a barrier between the gold source–drain material and the oligomer.

In addition, we also investigated TOC OFETs with source–drain contacts made from aluminium (work function 4.3 eV) and calcium (work function 2.9 eV). The LUMO of DCNDBQT measures –3.33 eV. Therefore, calcium forms an ohmic contact for electrons and a Schottky contact for holes. However, we were not able to observe any drain current for these transistor design structures, whether being n- or p-type. The direct investigation in situ without exposure to air before and after organic layer fabrication thus is mandatory und will be followed in our future work.

5.4 Conclusion

Novel oligomers based on β-substituted thiophene derivatives were synthesised with the aim to build-up a small molecule organic field-effect transistor (OFET). The developed material, α,ω-dicyano-β,β′-dibutylquaterthiophene (DCNDBQT), exhibits excellent thermal, optical and electrochemical stability.

Here, we investigated the properties of the applied films of DCNDBQT by different cast methods – spin-coating and vacuum sublimation.

The ultra-thin organic film formation on TiO_2 templates was effectively promoted through the specifi-cally designed, bifunctional self-assembly molecules (SAM) 5-cyano-2-(butyl(4-phosphonic acid))-3-butyl-thiophene (CNBTPA), whereas DBQT and DBST did not improve their layer structure when they were evaporated on the SAM layer (images not shown).

Different films were obtained depending on the cast method: Spin-coated films demonstrated thicknesses higher than 50 nm and did not form crystalline structures, which was proven by contact atomic force microscopy studies and X-ray diffraction measurements.

However, excellent structural properties were found for up to 9 DCNDBQT molecule thick films prepared through vacuum sublimation as investigated with UHV non-contact AFM and XRD. Both X-ray and AFM data indicate that the DCNDBQT molecules form a well-ordered terraced structure exhibiting step heights of 1.5 nm to 1.8 nm. Hence, the DCNDBQT molecules are linked to the functional SAM interface layer by H-bond interactions standing practically perpendicular to the TiO_2 template, and thus providing optimal orbital overlap between neighbouring thiophene rings. The vacuum deposited, ultra thin layers (in the range of 5 to 10 monolayers DCNDBQT) only remained stable under vacuum conditions. Note, that when exposed to air, the film properties changed (film thickness), which could be proven by the XRD measurements.

Finally, the electrical performance of an OFET-structure made of vacuum deposited DCNDBQT, which form a closed packed and dense molecular layer, was tested. The output characteristics of the OFET clearly show the behaviour of a p-type semiconductor. The resulting field mobilities of 10^{-5} cm^2/Vs reflect a high current density as interpreted in the framework of the ultra-thin molecular OFET structure. The sensitivity of the ultra thin material towards the environment necessitates that in the future, the OFET has to be produced and measured under vacuum-like conditions to the exclusion of air, which may be realised by appropriate passivation processes.

Acknowledgements

We would like to thank Dr. P. Friedel for *ab initio* calculation dimension of DCNDBQT. The authors are thankful for providing financial support by DFG (Project Schwerpunktprogramm 1121 OFET).

References

1. A. Facchetti, Mater. Today **10**, 28 (2007).
2. T. M. Barclay, A. W. Cordes, C. D. MacKinnon, R. T. Oakley, and R. W. Reed, Chem. Mater. **9**, 981 (1997).
3. A. Yassar, F. Demanze, A. Jaafari, M. El Idrissi, and C. Coupry, Adv. Funct. Mater. **12**, 699 (2002).
4. H. Wada, T. Taguchi, M. Goto, T. Kambyashi, T. Mori, K. Ischikawa, and H. Takezoe, Chem. Lett. **35**, 280 (2006).
5. A. Facchetti, M.-H. Yoon, C. L. Stern, G. R. Hutchison, M. A. Ratner, and T. J. Marks, J. Am. Chem. Soc. **126**, 13480 (2004).
6. S. E. Fritz, S. Mohapatra, B. T. Holmes, A. M. Anderson, C. F. Prendergast, C. D. Frisbie, M. D. Ward, and M. F. Toney, Chem. Mater. **19**, 1355 (2007).
7. J. Ackermanna, C. Videlota, P. Raynala, A. El Kassmia, and P. Dumas, Appl. Surf. Sci. Chem. Lett. **212/213**, 26 (2003).
8. E. Lim, B.-J. Jung, H.-K. Shim, T. Taguchi, B. Noda, T. Kambayashi, T. Mori, K. Ishikawa, H. Takezoe, and L.-M. Do, Org. Electron. **7**, 121 (2006).
9. M. Rittner, P. Baeuerle, G. Goetz, H. Schweizer, F. J. Balt Calleja, and M. H. Pilkuhn, Synth. Met. **156**, 21 (2006).
10. J. H. Kwon, J. H. Seo, H. Kang, D. H. Choi, and B. K. Ju, J. Appl. Phys. **101**, 064502 (2007).
11. N. Kiriy, A. Kiriy, V. Bocharova, M. Stamm, S. Richter, M. Plötner, W.-J. Fischer, F. C. Krebs, I. Senkovska, and H.-J. Adler, Chem. Mater. **9**, 4757 (2004).
12. P. T. Nguyen, U. Rammelt, W. Plieth, S. Richter, M. Plötner, W.-J. Fischer, N. Kiriy, K. Potje Kamloth, and H.-J. Adler, Electrochim. Acta **50**, 1757 (2005).
13. D. H. Kim, Y. D. Park, Y. Jang, H. Yang, Y. H. Kim, J. I. Han, D. G. Moon, S. Park, T. Chang, Ch. Chang, M. Joo, Ch. Y. Ryu, and K. Cho, Adv. Funct. Mater. **15**, 77 (2005).
14. V. Cuberos Guzman, R. Ponce Ortiz, M. C. Ruiz Delgado, R. Azumi, R. T. Oakley, J. Casado, V. Hernandez, and J. T. Lopez Navarrete, J. Mol. Struct. **744–747**, 403 (2005).
15. K. Haubner, E. Jähne, H.-J. Adler, A. A. Levin, and D. C. Meyer, in preparation.
16. S. A. Ponomarenko, S. Kirchmeyer, A. Elschner, N. M. Alpatova, M. Halik, H. Klauk, U. Zschieschang, and G. Schmid, Chem. Mater. **18**, 579 (2006).
17. H. J. Hug, B. Stiefel, P. J. A. van Schendel, A. Moser, S. Martin, and H.-J. Güntherodt, Rev. Sci. Instrum. **70**, 3625 (1999).
18. C. Loppacher, M. Bammerlin, F. Battiston, M. Guggisberg, D. Müller, H. R. Hidber, R. Lüthi, E. Meyer, and H. J. Güntherodt, Appl. Phys. A **66**, 215 (1998).
19. U. Zerweck, C. Loppacher, T. Otto, S. Grafström, and L. M. Eng, Phys. Rev. B **71**, 125424 (2007).

20. S. Sadewasser and M. Lux-Steiner, Phys. Rev. Lett. **91**, 266101 (2003).

21. H. Sirringhaus, N. Tessler, and R. H. Friend, Science **280**, 1741 (1998).

22. J. K. Herrema, Tuning of the luminescence in poly[(silanylene)thiophene]s (Rijksuniversiteit Groningen, Groningen, 1996), p. 161.

23. Private communication, R. Luschtinetz and Prof. G. Seifert, TU Dresden, 2007.

24. P. Hapiot, F. Demanze, A. Yassar, and F. Garnier, J. Phys. Chem. **100**, 8397 (1996).

25. A. Herasimovich, S. Scheinert, and I. Hörselmann, J. Appl. Phys. **102**, 054509 (2007).

26. G. Horowitz, Adv. Mater. **10**, 365 (1998).

27. T. Yasuda and T. Tsutsui, Chem. Phys. Lett. **402**, 395 (2005).

28. T. Koyanagi, M. Muratsubaki, Y. Hosoi, T. Shibata, K. Tsutsui, Y. Wada, and Yukio Furukawa, Chem. Lett. **35**, 20 (2006).

6
Organic Transistors Utilising Highly Soluble Swivel-Cruciform Oligothiophenes

Achmad Zen, Patrick Pingel, Dieter Neher, and Ullrich Scherf

6.1
Introduction

Many design strategies have been investigated in an attempt to achieve sufficient solubility and self-assembly of π-conjugated systems [1]. Although thin films fabricated from π-conjugated polymers such as poly(3-hexylthiophene) (P3HT) can be easily processed from solution, their charge carrier mobilities are highly influenced by the purity [2], regioregularity [3], and molecular weight [4–7]. Mobilities of up to 0.1 cm²/Vs at room temperature are observed for high regioregularity P3HT using optimised processing conditions [8]. Such samples exhibit a microphase-separated morphology with layers of regularly packed main chains separated by the disordered side chains. In fact, our earlier study on P3HT with different molecular weight has revealed a strong dependence of mobility on layer crystallinity [5]. It is well-known that oligothiophenes with a well-defined chemical structure crystallise well in a layered morphology in which the molecules stand almost upright on the substrate. This guarantees small intermolecular distances and efficient charge transport in the layer plane. Therefore, the study of oligothiophenes may provide us a better understanding on the relation between the crystallinity and field-effect mobility. Oligothiophenes also have been extensively investigated as components in OFETs and exhibit reasonable mobilities when deposited by thermal evaporation in vacuum ($\mu < 1$ cm²/Vs) [9–15]. It is well known that vacuum deposition techniques generally produce better-ordered layers with higher magnitude mobilities than those prepared from solution ($\mu < 0.1$ cm²/Vs) [16–20]. However, these processes can be costly and also tend to be wasteful of material. Therefore, there is demand for soluble thiophene oligomers.

Here, we summarise results obtained on soluble swivel-cruciform oligothiophenes. The term swivel-cruciform reflects the fact that the single bond connection of the two non-substituted oligothiophenes at the central thienylene moieties permits, in contrast to the corresponding spiro-type dimers [21], a certain degree of rotation of the arms. This leads to a good solubility even without the attachment of solubilising alkyl chains. Initially, a series of swivel-cruci-

Organic Electronics. Structural and Electronic Properties of OFETs. Edited by Christof Wöll

ISBN: 978-3-527-40810-8

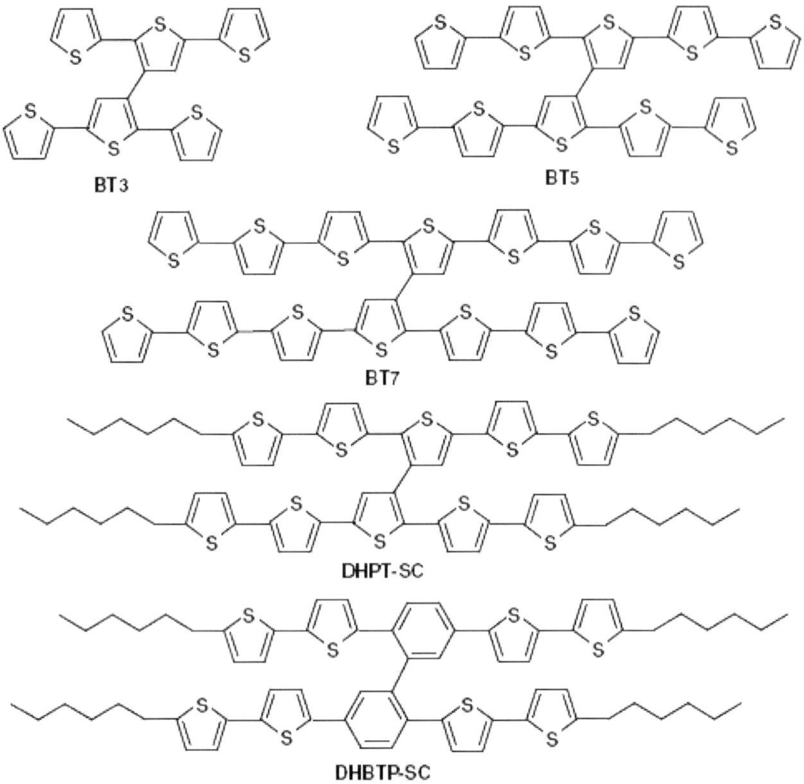

Figure 6.1 Molecular structure of the conjugated organic materials used in this work: bis(terthiophene) (BT3), bis(pentathiophene) (BT5), bis(heptathiophene) (BT7), α,α′-dihexylpentathiophene based swivel cruciform (DHPT-SC) and (dihexylbithiophene)$_2$-phenyl swivel cruciform (DHBTP-SC).

form oligothiophene dimers with increasing chain length (see Figure 6.1) – bis(terthiophene) (BT3), bis(pentathiophene) (BT5) and bis(heptathiophene) (BT7) – were used as semiconductor in organic field-effect transistors (OFETs). From this initial study, we have found that BT5 is the most prominent dimer for OFETs with a field-effect mobility of 3.7×10^{-5} cm^2/Vs and a current on/off ratio of $>10^3$. Therefore, two modifications were done to BT5, firstly an attachment of hexyl substituents to both terminals in order to obtain α,α′-dihexylpentathiophene swivel cruciform (DHPT-SC). Secondly, a modification to BT5 dimer-centre with biphenyl to get (dihexyl-bithiophene)$_2$-phenyl swivel cruciform (DHBTP-SC), see Figure 6.1. These modifications lead to increased solubility, crystallinity, better film forming properties and better field-effect mobilities. Note that the syntheses of these oligothiophene-based swivel cruciforms have been reported in our earlier publications [22–

24]. Using DHPT-SC as the semiconductor in OFETs, a field effect mobility of 0.012 cm^2/Vs and a current on/off ratio of $>10^5$ can be realised, which is among the highest OFET mobilities fabricated achieved with solution-processed oligothiophenes.

6.2 Optical and Thermal Properties

6.2.1 Optical Properties

As expected, the UV-Vis spectra of the swivel-cruciform oligothiophene dimers BT3, BT5 and BT7 exhibit a clear red-shift of the absorption maximum with increasing length of the oligothiophene in both solution (Figure 6.2a) and solid states (Figure 6.2b). The size of our most extended oligomer BT7 is still significantly below the effective conjugation length of oligothiophenes. Bäuerle et al. calculated the effective conjugation length of oligothiophenes by plotting the inverse chain length ($1/n$) versus the absorption energy. An extrapolation of the graph indicates no convergence of the optical properties until $n \sim 16$ [25–27].

The absorption maxima of the oligothiophene dimers in solution range from 361 nm (BT3) and 426 nm (BT5) to 451 nm (BT7). These values are similar to the optical properties of the corresponding unsubstituted linear oligothiophenes T3–T6 [28]. There is only a minor redshift of max of ca. 9 nm in BT3 compared to T3 and BT5 compared to T5, which is probably due to a weak intramolecular electronic interaction of the arms in BT3 and BT5. Comparison of the solution and solid state spectra (Figure 6.2a, b) shows a distinct red shift of the films spectra, which is more pronounced for the more extended dimers

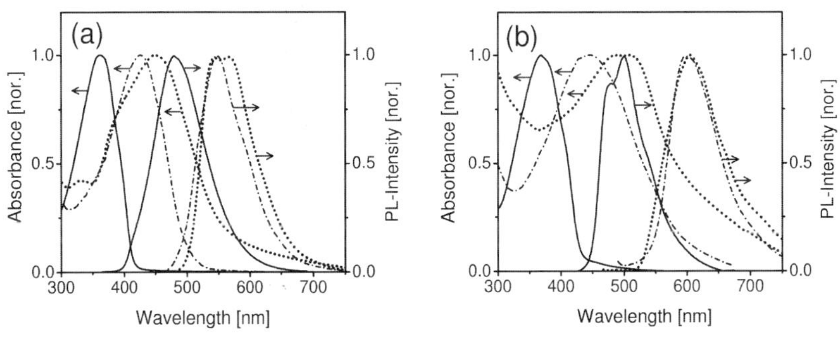

Figure 6.2 UV–Vis absorption and photoluminescence spectra of BT3 (solid lines), BT5 (dashed-dotted lines) and BT7 (dotted lines) recorded in chloroform solution (a) and as spin-coated thin films (b).

BT5 and BT7. The redshift is 7 nm for BT3, 20 nm for BT5 and 58 nm for BT7. As in poly(3-hexylthiophene)s (P3HT), this shift is explained by a transition into an intramolecularly higher ordered solid state structure [4]. In accordance with this explanation the appearance of a distinct vibronic structure is observed, especially for BT7 as also observed by other research group [29]. The PL emission maxima also show a bathochromic shift with increasing number of thiophene units. However, the PL spectra of BT5 and BT7 are very similar, due to essentially faster convergence of the emission properties. The solid state PL spectra reflect similar trends as described for the absorption properties.

The optical absorption and emission spectra of dilute chloroform solution of DHBTP-SC and DHPT-SC are shown in Figure 6.3a. The absorption spectra for both oligomers are broad and featureless with the all-thiophene cruciform DHPT-SC exhibiting the lower energy absorption as expected. Note that the absorption spectrum of DHPT-SC resembles that of BT5, naturally. Unlike the broad emission peak of DHPT-SC in solution (536 nm) the emission spectrum of DHBTP-SC exhibits two well-resolved maxima located at 461 nm and 484 nm. This may indicate that the emitting species adopts a more planar ordered structure in solution upon photoexcitation. Comparing the spectra recorded in solution and thin films (Figure 6.3a, b), the cruciforms exhibit no strong changes in absorption. This may indicate that the ground state conformation is quite similar in solution and solid state. The energy band gap (E_g), calculated from the onset of the solid state absorption spectra, taking into account an exciton energy binding of 300 meV, yields values of $E_g = 3.05$ eV and 2.65 eV for DHBTP-SC and DHPT-SC, respectively. For both cruciforms, the solid-state emission spectra exhibit a significant red shift and become broader compared to their spectra recorded in solution. This is a common feature of the optical spectra of swivel-cruciform structures and it is postulated that this is due to the formation of an intramolecular excimer upon photoexcitation.

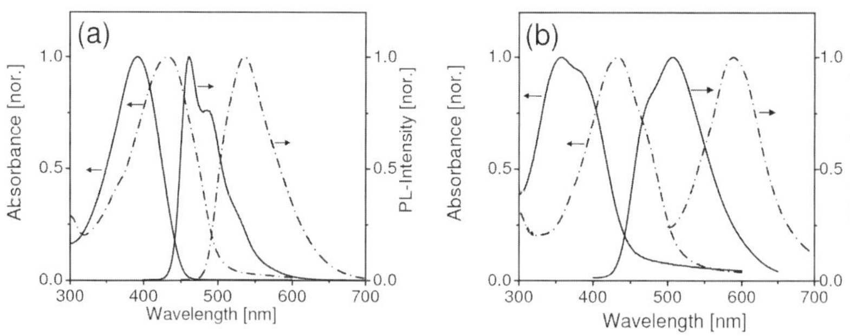

Figure 6.3 Absorption and photoluminescence spectra of DHBTP-SC (solid lines) and DHPT-SC (dashed-dotted lines) recorded in chloroform solution (a) and in thin films (b).

6.2.2
Thermal Properties

The thermal properties of all cruciform oligothiophenes were measured by differential scanning calorimetry (DSC) and are found to depend on the number of thiophene units and the presence of flexible side chains. Oligothiophene BT3 exhibits both a glass transition temperature T_g at 44 °C and a melting point T_m at 207 °C during heating (not shown here). In contrast, BT5 and BT7 display only a glass transition temperature at 107 °C and 146 °C, respectively. Melting points of BT5 and BT7 have not been observed and may be located above the decomposition temperature. The increase of the glass transition temperature indicates a higher morphological stability of the higher oligomers in the amorphous solid state. The DSC results for DHPT-SC and DHBTP-SC are shown in Figure 6.4. Both oligomers exhibit good thermal stabilities with melting and crystallisation temperatures above 150 °C, while the sharpness of melting and crystallisation peaks indicate good and homogeneous ordering within the materials. The endothermic melting enthalpies for DHBTP-SC and DHPT-SC were calculated from the integration of the melting curves and are 55.93 J/g and 50.90 J/g, respectively. They are ca. three times higher than the melting enthalpy measured for the highest molecular weight of P3HT as we published earlier [5].

6.3
Morphology Studies on Layers of Substituted Cruciforms

As we found that the OFETs performance of DHBTP-SC and DHPT-SC are far superior compared to BT3, BT5, and BT7 series, our focus was to study the layer morphology of these two prominent materials.

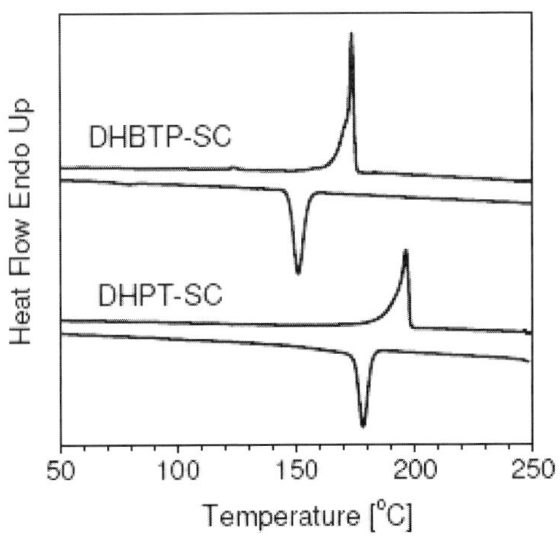

Figure 6.4 DSC thermograms of powder samples measured from DHBTP-SC and DHPT-SC.

6.3.1 XRD Studies

Figure 6.5a shows the reflectivity and XRD-GID (grazing incidence angle XRD) measured for thin films (40–55 nm) of DHBTP-SC. The films were prepared from chloroform solutions and were subsequently annealed at 120 °C for 5 minutes followed by slow cooling at a rate of 1.3 K/min. The first and second order Bragg peaks are clearly resolved in the reflectivity measurement as shown in the top curve (black line). The absence of any further prominent Bragg peaks suggests a layered structure. The *d*-spacing, calculated from the reflectivity curve is 3.2 nm and is attributed to the periodicity of DHBTP-SC within the lamellar structure and perpendicular to the substrate. In fact, this periodicity value is comparable with the length of a DHBTP-SC molecule. From the full width at half maximum (FWHM) of the Bragg peaks, the average domain size perpendicular to the substrate plane was estimated to be 30 nm. Further XRD-GID measurements were performed to investigate the surface and bulk crystallinity. Thin films from DHBTP-SC displayed first and second order Bragg peaks, independent of the incidence angles, indicating that the thin films of DHBTP-SC are uniformly crystalline throughout the layers (Figure 6.5a, coloured curves). In order to illustrate this further, reciprocal space maps were constructed from the measured data. In these maps, the scattering intensity is plotted as a function of the scattering angle (as a measure for the momentum transfer) and the incident angle (determining the penetration depth). Figure 6.5b shows the reciprocal space map for DHBTP-SC. In this figure, the intensity distribution of the first and second order Bragg peaks can be clearly identified as horizontal stripes. With increasing incidence angles, the intensity patterns at the first and second order Bragg peaks are homogeneous, which indicates that the crystalline domains are almost without any preferential orientation (there is no predominant orientation) throughout the layer.

Figure 6.5c shows the reflectivity (black curve) and XRD-GID data (coloured curves) obtained from a thin film of DHPT-SC. In the reflectivity curve, Kiessig peaks can be observed at low angles indicating that the DHPT-SC forms very smooth layers. From the distance between two Kiessig peaks one obtains an overall layer thickness of 52 nm. The *d*-spacing for the intermolecular distance within the lamellar structure is 3 nm, as calculated from the first and second order Bragg peaks. This value is slightly smaller than the length of a DHPT-SC molecule, which is ca. 3.4 nm and the one found in the films of DHBTP-SC. However, it is larger than the values for the related linear oligomers such as α,ω-dihexylquaterthiophene [30] or α-quinquethiophene [31]. From the FWHM of these peaks, we calculated the domain size in these films to be about 25 nm. The XRD-GID shows that the Bragg peaks are only discernible when the X-ray wave penetrates deeply into the film that is when the incidence angle α_i is larger than the critical angle of the total external reflection α_c. (Note that $\alpha_c = 0.16°$ and 0.22° for the oligomers and the silicon substrate, respectively). This clearly indicates that thin films from DHPT-SC

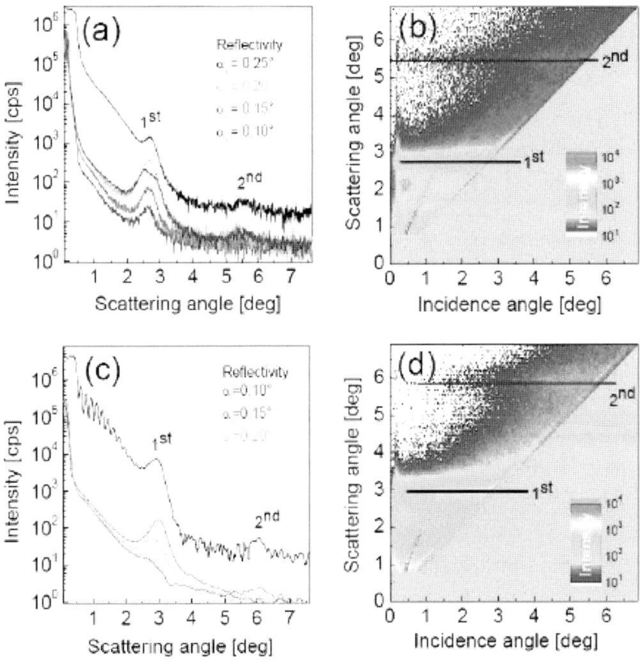

Figure 6.5 XRD patterns of DHBTP-SC and DHPT-SC that have been annealed at 120 °C for 5 minutes inside a N_2-filled glove box, followed by slow cooling to room temperature with a cooling rate of 1.3 K/min. (a) Reflectivity measurement (black line) and powder scans of DHBTP-SC with different angle of incidence (coloured), (b) reciprocal space map at small angles from powder scans of (a); the Bragg peaks can be clearly observed (horizontal stripes); the intensity at first and second order Bragg peaks are homogeneous with increasing incidence angle. (c) Reflectivity measurement (black line) and powder scans of DHPT-SC with different angle of incidence (coloured), (d) reciprocal space map at small angles from powder scans of (c); the first and second order of the Bragg peaks become more intense with increasing incidence angle.

are less crystalline in the surface compared to the bulk. The non-monotonic behaviour of the intensity of the first Bragg peak with the incident angle was due to the fact that thin films from DHPT-SC are less ordered in the surface compared to the bulk. Figure 6.5d shows the reciprocal space map for DHPT-SC. Here it is observed that when the angle of incidence is increased (deeper penetration to the films), the intensity patterns from the first and second order Bragg peaks are becoming more intense, which confirms that the crystalline domains are mainly in the bulk.

Moreover, the FWHM of the first order Bragg peak reduces slightly with increasing incidence angle indicating that the domain size at the sample surface is much smaller (ca. 10 nm) compared to the domain size in the bulk (ca. 25 nm).

6.3.2
AFM Studies

It is well known that annealing can have a pronounced influence on the morphology of organic thin films and thus affect the performance of OFETs. Therefore, we investigated the effect of annealing on the crystallinity of the oligomers by heating thin films (the same samples as investigated in OFETs) of both cruciforms at 120 °C for 30 minutes, 90 minutes and 120 minutes and comparing their AFM images to those of the as-prepared samples. Figure 6.6 displays the height images from thin films of cruciform DHBTP-SC as-prepared (Figure 6.6a) and after annealing at 120 °C for different time periods (Figure 6.6b–d). Annealing the layer for 30 minutes already leads to a significant roughening of the surface (see Table 6.1). It is attributed to the growth of large crystallites that protrude from the surface. After 120 minutes annealing, the roughness increases further. Note that the reciprocal space map as shown in Figure 6.5b revealed that thin films from DHBTP-SC are crystalline throughout the whole layer after annealing. Apparently, the crystallites seen at the surface extend deeply into the thin films and probably down to the substrate.

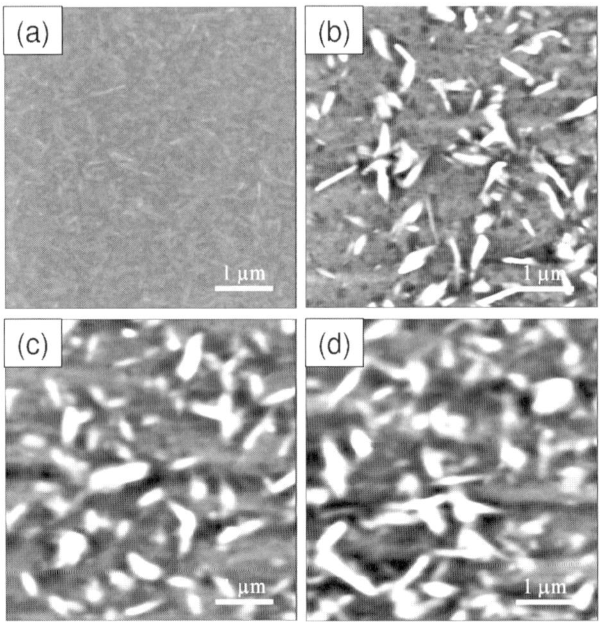

Figure 6.6 AFM height images (5 × 5 μm^2) of oligomer DHBTP-SC that has been treated as the following: (a) as-prepared, (b) annealed for 30 minutes, (c) annealed for 90 minutes and (d) annealed for 120 minutes. Annealing was performed at 120 °C inside N_2-filled glove box, followed by slow cooling to room temperature with a cooling rate of 1.3 K/min.

Table 6.1 Root-mean-square roughness (R_{RMS}) and the mean roughness (R_a) for both swivel cruciforms calculated using Veeco software and using AFM images shown in Figures 6.6a–d and 6.7a–d.

treatments	DHBTP-SC R_{RMS} (nm)	DHBTP-SC R_a (nm)	DHPT-SC R_{RMS} (nm)	DHPT-SC R_a (nm)
as prepared	1.4	1.1	2.3	1.8
annealed at 120 °C, 30 min.	10.1	6.8	2.1	1.7
annealed at 120 °C, 90 min.	10.6	7.4	2.3	1.8
annealed at 120 °C, 120 min.	12.7	9.6	4.5	3.4

On the other hand, DHPT-SC shows a rather different behaviour under the same conditions. If we compare the AFM images of the as-prepared sample (Figure 6.7a) to annealed samples (Figure 6.7b–d), it is difficult to observe any significant differences in the morphologies. Note that a significant increase in roughness occurred when the samples were annealed at 120 °C for 120 minutes. This is in agreement with the results from the X-ray measurements, which suggested that a short time of annealing does not initiate the growth of large

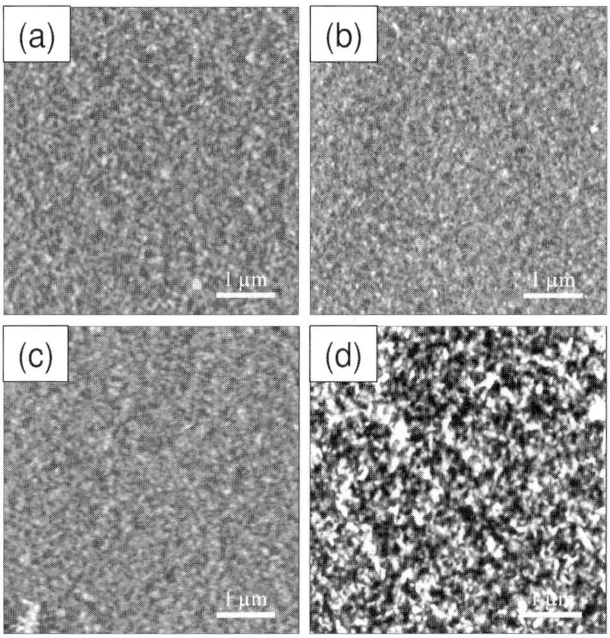

Figure 6.7 AFM height images ($5 \times 5\ \mu m^2$) of oligomer DHPTSC, treated as in Figure 6.6.

crystallites at the surface of DHPT-SC. Nonetheless, the reciprocal space map in Figure 6.5d clearly shows that the layer is highly crystalline in the bulk.

6.4 OFET Studies

OFETs were fabricated using bottom gate geometry on highly doped n-type silicon wafers which act as the gate electrode. A thermally grown 300 nm thick silicon dioxide layer with a capacitance of 11 nF/cm^2 served as the gate insulator. Additional oxygen plasma treatment and silanisation of the oxide with hexamethyldisilazane (HMDS) prior to the deposition of the oligothiophene dimers were done only for DHBTP-SC and DHPT-SC as other dimers showed dewetting on such treated substrates. BT3, BT5 and BT7 film layers were deposited by spin coating from hot chlorobenzene solution while DHBTP-SC and DHPT-SC are coated from chloroform solution. The solubility was very good for DHBTP-SC and DHPT-SC, good for BT3 and BT5 but fairly poor for BT7. 100 nm thick interdigitating Au electrode structures were evaporated on top of the thin organic semiconductor layer serving as source and drain electrodes (channel length 100 μm, total channel width 148.5 mm). All preparation processes and the characterisation of the devices were performed inside an N_2-atmosphere glove box.

Firstly, we investigated OFETs based on BT3, BT5 and BT7 as the semiconducting layers. The output characteristics from OFETs based on as prepared BT5 and BT7 layers are shown in Figure 6.8a and b, respectively. The OFETs exhibited negative amplification, which is typical of p-type semiconductors with well-defined linear and saturation regions. Although the film quality for BT3 was very good, we could not observe any reasonable transistor behaviour for the shortest dimers. The drain currents of the transistors from BT5 are distinctly higher than those of transistors made from BT7, indicating better transport properties (see Figure 6.9a). The poorer OFET performance of BT7 might

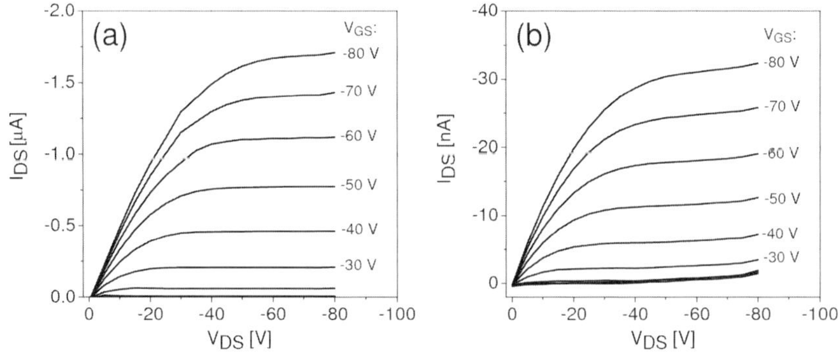

Figure 6.8 Output characteristics of OFET devices measured from as-prepared sample of (a) BT5 and (b) BT7.

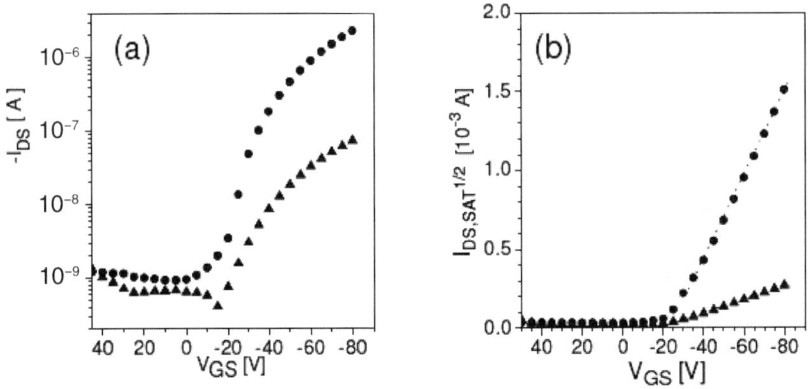

Figure 6.9 (a) Transfer characteristics of OFETs and (b) square root plot of $I_{DS,sat}^{1/2}$ versus V_{GS} (taken at a drain–source voltage $V_{DS} = -80$ V) of OFETs made from BT5 (circles) and BT7 (triangles) as the semiconducting layers.

be due to the poor solubility resulting in a lower quality of the organic films. Figure 6.9b shows square root plots of the drain current at saturation measured at $V_{DS} = -80$ V as a function of the gate voltage. The field-effect mobilities of the OFET devices were calculated from the slope of the linear fit of the data according to:

$$I_{DS,sat} = \frac{WC_i}{2L} \mu_{sat} (V_{GS} - V_0)^2 . \qquad (1)$$

Here, W and L are the channel width and length, respectively, and C_i is the capacitance per unit area of the SiO_2-insulator.

For OFETs made from BT5, we obtained a field-effect mobility of 3.7×10^{-5} cm²/Vs and a current on/off ratio of 2.5×10^3 at $V_{DS} = -80$ V. The device made from BT7 showed a reduced charge carrier mobility of 8.8×10^{-7} cm²/Vs with current on/off ratio of 1.1×10^2.

The output characteristics of transistors based on DHBTP-SC and DHPT-SC (as-prepared and annealed) show well resolved linear and saturation regions (Figure 6.10a, b). The absence of hysteresis during the measurement cycles indicates good stability and high purity of the materials. The current on/off ratio is $>10^4$ for both oligomers. Figure 6.11 shows the plot of the square root of the drain current in the saturation region $I_{DS,sat}^{1/2}$ versus the gate voltage V_{GS} from DHBTP-SC and DHPT-SC as the semiconducting layers. The field-effect mobilities at saturation region were calculated using the Eq. (1). For transistors with as-prepared thin films made from DHBTP-SC, we measured a field-effect mobility of ca. 10^{-3} cm²/Vs. With the same treatment, transistors from DHPT-SC exhibited a field-effect mobility of ca. 10^{-2} cm²/Vs.

For these oligomers, we found that annealing the thin films of both oligomers at 120 °C gave the optimum OFET performance. This is not surprising as it is well known that the annealing process affects the morphology of orga-

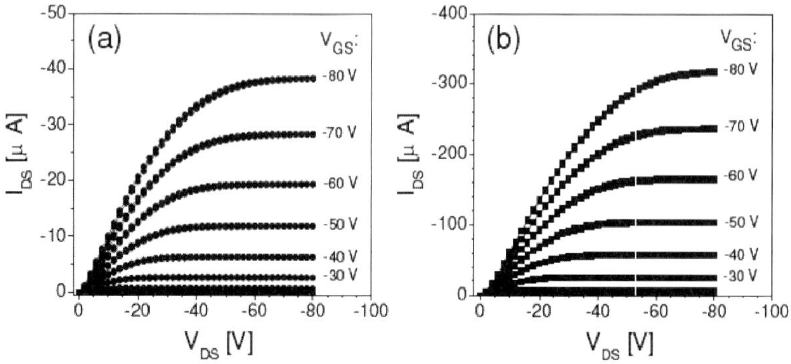

Figure 6.10 Output characteristics of OFET devices measured from as-prepared sample of (a) DHBTP-SC and (b) DHPT-SC.

nic materials. In order to study further the correlation between morphology and charge transport, we compared the OFET performances of the as-prepared layers to that of thin films that were annealed at 120 °C for 30 minutes, 90 minutes and 120 minutes prior to deposition of the Au source and drain electrodes (Figure 6.12). Interestingly, there is a four-fold increase in the mobilities of the OFETs prepared from DHBTP-SC layers after 30 minutes annealing. This might be explained by the significant change in the morphologies after the first annealing step (see Figure 6.6c, d). However, prolonged annealing does not increase the mobilities further. This indicates that the overall size of the crystals does not have a major impact on the transport properties. On the other hand, there is no significant difference in the OFET mobilities prepared from the as-prepared or the annealed films of DHPT-SC, even after 120 minutes of annealing. For all preparation conditions, the on/off current ratio (Figure 6.12) remains quite high indicating the overall good quality of all transistors.

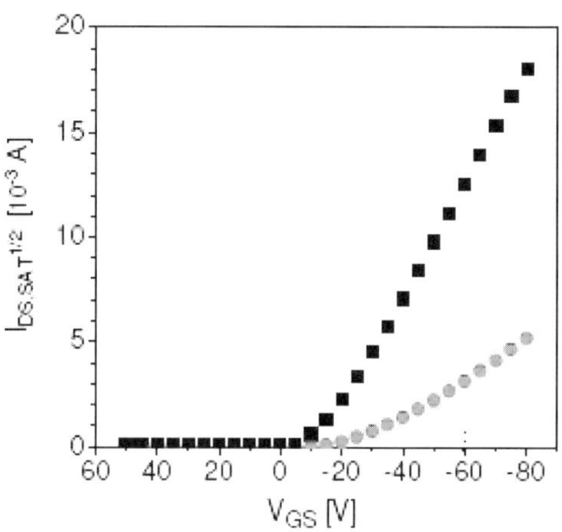

Figure 6.11 Square-root plot of $I_{DS,sat}^{1/2}$ versus V_{GS} from as-prepared OFET devices shown in Figure 6.10 with DHBTP-SC (red circles) and DHPT-SC (black rectangles) as the semiconductors.

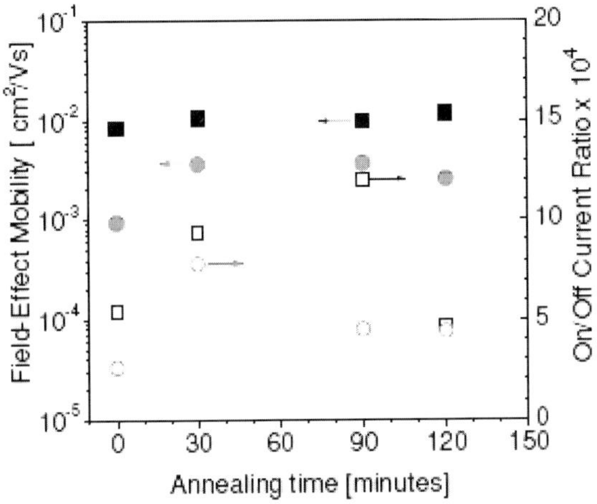

Figure 6.12 Field-effect mobilities and current on/off ratios of OFET devices annealed at 120 °C for different time intervals, using DHBTP-SC (circles) and DHPT-SC (rectangles) as the semiconductors. Open symbols depict the on/off current ratio whereas solid symbols the field-effect mobility.

6.5
Mobilities from Radiation Induced Conductivity Measurements

Pulse radiolysis time resolved microwave conductivity (PR-TRMC) measurements were performed for DHBTP-SC and DHPT-SC in order to obtain the charge carrier mobility at a probing frequency close to 30 GHz, using irradiation pulses of 10 ns. In fact, the TRMC mobility is not limited by the charge transport across grain boundaries [32]. Comparison of mobilities measured with PR-TRMC and in the OFET might thus provide valuable information on carrier trapping at grain boundaries in the device. The mobilities derived from the radiation induced conductivity measurements are shown as a function of temperature in Figure 6.13. At room temperature, the TRMC mobility in DHBTP-SC (6.9×10^{-3} cm^2/Vs) is somewhat higher than that for DHPT-SC (4.5×10^{-3} cm^2/Vs), in contrast to the OFET measurements where it was found that the mobility for DHPT-SC is generally higher. Note that the mobilities shown in Figure 6.13 are the minimum values, assuming that all the charge that is initially generated survives until after the pulse. Since charges may decay by recombination and trapping already during the irradiation pulse, the actual microwave mobility is expected to be significantly higher. Due to this complication, it is difficult to compare the microwave mobility values and the data obtained from OFET measurements quantitatively.

As shown in Figure 6.13 there are significant differences in the temperature dependent mobilities obtained for the swivel-cruciforms. Firstly, the tempera-

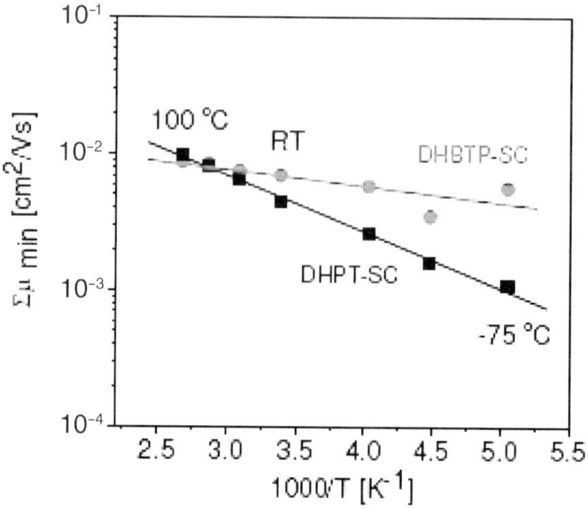

Figure 6.13 Temperature dependence of the charge carrier mobilities extracted from PR-TRMC measurements using DHBTP-SC (red circles) and DHPT-SC (black rectangles) powders.

ture dependence of DHBTP-SC is not as steep as for DHPT-SC. For DHBTP-SC, the activation energy is only 25 meV, which is significantly smaller than the value determined from the temperature-dependent OFET measurements (not shown here). In contrast, the microwave mobility of DHPT-SC exhibits a stronger temperature dependence and the activation energy is similar to the results obtained in the OFET measurements upon cooling. The lower activation energy for DHBTP-SC suggests that the structural order in this material is higher than in DHPT-SC, at least on a local scale, which is consistent with the higher microwave mobility at room temperature. The proposed higher structural order is further supported by the fact that the absorption spectrum of DHBTP-SC exhibits a vibronic progression while that of DHPT-SC is rather featureless. Also, our AFM studies on DHBTP-SC layers revealed the growth of large crystallites upon prolonged annealing, again indicating the ability of the molecules to form highly-ordered domains.

If the intrinsic (TRMC) mobilities of DHBTP-SC are higher because of a higher degree of ordering, it seems likely that the lower mobilities measured in transistors is due to inefficient transport across grain boundaries. In the cruciform DHBTP-SC highly ordered domains are formed with a high intrinsic mobility but with relatively inefficient charge transport between different domains. In DHPT-SC a more homogeneous structure is formed with a lower degree of order inside the grains, but with an apparent better charge transport between grains. This causes an overall higher mobility in the transistor measurement and rather similar activation energies in the OFET and PR-TRMC measurements.

6.6 Conclusions

A series of soluble swivel-cruciform oligothiophene dimers with increasing chain length – bis(terthiophene) (BT3), bis(pentathiophene) (BT5) and bis(heptathiophene) (BT7) – were used as semiconductor in organic field-effect transistors. Field-effect mobility of 3.7×10^{-5} cm^2/Vs and current on/off ratio of $>10^3$ was obtained with dimer BT5. By modifying BT5 into the highly soluble DHPT-SC (decoration with four terminal n-hexyls), we could obtain highly-crystalline layers with field-effect mobilities up to 10^{-2} cm^2/Vs. Films of DHPT-SC appeared to be morphologically stable upon annealing, while thermal treatment of the cruciform DHBTP-SC with a biphenyl core unit resulted in the growth of large crystallites. For both tetraalkylated swivel cruciforms (DHPT-SC, DHBTP-SC), the field-effect mobility was found to be temperature activated, with activation energy of ca. 80–90 meV upon heating and ca. 50 meV upon cooling. Interestingly, PR-TRMC measurements on the DHBTP-SC cruciform yielded a very weak temperature dependence of the microwave mobility. Compared to the mobility measurements in the OFET, PR-TRMC measures the transport of charges inside single grains of the material and the trapping at grain boundaries is insignificant. This result suggests a high structural order of this material on the microscopic level. We conclude that the electrical properties of these layers on macroscopic scales are ultimately determined by charge transport between crystalline (ordered) domains.

6.7 Experimental Section

For the absorption and photoluminescence measurements of the thin films, oligomers were dissolved into chloroform (10 mg/mL) and were spun onto glass substrates with a speed of 1500 rpm. For the XRD, AFM and OFET measurements, prior to deposition of the oligomers, the surface of the Si/SiO_2 substrates were carefully cleaned with several common solvents, activated with oxygen plasma and treated with hexamethyldisilazane (HMDS) for 26 h at 60 °C. Then, chloroform solutions from DHPT-SC and DHBTP-SC were spun onto those treated substrates. For BT3, BT5 and BT7, we used chlorobenzene and untreated Si/SiO_2. Note that we used the same substrates for the AFM and the OFET measurements. Optical absorption spectra were measured with a PerkinElmer Lambda 19 UV-Vis spectrometer. Photoluminescence emission spectra were recorded by using a PerkinElmer LS 55 Luminescence spectrometer. Differential scanning calorimetry (DSC) thermograms were measured using a PerkinElmer Thermal Analysis DCS-7 calibrated with Indium (99.99% purity). Atomic Force Microscopy (AFM) equipment from Digital Instruments, a Nanoscope IIIa working in tapping mode, was used to investigate the morphologies of the dry thin films on top of Si/SiO_2 substrates. X-ray diffraction was measured using coplanar X-ray diffraction (XRD) techniques at a home diffractometer equipped with a Goebel-mirror. Output and transfer characteristics of the OFET were measured using an Agilent 4155C semiconductor parameter analyzer from Hewlett Packard. Pulse Radiolysis Time Resolved Microwave Conductivity (PR-TRMC) measurements were carried out at the Delft University of Technology (DelftChemTech), The Netherlands.

Acknowledgement

We thank A. Bilge, F. Galbrecht, B. S. Nehls, and T. Farrell at the University of Wuppertal for the synthesis of the used materials. We would also like to acknowledge J. Grenzer (Institute of Ion Beam Physics and Materials Research in Dresden), W. Zhuang and J. P. Rabe (both Humboldt University Berlin) for performing the X-ray and AFM studies on our thin film samples. Microwave mobility measurements have been performed at the Interfaculty Reactor Institute, Delft University of Technology, NL with R. D. Abellon, F. C. Grozema, and L. D. A. Siebbeles. We further thank F. Jaiser (University of Potsdam) for fruitful discussions and assistance with the temperature dependent OFET measurements. This work was financially supported by the DFG under project SPP1121, the Fond der Chemischen Industrie, and the Ministerium für Wissenschaft, Forschung and Kultur of Brandenburg.

References

1. D. R. Gamota, P. Brazis, K. Kalyanasundaram, and J. Zhang (eds.), Printed Organic and Molecular Electronics (Kluwer Academic Publisher, 2004).

2. G. Dicker, M. P. de Haas, J. M. Warman, D. M. de Leeuw, and L. D. A. Siebbeles, J. Phys. Chem. B **108**, 17818 (2004).

3. Z. Bao, A. Dodabalapur, and A. J. Lovinger, Appl. Phys. Lett. **69**, 4108 (1996).

4. A. Zen, J. Pflaum, S. Hirschmann, W. Zhuang, F. Jaiser, U. Asawapirom, J. P. Rabe, U. Scherf, and D. Neher, Adv. Funct. Mater. **14**, 757 (2004).

5. A. Zen, M. Saphiannikova, D. Neher, J. Grenzer, S. Grigorian, U. Pietsch, U. Asawapirom, S. Janietz, U. Scherf, I. Lieberwirth, and G. Wegner, Macromolecules **39**, 2162 (2006).

6. R. J. Kline, M. D. McGehee, E. N. Kadnikova, J. Liu, and J. M. J. Frechet, Adv. Mater. **15**, 1519 (2003).

7. R. J. Kline, M. D. McGehee, E. N. Kadnikova, J. Liu, J. M. J. Frechet, and M. F. Toney, Macromolecules **38**, 3312 (2005).

8. H. Sirringhaus, P. J. Brown, R. H. Friend, M. M. Nielsen, K. Bechgaard, B. M. W. Langeveld-Voss, A. J. H. Spiering, R. A. J. Janssen, E. W. Meijer, P. T. Herwig, and D. M. de Leeuw, Nature **685**, 857 (1999).

9. M. Halik, H. Klauk, U. Zschieschang, G. Schmid, S. Ponomarenko, S. Kirchmeyer, and W. Weber, Adv. Mater. **15**, 917 (2003).

10. D. J. Crouch, P. J. Skabara, M. Heeney, I. McCulloch, S. J. Coles, and M. B. Hursthouse, Chem. Commun. **11**, 1465 (2005).

11. M. H. Yoon, S. A. DiBenedetto, A. Facchetti, and T. J. Marks, J. Am. Chem. Soc. **127**, 1348 (2005).

12. A. Facchetti, M. Mushrush, H. E. Katz, and T. J. Marks, Adv. Mater. **15**, 33 (2003).

13. A. Facchetti, J. Letizia, M. H. Yoon, M. Mushrush, H. E. Katz, and T. J. Marks, Chem. Mater. **16**, 4715 (2004).

14. H. Tian, J. Wang, J. Shi, D. Yan, L. Wang, Y. Geng, and F. Wang, J. Mater. Chem. **15**, 3026 (2005).

15. H. Tian, J. Shi, D. Yan, L. Wang, Y. Geng, and F. Wang, Adv. Mater. **18**, 2149 (2006).

16. J. G. Laquindanum, H. E. Katz, and A. J. Lovinger, J. Am. Chem. Soc. **120**, 664 (1998).

17. H. E. Katz, J. G. Laquindanum, and A. J. Lovinger, Chem. Mater. **10**, 633 (1998).

18. S. Ponomarenko, S. Kirchmeyer, A. Elschner, B. H. Huisman, A. Karbach, and D. Dreschler, Adv. Funct. Mater. **13**, 591 (2003).

19. M. Mushrush, A. Facchetti, M. Lefenfeld, H. E. Katz, and T. J. Marks, J. Am. Chem. Soc. **125**, 9414 (2003).

20. P. C. Chang, J. Lee, D. Huang, V. Subramanian, A. R. Murphy, and J. M. J. Frechet, Chem. Mater. **16**, 4783 (2004).

21. T. P. I. Saragi, T. Fuhrmann-Lieker, and J. Salbeck, Synth. Met. **148**, 267 (2005).

22. A. Bilge, A. Zen, M. Forster, H. B. Li, F. Galbrecht, B. S. Nehls, T. Farrell, D. Neher, and U. Scherf, J. Mater. Chem. **16**, 3177 (2006).

23. A. Zen, A. Bilge, F. Galbrecht, R. Alle, K. Meerholz, J. Grenzer, D. Neher, U. Scherf, and T. Farrell, J. Am. Chem. Soc. **128**, 3914 (2006).

24. A. Zen, P. Pingel, F. Jaiser, D. Neher, J. Grenzer, W. Zhuang, J. P. Rabe, A. Bilge, F. Galbrecht, B. S. Nehls, T. Farrell, U. Scherf, R. D. Abellon, F. C. Grozema, and L. D. A. Siebbeles, Chem. Mater. **19**, 1267 (2007).

25. P. Bäuerle, Adv. Mater. **4**, 102 (1992).

26. P. Bauerle, T. Fischer, B. Bidlingmeier, A. Stabel, and J. P. Rabe, Angew. Chem. **107**, 335 (1995).

27. P. Bäuerle, in: The oligomer approach, edited by K. Müllen and G. Wegner (Wiley-VCH, Weinheim, 1998).

Section III
Structural and Morphological Aspects

Organic Electronics. Structural and Electronic Properties of OFETs. Edited by Christof Wöll

ISBN: 978-3-527-40810-8

7
Chemical Approaches to the Deposition of Metal Electrodes onto Self-Assembled Monolayers – A Step Towards the Fabrication of SAM-Based Organic Field-Effect Transistors

Heidi Thomas, Jan Müller, and Andreas Terfort

7.1
Introduction

As a result of the increasing demand for "intelligent" consumer products, new approaches for the low-cost integration of logic circuitry are explored. One of the approaches is the use of field-effect transistors in which at least the semiconductor is an organic material. These organic field-effect transistors (OFETs) are expected to be cheaper and more versatile than the classical inorganic semiconductors, such as silicon or germanium, because the organic materials can be processed much more easily by gas-phase deposition or even wet-chemical technologies [1]. Additionally, organic materials have lower density and are often flexible, thus permitting relatively rough handling.

Since the channel height in (organic) field-effect transistors does not usually exceed a couple of nanometres, which thus lies within molecular dimensions, the ultimate semiconducting films need only to consist of a monomolecular layer. Self-assembled monolayers (SAMs) of chemisorbed molecules on solid surfaces present a technologically relevant form of organic films, since they are easily formed, often highly ordered, and relatively robust. If the SAM-forming molecules are semiconducting, in principle ultrathin OFETs should be achievable [2]. We consider transistors made of SAMs to be a practical alternative between today's bulk OFETs and the single-molecule transistors currently envisioned for the future, as they permit the minimisation of the amount of semiconducting material required, but on the other hand still permit the use of well-established methods of micro-fabrication.

A question remaining unsolved is how to make electrical contact to the organic semiconductors. While some efforts are also being made to manufacture the contacting electrodes (source, drain, gate) out of organic materials, it is still more common to contact the organic semiconductor with metal electrodes, such as aluminium, platinum, and in particular gold. A major problem with the methods for metal deposition used until now, such as physical vapour deposition [3–8], electrochemical deposition [9–12], soft-contact deposition [13], or nano-transfer printing [3, 14–16], is that they often lead to penetration of the

Organic Electronics. Structural and Electronic Properties of OFETs. Edited by Christof Wöll

ISBN: 978-3-527-40810-8

metal atoms into the organic material, thus destroying its desired electronic properties. Another problem with some of these methods is that the metal layers can often only be deposited uniformly, thus without the desired electrode pattern needed for the fabrication of transistors [9–12].

Here we want to present an approach for the deposition of metal contacts onto organic materials using very mild conditions. For our study we used SAMs, as they are not only suitable models for organic surfaces in general but can also be characterised using many methods of traditional surface science. The basic idea was to deposit a catalytic seed layer (in this case of metallic nanoparticles) onto the organic surfaces, which later would be "developed" to the desired metallic structures using relatively mild chemical processes, such as chemical vapour deposition (CVD) [17] or electroless deposition (ELD) [18, 19] (Figure 7.1).

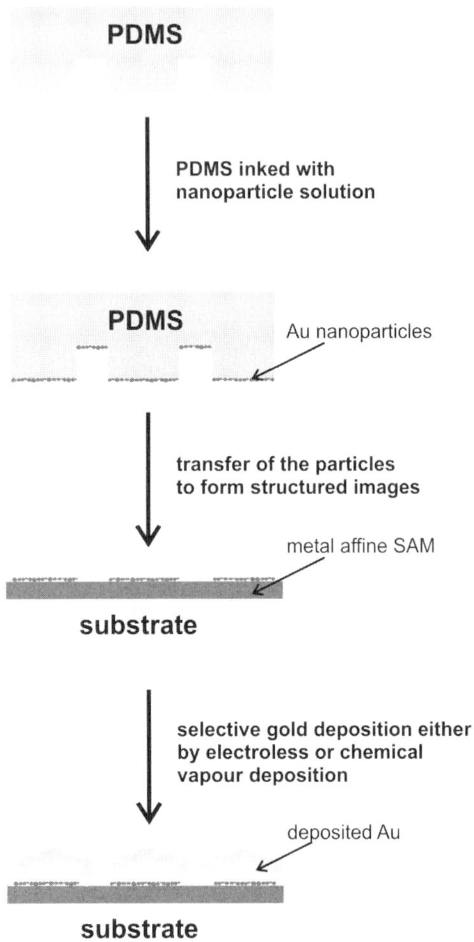

Figure 7.1 Schematic procedure for μCP of nanoparticles followed by selective gold deposition either by electroless deposition (ELD) or by chemical vapour deposition (CVD).
(see Colour plates p. LIII)

It should be pointed out that for the ELD method in particular, there exists a similarity to the photographic process, where initially formed Ag nanoparticles (the "latent picture") are used for the catalytic amplification in the development step. In our case, micro-contact printing (µCP) was used for the localised deposition of the seeds. This well-established, soft-lithographic method was originally developed by Whitesides and coworkers for the printing of alkanethiols [20–22] but was later extended *inter alia* to nanoparticles [23]. Such printed nanoparticles have already been used to seed the deposition of other metals, e.g. copper or silver [23, 24], but the quality of the metal layers obtained in this way was not satisfactory for application in micro-electronic devices. In addition, the problem of hetero-metallic contacts arose, as will be discussed later.

We therefore decided to perform an extended optimisation study involving the following parameters:

a) the nature of the SAM
b) the type of nanoparticles used for seeding (palladium vs. gold)
c) the stabilising layer of the nanoparticles
d) the amplification method (CVD vs. ELD)
e) the composition of the ELD bath.

For practical reasons, only gold was deposited in the development step. Gold has a high conductivity, is inert against air and most chemicals, and is ductile and therefore can cope with mechanical stress.

In the following we will discuss the influence of these parameters separately, but will present the results mostly in an integrated form, because due to the interplay of the various factors, trends are often blurred.

7.2 Results and Discussion

7.2.1 Nature of the SAM

A wide variety of SAMs has been prepared and described in recent decades. For the fabrication of OFETs, we want to focus mainly on those formed by the chemisorption of thiols onto gold, since these monolayers are usually dense and often highly ordered. Because the organic material should be semiconducting, the typically used aliphatic systems could be ruled out and aromatic systems had to be used instead.

We considered it advantageous to introduce functionalities into the respective organic molecules, which in principle could chemically interact initially with the nanoparticles and later with the metal electrodes. There are a few metal-affine groups that can work as headgroups: thiols [25, 26], amines [27, 28] or pyridines [29]. These should not only increase the adhesion of the metallic structures but also facilitate the charge-carrier injection from the electrodes into the organic material. The latter improves the performance of the resulting OFETs considerably, as the injection barrier represents a significant part of the electrical resistance found in real-life OFETs [30, 31]. The palette of molecules used in this study is depicted in Figure 7.2.

For comparison, an unsubstituted arenethiol, terphenylmethanethiol **1**, was used. This molecule is already known to form dense and ordered monolayers, but does not carry any functional headgroup that can interact with deposited metal (particles). It should be pointed out that the molecular structure is not the only factor determining the structure of the SAMs. In addition, the deposition conditions, in particular the solvent, can have a significant influence [32]. Therefore, the SAMs in this project were deposited either from ethanol or tetrahydrofuran (THF) solution (Table 7.1). Ethanol is usually considered advantageous for the monolayer formation, but THF is a much better solvent for the aromatic molecules, thus permitting higher concentrations.

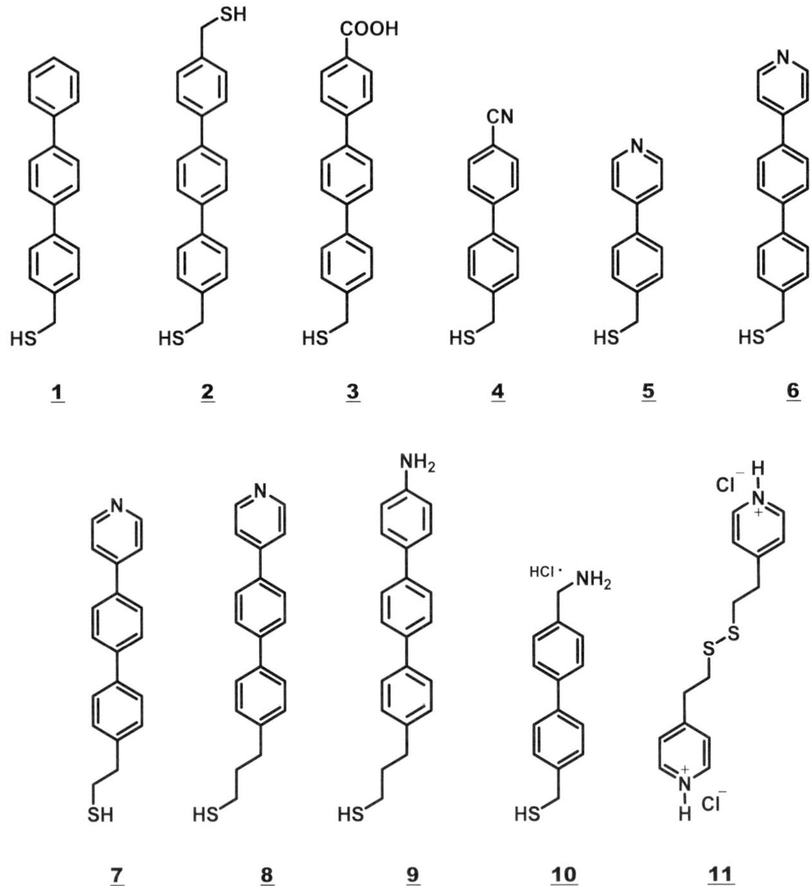

Figure 7.2 Aromatic thiols and disulfides used in this work for the monolayer formation.

Table 7.1 Calculated and measured layer thickness of the SAMs, as determined after immersion into solutions of the respective molecules for 24 h.

SAM	tilt angle	calculated layer thickness (Å)	determ. thickn. (Å) (from ethanol)	determ. thickn. (Å) (from THF)
1	19–28.5° (Refs. 33, 34)	14.5–15.6	13.5	19.9
2	19.3° (Ref. 35)	18.1	18.7	23.0
3	30° (estimated)	10.3	11.9	18.0
4	16–26° (Ref. 36)	16.5–17.6	19.2	32.2
5	30° (estimated)	8.8	5.2	17.0
6	30° (estimated)	13.3	14.5	22.5
7	25° (estimated)	15.4	10.9	17.9
8	30° (estimated)	15.3	15.4	21.0
9	30° (estimated)	17.0	18.7	28.4
10	30° (estimated)	12.8	15.6	26.3
11	30° (estimated)	12.0	13.9	16.3

One criterion for the quality of SAMs is the respective layer thickness, which, for example, can be determined by ellipsometry. As becomes clear in Table 7.1, the thicknesses of the monolayers deposited from ethanol are generally more consistent with the expected values. Most of the thiols used in this study are new, so the tilt angles could only be estimated based on experiences with other araliphatic systems [36]. Even if the estimated tilt angle for the molecules were to be lower, the calculated layer thicknesses would still be too low for the values determined for the THF-derived layers. The reason for this behaviour might be the formation of physisorbates. These layers were nevertheless used for the deposition study.

7.2.2 Seeding Material

In the original work by Whitesides and coworkers, palladium nanoparticles were used to initiate the electroless deposition of copper [23]. Palladium is renowned for its catalytic properties and the fabrication of its nanoparticles, which are considered to be air-stable, is well established. It was, therefore, the first choice in our study on exploring the metallisation of SAMs. This approach, however, bears a potential problem: if palladium nanoparticles are used and gold is deposited (Figure 7.3), a contact potential will build up at the interface. While this potential is usually small (μV to mV), it results in massive field gradients in the nanometre-scale, thus influencing the band structure in the adjacent organic material. For technological applications we thus considered it necessary to develop strategies based on gold nanoparticles as the seeding material.

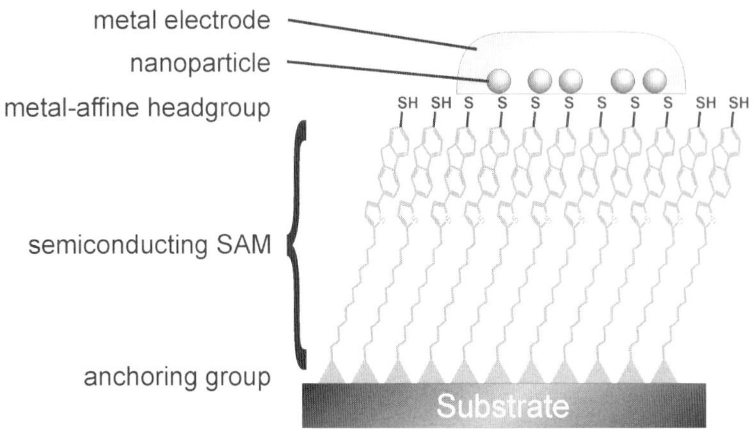

Figure 7.3 Schematic of the situation at the electrode/SAM interface: The use of palladium nanoparticles for the deposition of gold electrodes results in hetero-metallic interfaces at which band distortion in the organic material might build up due to field gradients. (see Colour plates p. LIV)

7.2.3
Stabilising Layer of the Nanoparticles

A very wide variety of procedures for the preparation of gold nanoparticles has been published. Most of these procedures follow the ones developed by Brust et al. [37] or Ulman and coworkers [38]: an Au(III)-salt is mixed with a thiol and reduced by sodium borohydride or superhydride. Most of the resulting gold nanoparticles obtained in this way are dispersed in organic solvents such as THF, toluene or methylene chloride [39–42]. Since these solvents are not compatible with poly(dimethylsiloxane) (PDMS), which is the material of the stamps used for μCP, we were looking for water-soluble gold nanoparticles instead [43]. The problem becomes more complicated by the fact that a suitable ligand should not only stabilise the respective nanoparticles over an extended period of time in the solution, but should, on the other hand, also permit access of the binding groups of the SAM after the stamping process, as well as permit the catalytic reaction of the metal deposition taking place. Thus typical multivalency approaches using polymers as stabilisers had to be ruled out. A lengthy study revealed that five different ligand systems provide the narrow set of properties required for the deposition process (Figure 7.4 and Table 7.2).

These hydrophilic nanoparticles were generally prepared by the reduction of tetrachloroauric acid, $HAuCl_4$, with borohydride in the presence of the ligand. Although this protocol was identical for all of the nanoparticles, their sizes and shapes varied considerably, as could be demonstrated by transmission electron microscopy (TEM, Figure 7.5).

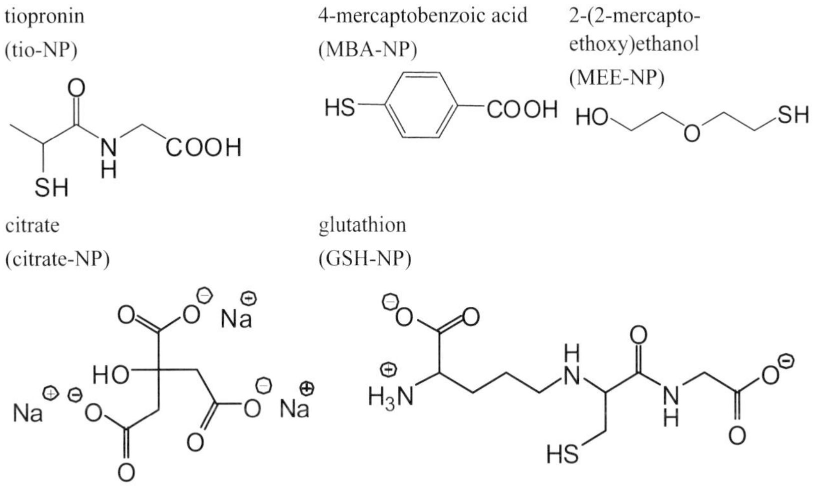

Figure 7.4 Structures of the ligands used for the stabilisation of water-soluble gold nanoparticles.

It therefore can be expected that the behaviour of the nanoparticles, for both the deposition step as well as the decomposition of gold precursors, differ, as will be discussed later.

7.2.4
Amplification Method (CVD vs. ELD)

A number of methods exist to deposit thin gold films onto surfaces. In particular, in the area of catalysis and microfabrication, the chemical vapour deposition (CVD) process became popular because it avoids the use of chemical baths, which in principle always carries the danger of contaminating the surface. For the CVD process, a volatile gold compound is required that decomposes upon heating or irradiation into metallic gold and volatile by-products. A well-established system is trimethylphosphanogoldmethyl ($(CH_3)_3PAuCH_3$)

Table 7.2 Water-soluble gold nanoparticles: synthesis, stability and colour.

ligand	synthesis	stability	colour
citrate	Ref. [44]	months	ruby
2-(2-mercaptoethoxy)-ethanol[a]	Ref. [45][b]	months[c]	dark purple
glutathione	Ref. [46]	days to months	auburn
4-mercaptobenzoic acid	Ref. [47]	days to months	clear yellow
tiopronin	Ref. [48]	days to months	yellow olive

[a] Preparation according to Ref. [49]. [b] 0.5-fold $NaBH_4$, 10-fold thiol. [c] The solution gets darker over time and has to be filtered after a few months.

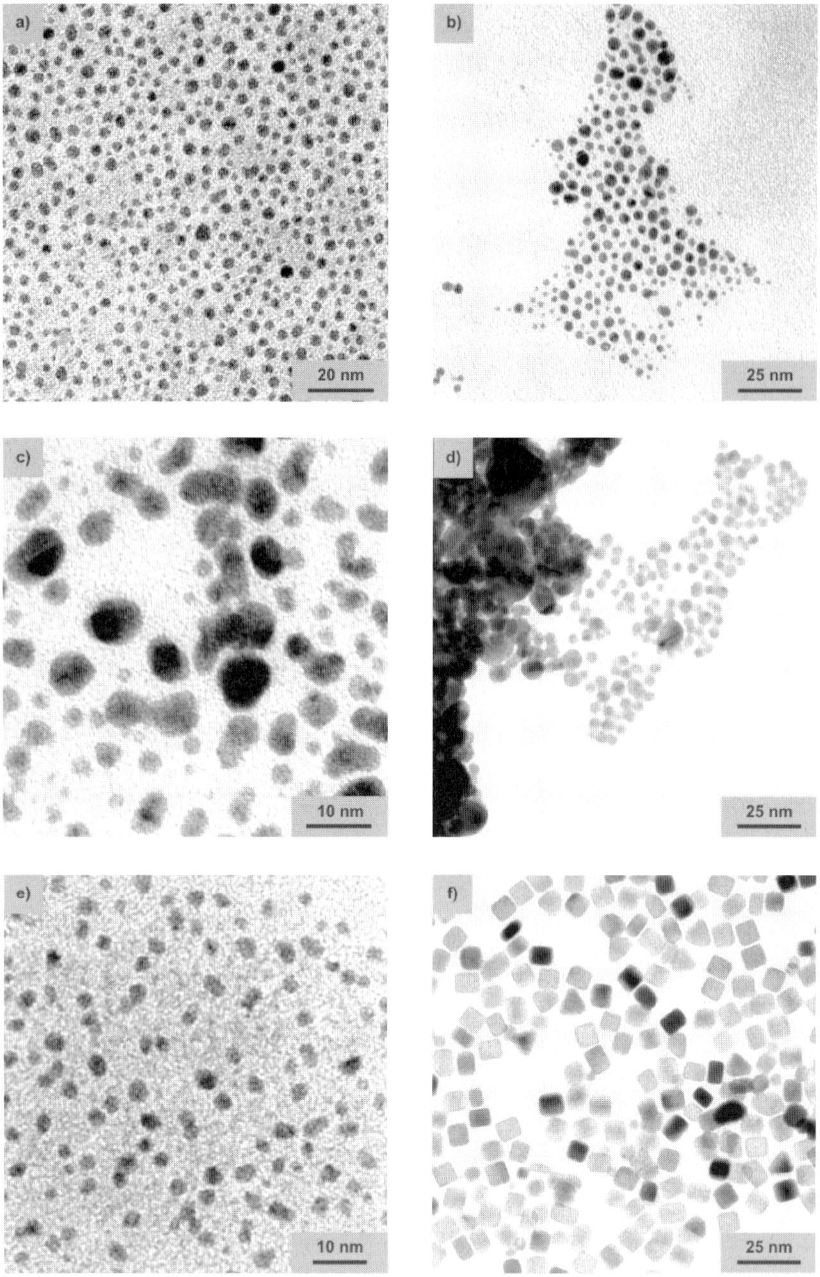

Figure 7.5 TEM micrographs of the nanoparticles used in this study: (a) citrate-NP (Au), (b) GSH-NP (Au), (c) MBA-NP (Au), (d) MEE-NP (Au), (e) tio-NP (Au) and (f) Pd/octylammonium-NP.

[50], which forms almost pure Au layers (see Eq. (1), decomposition of trimethylphosphanogoldmethyl). This process is autocatalytic and can also be catalysed by other metals, so that deposition temperatures as low as 70 °C can be reached. It is therefore possible to use selectively deposited nanoparticles as nuclei for the growth of laterally patterned gold layers.

$$(CH_3)_3PAuCH_3 \rightarrow Au^0 + (CH_3)_3P + \frac{1}{2}C_2H_6 \qquad (1)$$

In fact it turned out that palladium nanoparticles that have been stamped onto thiolated glass slides (using 3-mercaptopropyltrimethoxysilane, MPS, as a surface modifier) [51, 52], efficiently catalyse the deposition of gold (Figure 7.6).

To avoid the contact of two different metals in the vicinity of the organic semiconductor (as stated in 7.2.2), we shifted our attention to gold nanoparticles. Unfortunately, no selective deposition of gold could be achieved with any of the systems investigated, regardless of the protective layers used (Table 7.3), although the process should in this case be a truly autocatalytic one. In contrast, in some of the cases an inverse pattern was achieved, obviously due to some inhibition mechanism. We suggest that the stabilising reagent, which cannot diffuse away in a solvent-free environment, covers the active sites of the nanoparticles.

Figure 7.7 depicts representative results obtained for the "normal" pattern obtained using Pd nanoparticles and for a sample with inversed structure. The cross-sections in both cases indicate quite rough surfaces, a result of the formation of many very small grains grown together. In the case of the "normal" deposition, thicknesses of about 100 nm are generally obtained, while in the case of the inverse deposition the thickness difference can either be due to mass transport limitations in the CVD system or to a non-zero growth within

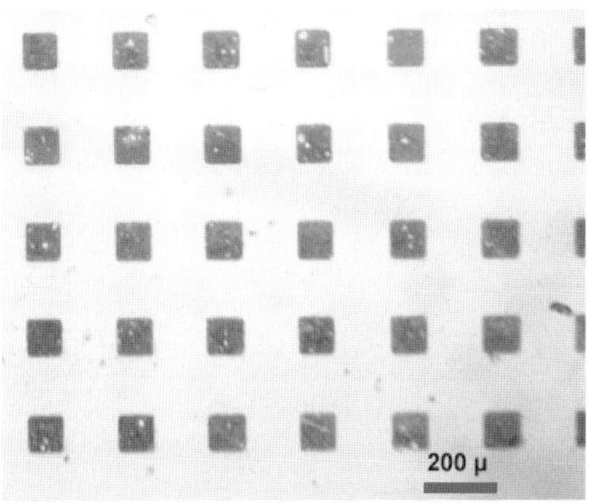

Figure 7.6 Optical micrograph of an array of gold squares deposited onto MPS-modified glass using micro-contact printing of Pd nanoparticles followed by CVD with ($(CH_3)_3PAuCH_3$) at 75 °C.
(see Colour plates p. LIV)

Table 7.3 Results of CVD experiments on selected SAMs using $((CH_3)_3P)AuCH_3$ (the solvent used for formation of the SAM is given in parentheses, x = no Au deposition).

SAM	citrate-NP	MBA-NP	tio-NP	MEE-NP	GSH-NP	Pd-NP
1 (EtOH)	x	inverse	inverse	unselective	x	unselective
2 (EtOH)	x	x	x	x	x	selective
3 (EtOH)	x	x	inverse	x	x	selective
4 (EtOH)	x	inverse	inverse	unselective	x	selective
5 (EtOH)	x	x	x	x	x	selective
10 (EtOH)	x	x	x	x	inverse	selective
11 (EtOH)	x	inverse	inverse	unselective	inverse	selective
1 (THF)	x	x	x	x	x	selective
2 (THF)	x	x	x	x	x	selective
3 (THF)	x	x	inverse	x	x	x
4 (THF)	x	x	inverse	x	x	x
5 (THF)	x	inverse	inverse	x	x	selective
10 (THF)	x	x	inverse	x	x	unselective
11 (THF)	x	x	x	x	x	x

the squares. Since the inverse systems did not seem useful for the contacting of OFETs, we did not put much effort into clarifying these questions.

Since the CVD process resulted in only limited success, alternative deposition processes were considered. Technically most important is the electrochemical approach, in which the surface to be covered is switched as the cathode in a bath containing Au ions in some form. Because this approach does need surfaces from which electrons easily can be transferred to the gold ions, thus discharging them, SAM-covered surfaces are most often not suitable. In addition, this process results in more or less homogeneously covered surfaces, although methods using masks have been reported [53]. An alternative source for the necessary electrons are chemical reductands, which can be homogeneously mixed into the gold salt solution. These so-called electroless deposition baths often deposit the metal at moderate temperature, in some systems even at room temperature. For the deposition of copper initiated by Pd nanoparticles, the previous results published in Ref. [23] were immediately successful, but copper is not a suitable metal for constructing stable devices, since it is readily oxidised in air.

For gold, a number of commercial systems exists, several of which were tested but did not show the desired selectivity in the process. Since the composition of these baths remains basically unknown, and thus cannot be rationally adjusted, we turned to gold ELD systems published in the literature, as described in the following.

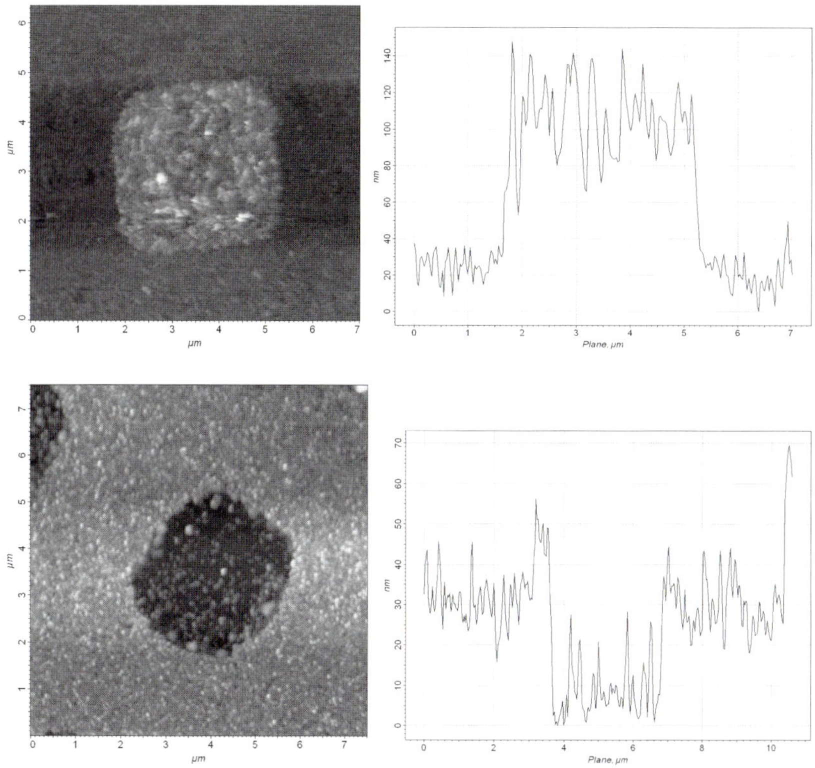

Figure 7.7 AFM micrographs and cross-sections of 3 µm × 3 µm squares obtained after the CVD of $((CH_3)_3P)AuCH_3$ onto printed nanoparticles. Upper: SAM of **10**, deposited from EtOH, printed with Pd nanoparticles; lower: SAM of **11**, deposited from EtOH, printed with GSH-NP (Au). (see Colour plates p. LV)

7.2.5 Composition of the ELD Bath

Basically three precursors are established for the deposition of gold from solution: tetrachloroauric acid ($HAuCl_4$), potassium gold cyanide ($KAu(CN)_2$), and sodium gold sulfite ($Na_3Au(SO_3)_2$). The cyanide was excluded from our investigation due to its toxicity and the tendency of the deliberated cyanide to dissolve/alter metals (including gold) already present on the samples. For the same reason, the baths could neither be extremely acidic nor basic. Several ELD baths have been published and were tested in this study with varying results (Table 7.4).

None of the baths were directly suitable for our purpose: while the baths based on $HAuCl_4$ showed some selectivity, the deposits were to thin and to rough. The gold layers obtained from $Na_3Au(SO_3)_2$ were denser and smoother, but not site selective. A series of alternative reducing reagents was tested with

Table 7.4 Results with previously published ELD baths. See also text for comments on alternative reductands.

gold source	other components	conditions	result	Ref.
$HAuCl_4$	citric acid, EDTA	RT–40/45 °C, 30 min	selective, but coating too thin and rough	54
$HAuCl_4$	$NH_2OH \cdot HCl$	RT, 30 min	selective, but coating too thin and rough	19, 55, 56
$Na_3Au(SO_3)_2$	Na_2SO_3, formaldehyde	RT	unselective	17, 19, 57–61
$Na_3Au(SO_3)_2$	$Na_2S_2O_3$, sodium citrate, ascorbic acid	RT–70/80 °C	no deposition	62

both precursors, such as hydrazine hydrate, glucose, propyleneglycol, glycerine, acetone, triethylamine, ethanol and glycine, but did not result in any gold deposition at all. Acetaldehyde and 2-ethoxyethanol gave unselective coatings. Hydroxylamine hydrochloride, as described in Refs. [19], [55] and [56], turned out to be the only suitable reductand available to hand, but only if the concentrations of the reactands was considerably increased. Dense Au layers were obtained with both $HAuCl_4$ and $Na_3Au(SO_3)_2$ under these conditions, but only for the sulfite complex, selectivity was obtained on surfaces patterned by μCP of nanoparticles. This modified bath (see Eq. (2), the assumed reduction process in the established ELD bath) was therefore used for the remainder of the study.

$$2NH_2OH + 4Na_3Au(SO_3)_2 \rightarrow 4Au^0 + N_2O + 3H_2O + 2SO_2 \quad (2)$$

It can be assumed that this process is not only autocatalytic because of the metallic gold, but is also self-enhancing: due to the liberation of SO_2 the pH value of the bath is lowered, which destabilises the sodium gold sulfite [63, 64]. Using the same representative monolayers as for the CVD process, an overview of the results as a function of the different SAM headgroups, the deposition conditions of the SAMs in addition to the type of nanoparticles, is given in Table 7.5.

To give an impression of the quality achievable in the deposition process, a selection of optical micrographs showing experiments resulting in more or less deposition is given in Figure 7.8.

Although there is still a number of defects visible in these micrographs, it should be pointed out that almost no gold was deposited outside the stamped area. This is the basic criterion for calling a deposition process "selective". Additionally, in most of these cases, the squares seem to be reproduced reliably, with only a few missing or severely deformed. A notable exception is the sample of GSH-NP on SAM **3**, the nitrile-terminated surface. We interpret the large blobs as being overgrown active sites, while most of the deposited nanoparticles are not very catalytically active, as suggested by their "washed-out" aspect. Since optical microscopy could not give any further information, we turned to AFM to look at the details (Figure 7.9).

Table 7.5 Results for the ELD experiments using the established bath on selected SAMs (the solvent used for formation of the SAM is given in parentheses, x = no Au deposition).

SAM	citrate-NP	MBA-NP	tio-NP	MEE-NP	GSH-NP	Pd-NP
1 (EtOH)	x	x	x	x	x	x
2 (EtOH)	x	x	x	x	unselective	unselective
3 (EtOH)	selective	x	x	x	unselective	x
4 (EtOH)	x	x	x	x	selective	unselective
5 (EtOH)	x	x	x	x	x	x
10 (EtOH)	x	x	x	selective	unselective	x
11 (EtOH)	x	x	x	x	x	x
1 (THF)	x	unselective	x	x	selective	unselective
2 (THF)	unselective	unselective	unselective	selective	selective	selective
3 (THF)	unselective	x	x	unselective	unselective	unselective
4 (THF)	x	x	x	x	x	x
5 (THF)	selective	x	x	unselective	x	selective
10 (THF)	x	x	x	x	x	x
11 (THF)	unselective	inverse	inverse	unselective	unselective	x

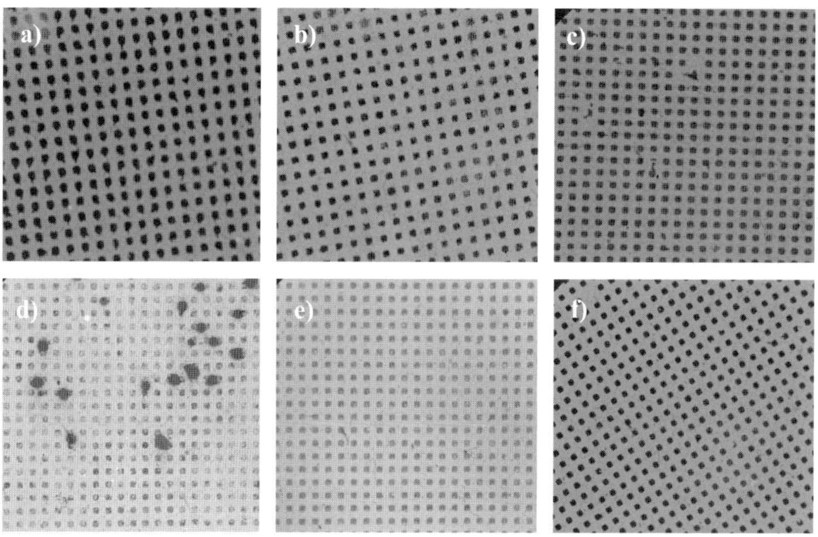

Figure 7.8 Optical micrographs of an array of 3 × 3 μm squares on different SAMs patterened by the μCP/ELD process: (a) SAM **2** (THF, MEE-NP), (b) SAM **2** (THF, Pd-NP), (c) SAM **10** (EtOH, MEE-NP), (d) SAM **3** (EtOH, GSH-NP), (e) SAM **3** (EtOH, citrate-NP) and (f) SAM **5** (THF, citrate-NP). (see Colour plates p. LVI)

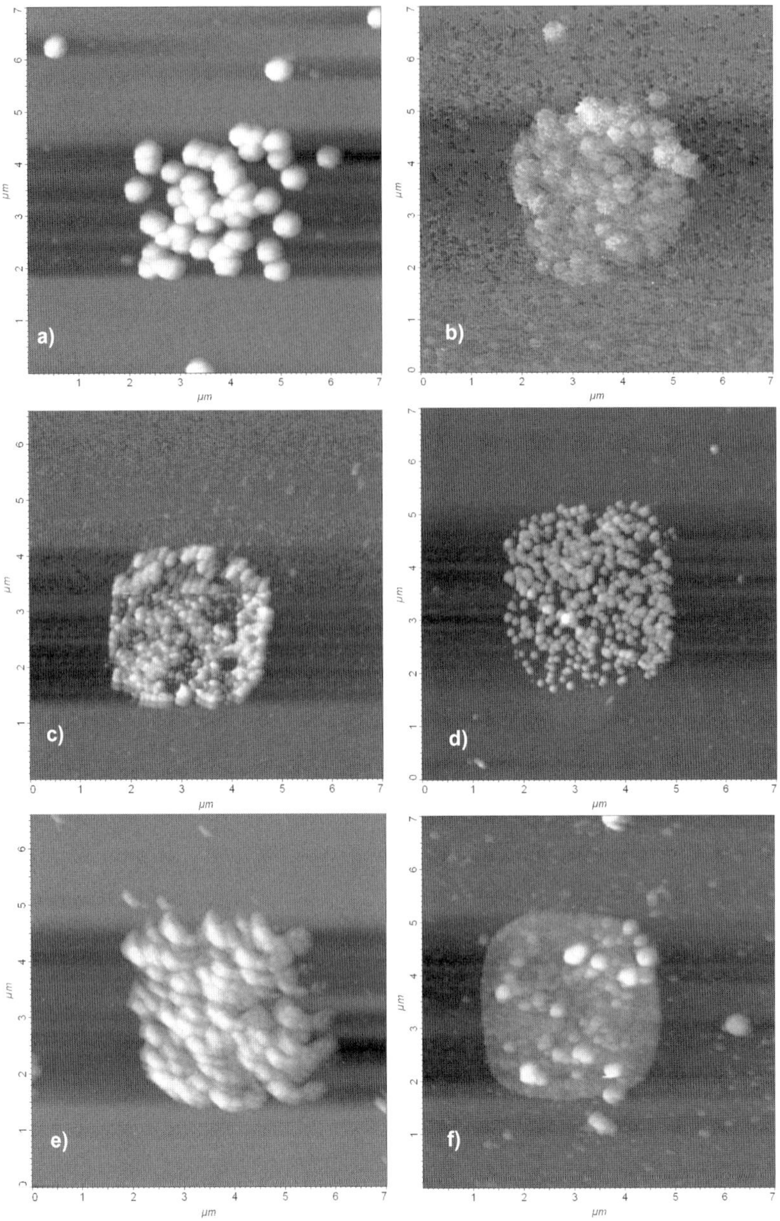

Figure 7.9 AFM micrographs of 3 μm squares deposited by the combined μCP/ELD process: (a) SAM **2** (THF, GSH-NP), (b) SAM **2** (EtOH, Pd-NP), (c) SAM **4** (EtOH, citrate-NP), (d) SAM **10** (EtOH, GSH-NP), (e) SAM **3** (THF, Pd-NP) and (f) SAM **5** (EtOH, MEE-NP). (see Colour plates p. LVII)

As the micrographs reveal, the squares are often not a solid gold film, but rather an assembly of tiny gold crystals. We assume that each gold crystal grew around a single nanoparticle. Two approaches should therefore solve this problem: (1) the growth process should be pushed further to permit the gold crystals to grow together; and (2) the area density of the seeding particles should become higher to reduce the distance between individual gold crystals. Since extension of the deposition time and further increasing the concentration of the reactands in the ELD bath did not result in closed films either, we focussed on the latter approach. The results given in Table 7.5, although somewhat erratic at first sight, gave hints on how to improve the transfer efficiency of the nanoparticles onto the SAMs. While on the SAM without any binding groups (the terphenyl thiol **1**) only one system resulted in selective deposition (if at all), the nitrogen-terminated SAMs, in particular **3**, **5** and **10**, gave the best selectivity together with closed Au islands, as shown by optical inspection. A typical example is the island shown in Figure 7.9f, and in Figure 7.10, where a cross-section is also depicted to give an idea of the quality achievable on a pyridine-terminated SAM (**5**). Although the film also looks grainy, the cross-section clearly shows that the island is a completely closed film of about 130 nm thickness.

Since the amino- and pyridine-terminated monolayers were considered optimal, four additional molecules (**6**–**9**) were synthesised based on the assumption that an extension of the aromatic system would result in better ordered monolayers. In addition, these molecules might even show the semiconducting properties required for the successful fabrication of SAM-based OFETs. The respective monolayers were again deposited from EtOH and THF: the resulting layer thicknesses have already been given in Table 7.1. For these SAMs, we used the established sulfite/hydroxylamine bath and restricted ourselves to the use of the citrate-NPs and MEE-NPs, because these nanoparticles were the most stable ones and, in addition, most often resulted in selective depositions. The approach turned out to be quite successful: in every case a good selectivity

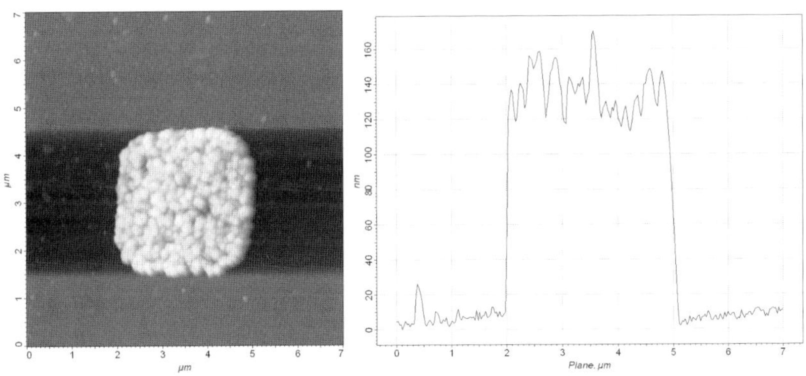

Figure 7.10 AFM images of an electrolessly deposited gold island on SAM **5** (THF) using citrate-NP as seeds. (see Colour plates p. LVIII)

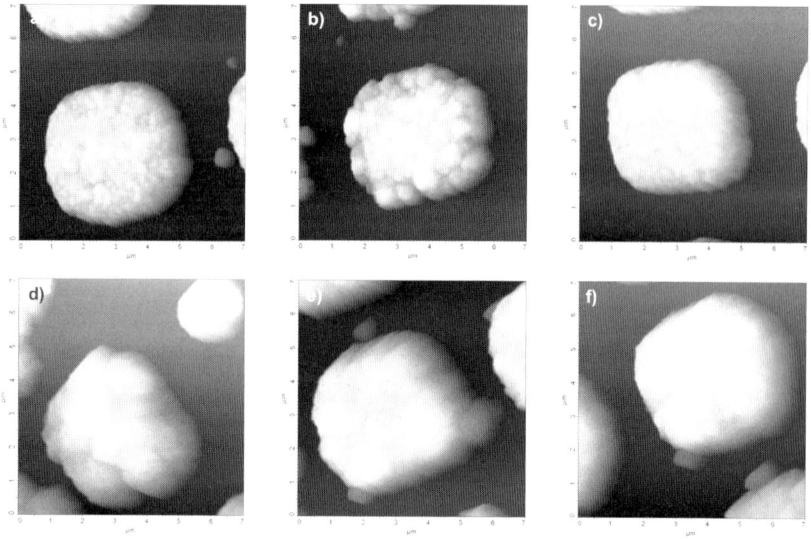

Figure 7.11 AFM micrographs of ELD substrates obtained by printing MEE-NPs onto: (a) SAM **7** (EtOH), (b) SAM **8** (EtOH), (c) SAM **10** (EtOH), (d) SAM **2** (THF), (e) SAM **9** (THF) and (f) SAM **6** (THF). (see Colour plates p. LVIII)

could be achieved, although some differences still turned up. While the islands grown from MEE-NPs were relatively thick (400–800 nm) and well defined (Figure 7.11), the ones obtained from citrate-NPs had a thickness of about 150–400 nm and consisted of an assembly of larger Au grains that blurred the edges of the squares. Obviously either a smaller number of nanoparticles were transferred in this case or the citrate-layer on the nanoparticles reduced their reactivity for the ELD process.

Because the use of MEE-NPs on nitrogen-terminated SAMs seemed to be the most successful for the μCP/ELD strategy, we decided to deposit an OFET electrode set-up onto a SAM-covered gold substrate to see whether this process permits the preparation of large-scale structures with a low defect density. For this, we devised an electrode set-up, in which the source and drain electrode interdigitate to maximise the channel width and thus the attainable currents. The active area was almost 10 mm^2 and thus posed a challenge to the reliability of the process. In fact, the respective pattern could be obtained reproducably by using the MEE-NP and the sodium gold sulfite/hydroxylamine bath (Figure 7.12).

To visualise the quality of the deposition, AFM was performed across the channels, which had a length of 10 μm and are therefore in the technologically relevant range (Figure 7.13). The micrographs reveal consistent electrodes with a thickness of about 220 nm, a low roughness on top, and – even more relevant – smooth edges. Only minimal deposits are visible within the channels,

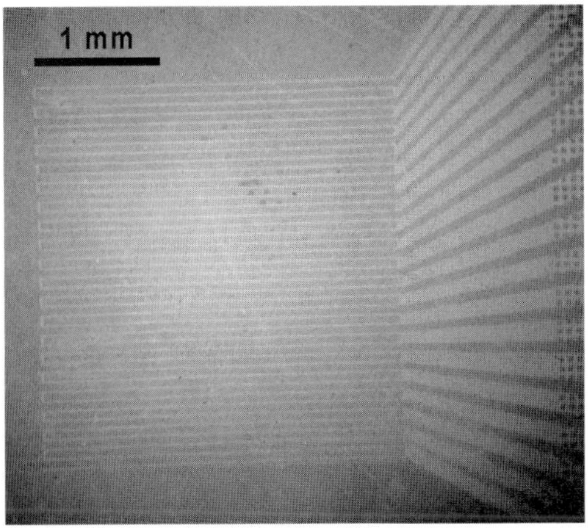

Figure 7.12 Optical micrograph of an OFET electrode get-up obtained by ELD using MEE-NP.
(see Colour plates p. LIX)

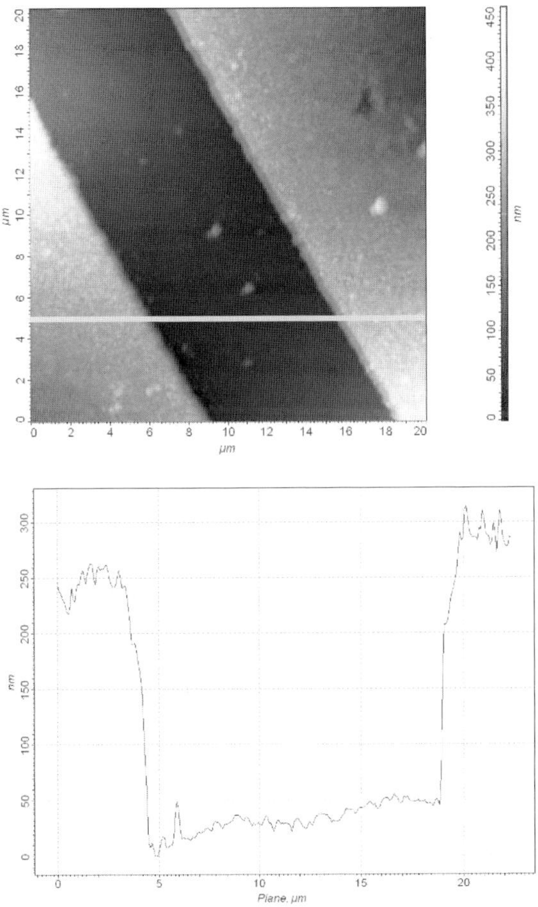

Figure 7.13 AFM detail on the OFET set-up shown in Figure 7.12. The deposition was made by ELD using MEE-NP printed onto a SAM of **6**. Top, a close-up on a channel within the transistor; bottom, cross-section along the channel.

nels, the nature of which remains unclear, as the AFM measurements were not performed in a dust-free room.

7.3 Conclusions

The electrical contacting of organic materials for their use in OFETs is a major problem to be solved. Based on low-temperature chemical processes that should not alter the organic materials, we performed an extensive study on different deposition methods based on the μCP of nanoparticles as seeds. The aim of this study was to use gold nanoparticles for the deposition of this same metal to avoid any problems due to hetero-metallic contacts. It turned out that the deposition by CVD did not result in the required selectivity for Au nanoparticles, while Pd nanoparticles performed better. The reason for the inability of the Au nanoparticles to catalyse their own growth might be the strong adherence of their stabilising layer, which cannot be removed in the vacuum system. In the case of the Pd seeds, a hand-over mechanism could be proposed, but no evidence for this assumption is available.

Better results can be obtained for the electroless deposition (ELD) in solutions. In this case the protective ligands can probably be dissolved in the liquid phase and thus diffuse away from the seeds. Of the several deposition baths considered, each consisting of a soluble gold salt and reducing reagent, only the one based on the sulfite complex of Au^{+} and hydroxylamine resulted in dense and relatively thick (up to 800 nm) gold layers. The reason for this behaviour remains unclear but there are hints that complex reactions are taking place in the autocatalytic deposition process.

It also turned out that there is a sensitive interplay between the stabilisation of the nanoparticles and the head group of the organic layers. The SAMs should carry coordinating headgroups to facilitate the attachment of the nanoparticles and thus increase the transfer yield from the stamps. As every nanoparticle is the seed for a gold crystallite, a less dense assembly of nanoparticles results either in non-closed gold layers or – after prolonged growth – in deformation of the printed shape.

A combination of nitrogen-terminated SAMs with MEE-stabilised nanoparticles turned out to be the optimal combination for the selective and dense deposition of gold layers. Through this, we were able to demonstrate that a large-scale (10 mm^2) electrode set-up could be deposited under very mild conditions.

We are currently in the process of determining the electrical properties of the gold islands deposited onto the SAMs to make sure that neither the μCP process nor the ELD chemistry results in destruction of the monlayers or the formation of shorts. Promising results have already been obtained in collaboration with Wöll's group in Bochum, which will be reported elsewhere.

7.4
Experimental

7.4.1
Nanoparticles

Sodium citrate trihydrate, tiopronin and 4-mercaptobenzoic acid were purchased from Sigma-Aldrich, sodium borohydride from Merck and acetic acid from Roth. Tetrachloroauric acid [65] and gold-nanoparticles (Table 7.1) were prepared according to the literature (see Table 7.2). After preparation, the nanoparticles were dissolved in water, filtered through syringe filters (PTFE, 0.2 μm) and stored at 4 °C. The Pd nanoparticles were prepared according to Ref. [23].

7.4.2
Substrate Preparation

Microscopy slides (ground edges 45°, Menzel-Gläser) were used after the following cleaning procedure: dust was removed by washing under flowing water. The slides were cleaned in piranha solution (H_2SO_4/H_2O_2, 30%; 3:1; Caution! Piranha solution is a corrosive and strongly oxidising agent) for 15 min, rinsed with deionised water, and blown dry with a stream of nitrogen. 50 nm of gold were deposited in ultra-high vacuum by e-beam evaporation at a rate of approx. 1 nm/s with an underlying layer of 2 nm of chromium as an adhesion promoter.

7.4.3
Plasma Cleaning [66]

Microwave induced plasma was generated by a Harrick Plasma Cleaner/Sterilizer PDG-32G. The operating pressure was achieved by a vacuum pump (Vacuubrand RD 4). Hydrogen was applied as a permanent stream, regulated by the integrated inlet valve and adjusted to approx. 0.7 mbar (determined by a MKS PDR 2000 Dual Capacitance Manometer). The gas flow was not determined.

7.4.4
Stamp Preparation

A poly(dimethylsiloxane) (PDMS, Sylgard 184®, Dow Corning) elastomer stamp was prepared by casting against a master mold. PDMS was mixed with the respective curing agent (10:1), poured over the master mold, and cured for 3 h at 60 °C. The elastomeric stamp was pealed off and the profile containing a 3 × 3 μm squares design and was cut out of the cast.

7.4.5
SAM Preparation

Gold substrates were dipped in a solution of 8.4 mmol/L hexadecanethiol in ethanol for 1 min, rinsed with ethanol, and cleaned by a hydrogen plasma. A 1 mM solution of each of **1**–**11** in absolute THF as well as absolute ethanol, was prepared under a stream of nitrogen. Substrates were stored in these solutions for 24 h in the absence of light, rinsed with ethanol, blown dry under a stream of nitrogen and stored in the dark until further use.

7.4.6
Ellipsometry

The thicknesses of the films were determined using an ellipsometer SE 400 (Sentech Instruments GmbH) under an incidence angle of 70° at a wavelength of 633 nm. Optical film thicknesses were determined assuming a complex refractive index $N = n - ik$ with a real part $n = 1.45$ and an imaginary coefficient $k = 0$. The parameters n and k of the gold substrates were obtained by ellipsometric measurements of the plasma-cleaned films before monolayer formation.

7.4.7
μCP of Nanoparticles

As illustrated in Figure 7.1, μCP was performed onto the SAM-coated gold films. The ultrasonically cleaned 3 × 3 μm squares stamp was treated with an air plasma for 20 s. Subsequently it was inked with the nanoparticle solution for 20 s and blown dry in a stream of nitrogen. Inking was repeated twice. The stamp was than brought into contact with the SAM-coated gold film for 30 s, generating a laterally structured nanoparticle layer. Only light pressure was applied.

7.4.8
Electroless Deposition of Gold

The plating bath consisted of 5 mL of a saturated hydroxylamine hydrochloride solution and 5 mL of a 1 wt% $Na_3Au(SO_3)_2$ solution (sodium gold sulfite was prepared according Ref. [67]). Both solutions were freshly mixed, stirred vigorously and the substrate was dipped into the mixture for 10 min at room temperature.

7.4.9
Chemical Vapour Deposition of Gold

The deposition was performed in a CVD chamber at a pressure of 10^{-1} mbar and 75 °C for 180 min. The precursor, Me_3PAuMe, was prepared according to Ref. [50].

7.4.10
AFM Measurements

Measurements were performed with a Solver-Pro instrument by NT-MDT.

Acknowledgements

We wish to thank the DFG focus program "Organic Field-Effect Transistors" (grant TE 247/4-3) and the DFG Graduate School 611 "Functional Materials" for financial support. We thank Carsten Brandt and Lisa Roeder for support with the experimental work.

References

1. H. Sirringhaus, Adv. Mater. **17**, 2411 (2005).
2. C. R. Kagan, A. Afzali, R. Martel, L. M. Gignac, P. M. Solomon, and A. G. Schrott, Ek B. Nano Lett. **3**, 119 (2003).
3. W. Jiang, N. Zhitenev, Z. Bao, H. Meng, D. Abusch-Magder, D. Tennant, and E. Garfunkel, Langmuir **21**, 8751 (2005).
4. Z. Zhu, T. A. Daniel, M. Maitani, O. M. Cabarcos, D. L. Allara, and N. Winogard, J. Am. Chem. Soc. **128**(42), 13710 (2006).
5. A. J. Wagner, G. M. Wolfe, and D. H. Fairbrother, Appl. Surf. Sci. **219**, 317 (2003).
6. G. Nagy, and A. V. Walker, J. Phys. Chem. C **111**, 8543 (2007).
7. B. Jaeckel, J. B. Sambur, and B. A. Parkinson, Langmuir **23**, 11366 (2007).
8. A. V. Walker, T. B. Tighe, B. C. Haynie, S. Uppili, N. Winograd, and D. L. Allara, J. Phys. Chem. B **109**, 11263 (2005).
9. V. Ivanova, T. Baunach, and D. M. Kolb, Electrochim. Acta **50**, 4283 (2005).
10. M. Manolova, V. Ivanova, D. M. Kolb, H.-G. Boyen, P. Ziemann, M. Büttner, A. Romanyuk, and P. Oelhafen, Surf. Sci. **590**, 146 (2005).
11. D. Qu, and K. Uosaki, J. Phys. Chem. B **110**(35), 17570 (2006).
12. T. Baunach, V. Ivanova, D. M. Kolb, H.-G. Boyen, P. Ziemann, M. Büttner, and P. Oelhafen Adv. Mater. **16**, 2024 (2004).
13. A. Vilan, and D. Cahen, Adv. Funct. Mater. **12**, 795 (2002).
14. E. Menard, L. Bilhaut, J. Zaumseil, and J. A. Rogers, Langmuir **20**, 6871 (2004).
15. Y.-L. Loo, R. L. Willett, K. W. Baldwin, and J. A. Rogers, Appl. Phys. Lett. **81**, 562 (2002).
16. N. A. A Hatab, J. M. Oran, and M. J. Sepaniak, ACS Nano **2**, 377 (2008).
17. R. J. Puddephatt, Polyhedron **13**, 1233 (1994).
18. Y. Jin, X. Kang, Y. Song, B. Zhang, G. Cheng, and S. Dong, Anal. Chem. **73**, 2843 (2001).
19. S. Hrapovic, Y. Liu, G. Enright, F. Bensebaa, and J. H. T. Luong, Langmuir **19**, 3958 (2003).
20. Y. Xia, and G. M. Whitesides, Angew. Chem. Int. Ed. **37**(5), 550 (1998).
21. A. Kumar, and G. M. Whitesides, Appl. Phys. Lett. **63**, 2002, (1993).
22. A. Kumar, H. A. Biebuyck, and G. M. Whitesides, Langmuir **10**, 1498 (1994).
23. P. C. Hidber, W. Helbig, E. Kim, and G. M. Whitesides, Langmuir **12**, 1375 (1996).

24. A. A. Mewe, E. S. Kooji, and B. Poelsema, Langmuir **22**, 5584 (2006).

25. M. T. Zin, H.-L. Yip, N.-Y. Wong, H. Ma, and A. K.-Y. Jen, Langmuir **22**, 6346 (2006).

26. S. Wakamatsu, J. Nakada, S. Fujii, U. Akiba, and M. Fujihira, Ultramicroscopy **105**, 26 (2005).

27. J. Wang, T. Zhu, M. Tang, S. M. Cai, and Z. F. Liu, Jpn. J. Appl. Phys. **35**, L1381 (1996).

28. J. Zheng, Z. Zhu, H. Cheng, and Z. Liu, Langmuir **16**, 4409 (2000).

29. T. Zhu, X. Zhang, J. Wang, X. Fu, and Z. Liu, Thin Solid Films **327–329**, 595 (1998).

30. C. Bock, D. V. Pham, U. Kunze, D. Käfer, G. Witte, and A. Terfort, Appl. Phys. Lett. **91**, 052110 (2007).

31. C. Bock, D. V. Pham, U. Kunze, D. Käfer, G. Witte, and Ch. Wöll, J. Appl. Phys. **100**, 114517 (2006).

32. D. Käfer, G. Witte, P. Cyganik, A. Terfort, and C. Wöll, J. Am. Chem. Soc. **128**, 1723, (2006).

33. A. Shaporenko, M. Brunnbauer, A. Terfort, M. Grunze, and M. Zharnikov, J. Phys. Chem. B **108**, 14462 (2004).

34. H.-J. Himmel, A. Terfort, and Ch. Wöll, J. Am. Chem. Soc. **120**, 12069 (1998).

35. Y. Tai, A. Shaporenko, M. Grunze, and M. Zharnikov, J. Phys. Chem. B **109**, 19411 (2005).

36. A. Shaporenko, M. Brunnbauer, A. Terfort, M. Grunze, and M. Zharnikov, J. Phys. Chem. B. **108**, 14462 (2004).

37. (a) M. Brust, M. Walker, D. Bethell, D. J. Schiffrin, and R. Whyman, Chem. J. Soc. Chem. Commun. 801 (1994); (b) M. Brust, J. Fink, D. Bethell, D. J. Schiffrin, and C. Kiely, J. Chem Soc. Chem. Commun. 1655 (1995).

38. C. K. Yee, R. Jordan, A. Ulman, H. White, A. King, M. Rafailovich, and J. Sokolov, Langmuir, **15**, 3486 (1999).

39. A. Moores, F. Goettmann, C. Sanchez, and P. Le Floch, Chem. Commun. 2842 (2004).

40. M. K. Corbierre, and R. B. Lennox, Chem. Mater. **17**, 5691 (2005).

41. N. R. Jana, and X. Peng, J. Am. Chem. Soc. **125**, 14280 (2003).

42. M. P. Rowe, K. E. Plass, K. Kim, C. Kurdak, E. T. Zellers, and A. Matzger, J. Chem. Mater. **16**, 3513 (2004).

43. C. J. Ackerson, P. D. Jadzinsky, and R. D. Kornberg, J. Am. Chem. Soc. **127**, 6550 (2005).

44. G. Frens, Nature Phys. Sci. **241**, 20 (1973).

45. S. F. Sweeney, G. H. Woehrle, and J. E. Hutchinson, J. Am. Chem. Soc. **128**, 3190 (2006).

46. T. G. Schaaff, G. Knight, M. N. Shafigullin, R. F. Borkman, and R. L. Whetten, Phys. Chem. B **102**(52), 10643 (1998).

47. S. R. Johnson, S. D. Evans, and R. Brydson, Langmuir **14**, 6639 (1998).

48. A. C. Templeton, S. Chen, S. M. Gross, and R. W. Murray, Langmuir **15**, 66 (1999).

49. G. H. Woehrle, M. G. Warner, and J. E. Hutchinson, Langmuir **20**, 5982 (2004).

50. H. Schmidbaur, and A. Shiotani, Chem. Ber. **104**, 2821 (1971).

51. J. J. Gooding, V. G. Praig, and E. A. H. Hall, Anal. Chem. **70**, 2396 (1998).

52. D. Losic, J. G. Shapter, and J. J. Gooding, Langmuir **17**, 3307 (2001).

53. J. Tien, A. Terfort, and G. M. Whitesides, Langmuir **13**, 5349 (1997).

54. European Patent Application EP 1319734A1, Plating method, Mariana Kanzler (2003).

55. F. Guan, M. Chen, W. Yang, J. Wang, S. Young, and Q. Xue, Appl. Surf. Sci. **240**, 24, (2005).

56. Y. Jin, X. Kang, Y. Song, B. Zhang, G. Cheng, and S. Dong, Anal. Chem. **73**, 2843 (2001).

57. R. Takahashi, and T. Ishiwatari, Chem. Commun. 1406 (2004).

58. A. Hilmi, and J. H. T. Luong, Anal. Chem. **72**, 4677 (2000).

59. J. Yan, Y. Du, J. Liu, W. Cao, X. Sun, W. Zhou, X. Yang, and E. Wang, Anal. Chem. **75**, 5406 (2003).

60. V. P. Menon, and C. R. Martin, Anal. Chem. **67**, 1920 (1995).

61. Z. Hou, N. L. Abbott, and P. Stroeve, Langmuir **14**, 3287 (1998).

62. United States Patent 5,935,306, A. H. Reed (1999).

63. P. Wilkinson, Gold Bull. **19**(3), 75 (1986).

64. P. C. Hydes, and H. Middleton, Gold Bull. **12**(3), 90 (1979).

65. G. Brauer, Handbuch der präparativen anorganischen Chemie, 3rd ed. (Ferdinand Enke Verlag, Stuttgart, 1975, p. 1014).

66. K. Raiber, A. Terfort, C. Benndorf, N. Krings, and H.-H. Strehblow, Surf. Sci. **595**, 56 (2005).

67. United States Patent, 3,966,880, G. Dietz Jr. (1976).

8
Growth Morphologies and Charge Carrier Mobilities of Pentacene Organic Field Effect Transistors with RF Sputtered Aluminium Oxide Gate Insulators on ITO Glass

M. Voigt, J. Pflaum, and M. Sokolowski

8.1
Introduction

Presently there exists a strong research interest in the understanding, development, and optimisation of organic field effect transistors (OFETs) [1, 2]. Two classes of semiconducting organic materials are considered, namely molecular materials which are processed into thin films by vacuum sublimation [1, 2], and polymers which are deposited onto substrates in the form of solutions, for instance by spin coating [3]. In this chapter we report on OFETs based on thin polycrystalline films of the molecular material pentacene (Pc) as the semiconducting material.

Different to many other investigations on OFETs, which utilise thermally oxidised Si (SiO_x) as the substrate [4], we prepared our OFETs on indium-tin-oxide (ITO) covered glass with an radio frequency (rf) sputtered film of aluminium oxide (AlO_x) as the gate insulator. OFETs of this type have recently also been reported by Lee and coworkers [5–8]. The attractive aspect of this type of OFET layout is that the OFET channel is semitransparent and thus allows optical functionality. In addition, such OFETs could be easily integrated into organic light emitting displays (OLEDs), which are usually also fabricated on ITO glass. However, since the morphology, the electrical properties (e.g., the relative permittivity), and the chemical reactivity of the AlO_x interface differs from that of SiO_x and other gate insulators, results obtained for other OFETs can be transferred only within a limited extent to this system. A detailed investigation of the Pc film growth on AlO_x and the resulting OFET characteristics is hence justified. In addition, effects due to the higher dielectric permittivity ε_r (e.g., interface charge trapping and high accumulated charge densities) can be expected for AlO_x based OFETs [9]. In particular concerning the aspect of the Pc film growth, our work extends that of Lee et al. [5–8], who demonstrated that Pc based OFETs can be realised on AlO_x sputtered onto ITO glass, but did not investigate the correlation between the structural/morphological and electrical properties in more detail. In addition, AlO_x was used as a gate insulator for Pc OFETs by Kalb et al. [10].

Organic Electronics. Structural and Electronic Properties of OFETs. Edited by Christof Wöll

ISBN: 978-3-527-40810-8

The growth scenario of Pc on the AlO_x substrate turned out to be complicated, since more than one structural phase of Pc forms, and since the structure and morphology of the Pc films are both strongly influenced by the preparation conditions. Concerning the charge transport along the OFET channel, the defect density of the substrate near thin film phase of Pc was found to be decisive. This defect density appears to be related to the detailed chemical termination of the AlO_x substrate and could not be directly measured by us. However, the volume fraction of the Pc bulk phase growing on top of the thin film phase was also found to be correlated to this defect density, and hence indicates the Pc/AlO_x interface quality. The field-induced charge transport has to be described by models which account for local trapping of the charge carriers along the OFET channel and scattering of the charge carriers in the Pc bulk and at the Pc/AlO_x interface. From a detailed investigation of the mobilities as a function of temperature, the structural quality of the Pc films, and the drain– and gate–source voltages we were able to conclude the relevance of the different mechanisms noted above.

The outline of this chapter is as following. The experimental details are reported in Section 8.2. The results are presented and discussed in Section 8.3. Conclusions are given in Section 8.4.

8.2
Experimental

The general layout of our OFETs is shown in Figure 8.1. As substrates we used commercially available ITO covered glass slides (25 mm × 25 mm). They were cleaned by ultrasonic treatments in acetone, H_2O_2 (37%), distilled H_2O, and subsequent drying in an argon flow. The sheet resistance and thickness of the ITO films were 50 Ω and 100 nm, as specified by the supplier (Deutsche Balzers GmbH). The rms roughness of the ITO was 4 nm, as determined by atomic force microscopy (AFM).

The AlO_x insulator layer was deposited on the ITO by rf magnetron sputtering (13.56 MHz) from a planar Al_2O_3 ceramic target of 99.99% purity in pure Ar (99.998%). The sputter gas pressure was 8×10^{-3} mbar. This yielded AlO_x deposition rates between 4.0 nm/min and 6.0 nm/min. The base pressure prior

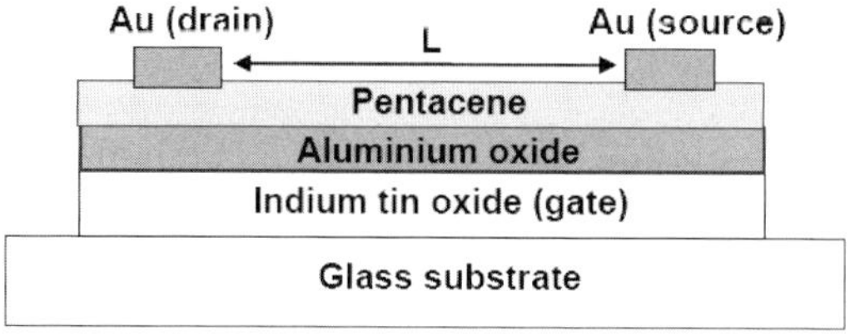

Figure 8.1 Schematic cross section of our pentacene (Pc) organic field effect transistors.

to the sputtering of AlO_x and the deposition of Pc varied in the range of 5×10^{-7} mbar to 5×10^{-6} mbar (for details see also Table 8.1). The thickness of the AlO_x films was 320 nm and monitored by a quartz microbalance that was calibrated by additional measurements with a mechanical profilometer. The electrical breakdown fields and relative permittivity ε_r of the AlO_x films were measured to be ~2.5 MV cm^{-1} and ~7, respectively [11]. The rms roughness of the AlO_x films was about 3–4 nm, as determined by AFM. More details of the preparation conditions and the electrical properties of the AlO_x films are reported in Ref. [11].

Pentacene (Pc) was obtained from Fluka in 98% purity and purified by temperature gradient sublimation under high vacuum. Characterisation of the purified Pc by mass spectroscopy revealed that the main impurities are *dihydropentacene*, *pentacenequinone* and *pentaceneol*. The sum of the concentrations of these impurities was estimated from the intensity ratio of the mass signals to be below 2%. In contrast to Jurchescu *et al.* [12], we find *dihydropentacene* (instead of *pentacenequinone*) as the main impurity.

The purified Pc was vacuum deposited by thermal evaporation onto the AlO_x films at different substrate temperatures T_S from −17 °C to 93 °C. The deposition rate of Pc was 5–7 nm/min (except for sample D, 1 nm/min). The shutter was closed during the heating of the evaporation cell in order to avoid deposition of low mass impurities, degassing from the Pc evaporation cell, onto the samples. The evaporation temperature of the cell was between 260 °C and 310 °C. From optical polarisation microscopy we derived that all Pc films were polycrystalline with no preferential lateral orientation. The thicknesses of the Pc films $d_{\mathrm{Pc\,AFM}}$ were determined from the averaged height AFM profiles and ranged from 90 nm to 155 nm (see Table 8.1). The film thicknesses of the Pc films could be verified independently by X-ray diffraction (XRD) measurements which are described below in detail.

Source and drain contacts were deposited on top of the Pc films by thermal evaporation of Au (purity 99.99%) through shadow masks. The layout of the shadow mask yielded 8 independent OFETs on one sample. Samples are noted as A to F in the following, and distinct OFETs are noted as A1, A2, etc. (see Tables 8.1 and 8.2). The thickness of the Au contacts was 80 nm and monitored by a quartz microbalance, also calibrated via mechanical profilometry. The channel width W of our OFETs was 2000 μm. The channel length L was between 35 μm and 70 μm. We note that the preparation chamber was vented after the deposition of the AlO_x, the Pc film and Au contacts for experimental reasons.

Current–voltage measurements of the OFETs were performed in a He cryostat under pressures below 4.0×10^{-5} mbar, absence of light, and at defined temperatures in the range from ~4 K to 300 K. We used a combination of two Keithley 6487 units, which allowed us to monitor the source drain current (I_D) and the leakage current across the gate insulator as a function of the drain (V_D) and gate voltage (V_G). The increase of the drain voltage (V_D) was typically 1.6–5.0 V s^{-1}. We observed hysteresis and memory effects of our OFETs which

Table 8.1 Overview on the preparation conditions for the Pc films A to F. T_s denotes the sample temperature during the Pc deposition. $d_{\mathrm{Pc\ AFM}}$ denotes the averaged Pc film thickness determined by AFM, $d_{\mathrm{Pc\ XRD}}$ the value determined from the X-ray diffraction results $d_{\alpha\mathrm{XRD}}$ and $d_{\beta\mathrm{XRD}}$. The $d_{\alpha\mathrm{XRD}}$ values were determined from the FWHM of the (001) spots; the $d_{\beta\mathrm{XRD}}$ values were determined from the Williamson–Hall plots. In both cases the instrumental broadening was subtracted, assuming Gaussian profile shapes. The $d_{\mathrm{Pc\ XRD}}$ values were calculated according the following formula: $d_{\mathrm{Pc\ XRD}} = d_{\beta\mathrm{XRD}}(1 + (I_{\alpha(001)}/I_{\beta(001)}))$, except for sample *F*. This calculation assumes that the β-phase forms a closed layer, as illustrated in Figure 8.6 (below).

	p_B^* (10^{-6} mbar)	T_S (°C)	$I_{\alpha(001)}/I_{\beta(001)}$	$d_{\mathrm{Pc\ AFM}}$ (nm)	$d_{\mathrm{Pc\ XRD}}$ (nm)	$d_{\alpha\mathrm{XRD}}$ (nm)	$d_{\beta\mathrm{XRD}}$ (nm)
A	4.4	−17	0.69	98 ± 9	53	30 ± 3	32 ± 3
B	3.1	23	0.08	91 ± 10	86	63 ± 14	80 ± 1
C	1.3	25	0.41	94 ± 6	110	44 ± 6	79 ± 3
D	1.6	28	0.17	90 ± 5	102	47 ± 7	88 ± 2
E	2.3	60	0.25	155 ± 17	218	77 ± 23	174 ± 9
F	3.2	93	only α-phase	131 ± 75	(136 ± 30)	136 ± 30	only α-phase

* p_B denotes the base pressure prior to the AlO_x deposition. For sample A, $d_{\mathrm{Pc\ XRD}}$ is underestimated due to structural disorder. For sample E the $_{\mathrm{Pc\ XRD}}$ value exhibits a larger systematic error due to experimental limitations.

caused variations in the drain source current of maximal 20%. Within these limits the I_D–V_D curves were stable for at least half an hour under current. However, after storage of the OFETs for about three weeks (in the dark and under vacuum) changes at small $|V_D|$ pointed to an increase of non ohmic contact resistances between the Au electrodes and the Pc.

Contact mode AFM and XRD in Bragg–Brentano geometry using Cu K_α radiation ($\lambda = 1.54$ Å [13]) were carried out in order to characterise the structural and morphological qualities of the Pc films. Table 8.1 summarises the most important parameters of the OFETs investigated in this work.

8.3 Results and Discussion

8.3.1 Structural and Morphological Properties of the Pc Films

8.3.1.1 X-Ray Diffraction

Figure 8.2 displays XRD scans of the Pc films of the different samples after the deposition of the Au electrodes. Two sets of diffraction spots can be identified and explained by two different phases of Pc. The set marked by α belongs to the (001) lattice planes of the so-called *Campbell* bulk phase (C-phase) of

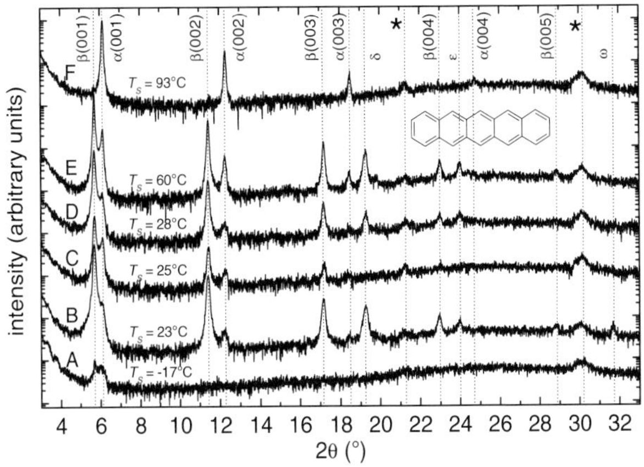

Figure 8.2 X-ray diffraction scans in Bragg–Brentano geometry obtained from the pentacene (Pc) films deposited on sputtered aluminium oxide at different growth temperatures (see Table 8.1). The reflections corresponding to the (001) planes of the α- and β-phase are numbered by their Bragg indices (001). The reflections marked by an asterisk belong to the underlying ITO layer which was proved by additional control measurements. The structural formula of Pc is shown in the inset. The spectra are vertically shifted for clarity.

Pc [14, 15]. The set marked by β originates from the (001) lattice planes of the *thin film* phase (TF-phase) of Pc which was characterised by several work groups [6, 16–20]. In both phases the long molecular axis are nearly *perpendicular* to the (001) lattice planes, and hence perpendicular to the substrate plane. Both phases are identified by their characteristic (001) lattice distances. The (001) lattice distance is 1.45 nm for the C-phase [15] and 1.54 nm for the TF phase [17], as calculated by using Bragg's law [13].

Three additional reflections at $2\theta \geq 19°$ are marked by δ, ε, and ω. Their intensity scales roughly with the intensity of the peaks of the β-phase. From Bragg's law the δ, ε, and ω reflections fit to the (110), (020), and (120) lattice planes of the orthorhombic TF phase which was first characterised by Drummy et al. [16]. Hence we assign these reflections to a small fraction of the TF phase with an orientation of the long molecular axis nearly *parallel* to the substrate plane, as it was found earlier [21, 22]. From additional control experiments, we observed that this fraction gained intensity after the deposition of the Au electrodes, likely due to the radiative heating of the Pc layer up to ~80 °C during the Au deposition. However, the fraction of these molecules must be small with respect to the fraction of molecules in the TF phase with a *perpendicular* orienta-tion to the substrate (β-phase), otherwise much higher intensities of the δ, ε, and ω reflections would be expected on the basis of the structure factors. (This was estimated using the structure factors of the C-phase, since atomic coordinates of the TF phase are not available.) We also note ahead that we were not able to identify morphological structures in the AFM images (see below) with a weight that consistently corresponded to the intensities of the δ, ε, and ω reflections. Hence we will not further consider this fraction of the TF phase in the following.

The main result from the XRD spectra is that the intensity ratio of the two phases α and β varies considerably with the substrate temperature T_S during Pc growth and the preparation conditions of the AlO_x (see Figure 8.1). The relative fractions of these two phases were estimated from the ratio of the 2θ-integrated intensities of the first order diffraction spots, i.e. $I_{\alpha(001)}/I_{\beta(001)}$, which were derived from Figure 8.3. We note that this is only correct under the condition that the rocking widths of the two phases are comparable.

We assume that this is the case, since the α-phase nucleates on top of the β-phase (as concluded below) and should hence exhibit the same mosaic spread with respect to the surface normal. In particular we find that the relative fraction of the α-phase is larger for higher T_S. E.g., at the extremely high growth temperature 93 °C, only the α-phase is observed, whereas around room temperature ($T_S = 23–28$ °C) the phase ratio varies from 10% to 40%.

The increase of the relative fraction of the bulk phase with higher T_S is in agreement with the observations of Bouchoms *et al.* [18] for Pc growth on SiO_x in a similar temperature range. The former result can be explained by a higher thermodynamic stability of the C-phase [20], which forms with larger probability at higher T_S since kinetic bar-riers presumably play a smaller role there. However, in contrast to Bouchoms *et al.* [18], a strong variation of the

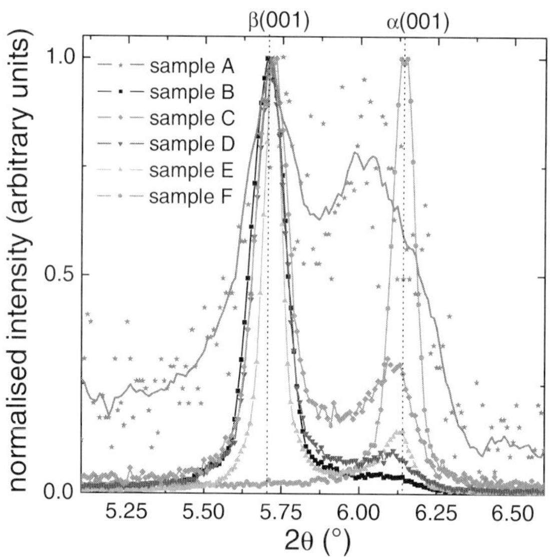

Figure 8.3 Profiles of the (001) reflections from Figure 8.2. The ideal Bragg positions of the α- and the β-phase are marked. All intensities were normalised to the maximum values.

intensity for the Pc films prepared at about room temperature is observed here, even if the T_S values are very similar (see Table 8.1, samples B, C, and D). This indicates that other details of the preparation process play a role. Below we will discuss that in particular the detailed chemical composition of the AlO_x may be important (see Section 8.3.3.). At low growth temperatures (−17°, cf. Table 8.1, sample A), a very large fraction of the α-phase is found, too, but there it is likely induced by a high density of growth defects resulting from kinetic limitations. We note that the growth of Pc films in these two different structural phases was observed in many experiments, and has been discussed controversially [18, 23].

Whereas the spot position of the β-phase is identical for all samples, a small shift of the (001) spot of the α-phase by about 0.05°–0.10° in 2θ to lower diffraction angles is observed for the Pc films in samples A and C (see Figure 8.3). The reason for this may be the presence of an additional Pc phase with a slightly larger (001) lattice constant (1.46 nm) compared to the C-phase (1.45 nm) which is unknown so far. For this larger lattice constant the presence of local asymmetric strain may play a role, too. The undiscovered presence of this additional phase may be the reason that a *small range* of different (001) lattice distances is reported in the literature for the so called *bulk* phase in thin Pc films [18, 23]. Finally, the thermodynamically most stable bulk phase of Pc, the so called H-phase [23], which was first observed by Holmes *et al.* [24] and which is characterised by a smaller (001) lattice distance of 1.41 nm, is not observed in our films.

8.3.1.2 **Scanning Force Microscopy**

Figure 8.4 shows AFM images of the Pc films. In all cases, except for the film F (see below), which was grown at very high T_S, we find that the Pc films consist of two morphological phases.

The phase at the vacuum side of the film is formed by small crystallites with lateral extensions of 0.4 μm to 1.8 μm which vary in their shape between more compact and dendritic, depending on the individual sample. E.g., compact crystallites are observed for samples A and F; dendritic or weakly branched crystallites are found for the samples B to E. The second phase which is closer to the AlO_x substrate is composed of more platelet-like, compact structures with lateral extensions of 1.3 μm to 1.8 μm, forming a dense film, as can be seen in particular for the samples B and D (dark areas between isolated dendritic crystallites). Correlating these results with the information obtained by the XRD, we assign the vacuum near phase to the bulk phase, which led to the XRD spots marked by α in Figure 8.2. The substrate near phase is assigned to the TF phase, causing the XRD spots marked by β in Figure 8.2. This assignment of the two phases is based on two independent arguments: (i) The averaged film thickness determined from the AFM images ($d_{\text{Pc AFM}}$) came out consistently with the average film thickness that was determined from the XRD data ($d_{\text{Pc XRD}}$) on the basis of this assignment (see below and caption of Table 8.1). (ii) The ratios of the α-phase to the β-phase determined by AFM

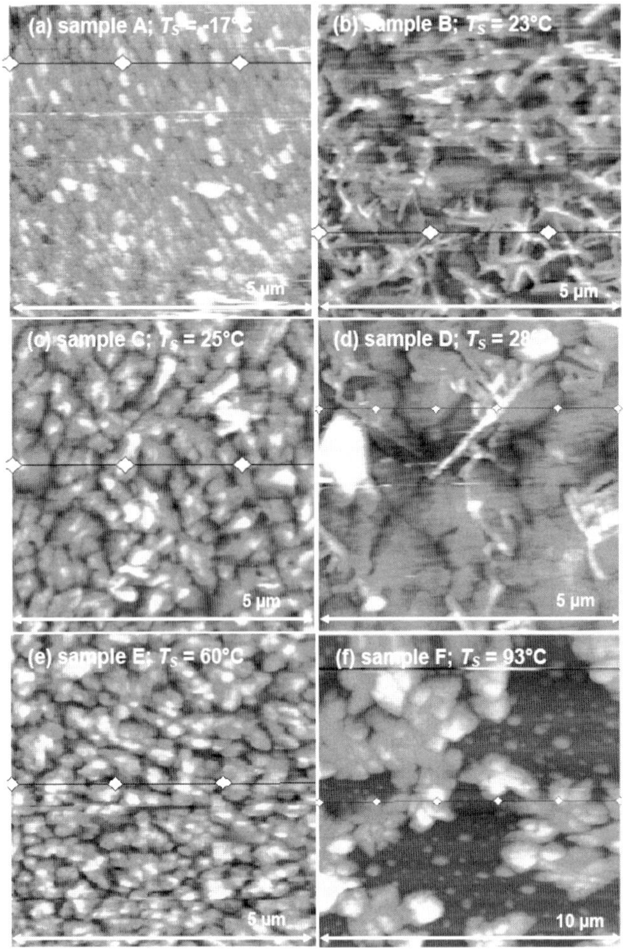

Figure 8.4 AFM images of the Pc films deposited on sputtered aluminium oxide (samples A to F of Table 8.1). For further details see text.

agreed with those estimated from the $I_{\alpha(001)}/I_{\beta(001)}$ ratios. These are listed in Table 8.1 and plotted in Figure 8.11 (below).

For the determination of the average thicknesses of the two phases from the diffraction data ($d_{\beta XRD}$), we performed a Williamson–Hall analysis to the Gaussian spot profiles [25]. The corresponding plot is given in Figure 8.5. The intention of this analysis was to separate the broadening effects due to micro lattice tensions and due to the finite dimension of the crystallites on the width of the diffraction spots. As it can be seen in Figure 8.5, we obtain a good linear dependence of the squared FWHM of the diffraction spots as a function of the squared momentum transfer perpendicular to the Pc films q_z^2, as expected for micro lattice tensions [25]. We note that this detailed analysis was only possible for the β-phase, since the higher order diffraction peaks of the α-phase

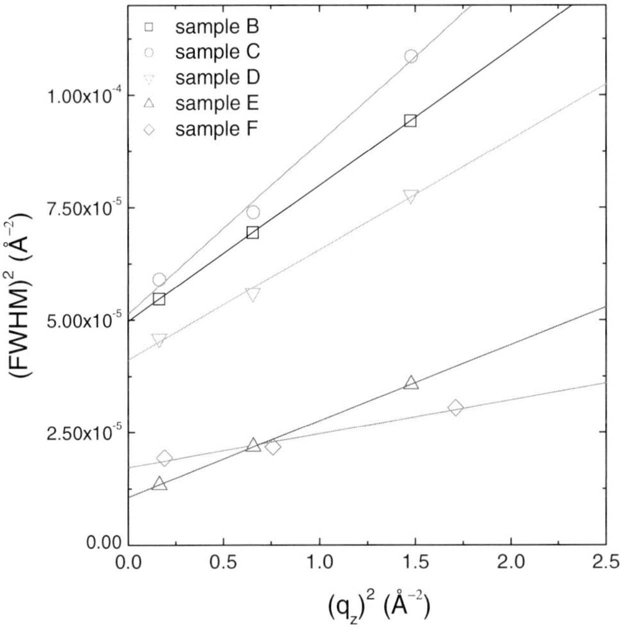

Figure 8.5 Williamson–Hall plots of all (001) reflections for the Pc films of the samples A to F (see Table 8.1). In all cases the reflections of the β-phase were evaluated, except for the sample F, for which the reflections of the α-phase were used. The FWHM values are corrected for experimental broadening.

were too weak to be analysed (with the exception of sample F). The average thicknesses $d_{\beta XRD}$ of the β-phase were obtained from the intersection of the linear fit with the ordinate via the Scherrer formula. The rms strain, σ_{rms}, of the (001) lattice planes was determined from the slope (m) of the curves in the Williamson–Hall plots ($\sigma_{rms} = \sqrt{m}$) [26]. The σ_{rms} values decrease with increasing values of T_S from 0.62% (sample C) to 0.27% (sample F). These values are comparable to those obtained for Pc films on polycrystalline Cu substrates [26]. From the (001) diffraction spots we also measured rocking curves of the β-phase. The full widths at half maximum of the rocking curves varied between 0.2° and 1.4°, indicating a good orientation of the domains parallel to the substrate, which is compatible with the value estimated from the rms roughness of the AlO_x and the lateral size of the grains of the β-phase.

Figure 8.6 gives a schematic drawing of the film morphology with the two phases. The only Pc film which was composed of only *one* phase was the one grown at the highest T_S (sample F). This Pc film is only composed of isolated crystallites of the α-phase, and does not show the substrate near β-phase. Evidently, there is the question why the relative ratio of the two phases varies for the different growth experiments and in particular, for Pc films which were grown at very similar T_S. As we will explain in the final discussion in Sec-

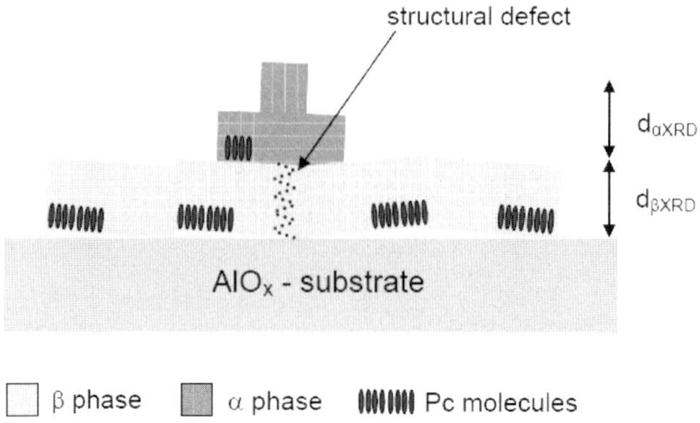

Figure 8.6 Proposed growth model for the Pc films deposited on sputtered aluminium oxide. The illustrated defect in the β-phase originates at the AlO_x interface. It is preserved throughout the growth of the β-phase and finally leads to the nucleation of a crystallite of the α-phase.

tion 8.3.3, there is evidence that the nucleation of the α-phase on top of the β-phase is supported by structural defects, which originate from the Pc/AlO_x interface. Such a defect is illustrated in Figure 8.6, and could be formed, e.g., by a screw or an edge dislocation. Defects of this type were observed by Nickel et al. for thin Pc films evaporated on SiO_x [27].

8.3.2 Analysis of the Electrical Characteristics

8.3.2.1 Overview of the I_D–V_D Characteristics

Figure 8.7(a) displays a typical set of original I_D–V_D curves for varying gate voltages (V_G). We find that larger negative values of V_G lead to increased negative source drain currents (I_D), as it is expected for hole accumulation in Pc. For the further evaluation, two corrections were applied to the I_D–V_D curves. First, *leakage currents* across the gate insulator which were measured independently in parallel were subtracted from the I_D values. In Figure 8.7(a), these leakage currents can be read at $V_D = 0$. Secondly, the I_D–V_D curves were corrected for the *ohmic currents* along the channel. These are possibly due to an unintentionally doping of the Pc films by impurities in combination with the large Pc film thickness. Hence they are independent of V_G, since they are not due to field-induced charge carriers. Such an ohmic contribution to I_D is responsible for the approximately linear $I_D - V_D$ curves which are obtained at positive values of ($V_G - V_D$), where no holes are accumulated at the Pc/AlO_x interface, e.g., at $V_G = 30$ V in Figure 8.7(a). This also demonstrates that no accumulation of mobile electrons occurs. Figure 8.7(b) and (c) display the $I_D - V_D$ curves of Figure 8.7(a) after these two corrections were applied stepwise. The final $I_D - V_D$ curves show clear saturation behaviour at high values of $|V_D|$ (see

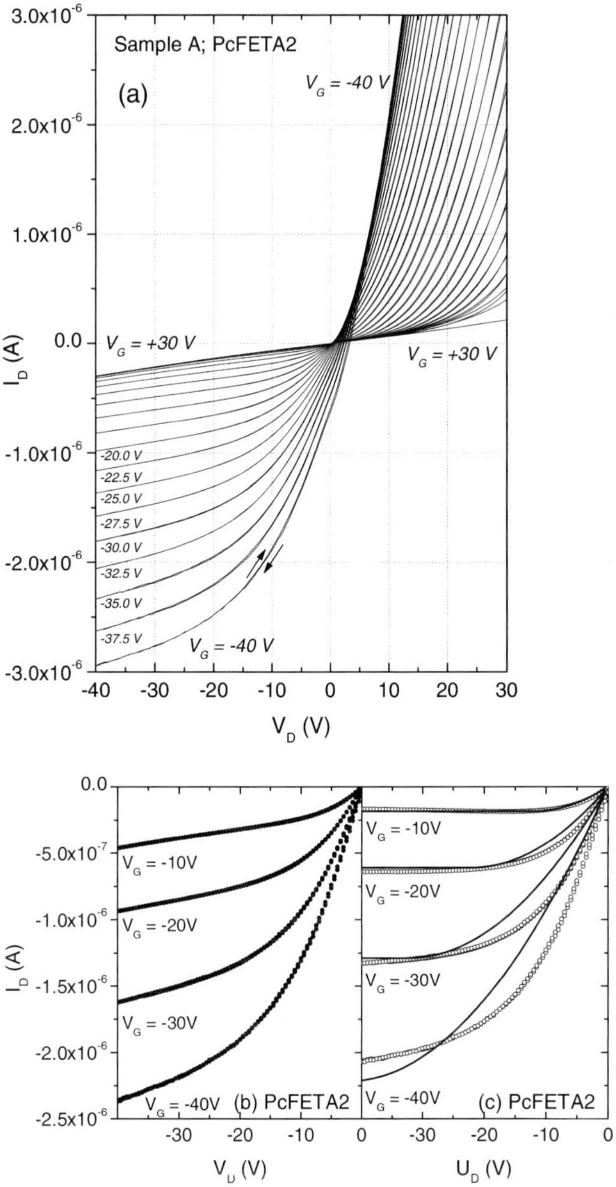

Figure 8.7 (a) Drain-current (I_D)–drain-voltage (V_D) characteristics of sample A measured at room temperature. (b) I_D–V_D curves from (a) corrected for leakage currents across the gate insulator. (c) as (b), but corrected for ohmic currents in addition. The lines in (c) represent simulated currents calculated on the basis of Eq. (2). V_G was reduced from 30 V to −40 V in steps of 2.5 V. V_D was varied from positive to negative values and back for each value of V_G in steps of 0.5 V, with a rate of 5 V/s.

Figure 8.7(c)). From these $I_D - V_D$ curves we determined the hole mobilities μ_+ by two different methods [28]. Either from the variation of the saturation current (I_D^{sat}) as a function of the gate voltage $\sqrt{|I_D^{sat}|} \sim V_G$ denoted as method A in the following, or from the linear increase of I_D versus V_G at constant V_D in the linear regime under the condition $|V_G| > |V_D|$:

$$\mu_+ = \frac{L}{Wc_i V_D} \times \left.\frac{\partial I_D}{\partial V_G}\right|_{V_D} . \quad (1)$$

In Eq. (1), c_i denotes the capacitance per unit area of the gate insulator ($c_i = 18.5$ nF/cm^2), W the channel width, and L the channel length. This second method is denoted as method B in the following. With respect to method A, it has the advantage that mobilities can be also determined in the region of small values of $|V_D|$.

Table 8.2 displays the values for μ_+ obtained by method A at 300 K for the different samples. The so determined mobilities vary between 0.005 cm^2/Vs and 0.20 cm^2/Vs, i.e. they span a range of more than *one* order of magnitude. The values of μ_+ are well comparable with those found for Pc based OFETs on other substrates [4]. However, we note ahead that the exact values of μ_+ have

Table 8.2 Field effect hole mobilities μ_+, threshold voltages V_0, and channel lengths L of the samples A to F (see Table 1). The μ_+ values were determined according to method A. E_a denotes the thermal activation energy of the field effect mobility determined from the slopes of the plots in Figure 8.8.

sample/transistor		L (μm)	V_0 (V)	μ_+ (cm^2/Vs)	E_a (meV)
A	PcFETA1	68 ± 5	−5	1.3×10^{-2}	107 ± 14
	PcFETA2	57 ± 5	0	5.6×10^{-3}	129 ± 29
	PcFETA3	57 ± 5	−5	6.3×10^{-3}	*n.d.*
	PcFETA4	60 ± 5	−7	1.1×10^{-2}	103 ± 10
B	PcFETB1	35 ± 5	−6	0.20	*n.d.*
	PcFETB2	43 ± 5	−5	0.14	*n.d.*
	PcFETB3	55 ± 5	−6	8.7×10^{-2}	129
C	PcFETC1	35 ± 5	−10	5.8×10^{-3}	67 ± 8
D	PcFETD1	10 −70	4	2.8×10^{-2} to 0.20	44 ± 7
E	PcFETE1	39 ± 5	−19	0.10	*n.d.*
	PcFETE2	43 ± 5	−2	3.0×10^{-2}	*n.d.*
	PcFETE3	47 ± 5	−14	3.5×10^{-2}	*n.d.*
F	PcFETF1	60 ± 5	*no field effect*	*no field effect*	*no field effect*

n.d. = not determined. For sample D, the channel length L was not well defined.

only a limited meaning, since they were determined using equations describing an ideal *trap free* charge transport [2]. This is not the case here, as we describe below. Hence for our samples, the mobilities should be better referred to as "effective mobi-lities". In addition, the threshold voltages V_0 were determined from the intersection of the linear fit of $\sqrt{I_D^{sat}}$ with the V_G axis according to $\sqrt{I_D^{sat}} \sim (V_G - V_0)$ [28] (not shown here).

The V_0 values are included in Table 8.2. Both the μ_+ and the V_0 values scatter significantly for OFETs that were prepared on the same sample, which illustrates the statistical variation of the data (see also Figure 8.11, below). In addition, we found that the μ_+ values also vary significantly for samples that have been prepared at very similar T_S (samples B–D, see Tables 8.1 and 8.2). As we will explain in Section 8.3.3, these variations are understandable from the different Pc film morphologies due to variations in the condition of the AlO_x surface prior to the Pc growth. For the sample F that was grown at the highest T_S, *no* source drain currents were measurable. This is understandable, since the pronounced cluster formation at this high T_S yields no closed transport paths between the source and drain electrodes.

8.3.2.2 **Temperature Dependence of the Mobilities**

From I_D–V_D curves measured at different temperatures down to ~4 K, mobilities were determined as a function of temperature using method A. The obtained μ_+ values are displayed in an Arrhenius plot in Figure 8.8. The mobilities are found to be temperature activated and decrease by about two orders of magnitude, when the temperature is reduced from 300 K to about 100 K. This is clear indication that trapping of the charge carriers plays a role here.

The data can be reasonably fitted by an Arrhenius type mobility $\mu_+ = \mu_0 \exp(-E_A/k_BT)$. For the activation energies E_A we find values in the range from 44 meV to 108 meV. Within the limitations of our data, no correlation of E_A and μ_+ values at 300 K is observed. We interpret this type of temperature dependence of μ_+ with a temperature dependent trapping of the field induced charge carriers. At low temperatures, the majority of the charge carriers is trapped and does not contribute to I_D; for elevated temperatures, thermal activation occurs and hence the fraction of the mobile charge carriers is increased. In principle, the prefactor μ_0 which describes the mobility of the mobile charge carriers in the Pc film, could also vary as a function of temperature, e.g., due to interaction with phonons, which would cause deviations from the linear behaviour in Figure 8.8. However, such deviations could not be discerned within the statistics of our data.

It is plausible, and will indeed be demonstrated for our data below, that the charge carrier traps are related to structural defects and/or disorder near the AlO_x/Pc interface. It is further conceivable that there exists some energetic distribution of the traps, and hence E_A has to be seen as a mean energy depth of the traps with respect to the valence band edge, or more generally with respect to some kind of a mobility edge. An often used model which is compatible

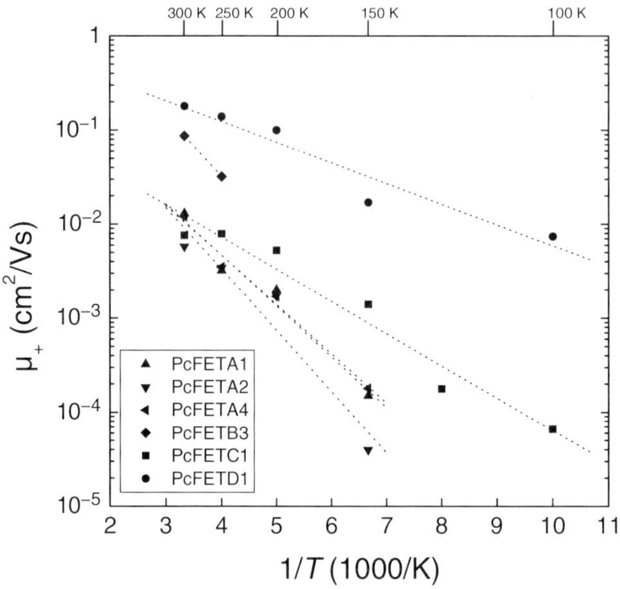

Figure 8.8 Temperature dependence of the field effect hole mobility (determined by method A, for the saturation regime) for different OFETs on the samples A to D (see Table 8.2). The linear fits correspond to Arrhenius behaviour; the slopes yield the effective activation energies required to release the charge carriers from traps.

with the here observed temperature activated mobility is the *multiple trapping and thermal release* (MTR) model [29, 30], which is used to explain the charge transport in structurally imperfect materials. The main assumption of this model is an equilibrium between trapped and mobile charge carriers which depends on the temperature and the total accumulated charge density (see below). In this model the majority of the field induced charge carriers is trapped and therefore not mobile. Mobile charge carriers are obtained by thermal activation into delocalised states. As we will show in the next section, the MTR model explains also the dependence of the effective mobilities on the gate and drain voltages in our OFETs.

8.3.2.3 Detailed Analysis of the Field Effect Mobilities as a Function of V_D and V_G

In order to demonstrate the noted role of traps for the I_D–V_D curves, we compare two OFETs based on Pc films prepared at very different growth temperatures (samples A and B). The Pc film in sample A was grown at −17 °C and exhibited a high density of structural defects. This can be seen from the corresponding XRD spectrum in Figure 8.2, which shows only the first order diffraction spots, since higher order spots are suppressed due to the structural defects. The Pc film in sample B was grown at 23 °C, and had a lower density of structural defects compared to sample A. Correspondingly, XRD spots can be observed up to the order of 5 for sample B in Figure 8.2.

To begin with, we demonstrate that I_D–V_D curves based on the ideal OFET equations according to Ref. [28]:

$$I_D = -\frac{W}{L} c_i \mu_+ \left[(V_G - V_0) V_D + \frac{V_D^2}{2} \right], \tag{2}$$

and a constant mobility, do *not* explain the measured curves. This is shown for the OFET A with a high density of defects in Figure 8.7(c), where we compare the experimental and simulated I_D currents. For the simulation, the μ_+ and V_0 values of Table 8.2 were used. We find that the simulated curves ($I_{D\,sim}$) deviate from the experimental curves (I_{Dexp}). In particular the experimental curves show a steeper increase of the drain current at low values of $|V_D|$ than the simulated curves. In Figure 8.9(a) the ratio of the experimental to the simulated currents ($I_{D\,exp}/I_{D\,sim}$) is analysed as a function of the total accumulated charge Q_{int} for different values of V_D and V_G. The value of $I_{D\,exp}/I_{D\,sim}$ describes the deviation of the experimental data from the ideal OFET characteristics calculated for a chosen constant mobility (see Figure 8.7(c)). Hence it describes the ratio of the *effective mobility* in the OFETs in relation to a constant mobility determined from the saturation currents. The total accumulated charge densities Q_{int} (integrated along the channel $Q_{int} = \int Q(x)\,dx$) were calculated from the local charge density $Q(x)$, which is given by the potential $V(x)$ by $Q(x) = -c_i(V(x) - V_G)$, whereby

$$V(x) = V_G + \sqrt{V_G^2 - 2\frac{x}{L}\left(V_G - \frac{1}{2}V_D\right)V_D}, \tag{3}$$

according to Ref. [31]. $V(x)$ is illustrated in Figure 8.9(b). Here x denotes the position along the channel in the direction from the source to the drain electrode. For an ideal OFET, without significant charge trapping, a constant value of $I_{D\,exp}/I_{D\,sim} = 1$ would be obtained, independent of V_D and V_G.

Two results can be derived from Figure 8.9(a). First, for a constant local charge density along the channel ($Q(x) = \text{const}$), which is expected to be present in the limit $V_D = 0$ V [31], the ratio $I_{D\,exp}/I_{D\,sim}$ increases significantly with Q_{int}. This demonstrates that the effective mobility depends on the total charge density and increases for higher values of Q_{int}. Such a behaviour is expected within the MTR model due to filling and saturation of the trap states at higher values of Q_{int} that makes more mobile charge carriers available.

Secondly, from Figure 8.9(a) we find that changes in Q_{int} due to V_D have a much stronger impact on $I_{D\,exp}/I_{D\,sim}$ than equivalent changes in Q_{int} due to V_G. This demonstrates that the distribution of the charge density along the channel ($Q(x)$) is also relevant, and not only the integrated value Q_{int}. The influence of V_D on $Q(x)$ occurs mainly near the drain electrode, where it causes a local depletion of the charge carrier density for higher values of $|V_D|$ (see Figure 8.9(b)).

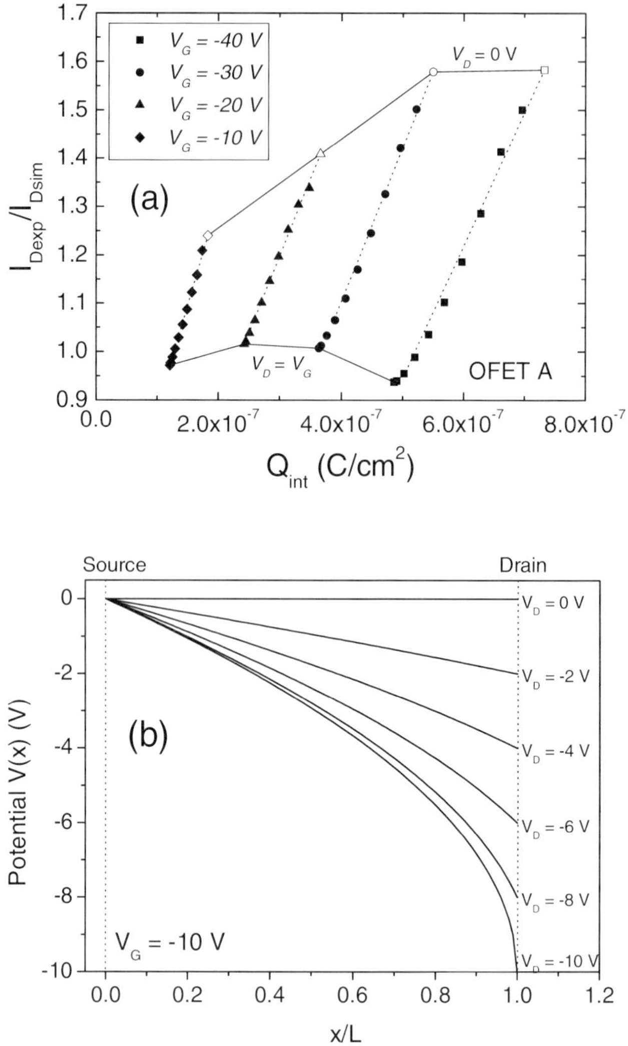

Figure 8.9 (a) Ratio of the experimental drain current to the simulated drain current ($I_{D\,exp}/I_{D\,sim}$) from Figure 8.7(c) as a function of the integrated charge density Q_{int} per area. We note that a value of 7.4×10^{-7} C/cm² of the charge density corresponds to a charge carrier density of 4.6×10^{12} cm^{-2}. For the limit $V_D = 0$ V, $I_{D\,exp}/I_{D\,sim}$ is not defined and was hence extrapolated. The solid lines mark the increase of $I_{D\,exp}/I_{D\,sim}$ as a function of Q_{in} for $V_D = 0$ V and $V_D = V_G$, and hence limit the linear region of the OFET. (b) Local potential along the OFET channel (hole accumulation) according to Eq. (3) for different drain voltages V_D at a constant gate voltage V_G.

Contrary, the influence of V_G on $Q(x)$ occurs more uniformly along the entire channel. Hence V_D induces larger absolute changes of $Q(x)$, compared to V_G, for equivalent changes in Q_{int}.

Due to the saturation of the traps, changes in $Q(x)$ have a super-linear effect on the local charge carrier mobilities. Since the charges have to pass the entire channel, $I_{D\,exp}/I_{D\,sim}$ and the effective mobility of the OFET are limited by the locally reduced mobility near the drain electrode, which explains the strong influence of V_D. Finally, in Figure 8.9(a), the dashed lines indicate the variation of $I_{D\,exp}/I_{D\,sim}$ as a function of Q_{int} at constant values of V_G by variation of V_D. It can be seen in Figure 8.9(a) that the slopes $(d(I_{D\,exp}/I_{D\,sim})/dQ_{int}$, $V_G = \text{const.})$ decrease, if larger values of Q_{int} are accumulated by an increase in V_G. This is also explained consistently in the above noted model by a reduced trapping of the charge carriers at larger values of Q_{int} which is related to the saturation of the traps.

We now turn to the comparison of the two OFETs prepared on samples A and B, from which we demonstrate the role of the structural quality on the above noted effects. For this purpose we plotted μ_+ as a function of V_G and V_D in Figure 8.10(a) and (b). These μ_+ values were determined at 300 K according to the method B.

We start with the μ_+ curves of the OFET A. These show three different regions (I, II, and III): In region I, the mobility increases by up to a factor of three on increasing values of $|V_G|$. In region II, the mobility saturates, and in region III the mobility slightly decreases on increasing values of $|V_G|$. In region I, this gate voltage dependence of μ_+ can be explained by the MTR model, as explained above due to the partial saturation of traps with increasing $|V_G|$. In region II, the observed limitation of the charge transport must be due to scattering at structural defects in the thin film phase. Finally, in region III, additional scattering at structural defects near the Pc/AlO_x interface is likely responsible for the small decrease in μ_+. This last effect is related to the concentration of the charge carriers at the organic/insulator interface at high $|V_G|$ as suggested by Refs. [2, 9]. In Figure 8.10(a) a marked decrease of μ_+ on increasing values of $|V_D|$ is observed. As discussed above, this can be explained by a *locally* reduced filling of the trap states near the drain electrode. This leads to a strongly reduced mobility in this region, which lowers the effective μ_+ of the entire OFET channel. As also discussed above, the effect of V_G on μ_+ becomes smaller for larger values of $|V_D|$, since the trap density near the drain electrode is mainly controlled by V_D and dominates the OFET characteristics. Hence the change in μ_+ with V_G becomes small for high values of $|V_D|$, as can be seen in Figure 8.10(a).

Similar to OFET A, OFET B also shows an increase of μ_+ with $|V_G|$ by a factor of 3 which is again due to the MTR of the charge carriers, although the drain currents of the OFET B are about a factor of 100 larger compared to the OFET A. However, in contrast to the OFET A, the OFET B does *not* show any saturation behaviour of the mobility in region II, which we explain by the lower density of structural defects of the Pc film in OFET B. Finally the de-

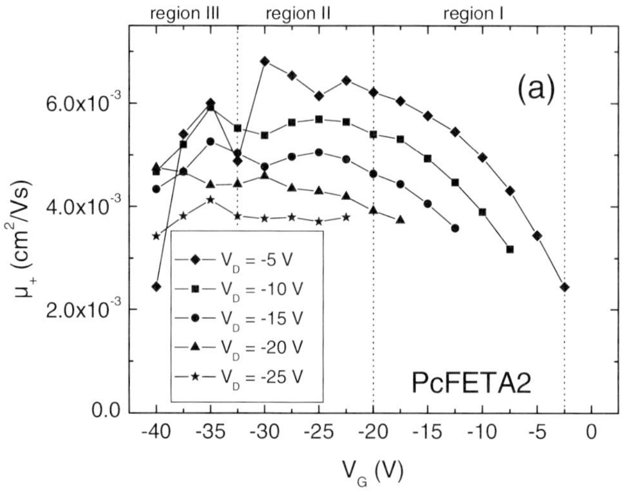

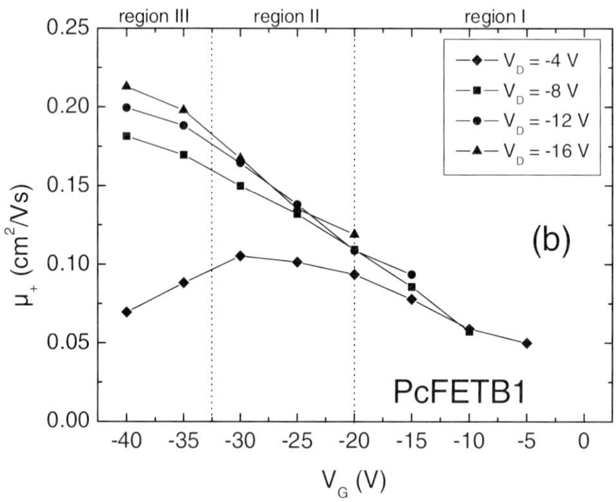

Figure 8.10 Gate-voltage and drain-voltage dependence of the field effect hole mobilities for two OFETs, which were prepared at very different growth temperatures (PcFETA2 and PcFETB1 of Table 8.2). Three regions (I–III) are marked for both OFETs which are discussed in the text.

crease of the mobility in region III is absent here, likely due to a higher quality of the Pc/AlO_x interface. The influence of V_D is opposite to that observed for OFET A and can be explained by limitations of I_D through non-ohmic contacts between the gold electrodes and Pc. The different influence of V_D on I_D demonstrates that the effect of V_D on I_D for OFET A cannot be explained by such non-ohmic losses at the contacts there, but is due to the noted carrier depletion.

8.3.3 Discussion and Conclusions

8.3.3.1 Correlation of the Electrical Transport Properties and the Film Morphology

As described so far, we found a significant variation of the charge carrier (hole) mobility and the growth morphology for samples grown under different conditions. From the detailed analysis of the $I_D - V_D$ curves there is evidence that trapping of the charge carriers in the Pc film is most relevant and limits the mobilities. From the locally accumulated charge density along the channel in combination with a partial saturation of the trap states it is possible to understand the $I_D - V_D$ curves of the OFETs in detail. Since the charge transport occurs in the first few organic layers [2, 9], the trap states must be located in the β-phase which grows on the AlO_x substrate. Evidently we can only speculate about the nature of the trap states. However, it is plausible that the trap states are related to structural defects in the β-phase. Such defects could be for instance grain boundaries between Pc domains of different azimuthal orientation, screw dislocations, or locally disordered regions.

When describing the different morphologies of the Pc films, we already noted that the probability for the nucleation of the α-phase on the substrate near the β-phase is plausibly related to the density of structural defects of the β-phase. The underlying idea is that the above noted structural defects form local sites at the vacuum/Pc growth interface which support the formation of nuclei of the thermodynamically more stable α-phase. Of course there is no reason that the structural defects which we discussed above in the context of charge carrier trapping are identical to those which play a role for the nucleation of the α-phase. However, from our data we found that there exists a correlation between the intensity ratio $I_{\alpha(001)}/I_{\beta(001)}$ and the mobility μ_+, which is shown in Figure 8.11. As can be seen the mobilities decrease with increasing values of $I_{\alpha(001)}/I_{\beta(001)}$ by about one order of magnitude. Hence, if we consider $I_{\alpha(001)}/I_{\beta(001)}$ to be proportional to the densities of structural defects in the β-phase, the data supports that these defects are not only relevant for the nucleation of the α-phase, but are also relevant for the trapping of the charge carriers and hence determine their mobilities. In addition, the correlation observed in Figure 8.11 is remarkable, since it reveals that the growth temperature of the sample is *not* the only important parameter for the resulting mobilities. Otherwise comparable mobilities would have to be expected for the samples B, C, and D, which were grown at very similar growth temperatures (see Table 8.1).

A correlation between the mobility and the phase ratio was also observed by Dimitrakopoulos et al. [19]. In contrast to our work, they interpret this correlation by grain lateral boundaries between the TF phase and the bulk phase at the Pc/insulator interface.

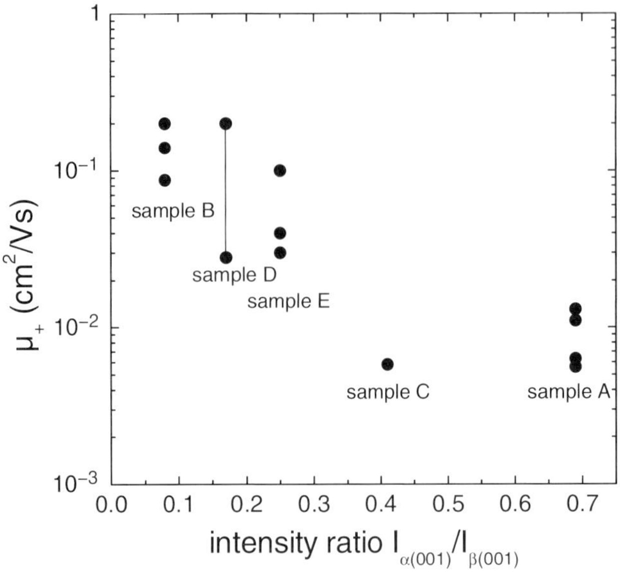

Figure 8.11 Field effect hole mobility as a function of the ratio of the integrated XRD intensities ($I_{\alpha(001)}/I_{\beta(001)}$) of the (001) reflections of the α- and β-phase in Figure 8.3. For sample D, the vertical bar indicates the uncertainty of μ_+ due to the uncertainty of L (see Table 8.2).

8.3.3.2 **Origin of the Structural Defects and Conclusions**

The above interpretation evidently asks for the origin of the proposed structural defects that influence both the charge transport and the growth morphology. Which is the relevant (key) parameter that needs to be controlled in order to achieve the highest mobilities? Within our experimental range of preparation parameters this would be equivalent to growth of Pc films consisting only of the β-phase, or at least a very small fraction of the α-phase.

The only parameter we found to correlate with the achieved $I_{\alpha(001)}/I_{\beta(001)}$ ratio was the chemical composition of the substrate, i.e. the AlO_x substrate. To explain this we note that we analysed the detailed stoichiometry of the AlO_x by elastic recoil detection of heavy ions (ERD). The detailed results of this investigation are described elsewhere [32]. Here we only concentrate on one specific finding, namely that we found significant concentrations of hydrogen in the AlO_x films in the order of 0.18 at% to 0.96 at%. The hydrogen must be built into the AlO_x films from the residual gas in the chamber during the rf sputter process and forms AlO_xH_y compounds [33]. Their concentration varied from sample to sample due to the variation of the background pressure of the residual gas, which was related to the day-to-day conditions of the sputter chamber. Interestingly, we observed that there exists a correlation between the background pressure of the residual gas during the AlO_x sputter process and the charge carrier mobilities μ_+, respectively the $I_{\alpha(001)}/I_{\beta(001)}$ ratio, as long as the growth temperatures for the Pc films were comparable. In particular, a higher background pressure was found to yield devices with higher field effect mobilities. We propose that the background pressure influences the resulting μ_+ via the stoichiometry of the AlO_x substrate. The idea is that an increasing H concentration of the AlO_x substrate makes the AlO_x surface more *hydrophilic* and supports the adsorption of a thin water film from the residual background pressure on the surface. This water film is similar to those found on SiO_x surfaces under high vacuum conditions [34]. It saturates defects on the AlO_x surface and hence leads to a smaller concentration of growth defects in the thin film phase (β), that is relevant for the charge transport. However, to clarify this role of the AlO_x, more controlled experiments, which use UHV conditions both for the AlO_x and for the Pc growth, are unavoidable.

8.4
Summary

OFETs based on thin evaporated Pc films as the semiconducting layer and rf sputtered AlO_x films as the gate insulator on ITO glass were successfully prepared and yielded field effect hole mobilities up to 0.2 cm^2/Vs. This finding agrees with those made recently also by Lee et al. [5–8]. Similar to the situation on SiO_x, Pc grows in two phases on AlO_x: a *thin film phase* on the gate insulator and a *bulk phase* which nucleates on top of the thin film phase. In addition, we obtained the following important new results: (i) The growth of the Pc bulk phase is supported by structural defects in the thin film phase. (ii) These structural defects originate at the Pc/AlO_x interface and are related to the growth temperature and the detailed preparation conditions of the AlO_x. (iii) These defects of the thin film phase also limit the charge carrier mobilities by charge trapping and scattering. (iv) The detailed temperature, gate- and drain voltage dependence of the field effect mobility is explained by multiple trapping and release of the charge carriers in combination with the local variation of the charge carrier density along the channel, i.e. the reduction of the charge carrier density near the drain contact that leads to increased local trapping and decreased local mobility. In the present work it was hence demonstrated that the detailed analysis of the charge transport limiting effects in this type of OFETs is possible and provides complementary information on the structural quality of the interface.

Acknowledgements

This work was supported by the Deutsche Forschungsgemeinschaft through the priority program "Organic field effect transistors" (SPP1121). We thank B. Klöter (Universität Bonn) for support with the AFM measurements on the Pc films and M. Böhmer for the support in programming the automatic data collection.

References

1. G. Horowitz, in: Organic Electronics, edited by H. Klauk (Wiley-VCH, Weinheim, 2006).
2. G. Horowitz, J. Mater. Res. **19**, 1946 (2004).
3. H. Sirringhaus, Adv. Mater. **17**, 2411 (2005).
4. A. Facchetti, M.-H. Yoon, and T. J. Marks, Adv. Mater. **17**, 1705 (2005).
5. J. Lee, J. H. Kim, and S. Im, Appl. Phys. Lett. **83**, 2689 (2003).
6. J. Lee, J. H. Kim, and S. Im, J. Appl. Phys. **95**, 3733 (2004).
7. J. Lee, J. H. Kim, S. Im, and D.-Y. Jung, J. Appl. Phys. **96**, 2301 (2004).
8. J. Lee, K. Kim, J. H. Kim, S. Im, and D.-Y. Jung, Appl. Phys. Lett. **82**, 4169 (2003).
9. M. Mottaghi and G. Horowitz, Org. Electron. **7**, 528 (2006).
10. W. Kalb, P. Lang, M. Mottaghi, H. Aubin, G. Horowitz, and M. Wuttig, Synth. Met. **146**, 279 (2004).
11. M. Voigt and M. Sokolowski, Mater. Sci. Eng. B **109**, 99 (2004).
12. O. D. Jurchescu, J. Baas, and T. T. M. Palstra, Appl. Phys. Lett. **84**, 3061 (2004).
13. H. Krischner and B. Koppelhuber-Bitschnau, Röntgenstrukturanalyse und Rietveldmethode (Vieweg, Braunschweig/Wiesbaden, 1994).
14. R. B. Campbell, J. M. Robertson, and J. Trotter, Acta Crystallogr. **14**, 705 (1961).
15. R. B. Campbell, J. M. Robertson, and J. Trotter, Acta Crystallogr. **15**, 289 (1962).
16. L. F. Drummy and D. C. Martin, Adv. Mater. **17**, 903 (2005).
17. A. C. Mayer, A. Kazimirov, and G. G. Malliaras, Phys. Rev. Lett. **97**, 105503 (2006).
18. I. P. M. Bouchoms, W. A. Schoonveld, J. Vrijmoeth, and T. M. Klapwijk, Synth. Met. **104**, 175 (1999).
19. C. D. Dimitrakopoulos, A. R. Brown, and A. Pomp, J. Appl. Phys. **80**, 2501 (1996).
20. T. Kakudate and N. Yoshimoto, Appl. Phys. Lett. **90**, 081903 (2007).
21. T. Minakata, H. Imai, M. Ozaki, and K. Saco, J. Appl. Phys. **72**, 5220 (1992).
22. M. Shtein, J. Mapel, J. B. Benziger, and S. R. Forrest, Appl. Phys. Lett. **81**, 268 (2002).
23. L. Farina, A. Brillante, R. G. D. Valle, E. Venuti, M. Amboage, and K. Syassen, Chem. Phys. Lett. **375**, 490 (2003).
24. D. Holmes, S. Kumaraswamy, A. J. Matzger, and K. P. C. Vollhardt, Chem. Eur. J. **5**, 3399 (1999).
25. G. K. Williamson and W. H. Hall, Acta Metall. **1**, 22 (1953).
26. M. Oehzelt, R. Resel, C. Suess, R. Friedlein, and W. R. Salaneck, J. Chem. Phys. **124**, 054711 (2006).
27. B. Nickel, R. Barabash, R. Ruiz, N. Koch, A. Kahn, L. C. Feldman, R. F. Haglund, and G. Scoles, Phys. Rev. B **70**, 125401 (2004).
28. G. Horowitz, R. Hajlaoui, H. Bouchriha, R. Bourguiga, and M. Hajlaoui, Adv. Mater. **10**, 923 (1998).
29. G. Horowitz, Adv. Mater. **10**, 365 (1998).
30. G. Horowitz, M. E. Hajlaoui, and R. Hajlaoui, J. Appl. Phys. **87**, 4456 (2000).
31. R. Schmechel, M. Ahles, and H. v. Seggern, J. Appl. Phys. **98**, 084511 (2005).
32. M. Voigt, A. Bergmaier, and M. Sokolowski, J. Vac. Sci. Technol. A 27(2), (2009).
33. J. M. Schneider, A. Anders, B. Hjörvarsson, I. Petrov, K. Macák, U. Helmersson, and J.-E. Sundgren, Appl. Phys. Lett. **74**, 200 (1999).
34. A. Mayer, G. Malliaras, B. Wang, Y. Wang, S. Wo, R. Headrick, and A. Kazimirov, Chess News Magazine (2005).

9
In Situ X-Ray Scattering Studies of OFET Interfaces

Alexander Gerlach, Stefan Sellner, Stefan Kowarik, and Frank Schreiber

9.1
Introduction

The advent of organic, molecular semiconductors for electronic and optoelectronic applications has opened new possibilities for manufacturing of devices with large area, flexible structure, low temperature processing and low cost [1–4]. All these devices crucially depend on the definition of interfaces between functional materials and the structural (crystalline) order within the organic semiconductor itself [5]. While for the growth of inorganic materials a certain level of understanding has been reached, the growth of organic molecules poses a range of new challenges due to the weaker van-der-Waals binding forces within this 'soft matter' and new degrees of freedom impact growth, such as molecular orientation and conformation [6, 7]. Here we review our work on controlling the growth of organic semiconductor as well as the interface formation between inorganic insulators and metals with organic thin films. As is obvious from a generic sketch of an organic field effect transistor (OFET), interfaces play a crucial role in its performance: the metal/organic semiconductor interface determines charge carrier injection, while the interface between gate insulator/organic semiconductor is crucial in the formation – or interruption – of a conducting channel. This structure-function relation is also illustrated in several other contributions to this book. Apart from the use as gate dielectric, insulators may also be used to encapsulate devices, which is crucial for device operation and preventing device breakdown as many organic semiconductors degrade by exposure to oxygen or humidity. According to these challenges in controlling growth and manufacturing this chapter covers the following systems: (1) thin films of organic semiconductors on different substrates and (2) organic heterostructures, i.e. metal contacts on top of organics and aluminium oxide capping layers which are the basic building blocks of OFETs. It does not cover the organic/organic interface which is relevant for bipolar devices such as solar cells or bipolar transistors, and also modification of growth by self-assembled monolayers will not be discussed here.

Organic Electronics. Structural and Electronic Properties of OFETs. Edited by Christof Wöll

ISBN: 978-3-527-40810-8

In the growth studies presented below emphasis is put on X-ray scattering techniques, which are well suited to measure the properties of organic thin films as they are non-invasive, i.e. they can be used for *in situ* studies in an ultra-high vacuum environment and they can be performed in real-time during organic molecular beam deposition (OMBD). We make extensive use of the capability of X-ray experiments to follow changes in sample structure, which allows one to study the growth mode during thin film deposition, but also the breakdown of encapsulated thin films can be directly measured. Also, *in situ* studies offer the advantage that post-growth sample changes such as oxidation or de-wetting do not obscure the results. From the X-ray data one can extract parameters such as film thickness and roughness, fractional coverage of individual layers, crystal structure, mosaicity, bonding distances between substrate and molecules, island sizes and correlation lengths as will be shown below. Of great importance for understanding the organic/inorganic interface is the ability of X-rays to penetrate into the sample and thereby provide access to microscopic properties also of buried interfaces.

This chapter is organised as follows: After giving a short introduction to X-ray scattering for the investigation of thin films (Section 9.2), we address some aspects of growth physics which are relevant for organic materials (Section 9.3). Focusing on the molecules shown in Figure 9.1 we then present three case studies of growth and interface formation for the systems pentacene on silicon oxide, diindenoperylene (DIP) on silicon oxide, and perylene-tetracarboxylicacid-dianhydride (PTCDA) on noble metals (Section 9.4). Increasing the complexity of the systems we finally cover structural properties of organic heterostructures, i.e. metal and insulator films on DIP (Section 9.5). These case studies are based mostly on our own work, but we provide numerous references to related studies in the literature.

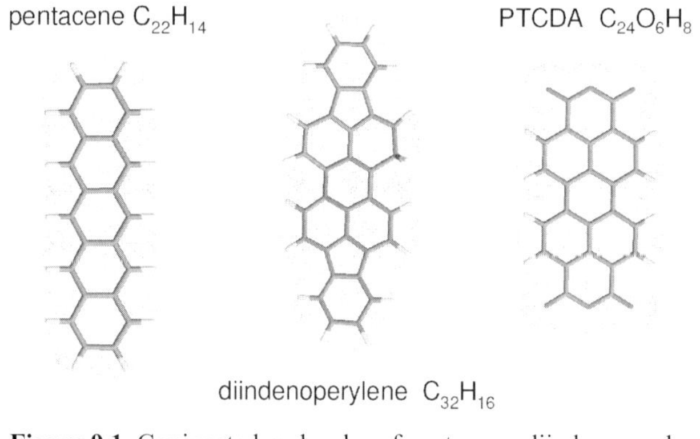

Figure 9.1 Conjugated molecules of pentacene, diindenoperylene (DIP), and perylene-tetracarboxylicacid-dianhydride (PTCDA).

9.2 X-Ray Scattering

Various X-ray scattering techniques for the investigation of bulk properties are well established. Due to the increased availability of synchrotron light and advanced detector systems X-ray studies of low dimensional systems, interfaces, and thin films – even for organic materials with their low scattering cross section – have become feasible. According to the principal X-ray scattering geometries shown in Figure 9.2 structural information can be derived both along the surface normal (out-of-plane) and parallel to the surface (in-plane). In the reciprocal space representation these measurements correspond to a momentum transfer along q_z and $q_{||}$, respectively. X-ray scattering data with high resolution and dynamic range taken in these geometries can be analysed quantitatively [8–11].

Since X-ray reflectivity (XRR) measurements on the specular path, i.e. with $q_{||} = 0$, probe the electron density along the surface normal, information about the film thickness D, surface and interface roughness σ, and the density ρ can be obtained [8]. With a layer model of the film the measured reflectivity curves can be analysed using either a kinematical approximation or the dynamical Parratt formalism [9]. Interference patterns with a periodicity of $2\pi/D$ in reciprocal space, so-called 'Kiessig fringes', which originate from the scattering at different interfaces (e.g., substrate-film and film-air), yield the parameters of the different layers. For a well-ordered film structure along the surface normal Bragg reflections observed at certain q_z-values give the spacing d_F of the individual layers shown in Figure 9.2. For rough surfaces a significant part of the reflected intensity is scattered off-specular. This diffuse part

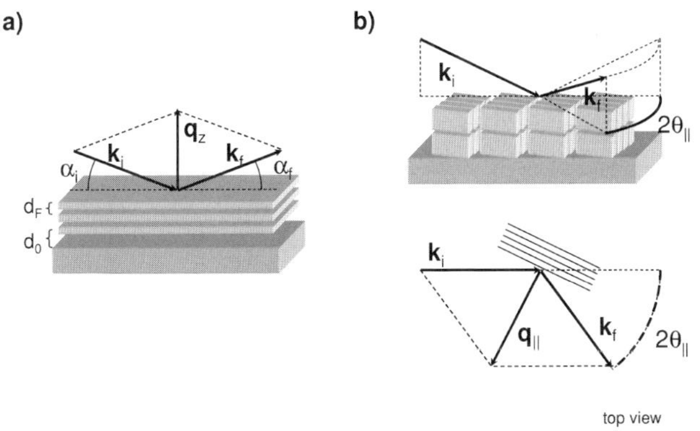

Figure 9.2 Scattering geometries with the corresponding initial and final wave vector $\boldsymbol{k}_i$ and $\boldsymbol{k}_f$. (a) Setup for specular X-ray reflectivity measurements which reveal the sample structure (roughness, lattice spacing) along the surface normal. (b) Setup for grazing incidence diffraction (GID), which measures lattice constants parallel to the sample surface. In both cases the initial and final wave vector $\boldsymbol{k}_i$ and $\boldsymbol{k}_f$ define the scattering plane.

contains information on the statistical properties of the surfaces and interfaces, such as the lateral correlation length ξ [10].

If the incident X-ray beam illuminates the surface at an angle smaller than the critical angle for total external reflection a so-called 'evanescent wave' is created, i.e. an electric field with exponentially decaying amplitude inside the sample [11]. This transmitted wave, which parallel to the surface still has oscillating character, can give rise to Bragg reflections from periodic structures in the plane with a corresponding momentum transfer $q_{\|}$. Using this technique of 'grazing incidence diffraction' (GID), the in-plane structure can be studied (down to $z \approx 50$ Å). These two scattering geometries will be applied in the following to study the structural evolution during and after thin film deposition.

9.3 Growth Physics

Growth of crystalline thin films is a rich subject with many different facets and theoretical approaches [12–15]. Here we only address some important aspects which are relevant for organic thin film growth [16], particularly the interface formation in the first monolayer and the different growth modes of organic molecules (see Figure 9.3).

9.3.1 Monolayer Deposition

It has been demonstrated that the first monolayer forms the crucial template for the growth of further molecular layers [16, 17], and the strength of the adsorbate-substrate interaction, the orientation of the molecules, their bonding distances to the topmost substrate layer determine to some extent the properties of the multilayer film. The bonding distance d_0 of the monolayer (Figure 9.2) is one of the central quantities in this context and can be determined, e.g., by

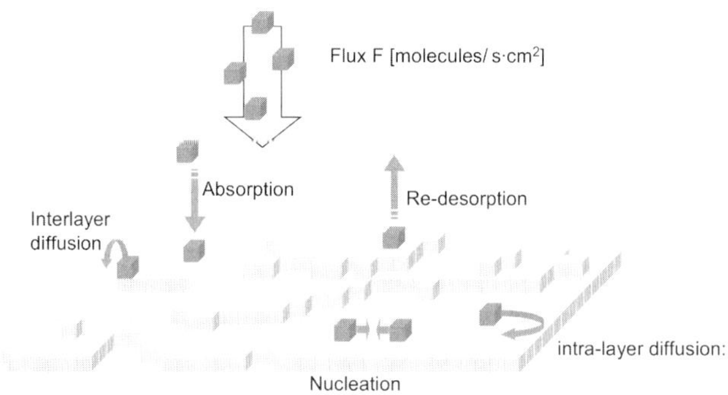

Figure 9.3 Schematic of atomistic processes relevant for OMBD.

X-ray scans on the specular path [18]. More precise and chemically resolved structural information, however, can be obtained from X-ray standing wave (XSW) experiments [19]. As will be discussed below XSW measurements of various organic molecules on metal substrates [20–27] show that even for molecules in a lying-down orientation the bonding distance depends strongly on the substrate-adsorbate interaction. The exact knowledge of bonding distance and molecular conformation becomes particularly important in the light of recent reports [28], that the electronic structure of the metal/organic interface depends on adsorption geometry of the first molecular layer. The injection barriers for electrons (and holes) from the electrode into the organic layer, which affect the performance of the organic device, is thereby related to the monolayer structure.

For growth on crystalline substrates, we also note that the strain induced by the lattice mismatch at the film-substrate interface is not only important in a crystallographic sense [29], but also influences the growth beyond the structure of the first monolayer. For example, it has been shown that DIP grown on NaCl single crystals exhibits herringbone type packing, but when DIP is grown on crystalline perylene thin films an unusual sandwich herringbone type packing is observed [30]. This suggests the possibility of controlling the stacking of epitaxial layers by careful selection of organic substrates beyond the first monolayer. For further details on organic epitaxy see the review by Hooks et al. [31].

9.3.2
Thin Film Growth and Dynamic Scaling

For multilayer coverage one can distinguish three growth modes: island (Volmer–Weber), layer-plus-islands (Stranski–Krastanov), and layer-by-layer (Frank–van-der-Merwe) growth. Thermodynamic reasoning can be used to relate the surface energies $\gamma_{\text{substrate}}$, γ_{film}, and $\gamma_{\text{interface}}$ to these different growth scenarios. The details of the growth, however, usually also depend on the deposition rate and the temperature of the sample. In order to change the growth behaviour specific modification of the substrate surface energy $\gamma_{\text{substrate}}$ is often desirable. This can be achieved, for example, by functionalising the substrate using self-assembled monolayers (SAMs) [32]. Several studies used alkanethiol SAMs on Au(111) to modify the growth of PTCDA films [33–35]. Similarly, SAMs have been used to modify the growth of pentacene thin films [36–39]. A more detailed discussion of this approach can be found in Ref. [32].

For a thorough description of a given growth scenario one might use specific growth models to quantify properties such as island size or film roughness and their evolution with film thickness. For island growth leading to roughening of the film surface rate equation models can be used [40]. In cases with little knowledge of atomic processes the roughness evolution can be more generally expressed in terms of growth exponents and dynamic scaling theories [41–43] which relate the growth mechanisms to a set of scaling exponents. Within this framework the film morphology can be described by three key parameters: the typical surface slope a, the correlation length ξ beyond which the heights at two points become uncorrelated, and the standard deviation of the film height σ (RMS roughness). These parameters scale with film thickness D according to

$$a \sim D^{\lambda}, \qquad \xi \sim D^{1/\tilde{z}}, \qquad \sigma \sim D^{\beta},$$

defining the steepening exponent λ, the dynamic exponent $\tilde{z}$, and the growth exponent β. Interestingly, both DIP on silicon oxide [44] as well as phthalocyanines [45] seem to display pronounced roughening, i.e. high β-values. Roughening parameters $\beta > 0.5$ indicate roughening faster than expected for random deposition in a 'hit-and-stick' model. Such 'rapid roughening' has been measured for DIP/SiO_2 ($\beta = 0.748 \pm 0.05$) [44], H_2Pc/glass ($\beta = 1.02 \pm 0.08$), and plasma polymer ($\beta = 0.7 \pm 0.10$) [45], but the effect is also found in inorganic materials [46].

9.3.3 Growth of Organic Molecular Materials

While the general considerations above apply both to organic and inorganic materials, there are several specific challenges related to the growth of organic materials. These may arise from the commonly encountered polymorphism (simultaneous occurrence of different structural phases), the complex epitaxial relations (different unit cell sizes of substrate and adsorbate), or the heterogeneous material properties at the interface (interdiffusion, different thermal expansion of the organic thin film and the substrate).

Moreover, the growth itself may be complicated by the additional internal degrees of freedom (DOF) characteristic for organic molecules. The vibrational DOF can affect the interaction with the substrate and also the thermalisation upon adsorption on the surface. Conformational DOF mean that the building block can change within the film, for example by bending to accommodate stress. Käfer et al. [7] found that the conformation of the organic semiconductor rubrene changes during growth, which may influence the film morphology [47]. Orientational DOF which are not included in conventional growth models can give rise to tilt domains and thereby an additional source of disorder, or may even entail the growth of competing 'lying down' and 'standing up' structures [6, 37] in the film as discussed below for the example of DIP.

The interaction between molecules and between molecules and non-metallic substrates is often dominated by weak van-der-Waals forces. It is important to emphasise that when integrated over all atoms within a molecule, the weak interaction energies add up and lead to substantial molecular binding energies. Nevertheless, the weaker interactions per atom lead to 'softer' materials and, for example, strain can be accommodated more easily. Due to the weaker interactions the thermal expansion coefficients (typically in the 10^{-4} 1/K range) are large when compared to inorganic materials, which possibly leads to higher thermally induced strain at film-substrate interfaces.

The size of the molecules and consequently the size of the unit cell is larger than that of inorganic materials. Therefore, the molecule-substrate interaction is averaged over a larger area with generally incommensurate (sub)structures. Due to this averaging the molecules experience a weaker specific interaction with the substrate. Also more translational and orientational domains for epitaxy on inorganic substrates are possible because of the different unit cell size. This introduces an additional source of disorder for organics.

9.4 Organic Thin Films

Below we present three case studies for OMBD which demonstrate how X-ray scattering techniques can be used to derive not only precise structural information, but also details of the growth on different substrates.

9.4.1 Pentacene on Silicon Oxide

Pentacene is attracting considerable attention as its charge transport properties are excellent [4, 48], and films of pentacene on silicon oxide are commonly used for thin film transistors in which the silicon oxide serves as gate dielectric.

Its thin film structure on silicon oxide as well as modified silicon surfaces has been studied extensively [51, 52]. Recently, the molecular arrangement within the unit cell has been solved using X-ray scattering [53, 54]. A dynamic-scaling analysis of the island distribution in sub-monolayer films shows that islands containing three or more molecules are stable [43, 55]. A modification of the hydrophobicity of the substrate has been shown to change the nucleation density of pentacene islands as well as the island size [38]. Under optimised conditions the island size in pentacene thin films can be as large as 0.1 mm [56]. In addition to varying substrate temperature and deposition rate, the kinetic energy of pentacene molecules has been varied in supersonic beam deposition [36, 57], providing an additional free parameter to influence the growth.

Using real-time techniques, the details of growth can be studied during deposition, yielding information about the dendritic island shape [56] as well as the coverage of individual layers [50, 58, 59]. From the X-ray reflectivity measurements shown in Figure 9.4, not only the crystal structure can be determined from the positions of the Bragg reflections, but also the evolution of the film roughness can be extracted. It was shown by Mayer et al. [60] that a

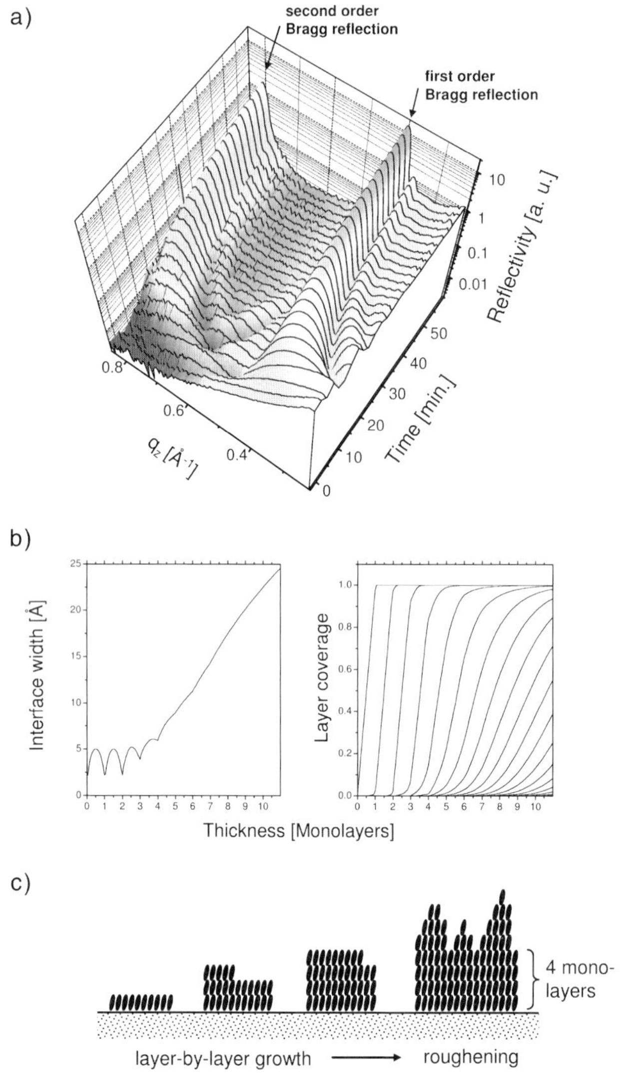

Figure 9.4 (a) Real-time measurements of the X-ray reflectivity in a wide q-range during pentacene deposition on silicon oxide have been performed with energy dispersive data acquisition [49] (deposition rate 3.5 Å/min, substrate temperature 50 °C). Two peaks corresponding to the first and second order Bragg reflection of the pentacene thin film phase can be seen to grow with increasing deposition time (from Ref. [50] with permission). (b) Using a diffusive growth model and the kinematic approximation in X-ray scattering the interface roughness of the pentacene film and the coverage of individual layers during growth can be extracted from the data-set. (c) From the evolution of the surface roughness it can be seen that layer-by-layer growth persists for the first four monolayers before roughening sets in.

second pentacene phase [51, 52, 61] nucleates already in the first pentacene monolayer as could be determined from following the Bragg reflections corresponding to the two phases.

From the data in Figure 9.4 as well as GID data [50], it can be seen that only the thin-film phase of pentacene [52] is growing for the conditions employed (out-of-plane lattice constant of $d_F = 15.6$ Å). Analysing not only the Bragg reflections, but also the evolution of the reflectivity between the Bragg reflections, we obtain additional information. Halfway between the Bragg reflections, i.e. $q_z = \pi/d_F$, at the so-called anti-Bragg point the interference of X-rays scattered from neighbouring layers interferes destructively, leading to an oscillating X-ray reflectivity when subsequent pentacene layers are filled. From these oscillations the number of pentacene monolayers that have been grown can be directly counted (oscillation period two monolayers). As can be seen from Figure 9.4 the X-ray reflectivity shows modulations not only at the anti-Bragg condition, but also at all q-values other than the Bragg condition. These growth oscillations at different q-points correspond to the several Fourier components of the real-space structure. Therefore, it is advantageous to measure the reflectivity in a wide q-space region (0.25–0.8 $Å^{-1}$ in this case) to get a precise measurement of the real-space structure and the film roughness [62].

For pentacene, the roughness data show a clear change in growth mode after four monolayers. While the growth oscillations in the beginning indicate that pentacene grows in a layer-by-layer fashion (with an oscillating surface roughness), after four monolayers the roughness starts to increase. This is due to an incomplete filling of individual layers which results in a higher film roughness and a reduced electron density at the interface. A change of the growth mode from layer-by-layer growth to roughening has also been observed in Refs. [58, 59], where the interlayer transport of pentacene molecules could be quantified. While the exact nature of this transition is unknown, several factors may contribute such as decreased interlayer transport (increasing Schwoebel barrier), faster nucleation, and decreased diffusivity on top of islands.

9.4.2
DIP on Silicon Oxide

Diindenoperylene (sometimes also referred to as periflanthene) is an organic semiconductor with both electron and hole conduction in single crystals, making this compound interesting for ambipolar electronics [63]. Thin film transistors with hole mobilities of up to 8×10^{-2} cm²/Vs have been reported [48, 64], and recent experiments show that mobilities beyond 10^{-1} cm²/Vs are feasible [65]. Further applications of DIP include usage for optical recording and in organic light emitting diodes [66, 67].

Growth of DIP molecules was studied in considerable detail [44, 68, 69]. When prepared under suitable conditions organic thin films of DIP deposited on silicon oxide exhibit high structural out-of-plane order [44]. The films form large flat terraces with a step height corresponding to the lattice spacing

$d_F = 16.5$ Å as derived from the position of the DIP Bragg peak. On silicon oxide at high substrate temperature (≥100 °C), DIP grows in a mostly 'standing upright' orientation (NEXAFS measurements give a tilt angle of 83 ± 5°). The coherent thickness as determined from the Laue fringes around the first order Bragg reflection at $q_z = 0.38$ Å^{-1} corresponds almost to the total film thickness as determined from Kiessig oscillations [69]. This demonstrates that the DIP films are coherently ordered over the entire thickness. Rocking widths of about 0.01° and lower measured for DIP films confirm the high order/fiber texture of the films perpendicular to the sample surface. Within the surface plane the films are polycrystalline with an isotropic distribution of crystallite orientations on silicon oxide.

To study the temperature dependent growth dynamics and to establish the growth mode, the X-ray reflectivity in a q-range from the anti-Bragg condition to the Bragg point has been measured in real-time during DIP deposition (Figure 9.5). For all three substrate temperatures studied the development of a strong Bragg reflection corresponding to standing upright molecules can be seen. Laue fringes develop and narrow with increasing film thickness next to the Bragg reflection. The Laue fringes (i.e. fringes along q at fixed time) – or equivalently the anti-Bragg oscillations (time dependency of reflectivity at $q_z = 0.19$ Å^{-1}) get damped for increasing film thickness as the growth mode changes from a 2D/layer-by-layer growth of the first monolayers to 3D/mound growth, which is characterised by simultaneous filling of different molecular layers. For growth at 130 °C the anti-Bragg oscillations get damped out after four oscillation maxima (see Figure 9.5b top panel), i.e. roughening sets in after ~8 monolayers as one full anti-Bragg oscillations corresponds to growth of two monolayers [6]. Compared to pentacene where roughening sets in after four monolayers DIP grows with smooth morphology for twice as long. For subsequent 3D growth, the DIP roughness has been found to increase faster than expected for random deposition of molecules. This rapid roughening has been followed up to film thicknesses of 10^4 Å [44] and has also been found in other organic systems [45]. When lowering the substrate temperatures the DIP anti-Bragg oscillations get damped out, and both for 30 °C and −10 °C the layer-by-layer growth breaks down after only ~2–3 monolayers (Figure 9.5b middle and bottom panel). Interestingly, the intensity at the Bragg-reflection is larger for lower growth temperatures, as the roughening leads to mounds with heights greater than the average film thickness, and therefore a larger number of layers scattering in phase. The reasons for the transition from layer-by-layer growth to roughening are not yet well understood for complex organic molecular materials. Strained growth in the first monolayers and strain relaxation may trigger a change in growth mode, but for organic molecules the molecular tilt angle and the molecular conformation may also change during growth. For DIP, it has indeed been found in real-time measurements that the in-plane lattice parameter changes during growth of the first three monolayers, but further work is needed to clarify the connection between growth mode and structural changes.

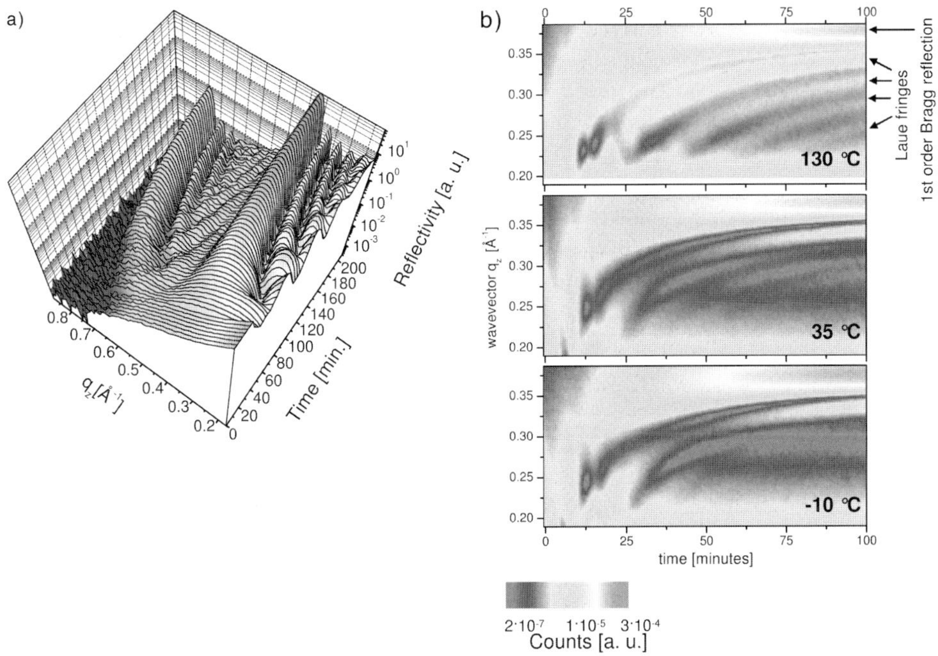

Figure 9.5 (a) Real-time measurements of the X-ray reflectivity in a wide q-range during DIP deposition on silicon oxide have been performed at beamline ID10B at the ESRF (deposition rate 0.6 Å/min, substrate temperature 130 °C). Two peaks corresponding to the first and second order Bragg reflection of DIP can be seen to grow with increasing deposition time. (b) Reflectivity of DIP thin films grown at 130 °C, 35 °C, and −10 °C substrate temperature with a rate of 3 Å/min on silicon oxide. The datasets comprise a region between the anti-Bragg and the Bragg condition in q, and range from time 0 min (i.e. bare substrate) to 100 min (corresponding to ~240 Å film thickness). In all three measurements a strong Bragg reflection at $q = 0.38$ Å^{-1} develops, showing that a DIP structure with standing upright molecules grows. Next to the Bragg reflections side maxima (Laue-fringes) develop and get narrower with time. At lower substrate temperatures the Laue fringes are damped more strongly indicating that 3D growth (roughening) sets in earlier.

Apart from the temperature dependent growth dynamics we observe a second structure of 'lying down' molecules (λ-structure) which at lower substrate temperatures competes with the growth of the 'standing upright' (σ-structure). While the λ-structure cannot be seen in the real-time measurements in Figure 9.5, GID data show Bragg reflections which correspond to DIP molecules in the 'lying down' phase [6, 70]. This change in molecular orientation does depend on substrate temperature as well as the type of substrate. For growth on silicon oxide at 35 °C, the λ-structure starts to nucleate after a critical thickness of 170 Å, i.e. the lying down structure grows on top of the 'standing upright' structure of DIP (see Figure 9.6a). In contrast, when growing DIP on top of the organic semiconductor rubrene no λ-structure nucleates within the thickness range studied. When using *A*-plane sapphire as a substrate, the nucleation of the lying phase occurs without threshold thickness. The substrate dependent occurence of the λ-structure can be rationalised by regarding the substrate interactions with DIP. Rubrene substrates have only weak van-der-Waals interactions that favour the standing upright structure. In contrast, the stronger effective interaction with sapphire due to the stepped sapphire surface and the slightly higher van-der-Waals interactions lead to molecules adopting a lying down orientation and therefore an early nucleation of the λ-structure. For a more detailed discussion regarding the influence of different substrates and the interaction of the molecules with the substrate see e.g. Ref. [17]. Figure 9.6b schematically summarises the connection between substrate interaction and temperature with molecular orientation in the thin film. For metal substrates the molecule-substrate interactions are even stronger and as expected the lying down structure has been observed to grow directly on top of polycrystalline gold substrates [71].

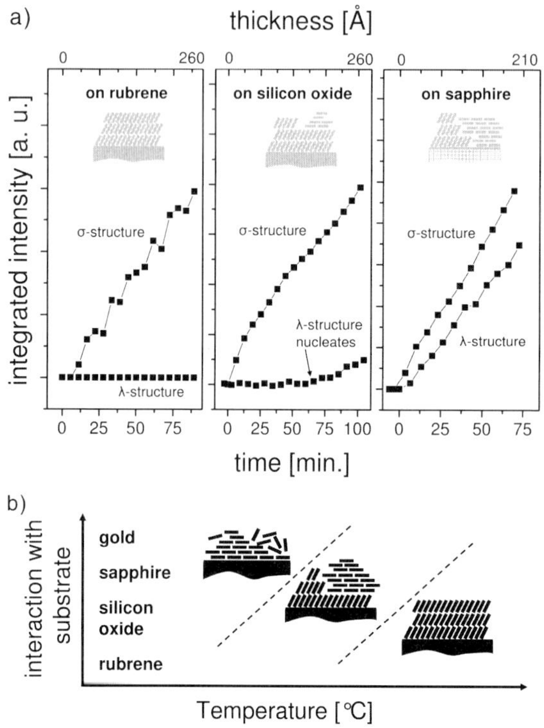

Figure 9.6 (a) Evolution of the characteristic reflections for the lying down structure (100) and the 'standing upright' structure (11) as a function of time (film thickness), for growth on rubrene (10 °C), silicon oxide, and stepped sapphire (both at 35 °C). Stronger interactions with the substrate promote earlier nucleation of the lying down structure. (b) Schematic showing influence of substrate temperature and strength of interaction with substrate on the orientational transition from lying down to standing upright structures. With permission from Ref. [6].

9.4.3
PTCDA on Ag(111), Cu(111), and Au(111)

Perylene-tetracarboxylicacid-dianhydride (PTCDA) is among the most thoroughly studied organic semiconductors [20–22, 72–84], both in the monolayer and multilayer regime.

One of the characteristic features of PTCDA is that it (almost) always grows in a lying-down configuration in contrast to pentacene and DIP, which is probably due to its layered crystal structure and molecular quadrupole moment. A strong interaction of PTCDA with the metal substrates ('chemisorption') is found for Ag(111) [73] and Cu(111) surfaces. Using the XSW technique a bonding distance of $d_0 = 2.86 \pm 0.01$ Å [20, 21] on silver and 2.61 ± 0.02 Å [20] on copper has been measured. A weaker interaction, however, is found for PTCDA on Au(111) with 3.27 ± 0.02 Å [22], see Figure 9.7. These differences reflect the complex bonding mechanism to the substrates where covalent and the van-der-Waals forces contribute to the interaction. Interestingly, there is a close correlation of these distances with the electronic properties of PTCDA monolayers on the different substrates [28] – an issue which is relevant for the alignment of energy levels at the interface and the charge carrier (electron or hole) injection from the metal contacts into the organic layer. We note that the XSW results of PTCDA/Ag(111) obtained in the monolayer regime agree with surface diffraction data obtained earlier from multilayer films of PTCDA [18]. This indicates that the growth of further layers does not influence the first layer and the XSW measurements of d_0 are indeed relevant beyond the monolayer coverage. Moreover, a substrate dependent distortion of the C = O bonds is found, see Figure 9.7: For PTCDA on Ag(111) the carboxylic oxygen atoms are located below the molecular plane ($d_0 = 2.68$ Å), but for PTCDA on Cu(111) these atoms are above the plane ($d_0 = 2.73$ Å). Similar adsorption induced distortions of organic molecules have been observed for NTCDA [23, 24] and F_{16}CuPc [25], respectively.

When grown at low substrate temperatures ($T < 50$ °C at a growth rate of 1 Å/min), PTCDA films exhibit a smooth morphology albeit with poor crystallinity. As is often observed for OMBD, higher substrate temperatures gives improved crystallinity, albeit with a rough morphology with separated crystals

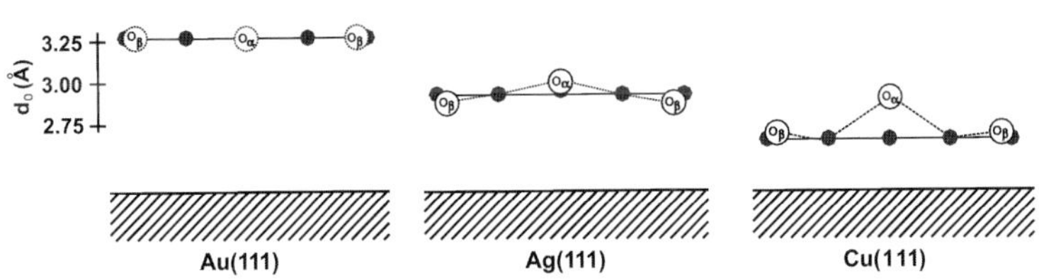

Figure 9.7 Different bonding distances and adsorption geometries of PTCDA on Au(111), Ag(111) and Cu(111) as measured by XSW [20–22]. From Ref. [28] with permission.

on a range of substrates such as PTCDA/InAs(001) [82], PTCDA/NaCl(001), KCl(001) and KBr(001) [81], Au(111) [83, 84], and Ag(111) [74–80]. It turned out that PTCDA exhibits very well-defined Stranski–Krastanov growth on Ag(111) as established in real-time X-ray experiments.

Figure 9.8 shows anti-Bragg oscillations for PTCDA deposition on Ag(111) at different substrate temperatures clearly demonstrating (1) the decay of crystallinity and therefore growth oscillations at low substrate temperatures, and (2) the damping of oscillations after deposition of two monolayers, indicating a transition from 2D to 3D growth (Stranski–Krastanov growth with two monolayers wetting) [76]. Again fits of the growth oscillations have been performed within the kinematic (single scattering) approximation of X-ray scatter-

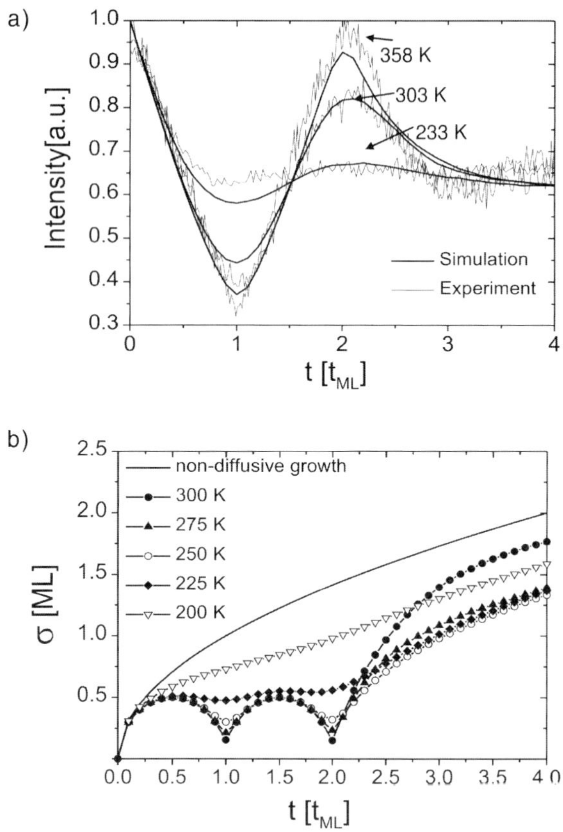

Figure 9.8 (a) Comparison of experimental anti-Bragg oscillations (substrate temperatures 233 K, 303 K, and 358 K) and Monte Carlo simulations (simulations at 200 K, 225 K, 250 K). (b) Interface roughness σ in units of monolayers (ML) as obtained from Monte Carlo simulations at different temperatures. A clear transition from 2D to 3D growth is visible after two monolayers, which occurs as early as after half a monolayer deposition for low temperatures (200 K). Figures by courtesy of B. Krause and from Ref. [76] with permission.

ing, but in this example kinetic Monte Carlo simulations [86] have been used to model the evolution of the fractional layer coverages $\theta_n(t)$. While temperatures for the calculations systematically lie below the real substrate temperature, indicating that the energy barriers in the calculations are slightly too low, the X-ray growth oscillations are fitted well. This allows a calculation of the film roughness σ from the simulated $\theta_n(t)$ as shown in Figure 9.8. At high substrate temperatures the first two layers grow in a well-defined layer-by-layer fashion with a pronounced transition to island growth after two monolayers. The layer-by-layer growth of the first layers is strongly temperature dependent and breaks down for lower substrate temperatures [76].

9.5 Organic Heterostructures

Below we discuss selected examples of more complex layer structures, which are essential for organic based devices.

9.5.1 Metal Capping Layers

Metal contacts are one obvious requirement for many applications of organic semiconductors. It turns out that the controlled deposition of metals on organics, e.g. as 'top electrode', is non-trivial. In order to reduce problems related to interdiffusion (and ultimately short-circuiting) and traps related to surface states, different strategies can be pursued:

– Deposition at low temperatures to 'freeze in' the interdiffusion.

– Deposition at (moderately) high rates with the idea that the metals are quickly forming larger aggregates which are then less mobile and diffuse less far into the organic film.

– Use of 'suitably reactive' metals and/or organics, so that a strong interaction at the top layer(s) of the organic material prevents interdiffusion.

– 'Soft deposition' by 'thermalising' or at least reducing the energy of the impinging metal atoms by 'baffling' these using a noble gas or other means.

– Miscellaneous other non-thermal deposition strategies including, e.g., electrochemical deposition may be attempted.

We performed studies of the deposition of gold, which is widely used as a hole injection material, onto well-defined DIP thin film surfaces to study the interdiffusion using X-ray reflectivity and transmission electron microscopy (Figure 9.9). The study followed the 'classical' approach without specific precaution against interdiffusion except for variation of the temperature and the rate [68, 71, 86–88]. The important result was that under rather typical deposition conditions near room temperature the metal interdiffusion was already significant, and the layers would exhibit electrical shorts, see Figure 9.9b. Only if the

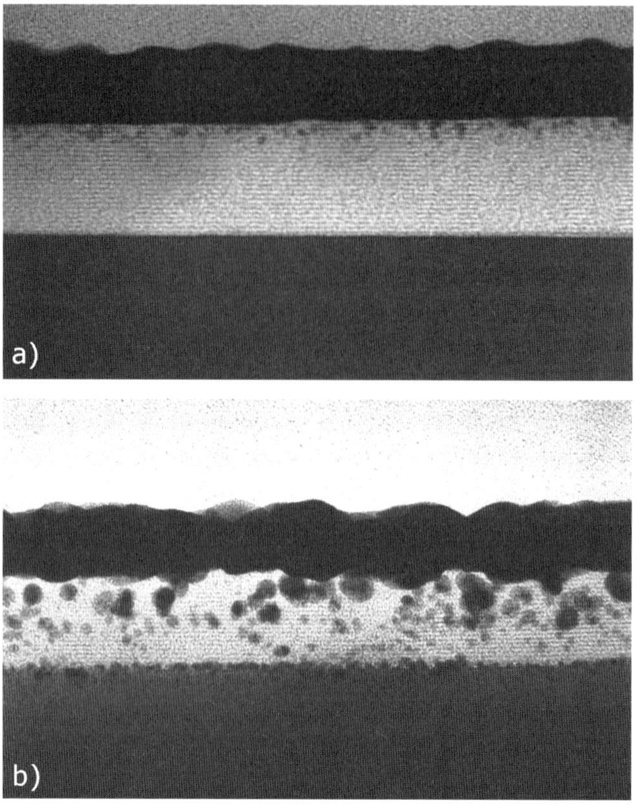

Figure 9.9 Cross-sectional transmission electron microscopy images of two Au/DIP/silicon oxide hetero-structures. While the Au contact prepared at (a) – 120 °C and a rate of 23 Å/min exhibits rather well-defined interfaces, the Au contact prepared at (b) 70 °C and a rate of 0.35 Å/min shows strong interdiffusion. Note that individual lattice planes of the DIP film can be resolved. Figures by courtesy of A. Dürr and from Ref. [86] with permission.

substrate is cooled, fairly well-defined interfaces could be obtained. Scharnberg et al. [89] found a very similar behaviour for silver capping layers on DIP and pentacene using the complementary approach of radio tracers.

9.5.2
Insulating Capping Layers

Organic devices eventually have to meet certain stability requirements to preserve their electrical and/or optical characteristics and to guarantee a longterm functioning.

9.5.2.1 **Degradation of Devices**

It is well known that organic semiconductor films exposed to ambient conditions may undergo alterations which significantly affect their optical and electrical properties. For amorphous rubrene films, for example, Käfer et al. [90] could show that exposure to ambient gases leads to the formation of rubrene-peroxide at the surface. In Ref. [91] we studied the oxidation and photo-oxidation of rubrene films and found it to be accompanied by a significant change of the optical properties. Other studies focused on the degradation processes of OLEDs and different sources could be identified (see also Ref. [92]) as for example: (1) structural changes of the organic films, i.e. crystallisation of initially amorphous films [93, 94], (2) oxidation/degradation of the top electrodes [95], and (3) gas evolution [96].

A number of recent articles also describe the influence of moisture and ambient gases on device characteristics of organic field-effect transistors (OFETs) [97, 98]. Pannemann et al. [99], for instance, studied the longterm effects of different gases such as oxygen and nitrogen on the performance of OFETs based on pentacene films. During a period of nine months they found a significant decrease in the maximum on-current from initially $-60.9\,\mu A$ to -187 nA while the charge carrier mobility decreased from $2.0 \times 10^{-3}\,cm^2/Vs$ to $1.2 \times 10^{-5}\,cm^2/Vs$. To preserve device characteristics it is obvious that either molecules have to be used which are stable against moisture and oxygen [100] or the devices have to be encapsulated [101]. In this section we will discuss some of our results in the context of recent progress in encapsulating organic devices. Our results focus on the preparation, growth and thermal stability of organic semiconductor films of DIP capped by sputtered aluminium oxide films.

9.5.2.2 **Encapsulation of Devices**

An encapsulation film for organic devices primarily has to fulfil the following requirements:

- protection of organic film and contacts against moisture or ambient gases,
- formation of well-defined interfaces with the organic film, i.e. no diffusion of the capping material into the organic film,
- stability at elevated temperatures,
- sufficient elasticity to be used on flexible substrates.

Different materials have been used to encapsulate organic devices and it has been shown that they can significantly enhance the lifetime of the devices [101–105]. Beyond protection of devices against ambient gases capping layers can also be attributed supplementary functions. Scharnberg et al. [106], for example, used a Teflon-based electret layer as a second gate and encapsulation material which could also be used to tune the threshold voltage of the device. Riel et al. [107] studied the tunability of the emission characteristics of top-emitting OLEDs by means of a dielectric capping layer. Furthermore, Peumans et al. [108] have shown that capping layers can also be used to effectively suppress substantial surface roughening during post-growth annealing of a blend of organic semiconductors.

9.5.2.3 Aluminium Oxide Capping Layers

Aluminium oxide has been successfully used as encapsulation material either as a pure AlO_x capping layer [109] or in combination with polyacrylate as a multilayer coating [110] for OLEDs. Ferrari et al. [111] have studied the effect of a capping layer of aluminium oxide on the electrical properties of thin film transistors (TFT) based on poly-3 hexylthiophene (P3HT). They could show that a P3HT-transitor capped with an Al_2O_3/PVP (poly-vinyl alcohol) layer resulted in almost the same charge carrier mobility as the uncapped transistor. While the uncapped TFT showed a high doping through moisture and oxygen adsorption resulting in an intolerable low on/off ratio the capped transistor was mostly unchanged upon exposure to air.

We prepared aluminium oxide films by radio frequency (r.f.) magnetron sputtering from an aluminium oxide target in a dedicated vacuum chamber. To study the growth and structure of these films deposited on silicon oxide and films of DIP we used X-ray reflectivity, cross-sectional transmission electron microscopy (TEM) and atomic force microscopy (AFM) in contact mode. For further details on the preparation of the aluminium oxide films we refer to Refs. [112, 113].

9.5.2.3.1 **Structure and Morphology of Aluminium Oxide Capped DIP Films** The aluminium oxide films are totally amorphous and consist of small grains. Using cross-sectional TEM, see Figure 9.10, we could show that the interface of the aluminium oxide and the DIP film is very well-defined and no significant penetration of aluminium oxide into the organic film could be observed. Furthermore, even individual layers corresponding to the length of upright standing DIP molecules could be observed indicating the high crystallinity of these organic semiconductor films.

Figure 9.11 shows typical atomic force microscopy (AFM) images (contact mode) of a crystalline DIP film of about 360 Å thickness deposited on silicon oxide (a), an aluminium oxide film (~174 Å thick) deposited on silicon ox-

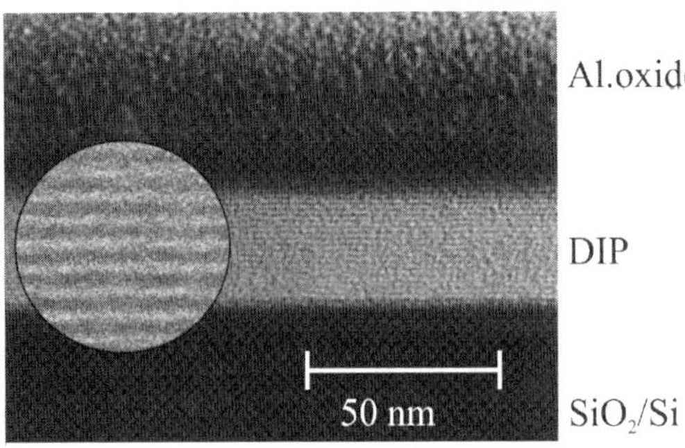

Figure 9.10 TEM image showing a well-defined heterostructure of aluminium oxide on DIP on silicon oxide together with the monolayered structure of the organic film (inset). With permission from Ref. [114].

ide (b), and an aluminium oxide film (681 Å) deposited on top of a DIP film (317 Å) (c). After capping DIP films with aluminium oxide the typical graininess of the aluminium oxide morphology on silicon oxide (see Figure 9.11b) can be recognised together with the typical DIP topography consisting of large flat terraces of monomolecular step height (upright standing molecules) as shown in Figure 9.11a.

9.5.2.3.2 **Growth of Aluminium Oxide Films on Silicon Oxide and Films of DIP** As was pointed out in Section 9.3.2 important information on the growth of thin films can be extracted from scaling theories and it was mentioned that the film roughness σ scales with film thickness D according to $\sigma \sim D^{\beta}$. From the growth exponent β information on the different growth processes involved can be extracted.

Thus, from measuring the aluminium oxide roughness for different film thicknesses the growth exponent β can be determined as the slope of a linear fit to a log–log plot of the aluminium oxide roughness versus its film thickness. To account for the roughness of the underlying substrate we used a renormalisation according to the relation [10]

$$\sigma_{\mathrm{Al_2O_3}}^{\mathrm{renom.}} = \sqrt{\sigma_{\mathrm{Al_2O_3}}^{2} - \sigma_{\mathrm{substr.}}^{2}}$$

with $\sigma_{\mathrm{substr.}} = 4$ Å as derived from measurements of the clean substrate.

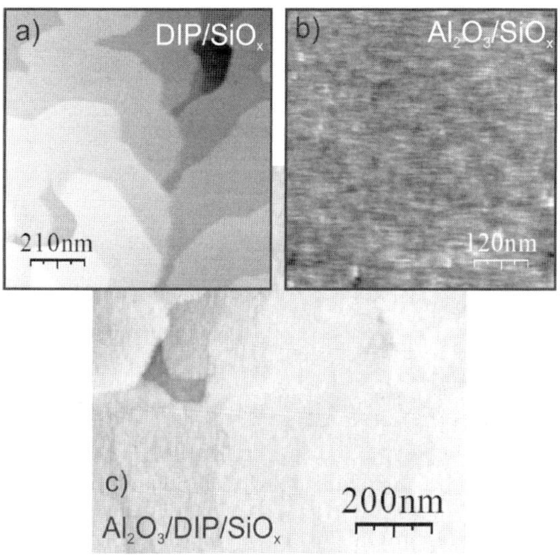

Figure 9.11 Topographical AFM images (contact mode) of (a) 360 Å DIP on silicon oxide, (b) 174 Å aluminium oxide on silicon oxide, and (c) 681 Å aluminium oxide on a DIP film. With permission from Ref. [112].

We have prepared aluminium oxide films on silicon oxide as well as on crystalline films of DIP with thicknesses ranging from 116 Å to ~6000 Å. From the analysis of X-ray reflectivity measurements on these samples we could determine the aluminium oxide surface roughness for different film thicknesses. In Figure 9.12 the renormalised aluminium oxide roughness is plotted as a function of its film thickness. For Al_2O_3/SiO_x we could extract a growth exponent of $\beta = 0.38$ and for Al_2O_3/DIP $\beta = 0.34$ [112]. It is quite surprising to find similar growth exponents considering the different chemical nature of the substrates especially the significantly lower surface energy of DIP compared to silicon oxide.

9.5.2.4 **Thermal Stability of Capped Organic Films**

Besides the obvious performance requirements, the devices also have to meet stability standards, which in some cases are actually the limiting factor of technological progress [2]. Indeed, stability at elevated temperatures, high electrical-field gradients, and against exposure to corrosive gases like oxygen is crucial for many commercial applications. Recently, it has been shown that thermal degradation of OLEDs is a serious problem for the lifetime of these devices [115].

To address this problem we have prepared highly crystalline films of the organic semiconductor DIP and capped them with r.f. magnetron sputtered aluminium oxide films.

We then performed *in situ* X-ray reflectivity and grazing incidence diffraction (GID) measurements to study the thermal stability of these samples. Figure 9.13 shows the reflectivity (a) with the first order DIP Bragg peak (b) for different temperature steps.

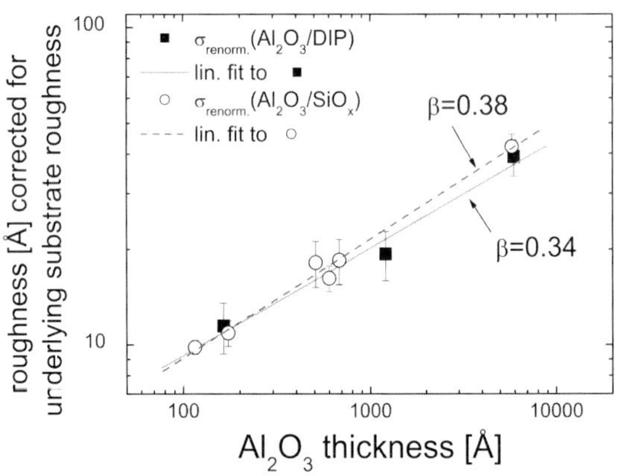

Figure 9.12 Renormalised roughness $\sigma_{Al_2O_3}^{renorm.}$ (DIP) for Al_2O_3/DIP (filled squares) compared to the roughness $\sigma_{Al_2O_3}^{renorm.}$ (SiO_x) of the Al_2O_3/SiO_x system (open circles). The scaling behaviour of aluminium oxide layers deposited on DIP and on SiO_x are in good agreement. With permission from Ref. [112].

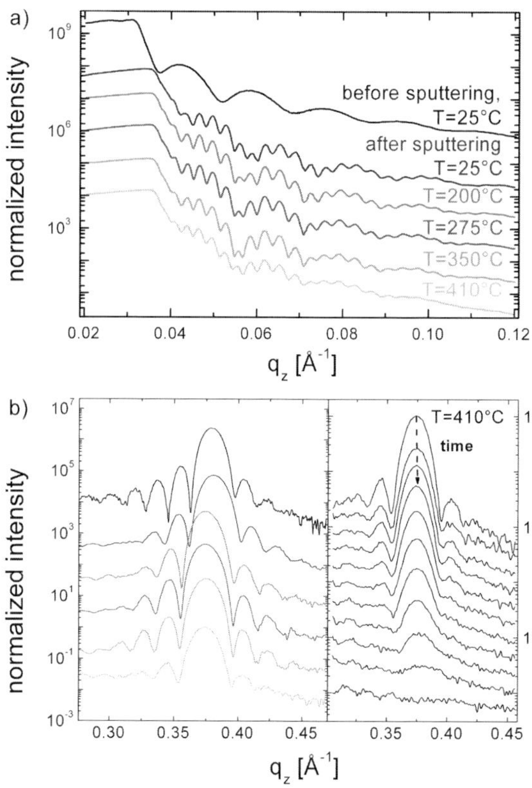

Figure 9.13 X-ray reflectivity data of the aluminium oxide/DIP multilayer with least-square fits at different temperatures. (a) By heating the sample the initially well-defined Kiessig fringes slowly degrade and the roughness of the DIP/aluminium oxide and aluminium oxide/vacuum interfaces increases with higher temperatures. (b) The first order Bragg peak with Laue oscillations remains visible up to $T = 460$ °C. For clarity the datasets are plotted with an offset. With permission from Ref. [113].

Surprisingly, the aluminium oxide/DIP multilayer structure did not show significant changes when heating the sample up to 350 °C. And only around 410 °C, i.e. 210 K above the desorption temperature of uncapped DIP films (on SiO_x) the multilayer structure broke down.

We further investigated the parameters influencing this strong enhancement of the thermal stability and we found that the 'breakdown' temperature depends on the heating rate (heating at lower rate leads to higher breakdown temperatures), the stoichiometry of the aluminium oxide capping layer (aluminium oxide films with higher metallic content were less effective in stabilising the DIP film), and the thickness of the capping layer [113]. The results are summarised in Figure 9.14 which shows the integrated intensity of the first order DIP Bragg reflection as a function of temperature. The integrated intensity

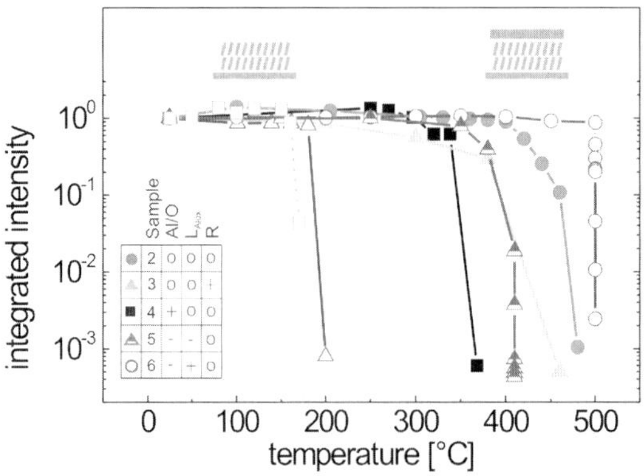

Figure 9.14 Comparison of the integrated Bragg intensity of different samples. Sample 1 (open triangle) had no capping layer, whereas samples 2–6 had an aluminium oxide capping layer. Variations of the Al/O stoichiometry, the capping layer thickness L_{Alox} and heating rate R (in comparison with sample 2 the symbols –, ○, and + in the inset indicating respectively smaller, similar, and higher values for the particular parameter) result in different breakdown temperatures. Sample 7 (open square) had a gold film instead of an aluminium oxide layer on top of DIP [86]. With permission from Ref. [113].

being proportional to the number of ordered DIP molecules in the film thus represents the degree of order/crystallinity of the organic film.

From *in situ* GID measurements we could extract the in-plane DIP unit cell parameters as a function of temperature and found that the thermal expansion of these lattice parameters shows a complex/non-linear behaviour. This suggests that due to the large difference in the thermal expansion of aluminium oxide ($6.5 \times 10^{-6}\,\mathrm{K}^{-1}$ [116]) and the in-plane lattice parameters of uncapped DIP (about an order of magnitude larger) thermal stresses eventually lead to the formation of cracks in the aluminium oxide capping layer which in turn allow for the desorption of DIP molecules through theses defects. This could also be confirmed by thermal desorption spectroscopy measurements. Furthermore, on a long term scale the crystallinity of DIP films was observed to slowly decrease possibly due to defects in the capping layer and consecutive desorption.

Figure 9.15a shows an optical micrograph of a capped DIP film after a thermal cycle. The DIP film was only deposited in the middle of the silicon oxide substrate and the whole sample was encapsulated with aluminium oxide to prevent desorption of the organic film from the sides. Clearly, a network of cracks formed on the part of the sample where DIP was below the aluminium oxide layer while the capping layer seems unaffected where it covers the silicon oxide. A possible scenario of this breakdown is illustrated in Figure 9.15b. Due to mostly thermally induced cracks in the aluminium oxide capping layer DIP is desorbing through the capping barrier. This leads to a decrease in the crystallinity of the organic film as a function of temperature and time. A second

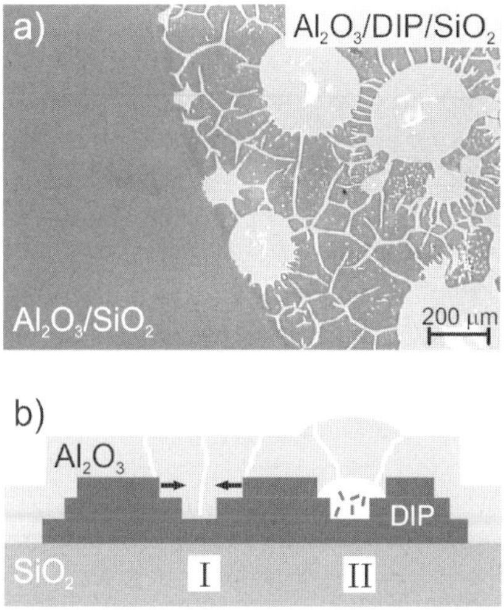

Figure 9.15 Optical micrograph of a heated Al_2O_3/DIP/SiO_x sample where the DIP film was only deposited in the middle of the sample. An extended network of cracks can be observed which is only limited to regions where the organic film was located underneath. With permission from Ref. [113].

possibility for the development of defects in the capping layer is due to argon inclusions which at elevated temperatures might lead to defects in the encapsulation.

In conclusion, we could show that the thermal stability of crystalline films of DIP can be strongly enhanced by aluminium oxide encapsulation [113, 114]. The important finding is that the crystallinity of these films could be preserved to temperatures up to 300 K above the desorption temperature of uncapped films.

The evidence of the enhanced thermal stability is of great practical importance and it offers a route for the stabilisation of compounds with sapor pressures so far considered too high for utilisation in organic-based devices. Capping layers also allow to measure for example the charge carrier mobility in organic TFTs at temperatures otherwise inaccessible. Meyer et al. [117] have built organic pentacene based TFTs capped with aluminium oxide films and poly-para-xylylene (PPX) and they could measure field-effect for transistors with both capping layers up to temperatures between 140–170 °C, i.e. 50 K above the desorption temperature of pentacene/SiO_x.

9.6 Conclusion

In this chapter we presented a selection of recent studies on organic thin films and organic-inorganic interfaces which demonstrate that X-ray scattering techniques can be used to investigate the various structural properties as well as the particular growth behaviour of organic molecules. The case studies of pentacene, DIP, and PTCDA deposition show how real-time experiments can be employed to derive detailed information about materials which are relevant for organic device application such as OFETs.

Acknowledgements

We wish to thank the DFG for financial support. We would like to acknowledge numerous fruitful interactions within the DFG focus program 1121 on organic field effect transistors. We also thank the synchrotron radiation sources ESRF (Grenoble), ANKA (Karlsruhe) and BESSY (Berlin) for providing excellent experimental support.

References

1. W. E. Brütting (ed.), Physics of Organic Semiconductors (Wiley-VCH, Weinheim, 2005).
2. R. Farchioni and G. Grosso, Organic Electronic Materials (Springer, Berlin, 2001).
3. G. Horowitz, D. Fichou, X. Peng, Z. Xu, and F. Garnier, Solid State Commun. **72**, 381 (1989).
4. C. D. Dimitrakopoulos and P. R. L. Malenfant, Adv. Mater. **14**, 99 (2002).
5. D. J. Gundlach, Y. Y. Lin, T. N. Jackson, S. F. Nelson, and D. G. Schlom, IEEE Electron Device Lett. **18**, 87 (1997).
6. S. Kowarik, A. Gerlach, S. Sellner, F. Schreiber, L. Cavalcanti, and O. Konovalov, Phys. Rev. Lett. **96**, 125504 (2006).
7. D. Käfer, L. Ruppel, G. Witte, and C. Wöll, Phys. Rev. Lett. **95**, 166602 (2005).
8. M. Tolan, X-Ray Scattering from Soft-Matter Thin Films (Springer, Heidelberg, 1999).
9. L. G. Parratt, Phys. Rev. **95**, 359 (1954).
10. V. Holý, U. Pietsch, and T. Baumbach, High-Resolution X-Ray Scattering from Thin Films and Multilayers (Springer, Berlin, 1999).
11. H. Dosch, Critical Phenomena at Surfaces and Interfaces, Springer Tracts in Modern Physics, Vol. 126 (Springer, Heidelberg, 1999).
12. J. Krug and T. Michely, Islands, Mounds and Atoms: Patterns and Processes in Crystal Growth Far From Equilibrium (Springer, Berlin, 2004).
13. A. Pimpinelli and J. Villain, Physics of Crystal Growth (Cambridge University Press, Cambridge, 1998).
14. I. V. Markov, Crystal Growth for Beginners: Fundamentals of Nucleation, Crystal Growth and Epitaxy (World Scientific, Singapore, 1994).
15. A. L. Barabási and H. E. Stanley, Fractal Concepts in Surface Growth (Cambridge University Press, Cambridge, 1995).
16. F. Schreiber, phys. stat. sol. (a) **201**(6), 1037 (2004).
17. G. Witte and C. Wöll, J. Mater. Res. **19**, 1889 (2004).
18. B. Krause, A. C. Dürr, F. Schreiber, H. Dosch, and O. H. Seeck, J. Chem. Phys. **119**, 3429 (2003).
19. J. Zegenhagen, Surf. Sci. Rep. **18**, 199 (1993).
20. A. Hauschild, K. Karki, B. C. C. Cowie, M. Rohlfing, F. S. Tautz, and M. Sokolowski, Phys. Rev. Lett. **94**, 36106 (2005).
21. A. Gerlach, S. Sellner, F. Schreiber, N. Koch, and J. Zegenhagen, Phys. Rev. B **75**, 045401 (2007).
22. S. K. M. Henze, O. Bauer, T. L. Lee, M. Sokolowski, and F. S. Tautz, Surf. Sci. **601**, 1566 (2007).
23. J. Stanzel, W. Weigand, L. Kilian, H. L. Meyerheim, C. Kumpf, and E. Umbach, Surf. Sci. Lett. **571**, 311 (2004).
24. C. Stadler, S. Hansen, A. Schöll, T. L. Lee, J. Zegenhagen, C. Kumpf, and E. Umbach, New J. Phys. **9**, 50 (2007).

25. A. Gerlach, F. Schreiber, S. Sellner, H. Dosch, I. A. Vartanyants, B. C. C. Cowie, T. L. Lee, and J. Zegenhagen, Phys. Rev. B **71**, 205425 (2005).

26. C. Stadler, S. Hansen, F. Pollinger, C. Kumpf, E. Umbach, T. L. Lee, and J. Zegenhagen, Phys. Rev. B **74**, 035404 (2006).

27. R. Woolley, C. Martin, G. Miller, V. Dhanak, and P. Moriarty, Surf. Sci. **601**, 1231 (2007).

28. S. Duhm, A. Gerlach, I. Salzmann, B. Bröker, R. L. Johnson, F. Schreiber, and N. Koch, Org. Electron. **9**, 11 (2008).

29. S. C. B. Mannsfeld and T. Fritz, Mod. Phys. Lett. B **20**, 585 (2006).

30. A. Hoshino, S. Isoda, and T. Kobayashi, J. Cryst. Growth **115**, 826 (1991).

31. D. E. Hooks, T. Fritz, and M. D. Ward, Adv. Mater. **13**, 227 (2001).

32. F. Schreiber, J. Phys.: Condens. Matter **16**, R881 (2004).

33. M. C. Gerstenberg, F. Schreiber, T. Y. B. Leung, G. Bracco, S. R. Forrest, and G. Scoles, Phys. Rev. B **61**, 7678 (2000).

34. F. Schreiber, M. C. Gerstenberg, H. Dosch, and G. Scoles, Langmuir **19**, 10004 (2003).

35. F. Schreiber, M. C. Gerstenberg, B. Edinger, B. Toperverg, S. R. Forrest, G. Scoles, and H. Dosch, Physica B **283**, 75 (2000).

36. A. S. Killampalli, T. W. Schroeder, and J. R. Engstrom, Appl. Phys. Lett. **87**, 033110 (2005).

37. D. Käfer, L. Ruppel, and G. Witte, Phys. Rev. B **75**, 085309 (2007).

38. R. Ruiz, B. Nickel, N. Koch, L. C. Feldman, R. F. Haglund, A. Kahn, and G. Scoles, Phys. Rev. B **67**, 125406 (2003).

39. H. Klauk, U. Zschieschang, J. Pflaum, and M. Halik, Nature **445**, 745 (2007).

40. C. E. Aumann, R. Kariotis, and M. G. Lagally, J. Vac. Sci. Technol. A **7**, 2180 (1989).

41. J. Krug, Adv. Phys. **46**, 139 (1997).

42. J. Krug, arXiv cond-mat/0403267 (2004).

43. R. Ruiz, B. Nickel, N. Koch, L. C. Feldman, J. R. F. Haglund, A. Kahn, F. Family, and G. Scoles, Phys. Rev. Lett. **91**, 136102 (2003).

44. A. C. Dürr, F. Schreiber, V. K. K. A. Ritley, J. Krug, H. Dosch, and B. Struth, Phys. Rev. Lett. **90**, 16104 (2003).

45. S. Yim and T. S. Jones, Phys. Rev. B **73**, 161305 (2006).

46. M. C. Lafouresse, P. J. Heard, and W. Schwarzacher, Phys. Rev. Lett. **98**, 236101 (2007).

47. S. Kowarik, A. Gerlach, S. Sellner, F. Schreiber, J. Pflaum, L. Cavalcanti, and O. Konovalov, Phys. Chem. Chem. Phys. **8**, 1834 (2006).

48. N. Karl, Charge carrier mobility in organic crystals, in: Organic Electronic Materials, edited by R. Farchioni and G. Grosso (Springer, Berlin, 2001).

49. E. Chason, T. M. Mayer, A. Payne, and D. Wu, Appl. Phys. Lett. **60**, 2353 (1992).

50. S. Kowarik, A. Gerlach, W. Leitenberger, J. Hu, G. Witte, C. Wöll, U. Pietsch, and F. Schreiber, Thin Solid Films **515**, 5606 (2007).

51. R. Ruiz, A. C. Mayer, G. G. Malliaras, B. Nickel, G. Scoles, A. Kazimirov, H. Kim, R. L. Headrick, and Z. Islam, Appl. Phys. Lett. **85**, 4926 (2004).

52. R. Ruiz, D. Choudhary, B. Nickel, T. Toccoli, K. C. Chang, A. C. Mayer, P. Clanc, J. M. Blakely, R. L. Headrick, S. Iannotta, and G. G. Malliaras, Chem. Mater. **16**, 4497 (2004).

53. K. I. H. Yoshida and N. Sato, Appl. Phys. Lett. **90**, 181930 (2007).

54. S. Schiefer, M. Huth, A. Dobinewski, and B. Nickel, J. Am. Chem. Soc. **129**, 10316 (2007).

55. S. Pratontep, M. Brinkmann, F. Nuesch, and L. Zuppiroli, Phys. Rev. B **69**, 165201 (2004).

56. F. J. Meyer zu Heringdorf, M. C. Reuter, and R. M. Tromp, Nature **412**, 517 (2001).

57. L. Casalis, M. F. Danisman, B. Nickel, G. Bracco, T. Toccoli, S. Iannotta, and G. Scoles, Phys. Rev. Lett. **90**, 206101 (2003).

58. A. C. Mayer, R. Ruiz, R. L. Headrick, A. Kazimirov, and G. G. Malliaras, Org. Electron. **5**, 257 (2004).

59. A. C. Mayer, R. Ruiz, H. Zhou, R. L. Headrick, A. Kazimirov, and G. G. Malliaras, Phys. Rev. B **73**, 205307 (2006).

60. A. C. Mayer, A. Kazimirov, and G. G. Malliaras, Phys. Rev. Lett. **97**, 105503 (2006).

61. I. P. M. Bouchoms, W. A. Schoonveld, J. Vrijmoeth, and T. M. Klapwijk, Synth. Met. **104**, 175 (1999).

62. S. Kowarik, A. Gerlach, S. Sellner, M. Skoda, and F. Schreiber, Eur. Phys. J. Special Topics 168, 11 (2009).

63. A. K. Tripathi and J. Pflaum, Appl. Phys. Lett. **89**, 082103 (2006).

64. M. Münch, Strukturelle Beeinflussung der elektrischen Transporteigenschaften dünner organischer Schichten, Ph.D. thesis, Universität Stuttgart (2001).

65. V. Pozdin, A. Papadimiraos, and G. Malliaras, private communication (2007).

66. H. E. Simmons, United states patent 4681834: optical recording element (1987).

67. H. Antoniadis and A. J. Bard, United states patent 6004685: LED doped with periflanthene for efficient red emission (1997).

68. A. C. Dürr, F. Schreiber, M. Kelsch, and H. Dosch, Ultramicroscopy **98**, 51 (2003).

69. A. C. Dürr, F. Schreiber, M. Münch, N. Karl, B. Krause, V. Kruppa, and H. Dosch, Appl. Phys. Lett. **81**(12), 2276 (2002).

70. A. C. Dürr, B. Nickel, V. Sharma, U. Täffner, and H. Dosch, Thin Solid Films **503**, 127 (2006).

71. A. C. Dürr, N. Koch, M. Kelsch, A. Rühm, J. Ghijsen, R. L. Johnson, J. J. Pireaux, J. Schwartz, F. Schreiber, H. Dosch, and A. Kahn, Phys. Rev. B **68**, 115428 (2003).

72. E. Umbach, Prog. Surf. Sci. **35**, 113 (1991).

73. Y. Zou, L. Kilian, A. Schöll, T. Schmidt, R. Fink, and E. Umbach, Surf. Sci. **600**, 1240 (2006).

74. L. Kilian, E. Umbach, and M. Sokolowski, Surf. Sci. **573**, 359 (2004).

75. B. Krause, A. C. Dürr, F. Schreiber, H. Dosch, and O. H. Seeck, Surf. Sci. **572**, 385 (2004).

76. B. Krause, F. Schreiber, H. Dosch, A. Pimpinelli, and O. H. Seeck, Europhys. Lett. **65**(3), 372 (2004).

77. B. Krause, A. C. Dürr, K. Ritley, F. Schreiber, H. Dosch, and D. Smilgies, Phys. Rev. B **66**, 235404 (2002).

78. B. Krause, A. C. Dürr, K. A. Ritley, F. Schreiber, H. Dosch, and D. Smilgies, Appl. Surf. Sci. **175/176**, 332 (2001).

79. B. Krause, Growth and Structure of the Organic Molecule PTCDA on Ag(111), Ph.D. thesis, Universität Stuttgart (2002).

80. M. Eremtchenko, J. Schaefer, and F. S. Tautz, Nature **425**, 602 (2003).

81. D. Schlettwein, A. Back, B. Schilling, T. Friz, and N. R. Armstrong, Chem. Mater. **10**, 601 (1998).

82. C. Kendrick and A. Kahn, J. Cryst. Growth **181**, 181 (1997).

83. L. Kilian, E. Umbach, and M. Sokolowski, Surf. Sci. **600**, 2633 (2006).

84. P. Fenter, F. Schreiber, L. Zhou, P. Eisenberger, and S. R. Forrest, Phys. Rev. B **56**, 3046 (1997).

85. S. Tan and P. M. Lam, Phys. Rev. B **60**, 8314 (1999).

86. A. C. Dürr, F. Schreiber, M. Kelsch, H. D. Carstanjen, and H. Dosch, Adv. Mater. **14**, 961 (2002).

87. A. C. Dürr, F. Schreiber, H. D. C. M. Kelsch, H. Dosch, and O. H. Seeck, J. Appl. Phys. **93**, 5201 (2003).

88. N. Koch, A. C. Dürr, J. Ghijsen, R. L. Johnson, J. J. Pireaux, J. Schwartz, F. Schreiber, H. Dosch, and A. Kahn, Thin Solid Films **441**, 145 (2003).

89. M. Scharnberg, J. Hu, J. Kanzow, K. Rätzke, R. Adelung, F. Faupel, C. Pannemann, U. Hilleringmann, S. Meyer, and J. Pflaum, Appl. Phys. Lett. **86**, 024104 (2005).

90. D. Käfer and G. Witte, Phys. Chem. Chem. Phys. **7**, 2850 (2005).

91. M. Kytka, A. Gerlach, F. Schreiber, and J. Kováč, Appl. Phys. Lett. **90**(13), 131911 (2007).

92. H. Aziz and Z. D. Popovic, Chem. Mater. **16**, 4522 (2004).

93. H. Aziz, Z. Popovic, S. Xie, A. M. Hor, N. X. Hu, C. Tripp, and G. Xu, Appl. Phys. Lett. **72**(7), 756 (1998).

94. E. M. Han, L. M. Do, N. Yamamoto, and M. Fujihira, Thin Solid Films **273**, 202 (1996).

95. L. M. Do, E. M. Han, Y. Niidome, M. Fujihira, T. Kanno, S. Yoshida, A. Maeda, and A. J. Ikushima, J. Appl. Phys. **76**(9), 5118 (1994).

96. H. Aziz, Z. Popovic, C. P. Tripp, N. X. Hu, A. M. Hor, and G. Xu, Appl. Phys. Lett. **72**(21), 2642 (1998).

97. S. Hoshino, M. Yoshida, S. Uemura, T. Kodzasa, N. Takada, T. Kamata, and K. Yase, J. Appl. Phys. **95**(9), 5088 (2004).

98. S. Ogawa, T. Naijo, Y. Kimura, H. Ishii, and M. Niwano, Appl. Phys. Lett. **86**(25), 252104 (2005).

99. C. Pannemann, T. Diekmann, and U. Hilleringmann, J. Mater. Res. **19**, 1999 (2004).

100. H. Tian, J. Shi, D. Yan, L. Wang, Y. Geng, and F. Wang, Adv. Mater. **18**, 2149 (2006).

101. P. E. Burrows, V. Bulovic, S. R. Forrest, L. S. Sapochak, D. M. McCarty, and M. E. Thompson, Appl. Phys. Lett. **65**(23), 2922 (1994).

102. C. Pannemann, T. Diekmann, and U. Hilleringmann, 5th International Conference on Polymers and Adhesives in Microelectronics and Photonics (2005), p. 63.

103. J. Lewis and M. Weaver, IEEE J. Sel. Top. Quantum Electron. **10**, 45 (2004).

104. K. Yamashita, T. Mori, and T. Mizutani, J. Phys. D, Appl. Phys. **34**(5), 740 (2001).

105. K. Tsukagoshi, I. Yagi, K. Shigeto, K. Yanagisawa, J. Tanabe, and Y. Aoyagi, Appl. Phys. Lett. **87**(18), 183502 (2005).

106. M. Scharnberg, V. Zaporojtchenko, R. Adelung, F. Faupel, C. Pannemann, T. Diekmann, and U. Hilleringmann, Appl. Phys. Lett. **90**(1), 013501 (2007).

107. H. Riel, S. Karg, T. Beierlein, W. Riess, and K. Neyts, J. Appl. Phys. **94**(8), 5290 (2003).

108. P. Peumans, S. Uchida, and S. R. Forrest, Nature **425**, 158 (2003).

109. S. H. K. Park, J. Oh, C. S. Hwang, J. I. Lee, Y. S. Yang, H. Y. Chu, and K. Y. Kang, ETRI J. **27**, 545 (2005).

110. A. B. Chwang, M. A. Rothman, S. Y. Mao, R. H. Hewitt, M. S. Weaver, J. A. Silvernail, K. Rajan, M. Hack, J. J. Brown, X. Chu, L. Moro, T. Krajewski, and N. Rutherford, Appl. Phys. Lett. **83**(3), 413 (2003).

111. S. Ferrari, F. Perissinotti, E. Peron, L. Fumagalli, D. Natali, and M. Sampietro, Org. Electron. **8**, 407 (2007).

112. S. Sellner, A. Gerlach, S. Kowarik, F. Schreiber, H. Dosch, S. Meyer, J. Pflaum, and G. Ulbricht, Thin Solid Films, **516**, 6377 (2008).

113. S. Sellner, A. Gerlach, F. Schreiber, M. Kelsch, N. Kasper, H. Dosch, S. Meyer, J. Pflaum, M. Fischer, B. Gompf, and G. Ulbricht, J. Mater. Res. **21**, 455 (2005).

114. S. Sellner, A. Gerlach, F. Schreiber, M. Kelsch, N. Kasper, H. Dosch, S. Meyer, J. Pflaum, M. Fischer, and B. Gompf, Adv. Mater. **16**, 1750 (2004).

115. Y. J. Lee, H. Lee, Y. Byun, S. Song, J. E. Kim, D. Eom, W. Cha, S. S. Park, J. Kim, and H. Kim, Thin Solid Films **515**, 5674 (2007).

116. R. G. Munro, J. Am. Ceram. Soc. **80**, 1919 (1997).

117. S. Meyer, S. Sellner, F. Schreiber, H. Dosch, G. Ulbricht, M. Fischer, B. Gompf, and J. Pflaum, Mater. Res. Soc. Symp. Proc. **965**, 0965-S06-13 (2007).

10
X-Ray Structural Studies of Low and High Molecular Weight Poly(3-hexylthiophene)

S. Joshi, S. Grigorian, and U. Pietsch

10.1
Introduction

In recent years great effort has been devoted to conjugated polymers as organic field-effect transistor (OFET) active layers [1]. An important parameter relevant to the operation of an OFET is the charge carrier mobility which determines the overall performance of any polymer-based devices.

Poly(3-hexylthiophene) P3HT is one of the most promising π-conjugated polymers used in polymer electronics and exhibits charge carrier mobility in the range of 10^{-3}–10^{0} cm^2/Vs [2–5]. Initially, P3HT reveals strong directional anisotropy of mobility: relatively easy charge carrier transport along backbones and in the π–π stacking direction and quite poor drift along side chains in response to an electric field. Therefore, the open question is the correlation between ingredients of micromobility of and macromobility of the whole device. Additionally, the packing and orientation of polymer chains in the transport layer may influence the transistor performance. Above all, the mobility seems to be critically dependent on the crystal orientation in the semicrystalline layers and the presence of a sufficient number of highly oriented crystallites might be the primary key for obtaining high mobility. Up to now correlation between morphology and charge carrier mobility has not been well established for P3HT-based OFETs devices and the understanding of the performance of polymer-based transistors with respect to the order and orientation of polymer chains is far from complete.

One of the important members of the P3HT family is regioregular poly(3-hexylthiophene) (RR-P3HT), with a rigid conjugated backbone and flexible alkyl side chains, introduced for greater solubility. Already in melts, this type of polymer exhibits microphase separation between the main chain and the alkyl side chains, leading to a layered liquid crystalline structure [6]. Upon crystallisation the layered structure is preserved, and the main chains as well as the side chains order to a common crystalline lattice. It was found that in such a class of material the length and bulkiness of the side chains strongly influence the charge carrier mobility [7–9]. There are several known parameters affect-

Organic Electronics. Structural and Electronic Properties of OFETs. Edited by Christof Wöll

ISBN: 978-3-527-40810-8

ing the morphology of P3HT thin films, mainly: the degree of regioregularity, the method of film deposition, annealing, molecular weights, nature of solvent and others. To date, various studies have been done on the structure of the film [10–12], the molecular weight dependence [12–14], preparation conditions [2, 10], pre-treatment of dielectric gate insulator [2, 9, 15, 16], and solvent used for the film preparation [17].

Kim et al. [3] used cross-sectional transmission electron microscopy and grazing incidence-angle X-ray diffraction (GID) to determine the internal structure of regioregular poly(3-hexylthiophene) (P3HT) on hydrophobised insulator substrates. It was found that these P3HT films are self-organised with layered molecular ordering and consist of nanometer-size domains with face-on orientations. They studied the layered molecular ordering in P3HT films grown on hydrophobised insulator substrates. After modifying the insulator substrates by an octyltrichlorosilane and annealing at 240 °C the nanocrystallites were completely reoriented into flat-on orientation.

The polymer chain/nanocrystal ordering in thin films of RR-P3HT and blends of P3HT with a soluble fullerene derivative was studied by Kim et al. [4]. A detailed analysis has been made based on GID data taken using synchrotron radiation. The results show that the P3HT chain ordering is strongly affected by regioregularity, and that thermal annealing improves chain ordering in the out-of-plane direction. They found a good agreement with high-resolution transmission electron microscope measurements of the film nanomorphology.

The first few monolayers close to the gate oxide are mostly responsible for the efficient charge flow from the source to drain inside the active channel region [5]. Therefore, it seems to be extremely important to know the particular film structure close to the interface to the gate insulator. There are several indications that the charge carrier mobility in OFETs is strongly dependent upon their surface and interface morphology [18–20]. Based on X-ray rocking curve measurements Kline et al. [21] suggested that highly oriented crystals of P3HT are mainly arranged at the insulator and film interface. Therefore, a systematic study of the crystalline structure as a function of the film thickness in P3HT is still desirable.

It has been reported that the mobility of field-effect transistors made from RR-P3HT increases strongly with molecular weight. Kline et al. [12] and Zen et al. [14] presented two different models for the explanation of this phenomenon. Kline et al. [21] concluded that the charge-carrier trapping at grain boundaries mainly hinders the charge transport in low molecular weight samples, while Zen et al. placed emphasis on the importance of conformation and packing of the polymer chains in thin layers of different molecular weight fractions of P3HT [11]. Moreover, recent studies propose that the transport properties of LMW and HMW thin films are largely determined by the crystallinity of the samples and not by the perfection of the packing of the chains in the individual crystallites [11].

In order to clarify these two hypotheses, one needs to understand the surface and interface induced ordering of the chains that ultimately enhances the crystallinity of the sample. Besides this, temperature-resolved measurements can help to indicate the influence of structural modifications on the charge-carrier mobility.

Our studies are focussed on X-ray investigations of the structure and the thermal behaviour of low molecular weight (LMW) and high molecular weight (HMW) P3HT films, in particular, the analysis of the structural properties and orientation of the ordered domains. X-ray measurements were done at a synchrotron radiation source in order to provide sufficient intensity in wide-angle regions and to resolve structural Bragg peaks appearing on the top of amorphous halos. Structural changes of the polymer films have been studied as a function of film thickness and temperature.

10.2 Sample Preparation

RR-P3HTs were prepared by the Grignard metathesis procedure according to McCullough and coworkers [22]. The raw polymer was fractioned by applying the solvent-extraction method [23] using ethyl acetate as a solvent for LMW and chloroform for HMW material yielding two polymer fractions with a polydispersity index between 1.3 and 1.4 respectively. The average molecular weights for LMW (2500 g/mol) as well as for HMW (30,000 g/mol) were determined by using gel permeation chromatography (GPC) with THF as the solvent (calibration with narrowly distributed polystyrene standards). Powders of all P3HT fractions have been precipitated from solvents into nonsolvents. Such powders are often highly crystalline.

Highly doped silicon crystal was used as a substrate covered by a thermally grown 300 nm thick silicon dioxide layer. Before spin coating, the surface of SiO_2 was treated with either hexamethyldisilazane (HMDS) or octadecyltrichlorosilane (OTS) with the aim of enhancing the molecular packing and ultimately the charge transport in P3HT thin films [2, 11]. In order to prepare the films of different thickness, t, the concentration of P3HT in solution was varied between 1 mg/mL ($t = 10$ nm) and 40 mg/mL ($t \approx 200$ nm).

All characterisation and sample preparation steps were executed in an inert N_2 atmosphere. The reported t values were measured by X-ray reflectivity and were found to vary from 10 nm to about 200 nm [24].

10.3 X-Ray Grazing-Incidence Diffraction Studies

For a detailed study of the structural properties of LMW and HMW fraction thin films ($t \sim 10$–200 nm) grazing-incidence diffraction techniques [25] have been employed. As mentioned above, P3HT films show mostly lamellae structure with possible edge-on and face-on orientation [21]. The scattering scheme is shown in Figure 10.1. This set-up allows measurement of a momentum trans-

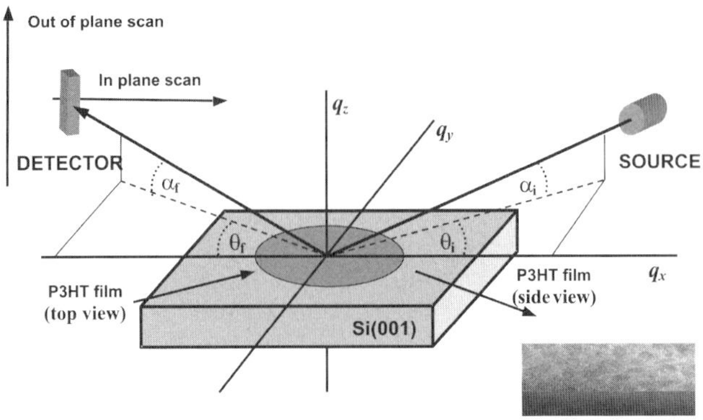

Figure 10.1 Experimental set-up and schematic view of P3HT film on the top of Si substrate.

fer in directions parallel and perpendicular to the surface normal. In the first case one changes the angle of exit, α_f with respect to the surface under conditions of fixed incidence angle α_i (out-of-plane diffraction – GOD [10, 21]), in the second case the in-plane diffraction angle $2\theta = \theta_i + \theta_f$ is varied for fixed α_i (in-plane diffraction – GID [14, 21]).

Both GID and GOD are realised at extremely reduced penetration depths of the probing X-ray within the sample. When the incident angle, α_i, is equal to, or smaller thans the critical angle of thin film, α_{cf}, the incident X-rays undergo total external reflection and penetrate into the sample as evanescent waves. For $\alpha_i < \alpha_{cf,}$ the penetration depth of the evanescent X-rays is of the order of few nanometers, only. Using a fixed α_i above the critical angle, α_{cf}, of the P3HT thin films but below the critical angle of the substrate, α_{cs}, one can reduce the scattering from the substrate relative to the scattering from the thin polymer films.

The scattering vector $\boldsymbol{Q}$ can be written as a function of the scattering angles and the X-ray wavelength in the following form:

$$Q_z = (2\pi/\lambda)\,(\sin\alpha_i + \sin\alpha_f)\,, \quad \text{for} \quad \text{GOD scans} \tag{1}$$

$$Q_{x-y} = (2\pi/\lambda)\,(\sin\theta_1 + \theta_f)\,, \quad \text{for} \quad \text{GID scans} \tag{2}$$

Our X-ray measurements were mainly performed at the European Synchrotron Radiation Facility (ESRF), Grenoble, France at the undulator beamlines ID01 and ID10B ("Trokia II")using wavelengths 1.54 Å and 0.92 Å, respectively, and an energy resolution of better than 10^{-3}. Additional measurements were performed at DELTA (University of Dortmund, Germany – see Acknowledgements). Considering the wavelengths used at both ESRF beamlines (ID01/ID10B) the critical angle of thin P3HT film and underlying substrate were $\alpha_{cf} = 0.14°$, $\alpha_{cs} = 0.23°$ and 0.09°, 0.14°, respectively.

The data were collected by a point detector or a 1D position-sensitive detector (PSD) oriented parallel to the surface normal. It could be scanned both horizontally and vertically with respect to the direct beam. For GID, a Soller slit was installed in front of the detector, providing an angular resolution of $\Delta\theta = 0.1°$. Due to alignment of the PSD in the vertical scanning direction a range of vertical scattering angles of about $\Delta\alpha_f \sim 6°$ could be captured by a single exposure.

GID profiles of films made from LMW and HMW fractions with film thickness of 40 nm and 48 nm are shown in Figure 10.2 (top and bottom), respecti-

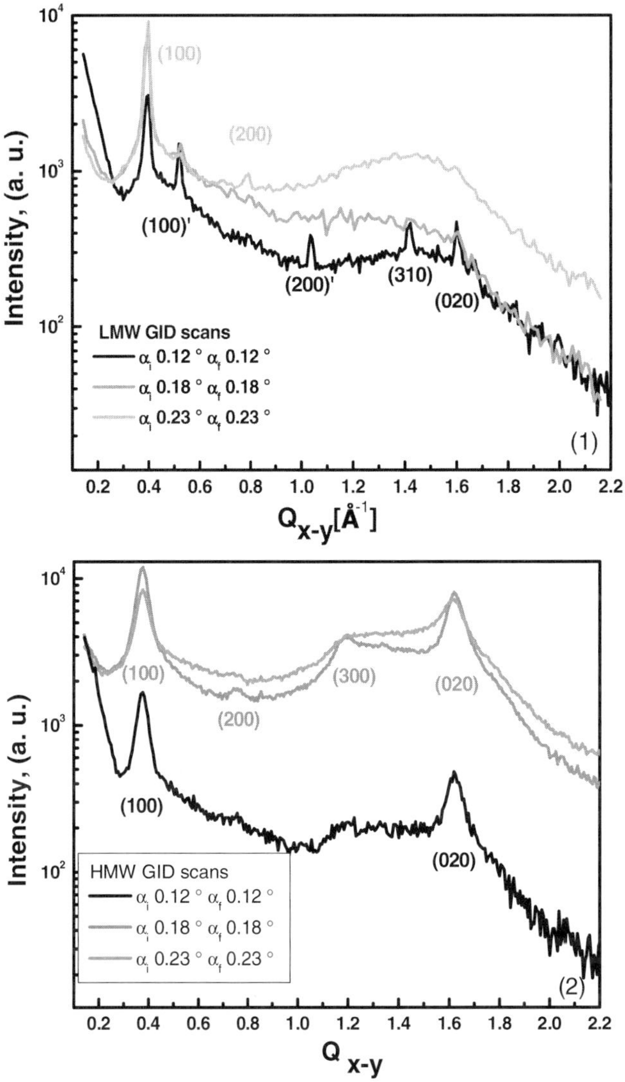

Figure 10.2 GID profiles for the low (top) and high (bottom) molecular weight fractions. Surface, subsurface and bulk profiles are shown by black and red and green curves, respectively.

vely. Depending on α_i the scattering signal was detected from polymer thin film only at $\alpha_i = 0.12°$ (black curves), the film and a few nanometers of substrate at $\alpha_i = 0.18°$ (red), and from the film and hundreds of nanometers of substrate ($\alpha_i = 0.23°$ green). Five peaks are present for the case of the LMW fraction at the smallest α_i. In comparison, the HMW fraction explicitly shows only two pronounced structural peaks marked by (100) and (020) reflections (see Figure 10.2. bottom). Increasing α_i structural peaks of higher order (Figure 10.2, green and red curves) appear only when $\alpha_i < \alpha_{cf}$. For higher incidence angle the diffuse halo close to $Q_{x-y} = 1.4$ Å^{-1} dominates on the diffraction pattern. Because strong amorphous scattering of underlying SiO_2 starts to dominate at higher angles the structural information of the polymer film can be probed only at very shallow angles (see black curve, Figure 10.2 top). The α_i dependence suggests that nanocrystallites of the LMW fraction mostly lie on the top of the thin film whereas for the HMW fraction such nanodomains are located in the entire film. Additionally, for the LMW fraction we found two different types of crystallisation (polymorphism) shown by black and green indexes in Figure 10.2 (top).

The diffraction intensity from planes parallel to the interface has been detected by grazing-incidence out-of-plane diffraction (GOD) (Figure 10.3). Here, α_i was fixed at the condition $\alpha_{cf} < \alpha_i < \alpha_{cs}$ and the detector angle α_f was scanned in a wide angular range.

Both molecular weight fractions display a well-pronounced periodicity normal to the surface due to alkyl chain stacking. The different interplanar distances for such stacked sheets were found to be ~15.5 Å and ~16.5 Å for low and high molecular weight fractions, respectively. This reflects the fact that

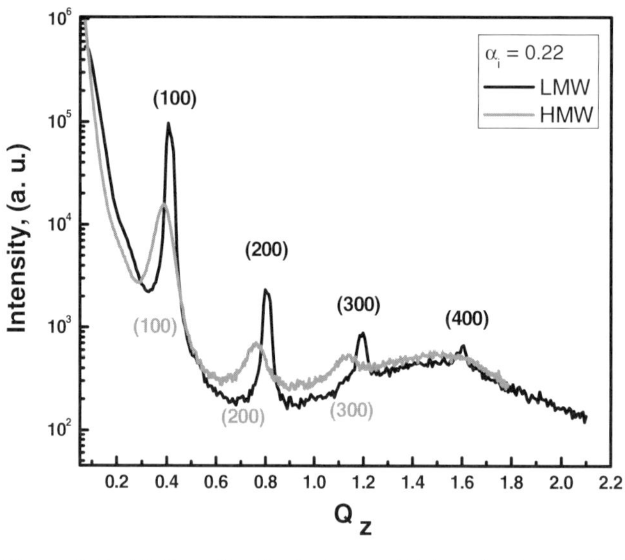

Figure 10.3 GOD profiles for the low and high molecular weight fractions, black and red curves, respectively.

the '*d*'-spacing might be a function of molecular weight. Both fractions show the same shape of amorphous halo centred at $Q_z \approx 1.4$ Å^{-1}; probably caused by the scattering from SiO_2.

The peak widths of the HMW material is broader than that of the LMW fraction, it reveals a smaller crystallite size compared with the LMW fraction material. The average crystal size was $D \sim 25$ nm and $D \sim 8$ nm for the LMW and the HMW fractions, respectively.

Due to the small crystallite size, the further Bragg peaks could not be resolved in the HMW fraction. In contrast to this, the morphology of the LMW films is more complex and provides better ordering of the nanocrystalline domains and existence of two polymorphs. Therefore, our further study was mostly focused on the investigations of the LMW P3HT films. Considering the angular dependent results shown in Figure 10.2, we have performed measurements at shallow incident angles (α_i^s) for further studies.

10.4
Structure Determination for LMW Fraction

GID and GOD data taken for very shallow angles, i.e. $\alpha_i = 0.10°$ and $\alpha_i = 0.11°$, have been used to determine the crystalline structure of nanocrystallites. Using GID at the high-resolution beamline ID10B, several crystalline peaks of the film become visible due to the strong reduction of the substrate scattering. As shown in Figure 10.4 at least 9 sharp reflections are visible for the thickest sample ($t \approx 200$ nm).

Following previous structure investigation of the same material, the low-order peaks of alkyl stacking indicated by (100), (200) and (300) are related to

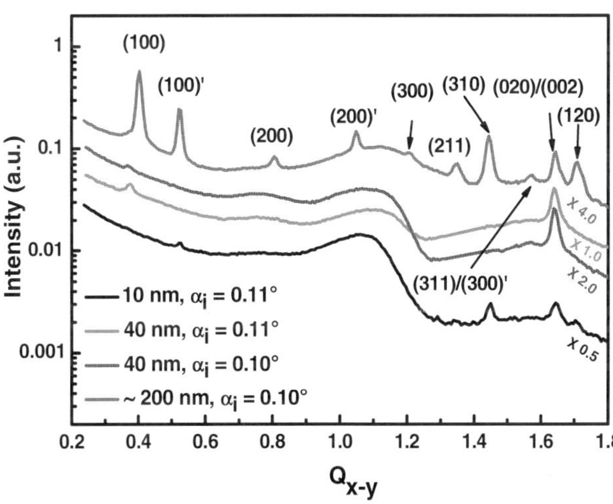

Figure 10.4 GID scans from samples with different thickness measured at $\alpha_i \leq \alpha_{cs}$. For better clarity all scans are separated by an offset factor.

the lamellar ordering of phase-separated structure. Further peaks appear at Q_{x-y} = 1.35, 1.44, 1.57, 1.64 and 1.70 Å^{-1}. The peak at Q_{x-y} = 1.64 A^{-1} is indexed by (020) [2, 11, 26] and is associated by the in-plane π−π stacking distance of 3.88 Å. Considering Prosa et al. [27], this distance is very close to the distance between two sulfur atoms of the thiophene backbones of 3.8 Å [16] indexed with (002). We suppose that both peaks are not resolved in our experiment. At the same time two amorphous humps were found. One is centred at about 1.1 Å^{-1} and is less dependent on film thickness and the second halo is centred at about 1.6 Å^{-1} and can be addressed to the amorphous scattering of disordered thiophene rings, which increases with layer thickness. It is observed that the low index (100) peaks decrease as the film thickness decreases. The peaks completely disappear for the 10 nm sample. At the same time, we have not found any pronounced thickness dependent behaviour of (020) peaks.

Figure 10.4 also shows the existence of two sets of reflections for the thickest sample: direct evidence of polymorphism. The second form indexed with (100)′ appears at higher Q_{x-y} = 0.525 Å^{-1} compared with (100) (Q_{x-y} = 0.405 Å^{-1}). This reflects an interplanar distance of 'd' = 12.0 Å compared with the main form with the 'd' = 15.6 Å. The appearance of higher-order peaks such as (200)′ and (300)′ in addition to (200) and (300) gives evidence for well-pronounced ordering of both polymorphs. Interestingly, such dual form is not observed for thinner films. Such polymorphism was also found in powder samples [17].

Figure 10.5 shows the GOD scans for two $\alpha_i < \alpha_{cf}$ of varying sample thickness. A number of peaks are visible for films of different thickness. All peaks appear independent of the film thickness but they vary in intensity. As in GID, intense (100) and (200) peaks are visible (modified by parasitic scattering of beamline components), which is a hint to the random orientation of nanocrystallites (powder rings). At the same time, further peaks are visible at Q_z = 1.28, 1.44, 1.57 and 1.70 Å^{-1} where the second one is most intense. These peaks are located in the region of the amorphous halo centred at $Q_z \approx 1.4$ Å^{-1}. Again we have observed the polymorphism in the out-of-plane scans only at (100) and (100)′ indexes for the thickest sample.

Based on data shown in Figure 10.5 we have calculated the interplanar distances, d_{100}, and d_{020}, and the size of crystals, D, along the vertical direction. d_{100} reveals a nearly constant value for all samples within ±0.02 error bars (d_{100} = 1.56 nm) and is smaller than the value obtained in previous powder measurements (d_{100} = 1.58 nm) [14]. This might be caused by a higher degree of interdigitations of the alkyl side chains for the present samples [26]. Also, d_{020} distances do not vary with increasing film thickness (d_{020} = 0.382 nm). Using the Scherrer formula [28] we have calculated the average size of crystals along the vertical direction for the samples of different thickness. Based on the width of (100) peaks the size of crystals varies from about 10 nm for the thinnest film (t = 10 nm) to 40 nm for the thickest film ($t \approx 200$ nm).

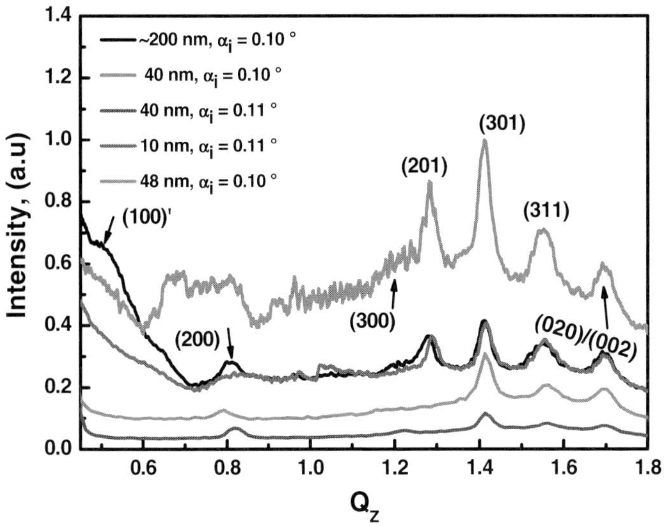

Figure 10.5 Out-of-plane GOD scans for samples of different thicknesses. The incidence angles are 0.10° and 0.11°, respectively. In low-q range the diffraction curves suffer from parasitic scattering of beamline components.

Using all angular positions of peaks measured in GOD and GID scans together we evaluated a suitable unit cell of the nanocrystallites within LMW P3HT samples. In contrast to the thin-films investigations by Kline et al. [12, 13] we observed additional peaks. These peak positions cannot be explained by the assumption of an orthorhombic cell as suggested by Tashiro et al. [29]. Multiples of (100) and the (020) or (002) have already been found by Kline et al. [12].

However, the number of structure peaks is still too small to evaluate a unique structure solution. The fitted unit cell can be triclinic or monoclinic. In the present case several unit cells match the experimental data. Of these, the two unit cells with parameters (1) $a_1 = 16.2$ Å, $b_1 = 7.6$ Å, $c_1 = 7.1$ Å, $\beta = 105.5$ and $\alpha = \gamma = {\sim}90°$ and (2) $a_2 = 15.9$ Å, $b_2 = 7.5$ Å, $c_2 = 7.6$ Å, $\beta = 100.7$ and $\alpha = \gamma = {\sim}90°$ provided the lowest residuum. Both our cells are approximately monoclinic and close to the structure solution suggested by Prosa et al. [26] and Brinkmann et al. [30]. All the peak indexing has been done based on our second solution of the proposed monoclinic unit cell. A scheme of the unit cell and its orientation with respect to the film is shown in Figure 10.6. The indexing of Bragg peaks used in this chapter corresponds to a unit cell as depicted in Figure 10.6.

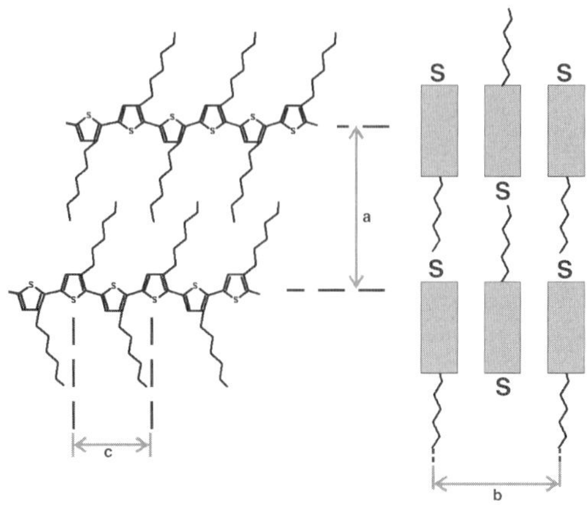

Figure 10.6 Monoclinic unit cell of LMW P3HT fraction.

10.5 Temperature-Dependent Measurements

All X-ray measurements were performed under vacuum conditions (~10^{-3} mbar) using the DHS 900 domed hot stage provided by Anton Paar GmbH, Graz, Austria. This was essential in order to avoid radiation damage of the polymer sample and to reduce background scattering. For temperature resolved measurements the temperature has been increased in steps of 10 °C from room temperature to the melting point of the individual sample with an accuracy of ±0.5 °. Data were taken after stabilisation of the respective temperature at the sample surface.

Figure 10.7 shows the effect of annealing on the structure of films made from the LMW fraction that initially has shown polymorphism. After annealing the second polymorph (100)′ disappears, showing that the first one (100), which is same as for the other films, is thermodynamically the more stable one. At the same time, the decrease of (100) peak intensity after heating indicates the reduction of insulating chains along the $\pi - \pi$ conjugation direction, i.e. it indicates the increase of ordering of the crystalline domain after the heat treatment.

Figures 10.8 and 10.9 show synchrotron measurements of the melting behaviour of films made from the LMW fraction. The temperature dependence of both (100) and (020) reflections was measured in the GOD and GID geometry, respectively. From Figure 10.8 one can see that as the temperature increases the angular position of the (100) reflections shift to lower Q_{x-y}-values. This corresponds to an increase of the d_{100} distance indicating expansion of the distance of the alkyl chain stacking. The lattice expansion is accompanied by a decrease in the peak intensity. This trend is continuous up to the complete disappearance of the peak. The temperature of complete disappearance of the peak can be associated with the melting point of the ethyl acetate fraction (LMW) P3HT. At the same time, Figure 10.9, the (020) peak position is shif-

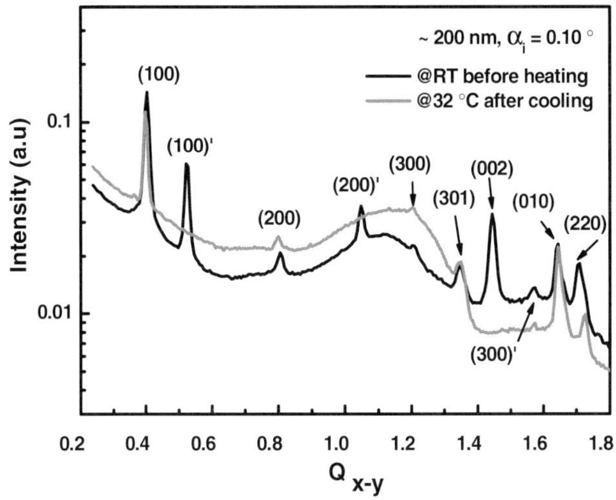

Figure 10.7 Annealing study of sample made from LMW fraction at $\alpha_i = 0.10°$. It shows polymorphism for the pristine sample, but the second form disappears after a heating cycle.

ting towards higher Q_{x-y} values as the temperature increases, indicating a decreasing d_{020}, i.e. a reduction of the $\pi-\pi$ distance. Combining both directions, an increase of temperature gives rise to a stretching of nanocrystals along the axis of the alkyl chains accompanied by shrinking of $\pi-\pi$-distance. A qualita-

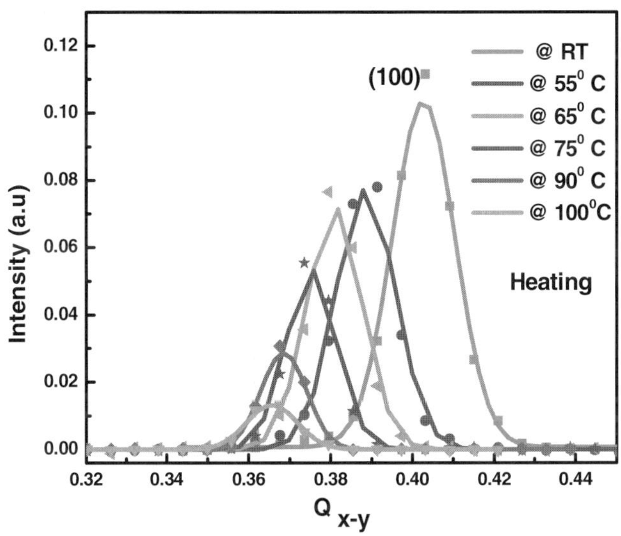

Figure 10.8 Temperature dependent 1st order (100) peak study of ~200 nm thick sample at $\alpha_i = 0.10°$. The shift of the (100) peak towards low Q_{x-y} values indicates the increase of the '*d*' spacing along the normal direction.

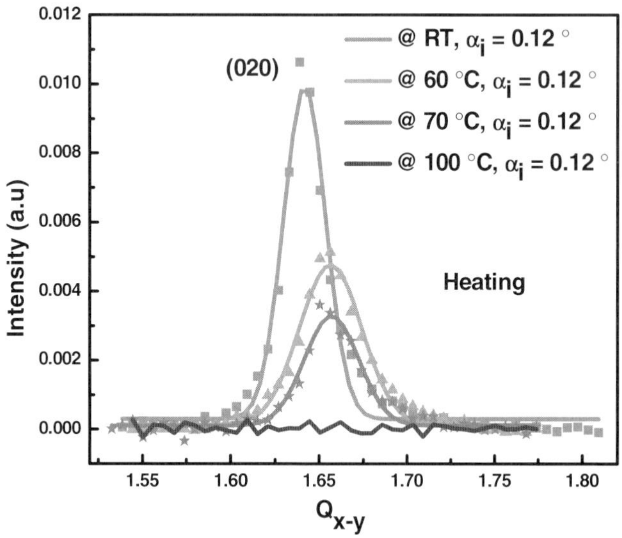

Figure 10.9 Temperature dependent (020) peak study of sample 3 at $\alpha_i = 0.12°$. The shift of the (010) peak towards higher Q_{x-y} values indicates the decrease of '*d*' spacing lateral direction.

tively similar effect for poly(3-dodecylthiophene) was also observed by Prosa et al. [27] where the interlamellar '*d*' spacings also increase with increasing temperature. Comparing both molecules the variation in the corresponding '*d*' spacing as well as in the size of crystallites corresponds to the different length of the molecules.

Figure 10.10 shows the temperature variation of the (100) peak intensity measured by GOD for three samples of different thickness. It shows a continuous decrease in intensity versus temperature for samples of thickness above 200 nm and 48 nm but a nearly constant intensity for the thinnest sample of thickness 11 nm up to about 70 °C followed by a sharp drop. The same behaviour was also found for a 17 nm thick sample but in a different experiment performed at DELTA (not shown here). This much thinner sample ($t = 10$ nm) exhibits a sharp drop of (100) intensity at 80 °C. This behaviour can be interpreted by a strong substrate–layer interaction becoming remarkable for thin films only. The interaction also modifies the character of phase transition. It appears as a continuous drop of intensity for thick films but a sharp drop of intensity was found for thin films. Similarly, the inter-planar distances 'd_{100}' calculated for all three samples of thickness 11 nm, 48 nm and 200 nm at different elevated temperatures indicate that all the curves of the thin and thick films show an increase of d_{100} up to the temperature of the intensity drop. The inter planar distances ('*d*') values are reversible as long as it is not heated beyond the melting point, while the intensity was always found to decrease after cool-

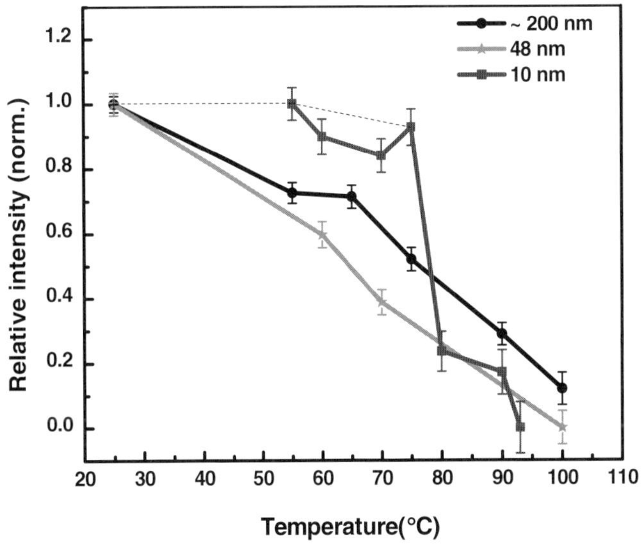

Figure 10.10 Temperature dependent measurements on three samples with different thickness measured at ESRF.

ing back to room temperature and the fraction of intensity recovery depends on the film thickness.

The HMW fraction has shown another temperature-dependent behaviour: already laboratory measurements revealed a significant increase of peak intensities corresponding to formation of a lamellar phase (see Figure 10.11). After

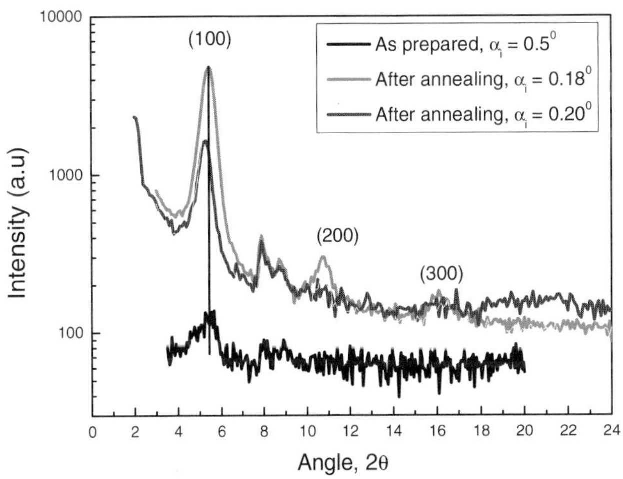

Figure 10.11 Increase of crystalline fraction after annealing of HMW thin films up to 180 °C, black and red, blue curves correspond to initial and temperature treated samples, respectively.

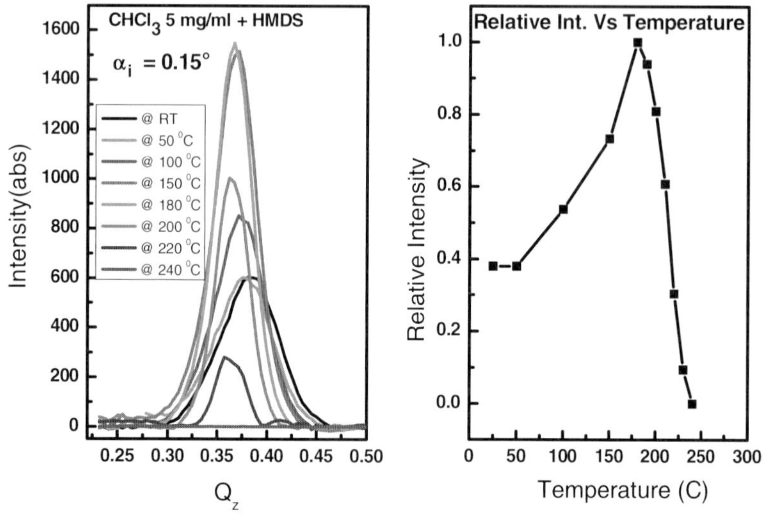

Figure 10.12 Temperature resolved (100) peak of P3HT film made from HMW fraction for $\alpha_i = 0.15°$.

2 h annealing at about 180 °C (50 K below the melting point) the (100) peak increased by a factor of 10. This process was observed in more detail using synchrotron radiation. Here, we raised the temperature in steps of 20 °C and measured the intensities of the (100) and (020) peaks at elevated temperatures. The (100) peak (see Figure 10.12) shows a continuous increase in intensity up to $T = 180$ °C and decreases upon further heating accompanied by a small increase of lattice spacing. At the same time the peak position remained essentially unchanged. This result is in contrast to the behaviour of in-plane peaks (not shown). Here, the temperature seems not to have an influence on the in-plane lattice parameter (b). In addition, the (020) peak remains unchanged with temperature. Both findings suggest the crystal growth in the direction of the surface normal that is linked with an improved lamellar alignment of alkyl chains in this direction. Recent experiments have shown that this behaviour also seems to be thickness dependent. The detailed results of this study are to be publish in upcoming publication [31].

10.6 Discussion

For the first time we have shown that the crystalline order of films made of LMW and HMW P3HT changes as a function of film thickness. For film thicknesses above 20 nm the film structure consists of randomly oriented nanocrystallites of 25–40 nm size diluted in an amorphous matrix. Using the ratio between the total amount of scattering intensity under Bragg peaks and that under amorphous halos one can make an estimate of the degree of crystallinity. In contrast to powder samples [32] an estimate for thin films contains several uncertainties. Presently we cannot provide absolute numbers but we can compare relative numbers from different samples. We have found a decrease in the degree of crystallinity for decreasing film thickness.

A dramatic change in the temperature dependence of the peak position and peak intensity was realised for the thinnest films below 20 nm. The disappearance of (100) peak in GID can be interpreted by a preferential alignment of nanocrystals at the film/substrate interface. Considering the average crystal size, the pronounced change in alignment appears when the domain size becomes of the order of the film thickness. The effect is known from liquid crystals where a characteristic length of interface ordering decays exponentially from the interface towards the bulk.

For LMW material the structure of nanocrystals can be defined as monoclinic. Considering the stereochemical arguments given by Prosa et al. [10, 26], Tashiro et al. [29], and Meille et al. [33], the molecular packing is such that the thiophene rings are arranged more or less parallel to each other defining the lattice parameter b. The side chains interpenetrate each other where each second chain attached to a thiophene ring aligns into same direction. The degree of interdigitation defines the stacking distance (lattice parameter a). The distance along a single thiophene backbone (lattice parameter c) is very close to the distance to the neighbouring thiophene backbones. Due to the alternative arrangement of chains, c is the distance between two second consecutive thiophene rings (Figure 10.6). Similar arguments hold for b, it measures the distance between two subsequent backbones assuming that two neighbouring thiopene rings are shifted by $c/2$ to each other. Following the model of Prosa et al., the aliphatic chains are tilted by a certain angle (β angle) with respect to the thiophene rings. A local disorder of side-chain alignment may result in the appearance of the repetition unit along the polymer backbone and it might also change the alignment of thiophene rings with respect to each other, as seen by a change in angles between the axes and the diffuse scattering hump at $Q_{x-y} = 1.6$ Å^{-1}.

The refined crystal structure is assumed to be an average valid over the whole sample. The lattice parameters might vary slightly among the crystals and probably change at the film/substrate interface. The latter could explain the observation that a few peaks appeared in-plane or out-of plane only. This is supported by the polymorphism that appeared at thick films of pristine films only. It reflects the effect of small fluctuations of crystal growth energy during the preparation process.

Temperature-resolved measurements were revealed to be very instructive for the characterisation of the interface interaction. For LMW material we detected a complete change of crystal unit cell with increasing temperature. Moreover, the phase transition between solid and melt changes as a function of temperature. A continuous drop of intensity was found for thick films but there was a sharp drop of intensity for the thinnest films. The film thickness gives a rough estimate of the decay length of interface interaction towards the bulk. Because the peak width remains constant, while decreasing in intensity, the number of crystallites decrease with increasing temperature for film thickness above the interface decay length. The crystals dissolve and the amorphous film becomes molten as a whole. For thin films the melting behaviour is different because the crystals remain pinned at the film/substrate interface up to the melting temperature. Due to the interface interaction the latter is expected to change as a function of film thickness.

For HMW material we found a maximum for the stacking of alkyl chains at 180 °C, which is 60 °C below the melting temperature. Also, here the interface induces the direction of ordering, i.e. stacking along the film normal. Because the peak width is unchanged, the increased intensity of the (100) peak is caused by the increased number of ordered crystalline domains. Stepwise heating and further subsequent cooling reduces the peak intensity, i.e. the number of stacked nanocrystals in the HMW material. One may suggest that the additional nanocrystals redistributed into a preferential order at elevated temperature and this behaviour is maximised at 180 °C, as further heating can only reduce the intensity and suggests distortion of such alkyl side-chain staking.

In general, we have shown that the thickness dependence is essential for the understanding of structural properties of the thin layers and temperature treatment plays an important role for the optimisation of P3HT films.

Acknowledgements

The authors thank Dieter Neher (University of Postdam) for electric measurements and fruitful discussions, Ullrich Scherf (University of Wuppertal) for providing of polymers. In addition, we thank DELTA (Dortmund) and ESRF (Grenoble) synchrotron radiation sources for experimental support. This work was financed by the Priority program SPP-1121 of the German Science Foundation.

References

1. Z. Bao, A. Dodabalapur, and A. Lavinger, J. Appl. Phys. Lett. **69**, 4108 (1996).
2. H. Sirringhaus, P. J. Brown, R. H. Friend, M. M. Nielsen, K. Bechgaard, B. M. W. Langeveld-Voss, A. J. H. Spiering, R. A. J. Janssen, E. W. Meijer, P. T. Herwig, and D. M. de Leeuw, Nature **401**, 685 (1999).
3. D. H. Kim, Y. Jang, Y. D. Park, and K. Cho, Macromolecules **39**, 5843 (2006).
4. Y. Kim, J. Nelson, J. R. Durrant, D. D. C. Bradley, K. Heo, J. Park, H. Kim, I. Mcculloch, M. Heeney, M. Ree, and C. S. Ha, Soft Matter **3**, 117 (2007).
5. C. Tanase, E. J. Meijer, P. W. M. Blom, and D. M. de Leeuw, Org. Electron. **4**, 33 (2003).
6. M. Ballauff, Angew Chem., Int. Ed. Engl. **28**, 253, (1989).
7. Y. N. Gartstein, and E. M. Conwell, Chem. Phys. Lett. **245**, 351 (1995).

8. D. H. Dunlap, P. E. Parris, V. M. Kenkre, Phys. Rev. Lett. **77**, 542 (1996).

9. S. V. Novikov, D. H. Dunlap, V. M. Kenkre, P. E. Parris, and A. V. Vannikov, Phys. Rev. Lett. **81**, 4472(1998).

10. T. J. Prosa, M. J. Winokur, J. Moulton, P. Smith, and A. J. Heeger, Macromolecules **25**, 4364 (1992).

11. A. Zen, M. Saphiannikova, D. Neher, J. Grenzer, S. Grigorian, U. Pietsch, U. Asawapirom, S. Janietz, U. Scherf, I. Lieberwirth, and G. Wegner, Macromolecules **39**, 2162, (2006).

12. R. J. Kline, M. D. McGehee, E. N. Kadnikova, J. Liu, J. M. J. Frechet, and M. F. Toney, Macromolecules **38**, 3312 (2005).

13. R. J. Kline, M. D. McGehee, E. N. Kadnikova, J. Liu, J. M. J. Frechet, and M. F. Toney, Adv. Mater. **15**, 1519 (2003).

14. A. Zen, J. Pflaum, S. Hirschmann, W. Zhuang, F. Jaiser, U. Asawapirom, J. P. Rabe, U. Scherf, and D. Neher, Adv. Funct. Mater. **14**, 757 (2004).

15. J. A. Merlo, and C. D. Frisbie, J. Polym. Sci. Part B: Polym. Phys. **41**, 2674 (2003).

16. T. C. Gorjanc, I. Levesque, and M. D. Lorio, Appl. Phys. Lett. **84**, 930, 2004.

17. J. Chang, B. Sun, D. W. Breiby, M. M. Nielsen, T. I. Solling, M. Giles, I. McCulloch, and H. Sirringhaus, Chem. Mater. **16**, 4772 (2004).

18. R. Ruiz, A. Papadimitratos, A. C. Mayer, and G. G. Malliaras, Adv. Mater. **17**, 1795 (2005).

19. A. Dodabalapur, L. Torsi, and H. E. Katz, Science **268**, 270 (1995).

20. E. L. Granstrom, and C. D. Frisbie, J. Phys. Chem. B **103**, 8842 (1999).

21. R. J. Kline, M. D. McGehee, and M. F. Toney, Nature Mater. **5**, 222 (2006).

22. R. S. Loewe, S. M. Khersonsky, and R. D. Mc Cullough, Adv. Mater. **11**, 250 (1999).

23. M. Trznadel, A. Pron, M. Zagorska, R. Chrzaszcz, and J. Pielchowski, Macromolecules **31**, 5051 (1998).

24. S. Joshi, S. Grigorian, U. Pietsch, P. Pingel, A. Zen, D. Neher, and U. Scherf, Macromolecules, to be published.

25. U. Pietsch, V. Holy, and T. Baumbach, High-resolution X-ray Diffraction from Thin Films and Lateral Nanostructures, (Springer, Berlin, 2005).

26. T. J. Prosa, M. J. Winokur, and R. D. McCullough, Macromolecules **29**, 3654 (1996).

27. T. J. Prosa, J. Moulton, A. J. Heeger, and M. J. Winokur, Macromolecules **32**, 4000 (1999).

28. G. Boder, Structural Investigation of Polymers (Ellis Horwood Ltd., Chichester, West Sussex, England, 1991).

29. K. Tashiro, O. Keiko, M. Yasuhisa, K. Masamichi, K. Tsuyoshi, and Y. Katsumi, J. Polym. Sci.: Part B: Polym. Phys. **29**, 1223 (1991).

30. M. Brinkmann, and P. Rannou, Adv. Funct. Mater. **17**, 101 (2007).

31. S. Joshi, P. Pingel, S. Grigorian, Tobias Panzer, U. Pietsch, D. Neher, Michael Forster, and U. Scherf, Macromolecular, submitted (2009).

32. W. Ruland, Acta Crystallogr. **14**, 1180 (1961).

33. S. V. Meille, V. Romita, and T. Caronna, Macromolecules **30**, 7898 (1997).

11
Molecular Beam Deposition and Characterisation of Thin Organic Films on Metals for Applications in Organic Electronics

G. Witte and Ch. Wöll

11.1
Introduction

Despite many fundamental differences organic semiconductors (OSCs) are similar to inorganic semiconductors (Si, Ge, GaAs) insofar as many of their key properties depend critically on the presence of disorder (like grain boundaries, dislocations) and impurities. In addition, for both types of semiconductors the injection of charge carriers, both electrons and holes, requires a proper electronic level alignment at the electrode/semiconductor interface. Since the resulting characteristic of organic electronic devices is frequently limited by extrinsic properties (such as defects and contacts) it is often difficult to compare the intrinsic properties (e.g. mobility) of different organic materials. This is an unfortunate situation since presently many different molecules are being considered for applications in organic electronics, both polymers and small organic molecules, and these problems make a direct comparison rather difficult.

One of the most critical parameters of organic semiconductors is the charge carrier mobility which has been shown in previous work to depend critically on contaminations and structural defects [1, 2].

Presently, the most important active device in the area of organic electronics is an organic field effect transistor (OFET) [3, 4], the two most common types of OFETs, with top or bottom contact geometry, are shown schematically in Figure 11.1. Such devices constitute a key component for the fabrication of radio frequency identification (RFID) tags which are likely to become the first mass product entering the market. The reading distance of such tags increases with the transmitter frequency for which only a limited number of frequency bands are available (125 kHz, 13.56 MHz, or 900 MHz). Since the maximum switching frequency of an OFET scales proportional to $f_{max} \sim \mu U/L^2$ the fairly high value of the frequencies puts a stringent requirement on the charge carrier mobility μ. For example, an organic RFID tag operating at 13.56 MHz which has been demonstrated recently requires a charge carrier mobility of about 0.02 cm^2/V s when using transistors with a channel length of $L = 4$ µm [5].

Organic Electronics. Structural and Electronic Properties of OFETs. Edited by Christof Wöll

ISBN: 978-3-527-40810-8

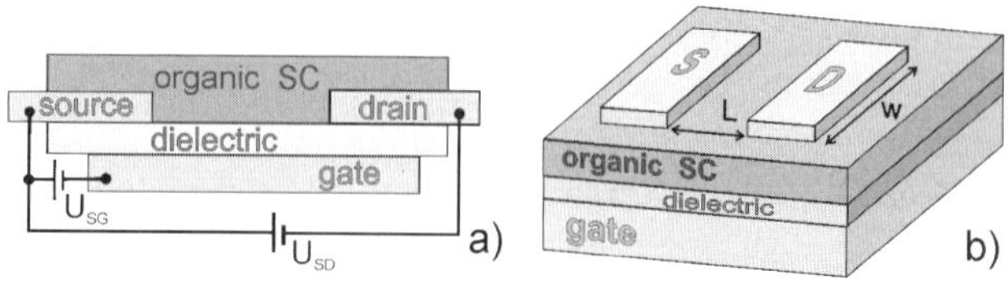

Figure 11.1 Schematic layout of organic field effect transistors with top (a) and bottom contacts (b).

The traditional method to determine the intrinsic charge transport and carrier mobility of crystalline organic semiconductors is the time-of-flight technique, originally introduced by Kepler [6] and Leblanc [7]. While this technique is rather accurate it requires high purity macroscopic-sized single crystals of the respective organic material. Since such mm-sized organic single crystals are often difficult to grow they have been produced and studied only for a small set of materials [1]. At present, the most common way to determine the charge carrier mobility of a given organic material is to extract this value from the device characteristics of an OFET. It should be noted, however, that the derived values depend critically on the details of the analysis (e.g. low voltage slope vs. saturation current, consideration of contact resistances). Moreover, it has been demonstrated that mobilities derived from field effect or *I*/*V*-measurements can differ substantially from the time-of-flight mobilities. This fact reflects the influence of extrinsic properties mostly related to trap-states at the electrodes which have to be taken into consideration when determining intrinsic material properties [8].

In the present chapter we will briefly review recent efforts in trying to identify the origins of extrinsic effects in organic electronic devices by first considering the electronic structure at the electrode/organics interface, then by unravelling fundamental principles in organic molecular beam deposition (OMBD) and third by fabricating an "ideal" organic electronic device where the presence of extrinsic defects can be strictly excluded.

11.2
Electronic Level Alignment at the Metal/Organics Interface

An important issue for the performance of an organic electronic device like an OFET is the injection of charge carriers, electrons or holes, from the electrode into the organic material. In case of the commonly used metal electrodes an efficient electron injection is possible only if the Fermi level of the metal and the energy of the lowest unoccupied molecular orbital (LUMO) of the organic material differs by a small amount only. A similar statement applies for hole injection, in this case the position of the highest occupied molecular orbital (HOMO) has to match with the position of the Fermi level. When noble metals, in particular Au, are being used for an electrode one may naively assume that with regard to the electronic level alignment the so called common vacuum level (CVL) approximation applies [9]. This assumption, which dates back to Mott [10] and Schottky [11] can be used to predict the relative positions of the electronic states at the metal/organics interface by simply assuming that – because of the absence of any chemical interaction – the electronic states of the molecule and the substrate are unchanged relative to the vacuum level which is then set to be the same value for the metal and the molecules. This situation is schematically illustrated in Figure 11.2a. A simple consequence of the presence of such a common vacuum level is that there is no change in work function of the metal substrate upon adsorption of the molecule.

As first noted by Seki and coworkers [12] and later confirmed by Kahn and coworkers [13] even in the most simple cases this approximation does not hold. When saturated hydrocarbons, which are among the least reactive molecules in organic chemistry, are deposited onto the most inert metal, Au, a reduction of the work function on the order of 1 eV can be seen. It is interesting to note that this work function lowering has about the same amount encountered when an electro-positive element like Cs is deposited on a metal surface. These changes of the work-function even in the absence of any chemical interaction (like charge transfer or bond formation) are attributed to the formation of an interface dipole located between the molecule and the metal substrate.

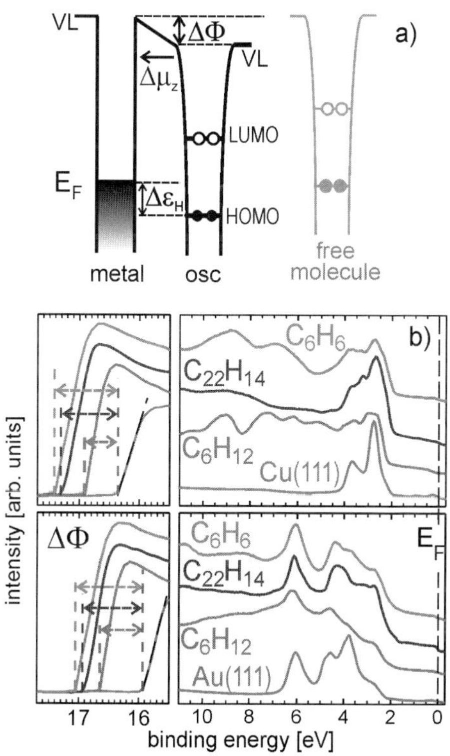

Figure 11.2 (a) Schematic representation of energy level positions at the metal/molecule interface. In contrast to a simple vacuum level (VL) alignment a realistic scheme has to take the interface dipole $\Delta\mu$ into account. (b) HeI UPS data showing the high binding energy cutoff and the valence band region for a monolayer of benzene, pentacene and cyclohexane on Cu and Au surfaces.

In the past a number of different explanations have been proposed for these unexpected changes (see e.g. [14]) but a thorough theoretical description of the physics governing the formation of these interface dipoles was only possible after *ab initio* electronic structure calculations became available [9, 15].

In the case of a Cu(111)-substrate these calculations carried out for saturated and unsaturated hydrocarbon adsorbates provided theoretical evidence that the adsorption of organic molecules on metals is always accompanied by a lateral displacement of electronic charge at the metal surface [15, 16]. The origin of this effect is a quantum-mechanical phenomenon referred to as exchange repulsion which is also the basis for Pauli-repulsion. Although these early calculations have made it possible to unravel the origin for the interface dipole in a simple case with regard to understanding the charge injection properties in real organic electronic devices the particular molecules used as OSC (which are typically significantly larger than benzene) have to be considered.

At present one of the most promising OSCs is pentacene ($C_{22}H_{14}$) because this material exhibits a remarkably high charge carrier mobility in the crystalline phase and reveals an ability to form highly ordered (poly-)crystalline films allowing the fabrication of thin film OFETs with very high charge carrier mobilities [17, 18]. Previous experimental work [19] has demonstrated that also for pentacene deposited onto various metal substrates interface dipoles are present and exhibit values of about the same size than seen for saturated hydrocarbons. One important consequence is the presence of contact barriers which sensitively affects the charge carrier injection [20]. Unfortunately, however, the precise *ab initio* wave-function-based calculations required for a reliable theoretical analysis of the pentacene/metal interaction are by far too expensive. Theoretical approaches employing density functional theory (DFT), on the other hand, which can be used to describe larger systems and which for that reason, are presently the most popular approach to carry out electronic structure calculations, have limitations in situations where the *van der Waals* interaction contributes significantly to the interaction. To overcome the limitations resulting from the fairly large size of pentacene we have followed a two step approach. First, it was demonstrated experimentally by using ultraviolet photoelectron-spectroscopy (UPS), that the work function shift observed for a monolayer of benzene (C_6H_6) and pentacene was equal to within 0.15 eV, when adsorbed on Au(111) or Cu(111), respectively (see Figure 11.2b). This observation together with the fact that both, benzene and pentacene, adsorb with their molecular plane parallel to the metal surface allows us to interrogate high quality wave function based calculations for the case of benzene adsorbed on metal clusters in order to understand the complex interplay between interface dipole formation, charge transfer and possibly chemical interactions between pentacene and the supporting metal.

In Table 11.1 we show a compilation of the theoretical and experimental results obtained for benzene and cyclohexane, a saturated hydrocarbon, deposited on Au and Cu substrates. The change of the work function induced by benzene adsorption has been calculated using the Helmholtz equation by as-

Table 11.1 Comparison of work function changes induced by adsorption of molecular monolayer on either Cu or Au.

	Cu(111)		Au(111)	
	benzene	cyclohexane	benzene	cyclohexane
$\Delta\Phi_{exp}$	−1.05 eV	−0.50 eV	−1.10 eV	−0.70 eV
$\Delta\Phi_{calc}$	−1.08 eV	−0.43 eV	−0.87 eV	−0.54 eV

suming the presence of a dense monolayer of adsorbed molecules. Comparison of the respective experimental and theoretical values reveals a rather satisfying agreement and makes it possible to analyse the microscopic origin for the fairly large changes in work function caused by benzene on these metals. The results [9, 15, 16] allow to conclude that the "cushion" of electronic charge at a metal surface is impressed by the molecule, which results in a fairly substantial charge decrease in the area between molecule and surface. In all cases studied the electrostatic field arising from this charge depletion provides the major contribution to the interface dipole and in turn to the change in work function. The calculations, however, also allowed identifying other contributions. As one may have expected in the case of benzene there is also a small charge transfer from the metal to the molecule, which is smaller for Au than for Cu. This contribution explains why the resulting interface dipole is smaller on Cu than on Au. Recent theoretical results reported for Ag(111) fully support the above findings and reveal that the adsorption of benzene on Ag(111) is more similar to Au than to Cu, as expected [21].

These results on physisorption-induced work function changes on metals are also interesting with regard to the common assumption on metal/molecule hetero-interfaces, namely that metals are the "hard" and molecules are the "soft" component. The theoretical results clearly demonstrate that with respect to the charge density the metal acts as "soft" and the molecule as the "hard" partner, since the – essentially undisturbed – organic molecule imprints its shape in the charge distribution on the metal surface.

The results presented in the above case are rather general and allow to establish the rule that molecular adsorption on a metal should normally be accompanied by a significant reduction of the work function. When no substantial reduction of work function is seen or when the work function is even observed to increase this can be taken as a strong indication for the presence of chemical interactions, i.e. transfer and chemical bond formation.

11.3 Structural Properties at the Metal/Organic Interface

After we have shown in the previous paragraph that the electronic structure at the metal/organic interface can – in the absence of chemical interactions – be predicted by assuming that the respective electronic structure remains unchanged but are shifted with regard to each other by a fixed amount (interface dipole) the simplest way to realise a metal/organic contact would to bring a clean metal into direct contact with an organic single crystal. Under ambient conditions metal surfaces are not free of contaminations and, as a result, the bottom contact approach where first metal electrodes are evaporated onto a substrate and then the OSC is deposited by spin coating or OMBD very often are problematic. For this reason frequently top contacts are used which are prepared by evaporating metals onto the organic material. A closer inspection, however, reveals a significant amount of inter-diffusion [22] thus demonstrating that such metal/organic interfaces are in fact not well defined. (For a more detailed discussion of this topic see also Chapter 9 by Gerlach et al. and Chapter 19 by Scharnberg et al.). An alternative approach to contact the rather fragile organic single crystals utilises gold coated elastomeric contact stamps which are gently laminated to the OSC. This method which frequently is referred to as "flip crystal" technique has been employed by Sundar et al. to build a functioning single crystal OFET on the basis of rubrene single crystals [23].

Unfortunately, rubrene – as well as other organic molecules used in organic electronics – is rather sensitive to O_2 and one always has to consider that the surface of these organic single crystals contains oxidation products. Since the very surface of the rubrene single crystal pressed against the laminated gate dielectric is actually the region where charge transport in an organic field effect transistor occurs [24] one cannot rule out the possibility that the devices fabricated using the "flip-crystal" technique are affected by doping the organic material with oxidation products. In order to investigate the importance of such contributions, an analysis of the surface composition of rubrene single crystals has been carried out using laser desorption ionisation time-of-flight mass spectroscopy (LDI-TOF MS) [25]. These experiments clearly demonstrated that on the surface of rubrene single crystals concentrations of rubrene-peroxide exceeding 10% can be observed even after fairly short contact with air, thus challenging the interpretation of the data obtained by Sundar et al. [23] in that the measured mobility is an intrinsic property of rubrene.

Because of this rather high sensitivity with regard to oxidation a better approach to determine intrinsic properties of OSC and to construct devices which are not limited by extrinsic properties appears to be organic molecular beam deposition (OMBD) where organic molecules are deposited from the gas phase onto clean, contamination-free metal substrates exhibiting a high structural quality.

11.4
General Principles Governing Organic Molecular Beam Deposition (OMBD) on Metal Substrates: Case Studies for Rubrene, Perylene and Pentacene

Even a brief look at the literature demonstrates that organic molecular beam deposition is a very complex field with different organic molecules actually showing rather different properties. There are cases where almost perfect epitaxial growth can be achieved, for example in the case of PTCDA deposited silver substrates [26]. In other cases the presence of different phases in the deposited organic thin films was observed, some of which occur only at the surface of the (inorganic) substrates e.g. the thin film phase for pentacene [27, 28] or columnar structures in thin films of hexabenzocoronene and its derivatives [29] (for details on the latter system see also Chapter 4 by Tsao et al.).

In the following we will present exemplarily three case studies for rubrene, perylene and pentacene. For a more thorough survey of other work see the recent review articles by Witte and Wöll [30], Schreiber [31] and Forrest [32].

11.4.1
Rubrene Deposition on Au(111)

In the first case study we will focus on rubrene and illustrate a fundamental problem in organic molecular beam deposition, namely the importance of the flexibility of the molecular entities. Molecules can, in contrast to inorganic materials, adopt different conformations with distinctly different geometries but rather similar energies. As a result, the precise geometry of the molecules in the bulk is not necessarily identical to that of the free molecules or molecules adsorbed on a substrate, even if that is a noble metal. By using high resolution X-ray absorption fine structure (NEXAFS) spectroscopy we could demonstrate that rubrene molecules ($C_{42}H_{28}$) in monolayers deposited on metal or oxide surfaces adopt a molecular conformation which is different from that in the crystalline phase [33]. In the crystal the tetracene backbone of the molecule (see Figure 11.3b, c) is essentially planar, whereas for the free molecule (gas phase) and the molecule present in a monolayer a torsion of the two outermost aromatic rings by about 42° with respect to each other is seen (Figure 11.3a). This large conformational change can be understood by considering the fairly large steric repulsion between the phenyl side groups attached to the tetracene backbone of the rubrene molecule. The energy lowering for the free molecule relative to the structure in the bulk amounts to 210 meV, this energy is in the bulk compensated by the gain in lattice energy during crystallisation. When depositing the molecule on an inert substrate, the neighbours present in the crystal which stabilise the planar conformation are missing and it is energetically more favourable for the molecule to realise the structure with the twisted tetracene unit. High resolution data from scanning tunnelling microscopy (STM) recorded for single rubrene molecules deposited on Au(111) recently reported by Schneider and coworkers [34] have nicely corroborated this finding.

Apparently the presence of such twisted molecules on the substrate leads to a strongly delayed nucleation and is not compatible with the growth of crystalline rubrene organic thin films. In fact it has been shown by using X-ray diffraction that rather smooth but amorphous rubrene films are formed upon room temperature deposition onto SiO_2 [35]. These observations explain the lack of

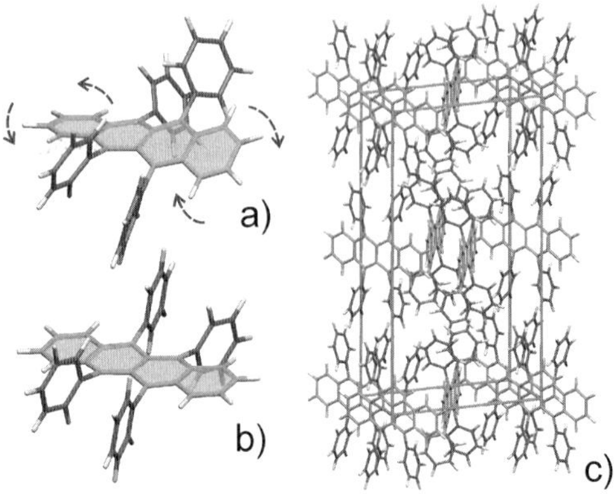

Figure 11.3 Molecular conformation of rubrene in (a) the gas phase and (b) crystalline phase together with the crystal structure (c).

work where thin film rubrene OFETs with satisfying electronic properties could be produced by OMBD. The situation for this organic molecules is thus in pronounced contrast to the case of pentacene, where high-performance devices with large charge carrier mobility have been fabricated by a number of different groups [17, 18]. A detailed description of OFET device properties and their relation to the microstructure can be found in Chapter 20 by Scholz et al., Chapter 15 by Nickel et al. and Chapter 8 by Voigt et al.

One approach to resolve the problems encountered during OMBD of this particular non-planar organic molecule is to use a capsulated (so called "hot wall") evaporator. In that case the enhanced rubrene vapour pressure above the substrate makes a deposition at elevated temperatures close to the sublimation temperature possible. Such a film growth under thermodynamic rather than kinetic control makes the growth of highly ordered films consisting of laminar rubrene crystallites possible [25].

That the presence of conformational freedom affects the growth of thin organic layers has also been demonstrated for paraphenylenes ($C_6H_5(C_6H_4)_{n-2}$ C_6H_5, p–nP). Using STM, UPS and NEXAFS it was shown that in the initial stage of growth p–nP molecules do not adopt a planar conformation as in the bulk structure but instead exhibit a twisting of the individual phenyl units as expected for the free molecule [36–38].

11.4.2
Adsorption-Induced Restructuring of Metal Substrates: Perylene on Cu(110)

In the previous paragraph it has been shown that the flexibility of organic molecules allows for different molecular conformations which can give rise to unexpected phenomena in organic thin film growth. In this section we will demonstrate that also the metal substrate, which is usually regarded as a mechanically "hard" material, not always remains unaffected by the molecular adsorption of larger organic molecules.

When depositing perylene, a planar aromatic molecule (see inset of Figure 11.4a), on a Cu(110) surface, data obtained from STM indicate the presence of a regular nanoscopic pattern of stripes with a distance of 18 Å [39]. Figure 11.4 displays a typical STM micrograph of this phase which spontaneously starts to develop after deposition of a close packed saturated monolayer and is completed after gently annealing the substrate at 350–450 K. Whereas at first sight one may assume that the stripes visible in the STM micrographs correspond to molecular rows separated by uncovered parts of the surface, a closer inspection of the area in between the rows reveals the presence of perylene molecules not only at the rim but also at the bottom of the troughs (see line scans in Figure 11.4b). By combining the results of various experimental techniques (scanning tunnelling microscopy (STM), He-atom scattering, thermal desorption spectroscopy (TDS), and near edge X-ray absorption spectroscopy (NEXAFS)) is was possible to identify this *stripe*-phase as a commensurate (5 × 5) phase where all molecules are uniformly aligned while the row pattern is actually caused by a restructuring of the copper substrate [39, 40].

The structural model derived form the experimental data is depicted in Figure 11.5 and shows a periodic stripe pattern defined by alternating "up" and "down" atomic-height copper steps which are not present for the saturated monolayer where the molecules adopt a typical herring bone packing motive yielding a closed packing with a (4.5 × 5) structure (see Figure 11.5a). On the basis of the precise structural analysis of the stripe phase it was further possible to identify the driving force for this phenomenon.

While the molecular coverage of the stripe phase is 10% lower than in the saturated monolayer the uniform alignment of molecules makes a commensurate registry of the molecular carbon frame with respect to the underlying copper lattice possible, resulting in larger adsorption energy. Using effective me-

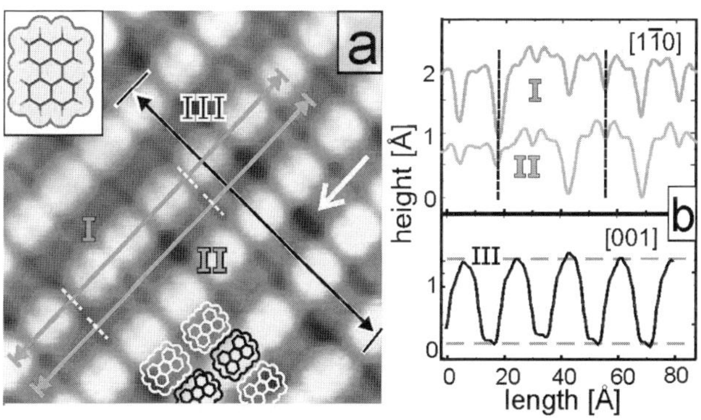

Figure 11.4 STM micrograph of the (5 × 5) monolayer phase of perylene on Cu(110) (a) together with corresponding line profiles along the rim (I) and the bottom (II) of the troughs, and perpendicular to the rows (III) (b). The inset in (a) shows the molecular structure of perylene ($C_{20}H_{12}$).

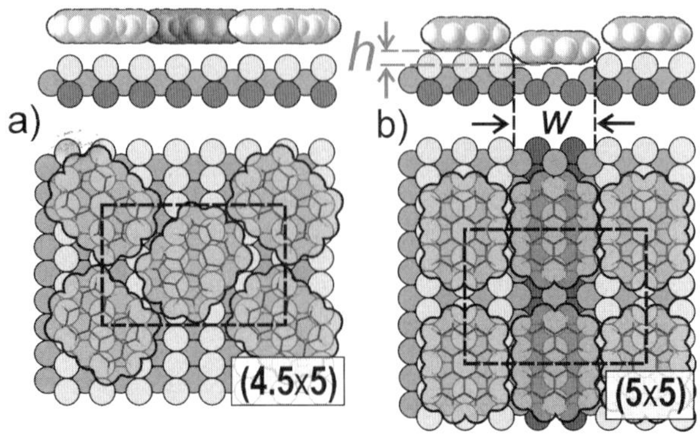

Figure 11.5 Structure models of (a) the saturated perylene monolayer and (b) the (5 × 5) stripe phase formed on the reconstructed Cu(110) substrate.

dium theory an energy of 0.1 eV per molecule is estimated for the formation of such alternating substrate steps [39], an amount which has to be compensated by the gain in perylene adsorption energy.

When heating the Cu-substrate for longer times a rather pronounced faceting of the surface is seen [39]. The individual facets have a structural motif similar to that of the (5 × 5) phase and also consist of narrow copper terraces. In this case, however, the steps form a staircase instead of a periodic "up" and "down" arrangement. This structure is again driven by the gain of adsorption energy while the mass transport required for the formation of such substrate facets is made possible by thermally activated diffusion. Similar effects have also been seen for molecules deposited on other metals [41–43] and it can be concluded that it is not always the metal which is the (mechanical) hard partner in the deposition of organic molecules on metal substrates.

11.4.3
Organic Molecular Beam Deposition of Pentacene on Clean Metal Surfaces

In this section we describe the results of a systematic study on the growth of pentacene on different types of metal surfaces. The results were obtained by combining a fairly large number of complementary techniques, including STM, atomic force microscopy (AFM), scanning electron microscopy (SEM), NEXAFS, and X-ray diffraction (XRD). These studies reveal a rather complex growth scenario which not only depends on the type of metal but also on its cleanliness and roughness.

Generally, in the first monolayer on (coinage) metals pentacene and other planar aromatic molecules were found to adsorb with their aromatic ring plane parallel to the substrate. A detailed discussion of the molecule/metal interaction probed by means of low temperature STM can be found in Chapter 12 by Soubatch et al. The preference for such a planar adsorption geometry can be understood by first considering the case of benzene. As demonstrated by accurate *ab initio* calculations, for all close-packed (111)-surfaces of the coinage metals, the benzene/metal interaction is governed by fairly strong dispersion (or *van der Waals*) forces acting between a planar organic molecule and the metal substrate. This interaction is stronger for a planar than for a side-on orientation because of the closer proximity in the former cases. For the larger pentacene the same reasoning can be used if chemical interactions are rather weak, e.g. in the case of pentacene/Au(111). In the case of Cu(111) chemical interactions arising from an interaction of the p_z-orbitals of the C-atoms with the metal d-orbitals may come into play [44], which also favour a planar orientation.

In all cases investigated the orientation of pentacene in the first layer is not affected by molecules adsorbed in the second, third or further layers since the interaction between the planar organic molecule and the metal substrate is considerably stronger than the lattice energy of this aromatic hydrocarbon in the corresponding organic single crystal. In the case of pentacene the sublimation enthalpy amounts to 157 kJ/mol leading to an onset of evaporation at temperatures of about 390 K under vacuum conditions while monolayer films chemisorbed on such metal surfaces remain stable up to much higher temperatures (Au: 480 K, Ag: 500 K, Cu: >700 K [45, 46]).

Note, that in the case of bismuth the growth of upright oriented pentacene molecules has been observed – even in the very first monolayer [47]. This orientation has been explained [47] by the rather small density of states near the Fermi level of this semi-metal leading to a rather small molecule-substrate interaction which becomes comparable or even weaker than the mutual molecular interaction and thus parallels the situation observed for pentacene deposited on inert substrates such as SiO_2 [48].

As regards the lateral structure of pentacene monolayers adsorbed on the close packed surfaces of gold and silver, ordered monolayer films were identified [45, 49–51], where the flat lying molecules adopt a dense packing with intermolecular distances given by their *van der Waals* dimensions. The long range ordering of such films is limited by the presence of rotational domains due to the symmetry of the substrate.

In order to exclude the appearance of rotational domains in the initial stage of growth we have used a Cu(110) substrate and have studied the growth of pentacene films on this substrate quite extensively. In fact, a highly ordered submonolayer and a saturated monolayer phase are found which exhibit a uniform alignment of the flat lying molecules with their long axes orientated parallel to the $\langle 1\bar{1}0 \rangle$-azimuth direction of the substrate [46, 52].

For the subsequent growth of multilayer pentacene films on this surface, however, a rather complex scenario has been observed. By combining low energy electron diffraction (LEED) and He atom scattering (HAS) with NEXAFS, we have shown that already from the second layer on the molecules grow in a tilted fashion comprising a rotation around their long axis which remains parallel to the surface. These films still exhibit high degree of lateral order. Surprisingly, this phase is not retained upon continued deposition. When the thickness of the films exceeds a value of about 2 nm the molecular orientation changes and the pentacene molecules continue to grow in an upright orientation (see Figure 11.6) and form laterally close packed films with a rectangular unit cell which bears close similarity to the (001)-plane of the thin film phase of pentacene [27]. Coverage dependent XPS measurements indicate further that this second reorientation is accompanied by a substantial increase of film roughness [46].

Film roughness is a severe problem encountered in OMBD of pentacene and becomes particularly pronounced when growing pentacene on the close packed surfaces of silver or gold. In that case film growth proceeds by formation of tall crystalline islands on top of the chemisorbed first monolayer [45, 53]. This is illustrated in Figure 11.7a which reveals a typical SEM micrograph recorded for a 2 nm pentacene film deposited at room temperature onto a clean Au(111)/mica substrate. Using STM it was possible to also image the pentacene monolayer in the regions between the islands (see Figure 11.7b). The lateral structure observed by STM within the monolayer is in close agreement with the previously reported monolayer structure [49].

In contrast, unfortunately, it was not possible to achieve stable tunnelling conditions for the STM which would allow imaging the top of such islands, even for tunnelling currents as low as 15 pA. For that reason these islands

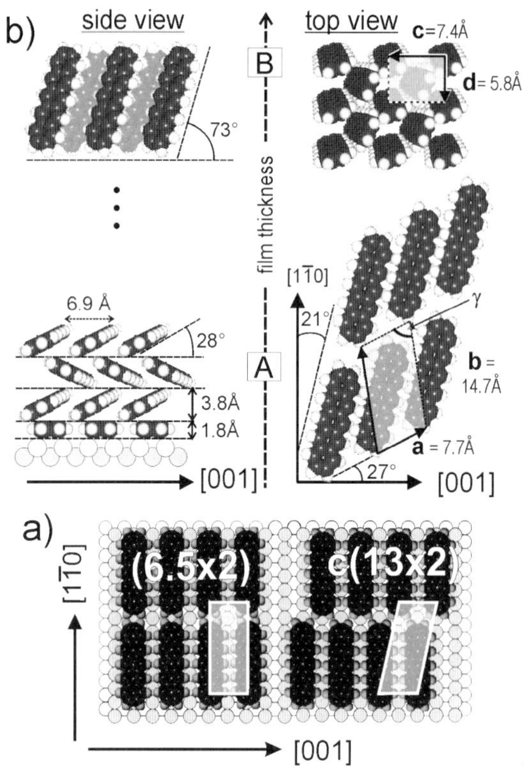

Figure 11.6 Evolution of pentacene films on Cu(110). After completion of the first monolayer revealing a predominant (6.5 × 2) phase and occasionally a coexisting *c*(13 × 2) phase (a) an intermediate phase A is formed whereas for thickness above 2 nm the molecules continue in an upright orientation adopting the well known *thin film phase* B (b).

had to be characterised by *ex situ* AFM measurements which yielded a typical island height of more than 50 nm. This value is surprisingly large considering that the nominal deposition (as measured by a quartz crystal micro balance) amounted to only 2 nm, hence demonstrating the presence of a pronounced and distinct Stranski–Krastanov growth mode. This pronounced dewetting giving rise to tall, isolated islands was observed for a wide range of growth conditions and is still present for thicker films as shown in Figure 11.7c for a 30 nm film (grown at a rate of 15 nm/min). Note, that due to the large aspect ratio of the tall islands there are severe artifacts in the AFM-data because of the convolution of the AFM tip shape with that of the huge pentacene islands.

Using X-ray diffraction (XRD) it was found that these islands reveal an excellent crystalline ordering. By comparing the recorded XRD pattern with the powder-patterns of the different polymorphism reported for pentacene [54] it could be demonstrated that pentacene films grown by OMBD on Ag(111) and on Au(111) adopt the bulk phase reported by Siegrist et al. [55]. Moreover, it was found, that the crystalline islands adopt distinct orientations which reflect the presence of characteristic contact planes, (022), (121), (122), $(1\bar{2}\bar{1})$, and (123), with respect to the (111)-surface of both metal substrates [45, 53]. Interestingly, all these crystalline planes exhibit molecules whose molecular plane is typically tilted but their long axis is still orientated parallel to the substrate surface (as shown schematically in Figure 11.7d) and thus can be considered as a close match between a crystalline bulk plane and the chemisorbed monolayer. Because such orientations of crystallites with the pentacene bulk structure still imply a significant lattice mismatch between the crystal lattice and the wetting layer we attribute the observed dewetting to a minimisation of the strain at the pentacene bulk/pentacene monolayer interface.

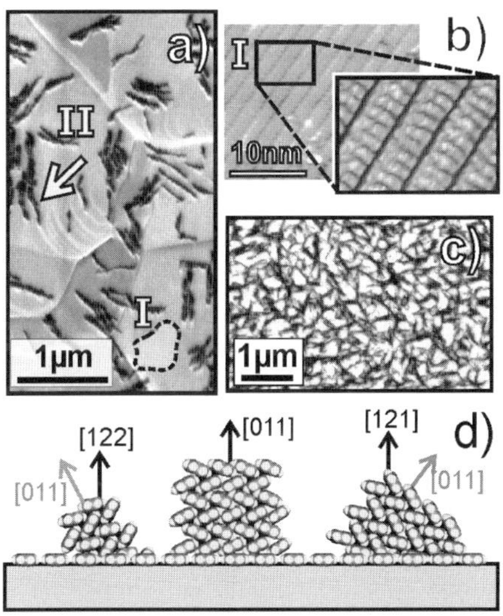

Figure 11.7 Structure and morphology of pentacene films grown on Au(111). (a) SEM micrograph of 2 nm grown at room temperature together with (b) STM data ($U = -2$ V, $I = 15$ pA) showing the structure of the first monolayer (indicated by I) between the islands formed (indicated by II). A similar morphology was also obtained for films grown at various conditions, (c) 30 nm (15 nm/min). The resulting islands reveal characteristic orientations relative to the substrate as shown schematically in (d).

We would like to note that the conclusions on the pentacene growth on Au substrates presented here are not consistent with the results of a previous STM study by Kang and Zhu [56]. These authors interpreted their STM-data so as to indicating the formation of highly ordered pentacene multilayers on Au(111) exhibiting a cofacial π-stacking with all the pentacene molecular planes orientated parallel to the substrate surface. Such a packing is neither present in any of the known pentacene bulk structures nor in the thin film phases. We attribute these apparent discrepancies to our findings to the fact that STM can cause severe artifacts when imaging tall islands of a poorly conducting material (see also next section). Moreover, the extreme dewetting results in a large fraction of the Au-substrate where the first pentacene monolayer still remains uncovered – even after deposition of films with a nominal thickness of several nanometers.

It is important to note that the structure depicted in Figure 11.7d implies a structural transition between the first monolayer and the organic thin film which is grown on this wetting layer. This is an important fact which has to be considered when analysing the charge carrier injection from a metal electrode into the organic thin film. Since this significant change in packing will act as a grain boundary it will obviously affect the charge injection from the metal electrode into organic (multilayer) films in organic thin film devices.

Frequently, the occurrence of a pronounced island growth mode is attributed to the presence of diffusion barriers (e.g. Schwoebel barriers) at step edges which can cause such growth instabilities during molecular beam deposition. A thorough characterisation and discussion of the increasing roughness upon organic film growth can be found in Chapter 9 by Gerlach et al. We would like to note, however, that this dewetting in the case of pentacene OMBD is observed to even continue after the end of the deposition. In several cases a corresponding change in the photoelectron spectroscopic data has been observed for times as large as twelve hours after deposition of pentacene films prepared and stored at room temperature in UHV. This is an important finding and demonstrates that post-deposition dewetting has to be considered in the fabrication of organic electronic devices [57]. A more detailed description of this phenomenon will be presented in the next section.

11.5 Organic Molecular Beam Deposition of Perylene

The second planar aromatic hydrocarbon for which a more extensive investigation of adlayers fabricated by OMBD has been carried out is perylene. Especially the deposition on Cu(110) substrates is of particular interest since this low symmetry surface may provide a useful template which rules out the formation of rotational domains in the initial stage of growth (as found for the case of pentacene on this substrate [52]).

Moreover, in a previous growth study of perylene on Cu(110), Chen et al. reported the formation of highly ordered organic multilayers revealing a new, substrate stabilised crystalline phase with a homogeneous orientation of the molecular planes parallel to the surface [58]. As in the case of pentacene (see previous paragraph) such a structure with an all-planar stacking had not been seen before. Whereas our data recorded for the various (sub)monolayer phases [40] were consistent with the result of the previous work [58], for the structures obtained after the saturated monolayer was heated to yield a (5×5) *stripe*-phase (see Figure 11.5) our results revealed a different, non-planar packing. On the basis of NEXAFS measurements which allow for a precise determination of the orientation of planar aromatic molecules it could be demonstrated that the molecular planes are orientated parallel to the surface only in the first monolayer while this orientation is not transferred to multilayers [59]. As described in a previous paragraph this pronounced striped phase is caused by a restructuring of the Cu(110) substrate and not by structures within the molecular adlayer. The molecular reorientation upon multilayer growth has also been observed directly during bi-layer formation of perylene on Au(111) by using low temperature STM [60].

Detailed measurements using a number of different techniques revealed that the subsequent multilayer growth on top of the reached copper structure substrate is accompanied by a pronounced dewetting leading to the formation of tall 3D islands with a bulk-like structure on top of the chemisorbed wetting layer [61].

The pronounced Stranski–Krastanov growth mode which is commonly observed upon deposition of various organic materials on clean metal surfaces is nicely illustrated by the *in situ* XPS measurements shown in Figure 11.8a. There the attenuation of the Cu 2p substrate signal together with the increase of the C 1s signal is shown as a function of the nominal film thickness as accurately determined by a quartz crystal microbalance. At low temperature (200 K) the diffusion is kinetically hindered and the growth of homogeneous films is observed, while for room temperature deposition a pronounced islanding takes place. Additional evidence for the importance of post-deposition dewetting is provided by measuring the temporal evolution of the Cu 2p to C 1s intensity ratio for perylene films that had been deposited at low temperature and were allowed to warm back to room temperature (see Figure 11.8b). Note, that desorption can be safely excluded in these experiments since the onset of multilayer desorption starts at temperatures above 360 K. The morphological peculiarities caused by the post-deposition diffusion and island formation is nicely demonstrated by tapping-mode AFM measurements which clearly reveal the presence of tall nano-crystalline islands – typical island heights of more than 50 nm were found although only a film of 4 nm was deposited. As a consequence of this dewetting only a rather small part of the surface is covered by islands (cf. also Figure 11.7a) which has to be taken into account when assigning features in spectroscopic data acquired for films with a nominal thickness in the multilayer regime.

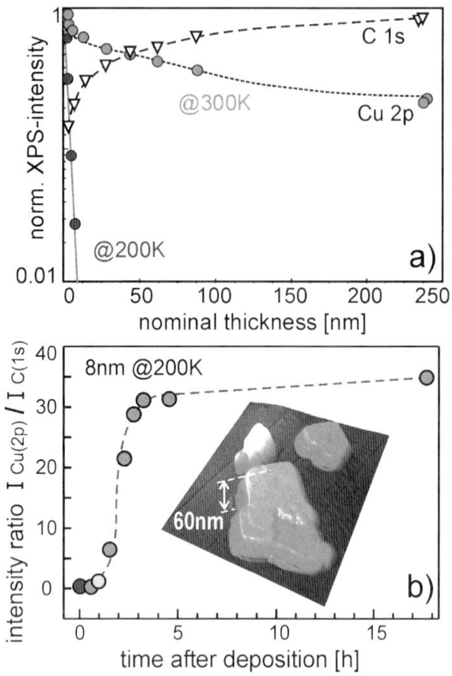

Figure 11.8 (a) Variation of the normalised C 1s and Cu 2p XPS signals as a function of the nominal thickness of perylene films growth on Cu(110) at different temperatures and (b) temporal evolution of the Cu 2p to C 1s intensity ratio for a 8 nm perylene film after deposition at 200 K and raising the temperature to room temperature. The inset shows the typical morphology (AFM tapping mode) of a 4 nm perylene film grown at low temperature and keeping 12 h at room temperature.

When attempting to image multilayer films by STM frequently severe problems are encountered which are nicely illustrated in Figure 11.9 for a thick perylene film. If the conductivity of the organic multilayer becomes too low the STM tip will penetrate into the organic multilayer in order to achieve the pre-determined tunnelling current. In order to do so the feedback control will push the STM-tip inside the molecular material, thus causing indentation, a severe mechanical damage of the organic thin film. When the tip is then scanned (i.e. moved laterally) parallel the surface organic material in multilayers is

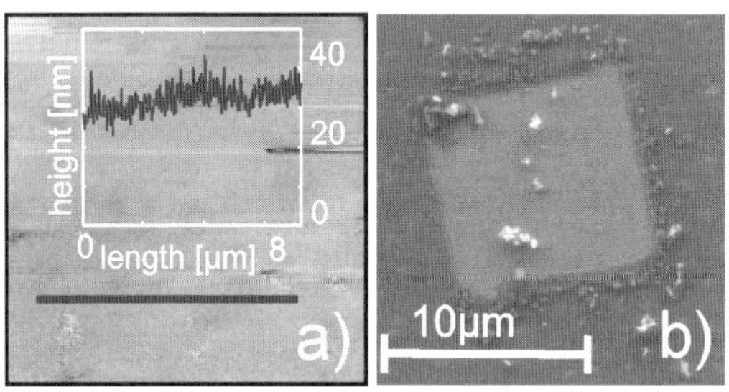

Figure 11.9 Comparison of (a) STM and (b) SEM data recorded for a 100 nm perylene film deposited at low temperatures on Cu(110). While the STM data (10 × 10 μm, $U = -2$ V, $I = 50$ pA) suggest a rather smooth film subsequently recorded SEM data clearly reveal a tip-induced removal of perylene.

pushed sideways by the tip, leaving behind the wetting layer consisting of a single adlayer of molecules bound rather strongly to the metal surface. Eventually, fairly clear images can be obtained in the STM for the wetting layer below, while all other organic material has been removed [61].

These observations clearly reveal that (room temperature) STM is not the method of choice for the investigation of organic multilayers, especially in the presence of islanding, and also explain the discrepancy between our data and the STM-results reported for the growth of thin films of pentacene and perylene on clean metal surfaces [56, 58].

11.6
Growth of Other Molecules of Interest for Organic Electronics on Metal Substrates

A more complete review of all previous experimental work on the structure and orientation of organic adlayers grown by OMBD on solid substrates is beyond the scope of this chapter. The literature up to the year 2004 has been reviewed previously in the paper by Witte and Wöll [30] as well as in the papers by Schreiber [31] and Forrest [32].

In the following we will briefly sketch the most important principles governing organic molecular beam deposition of organic molecules on metallic substrates.

A comparison among the various molecular adlayers shows that the growth studies described above for pentacene and perylene on noble metals are rather typical for planar aromatic molecules which do not exhibit an orientational precursor in their bulk structure. Here, we define an orientational precursor [30] as a plane of the crystalline bulk structure which consists only of non-tilted, densely packed molecules. In the bulk structures of pentacene and perylene and for most other planar aromatic molecules such an "all-planar" plane does not exist due to the typical face-on-edge herring-bone packing motive which implies that any crystalline plane contains molecules with different tilt angles relative to this lattice plane. Since on coinage metals planar organic molecules form close packed "all-planar" monolayer films with an orientation of their aromatic plane parallel to the substrate to maximise the adsorption energy (which is typically dominated by dispersion or *van der Waals* forces [15]) a true epitaxial growth on these metallic supports is not possible.

There are, however, a few planar aromatic molecules with different bulk structures which do show the existence of such orientational precursors. One of these exceptions is PTCDA (3,4,9,10-perylenetetracarboxylic dianhydride) which has been extensively studied in the past and is commonly considered as a prototype system for organic thin film growth [62, 63]. Especially on Ag(111) an epitaxial growth of well-ordered PTCDA films was observed where molecules adopt the same arrangement as in the (102)-plane of the bulk structure of PTCDA [62]. An examination of this (102)-plane reveals that this effect is an orientational precursor, i.e. molecules in this plane are co-planar with the plane. In the initial stage of PTCDA deposition well ordered films with thicknesses of up to several monolayers can be grown. Also in this case, however, a significant dewetting and islanding was observed with increasing film thickness [64] which finally results in a rather rough film morphology. Studies reported later also revealed a substantial post-deposition dewetting and island formation upon annealing of PTCDA films that were grown at room temperature on Ag(111) [65]. In part this has been attributed to a release of strain within the organic films which results from a small but noticeable lattice mismatch between the PTCDA and the metal and a different thermal expansion coefficient of both materials which was also reported for the growth of PTCDA on Au(111) [66].

Comparing the situation for PTCDA with the above discussed cases of perylene and pentacene which exhibit a large diffusion already at room temperature (due to their smaller sublimation enthalpy) thus indicates that the film structure of large molecules such as PTCDA prepared at room temperature may not represent the thermodynamic equilibrium structure. Finally, we note that PTCDA reveals only a rather small charge carrier mobility (for thin films values of less than 3×10^{-2} cm^2/V s were reported [67]), and thus is not well suited for the fabrication of OFETs.

11.7 Growth of Pentacene on Modified Gold Surfaces

Since the results described above for the orientation, structure and morphology of pentacene thin films deposited on metal substrates are incompatible with the goal to grow perfectly oriented (preferably) single crystal adlayers of pentacene on a metal electrode, a different approach was taken. In order to avoid the presence of a wetting layer directly on top of the metal with the molecules all oriented with their aromatic-plane parallel to the metal surface – an orientation, which is incompatible with the bulk structure of pentacene and most other planar organic molecules of interest for organic electronics – the gold substrate has been modified by the adsorption of organothiols. When immersing gold substrates into ethanolic solutions of organothiols, monomolecular films are formed spontaneously which are generally referred to as self-assembled monolayers (SAMs) [68–70]. Several studies have shown that when prepared under proper conditions the structural quality of these SAMs is essentially limited by that of the corresponding gold surface. In the case of gold single crystals, self-assembled monolayers can be prepared which exhibit a very high degree of long range ordering (see e.g. [71]). Since a large variety of differently ω-functionalised organothiols is available for the preparation of SAMs this provides a rather versatile route to fabricate organic surfaces of different chemical terminations exposing e.g. carboxylic acid (COOH), hydroxy (OH), or amino (NH_2) groups [69].

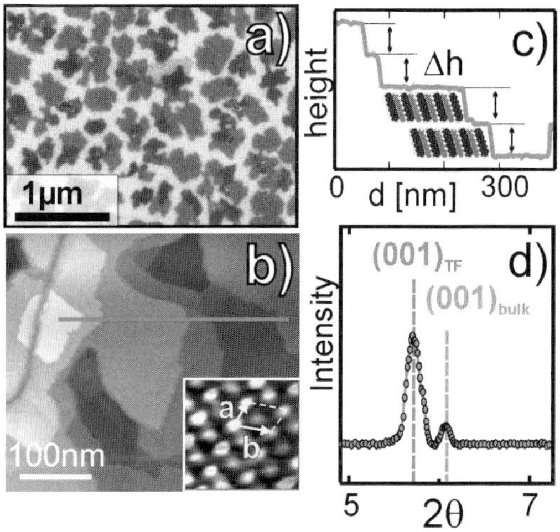

Figure 11.10 Morphology and structure of thin pentacene films on a SAM pre-covered Au(111) substrate: (a) SEM micrograph of a 2 nm film together with corresponding STM data (b), (c) showing a layered arrangement of upright standing molecules. XRD data recorded for a 30 nm film (d) clearly reveal the presence of (001) oriented films revealing the *thin film* phase while at larger thickness the bulk phase is adopted.

In fact a distinctly different growth scenario was observed when pentacene was deposited using OMBD on SAM-covered gold surfaces. Combining STM, SEM and NEXAFS measurements it was possible to characterise the structure and morphology of such films in considerable detail. As shown in Figure 11.10 on SAM covered Au(111) substrates the formation of a wetting layer with a planar orientation is completely suppressed and pentacene grows in a quasi layer-by-layer fashion with an upright orientation of the long molecular axis, leading to the formation of rather smooth films [53]. The same upright orientation of the large molecular axis has also been observed for pentacene growth on SAMs exhibiting a different organic surface terminated by e.g. CH_3, CF_3 or COOH end groups [53]. It is interesting to note, that the same growth mode has also been reported before by Hu et al. for SAM covered polycrystalline gold substrates [72].

Frequently, also the growth of pentacene on SiO_2 surfaces coated with silane-based SAMs have been investigated [73]. Since silane-based SAMs and the organic surfaces exposed by them are inferior to thiol-based SAMs as regards their structural (defects, pin holes) and chemical (presence of monomers on top of the SAM, etc.) properties, a direct comparison is difficult. A thorough discussion of the growth phenomena observed on silane-based SAMs is therefore outside the scope of this chapter and can be found elsewhere [53].

Although it was possible to image the lateral arrangement of the upstanding pentacene molecules on top of the SAM-modified Au substrate by means of STM even in multilayer films by using tunnelling current as low as 15 pA, the presence of image distortions due to thermal drift and/or piezo creep hampered a precise determination of the crystalline phase. Additional XRD measurements demonstrated unambiguously that the thin pentacene films initially adopt the so-called thin film phase [27] while with increasing film thickness a transition towards the bulk Campbell-phase occurs [53]. This scenario thus parallels the situation observed for the growth of pentacene on other non-metallic surfaces such as SiO_2 [74].

We note further, that the reduction of dewetting and islanding enables in case of the SAM-modified Au substrates in particular the preparation of crystalline thin films of well defined thickness on single crystalline metal surfaces which permits the use of tunnelling spectroscopic measurements to study charge transport properties as described in the next section.

The use of modified gold substrates carries a large potential with regard to application in organic electronic devices, since the presence of SAMs not only affects the growth properties in OMBD but also changes and in many cases improves the charge injection properties. Using SAMs of aromatic organothiols such as anthracene-2-thiol [75] as contact primers for electrodes of bottom contact pentacene-based OFETs, significantly improved device characteristics were demonstrated such as e.g. an enhancement of the on/off-ratio by a factor of 10^4 [76]. More recent experiments further demonstrated that this improvement cannot be only related to the improved growth of the OSC at the SAM-treated electrodes since the contact resistance is barely affected while the sheet resistance reveals significant differences [77]. These findings indicate a more complex interplay of nucleation and film growth on top of the electrodes and the gate dielectrics which need to be further explored.

11.8
Realisation of an "Ideal" Diode-like Organic Electronic Device

As pointed out in the introduction the performance of organic electronic devices depends critically on the structural quality of the organic semiconductor (OSC), and for that reason it is highly desirable to realise model devices where the presence of disorder and impurities can be strictly excluded. Such a device would allow to determine true intrinsic properties of organic semi-conductors like the mobility of the two different charge carriers, electrons and holes, as well as to derive conclusions about the conduction mechanism. In this context the results recently published by Friend and coworkers [78] where it has been reported that the typical unipolar behaviour (for pentacene based OFETs usually p-conduction is dominating) changes to an ambipolar behaviour if the presence of hydroxyl groups at the gate dielectric can be strictly excluded. Such OH-species act as electron traps and thus virtually eliminate n-type transport in the regions of the OSC in close proximity to the interface.

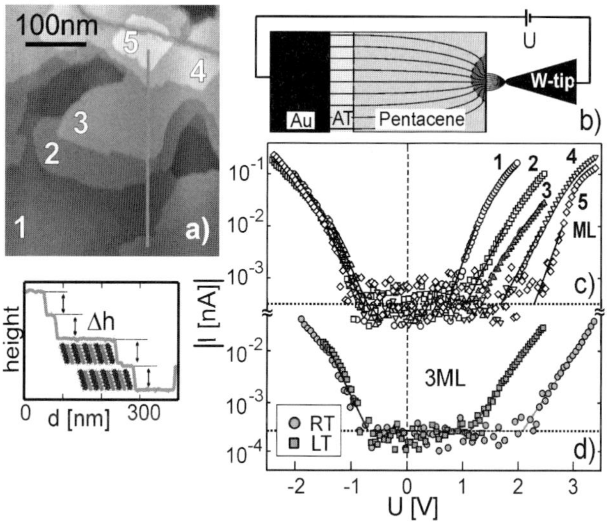

Figure 11.11 (a) STM micrograph showing the layered structure of crystalline pentacene islands grown on a SAM-covered Au(111) surface which enables the recording of tunnelling spectroscopy *I*/*V*-curves as a function of film thickness. A series of typical *I*/*V*-curves recorded at low temperature (LT = 70 K) are displayed in (c) and a comparison of *I*/*V*-curves for a 3ML film at different temperatures in (d), together with a schematic setup of the model diode (b).

Based on the knowledge acquired during the work described above we have built a structurally rather perfect two-terminal model device [79]. By first precoating an Au(111) substrate with an alkanethiol SAM the subsequent OMBD of pentacene results in the formation of thin, (001)-oriented crystalline islands with lateral extensions of several 100 nm. These islands were observed to grow in a quasi layer-by-layer fashion. As demonstrated by XRD this organic thin film contains pentacene islands with the so called "thin film phase" for such crystalline pentacene islands [53]. When using the STM-tip as a second electrode the electrical characteristics for such a diode-like device with two tunnelling contacts (shown schematically in Figure 11.11b) could be determined as a function of the layer thickness. By using an SEM it could be demonstrated that for layer thicknesses up to 5 layers a non-destructive imaging with the STM tip could be obtained. This is an important demonstration since, as discussed above, when using STM to characterise organic thin layers special care needs to be exercised.

A set of typical *I*/*V*-curves recorded at low temperature (70 K) on top of pentacene islands of various thickness are summarised in Figure 11.11c. Interestingly, they are quite asymmetric and reveal different behaviour at negative and positive sample bias which also depends on the film thickness. While for negative sample bias an almost thickness-independent, constant onset voltage of –1 V is observed the onset voltage at positive bias was found to increase with the number of layers. For thicknesses exceeding 5 layers no stable imaging of the pentacene islands could be achieved and as a result no *I*/*V*-curves could be recorded. This surprising asymmetry and the dependence on the layer thickness suggests the presence of space charge effects caused by the rather high charge density in the region of the OSC where charge carriers (electrons or holes) are injected from the STM-tip. In order to test this hypothesis we have additionally determined the temperature dependence of these *I*/*V*-curves. As displayed in Figure 11.11d for a film of 3 monolayers the asymmetry of the *I*/*V*-curves is significantly reduced when lowering the temperature strongly. Since tunnelling is not temperature dependent, this observation indicates that transport properties determine the *I*/*V*-curves. Because at lower temperatures the curves become more symmetric, we conclude that the space charge effects are reduced at low temperatures, pointing towards a band-like transport mechanism [2] within the crystalline pentacene layers.

When attempting to simulate the device characteristics using commercial software packages, where the charge transport through organic semiconductors is described in a self-consistent way we found it impossible to reproduce the experimental asymmetry without assuming the presence of n-type conduction [79].

This is of particular interest since for all pentacene-based thin film transistors using high work function electrodes (such as gold) only p-type transport has been found although pentacene is intrinsically an ambipolar semiconductor. Note, that previously n-type transport has only been observed in connection with low work function electrodes such as calcium [80] (see also Chapter 24 by Benson et al.). In accord with previous findings by Friend and coworkers [78] we propose that we can observe n-conduction in our model device because the presence of any OH-groups at the SAM/pentacene interface can be safely excluded by the use of ultra-clean electrodes.

Although it has not yet been possible to fully analyse the temperature behaviour and, in particular, to determine the mobility from the device characteristics of this "ideal" OSC-device, we believe that such model devices, where structural defects such as domain boundaries and reorientation layers as well as contamination are largely absent will be very important for unravelling the physics governing charge transport in organic materials.

Acknowledgement

This work was supported by the DFG (focus program SPP 1121 OFET). Valuable discussions with the other partners in the frame of this focus program are gratefully acknowledged. We thank D. Käfer and Dr. Z. Wang for a careful reading of the manuscript.

References

1. N. Karl, Charge-Carrier Mobility in Organic Crystals, in: Organic Electronic Materials, Springer Series in Materials Science, edited by R. Farchioni and G. Grosso, Vol. 41 (Berlin, Springer Verlag, 2001), p. 283.
2. N. Karl, Charge carrier transport in organic semiconductors, Syn. Met. **133**, 649–657 (2003).
3. G. Horowitz, Organic field-effect transistors, Adv. Mater. **10**(5), 365–377 (1998).
4. C. D. Dimitrakopoulos and P. R. L. Malenfant, Organic thin film transistors for large area electronics, Adv. Mater. **14**(2), 99–117 (2002).
5. E. Cantatore, T. C. T. Geuns, G. H. Gelinck, E. van Veenendaal, A. F. A. Gruijthuijsen, L. Schrijnemakers, S. Drews, and D. M. de Leeuw, A 13.56-MHz RFID system based on organic transponders, IEEE J. Solid-State Circuits **42**(1), 84–92 (2007).
6. R. G. Kepler, Charge Carrier Production and Mobility in Anthracene Crystals, Phys. Rev. **119**(4), 1226–1229 (1960).
7. O. H. Leblanc, Hole and Electron Drift Mobilities in Anthracene, J. Chem. Phys. **33**(2), 626–626 (1960).
8. R. W. I. de Boer, M. Jochemsen, T. M. Klapwijk, A. F. Morpurgo, J. Niemax, A. K. Tripathi, and J. Pflaum, Space charge limited transport and time of flight measurements in tetracene single crystals: A comparative study, J. Appl. Phys. **95**(3), 1196–1202 (2004).
9. G. Witte, S. Lukas, P. S. Bagus, and C. Wöll, Vacuum level alignment at organic/metal junctions: Cushion effect and the interface dipole, Appl. Phys. Lett. **87**, 263502 (2005).
10. N. Mott, On the contact between a metal and an insulator or semiconductor, Proc. Camb. Philos. Soc. **34**, 568 (1938).
11. W. Schottky, Vereinfachte und erweiterte Theorie der Randschichtgleichrichter, Z. Phys. **118**, 539 (1942).
12. S. Narioka, H. Ishii, D. Yoshimura, M. Sei, Y. Ouchi, K. Seki, S. Hasegawa, T. Miyazaki, Y. Harima, and K. Yamashita, The Electronic-Structure and Energy-Level Alignment of Porphyrin/Metal Interfaces Studied by Ultraviolet Photoelectron-Spectroscopy, Appl. Phys. Lett. **67**(13), 1899–1901 (1995).
13. I. G. Hill, A. Rajagopal, A. Kahn, and Y. Hu, Molecular level alignment at organic semiconductor-metal interfaces, Appl. Phys. Lett. **73**(5), 662–664 (1998).
14. H. Ishii, K. Sugiyama, E. Ito, and K. Seki, Energy level alignment and interfacial electronic structures at organic metal and organic organic interfaces, Adv. Mater. **11**(8), 605 (1999).
15. P. S. Bagus, V. Staemmler, and C. Wöll, Exchange-like effects for closed-shell adsorbates: Interface dipole and work function, Phys. Rev. Lett. **89**(9), 0961041-3 (2002).
16. P. S. Bagus, K. Hermann and C. Wöll, The Interaction of C_6H_6 and C_6H_{12} with Noble Metal Surfaces: Electronic Level Alignment and the Origin of the Interface Dipole, J. Chem. Phys. **123**, 184109 (2005).
17. D. J. Gundlach, Y. Y. Lin, T. N. Jackson, S. F. Nelson, and D. G. Schlom, Pentacene organic thin-film transistors – Molecular ordering and mobility, IEEE Electron Device Lett. **18**(3), 87–89 (1997).
18. H. Klauk, M. Halik, U. Zschieschang, G. Schmid, W. Radlik, and W. Weber, High-mobility polymer gate dielectric pentacene thin film transistors, J. Appl. Phys. **92**(9), 5259–5263 (2002).
19. N. J. Watkins, L. Yan, and Y. L. Gao, Electronic structure symmetry of interfaces between pentacene and metals, Appl. Phys. Lett. **80**(23), 4384–4386 (2002).
20. L. Diao, C. D. Frisbie, D. D. Schroepfer, and P. P. Ruden, Electrical

characterization of metal/pentacene contacts, J. Appl. Phys. **101**(1), 014510 (2007).

21. R. Caputo, B. P. Prascher, V. Staemmler, P. S. Bagus, and C. Wöll, Adsorption of benzene on coinage metals: A theoretical analysis using wavefunction based methods, J. Phys. Chem. A **111**, 12778 (2007).

22. A. C. Durr, F. Schreiber, M. Kelsch, H. D. Carstanjen, and H. Dosch, Morphology and thermal stability of metal contacts on crystalline organic thin films, Adv. Mater. **14**(13/14), 961–963 (2002).

23. V. C. Sundar, J. Zaumseil, V. Podzorov, E. Menard, R. L. Willett, T. Someya, M. E. Gershenson, and J. A. Rogers, Elastomeric transistor stamps: Reversible probing of charge transport in organic crystals, Science **303**(5664), 1644–1646 (2004).

24. T. Muck, V. Wagner, U. Bass, M. Leufgen, J. Geurts, and L. W. Molenkamp, In situ electrical characterization of DH4T field-effect transistors, Synth. Met. **146**(3), 317–320 (2004).

25. D. Käfer and G. Witte, Growth of crystalline rubrene films with enhanced stability, Phys. Chem. Chem. Phys. **7**, 2850–2853 (2005).

26. E. Umbach, K. Glockler, and M. Sokolowski, Surface architecture with large organic molecules: interface order and epitaxy, Surf. Sci. **404**(1–3), 20–31 (1998).

27. S. Schiefer, M. Huth, A. Dobrinevski, and B. Nickel, Determination of the Crystal Structure of Substrate-Induced Pentacene Polymorphs in Fiber Structured Thin Films, J. Am. Chem. Soc. **129**, 10316–10317 (2007).

28. H. Yoshida, I. Katsuhiko, and N. Sato, X-ray diffraction reciprocal space mapping study of the thin film phase of pentacene, Appl. Phys. Lett. **90**, 181930 (2007).

29. P. Ruffieux, O. Groning, M. Bielmann, C. Simpson, K. Müllen, L. Schlapbach, and P. Groning, Supramolecular columns of hexabenzocoronenes on copper and gold(111) surfaces, Phys. Rev. B **66**(7), 073409 (2002).

30. G. Witte and C. Wöll, Growth of aromatic molecules on solid substrates for applications in organic electonics, J. Mater. Res. **19**(7), 1889–1916 (2004).

31. F. Schreiber, Organic molecular beam deposition: Growth studies beyond the first monolayer, phys. stat. sol. (a) **201**(6), 1037–1054 (2004).

32. S. R. Forrest, Ultrathin organic films grown by organic molecular beam deposition and related techniques, Chem. Rev. **97**(6), 1793–1896 (1997).

33. D. Käfer, L. Ruppel, G. Witte, and C. Wöll, Role of molecular conformations in rubrene thin film growth, Phys. Rev. Lett. **95**, 166602 (2005).

34. M. C. Blum, E. Cavar, M. Pivetta, F. Patthey, and W. D. Schneider, Conservation of chirality in a hierarchical supramolecular self-assembled structure with pentagonal symmetry, Angew. Chem., Int. Ed. **44**(33), 5334–5337 (2005).

35. S. Kowarik, A. Gerlach, S. Sellner, F. Schreiber, J. Pflaum, L. Cavalcanti, and O. Konovalov, Anomalous roughness evolution of rubrene thin films observed in real time during growth, Phys. Chem. Chem. Phys. **8**(15), 1834–1836 (2006).

36. K. F. Braun and S. W. Hla, Probing the conformation of physisorbed molecules at the atomic scale using STM manipulation, Nano Lett. **5**(1), 73–76 (2005).

37. N. Koch, G. Heimel, J. S. Wu, E. Zojer, R. L. Johnson, J. L. Bredas, K. Müllen, and J. P. Rabe, Influence of molecular conformation on organic/metal interface energetics, Chem. Phys. Lett. **413**(4–6) 390–395 (2005).

38. S. Müllegger, K. Hänel, T. Strunskus, C. Wöll, and A. Winkler, Organic molecular beam deposition of oligophenyls on Au(111): A study by X-ray absorption spectroscopy, Chem. Phys. Chem. **7**, 2552–2558 (2006).

39. G. Witte, K. Hänel, C. Busse, A. Birkner, and C. Wöll, Molecules coining patterns into a metal: the hard core of soft matter, Chem. Mater. **19**, 4228 (2007).

40. S. Söhnchen, K. Hänel, A. Birkner, G. Witte, and C. Wöll, Molecular Beam Deposition of Perylene on Copper: Formation of Ordered Phases, Chem. Mater. **17**, 5297–5304 (2005).

41. Q. Chen and N. V. Richardson, Surface facetting induced by adsorbates, Prog. Surf. Sci. **73**(4–8), 59–77 (2003).

42. P. Guaino, D. Carty, G. Hughes, O. McDonald, and A. A. Cafolla, Long-range order in a multilayer organic film templated by a molecular-induced surface reconstruction: Pentacene on Au(110), Appl. Phys. Lett. **85**(14), 2777–2779 (2004).

43. F. Evangelista, A. Ruocco, D. Pasca, C. Baldacchini, M. G. Betti, V. Corradini, and C. Mariani, Au(110) induced reconstruction by π conjugated molecules adsorption investigated by photoemission spectroscopy and low energy electron diffraction, Surf. Sci. **566**, 79–83 (2004).

44. R. Hoffmann, Solids and Surfaces: A Chemist's View of Bonding in Extended Structures (VCH Publishers Inc., 1988).

45. D. Käfer and G. Witte, Evolution of pentacene films on Ag(111): Growth beyond the first monolayer, Chem. Phys. Lett. **442**, 376–383 (2007).

46. A. Söhnchen, S. Lukas, and G. Witte, Epitaxial growth of pentacene films on Cu(110), J. Chem. Phys. **121**(1), 525–534 (2004).

47. J. T. Sadowski, T. Nagao, S. Yaginuma, Y. Fujikawa, A. Al-Mahboob, K. Nakajima, T. Sakurai, G. E. Thayer, and R. M. Tromp, Thin bismuth film as a template for pentacene growth, Appl. Phys. Lett. **86**(7), 073109 (2005).

48. R. Ruiz, D. Choudhary, B. Nickel, T. Toccoli, K. C. Chang, A. C. Mayer, P. Clancy, J. M. Blakely, R. L. Headrick, S. Iannotta, and G. G. Malliaras, Pentacene thin film growth, Chem. Mater. **16**(23), 4497–4508 (2004).

49. P. G. Schroeder, C. B. France, J. B. Park, and B. A. Parkinson, Energy level alignment and two-dimensional structure of pentacene on Au(111) surfaces, J. Appl. Phys. **91**(5), 3010–3014 (2002).

50. M. Eremtchenko, R. Temirov, D. Bauer, J. A. Schaefer, and F. S. Tautz, Formation of molecular order on a disordered interface layer: Pentacene/Ag(111), Phys. Rev. B **72**(11), 115430 (2005).

51. O. McDonald, A. A. Cafolla, Z. S. Li, and G. Hughes, Synchrotron photoemission studies of pentacene films on Cu(110), Surf. Sci. **600**(9), 909–1916 (2006).

52. S. Lukas, G. Witte, and C. Wöll, Novel mechanism for molecular self-assembly on metal substrates: Unidirectional rows of pentacene on Cu(110) produced by a substrate-mediated repulsion, Phys. Rev. Lett. **88**(2), 028301 (2002).

53. D. Käfer, L. Ruppel, and G. Witte, Growth of pentacene on clean and modified gold surfaces, Phys. Rev. B **75**, 085309 (2007).

54. R. G. Della Valle, A. Brillante, E. Venuti, L. Farina, A. Girlando, and M. Masino, Exploring the polymorphism of crystalline pentacene, Org. Electron. **5**(1–3), 1–6 (2004).

55. T. Siegrist, C. Kloc, J. H. Schön, B. Batlogg, R. C. Haddon, S. Berg, and G. A. Thomas, Angew. Chem., Int. Ed. **40**, 1732 (2001).

56. J. H. Kang and X. Y. Zhu, π-stacked pentacene thin films grown on Au(111), Appl. Phys. Lett. **82**(19), 3248–3250 (2003).

57. G. Beernink, T. Strunskus, G. Witte, and C. Wöll, Importance of dewetting in organic molecular beam deposition:

pentacene on gold, Appl. Phys. Lett. **85**(3), 398–400 (2004).

58. Q. Chen, T. Rada, A. McDowall, and N. V. Richardson, Epitaxial growth of a crystalline organic semiconductor: Perylene/Cu(110), Chem. Mater. **14**(2), 743–749 (2002).
59. K. Hänel, S. Söhnchen, S. Lukas, G. Beernink, A. Birkner, T. Strunskus, G. Witte, and C. Wöll, Organic molecular beam-deposition of perylene on Cu(110) Results from near-edge x-ray absorption spectroscopy, X-ray photoelectron spectroscopy, and atomic force microscopy, J. Mater. Res. **19**(7), 2049–2056 (2004).
60. L. Gao, J. T. Sun, Z. H. Cheng, Z. T. Deng, X. Lin, S. X. Du, and H. J. Gao, Structural evolution at the initial growth stage of perylene on Au(111), Surf. Sci. **601**(15), 3179–3185 (2007).
61. G. Witte, K. Hänel, S. Söhnchen, and C. Wöll, Growth and morphology of thin films of aromatic molecules on metals: the case of perylene, Appl. Phys. A **82**, 447–455 (2006).
62. E. Umbach, M. Sokolowski, and R. Fink, Substrate-interaction, long-range order, and epitaxy of large organic adsorbates, Appl. Phys. A, Mater. Sci. Process. **63**(6), 565–576 (1996).
63. Y. Hirose, S. R. Forrest, and A. Kahn, Quasiepitaxial Growth of the Organic Molecular Semiconductor 3,4,9,10-Perylenetetracarboxylic Dianhydride, Phys. Rev. B **52**(19), 14040–14047 (1995).
64. B. Krause, A. C. Durr, K. Ritley, F. Schreiber, H. Dosch, and D. Smilgies, Structure and growth morphology of an archetypal system for organic epitaxy: PTCDA on Ag(111), Phys. Rev. B **66**(23), 23404 (2002).
65. B. Krause, A. C. Durr, F. Schreiber, H. Dosch, and O. H. Seeck, Thermal stability and partial dewetting of crystalline organic thin films: 3,4,9,10-perylenetetracarboxylic dianhydride on Ag(111), J. Chem. Phys. **119**(6), 3429–3435 (2003).
66. P. Fenter, F. Schreiber, L. Zhou, P. Eisenberger, and S. R. Forrest, In situ studies of morphology, strain, and growth modes of a molecular organic thin film, Phys. Rev. B **56**(6), 3046–3053 (1997).
67. N. Karl, K. H. Kraft, J. Marktanner, M. Munch, F. Schatz, R. Stehle, and H. M. Uhde, Fast electronic transport in organic molecular solids?, J. Vac. Sci. Technol. A **17**(4), 2318–2328 (1999).
68. A. Ulman, Formation and structure of self-assembled monolayers, Chem. Rev. **96**(4), 1533–1554 (1996).
69. F. Schreiber, Structure and growth of self-assembling monolayers, Prog. Surf. Sci. **65**(5–8), 151–256 (2000).
70. J. C. Love, L. A. Estroff, J. K. Kriebel, R. G. Nuzzo, and G. M. Whitesides, Self-assembled monolayers of thiolates on metals as a form of nanotechnology, Chem. Rev. **105**(4), 1103–1169 (2005).
71. P. Cyganik, M. Buck, T. Strunskus, A. Shaporenko, J. Wilton-Ely, M. Zharnikov, and C. Wöll, Competition as a design concept: Polymorphism in self-assembled monolayers of biphenyl-based thiols, J. Am. Chem. Soc. **128**(42), 13868–13878 (2006).
72. W. S. Hu, Y. T. Tao, Y. J. Hsu, D. H. Wei, and Y. S. Wu, Molecular orientation of evaporated pentacene films on gold: Alignment effect of self-assembled monolayer, Langmuir **21**(6), 2260–2266 (2005).
73. T. W. Kelley, L. D. Boardman, T. D. Dunbar, D. V. Muyres, M. J. Pellerite, and T. P. Smith, High-performance OTFTs using surface-modified alumina dielectrics, J. Phys. Chem. B **107**, 5877 (2003).
74. I. P. M. Bouchoms, W. A. Schoonveld, J. Vrijmoeth, and T. M. Klapwijk, Morphology identification of the thin film phases of vacuum evaporated pen-

tacene on SiO_2 substrates, Synth. Met. **104**(3), 175–178 (1999).

75. D. Käfer, G. Witte, P. Cyganik, A. Terfort, and C. Wöll, A comprehensive study of self-assembled monolayers of anthracenethiol on gold: Solvent effects, structure, and stability, J. Am. Chem. Soc. **128**(5), 1723–1732 (2006).

76. C. Bock, D. V. Pham, U. Kunze, D. Käfer, G. Witte, and C. Wöll, Improved morphology and charge carrier injection in pentacene field-effect transistors with thiol-treated electrodes, J. Appl. Phys. **100**, 114517 (2006).

77. C. Bock, D. V. Pham, U. Kunze, D. Kaefer, G. Witte, C. Wöll, and A. Terfort, Influence of anthracene-2-thiol treatment on the device parameters of pentacene bottom-contact transistors, Appl. Phys. Lett. **91**, 052110 (2007).

78. L. L. Chua, J. Zaumseil, J. F. Chang, E. C. W. Ou, P. K. H. Ho, H. Sirringhaus, and R. H. Friend, General observation of n-type field-effect behaviour in organic semiconductors, Nature **434**(7030), 194–199 (2005).

79. L. Ruppel, A. Birkner, G. Witte, C. Busse, T. Lindner, G. Paasch, and C. Wöll, A defect-free thin film pentacene diode: Interplay between transport and scanning tunneling microscope tip tunneling injection, J. Appl. Phys. **102**, 033708 (2007).

80. R. Schmechel, M. Ahles, and H. von Seggern, A pentacene ambipolar transistor: Experiment and theory, J. Appl. Phys. **98**(8), 103702 (2005).

12
Fundamental Interface Properties in OFETs: Bonding, Structure and Function of Molecular Adsorbate Layers on Solid Surfaces

S. Soubatch, R. Temirov, and F. S. Tautz

12.1
Introduction

Organic electronics, i.e. electronics on the basis of organic semiconductors, is generally believed to be a key technology for the future of electronics as a whole. The reason is twofold. On the one hand, organic semiconductors open up new low-cost application areas for electronics (e.g. printed electronics, electronic paper, lighting applications). On the other hand, organic electronics are particularly amenable to miniaturisation, the grand challenge of all electronics: Making organic electronic devices smaller and smaller, one naturally arrives at the concept of molecular electronics, where the full functionality of an electronic device such as a transistor is embedded in the chemical structure of a single molecule. Following this idea, one would first create an essentially unlimited number of such elementary "device molecules" by conventional chemical synthesis in the test tube and then use bottom-up structuring techniques (such as self-assembly) to create large-scale architectures for computation. Of course, up to now this concept is a mere vision which may or may not become true in the distant future. However, the electrical properties of single molecules are currently a topic of active research and, quite apart from any potential applications, a lot of interesting physics is discovered through these efforts.

As opposed to single molecule devices, organic field effect transistors (OFETs) are a reality today; a few applications are already close to being marketed commercially. A schematic diagram of an OFET is shown in Figure 12.1. It resembles an ordinary field effect transistor, with the exception that the active layer is made from an *organic* semiconductor. In some cases also the other components, such as electric contacts and the dielectric, are made from organics. However, this is not necessary and many of the devices made for the purposes of research contain conventional metals and inorganic insulators.

Figure 12.2 displays some of the molecules which may be used in OFETs, e.g. Refs. [1–5]. All of them are polycyclic aromates. Indeed, the extended systems of π-electrons bear the semiconducting functionality of these mole-

Organic Electronics. Structural and Electronic Properties of OFETs. Edited by Christof Wöll

ISBN: 978-3-527-40810-8

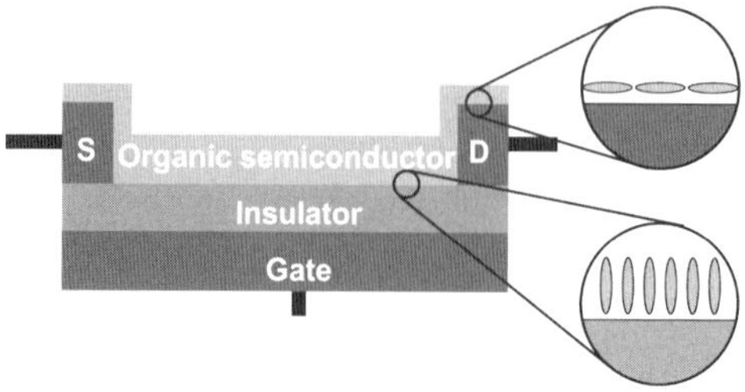

Figure 12.1 Schematic of an organic field effect transistor with its relevant interfaces. The molecules are shown in green. (see Colour plates, p. LIX)

cules. But also their chemical and optical properties are closely coupled to the π-electrons, as we will see below.

Looking at the scheme of an OFET, one immediately notices that interfaces are the elements which bestow the functionality to the device. At the *metal-organic interface*, charge injection from the source and drain contacts into the transistor channel takes place. The orientation and electronic structure of molecules in direct binding contact to the electrodes is therefore crucial for the quality of the device [6]. At the *interface between the semiconductor and the dielectric*, the injected charge accumulates and is transported when a source-drain voltage is applied. Because it is the first monomolecular layer which bears the brunt of charge transport [7, 8], molecular orientation [9, 10, 11], structural order [6, 12, 13] and electronic defect structure [14, 15, 16] in the interface region determine the overall performance of the device.

It is therefore clear that the physics and chemistry at OFET interfaces must be understood before device properties can be optimised efficiently. Two approaches are conceivable. Firstly, one can investigate technical interfaces in real devices, e.g. the interface of polycrystalline gold contacts with a textured

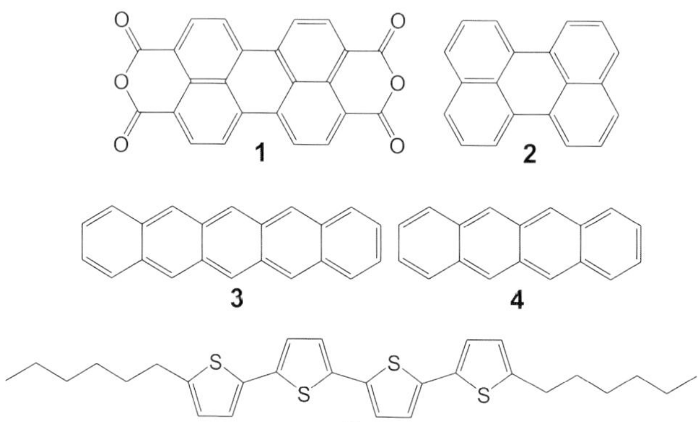

Figure 12.2 Semiconductor molecules discussed in this paper. 1 perylene-3,4,9,10-tetracarboxylic-dianhydride (PTCDA), 2 perylene, 3 pentacene, 4 tetracene, 5 α,ω-dihexyl-quarterthiophene (DH4T).

thin film of an organic semiconductor. While this is an important task for the device engineer, it is not the approach taken in the present work. Rather, we employ a methodology based on highly ordered, crystalline model interfaces. Usually these are prepared by organic molecular beam deposition (OMBD) which allows ultimate control of the deposition process. The extremely thin adsorbate layers (often only a molecular monolayer thick) and the large degree of order obtained by OMBD allow the deployment of the powerful armoury of highly sensitive surface analytical methods, with the result that even minute details of bonding, structure and function can be investigated microscopically, spectroscopically and by diffraction methods. Of course, transferring the results of these model experiments back to real devices must be done with care, but the insight into fundamental properties and processes gained in this way is indispensable for the field. As an intermediate approach it is also fruitful to create single crystalline model devices which can serve as a benchmark for more realistic OFETs [6, 17].

In the present chapter we report on *bonding* and *structure* of organic adsorbate layers on crystalline metals and insulators and their *function*, following the model-system philosophy outlined above. Evidently, the substrate-bonding and the bonding between neighbouring molecules both influence the structure of the interface. Together, bonding and structure determine the electronic properties of the interface (charge transfers, energy level alignments, and band dispersion, if applicable). These in turn govern interface functionality, most notably charge carrier injection at the contacts and the mobility in the transistor channel along the interface to the gate dielectric.

We have placed the focus on metal/organic interfaces, because contacts remain one of the largest challenges not only in organic but also in molecular electronics. Although bonding, structure and function are interrelated, the chapter is divided into three sections, each of which emphasises one of these aspects separately. In Section 12.2 (bonding) we will see that in many cases the bond between polycyclic aromates and metal surfaces may contain a chemical component in which the π-electrons of the molecule are implicated. Section 12.3 (structure) focuses on the role of intermolecular interactions in the formation of structures. Especially for molecules without additional functional groups this interaction may be of similar size as the substrate interaction, which can lead to the evolution of rather complex interface structures and/or phase diagrams.

In Sections 12.2 and 12.3 which deal with bonding and structure, interface properties are first discussed on a fundamental level, before the results are evaluated in the context of OFETs, in a section entitled "what can be learned for OFETs". Because we do not present data on actual devices in this chapter, this discussion must remain rather general. However, it is our firm conviction that the understanding of fundamental interface properties of the type discussed here will eventually become beneficial for device engineering, too.

Most of our work is concerned with interface microscopy and spectroscopy (Sections 12.2 and 12.3), but at the end of the paper (Section 12.4) we also report charge transport experiments. In line with our general philosophy to measure under circumstances of optimum control, these experiments are carried out on a single molecule. We will show that at the single-molecule level the contact chemistry can have a major influence on transport properties, essentially switching transport from the single-particle to the many-body regime.

12.2 Bonding

To discuss the bonding of large π-conjugated molecules on metal surfaces, we turn to the molecule 3,4,9,10-perylenetetracarboxylic dianhydride (PTCDA) (cf. **1** in Figure 12.2). It is a dye molecule of deep red colour. Its thermal stability makes it easy to purify by sublimation and very well-suited for OMBD. As opposed to e.g. pentacene (**3** in Figure 12.2) and tetracene (**4** in Figure 12.2), the bulk crystal structure of PTCDA has a lattice plane (the [102] plane) in which the molecules are nearly co-planar. This layered crystal structure of PTCDA has led to the original idea of growing bulk PTCDA epitaxially on metal substrates (thereby obtaining an optimum interface) [18–22], because a flat orientation of π-conjugated molecules on metals is often observed.

In addition to its π-electron system, PTCDA possesses a second functionality, namely the carboxylic oxygen atoms in the four corners of the molecule. Their existence creates a quadrupole moment in the molecule, allows hydrogen bonding between molecules and offers an additional interaction channel with any substrate. The interplay of all these factors makes PTCDA an interesting model molecule in the context of molecular adsorbate layers. As a consequence, PTCDA is still a "fruit fly", although it is not used any more to fabricate electronic devices such as OFETs, in spite of early reports of an organic heterojunction device based on PTCDA [23, 24].

Most surface science work on PTCDA has been carried out on noble metal surfaces. Generally, the decreasing order of reactivity from Cu to Au is also observed for PTCDA adsorption. Extreme cases are Cu(100), where dissociative adsorption of PTCDA has been reported [25], and Au(111), where the interaction is purely physisorptive [26], as we will see below. The possibility of organic epitaxy has been investigated on a surface of intermediate reactivity, namely Ag(111), where PTCDA forms a commensurate monolayer [18, 27]. Unfortunately, it has turned out that epitaxial growth does not extend beyond 2nd monolayer [28], in spite of relatively low misfit of 2% between Ag(111) and the (102) bulk plane of PTCDA. Nevertheless, the PTCDA/Ag(111) adsorption system has revealed a lot of interesting physics, including a strongly non-adiabatic coupling between molecular vibrations and metal electrons [29] as well as a two-dimensional interface state with an anomalously large dispersion [30]. Very recently, PTCDA has been shown to be well-suited for single molecule transport experiments [31], because of its special bonding to Ag(111). In the following discussion of PTCDA bonding we mainly focus on our own work. In a recent review article [32], a more complete account of the field is given, including the work of other authors.

Let us begin our discussion with an image of the molecule as it appears in scanning tunnelling microscopy on the Ag(111) surface (Figure 12.3). The extreme resolution of this image is only achieved if a functionalised tip is employed [27].

A clean metal tip produces the same kind of pattern, but because of the lower resolution the two outer lobes at the long edges of the molecules merge into one, and the inner lobes are barely distinguishable. The STM imaging process of PTCDA on Ag(111) has been studied in detail in Refs. [27, 33]. Irrespective of the tip state, the images which are observed from PTCDA for small positive and negative biases clearly resemble the charge distribution in the lowest unoccupied molecular orbital (LUMO) of the free PTCDA molecule, as a simple quantum chemical calculation shows [34].

The extremely high resolution of the images has allowed us to determine the actual adsorption site of the molecule [27]. The existence of a definite adsorption site must not be taken for granted, because one may rightly ask what makes a large molecule which averages over many periods of the substrate corrugation potential adsorb site-specifically, especially if the molecule-substrate interaction is, as we will see below, connected to the extended π-electron system of the molecule. On Au surfaces, e.g., PTCDA and many other molecules form incommensurate overlayers, which even overgrow the herringbone reconstruction of the Au(111) surface without being influenced by the complex substrate structure. But on Ag(111) the interaction is stronger than on Au(111) and a definite adsorption site ensues. It turns out that molecules of type A and B both adsorb in a bridge sites. Their different contrast in the image (cf. Ref. [27]) is therefore not related to distinct sites, but to unlike in-plane orientations of the molecules relative to the high symmetry directions of the substrate and possibly to different intermolecular interactions (see below) [27, 33]. The bridge site is also predicted by density functional calculations, although the calculated binding energies are too large, because of general problems with DFT on weakly interacting systems [33]. Experimentally, the ad-

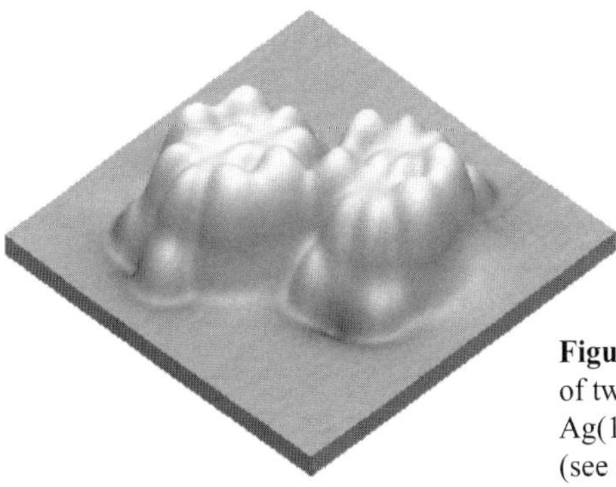

Figure 12.3 Submolecular STM contrast of two PTCDA molecules adsorbed on the Ag(111) surface.
(see Colour plates, p. LX)

sorption energy per molecule is not known, because the molecule does not desorb intact from the surface [36], such that thermal desorption spectroscopy is not available. Differential micro-calorimetry experiments would be required to measure the adsorption energies.

Coming back to Figure 12.3, it is remarkable that an originally empty orbital is imaged at negative bias voltages. This can only result from a substantial charge transfer into the molecule. The STS spectra (red and blue) of Figure 12.4 are dominated by three features and indeed confirm this charge transfer, in the sense that a state appears just below the Fermi level at –0.3 eV binding energy. Judging from its appearance in STM and STS images (Figure 12.3), this state must be the former LUMO of the molecule. We also note that this state is much broader than the peak at –1.7 eV, which by spectroscopic imaging is revealed as the former HOMO [27]. The structure at 0.7 eV will be discussed below.

Both the charge transfer into the LUMO and its broadening are indicative of chemical bonding, as the scheme in Figure 12.5 indicates. Since for any chemical bond the bond length is a valid parameter to characterise the bond strength, we have measured the vertical distance of PTCDA above the Ag(111) surface by the NIXSW technique [37, 38]. Additionally, the technique is in principle able to measure the internal distortion of the molecule. The following is found. Firstly, the bonding distance of PTCDA to Ag(111) is substantially shorter (2.86 Å) than the interlayer distance in bulk PTCDA crystals (3.22 Å), which is commonly interpreted as a van der Waals bonding distance. Secondly, the molecule is distorted, with the carboxylic oxygen atoms 0.18 Å closer to the silver than the average carbon. It thus seems as if both the π-electron system

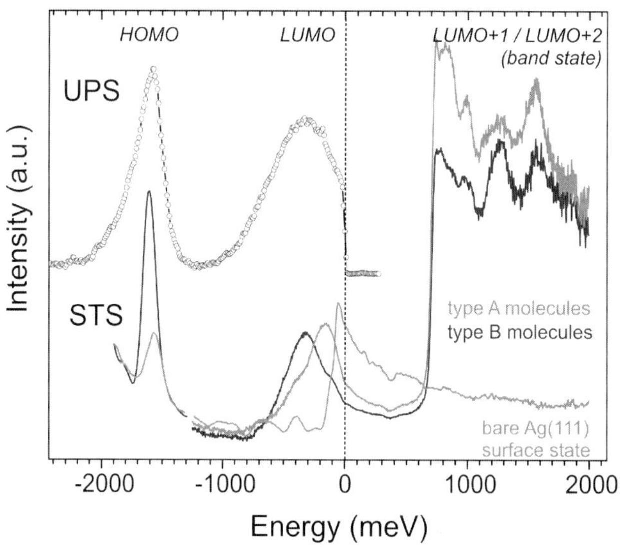

Figure 12.4 UPS and STS spectra of PTCDA/Ag(111) [27, 30, 35]. (see Colour plates p. LX)

80
46

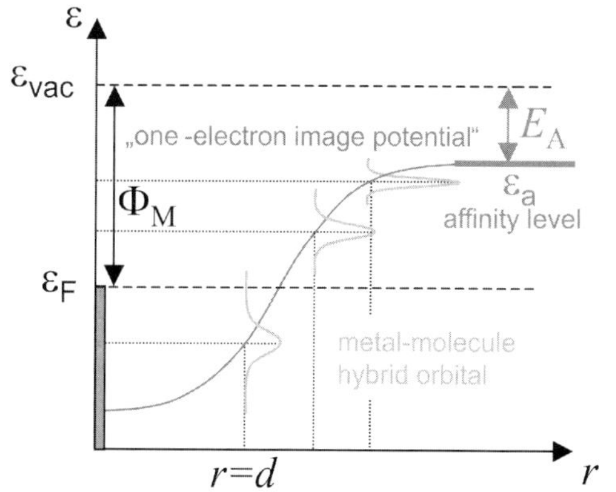

Figure 12.5 Schematic representation of the chemisorption process of an electron-accepting molecule. *r* is the distance coordinate between molecule and surface, *d* the equilibrium, after [42, 43].
(see Colour plates p. LXI)

system and the carboxylic oxygen atoms are involved in the bonding to the substrate. Note that the lowered oxygen atoms are just one aspect of the molecular distortion, which is shown fully in Figure 12.6.

To see whether these two observations really support the scenario of a chemical interaction, we have repeated the NIXSW experiment on Au(111), where all spectroscopic data reveal an essentially physisorptive bonding. And indeed, properly accounting for the reconstruction and relaxation of the Au(111) surface, we find a much larger bonding distance of 3.27 Å [26], close to the van der Waals contact distance between carbon and gold. For technical reasons, the position of the oxygen atoms could not be measured on Au(111). We note in passing here that the larger contact distance and electronic structure on Au(111) agree well with each other, because it is found by STS on PTCDA on Au(111) that the Fermi level is located in the gap between HOMO and LUMO of the molecule [39–41].

In the light of the PTCDA/Au(111) contact distance, the PTCDA/Ag(111) contact really looks like a chemisorptive one. To analyse the bonding more closely, the results of density functional calculations, of which several are

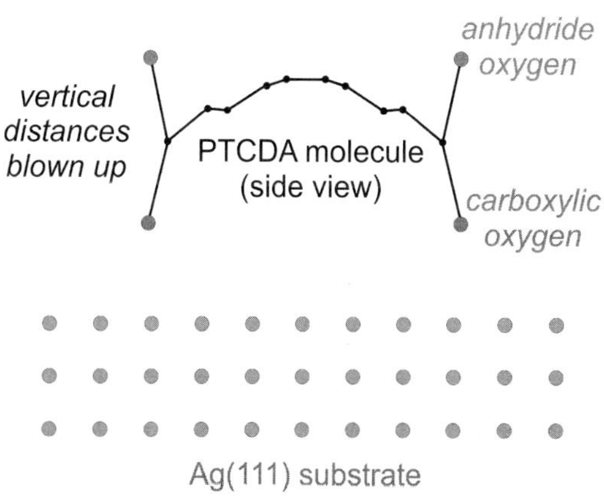

Figure 12.6 Schematic representation of the distortion of PTCDA on Ag(111). The structure has been calculated by density functional theory in the local density approximation [33].
(see Colour plates p. LXI)

available in the literature [37, 38, 44–47], are most welcome. Unfortunately, their results are in disagreement with each other and partly also with experiment [33]. Some calculations do not reproduce the charge transfer which experiments unambiguously prove, while others do not find the molecular distortion. The calculated results depend strongly on the employed density functional. In state-of-the-art generalised gradient approximations the bonding distance comes out far too large (3.22 to 3.4 Å) [37], while LDA gives a distance which is slightly too short [33]. Nevertheless, LDA still yields the best overall agreement with experiments [27, 33].

Why do the calculations get the physics of the PTCDA/Ag(111) interface wrong? It is generally believed that at the interface of large aromatic molecules with metal surfaces the van der Waals interaction plays a significant role [22], but this many-body interaction is not systematically included in the exchange correlation functionals employed widely today. Quantities which scale with the total interaction strength, such as the adsorption energy or the adsorption distance, suffer from this systematic error. On the other hand, we have seen that the PTCDA/Ag(111) interaction also contains a chemical contribution which should be well-described by density functional theory.

Since the van der Waals interaction is usually unspecific, any relative comparison, e.g. between adsorption energies at two different adsorption sites, should be determined by the chemical interaction and thus more or less reliably predicted by DFT. In this way it can be justified that we have used the results of DFT to confirm the experimentally found bridge site adsorption. Indeed, both LDA and GGA predict the bridge site to be the lowest energy adsorption site, but the adsorption energies themselves are vastly different (E_b (LDA) = –5.99 eV/cell; E_b (GGA) = 150 meV/cell) [33]. Other "relative properties" which can be rationalised on the basis of DFT are the site-specific electronic structure or the distortion of the molecule (i.e. relative molecule-substrate interaction strength between different parts of the molecule). Moreover, STM images have been shown to be reliably reproduced by a Tersoff-Haman approach based on DFT-LDA calculations [33].

On the basis of experiment and calculations, we can draw up the following picture of the PTCDA/Ag(111) interaction (Figure 12.7). The molecule interacts with the Ag surface via its two functionalities. (1) The π-system hybridises with states from the Ag 5s band and accepts charge from the metal. This is an extended interaction. (2) The carboxylic oxygen atoms form local bonds

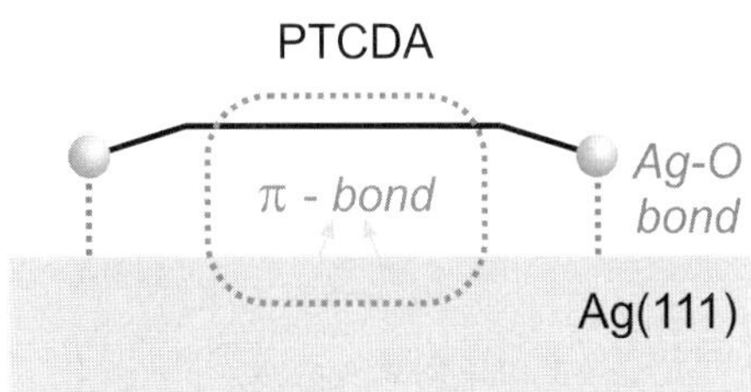

Figure 12.7 Schematic representation of the bonding interaction of PTCDA with Ag(111) [27, 33]. (see Colour plates p. LXI)

which certainly contain an electrostatic but possibly also a weak covalent component. At first sight it may seem as if these two interaction channels are independent of each other. However, if the influence of both interactions on the carbon skeleton is analysed, one finds them to be synergetic in the sense that they contract and expand the same C–C bonds throughout the molecule [33].

Concerning the extended interaction, two questions arise. Firstly, is this interaction a bonding one? One could argue that the larger distance of the central part of the molecule results from a repulsive interaction. However, since the behaviour of the former LUMO resembles the (bonding) Newns-Anderson scheme in Figure 12.5 quite closely and since the DFT-LDA calculation finds a build-up of charge between the molecule and the silver surface [33], the interaction seems to be bonding rather than anti-bonding. In a simple model of an anti-bonding interaction, one would expect a wave function (and thus charge density) node between the two partners. This is definitely not observed in the calculation, which – in spite of its problems – models the experimentally observed spectroscopic signature of the interface quite well. But it is nevertheless true that distinct optimum distances in both bonding channels require a compromise with respect to the vertical distances, with the result that the distance of the central molecule may well be shorter than its optimum while the opposite is true for the oxygen atoms. In this sense, the oxygen atoms may actually pull the carbon skeleton into a repulsive part of the interaction potential.

The second question regarding the extended interaction is whether it can be responsible for a site-specific interaction? In principle, the answer is yes, because the LUMO consists of lobes of a lateral size which is comparable to the characteristic length scale of the atomic corrugation potential. Indeed, in our DFT analysis [33] we find that the accumulated charge between molecule and metal (the "bond") does not have the same lateral distribution of the LUMO but exhibits contributions of the LUMO *and* the silver surface [33]. There is no reason why such a structured bond should not be site specific. We can thus conclude that it is not necessarily the local oxygen bonds alone which are responsible for site specificity.

Finally, we stress the twofold role of the oxygen atoms for the substrate interaction. On the one hand, direct bonds with the substrate are formed. On the other hand, the electronegative anhydride groups increase the electron affinity of the π-electron system towards metal electrons, as the comparison with the perylene molecule (**2** in Figure 12.2) shows: perylene interacts much more weakly with Ag(111) and at room temperature forms a orientationally disordered layer [48]. A third role of the oxygen atoms, namely their participation in intermolecular interactions, will be discussed briefly in Section 12.3.

12.2.1
Bonding: What can be Learned for OFETs?

As pointed out above, metal-organic interfaces are important because charge injection takes place across them. In any OFET, one therefore tries to optimise their properties for injection. Charge transport in organic semiconductors is highly anisotropic. Good charge carrier mobilities are only achieved in the π-stacking direction. For good injection properties, it is therefore also desirable that molecular π-systems overlap with the metal. Fortunately, this is usually the case. On many metal surfaces aromatic molecules adsorb flat [6, 16]. But flat orientation cannot be taken for granted on all surfaces. We will see below that for non-metallic surfaces upright adsorption is often more stable, and even for a metal we have found an example where the interaction with the substrate is so weak that under the influence of the interaction with the second layer molecules in the first layer tilt out of the surface plane [49].

What is the origin of flat adsorption on metals? PTCDA adsorbs flat on silver surfaces because the π-system – its reactivity being enhanced by the presence of anhydride functional groups – dominates the chemistry between molecule and substrate. The result that π-systems can interact chemically with metal surfaces is in fact one of the most important outcomes of our study of PTCDA on Ag(111). But other examples (e.g. PTCDA/Au(111)) show that the flat orientation which is commonly observed on metals is not always the consequence of a chemical interaction; it can also result from van der Waals interactions with metal surfaces, since the latter usually have a large electronic polarisability. At any rate, on most metals organic semiconductor molecules lie flat, and this is certainly helpful if the molecules are deposited onto the metal, as in OFETs with bottom contact geometry. If the organic layer is metallised (top contacts), the structural evolution at the interface is more complicated, because intermixing of organic material and metal usually occurs [50, 51].

From the point of view of fundamental quantum transport across interfaces, any chemical interaction (in the sense of wave function overlap) at the source/drain contacts should aid the charge carrier injection. But whether holes or electrons can be transported efficiently across the interface depends also on the exact energy level alignment. Extensive studies of energy level alignments in OFET structures have been carried out (e.g. [6, 52–54]). An interesting outcome of our work in this context is the finding that in special cases the metallicity of the surface can even extend into the molecule, as observed for PTCDA on Ag(111). Furthermore, the energy level alignment may even generate many-body signatures in the transport across the interface, as we will see in Section 12.3. Such effects, however, will be more important for the metal-molecule contacts in single-molecule devices than in thin film organic electronics.

Comparison with other flat-lying molecules shows that the influence of the metal on the electronic structure of the molecule is not always as strong as for PTCDA on Ag(111). For example, in the case of tetracene on Ag(111) (Figure 12.8) we again observe both a substantial broadening of the LUMO and a shift to lower energies, but no population, i.e. the LUMO remains above the Fermi level. Apparently the lack of positive polarisation in the molecule reduces the tendency to attract electrons from the metal. However, binding energy shifts and level broadenings (which seem to be a general occurrence) are

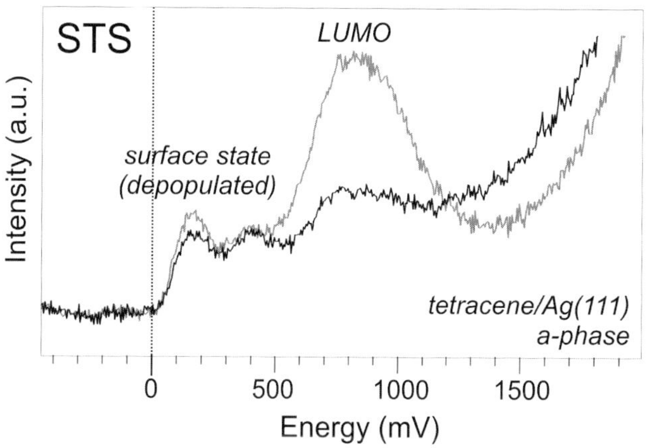

Figure 12.8 STS spectra of tetracene/Ag(111), α-phase, recorded at two different positions inside the molecule. (see Colour plates p. LXII)

also the result of a chemical interaction, as the scheme in Figure 12.5 illustrates.

The example of PTCDA also illustrates that multiple interaction channels between one molecule and a surface may exist if the former contains functional groups with specific reactivities [cf. also 55]. We have seen that they may influence the π-metal interaction (as in PTCDA), but if needed groups can be introduced into the molecule which form covalent bonds with the metal by themselves and thus enhance the molecule-metal coupling. The best-known example of this type is probably the thiol bond which is often used to graft molecules to metal surfaces [56, 57]. Such bonds can transport an electrical current very effectively, as many single-molecule experiments show. However, the example of PTCDA also shows that as soon as the molecule has more than one reactive group, the bonding to the metal can distort its structure, and in some cases this may modify the molecular properties, among them the electrical ones.

Good injection does not only require transfer of electrons or holes from the metal into molecules in direct contact with the metal; eventually the current has to pass from there into the "bulk" of the semiconductor. In the context of our model interfaces, this calls for a good coupling of molecules in the first monolayer to those in higher layers; the single most important factor is the correct orientation of molecules in higher layers. However, even the example of PTCDA/Ag(111) shows that extending the structure of the monolayer into higher layers is tricky: Strictly epitaxial growth only continues into the second layer [58]. But in this example the molecular orientation stays flat at least. In other cases, e.g. pentacene on single-crystalline and polycrystalline gold [59], the molecular orientation changes from flat to tilted to upright as the film thickness increases, and because of the transport anisotropy such changes of orientation have the potential to disrupt the carrier transport.

The example of PTCDA adsorption on low index silver surfaces shows that the adsorption properties are strongly dependent on surface orientation [18, 19, 60]. Moreover, at surface defects such as steps or kinks, preferential adsorption may occur, and the contact properties will again be different [18]. Since in most OFETs one usually has polycrystalline electrodes, the actual injection characteristics will be determined by a complex average over many interface configurations, or even be dominated by a small number of special configurations. This may introduce spatial inhomogeneities into the injection process, which could become relevant for small devices. Also, real interfaces may be locked in non-equilibrium, while model experiments are usually carried out on equilibrium structures. Finally, the fact that organic adsorbates may restructure metal surfaces extensively if enough thermal energy is provided complicates the matter even further. The latter phenomenon has been studied in detail for PTCDA/ Ag(111) [61].

The discussion of the previous paragraphs shows that it is not enough to consider the interaction of individual molecules with the surface; the overall structure of the metal-organic interface region also has a strong influence on its injection properties. This brings us to our next topic.

12.3 Structure

Interface structures are determined by adsorbate-substrate interactions *and* interactions between adsorbates, *intra*layer as well as *inter*layer. In the previous section the emphasis was placed on the molecule-substrate interaction. The present section focuses on molecule-molecule interactions and their competition/cooperation with the molecule-substrate interaction, because the wealth of structures observed at organic-inorganic interfaces stems from their variable interplay. Disentangling the systematics of this interplay requires the consideration of a range of different molecules and substrates. Nevertheless, let us start again with the case of PTCDA on Ag(111).

Although the structure of the PTCDA/Ag(111) interface is dominated by the molecule-substrate interaction (cf. the commensurate interface structure), the interaction between molecules does play an important role for the structural details and the energetics at the interface. On the one hand, there is the attractive *electrostatic* interaction between molecules. Because of this interaction PTCDA molecules always cluster in two-dimensional islands.[1] However, at surface temperatures below 150 K, these islands do not yet exhibit the familiar herringbone structure [35]; since the electrostatic interaction does not exhibit sufficient directional specificity, a considerable degree of structural disorder prevails, in spite of a clear propensity of the molecules to arrange in a T-like

[1] It turns out that even at low substrate temperatures of 100 K, the effective surface diffusion length is large enough that all molecules are able to reach an island [35].

pattern to minimise the electrostatic interaction. The formation of the herringbone structure in fact requires the establishment of *specific intermolecular hydrogen bonds* which go beyond the unspecific electrostatic attraction [35]. As shown schematically in Figure 12.9, these H-bonds form once the Ag–O bonds have been weakened and the whole molecule has lifted up from the surface. Below temperatures of 150 K, thermal energy is insufficient to overcome the required activation energy, and the herringbone structure does not form.

The above example illustrates the complex interplay between adsorbate-substrate and adsorbate-adsorbate interactions for PTCDA/Ag(111). On the one hand, the ordering implies the formation of intermolecular hydrogen bonds and thus requires the weakening of the molecule-substrate bond, but on the other hand it also enables *all* the molecules to adsorb at their preferred bridge sites, i.e. it optimises the molecule-substrate interaction.

Intermolecular interactions must depend strongly on the chemical structure of the molecule. Aromatic molecules with functional groups should behave very differently from those without. We therefore now turn to the non-functionalised molecules pentacene (**3** in Figure 12.2) and tetracene (**4** in Figure 12.2), again on the Ag(111) surface. In the context of this paper these molecules are indeed of special relevance because some of the best OFETs have been made with them [2, 62, 63]. Pentacene poses some challenges with respect to chemical stability [64], while tetracene is more stable and easier to crystallise, but has the slightly worse charge carrier mobilities of the two [65].

On the basis of what we have learned about the molecule-substrate interaction of PTCDA with Ag(111), we expect that the adsorbate-substrate interactions of tetracene and pentacene are smaller than that of PTCDA – both the direct bonding of the oxygen atoms and their enhancing influence on the π-bonding are missing. For pentacene a large number of structures has been reported, depending on preparation conditions [66–68]. This already indicates an indecisive balance between the weak molecule-substrate interaction and the

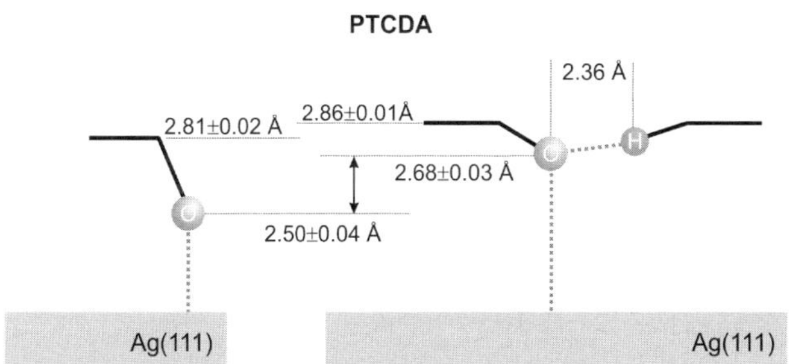

Figure 12.9 Schematic representation of structural changes of PTCDA on ordering in the herringbone phase [35]. Left: disordered low temperature phase. Right: herringbone phase. (see Colour plates p. LXII)

intermolecular interaction. Tetracene forms a point-on-line registry with the Ag(111) substrate [69]. But this is not the whole story: Up to a coverage of approx. 0.7 monolayers no order at all is observed, as the schematic phase diagram in Figure 12.10 shows.[2]

The lack of ordering is not at all surprising since the interaction between adsorbed tetracene molecules seems to be *repulsive*. This is in no contradiction with the bulk structure of tetracene, because in the three-dimensionally ordered structure the *attraction* occurs between molecules which are tilted with respect to each other (a so-called herringbone structure), such that hydrogen-terminated edges of one molecule point toward the π-system of the other [70, 71]. On the surface the molecules are forced into a co-planar orientation, with the result that their interaction becomes repulsive; while it is not clear whether the repulsion is due to *direct* or *substrate-mediated* intermolecular interaction, it is clearly related to co-planarity and the absence of functional groups.

If the intermolecular interaction is indeed repulsive, one may wonder why close to a coverage of $\Theta = 1$ the monolayer orders at all. On the one hand, the ordering may simply result from the fact that in an ordered layer molecules can be packed more densely, and thus a large number of molecules can be in contact with the metal surface. The driving force for ordering would then be an energetic one, i.e. the maximisation of molecule-substrate bonding. But on the other hand there is a second mechanism which favours the emergence of order, irrespective of the molecule-substrate interaction. A clue to this mechanism is provided by the STM images in Figure 12.11. At coverages below the ordering threshold, molecules spread evenly on the surface, randomly pointing in one of three possible directions along the close-packed atom rows of the substrate. If the coverage of this disordered phase is increased, molecules start to interlock

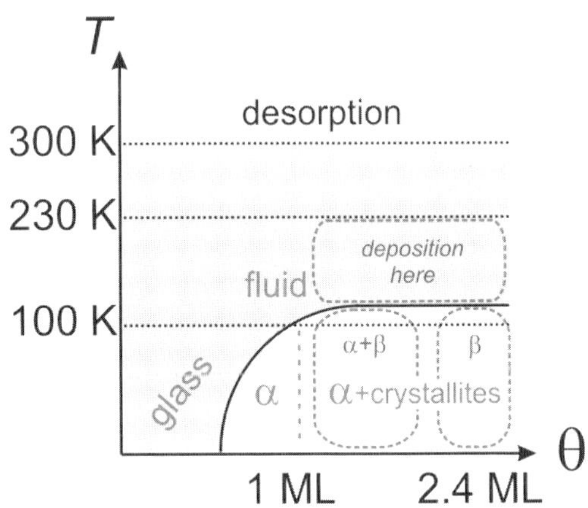

Figure 12.10 Schematic phase diagram of tetracene on Ag(111) (after Langner et al. [69]). Stable phases are shown in red, metastable phases and their preparation parameters in green.
(see Colour plates p. LXIII)

[2] Incidentally, our results on the pentacene/Ag(111) system at room temperature are consistent with the behaviour found for tetracene/Ag(111).

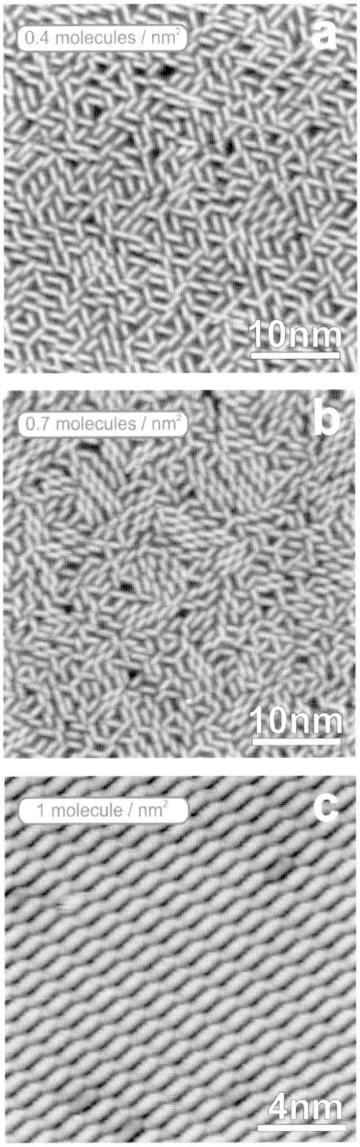

Figure 12.11 STM images of tetracene on Ag(111) at various coverages as indicated. Parameters: (a) 40 × 40 nm^2, 16 pA, 1.3 V. (b) 40 × 40 nm^2, 22 pA, 0.8 V. (c) 16 × 16 nm^2, 0.1 nA, 1.5 V. (see Colour plates p. LXIV)

and hinder each other, whence the layer looses its "*free area entropy*". In an ordered phase of the same coverage, on the other hand, molecules would still have free area around them which they could explore in their thermal motion. This yields a positive contribution to the entropy. Under certain circumstances, the loss of *configurational entropy* due to ordering is more than offset by the gain of free area entropy – and in such cases a disorder-to-order phase transition into a new equilibrium state, entirely driven by entropy, will occur. At first glance this sounds counterintuitive, because entropy is usually associated with disorder. However, if the definition of entropy as $S = k_B \ln \Omega$ is considered and *all* possible microstates are properly accounted for (including those generated by translating and rotating molecules in the free area surrounding them), there is nothing strange about "order-through-entropy". As a matter of fact, entropy-driven crystallisation has been observed in colloidal suspensions of hard spheres above a threshold density [72], and an analogous effect was predicted by Onsager already in 1949 for hard rods [73]. Apparently, tetracene on Ag(111) may be a two-dimensional realisation of the Onsager model system.

If a phase is entropically stabilised, then heating it to higher temperatures must stabilise it even more, because the entropy term $-TS$ in the free energy becomes more negative. Yet, for the tetracene/Ag(111) system an order-to-disorder transition is observed as the temperature at constant coverage is increased (phase diagram in Figure 12.10). This behaviour is explained naturally if it is assumed that close to the transition temperature the thermal energy becomes large enough to overcome the molecules' strict confinement to the surface. Then molecules can transiently tilt out of the surface plane. This breaks the two-dimensionality of the system, and the exploration of the third dimension provides enough additional entropy to tip the balance towards disorder. We can thus conclude that important features of the phase diagram of tetracene on Ag(111) may be explained by a hard rod model with repulsive interaction. Whether the energetic or the entropic mechanism is the dominant force behind the ordering of tetracene on Ag(111) needs to be clarified by further experiments.

Apart from the (disordered and ordered) monolayer phases of tetracene/Ag(111), a second ordered phase is observed at higher coverage. This so-called β-phase [69] is a bilayer with an extremely complex structure which is discussed elsewhere [49]. There are two notable facts about this phase. Firstly, the first layer of the bilayer is *not* the flat lying monolayer phase (α-phase). Rather, under the influence of the *attractive* intermolecular interactions with molecules in the second layer, the first layer re-orders and (partly) tilts up. This behaviour is markedly different from PTCDA, where the first layer forces the second layer into its epitaxial structure. This disparity indicates once more that for the two systems the weighting between intermolecular and interfacial interactions is different. Secondly, a detailed analysis of single-molecule spectra in the β-phase shows that the molecular environment has a very strong influence on the electronic properties of individual molecules, even for molecules far away from the metal.

We note in passing that pentacene, although interacting slightly more strongly with Ag(111) because of its larger size, exhibits a behaviour which is similar to tetracene on Ag(111), especially at sub-monolayer coverage. For both systems it has been conjectured that ordered structures can nucleate *at room temperature* on disordered monolayers [67, 69]. Presumably, the order in higher layers drives the first layer to crystallise, although no direct evidence for this is available. Also, the ordered layers reported in Ref. [67] show some similarities with the β-phase of tetracene. On the other hand, it has been shown in Ref. [68] that if additional layers of pentacene are grown on top of a monolayer, the molecules in the latter stay parallel to the surface. It is conceivable that the slightly stronger substrate interaction is in this case just large enough to permanently confine the molecules in the surface plane.

Pentacene and tetracene still tend to lie flat on metal surfaces, in spite of their weak substrate interaction (as compared with PTCDA). What happens if the substrate interaction is further reduced? Can we expect upright molecules on insulating layers, as would be desirable for OFETs [13]? In many cases, an upright geometry is indeed observed, albeit in a different structure from the bulk. An example is provided by the so-called thin film phase of pentacene on SiO_2 [74, 75]. However, other results show that this cannot be taken for granted. If tetracene is deposited onto epitaxial aluminium oxide films on $Ni_3Al(111)$ at low temperature, molecules initially adsorb *flat*, as the quenching of optical luminescence from the molecules proves, among other observations [76, 77]. The luminescence quenching on the oxide is an interesting finding in itself, because the double layer of oxide should in principle provide an efficient separation of the molecular π-electrons from the metal. After annealing, the molecules do in fact stand up and form three-dimensional clusters. This is coupled to the re-emergence of the luminescence.

Up to this point we have seen that the chemical structure (i.e. presence or absence of functional groups with their specific molecular interaction potentials) has a strong impact on the interface structure. Of course, also the steric properties of molecules may influence the film structure. As an example, we consider molecules which exhibit a considerable structural flexibility. In their bulk states, such molecules often form liquid-crystalline phases [78, 79]. For OFET applications, mesogenic molecules sometimes offer a benefit. For example, it was shown that the field effect mobility of oligomers such as quaterthiophene (4T) increases if two hexyl chains are grafted to either end of thiophene backbone [78, 80–82]. These chains induce a mesogenic behaviour of the resulting molecules (DH4T, **5** in Figure 12.2).

The mesogenic property of an organic semiconductor may indeed influence the evolution of its interface structure, as the example of DH4T on Ag(111) shows [83]. At room temperature, submonolayers of DH4T do not exhibit any long range order, but a reversible disorder-to-order transition is found at approx. 273 K on cooling. Notably, at temperatures slightly above the melting point the DH4T phase displays characteristics of a two-dimensional liquid crystalline structure. On Ag(111), DH4T monolayers thus exhibit a similar type of thermal behaviour as the bulk material, i.e. a succession of liquid, liquid crystalline and ordered phases with decreasing temperature. However, the actual temperatures are changed by the presence of the substrate and/or the confinement to two dimensions. The 2D crystallisation temperature, e.g., is much smaller than that of the bulk [78].

The example of DH4T/Ag(111) also demonstrates that even the structure of the ordered phase is affected by the mesogenic property of the molecules [83].

STM images of the crystalline phase reveal a considerable variation of the 4T backbone- and hexyl-chain orientations from unit cell to unit cell, although Fourier transformed images clearly confirm a good long-range order that corresponds to the observed diffraction patterns. Hence, DH4T displays a very peculiar type of ordering in which short range order is weak while long-range order is preserved. That this feature of the ordered phase must indeed be related to the flexibility of the molecule is clearly demonstrated by the properties of binary phases, in which the flexible molecule is mixed with a more rigid species of well-adjusted size and shape – for example tetracene [83]. The rigid molecule serves as an internal template and induces a much higher degree of short range order. Remarkably, we also find that the presence of the achiral tetracene molecule induces chiral recognition between the pro-chiral DH4T molecules, again by stabilising the structure of the latter. Because of the poor short range order in pure DH4T phases on Ag(111), no chiral recognition was observed there.

12.3.1
Structure: What can be Learned for OFETs?

As pointed out before, interface structure in OFETs is important for both charge injection (metal-semiconductor interface) and charge transport (semiconductor-dielectric interface). The discussion of the previous section demonstrates that even for the small number of model systems considered here a large variety of structures is observed. Because many factors influence the interface and film structures, it is difficult to establish general rules. However, a few important observations regarding the structure-forming factors can be kept hold of:

1) The presence of the metal or insulator does not only add the molecule-substrate interaction as a formative influence, but can also alter the effective intermolecular interactions. For example, whereas the crystallisation of bulk tetracene is governed by the *attractive* interaction between molecules in a particular relative orientation, the surface-confined molecules (on Ag(111)) *repel* each other. The modification of the effective intermolecular interaction may originate both from substrate-mediation and from the intrinsically anisotropic molecular interaction potentials. As the possibly entropy-driven ordering of tetracene on Ag(111) shows, the modified interactions may introduce new ordering mechanisms at the interface.

2) As a result of the above, and of the direct competition between molecule-substrate and intermolecular interactions, the presence of the metal or insulator can induce interface polymorphs which do not exist in the bulk. Examples for this are the specific thin film phases of pentacene on insulators [16, 74, 75] or metals (e.g., Cu(110) [16]), the α- and β-phases of tetracene on Ag(111) [69], or the square phases of PTCDA on Ag(111) [30] and Au(111) [84]. It is evident that the charge carrier mobilities of organic semiconductors will depend on the crystal phase.

3) The detailed study of electronic properties of individual molecules in thin films of organic semiconductors shows that their electronic structure can vary substantially depending on the environment. This effect is not limited to molecules in contact with the metal (an example would be the anomalously strong dispersion of PTCDA on Ag(111), see below), but also extends to molecules in molecular environment (cf. the β-phase of tetracene on Ag(111), where significant orbital energy shifts are observed [49]). This means that injection barriers at the metal-organic interface not only depend on the metal and the molecular species as such, but also on the actual structure at the immediate interface and in the wider interface region.

4) Finally, the example of DH4T demonstrates a wide scope for structural engineering of molecular interfaces on the basis of building blocks with different chemical and steric properties. This scope can be extended even further by considering binary or multinary phases. Evidently, the mixing of several components will not only change the structure of the organic material but also its electronic properties. Currently, we are still at the very beginning of exploring the properties of such newly created artificial materials.

An interesting aspect of intermolecular interactions which is relevant for charge transport (because of charge delocalisation) is the formation of an electronic band structure in the organic semiconductor. Data on some extremely pure organic crystals reveal a mobility that rises with decreasing temperature [84]. Because this is a hallmark of band transport, the study of dispersion in organic materials has become an important topic. Evidently, energy band dispersion is strongly related to structure. Quite generally, even in bulk organic semiconductors the issue is still debated [85], because dispersion measurements on organic materials are fraught with difficulty [86]: organic single crystals are usually small, poorly conducting, and limited to surfaces orientations along cleavage planes. Therefore, the preparation of highly ordered thin films is an important instrument for learning about intrinsic charge transport in organic semiconductors. However, one has to be careful here, because electronic interface states in ultra-thin films may also show different dispersion from the bulk.

Dispersion measurements on PTCDA [30, 87], pentacene [88] and the rod-like molecule of sexiphenyl [86] have been reported in the literature. Because the intermolecular interaction between organic semiconductor molecules is anisotropic, one expects the dispersion to be anisotropic as well, strong in the direction of π-stacking and much weaker or absent in directions where molecules touch each other edge-on. This is nicely illustrated in the case of well-oriented 6P films on Cu(110) [86], for which in the directions of the molecules a quasi-one-dimensional *intra*molecular dispersion of 0.2 eV is observed, while perpendicular to the molecules, in the direction of π-overlap, a large 0.7 eV *inter*molecular dispersion is found (intermolecular band formed by non-bonding orbitals HOMO-3 to HOMO-8).

The data on 6P have been recorded on 200 Å thick films, and the measured dispersion is thus a property of the bulk. In contrast, the case of PTCDA reveals that intermolecular dispersion can indeed be strongly influenced by the presence of a substrate. For bulk PTCDA (supported on MoS_2), Yamane et al. [87] have measured a 0.2 eV dispersion of the highest occupied molecular orbital along the π-stacking direction (perpendicular to the sample surface), yielding a moderate effective mass of $5.28m_e$. Because of the above mentioned anisotropy, one would expect the dispersion with surface-parallel wave vectors, where molecules are oriented edge-to-edge, to be much smaller, and indeed, thick films do not display any dispersion at all in this direction. If, however, experiments are performed on PTCDA monolayers on Ag(111), a very different picture emerges. Now a strongly dispersive band with width 0.27 eV and with an effective mass $m^* = 0.48m_e$, i.e. *smaller* than for the π-stacking directions is recorded [30]. The similarity with the effective mass of the silver surface state suggests that the strong dispersion is caused by a *substrate-mediated* intermolecular interaction, rather than by a direct intermolecular interaction between the planar molecules via their C–H edges. On the other hand, the position of the dispersing band and its dispersion itself depends sensitively on the intermolecular order, being different for the herringbone and the square phases of PTCDA on Ag(111). From these observations one may infer that the wave functions of the dispersing interface state are a composite of Shockley-like surface state wave functions inside the substrate and molecular states (LUMO+1/LUMO+2) outside. In this picture, the surface state of the bare Ag(111) surface is depopulated by the general charge transfer from the metal into the LUMO and shifted up in energy as a consequence, whence it becomes degenerate with the LUMO+1 and the LUMO+2 of the molecule.

A similar in-plane monolayer dispersion has been reported for pentacene on Cu(110) [88], and indeed much earlier already for flatly adsorbed benzene on metal surfaces [89–91]. In the latter cases, the dispersion was ascribed to a direct molecule-molecule interaction of co-planar molecules, enhanced by the compression of the molecular lattice as enforced by the registry with the substrate. In the former case, neither a layer compression nor a Shockley surface state are present, leaving the origin of the strong dispersion unexplained. It must therefore be concluded that the origins of interfacial dispersion of π-conjugated molecules is not yet fully understood in general terms, especially since in all cases where dispersion was observed so far it occurred selectively for one molecular state only, while the other states do not show any dispersion. Obviously, this issue could be addressed by *ab initio* calculations. However, to our knowledge there is currently no such calculation which would explain the origin of the strong dispersion observed in any of these interface systems. Finally, the interest in electronic band dispersion is highlighted by the intriguing observation of a temperature dependent dispersion in (bulk) pentacene on graphite [92].

12.4 Function

Charge transport is the functionality of predominant importance for organic semiconductors. If one wants to measure transport under circumstances of maximum control in order to exclude extrinsic effects, and thus establish reference standards for real devices, two approaches may be taken. Firstly, one may seek to prepare a conventional OFET in which organic layer and gate dielectric are single-crystalline and the metal electrodes have well-defined contacts to the organic. This approach is of particular relevance for standard organic electronics and is being actively pursued in the present priority program [6, 17].

Secondly, one may attempt to measure current transport through a single molecule. This line of attack pertains to molecular electronics, the single-molecule limit of organic electronics. Both approaches are in fact quite demanding. For the single crystalline OFET the preparation is the main challenge, while for the single-molecule transistor measurements under controllable conditions are hard to achieve.

Here we report on the second approach. To conduct an experiment as the one sketched in Figure 12.12a would be the dream of every experimentalist in the field. Indeed, many mechanical break junction experiments have been reported which come quite close to this ideal [93–95]. However, a major unsolved problem of all these experiments is that there are currently no robust methods to image and determine the precise adsorption site and conformation of the molecule in the break junction [96]. But at the same time it is known that the connection between molecule and electrode greatly affects the current–voltage characteristics, sometimes even more than the properties of the molecule itself [97]. This poses a serious problem for the interpretation of single molecule transport data.

As a way out of this dilemma, we have developed a two-step approach based on commensurate, highly ordered molecular layers on metal surfaces. In step 1, the powerful armoury of surface science is employed to characterise the structural and electronic properties of the metal-molecule contact. For the model molecule PTCDA, some of the results of these experiments have in fact been described in Sections 12.2 and 12.3. Once step 1 has been completed, the tip of a low-temperature STM is used to establish a covalent contact with an individual molecule of the highly ordered monolayer (step 2). Because of the excellent imaging properties of the STM, it is in fact possible to select the part of the molecule which is contacted with very high accuracy.

The resulting geometry of the transport measurement is shown in Figure 12.12b. It is apparent from this schematic that PTCDA on Ag(111) is well suited for this experiment, because the carboxylic oxygen atoms offer a suitable "terminal" at which the molecule can be contacted covalently. The figure has been drawn as if one of the carboxylic oxygen atoms has jumped up from the surface into contact with the tip. The experimental data in fact prove this scenario [31]. Once the tip-oxygen contact has been formed, a single-molecule

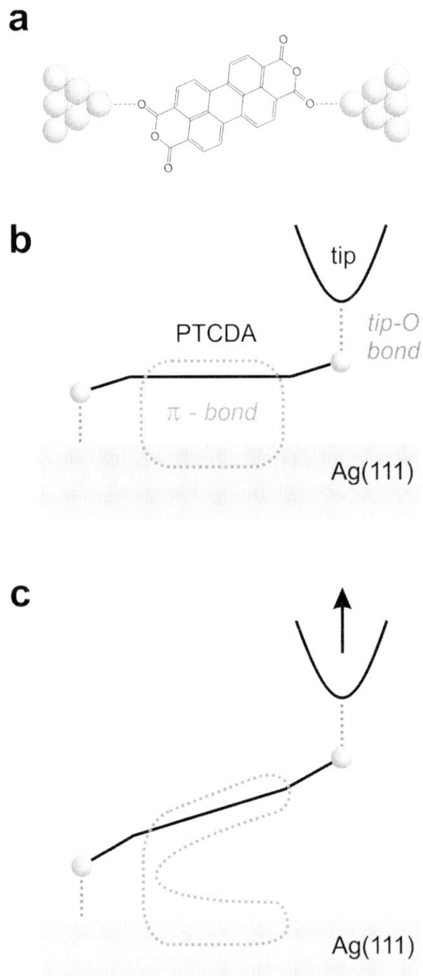

Figure 12.12 (a) Idealised single-molecule transport experiment. (b) Single-molecule transport experiment with optimal structural control based on a (1) epitaxial molecular monolayer with well-characterised adsorbate-substrate bond and (2) a specific, chemically well-defined tip-molecule contact. (c) Mechanically gated single-molecule wire based on experiment in b and STM tip retraction. (see Colour plates p. LXV)

wire is established; most importantly, its two contacts are precisely defined at the atomic scale. The epitaxial contact to the metal substrate has been characterised in step 1, and the contact to the tip is at least chemically well-defined as a covalent Ag–O contact.[3] Transport measurements on this wire have been performed [31], and they allow a meaningful comparison to first-principles calculations, because experimental and theoretical data are generated on wire junctions with the same well-defined atomic structure.

---

3 Since the structure of the tip apex is not known, the precise structure of the tip-molecule contact is unclear. The oxygen atom could, e.g., be bonded to a single adatom, an edge site between two facets, or a corner site between three facets. This residual uncertainty notwithstanding, the structural control in the present experiment is still far better than in standard break junction experiments.

However, optimal structural control is not the only benefit of our approach; it is also able to address the second quandary of single molecule experiments, namely device tuneability. Measuring transport reproducibly in a two terminal single-molecule device is certainly a valuable achievement, but interesting transport physics is more likely to be accessed once the wire is gated by a third electrode. In this way, the ultimate OFET, a single molecule transistor, can be realised. Break junction experiments combined with a gate electrode have been reported in the literature [98, 99]. But while these display excellent tuneability through various transport regimes, they again lack the structural control of the experiment sketched in Figure 12.12b. Ideally, the two important objectives *tuneability and atomic structure control* should be realised in one and the same experiment, because this combination would provide an optimal interface to *ab initio* transport calculations.

We have argued that STM experiments allow unparalleled structural control, but it is very difficult, if not impossible, to provide a third electrode besides tip and sample that is sufficiently close to the molecular wire. However, it is possible to use the manipulation capability of the STM to achieve a *mechanical* gating of the molecular wire [31]. The idea of this experiment is shown in Figure 12.12c. By retracting the tip from the surface, the wire junction is stretched and the molecule is peeled off the surface as a consequence. In this way, it is in fact possible to pull individual molecules into an upright orientation and even remove them from the surface. The detailed analysis of the experimental retraction spectra and first principle DFT simulations of the tip retraction process confirms this scenario.

The effect of stretching the junction on the current voltage characteristic is shown in Figure 12.13. If the tip position is identified with a gate voltage, the graph indeed looks like the output characteristics of a field effect transistor. Note that the $I-V$ curves have been measured on a *single* PTCDA molecule in

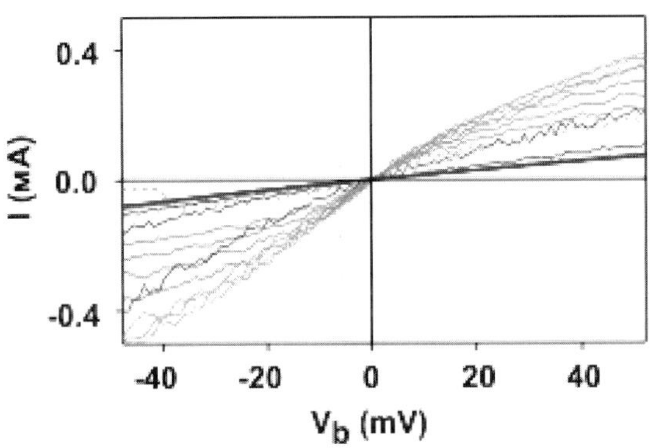

Figure 12.13 Current voltage characteristic of the device in Figure 12.12c [31]. Curves of different colours refer to different vertical tip positions and hence “streching states” of the junction. (see Colour plates p. LXVI)

the STM junction – in other words, a structurally well-controlled single molecule "transistor" has been realised.

But why does pulling the molecule gate the transport? This has to do with the specific properties of PTCDA/ Ag(111) that have been discussed in Section 12.2. As the molecule is peeled off the surface, the metal-molecule π-bond is gradually cleaved. The schematic in Figure 12.5 shows that this dehybridisation must shift the former LUMO back up in energy and sharpens it. Resonant transport then enhances the slope of the $I-V$ curves at low biases and thus yields the transistor-like output curves of Figure 12.13.[4]

On further pulling, the former LUMO eventually reaches the Fermi level. At this point, electron correlation effects in the transport orbital become important and the transport through the wire is strongly affected by the non-equilibrium Kondo effect [31]. In the Kondo regime, transport through the wire is highly correlated: as soon as an electron leaves the molecule into the positive lead, another one from the negative electrode fills the vacancy. At low bias voltages (and low temperatures), this correlated transport channel is very efficient, which further enhances the differential conductance at zero bias and therefore the gating effect. It is remarkable that a non-magnetic molecule such as PTCDA exhibits the Kondo effect which is usually observed for bulk magnetic impurities or magnetic adsorbates. But the reason is clear: by tuning the chemisorption bond, single occupancy of the former LUMO is achieved, making the wire magnetic.

The device of Figure 12.12c as such is of course not suitable for any real application. But its merit is the provision of a good testing system for exploring the transport physics in single molecules. In fact, although some aspects of the experiment discussed here are specific to PTCDA/Ag(111), the technique itself is much more general. Experiments on PTCDA/Au(111) revealed [41] that the molecule can be contacted and lifted up in essentially the same way as on Ag(111). However, since the LUMO of PTCDA remains unoccupied upon adsorption in this case, no gating effect can be achieved by removing the molecule from the surface.

The great advantage of our experiments is that they eliminate the need for large-number statistics in single-molecule transport experiments and offer fascinating opportunities for the understanding of the transport at the *ab initio* level, because theorists know exactly which geometry and electronic contact properties generate the observed transport. In fact, the transport simulations of our wire geometry have already begun, see Figure 12.14. We are convinced that the utilisation of the potential of surface science is a key for further progress in molecular electronics, and our experiments are a first step in this direction.

---

4 Note, however, that the mechanical gating employed here simultaneously changes the coupling constant Γ of the molecule to the surface.

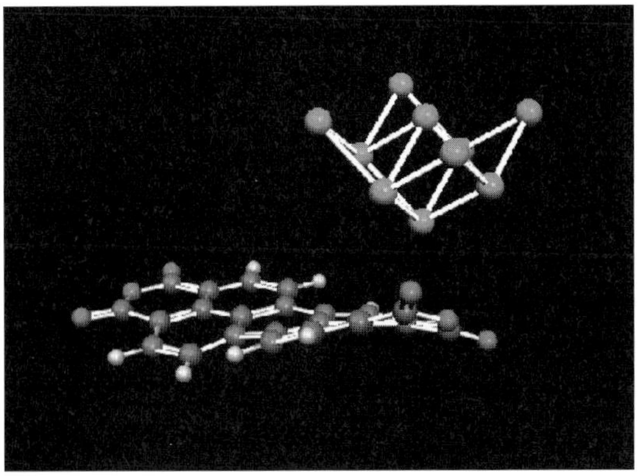

Figure 12.14 Simulation of the PTCDA/Ag(111) bond cleavage by tip retraction (calculated by local density approximation of density functional theory) [41]. (see Colour plates p. LXVI)

12.5 Conclusion

We have argued in this chapter that well-devised model systems can be used to lay the foundations of a deeper understanding of organic field effect transistors, by studying the fundamental principles of bonding and structure at relevant interfaces. Ultimately, it is even possible to construct a mechanically gated single molecule transistor that can serve as benchmark for describing transport physics through molecules.

Of course, it would be wrong to transfer the observations from the model interfaces or devices directly to their technological counterparts. But on the other hand, a number of effects that point to interesting physics have been discovered, for example the possibly entropy-stabilised ordered phase at the tetracene/Ag(111) interface, the unexpectedly strong dispersion of an unoccupied PTCDA/Ag(111) interface state, the remarkably strong dependence of the electronic structure of individual molecules on the molecular environment (tetracene β-phase), or the Kondo physics in a non-magnetic molecular wire.

In principle, all of these effects could also be present in real devices and have an impact on their performance. In this sense, experiments on model systems offer the unique opportunity to unearth potential contributions to the physics of OFETs that are hidden behind the average behaviour of technical, often non-ideal, devices. But as OFETs are getting smaller, some of these interesting effects may even become decisive for their performance.

Acknowledgements

The work reported here was carried out during the six years of the DFG priority programme 1121 "Organic field effect transistors: structural and dynamic properties" and the project TA244.

References

1. P. V. Necliudov, M. S. Shur, D. J. Gundlach et al., Solid-State Electron. **47**, 259 (2003).
2. A. Hepp, H. Heil, W. Weise et al., Phys. Rev. Lett. **91**, 157406 (2003).
3. T. Muck, V. Wagner, U. Bass et al., Synth. Met. **146**, 317 (2004).
4. J. R. Ostrick, A. Dodabalapur, L. Torsi et al., J. Appl. Phys. **81**, 6804 (1997).
5. M. Kotani, K. Kakinuma, M. Yoshimura et al., Chem. Phys. **325**, 160 (2006).
6. G. Witte and Ch. Wöll, Phys. Stat. Solidi A **205**, 497 (2008).
7. T. Muck, J. Fritz, and V. Wagner, Appl. Phys. Lett. **86**, 232101 (2005).
8. A. Hoppe, T. Balster, T. Muck, and V. Wagner, Phys. Stat. Solidi A **205**, 612 (2008).
9. D. Schlettwein, K. Hesse, H. Tada et al., Chem. Mater. **12**, 989 (2000).
10. Z. Bao, A. J. Lovinger, and J. Brown, J. Am. Chem. Soc. **120**, 207 (1998).
11. A. Gerlach, S. Sellner, S. Kowarik, and F. Schreiber, Phys. Stat. Solidi A **205**, 461 (2008).
12. H. Moon, R. Zeis, E. J. Borkent et al., J. Am. Chem. Soc. **126**, 15322 (2004).
13. M. Voigt, J. Pflaum, and M. Sokolowski, Phys. Stat. Solidi A **205**, 449 (2008).
14. J. Zaumseil, R. H. Friend, and H. Sirringhaus, Nature Mater. **5**, 69 (2006).
15. N. Benson, Ch. Melzer, R. Schmechel, and H. von Seggern, Phys. Stat. Solidi A **205**, 475 (2008).
16. G. Witte and Ch. Wöll, J. of Mat. Res. **19**, 1889 (2004).
17. S. Sachs, M. Paul, A. Schöll et al., Verhandlungen der DPG, Europhysics Conference Abstracts, Dresden, 2006, Vol. 30A.
18. K. Glöckler, C. Seidel, A. Soukopp et al., Surf. Sci. **405**, 1 (1998).
19. E. Umbach, K. Glockler, and M. Sokolowski, Surf. Sci. **404**, 20 (1998).
20. A. C. Hillier and M. D. Ward, Phys. Rev. B **54**, 14037 (1996).
21. C. Seidel, C. Awater, X. D. Liu et al., Surf. Sci. **371**, 123 (1997).
22. S. R. Forrest, Chem. Rev. **97**, 1793 (1997).
23. S. R. Forrest, J. Phys.: Condens. Matter **15**, S2599 (2003).
24. S. R. Forrest, M. L. Kaplan, P. H. Schmidt et al., Appl. Phys. Lett. **41**, 90 (1982).
25. A. Schmidt, T. J. Schuerlein, G. E. Collins et al., J. Phys. Chem. **99**, 11770 (1995).
26. S. K. M. Henze, O. Bauer, T. L. Lee et al., Surf. Sci. **601**, 1566 (2007).
27. A. Kraft, R. Temirov, S. K. M. Henze et al., Phys. Rev. B **74**, 041402(R) (2006).
28. L. Kilian, E. Umbach, and M. Sokolowski, Surf. Sci. **573**, 359 (2004).
29. F. S. Tautz, M. Eremtchenko, J. A. Schaefer et al., Phys. Rev. B **65**, 125405 (2002).
30. R. Temirov, S. Soubatch, A. Luican et al., Nature **444**, 350 (2006).
31. R. Temirov, A. C. Lassise, F. Anders et al., arxiv.org/abs/ cond-mat/ 0612036; Nanotechnology **19**, 065401 (2008).
32. F. S. Tautz, Prog. Surf. Sci. **82**, 479 (2007).
33. M. Rohlfing, R. Temirov, and F. S. Tautz, Phys. Rev. B **76**, 115421 (2007).
34. M. Eremtchenko, D. Bauer, J. A. Schaefer et al., New J. Phys. **6**, 4 (2004).
35. L. Kilian, A. Hauschild, R. Temirov et al., Phys. Rev. Lett. **100**, 136103 (2008).
36. L. Kilian, University of Würzburg (2002).
37. A. Hauschild, K. Karki, B. C. C. Cowie et al., Phys. Rev. Lett. **95**, 209602 (2005).
38. A. Hauschild, K. Karki, B. C. C. Cowie et al., Phys. Rev. Lett. **94**, 036106 (2005).

39. N. Nicoara, E. Roman, J. M. Gomez-Rodriguez et al., Org. Electron. **7**, 287 (2006).
40. O. Neucheva, R. Temirov, S. Soubatch, and F. S. Tautz, to be published.
41. F. Pump, R. Temirov, O. Neucheva et al. submitted.
42. D. M. Newns, Phys. Rev. **178**, 1123 (1969).
43. J. K. Norskov, Rep. Prog. Phys. **53**, 1253 (1990).
44. S. X. Du, H. J. Gao, C. Seidel et al., Phys. Rev. Lett. **97**, 156105 (2006).
45. W. Ji, Z.-Y. Lu, and H. Gao, Phys. Rev. Lett. **97**, 246101 (2006).
46. S. Picozzi, A. Pecchia, M. Gheorghe et al., Phys. Rev. B **68**, 195309 (2003).
47. R. Rurali, N. Lorente, and P. Ordejon, Phys. Rev. Lett. **95** (2005).
48. M. Eremtchenko, D. Bauer, J. A. Schaefer et al., J. Mater. Res. **19**, 2028 (2004).
49. S. Soubatch, R. Temirov, and F. S. Tautz, submitted.
50. M. Scharnberg, J. Hu, J. Kanzow et al., Appl. Phys. Lett. **86**, 024104 (2005).
51. M. Scharnberg, J. Hu, J. Kanzow et al., Defects and Diffusion Forum **237–240**, 993 (2005).
52. D. Cahen, A. Kahn, and E. Umbach, Mater. Today **8**, 32 (2005).
53. I. G. Hill, A. Kahn, J. Cornil et al., Chem. Phys. Lett. **317**, 444 (2000).
54. K. Sugiyama, D. Yoshimura, T. Miyamae et al., J. Appl. Phys. **83**, 4928 (1998).
55. L. Romaner, G. Heimel, J.-L. Brédas et al. Phys. Rev. Lett. **99**, 256801 (2007).
56. K. Y. Kwon, K. L. Wong, G. Pawin et al., Phys. Rev. Lett. **95**, 166101 (2005).
57. L. Bartels, B. V. Rao, and A. Liu, Chem. Phys. Lett. **385**, 36 (2004).
58. L. Kilian, E. Umbach, and M. Sokolowski, Surf. Sci. **573**, 359 (2004).
59. D. Käfer, L. Ruppel, and G. Witte, Phys. Rev. B **75**, 085309 (2007).
60. F. S. Tautz, M. Eremtchenko, J. A. Schaefer et al., Surf. Sci. **502**, 176 (2002).
61. X. Ma, H. L. Meyerheim, J. Barthel et al., Appl. Phys. Lett. **84**, 4038 (2004).
62. O. D. Jurchescu, J. Baas, and T. T. M. Palstra, Appl. Phys. Lett. **84**, 3061 (2004).
63. H. Klauk, M. Halik, U. Zschieschang et al., J. Appl. Phys. **92**, 5259 (2002).
64. M. Yamada, I. Ikemoto, and H. Kuroda, Bull. Chem. Soc. Jpn. **61**, 1057 (1988).
65. C. Goldmann, S. Haas, C. Krellner et al., J. Appl. Phys. **96**, 2080 (2004).
66. L. Casalis, M. F. Danisman, B. Nickel et al., Phys. Rev. Lett. **90**, 206101 (2003).
67. M. Eremtchenko, R. Temirov, D. Bauer et al., Phys. Rev. B **72**, 115430 (2005).
68. D. Käfer and G. Witte, Chem. Phys. Lett. **442**, 376 (2007).
69. A. Langner, A. Hauschild, S. Fahrenhoz et al., Surf. Sci. **574**, 153 (2005).
70. R. B. Campbell and J. M. Robertson, Acta Crystallogr. **15**, 289 (1962).
71. J. Robertson, J. Trotter, and V. C. Sinclair, Acta Crystallogr. **14**, 697 (1961).
72. J. Baumgartl, R. P. A. Dullens, M. Dijkstra et al., Phys. Rev. Lett. **98**, 198303 (2007).
73. L. Onsager, Ann. New York Acad. Sci. **51**, 627 (1949).
74. I. P. M. Bouchoms, W. A. Schoonveld, J. Vrijmoeth et al., Synth. Met. **104**, 175 (1999).
75. D. J. Gundlach, T. N. Jackson, D. G. Schlom et al., Appl. Phys. Lett. **74**, 3302 (1999).
76. A. Langner, Y. Su, and M. Sokolowski, Phys. Rev. B **74**, (2006).
77. S. Soubatch, R. Temirov, M. Weinhold et al., Surf. Sci. **600**, 4679 (2006).
78. F. Garnier, R. Hajlaoui, A. El Kassmi et al., Chem. Mater. **10**, 3334 (1998).
79. K. R. Amundson, H. E. Katz, and A. J. Lovinger, Thin Solid Films **426**, 140 (2003).
80. J. Ackermann, C. Videlot, P. Raynal et al., Appl. Surf. Sci. **212**, 26 (2003).

81. C. D. Dimitrakopoulos, B. K. Furman, T. Graham et al., Synth. Met. **92**, 47 (1998).
82. F. Garnier, A. Yassar, R. Hajlaoui et al., J. Amer. Chem. Soc. **115**, 8716 (1993).
83. S. Soubatch, R. Temirov, and F. S. Tautz, Langmuir **22**, 9572 (2006).
84. N. Karl, Synth. Met. **133**, 649 (2003).
85. Y. M. Sun, Y. Q. Liu, and D. B. Zhu, J. Mater. Chem. **15**, 53 (2005).
86. G. Koller, S. Berkebile, M. Oehzelt et al., Science **317**, 351 (2007).
87. H. Yamane, S. Kera, K. K. Okudaira et al., Phys. Rev. B **68**, 033102 (2003).
88. H. Yamane, D. Yoshimura, E. Kawabe, R. Sumii, K. Kanai, Y. Ouchi, N. Ueno, and K. Seki, Phys. Rev. B **76**, 165436 (2007).
89. H. H. Graen, M. Neuber, M. Neumann et al., Europhys. Lett. **12**, 173 (1990).
90. F. P. Netzer, Vacuum **41**, 49 (1990).
91. M. G. Ramsey, D. Steinmuller, F. P. Netzer et al., Surf. Sci. **251**, 979 (1991).
92. N. Koch, A. Vollmer, I. Salzmann et al., Phys. Rev. Lett. **96**, 156803 (2006).
93. C. Kergueris, J. P. Bourgoin, S. Palacin et al., Phys. Rev. B **59**, 12505 (1999).
94. J. J. Parks, A. R. Champagne, G. R. Hutchison et al., arXiv:cond-mat/0610555v2.
95. M. A. Reed, C. Zhou, C. J. Muller et al., Science **278**, 252 (1997).
96. C. Joachim and M. A. Ratner, Proc. Natl. Acad. Sci. USA **102**, 8800 (2005).
97. A. Nitzan and M. A. Ratner, Science **300**, 1384 (2003).
98. A. R. Champagne, A. N. Pasupathy, and D. C. Ralph, Nano Lett. **5**, 305 (2005).
99. H. S. J. van der Zant, Y. V. Kervennic, M. Poot et al., Faraday Discuss. **131**, 347 (2006).

13
Metal/Organic Interface Formation Studied *In Situ* by Resonant Raman Spectroscopy

G. Salvan, B. A. Paez, D. R. T. Zahn, L. Gisslen, and R. Scholz

13.1
Introduction

In recent years, enormous progress was made in the understanding of organic semiconductors. The research interest in these materials is driven by the various applications ranging from organic light emitting devices (OLEDs) over organic field effect transistors (OFETs) to organic photovoltaic cells (OPVCs). Experimental techniques such as photoemission spectroscopy which has been one major surface science technique in the field of inorganic semiconductor research for decades has also been successfully applied to study organic semiconductors and their interfaces for quite some time. Other experimental techniques such as Raman spectroscopy are less widely used for studying organic semiconductor interfaces. However, also this optical spectroscopy technique, which probes the vibrational modes and may thus also be called vibrational spectroscopy, can provide very valuable information about interface properties such as geometric structure, band bending, and interfacial chemistry. It is the intention of this chapter to illustrate that Raman spectroscopy can contribute significantly to the field of organic interface characterisation by providing an insight into the chemistry and structural aspects of the organic interface formation.

13.2
Methods

13.2.1
Sample Preparation and Characterisation

The organic layers were grown onto sulfur-passivated GaAs(100) substrates by organic molecular beam deposition (OMBD) in an ultra high vacuum chamber (UHV) with 7×10^{-10} mbar base pressure. The passivation procedure is described in Ref. [1].

Organic Electronics. Structural and Electronic Properties of OFETs. Edited by Christof Wöll

ISBN: 978-3-527-40810-8

The molecular materials obtained from Syntec GmbH Wolfen were pre-purified by a two-step sublimation prior to the deposition. The organic materials and the metals were evaporated from Knudsen cells kept at 280 °C for PTCDA, 270 °C for DiMe-PTCDI, 930 °C for Ag, 830 °C for In, and at 375 °C for Mg, resulting in deposition rates of 0.3 nm/min for both perylene derivatives, 1.6 nm/min for Ag and 2 nm/min for In and Mg. During the growth of these films, the substrates were kept at room temperature.

The films consist of islands (see AFM images in Figure 13.1) the crystalline character of which was proven by Raman spectroscopy [2].

For the *in situ* Raman measurements, the UHV system is optically aligned with a triple monochromator Raman spectrometer (Dilor XY) equipped with a CCD camera for multichannel detection [1]. The samples were excited with the 488 nm (2.54 eV) Ar^+ laser line that lies in the first absorption maximum of both organic molecules and thus ensures resonance conditions for the Raman process.

13.2.2 Theoretical Methods

The ground state geometries of the molecules are optimised with density functional theory (DFT), and the geometries in the excited states with time-dependent density functional theory (TD-DFT). As discussed elsewhere, the deformation patterns can be understood from the changes of the node patterns of the two electronic orbitals involved in the dipole-active transition, in our cases the highest occupied molecular orbital (HOMO) and the lowest unoccupied orbital (LUMO) [3, 4]. For each compound, the deformation pattern between the geometries in the relaxed excited state and in the ground state is projected onto the vibrational eigenvectors obtained on the potential energy surface of the ground state. All calculations are performed with the program packages Gaussian98 [5] and turbomole5.7 [6], using the hybrid functional B3LYP and localised basis sets.

13.3 Results and Discussion

13.3.1 Chemistry of Metal/Organic Interfaces

Two spectral regions were recorded upon step-wise metal deposition. The region between 25 cm^{-1} and 650 cm^{-1} contains a mode corresponding to the breathing of the whole molecule and C–C stretching modes [7]. The frequency region between 1200 cm^{-1} and 1650 cm^{-1} contains internal modes with C–H and C–C character. Figure 13.1 shows the Raman spectrum of a bare 15 nm PTCDA film and its evolution upon the step-wise deposition of indium.

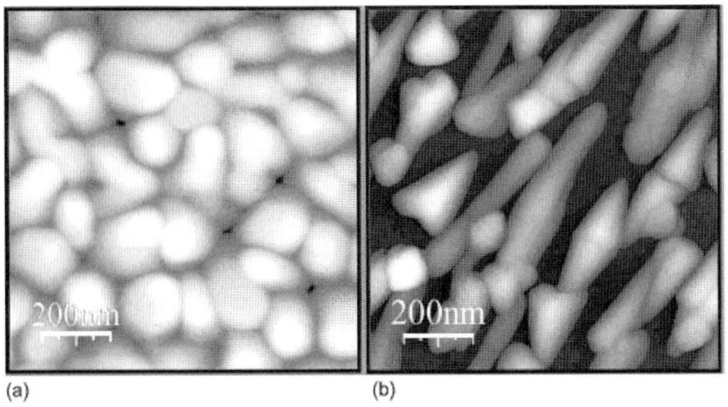

Figure 13.1 AFM images of PTCDA (a) and DiMePTCDI (b) films with a nominal thickness of 20 nm grown on S-GaAs(100). (see colour plates p. LXVII)

The overall signal intensity increases with the In thickness up to 15 nm and decreases for higher metal coverages as shown by the normalisation factors in Figure 13.2. The same behaviour is observed when Ag is deposited onto PTCDA and for both metals deposited onto DiMe-PTCDI [8, 9].

This effect is known as surface enhanced Raman scattering (SERS) [10–12]. The SERS effect has been widely investigated for various molecules adsorbed on rough metallic surfaces or on metallic clusters in colloids. Reviews on this topic can be found in Refs. [12–15]. The enhancement of normally Raman-active modes is a consequence of the enhancement of the electric field of the incoming and scattered radiation in the vicinity of a rough metal film upon coupling with the dipolar plasmon resonances in the metal clusters. This enhancement affects molecules located up to even 10 nm away from the metal surface [11, 12]. The enhancement factors are essentially determined by the electronic properties of the metal and by the morphology of the metal film.

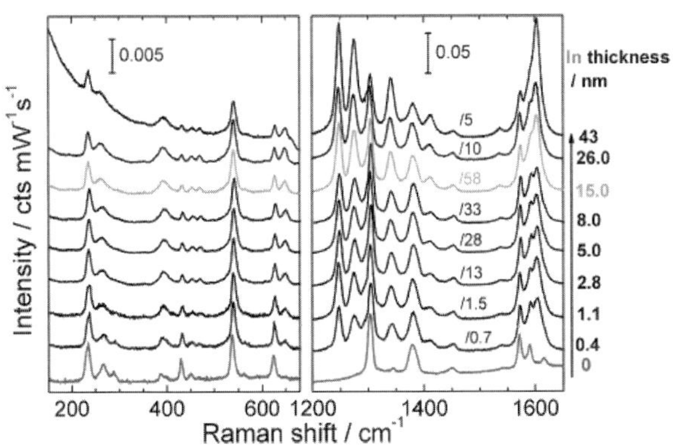

Figure 13.2 Raman spectra of PTCDA acquired upon successive deposition of In onto a 15 nm PTCDA film. The spectra normalisation is done with respect to the intensity of the C–C stretch modes (1572 cm^{-1}). (see colour plates p. LXVII)

In addition to the enhancement of the totally symmetric Raman active modes, already the deposition of only 0.4 nm In leads to the appearance of some modes in the higher frequency region. The same modes were also observed when depositing In or Ag onto a ML of PTCDA on S-GaAs, only with different relative intensities (Figure 13.3). They are therefore a signature of the molecules in direct contact or in the very near vicinity of a metal surface. In general, such a break-down of the selection rules accompanies the SERS effect and can be induced by several mechanisms: structural deformation of the molecule, charge transfer from the molecule into the metal or *vice versa*, or formation of new chemical bonds. Thus the spectral changes induced by SERS can be used to extract information about chemical reactions at the interface, as well as about the morphology of the metal film.

The band at 1338 cm^{-1} was identified in Ref. [16] to be a B_{3g} band based on its frequency and intensity in the crystal spectra, while a band at 1292 cm^{-1} is likely to be a shifted variant of the C–H deformation A_g mode at 1303 cm^{-1} in the single crystal [17]. The other bands correspond to modes which normally show infrared activity (see Figure 13.4). Considering that all the modes occurring upon In and Ag deposition are normal modes of the PTCDA molecule, the observed break-down of the Raman-infrared selection rules was proposed to originate from a weak charge transfer between the molecules and the metal surface mediated by molecular vibrations [9].

This is in contradiction to previous findings of Hirose et al. [18] and Kera et al. [19] who proposed the formation of a In_4PTCDA complex.

In order to assess the effect of a metal-PTCDA complex formation on the vibrational frequencies, theoretical calculations were preformed with the Gaussian'98 package on the In_4PTCDA complex using the B3LYP functional and

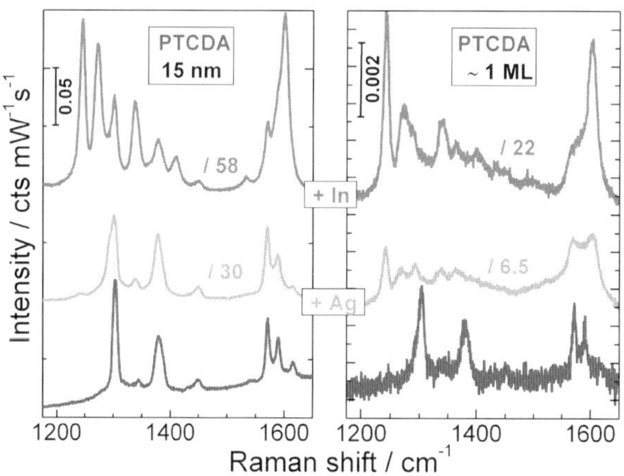

Figure 13.3 Raman spectra of a 15 nm (left) and of a ML (right) PTCDA film clean (bottom) and covered with nominally 15 nm of silver (middle) and indium (top) each. (see colour plates p. LXVIII)

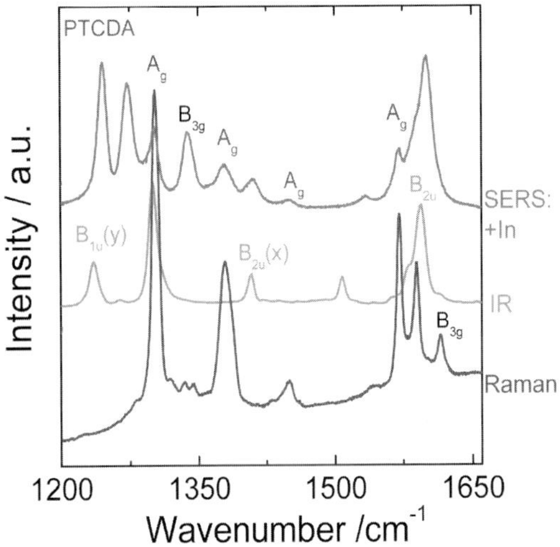

Figure 13.4 Example of the activation of infrared active modes in the Raman spectrum for the case of indium deposition onto PTCDA. (see colour plates p. LXVIII)

the basis set LANL2DZ which takes into account possible relativistic effects due to the presence of the heavy metal atoms [5]. Four In atoms were assumed to interact with the PTCDA molecule via the O atoms in the anhydride groups according to Ref. [19]. The optimised geometry, the electronic levels and the charge distribution over the complexes are similar to those reported in Ref. [19]. Figure 13.5 shows the spectrum of a PTCDA single crystal together with the calculated frequencies for a single PTCDA molecule (rhombus) and for an In_4PTCDA complex (triangles).

The most dramatic effect of the complex formation resides in the two-fold splitting of the breathing mode at 233 cm^{-1} (see the elongation patterns in Fig-

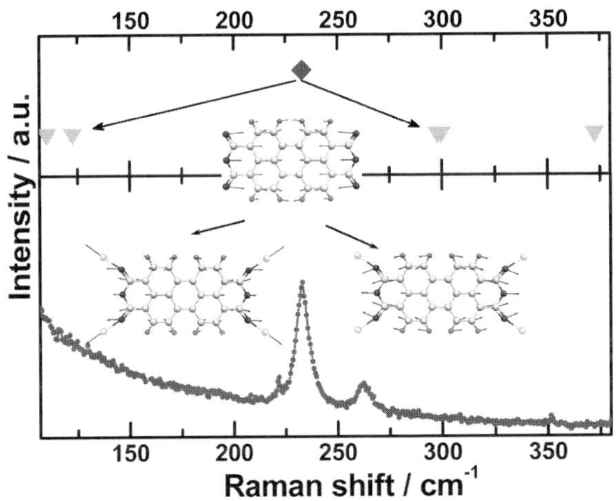

Figure 13.5 Raman spectrum of a PTCDA crystal along with calculated frequencies for a single molecule (rhombus) and for a In_4PTCDA complex (triangles). (see colour plates p. LXIX)

ure 13.5). A first component originates from a breathing of the whole complex and the significant increment of mass compared to the single PTCDA molecule results in a dramatic shift towards lower frequencies (110 cm^{-1}). For the second component the In atoms are fixed and constrain the breathing of the PTCDA molecule increasing its frequency to 299 cm^{-1}. In the experimental Raman spectra (Figure 13.1) of the In/PTCDA heterostructure, the molecular breathing mode is still observed at 233 cm^{-1}, ruling out the possibility of this chemical reaction. This conclusion is further supported by infrared spectroscopy studies [9] and by recent high resolution Photoemission Spectroscopy (PES) results [2].

In Figure 13.6 the Raman spectra of 15 nm films of PTCDA are shown for metal coverages of 5 nm In, 4.5 nm Ag, and 5 nm Mg. The spectra in the low frequency windows are normalised to the height of the molecular breathing mode at 233 cm^{-1}. The normalisation in the high frequency region is again performed with respect to the C=C stretching mode (1572 cm^{-1}).

The Raman spectra of the (5 nm) Mg/(15 nm) PTCDA system also exhibit the break-down of selection rules, with the occurrence of the modes observed in the other two metal/organic heterostructures (see Figure 13.6). In addition, several modes with significant intensity (marked with asterisks in Figure 13.6) appear at: 307 cm^{-1}, 502 cm^{-1}, 598 cm^{-1}, 696 cm^{-1}, 1088 cm^{-1} as well as at 1225 cm^{-1} and 1433 cm^{-1}. The frequency of the mode at 598 cm^{-1} is very close to the calculated value of 592 cm^{-1} for a B_{3g} mode of an isolated PTCDA molecule [4].

Frequency calculations performed with the same basis set and density functional methods in Gaussian'98 as in Ref. [4] but for a modified PTCDA mole-

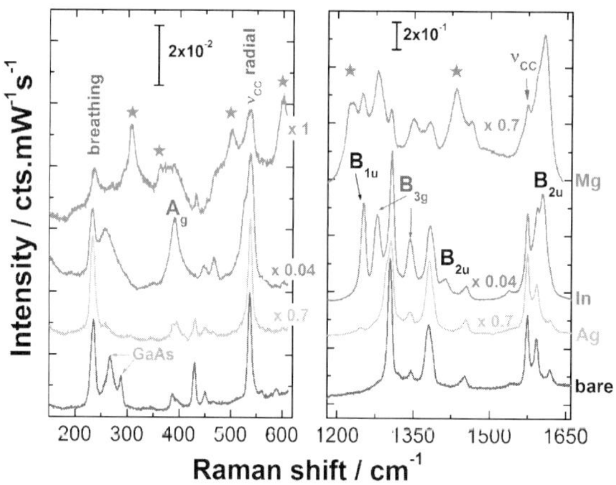

Figure 13.6 Raman spectra for In (5 nm), Ag (4,5 nm) and Mg (5 nm) coverages on 15 nm thick PTCDA films, compared with the spectrum of the bare PTCDA film in the spectral region of the internal breathing mode (left) and in the spectral region of C–H deformation and C=C stretching modes (right). (see colour plates p. LXIX)

cule, in which the central O atom in the anhydride group is removed, deliver several frequencies that may be candidates for the assignment of the experimentally observed modes: 308 cm^{-1}, 500 cm^{-1}, 581 cm^{-1}, 702 cm^{-1}, 1090 cm^{-1}. Raman active modes in MgO microcrystals [20] were observed at 595 cm^{-1}, 719 cm^{-1} and 1096 cm^{-1}. Thus the modes observed in the present work at 598 cm^{-1}, 696 cm^{-1} and 1088 cm^{-1} are very likely to indicate the formation of MgO as a result of the interaction between Mg and PTCDA. No modes of PTCDA or the modified molecule are found in the vicinity of 1225 cm^{-1}.

Whatever the final assignment of the new modes is, they are not activated in the molecules in contact with either Ag or with In. Therefore it can be concluded that the model of weak charge transfer is not sufficient to describe the interaction at the Mg/PTCDA interface.

To validate the model deduced from the results of these Raman spectroscopy experiments detailed investigations of the Mg/PTCDA system by means of other methods that are highly sensitive with respect to the changes of the chemical environment and charge redistribution such as photoemission spectroscopy were performed [2].

In Figure 13.7 the spectra of 15 nm DiMe-PTCDI films for metal coverages of 5 nm In, 4.5 nm Ag and 5 nm Mg are shown. The spectra in the low frequency windows are normalised to the height of the breathing mode at 221 cm^{-1}. The normalisation in the high frequency region is performed with respect to the C–C stretching mode (1570 cm^{-1}).

In the case of DiMe-PTCDI all the investigated metals, i.e. Ag, In and Mg, lead to the break-down of selection rules with the occurrence of normally infrared active modes at 1246 cm^{-1} and 1606 cm^{-1}. The breathing mode at 221 cm^{-1} survives with increasing metal coverage. Thus a chemical reaction

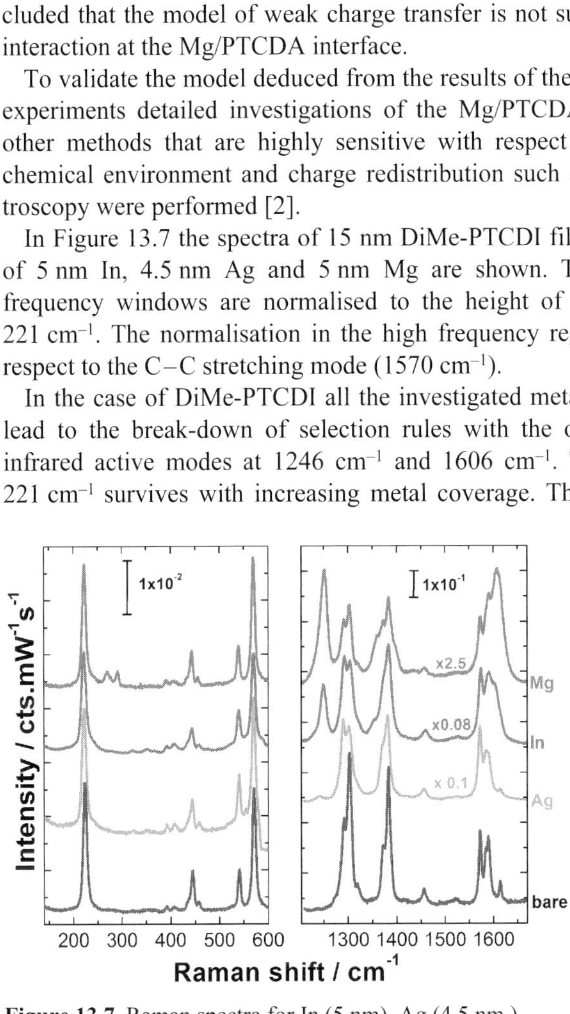

Figure 13.7 Raman spectra for In (5 nm), Ag (4,5 nm) and Mg (6 nm) coverages on 15 nm thick DiMe-PTCDI films, compared with the spectrum of the bare DiMe-PTCDI film. (see colour plates p. LXX)

between these metals and the O atoms of DiMe-PTCDI molecules can again be ruled out. Interestingly, the features potentially assigned to MgO phonons do not appear in the spectra even for higher coverages of Mg. It can thus be concluded that the methylimide group in the DiMe-PTCDI molecule is less reactive compared with the O atoms in the anhydride group of PTCDA.

13.3.2
Morphological Properties and Indiffusion of Metals at the Interfaces with Organic Semiconductors

Besides the occurrence of internal modes related to molecules in direct contact with the metal the totally symmetric modes are also enhanced (as shown by the normalisation factors in Figure 13.8 (a) and (b)) in the spectra of Ag, In and Mg on 15 nm thick PTCDA and DiMe-PTCDI films.

The latter effect originates from the coupling between the incident and scattered radiation with localized and/or collective plasmon resonances in the rough metal film. Therefore the intensity of the totally symmetric A_g modes is very sensitive to the morphology of the metal film.

For a quantitative determination of the enhancement factors curve fitting of each set of spectra recorded during silver, indium and magnesium deposition onto PTCDA and DiMe-PTCDI was performed using Lorentzian peaks. The dependence of relative area on metal coverage is plotted in Figure 13.6 for a representative totally symmetric mode and for a normally infrared active mode of each organic material. The relative intensities were calculated by dividing the intensities in the spectra at a given coverage to the initial intensities in the spectrum where the mode occurs for the first time. For example, the reference spectrum for the totally symmetric Raman band is that of the pure organic film,

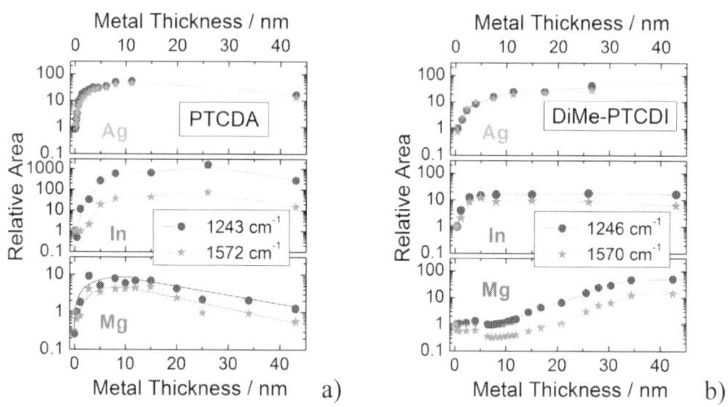

Figure 13.8 Enhancement factors of the B_u mode ($1243\ cm^{-1}$ in PTCDA and $1246\ cm^{-1}$ in DiMe-PTCDI) and of the C-C stretching A_g mode ($1572\ cm^{-1}$ in PTCDA and $1570\ cm^{-1}$ in DiMe-PTCDI) for PTCDA (a) and DiMe-PTCDI (b) as a function of the metal coverage. (see colour plates p. LXX)

film, while the reference spectrum for normally infrared active band is that taken after the first metal deposition.

The intensities of the A_g modes initially increase upon Ag and In deposition reflecting an increase in number and size of metal clusters as their plasmon energy approaches the energy of the laser electromagnetic field. When Mg is deposited onto PTCDA the intensities initially decrease, reflecting a reduction in number of Raman active PTCDA molecules. This is in good agreement with the conclusion drawn in the previous section regarding the disruption of PTCDA molecular structure upon reaction with Mg. In the next deposition step, i.e. at 2.8 nm Mg nominal coverage, the A_g Raman modes start to be enhanced indicating the formation of metallic clusters. The enhancement of the totally symmetric modes of DiMe-PTCDI occurs only above 15 nm nominal Mg thickness.

The PES studies (see Ref. [2]) showed that the metallic character of Mg clusters occurs at coverages above 1.6 nm for PTCDA and between 9 nm and 15 nm for DiMe-PTCDI. The large difference in the nominal thickness for which the metallic character of Mg clusters is formed on the two molecules is probably related to the different morphology of the underlying organic layer. The DiMe-PTCDI films have very large voids between the organic islands, while the PTCDA films are much more compact (see Figure 13.1).

The maximum enhancement of PTCDA modes for the Ag/PTCDA (15 nm) system is observed around 11 nm nominal Ag coverage (Figure 13.8 (a)) and corresponds to the optimum cluster size for the dipolar plasmon resonance. The In film thicknesses yielding the maximum enhancement for PTCDA and DiMe-PTCDI films are 26 nm and 5 nm, respectively. Further increase in the metal thickness leads to increasing size of the metal clusters associated with decreasing strength of the plasmon coupling with the incident and scattered radiation. Furthermore, the absorption in the metal film also plays an important role in decreasing the Raman signal for higher nominal coverages, when the clusters start to percolate.

The signal from PTCDA and DiMe-PTCDI internal modes remains visible even for a metal coverage of 43 nm, with higher intensity compared to the pure organic film. For Ag deposition onto DiMe-PTCDI no saturation of the signal intensity was observed up to a coverage of 263 nm.

Considering that I_0 is the intensity of the light incident on the sample, d is the nominal thickness of the metal coverage and δ is the light penetration depth in the metal, the light intensity I scattered by the sample can be described by:

$$I \propto I_0 \cdot e^{-2\frac{d}{\delta}} \qquad (1)$$

A summary of the values obtained from the fitting of the experimental decay of the enhancement factors for the totally symmetric C=C stretching mode in all investigated heterostructures is given in Table 13.1. The obtained values are much larger compared with the penetration depth of 488 nm light into smooth closed metal films. This is a clear indication that In and Ag films grown on

Table 13.1 Skin depth of smooth metallic films, apparent penetration depth of 488 nm light in In, Ag and Mg films grown on DiMe-PTCDI and PTCDA.

	In	Ag	Mg
$\delta_{\lambda = 488\ \mathrm{nm}}$ smooth metal film	8 nm	2.5 nm	14 nm
$\delta_{\lambda = 488\ \mathrm{nm}}$ (DiMe-PTCDI)	49 nm	50 nm	15 nm
$\delta_{\lambda = 488\ \mathrm{nm}}$ (PTCDA)	98 nm	–	24 nm

PTCDA and DiMe-PTCDI are not closed and have a high degree of roughness. The apparent light penetration depth in Mg films grown on PTCDA and DiMe-PTCDI estimated from the decrease in intensity of the C=C stretching mode has values comparable with the penetration depth in a closed smooth Mg film. This indicates that the Mg film is smoother and that the efficiency of the 488 nm radiation in exciting dipolar resonances is lower for Mg. The AFM topographic images in Figure 13.9 confirm the higher roughness of In compared with that of Mg films.

To recall, the A_g modes are enhanced via the long range electromagnetic effect, while the activation of B_u modes is characteristic for the molecules in intimate contact with the metal. Therefore the intensity of B_u modes relative to that of A_g modes will be considered in the following to extract the metal diffusion depth into the organic films.

In the case of Ag/PTCDA and Ag/DiMe-PTCDI the intensity of B_u modes is relatively low, indicating that only few molecules have intimate contact with Ag. This leads to the conclusion that Ag atoms diffuse very little into PTCDA layers. On the other hand, the B_u bands are stronger compared with the A_g modes in the spectra of In/PTCDA. This suggests that a large number of PTCDA molecules have intimate contact with the metal indicating a strong diffusion of In into PTCDA layers. In/DiMe-PTCDI represent an intermediate case between Ag/PTCDA and In/PTCDA.

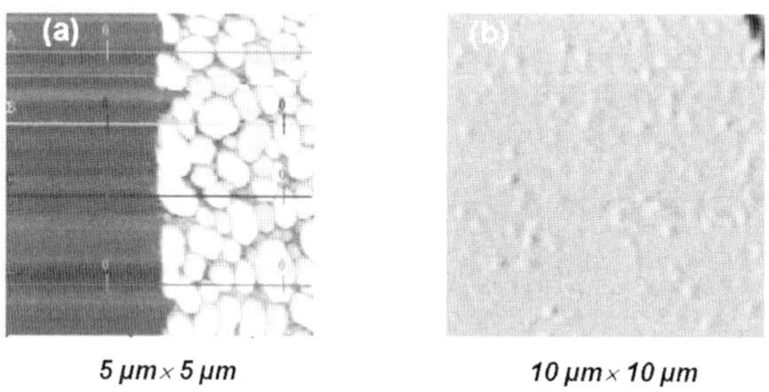

Figure 13.9 AFM topographic images of a 30 nm thick In film on PTCDA (a) (the left part of the image corresponds to the substrate and the right part of the image corresponds to the In film grown on PTCDA) and of a 113 nm thick Mg film on PTCDA (b). (see colour plates p. LXXI)

The ratio between the area of the B_u mode at 1243 cm^{-1} (1246 cm^{-1}) and that of the A_g mode at 1572 cm^{-1} (1570 cm^{-1}) in PTCDA (DiMe-PTCDI) is shown as a function of metal thickness in Figure 13.10. In the case of Ag/DiMe-PTCDI the maximum value is observed for the first Ag deposition, i.e. 0.4 nm Ag, whereas for PTCDA it increases up to a 1.3 nm nominal coverage of Ag. For In deposition onto both organics this ratio shows a saturation tendency only above 15 nm nominal In coverage, but its value is lower for In/DiMe-PTCDI. It is proposed that a maximum in the above defined ratio can be directly related to the metal diffusion length in the organic film. Thus the Ag atoms arriving at the organic film surface diffuse into the PTCDA islands up to a nominal Ag coverage of 1.3 nm.

For metal coverages at which the enhancement of the Raman signal due to dipolar plasmon resonances is observed, the B_u bands at around 1606 cm^{-1} (in both PTCDA and DiMe-PTCDI) become asymmetric towards the low frequency side in all the investigated systems except for Mg/PTCDA. This lineshape asymmetry is likely to be related to a Fano resonant coupling between the molecular electronic levels and the plasmons in the metallic clusters modulated by the molecular vibrations [21]. Interestingly, the mode at 1606 cm^{-1} stems from a breathing mode of the carbon rings.

This observation reinforces the conclusion that the interaction between the considered metals and PTCDA (DiMe-PTCDI) takes place via the perylene core.

The conclusions regarding diffusion of the metal atoms into the organic polycrystalline layers drawn from the enhancement factors of the internal modes are further confirmed by the spectral changes in the region of external modes of the organic layers. The external modes of PTCDA disappear almost completely already after depositing 0.4 nm In, while they only get broader and decrease in intensity after Ag and Mg deposition (Figure 13.11).

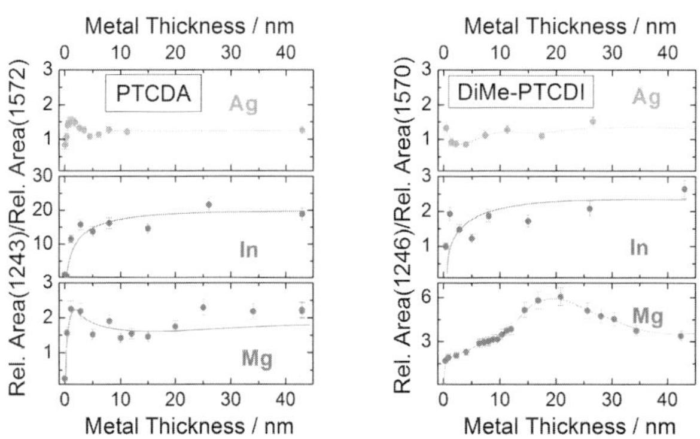

Figure 13.10 Ratio between enhancement factor of the B_u mode (1243 cm^{-1} in PTCDA and 1246 cm^{-1} in DiMe-PTCDI) and of the C–C stretching A_g mode (1572 cm^{-1} in PTCDA and 1570 cm^{-1} in DiMe-PTCDI) as a function of the metal coverage. (see colour plates p. LXXI)

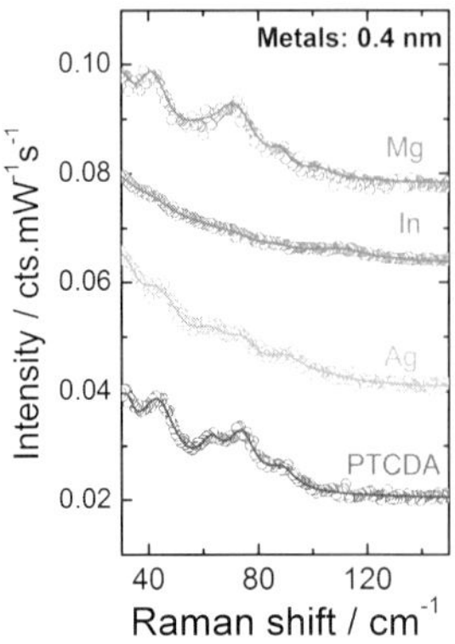

Figure 13.11 Raman spectra of 15 nm thick PTCDA films covered with Ag, In and Mg in the region of the external modes. (see colour plates p. LXXI)

In Figure 13.12 the spectra of the external modes are shown upon step-wise metal deposition onto 15 nm of PTCDA. For a quantitative evaluation the spectra of Ag/PTCDA and Mg/PTCDA were fitted using Lorentzian functions.

The evolution of the FWHM as a function of Ag and Mg thickness is plotted in Figure 13.13 for the external mode at 41 cm^{-1}. This mode is fairly well separated from its neighbours and hence the fitting parameters of the corresponding Lorentzian function are less correlated. As the metal thickness increases, the FWHM of the external modes increases faster in Ag/PTCDA compared with Mg/PTCDA.

For Mg the external modes are still visible at 12 nm coverage, whereas they are almost completely smeared out at 1.3 nm of Ag. This is a clear indication that the crystalline structure of the organic layers is less affected by the Mg deposition compared with Ag. However, it should be noted that curve fitting of

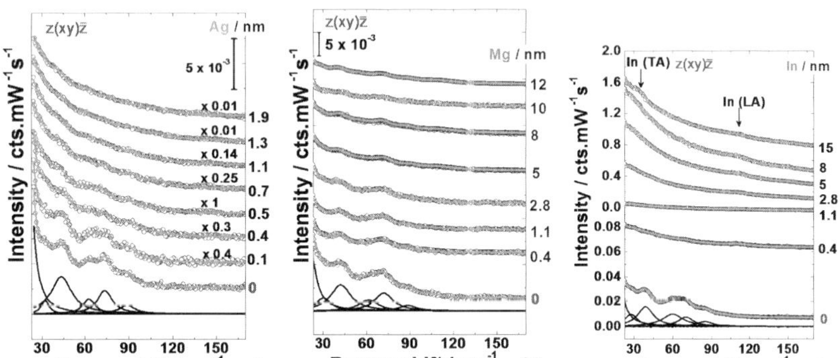

Figure 13.12 Raman monitoring in the external mode region upon metal deposition: Ag (a), Mg (b), and In (c). The experimental spectra are shown by open symbols and the fitted spectra by red lines. The Lorentzian functions used for curve fitting of the Raman spectrum of the pure PTCDA film are shown by lines in the lower part of the Figs. The spectra are normalised for Ag/PTCDA for a better observation of the phonons. (see colour plates p. LXXII)

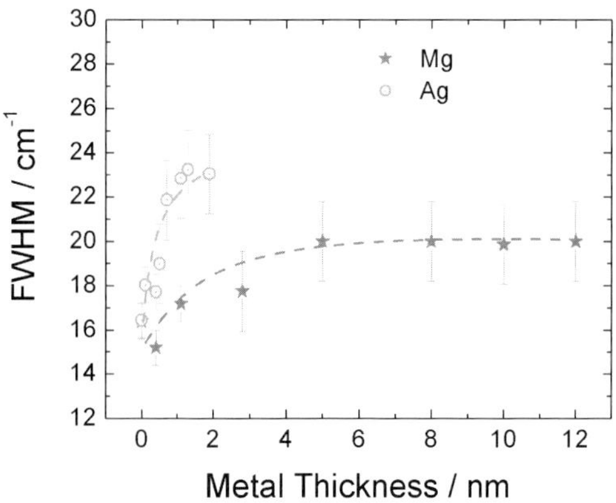

Figure 13.13 Evolution of the FWHM of the external mode in PTCDA at 41 cm^{-1} as a function of the metal coverage relative to the initial values before the metal deposition: for Ag (a) and Mg (b). The dashed lines are guidelines for the eyes. (see colour plates p. LXXII)

the spectra in the case of Ag/PTCDA is complicated by the significant increase in the low frequency background (which was already subtracted in Figure 13.11). The background evolution reflects an increasing degree of roughness, which is consistent with an increasing number of metallic clusters that diffusely scatter the light. A strong increase in the low frequency background is also observed for the case of In deposition onto PTCDA, while it hardly affects the spectra of Mg/PTCDA, supporting the conclusion that the roughening due to Mg is lower compared with that of the Ag and In.

While the external molecular modes disappear already in the first deposition stages for In/PTCDA, two new modes develop at 33 cm^{-1} and 112 cm^{-1} above In coverage of 2.8 nm. They may correspond to the transverse acoustic and longitudinal acoustic phonons located at 34 cm^{-1} and 114 cm^{-1}, respectively, in bulk indium [22, 23] activated due to the low dimension of the In clusters. This observation corroborated by the concomitant increase in the low frequency background indicates the formation of metallic In clusters. Moreover, the enhancement of the internal modes also increases dramatically above 2.8 nm In, supporting the conclusion of metallic cluster formation.

In the case of Mg deposition onto DiMe-PTCDI the external modes are attenuated only for a Mg nominal coverage above 21 nm (Figure 13.14), i.e. similar to the case of Mg/PTCDA, indicating that the Mg atoms do not protrude into the crystalline molecular islands. However, the ratio between the area of the B_u mode at 1246 cm^{-1} and that of the A_g mode at 1570 cm^{-1} (Figure 13.10) continuously increases up to about the same coverage. These two observations might seem contradictory when recalling the previous discussion on the correlation between the evolution of this ratio and the diffusion of the metal atoms. On the other hand, the relative intensities of the GaAs related features change already for low Mg coverages (Figure 13.14) reflecting a change in the GaAs band bending. This can only occur if the Mg atoms reach the GaAs surface in the large voids between the DiMe-PTCDI islands and react with it as also learned from the PES results [2].

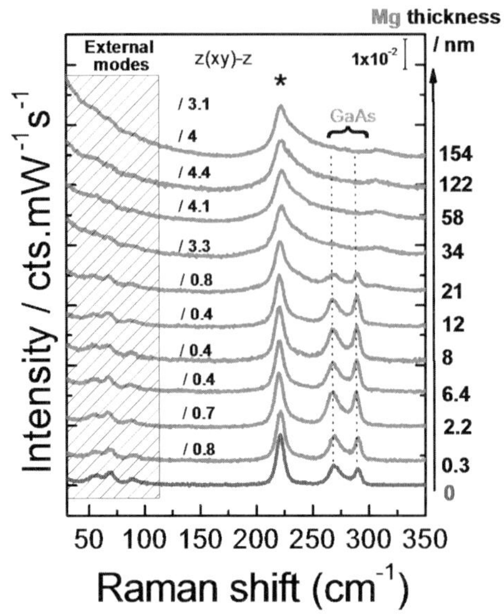

Figure 13.14 Raman spectra of DiMe-PTCDI upon Mg deposition in the region of external modes and GaAs phonons. The spectra are normalised with respect to the intensity of the breathing mode at 221 cm^{-1}. (see colour plates p. LXXIII)

13.3.3
Assignment of Raman Intensities with DFT Calculations

As discussed in Section 13.2.2, the deformation between the geometries in the relaxed excited state of the molecule and the ground state is the key to an understanding of the elongations of different internal vibrations. The projection of this deformation pattern onto the vibrational eigenvectors yields the respective Huang-Rhys factors S_j of the mode at the frequency ω_j. Concerning linear absorption, the intensities of the vibronic subbands are governed by a Poisson progression, $P_n = e^{-S_j} \cdot S_j^n/n!$, as observed e.g. for monomers in the gas phase or in weakly interacting surroundings like superfluid He [24, 25]. The intensities of vibronic bands involving several internal vibrations can be calculated from the product of their respective Poisson progressions.

The resonant Raman intensities are determined by the same Huang-Rhys factors as $I_j \propto S_j\omega_j^2$ to leading order. Therefore, when comparing two internal modes with similar Huang-Rhys factors, the mode at the higher frequency looks more prominent in the resonant Raman spectra, compare e.g. the PTCDA breathing modes at 233 and 1303 cm^{-1}. The Huang-Rhys factors calculated with turbomole5.7 using the B3LYP functional and a double-ζ (DZ) basis are shown in Figure 13.15 for PTCDA, DiMePTCDI, and Mg_2PTCDA_2 in the geometry visualised in Figure 13.16.

For PTCDA, the modes with the largest calculated Huang-Rhys factors compare favourably with the observed resonant Raman spectra. More specifically, we found the following correspondence between the most prominent calculated (observed) breathing modes: 233 (233) cm^{-1}, 540 (537) cm^{-1}, 627 (624) cm^{-1}, 1347 (1303) cm^{-1}, 1369 (1347) cm^{-1}, 1412 (1379) cm^{-1}, 1612 (1572) cm^{-1},

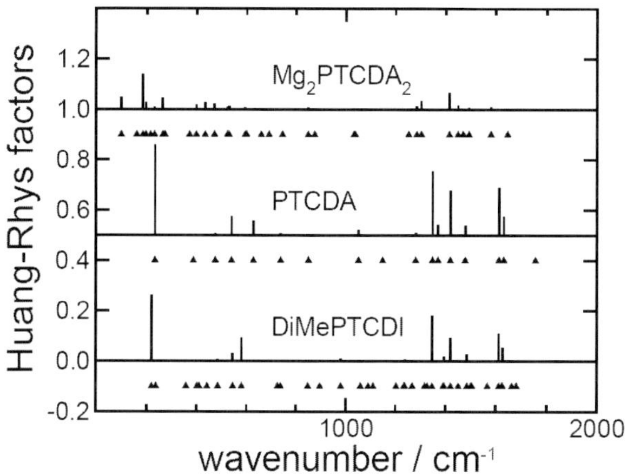

Figure 13.15 Calculated Huang-Rhys factors for DiMePTCDI, PTCDA, and a nonplanar Mg_2PTCDA_2 compound visualised in Figure 13.16. In all cases, the geometries have been optimised with B3LYP/DZ, and the Huang-Rhys factors are based on excited state geometries obtained with time-dependent DFT at the same level. For each compound, the positions of all breathing modes are indicated by triangles. The reference lines for the Huang-Rhys factors are shifted for clarity.

and 1632 (1492) cm^{-1}. As expected, the B3LYP functional gives very precise values for the low frequency modes because the harmonic approximation used for the diagonalisation of the dynamical matrix has only a minor influence. For high frequency modes, on the other hand, the harmonic approximation leads to an overestimate by about 3 percent with respect to the observed frequencies, a deficiency which is usually eliminated by scaling the calculated mode frequencies by a factor of about 0.97.

For DiMePTCDI, formally the smaller point group C_{2h} allows 46 breathing modes. However, when comparing PTCDA and DiMePTCDI, the most strongly elongated modes have a clear correspondence, e.g. for the lowest breathing mode at 220 (221) cm^{-1} in DiMePTCDI with respect to 233 (233) cm^{-1} in PTCDA. In general, for DiMePTCDI, the correspondence between calculated and observed intensities is again quite remarkable, with few exceptions, e.g. for pairs of vibrations very close in frequency, like the two observed modes at 1290 cm^{-1} and 1301 cm^{-1} where the calculated Huang-Rhys factor for the lower of the two modes is much too small. However, such a borrowing of intensities between modes close in frequency could also be the consequence of inter-molecular interactions in the crystalline phase, so that the wrong relative intensity of the two modes does not necessarily indicate a principal failure of the TD-DFT approach.

In the search for useful compounds based on PTCDA and Mg, we have investigated a planar object with 4 Mg atoms surrounding the molecule in a fashion resembling Figure 13.5 for In_4PTCDA, and a molecule with one Mg atom attached to each end. However, none of these candidates showed a substantial

(a)
(b)
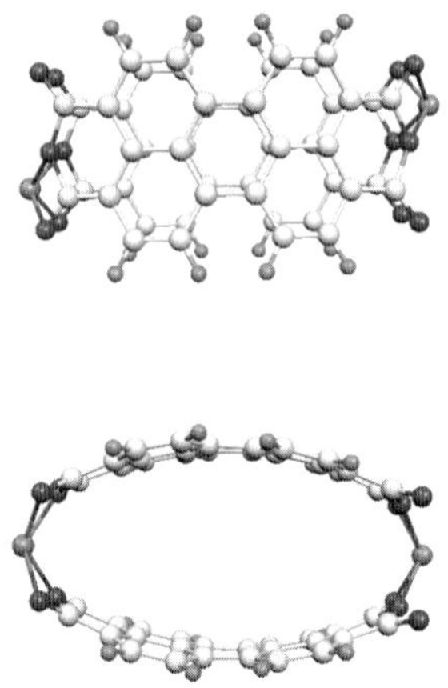

Figure 13.16 Model geometry of intercalated Mg ions, optimised at the B3LYP/DZ level, (a) perspective view along the molecular normal, (b) view along a crystal plane. This geometry was used for the calculation of the Huang-Rhys factors visualised in Figure 13.15. (see colour plates p. LXXIII)

binding energy, so that we presume that they do not play any role. Moreover, due to steric hindrance between neighbouring coplanar molecules and additional Mg atoms, such candidates are highly improbable.

Instead, we have realised that pairs of intercalated Mg atoms produce a substantial binding energy of –3.9 eV in the geometry visualised in Figure 13.16, and from the Mulliken charges we conclude that the net charge on the Mg sites is close to Mg^{+1}, compensated by oppositely charged $PTCDA^{-1}$. As the B3LYP functional does not account for dispersion interactions, the Pauli repulsion and the electrostatic repulsion between negatively charged PTCDA sheets are not counterbalanced by the van der Waals attraction, so that they repel each other substantially. However, in an extended model of this kind, the Ewald sums over the electrostatic interactions would still be favourable, and the Mg ions could then also connect adjacent PTCDA stacks, as proposed recently for potassium [26]. Due to the deficiencies of the B3LYP functional, we do not consider our model geometry to be realistic, but it merely represents the smallest finite size object reproducing the main features of the chemical binding of intercalated metal ions to pairs of stacked molecules.

From the Huang-Rhys factors of this Mg_2PTCDA_2 compound shown in Figure 13.15, we expect substantial resonant Raman intensities in regions where the PTCDA molecules do not show any breathing modes. This reproduces the observed trends, so that despite the limitations of the DFT geometry, our finite size approximant to Mg intercalation demonstrates that chemical interaction is necessary to produce substantial changes in the vibrational signature observed in resonant Raman spectra.

13.4 Conclusion

The deposition of several metals with different reactivity onto thin films of perylene-derivatives was monitored *in situ* by means of resonant Raman spectroscopy. Theoretical calculations of the resonant Raman spectra describe well the observed experimentally observed frequencies of the perylene-derivatives and enable thus a reliable assignment of the Raman modes, which is essential for the understanding of the chemical aspects of the interface formation. Raman spectroscopy has shown that Ag and In behave similarly in terms of chemical reactivity when deposited onto PTCDA and DiMe-PTCDI. Both metals induce a break-down of selection rules which was interpreted as a result of a dynamical charge transfer process. The vibrational signature of Mg deposited onto PTCDA is very different compared with the other metals, indicating that a strong chemical interaction between the two partners takes place. From investigations involving one or two PTCDA molecules and two to four Mg atoms, we found that intercalated Mg ions have a favourable binding energy. In such a model for intercalation, relatively large Huang-Rhys factors in regions where PTCDA does not have any breathing modes indicate that chemical interactions are a necessary condition for the dramatic changes of the resonant Raman intensities observed after evaporation of Mg onto PTCDA films.

Upon the Ag and In deposition the totally symmetric modes are initially strongly enhanced. Subsequently the signal is attenuated exponentially with an exponent that is much smaller than the penetration depth of the incident radiation in a smooth closed metal film, reflecting a high level of roughness of the metal overlayer. On the other hand, the intensity of the normally infrared active modes relative to the Raman modes provides information on the metal diffusion depth in the organic films.

Acknowledgements

This work was financially supported by the Deutsche Forschungsgemeinschaft within the OFET-SPP 1121 (Projects Za 146/14-2 and Scho 521/5-2).

References

1. G. Salvan, D. A. Tenne, T. U. Kampen, R. Scholz, G. Jungnickel, T. Frauenheim, and D. R. T. Zahn, Appl. Surf. Sci. **179,** 113 (2001).

2. D. R. T. Zahn, G. N. Gavrila, and G. Salvan, Chem. Rev. **107,** 1161 (2007).

3. R. Scholz, A.Yu. Kobitski, T. U. Kampen, M. Schreiber, D.R.T. Zahn, G. Jungnickel, M. Elstner, M. Sternberg, and T. Frauenheim, Phys. Rev. B **61**, 13659 (2000).

4. R. Scholz, *Organic Semiconductors*, in: *Encyclopedia of Condensed Matter Physics*, ed. by G. Bassani, G. Liedl, and P. Wyder, Elsevier (Oxford 2005).

5. Gaussian'98 (Revision A.1), M. J. Frisch, G. W. Trucks, H. B. Schlegel et al., Gaussian, Inc., Pittsburgh PA, **1998**.

6. R. Ahlrichs, M. Bär, M. Häser, H. Horn and C. Kölmel, Chem. Phys. Lett. **162**, 165 (1989).

7. A. Yu. Kobitski, R. Scholz, and D. R. T. Zahn, J. Molec. Struct. (Theochem). **625,** 39 (2003).

8. G. Salvan, D. A. Tenne, T.U. Kampen, R. Scholz, G. Jungnickel, Th. Frauenheim, and D. R. T. Zahn, Appl. Surf. Sci. **179**, 113 (2001).

9. D. R. T. Zahn, G. Salvan, B. A. Paez, and R. Scholz, J. Vac. Sci. Tech. A **22,** 1482 (2004).

10. F. Brouers, S. Blacher, A. N. Lagarkov, A. K. Sarychev, P. Gadenne and V. M. Shalaev, Phys. Rev. B **55**, 13234, (1997).

11. J. A. Creighton, Spectroscopy of Surfaces, edited by R. J. Clark and R. E. Hester, John Wiley and Sons Ltd. (1988).

12. A. Otto, I. Mrozek, H. Graborn, W. Akemann, J. Phys.: Condens. Matter. **4**, 1134 (1992).

13. K. Kneipp, H. Kneipp, I. Itzkan, R. R. Dasari, M. S. Feld, Chem. Rev. **99**, 2957 (1999).

14. A. Otto, phys. stat. sol. (a), **188–189**, 1455 (2001).

15. A. Otto, I. Mrozek, H. Graborn, W. Akemann, J. Phys.: Condens. Matter. **4**, 1134 (1992).

16. A. Yu. Kobitski, G. Salvan, R. Scholz, D. Tenne, T. U. Kampen, H. P. Wagner, D. R. T. Zahn, Appl. Surf. Sci. **190**, 386 (2002).

17. G. Salvan and D. R. T. Zahn, Europhys. Lett. **67**, 827 (2004).

18. Y. Hirose, A. Kahn, V. Aristov, P. Soukiassian, V. Bulovic, S. R. Forrest, Phys. Rev. B **54**, 13748 (1996).

19. S. Kera, H. Setoyama, M. Onoue, K. Okudaira, Y. Harada, N. Ueno, Phys. Rev. B **63**, 115204 (2001).

20. H. K. Böckelmann and R. G. Schlecht, Phys. Rev. B. **10**, 5225 (1974).

21. B.A. Paez, G. Salvan, R. Scholz, D. R. T. Zahn, phys. stat. sol. (c). **2**, 4048 **(2005)**.

22. K. Fleischer, S. Chandola, N. Esser, W. Richter, and J. F. McGilp, Phys. Rev. B **67**, 235318 (2003).

23. A. S. Bakulin, A. W. Overhauser, H. Kaiser, S. A. Werner, J. A. Fernandez-Baca, H. G. Smith, Phys. Rev. B **63**, 052301 (2001).

24. M. Wewer and F. Stienkemeier, Phys. Rev. B. **67**, 125201 (2003).

25. R. Scholz and M. Schreiber, Chem. Phys. **325**, 9 (2006).

26. C. Zazza, S. Meloni, A. Palma, M. Knupfer, G. G. Fuentes, and R. Car , Phys. Rev. Lett. **98**, 046401 (2007).

14
Development of Single-Crystal OFETs Prepared on Well-Ordered Sapphire Substrates

S. Sachs, M. Paul, F. Holch, J. Pernpeintner, P. Vrdoljak, M. B. Casu, A. Schöll, and E. Umbach

14.1
Introduction

Field effect transistors (FETs) of inorganic materials, e.g. GaAs, can be optimised to demonstrate extremely good properties such as high charge carrier mobilities beyond 10^7 cm^2/Vs. Organic thin film FETs are considered to be easily and cheaply producible and to be useful for all-organic devices. They perform well if their carrier mobilities are beyond 1 cm^2/Vs. Organic single crystal mobilities may reach 100 cm^2/Vs at low temperatures if the mobilties are measured contactless [1, 2]. Thus the question arises, how organic thin film FETs can be optimised and which properties, e.g., charge carrier mobilities can be obtained at best.

In the present work we make an attempt to approach the "high end" of organic FETs (OFETs). For this purpose various aspects have to be considered. For high mobilities the active medium should be single crystalline; in particular the layers at the interface to the gate dielectric should be defect free, since the charge transport mainly occurs in this region. This can be accomplished by, e.g., growing a macroscopic single crystal and subsequently attaching metal (source and drain) and insulator (gate) contacts. The handling of organic single crystals and the deposition of metal contacts, however, are difficult tasks. The latter almost indispensably results in structural defects, morphological roughening of the important interface region, and possibly uncontrolled chemical reactions at the interface due to diffusion of oxygen and water from the air. These influences are particularly problematic for the gate insulator since the possible methods to deposit the insulator on the organic material either include wet chemical processes, which cannot be utilised in a clean UHV environment and thus inevitably imply contamination problems or, like sputtering, involve high energies of the respective atoms yielding substantial damage at the organic interface. In particular, the most important organic layers in the device, namely those in which the charge transport occurs, are in direct contact to the gate insulator and hence will be most influenced by the application of the gate insulator.

Organic Electronics. Structural and Electronic Properties of OFETs. Edited by Christof Wöll

ISBN: 978-3-527-40810-8

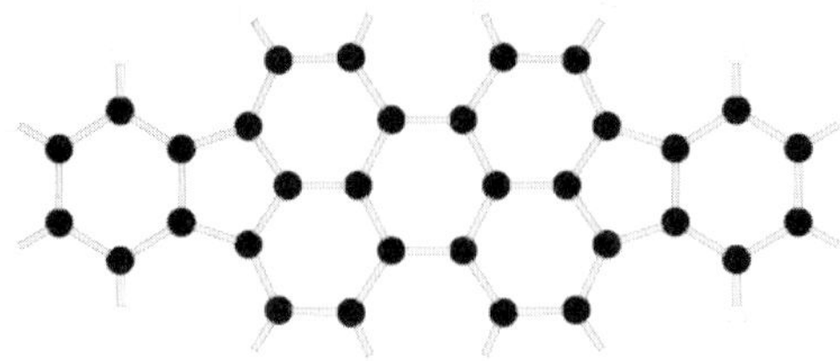

Figure 14.1 Sketch of the molecule diindenoperylene (DIP)

In this work we thus choose an alternative approach. We try to grow polycrystalline layers with micro-crystallites of the active material on a defect free, single crystalline, and atomically flat insulator substrate which is prepared according to the recipes and characterisation methods provided by surface science. The substrate of choice is a sapphire $(11\bar{2}0)$ surface, and diindenoperylene (DIP) (see Figure 14.1) has been selected as organic material. We note that other groups have tried alternative approaches, for instance depositing single crystallites on bottom contacts [3–5].

14.1.1
The Present Micro-OFET Concept

The present (micro-)OFET concept is shown in Figure 14.2 and based on the common approach: gate insulator (with gate electrode underneath) as substrate, then deposition of organic material, then application of source and drain contacts, and finally connection of the OFET to three leads. In the present approach we concentrate on drastically improving a) the interface between gate electrode and organic layer, b) the structural and defect quality of the organic material, and c) the interfaces to the metal contacts.

High structural quality and minimisation of defects demand for an *in situ* UHV preparation of the substrate surface, deposition of the active organic layer, and application of the metal contacts without intermediate exposure to

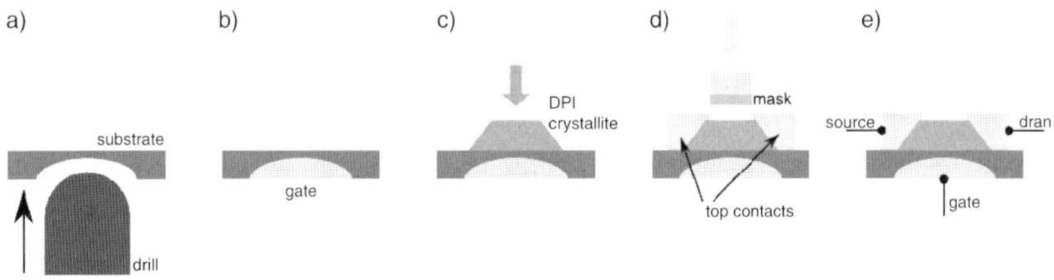

Figure 14.2 Illustration of our micro-OFET concept: (a) Thinning of the sapphire substrate from the backside to achieve a thin gate dielectric; (b) mounting of the metal gate electrode by thermal evaporation; (c) deposition of the active DIP layer by OMBD; (d) application of the source and drain top contacts by Au deposition using a lithographic mask; e) electrical characterization. (see colour plates p. LXXIII)

ambient conditions. However, not all possible (gate) insulators, mainly oxides, can be handled accordingly. In various cases the UHV cleaning by Ar-sputtering and annealing leads to an oxygen deficiency at the surface and consequently to structural defects. Sapphire (α-Al_2O_3) is known to be an exception in this respect, and recipes for the UHV treatment exist in literature, e.g. for the $(11\bar{2}0)$ surface [6]. Prior to the UHV treatment the substrate has to be thinned out (Figure 14.2a) to achieve an as thin as possible gate insulator. After the sputter-annealing of the top surface, the gate electrode is deposited by *in situ* thermal evaporation of metal (Figure 14.2b) from the backside.

Various organic molecules can be deposited by organic molecular beam deposition (OMBD) in the UHV yielding films almost free of impurities. We here use diindenoperylene (DIP; see sketch in Figure 14.1) as model material since it has shown rather high charge carrier mobilities of up to 10^{-1} cm^2/Vs in thin films [7] and forms a morphologically relatively flat contact interface with Au with only little interdiffusion [8, 9]. Given that the preparation conditions are optimised accordingly, layers with (micro-)crystalline structures can be achieved (Figure 14.2c) [10]. In the next step one large enough (micro-) crystallite is selected and is then contacted by thermal deposition of Au using an optical microscope together with micromanipulation of a shadow mask (Figure 14.2d). Finally, the leads can be connected and the electrical device characteristics can be determined by standard methods (Figure 14.2e).

After a brief introduction into the experimental characterisation methods (Section 14.2) we will in the following (Section 14.3) report on our progress in realising an optimised micro-OFET. The effort is split into four sections, which address the preparation of the sapphire substrate (Section 14.3.1.1), the growth of DIP layers on sapphire (Section 14.3.1.2), the application of Au-contacts (Section 14.3.1.3), and the mounting of the gate electrode (Section 14.3.1.4). In addition, many other important aspects, which are directly related to OFETs and the present approach, were also investigated within this project but these cannot be addressed in the present publication and hence are, or will be, published elsewhere. Amongst these aspects are the growth dynamics of organic layers [11], the orientational behaviour and interface properties of various organic molecules suited for OFETs on insulator surfaces [12–14], the surface or interface polarisation behaviour, and hence charge energetics, of organic molecules [15], and the determination of the charge transport levels [16].

14.2 Experimental

Organic films on sapphire were prepared in various UHV chambers with base pressures of less than 5×10^{-10} mbar. Chemical composition and cleanliness of all samples were determined with XPS using Mg-Kα radiation from a laboratory source and a hemispherical electron analyser. The preparation and characterisation of the sapphire substrates are discussed in Section 14.3.1.1.

The organic material was evaporated onto the samples with a home-made Knudsen-cell type evaporator at temperatures between 580 and 640 K resulting in growth rates of 0.4 to 1.0 nm/min. The film thicknesses were established by calibrated evaporators (accuracy 10%) and by measuring the attenuation of the substrate signal by the adsorbed films using XPS, applying an energy dependent inelastic electron mean free path for PMMA [17], and an exponential decay model for the attenuation. Monolayer films of DIP on Au were prepared by thermal desorption of a multilayer film at 570 K.

The Au films were evaporated from different resistively heated sources as well as from an electron beam evaporator. The Au film thicknesses and the growth rates were determined during growth by a quartz micro balance.

The synchrotron experiments were performed at beamline UE52-PGM at BESSY, Berlin. Substrates and films were prepared at the end station in a UHV preparation chamber at a base pressure of 5×10^{-10} mbar and transferred into the measurement chamber without exposure to air. The NEXAFS spectra on sapphire substrates were recorded by measuring the partial electron yield with a dedicated channeltron detector, the spectra on Au substrates were measured using the sample current. Energy calibration and intensity normalisation were carefully done according to [18].

The AFM experiments were performed under ambient conditions with a "Topometrix Explorer" AFM. The sapphire substrate samples were investigated in contact mode. The organic films were examined in tapping mode in order to decrease the interaction of the tip with the sample. The resolution of the AFM is 50 nm in lateral and less than 1 nm in vertical direction.

14.3 Results and Discussion

14.3.1 Realisation of the Micro-OFET Concept

14.3.1.1 Sapphire Substrate

Sapphire $(11\bar{2}0)$-oriented single crystals with a size of $10 \times 10 \times 1$ mm^3 and a nominal miscut of less than 1° were purchased from "Mateck GmbH", Jülich. Following the preparation procedure of [6] the samples were initially cleaned with acetone and ethanol in an ultrasonic bath and subsequently heated for 5 hours at 1550 K under ambient conditions. An XPS spectrum of the sample after the ambient treatment and before any further UHV cleaning is displayed in the lower part of Figure 14.3a. The spectrum clearly shows the surface contamination by carbon containing species (most likely adsorbed hydrocarbons) which result in a marked C 1s signal at about 290 eV binding energy. After transfer into the UHV the substrates were sputtered with 1 keV Ar^+-ions for 1 h at a constant sample temperature of 800 K and annealed for 15 min at 1150 K. This procedure leads to an almost perfect removal of carbon contaminants as can be observed in the spectrum recorded after sputter/annealing in the upper part of Figure 14.3a. Moreover, the stoichiometry of the surface is not altered significantly, which can be deduced from the intensity ratio of the Al 2p and O 1s signals. The O 1s/Al 2p ratio increases by about 5% after sputter/annealing, which is within the error bar of the data evaluation. This shows that oxygen depletion at the sapphire surface, which usually accompanies excessive heat treatment of many oxide materials in an oxygen free atmosphere, does not occur in the present case.

Figure 14.3b presents a LEED picture of the sapphire $(11\bar{2}0)$ surface after the UHV sputter/annealing treatment. Note that charging does complicate the

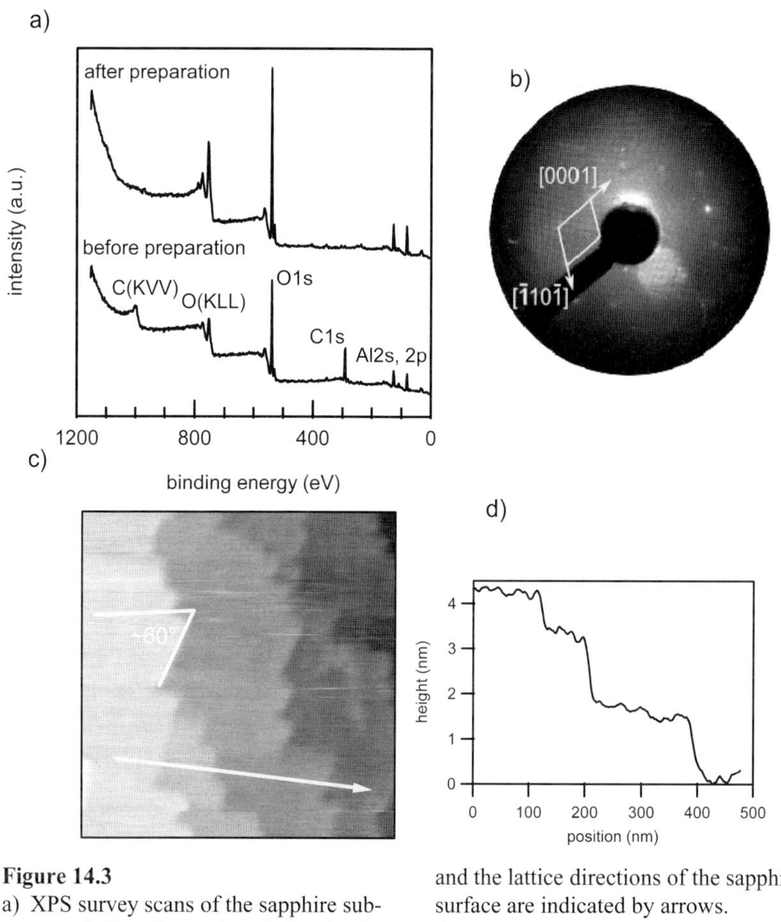

Figure 14.3
a) XPS survey scans of the sapphire substrate before (bottom) and after (top) UHV preparation by sputtering and annealing. The most prominent signals are denominated.
b) LEED picture of the sapphire substrate after UHV preparation. The unit cell and the lattice directions of the sapphire surface are indicated by arrows.
c) Contact mode AFM picture of a sapphire substrate after UHV preparation (window size: 500 nm × 500 nm).
d) Height profile along the line indicated by the arrow in Figure 14.3c.
(see colour plates p. LXXIV)

characterisation of the sample by conventional electron diffraction. This leads to high background intensity particularly if energies below 80 eV and normal incidence geometry are being used. For the present LEED picture the sample was thus rotated by 15° off normal direction; an electron energy of 126 eV was applied in order to achieve the best signal to background ratio. Thus we succeeded to obtain LEED pictures, though with low intensity. Nevertheless, the (preserved) long range order can be demonstrated by the relatively sharp diffraction spots, which are clearly distinguishable. The unit cell and the direction of the [0001] and $[\bar{1}10\bar{1}]$ unit cell vectors are depicted on the left and reflect the expected sapphire $(11\bar{2}0)$ surface geometry. No indication of a possible surface reconstruction was found.

Since an atomically flat surface is a prerequisite for the desired single crystal growth the surface morphology was checked by AFM. A typical picture of a 500 nm × 500 nm sampling spot is displayed in Figure 14.3c and shows a characteristic stepped morphology. A typical line scan as displayed in Figure 14.3d along the direction marked by an arrow in Figure 14.3c indicates that the surface consists of atomically flat terraces separated by steps. The preferential step directions are illustrated in Figure 14.3c; the indicated angle of ~60° fits well to the expected angle of 58.5° between the [0001] and $[\bar{1}10\bar{1}]$ directions, thus suggesting a preferential step orientation perpendicular to the [0001] and $[\bar{1}10\bar{1}]$ direction. The terrace widths are 100–200 nm. The heights of the steps that separate the terraces are in the order of 1–2 nm for this sample. The present step and terrace morphology is caused by the miscut angle of the polished sapphire substrate, which was determined from the AFM data to be 0.6° for this sample. Employing substrates with a lower miscut angle should thus allow the preparation of much larger terraces.

14.3.1.2 **Growth of DIP on Sapphire**

In the case of planar aromatic molecules the charge carrier mobility is known to be strongly anisotropic. The preferred electron (or hole) transport occurs in the direction perpendicular to the molecular plane, since the hopping barrier between the molecules is reduced in this direction due to the $\pi-\pi$-interaction of the aromatic system. Thus an upright molecular orientation is desired for a bottom-gate OFET geometry, since it helps to maximise the mobility in the direction parallel to the substrate surface where the conductive channel between source and drain is established when an electric field is built up by applying a voltage to the gate electrode. In order to achieve this kind of molecular orientation the growth parameters during OMBD have to be optimised accordingly.

Figure 14.4a shows some NEXAFS spectra from samples prepared at constant growth rate (0.9 nm/min) but at three different sample temperatures of 101, 303, and 373 K. The spectra were recorded with s- and p-polarisation of the incoming light; the respective experimental geometries are depicted in Figure 14.4d. The spectra are dominated by sharp pre-edge resonances at about 285 eV, which can be associated with transitions from C 1s core states into the lowest unoccupied molecular π^*-orbitals. The average molecular orientation,

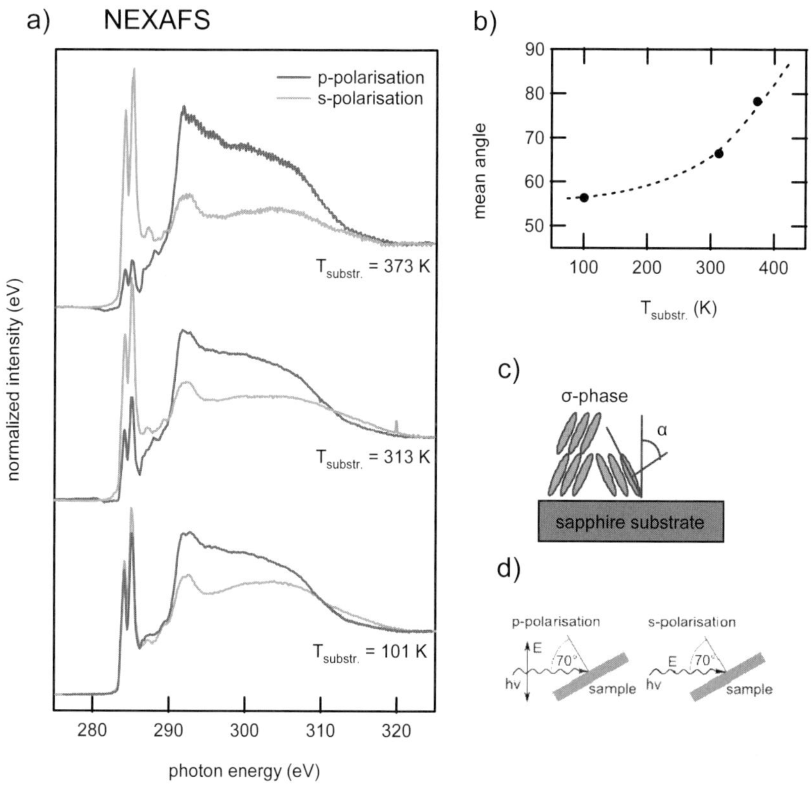

Figure 14.4
a) C K-NEXAFS spectra of DIP samples prepared at different substrate temperatures (373, 313, and 101 K), recorded in the partial electron yield (PEY) mode with p- (blue) and s-polarisation (red) of the incident X-rays.
b) Plot of the average molecular tilt angle versus substrate temperature deduced by evaluation of the NEXAFS dichroism from Figure 14.4a. The red curve is a guideline to the eye.
c) Illustration of the molecular tilt angle for growth in the σ-phase (from [24]).
d) Experimental geometry of the NEXAFS experiment. (see colour plates p. LXXV

i.e. the tilt angle α of the molecular planes relative to the surface, can now be determined from the NEXAFS dichroism, i.e. the polarisation dependent intensity of the π*-resonances, utilising the well-established selection rules and simple evaluation formulae of the matrix elements [19, 20]. The respective values are plotted versus the substrate temperature in Figure 14.4b. Although the number of data points does not allow a detailed statement regarding the exact form of the resulting curve – the exponential curve plotted in Figure 14.4b is intended to serve as a mere guideline to the eye – a clear increase of the average molecular tilt angle α is observed if the substrate temperature is increased.

This behaviour is expected, since with increasing substrate temperature the molecular mobility on the surface is increased. Since the molecules interact only weakly with the relatively inert sapphire surface, the mobility leads to the energetically preferred molecular arrangement. The interaction with the substrate is – according to the desorption temperatures – even weaker than that between the molecules because the latter involves an optimised intermolecular $\pi-\pi$-interaction leading to micro-crystallites. For many planar (homo)aromatic molecules such as DIP the crystalline structure is such that the molecules are arranged in a herringbone structure with two or four molecules per unit cell standing roughly perpendicular to cleavage planes. Since the cleavage plane of a large enough crystallite is preferentially oriented parallel to the surface of a hardly or non-interacting substrate, the molecules hence take the preferred perpendicular orientation with respect to the substrate leading to an enhanced π-overlap and hence enhanced charge mobility parallel to the surface.

A similar orientation behaviour was also found for other aromatic compounds on oxide surfaces, e.g. for phthalocyanines on sapphire [21, 22]. To understand the orientational ordering of aromatic molecules in more detail we have systematically studied the growth of perylene on $Al_2O_3/Ni_3Al(111)$. This substrate system serves as a model for sapphire since the highly-ordered thin Al_2O_3 film that can be produced by controlled oxidation of a $Ni_3Al(111)$ crystal resembles very much a perfect sapphire surface and can much better be utilised for surface studies using, e.g., electron spectroscopies because no charging problems occur. Also in this investigation, which will be published elsewhere [23], a preferred upright orientation of the molecules was found. Moreover, the observed average tilt angle of about 75° for DIP/sapphire is very similar to the σ-modification of DIP on SiO_2 (see illustration in Figure 14.4c and Ref. [24], which has shown high structural quality. This finding corroborates the present interpretation since the substrate interaction in the latter case is similarly weak and hence a similar film structure is likely.

We briefly note that annealing to 430 K of thin DIP films (few nm thick) deposited at room temperature leads to an average tilt angle of about 54° as determined by NEXAFS. Since this is the so-called magic angle this result most likely means that the crystallites formed by annealing have random orientation. Thus annealing not always leads to optimum results, at least not in terms of molecular orientation.

Figure 14.5 shows a typical AFM picture of a nominally 2 nm thick DIP film prepared at a substrate temperature of 328 K. The sampling area is 2 μm × 2 μm. The picture shows a competing growth of layers and 3D islands, which is illustrated by the height profiles plotted next to the 2D-picture in Figure 14.5. While height profile 2 cuts through an island with a height of ~ 25–30 nm profile 1 documents a relatively smoothly covered surface region. The latter scan shows the profile across a large flat island (light bright area) of 200 nm × 200 nm size with a height variation of less than a 1 nm corresponding to the morphology of the sapphire substrate. At the edges of the profile the height of the island can be measured which amounts to about 1.5 nm which fits

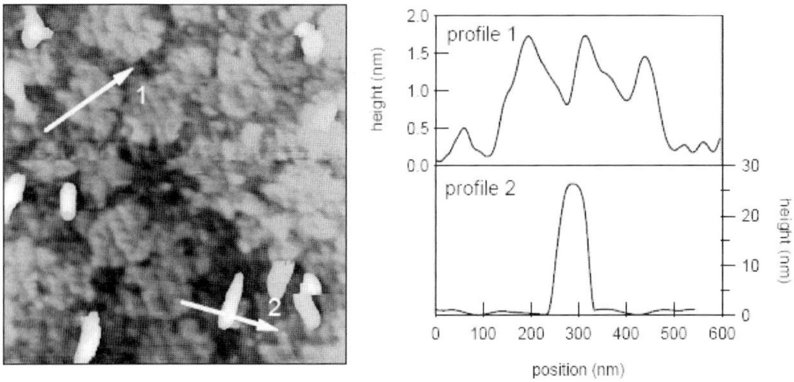

Figure 14.5 Tapping mode AFM picture (window size: 2 μm × 2 μm) of a nominally 2 nm thick DIP film deposited on a sapphire surface at a substrate temperature of 330 K. The profiles were extracted along the lines indicated by arrows in the AFM picture. (see colour plates p. LXXVI)

very well to the height of a DIP monolayer in the σ-phase (nearly upright standing molecules) of 1.65 nm [10]. Thus it can be concluded that the sapphire surface is partly covered by (3-dimensional) micro-crystallites (profile 2) and partly by monolayer domains that grow even across sapphire terrace steps. From X-ray diffraction experiments [24], it can be derived that the 3D islands are grown in the λ-phase. These two competing phases can nicely explain the observed NEXAFS data: the average orientation angle of molecules grown at room temperature is the average of two phases with σ- and λ-orientation. For higher growth temperatures the σ-phase is favoured and hence the mean orientation angle α is increased.

14.3.1.3 **Contacts – the Au/DIP Interface**

The fabrication of metal contacts to an organic film or single crystal is generally a demanding task. However, the electronic structure and morphology at the contact interface are of prime importance for the charge carrier injection. Moreover, the interaction at the interface may lead to the appearance of new interface states, which can substantially influence the injection and hence the transport properties [25, 26]. This is particularly the case for chemically interacting systems [27] for which the interaction has a crucial influence on the molecular frontier orbitals at the interface and consequently on the charge transport through this interface. Especially reactive metals may (partially) dissociate the adjacent molecules and form new metal-organic compounds which can have a negative influence on the electronic properties, e.g. by forming a barrier for the charge carriers, and which may be unstable under ambient conditions.

To avoid such effects relatively inert metals such as Au are often applied as contacts. However, also in this case covalently interacting organic adsorbates are reported. For the systems DIP/Au and Au/DIP former studies revealed a weak interaction (physisorption) between molecules and metal, which was somewhat influenced by the contact morphology [28]. In order to characterise the bonding mechanism between DIP and Au we started with XPS and NEXAFS measurements on DIP layers deposited on a Au(111) single crystal substrate. We employed this well-defined, morphologically flat model system with high structural order and no spurious influence of interdiffusion at the interface in order to get clear and unambiguous data.

The resulting spectra are displayed in Figure 14.6. Figure 14.6a shows the energy regime of the π^*-resonances in the C 1s NEXAFS spectra of a 10 nm thick film (top) compared with a single monolayer spectrum (bottom). Both sets were recorded with p-polarisation of the incident synchrotron light. Although the relatively low signal and the consequently demanding normalisation at the C K-edge [18] leads to some fluctuations in the monolayer data the spectroscopic signature is evidently not altered if the molecules are in direct contact to the Au substrate. The observation is a strong argument against a significant chemisorptive contribution to the interface interaction, since the effect of wave function overlap and/or static charge transfer involved in the covalent interaction generally leads to marked changes to the NEXAFS data [27, 29]. Nevertheless, the fact that a monolayer can be achieved by thermal desorption of a multilayer (see Experimental, Section 14.2) means that the molecules interact stronger with the Au surface than amongst themselves.

The XPS spectra of the multi- and monolayer samples are compared in Figure 14.6b. Each spectrum is dominated by one peak, which is centred at 284.4 eV in the multilayer (top spectrum) and at 284.0 eV in the monolayer data (bottom spectrum). These photoemission signals stem from the DIP C 1s levels and generally comprise the contributions from 9 symmetrically inequivalent carbon atoms. The chemical shifts of the different C 1s sites, however, are apparently very small and can thus not be resolved considering the resolution limiting effects of natural line width, experimental contributions (source, spectrometer), vibronic and inhomogeneous broadening. The overall shift of about 0.4 eV is due to a better polarisation screening of the C 1s core hole in the vicinity of the Au surface ("image potential screening"), which leads to a reduced binding energy of the photoelectrons from the molecules in the first layer.

Also in the photoemission case the signature of the multilayer spectrum is essentially conserved at the interface, but in this case minor changes occur. The peak at 286.2 eV in the multilayer spectrum, which is due to a shake up excitation, is almost completely quenched for the monolayer data. Moreover, the symmetric line shape of the multilayer C 1s line is slightly changed in the monolayer; it becomes more asymmetric with additional spectral weight on the low energy side. At the same time the line width is slightly reduced from 0.7 eV to about 0.6 eV. While the latter observation could be explained by a

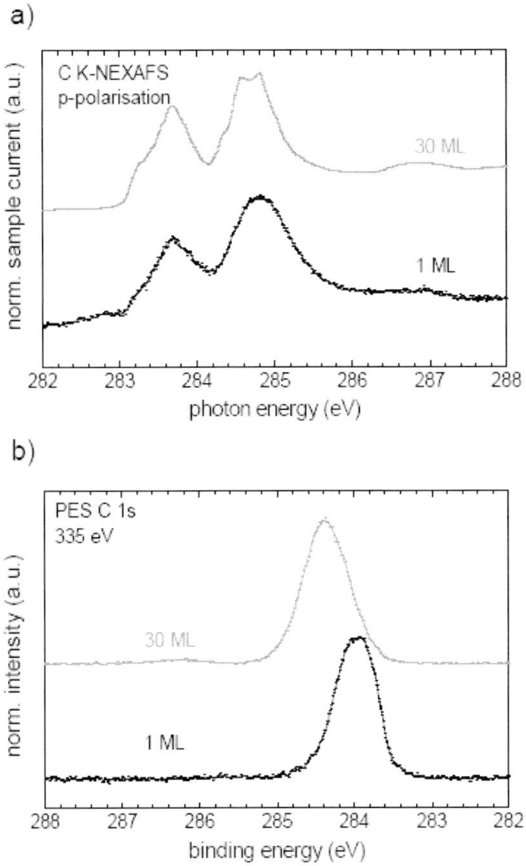

Figure 14.6
a) Pre-edge regime of high-resolution C K-NEXAFS spectra of a 30 ML (top, red) and a 1 ML thick (bottom, black) DIP film on Au(111). The spectra were recorded by measuring the sample current with p-polarisation of the incident X-rays.
b) High-resolution C 1s XPS spectra of a 30 ML (top, red) and a 1 ML thick (bottom, black) DIP film on Au(111). The spectra were recorded using a photon energy of 335 eV in normal emission geometry. (see colour plates p. LXXVII)

reduction of the layer inhomogeneity in the monolayer case, the change of the line profile demands for a more comprehensive explanation. Since furthermore a chemical interaction at the interface is not plausible as implied by the NEXAFS data, the changes can likely be understood as being due to a change of the vibronic envelope of the C 1s photoemission line. This is corroborated by the NEXAFS data. Also in the NEXAFS spectra the vibronic fine structure, which can be clearly observed as a series of shoulders to the 2nd resonance at 284.6 eV, is quenched in the case of the monolayer. A possible explanation for this effect is that the screening by the metal electrons drastically changes the vibronic coupling in the monolayer molecules and hence reduces the intensity of the vibronically excited states, an effect that is often more or less directly observed for physisorbed molecules on (weakly or not interacting) metallic substrates.

The investigation of an idealised interface, like DIP on a Au(111) single crystal surface, is a mandatory prerequisite to understand the fundamental processes involved in the interaction. However, it may not be representative for the situation at a real device contact. The evaporation of metal atoms onto an organic layer generally leads to a morphologically rough interface with unavoidable diffusion of metal atoms into the film. Figure 14.7a shows an AFM picture of a gold contact (nominal thickness 10 nm) prepared on a 40 nm DIP film by thermal evaporation. Note that the substrate was SiO_2 in this case and the DIP film was grown at room temperature. The line scan (Figure 14.7b) clarifies the morphology: islands with lateral dimensions on the μm scale and heights of the order of 10–20 nm are observed.

The investigation of the interface of a real device contact with surface sensitive techniques is generally not possible, since the interface is buried and these techniques "penetrate" only a few atomic layers. However, an interface analysis with, e.g., electron spectroscopic techniques would be highly desirable since these methods provide valuable (semi-) quantitative information on the electronic structure and chemical interaction in the contact regime. To tackle this problem, i.e. to access the buried interface, we have applied a lift-off technique, which is illustrated in Figure 14.8a and was previously developed by our group [30]. Therefore, we prepared a sample sandwich which consists of the actual model OFET, i.e. a gold film deposited on a DIP film that itself

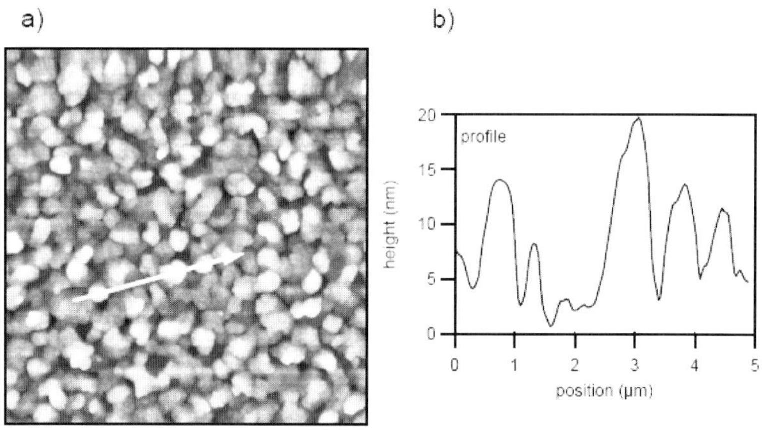

Figure 14.7
a) Contact mode AFM picture (window size: 10 μm × 10 μm) of an Au contact with 10 nm thickness evaporated on a DIP film.
b) Height profile along the direction indicated by the arrow in Figure 14.7a. (see colour plates p. LXXVIII)

was deposited on a gate insulator (in this case SiO_2). The SiO_2 substrate was then glued to a metal handle. On the top side, another SiO_2 pad was attached to the Au contact using a very thin layer of UHV compatible glue. Also the top SiO_2 pad was equipped with a glued-on metal handle.

By tearing the metal handles apart the sample sandwich can now be separated. Provided this process is done properly, the sample separates within the organic layer, and both parts of the sample can be mounted for UHV analysis. Figure 14.8b shows a photograph of a lifted-off Au contact after mounting on a sample holder. The XPS spectrum (Figure 14.8c) after transfer into the UHV is dominated by a C 1s peak at 284.4 eV. Already in this spectrum the Au contact can be observed as demonstrated by the Au 4f, Au 4d, and Au 4p lines. This is most likely explained by an inhomogeneous film thickness after separation. Though the sample was exposed to ambient conditions during the separation and subsequent sample mounting only minor traces of contaminants (mainly H_2O and other O containing species) are detected as indicated by the O 1s signal. Figure 14.8c shows the XPS data of the lifted-off Au contact after careful annealing (10 min at 470 K), which was performed to thin out the DIP film. The spectrum shows a strong increase of the Au signal while the C 1s line is

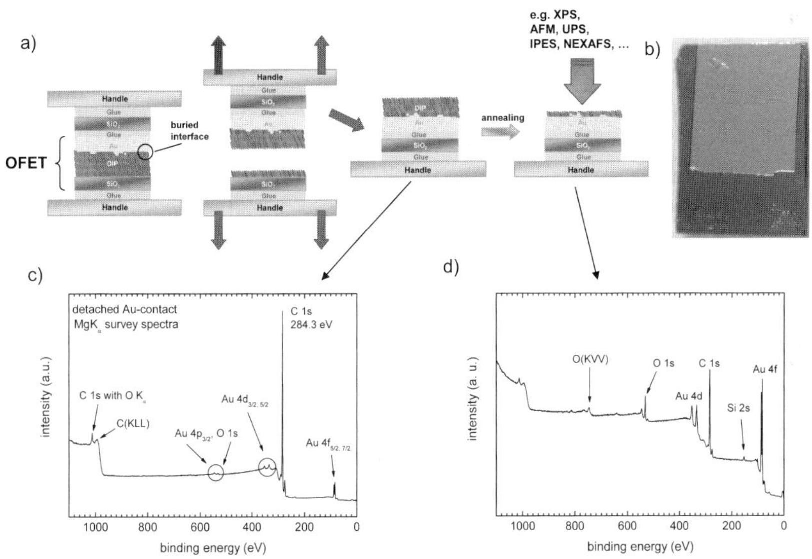

Figure 14.8

a) Illustration of the lift-off technique applied to investigate the buried Au/DIP interface; for details see text.

b) Picture of a separated Au top contact mounted on a sample holder for spectroscopic analysis.

c) XPS survey scan of the separated Au top contact recorded with a MgKα laboratory source. The most prominent signals are indicated.

d) XPS survey scan of the separated Au top contact after annealing at 470 K which leads to a partial desorption of the DIP layer. (see colour plates p. LXXVIII) reduced. Moreover, a clear contribution of the SiO_2 substrate is now visible, as indicated by the Si 2s line at 150 eV and the increased O 1s intensity. The SiO_2 signal may be the result of the very rough and not completely closed Au layer. The thus prepared surface can now be analysed in detail using, e.g., various electron spectroscopic techniques with high resolution as demonstrated by Figures 14.4 and 14.6.

In the present case after a rough test no major deviations from the ideal DIP/Au model system could be detected but further, more detailed studies using high-resolution electron spectroscopies with synchrotron radiation and additional morphological techniques such as AFM may reveal subtle differences. These will be performed after optimisation of the lift-off technique which is presently under way and will be applied completely under UHV conditions, thus allowing the minimisation of spurious influences. Then further investigations on buried interfaces combining XPS, UPS, IPES, NEXAFS, and AFM will be performed on the present and several other systems which are a matter of present research.

14.3.1.4 Gate Electrode

The attachment of the gate electrode is also a demanding task mainly from a technical point of view. In order to enable low switching voltages and to avoid short channel effects the gate dielectric has to be very thin. As indicated by Figure 14.2a, we utilised a thin-out technique, i.e. we drilled a large and sufficiently deep hole into the back side of the sapphire. This is particularly complicated for sapphire due to its mechanical hardness and chemical inertness. The desired thickness, which should at least be in the range of a few μm, can generally be achieved by a combination of different methods like, e.g., drilling with diamond drills and wet chemical etching. In this work mechanical drilling and simultaneous thickness measurements have been pursued. Figure 14.9a shows a sketch of the applied experimental set up. The sapphire sample is attached to an insulating PVC sample mount and fixed by a PVC ring. A high-precision diamond drill with micrometer drive was then approached to the sample from the top. During drilling the residual sapphire thickness was probed by the sample capacitance, which is measured between the drill and a thin aluminium foil attached to the front of the sapphire sample.

Figure 14.9b shows the profilometer measurements of three different blind holes applied to a sapphire sample with this set-up. The used drill had a diameter of 1 mm and a hemispherical tip with a radius of 0.5 mm. The profilometer scans show that holes with a relatively flat base with a roughness in the range of only a few μm are possible. The accuracy of the capacitance measurement should generally be in the order of 100 nm if the film thickness is in the μm range and should thus not constitute a technical limitation. A final plasma etching step according to [31] should allow to remove material with high accuracy (though very low rates) and can thus provide the desired gate insulator thickness. Subsequently, the gate electrode is deposited by thermal evaporation. After this successful proof of principle the method can be further improved thus

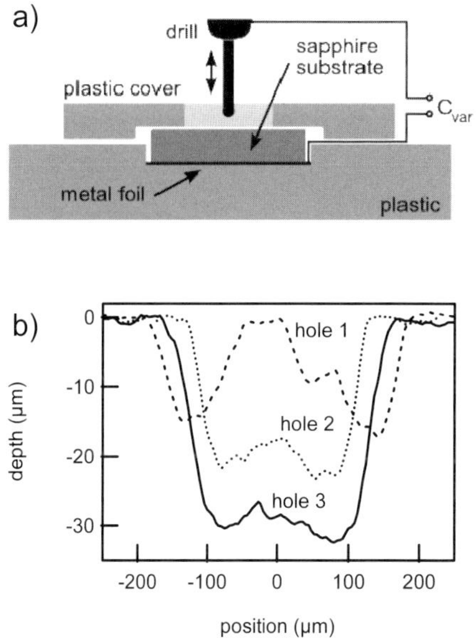

Figure 14.9
a) Illustration of the set-up used for high-precision drilling and thickness monitoring of the blind hole for the gate electrode; for details see text.
b) Profilometer scans of 3 different blind holes established in sapphire with the set-up sketched in Figure 14.9a. (see colour plates p. LXXIX)

approaching gate insulator thicknesses fairly below 1 µm which should reduce the required gate voltages to values in the 10 V range.

14.3.1.5 *In Situ* Device Characterisation

In order to enable the complete device fabrication and the electrical characterisation completely under UHV conditions without intermediate exposure to ambient conditions, a dedicated experimental set-up has been developed within this project. The system includes a transferable sample holder which has been equipped such that the mounting of additional contacts is possible. The sample holder set-up is displayed in Figure 14.10a (with a CuPc sample and Au fingers on a transparent glass sample). With the present set-up the sample can be heated by resistive heating elements inserted into the copper brackets. Cooling is possible with a LHe cryostat, and the temperature at the sample can be measured with a K-type thermocouple and stabilised by a PID controller. The accessible temperature range is thus 150–500 K depending on the thermal conductivity of the substrate.

For metal deposition different shading masks can be applied by manipulation in the UHV. Figure 14.10b shows a mask to deposit a finger structure with 0.3 mm finger width and 0.2 mm finger distance. First IV-measurements were performed using a standard Keithley source meter. They have nicely shown the feasibility of *in situ* preparation and electrical measurements. Moreover, the sample holder is compatible with various UHV spectrometers and can be transferred under UHV conditions using a home-made load lock system, thus allowing the combination of spectroscopic tools such as UPS, IPES, XPS, and NEXAFS with electrical measurements. Detailed investigations combining

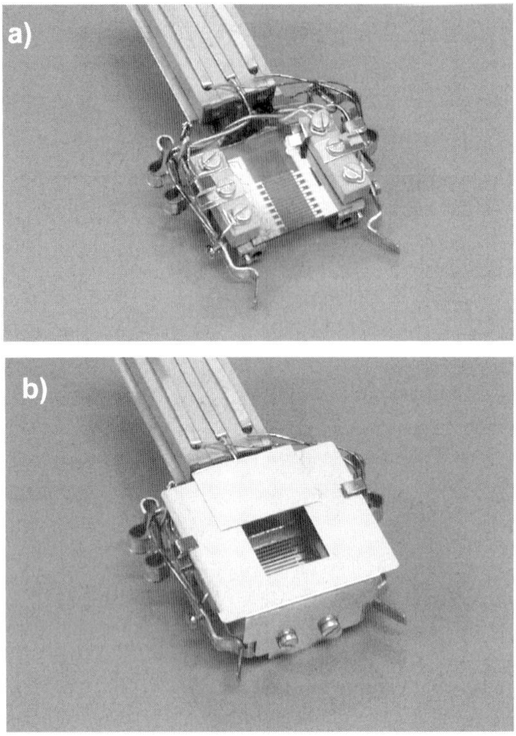

Figure 14.10 Picture of the transferable sample holder dedicated for *in situ* device fabrication, IV-measurements, and surface analysis in UHV without (Figure 14.10a) and with (Figure 14.10b) a mask to evaporate a metal finger structure. In Figure 14.10a a CuPc film contacted by a Au finger structure is mounted on the holder. (see colour plates p. LXXIX)

UHV preparation, electrical measurements, and surface analysis are underway and will be published later.

14.4 Conclusions

In the present publication we have shown that the micro-OFET concept generally works. The various successful steps are:

- preparation of a suitable sapphire substrate with an atomically flat, long-range ordered and impurity-free surface,
- optimal DIP growth yielding large crystallites on top of the well-prepared sapphire surface with upright standing molecules (required for optimum transport parallel to the surface),

- weak interaction between DIP and the Au interface thus yielding a little (or un-)perturbed first layer of organic molecules at the DIP/Au interface with little inter-diffusion (from previous results),
- preparation of a thin gate insulator with attached gate electrode (technically demanding task),
- characterisation of buried interfaces by surface sensitive techniques, and
- complete UHV approach, i.e. preparation of substrate surface, organic layer, and device as well as electrical characterisation and surface analysis of OFETs completely within UHV environment without transfer through air.

We note that all of these achievements are little breakthroughs on the way to a perfect micro-OFET device. However, we also admit that we could not yet demonstrate that this approach leads to OFET devices with improved properties as compared with those prepared in a conventional way using less well defined interfaces, active layers and contacts. Such demonstration experiments are underway and will be published later. We further plan to optimise the contacts by a "soft landing" technique for metal deposition, to implement micromanipulation for contacting selected single crystallites, and to do a careful electrical characterisation of several such prepared micro OFETs.

Acknowledgements

We highly appreciate the input of several collaborators in the DFG Schwerpunktprogramm SPP1121, particularly C. Heske, J. Pflaum, M. Sokolowski, and W. Brütting, and thank for fruitful discussions and valuable information. Especially, Clemens Heske deserves much credit because he managed the project at its beginning and introduced several good ideas. We also gratefully acknowledge financial support by the DFG in the framework of the SPP1121 under contracts Um 6/8-1 through Um 6/8-3 and by the BMBF under contract 05KS4WWC/2. One of us (E.U.) likes to thank the Fonds der Chemischen Industrie for support.

References

1. N. Karl, Synthetic Metals **133**, 649 (2003).
2. N. Karl, Journal of Crystal Growth **99**, 1009 (1990).
3. M. E. Gershenson, V. Podzorov and A. F. Morpurgo, Reviews of Modern Physics **78**, 973 (2006).
4. E. Menard, V. Podzorov, S. H. Hur, A. Gaur, M. E. Gershenson and J. A. Rogers, Advanced Materials **16**, 2097 (2004).
5. G. Horowitz, F. Garnier, A. Yassar, R. Hajlaoui and F. Kouki, Advanced Materials **8**, 52 (1996).
6. T. Becker, A. Birkner, G. Witte and C. Woll, Physical Review B **65** (2002).
7. N. Karl, Synthetic Metals **133**, 649 (2003).

8. A. C. Dürr, F. Schreiber, M. Kelsch, H. D. Carstanjen, H. Dosch and O. H. Seeck, Journal of Applied Physics **93**, 5201 (2003).

9. M. Scharnberg, R. Adelung and F. Faupel, Physica Status Solidi a-Applications and Materials Science **205**, 578 (2008).

10. A. C. Dürr, F. Schreiber, M. Munch, N. Karl, B. Krause, V. Kruppa and H. Dosch, Applied Physics Letters **81**, 2276 (2002).

11. H. Marchetto, U. Groh, T. Schmidt, R. Fink, H. J. Freund and E. Umbach, Chemical Physics **325**, 178 (2006).

12. G. Tzvetkov, G. Koller, Y. Zubavichus, O. Fuchs, M. B. Casu, C. Heske, E. Umbach, M. Grunze, M. G. Ramsey and F. P. Netzer, Langmuir **20**, 10551 (2004).

13. S. Kera, M. B. Casu, K. R. Bauchspiess, D. Batchelor, T. Schmidt and E. Umbach, Surface Science **600**, 1077 (2006).

14. M. B. Casu, B., P. Cosseddu, D. Batchelor, A. Bonfiglio and E. Umbach, Journal of Chemical Physics **128** (2008).

15. M. B. Casu, Y. Zou, S. Kera, D. Batchelor, T. Schmidt and E. Umbach, Physical Review B **76** (2007).

16. S. Krause, M. B. Casu, A. Schöll and E. Umbach, New Journal of Physics**,** 085001 (2008).

17. S. Tanuma, C. J. Powell and D. R. Penn, Surface and Interface Analysis **21**, 165 (1994).

18. A. Schöll, Y. Zou, T. Schmidt, R. Fink and E. Umbach, Journal of Electron Spectroscopy and Related Phenomena **129**, 1 (2003).

19. J. Stöhr and D. A. Outka, Journal of Vacuum Science & Technology a-Vacuum Surfaces and Films **5**, 919 (1987).

20. J. Stöhr and D. A. Outka, Physical Review B **36**, 7891 (1987).

21. S. Tabuchi, H. Tabata and T. Kawai, Surface Science **571**, 117 (2004).

22. M. I. Alonso, M. Garriga, J. O. Osso, F. Schreiber, E. Barrena and H. Dosch, Journal of Chemical Physics **119**, 6335 (2003).

23. M. B. Casu, A. Schöll, K. R. Bauchspieß, D. Hübner, T. Schmidt, C. Heske and E. Umbach, submitted.

24. A. C. Dürr, B. Nickel, V. Shan-Fia, U. Taffner and H. Dosch, Thin Solid Films **503**, 127 (2006).

25. D. Cahen and A. Kahn, Advanced Materials **15,** 271 (2003).

26. A. Kahn, N. Koch and W. Y. Gao, Journal of Polymer Science Part B-Polymer Physics **41**, 2529 (2003).

27. Y. Zou, L. Kilian, A. Schoell, T. Schmidt, R. Fink and E. Umbach, Surface Science **600**, 1240 (2006).

28. A. C. Dürr, N. Koch, M. Kelsch, A. Ruhm, J. Ghijsen, R. L. Johnson, J. J. Pireaux, J. Schwartz, F. Schreiber, H. Dosch and A. Kahn, Physical Review B **68** (2003).

29. D. Gador, C. Buchberger, R. Fink and E. Umbach, Journal of Electron Spectroscopy and Related Phenomena **96**, 11 (1998).

30. L. Weinhardt, O. Fuchs, A. Peter, E. Umbach, C. Heske, J. Reichardt, M. Bär, I. Lauermann, I. Kotschau, A. Grimm, S. Sokoll, M. C. Lux-Steiner, T. P. Niesen, S. Visbeck and F. Karg, Journal of Chemical Physics **124** (2006).

31. Y. J. Sung, H. S. Kim, Y. H. Lee, J. W. Lee, S. H. Chae, Y. J. Park and G. Y. Yeom, Materials Science and Engineering B-Solid State Materials for Advanced Technology **82**, 50 (2001).

Section IV
Device Performance and Characterisation

Organic Electronics. Structural and Electronic Properties of OFETs. Edited by Christof Wöll

ISBN: 978-3-527-40810-8

15
Pentacene Devices: Molecular Structure, Charge Transport and Photo Response

Bert Nickel

15.1
Introduction

Today, the field of organic electronics is wide open [1, 2]. While organic light emitting diodes (OLEDs) have already successfully entered the market, the roadmap for integrated circuits is still not settled. Mobilities μ for pentacene organic thin film transistors (OTFTs) commonly ranging between 0.1 cm/V s and 1 cm/V s have been reproduced by many groups world wide [3–6]. In spite of the enormous activities in synthesising and screening for new materials for transistor applications, pentacene has successfully defended its leading position for the production of OTFTs.

The reason for pentacene being superior for the production of TFT devices [7, 8] when compared with other molecules [9] is still not obvious. In this chapter, we will discuss to what extent the peculiar growth properties [10] of pentacene on metallic contacts and gate dielectrics contribute to the device performance. For this purpose, first the early growth state of pentacene films and the molecular structure of the so called thin film phase is reviewed. Then, major sources of crystal defects in thin films as determined by advanced synchrotron diffraction techniques are discussed. The relation of these defects to the frequently discussed electronic traps that strongly influence transport properties of TFTs [6, 11, 12] is indicated. Finally, the spatially resolved photo response of pentacene OTFTs will be discussed in the context of injection barriers and contact homogeneity.

15.2
Pentacene Thin Films

15.2.1
Film Formation on Inert Surfaces

Pentacene can be evaporated quite conveniently by resistive heating of a tantalum crucible filled with purified pentacene powder. This procedure is ultra

Organic Electronics. Structural and Electronic Properties of OFETs. Edited by Christof Wöll

ISBN: 978-3-527-40810-8

high vacuum compatible and a molecular beam with typical deposition rates ranging from below 0.01 nm/s up to 1 nm/s can be easily achieved [13], as verified by a quartz micro balance at the sample position. Also alternative methods such as laser evaporation have been reported [14]. Growth temperatures typically range from room temperature up to 50 °C. At higher temperatures, a much higher deposition rate is needed to compensate for desorption. Thus, with MBE techniques, the maximum growth temperature is limited to about 75 °C. A typical example for a pentacene film grown on a thermal oxide with Au bottom contacts is shown in Figure 15.1a. If substrate and growth conditions are properly chosen, a continuous film forms on SiO_2 (Figure 15.1a) with grains of the order of several microns which exhibit step heights comparable to the length of the pentacene molecule (Figure 15.1b). On the gold contacts, the film morphology is much more rugged (Figure 15.1a). The reason for the profound difference in growth behaviour on SiO_2 and Au is discussed below.

The initial stage of pentacene thin film growth on SiO_2 can be described by diffusion limited aggregation (DLA) [15]. In this growth mode, incoming mole-

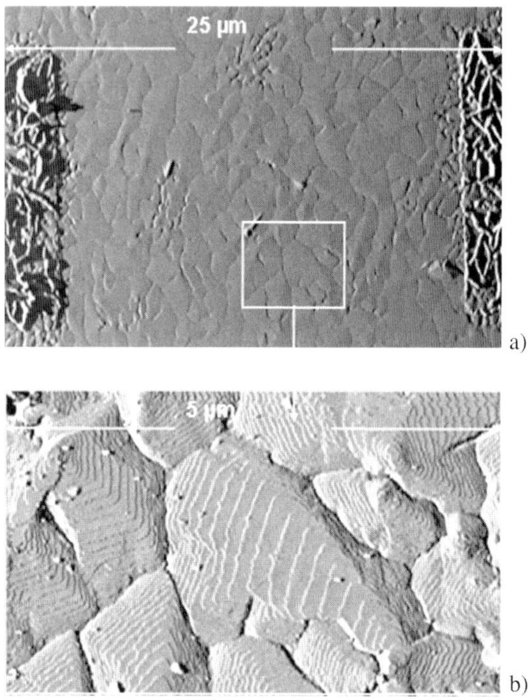

Figure 15.1 AFM amplitude micrograph of a pentacene OTFT structure. (a) Gold contacts on SiO_2 have been covered by a 50 nm pentacene film. (b) A zoom reveals a grainy structure and terraces with step heights of ca. 15.4 Å.

cules initially diffuse across the surface. Once they meet a critical number of molecules, they form a stable nuclei, which subsequently grows in area during deposition. The submonolayer islands stemming from the growth of such nuclei exhibit a fractal shape, which is the fingerprint of this growth mode. Pentacene submonolayers grown on bare Si surfaces show such fractal shapes, as revealed first by *in situ* photo electron emission microscopy experiments [16]. Subsequent AFM studies have shown that the diffusion length on silicon oxide is significantly reduced [17] compared with e.g. H-terminated Si. A detailed analysis of the island size and island density as a function of coverage by AFM and synchrotron experiments fully confirmed scaling predictions of the DLA theory [18, 19] and allowed to determine the stable nuclei size to four pentacene molecules [20]. If the growth temperature is lowered to 0 °C, the diffusion length is reduced by a factor ~ 4 [20].

Pentacene submonolayer islands exhibit a layer thickness of 15.4 Å which has been measured by AFM and by X-ray reflectometry [17]. This layer thickness implies that the molecules are oriented in an upright configuration, i.e. the long molecular axis is oriented predominantly along the surface normal, see Figure 15.2a. Multilayer films exhibit different crystalline phases, which are usually identified by their d_{001} spacing along the surface normal. The substrate-induced $d_{001} = 15.4$ Å polymorph, which is commonly termed *thin film phase*, is the most relevant for OTFT applications. Grazing incidence in plane diffraction experiments [21, 22] confirmed that the grains of the thin film phase are (001) oriented. The lateral dimension of the two dimensional unit cell inferred from these experiments, as well as the observed selection rules, suggest a unit cell with two non-equivalent molecules in a Herringbone arrangement, similar to the bulk pentacene ordering motive [23, 24].

In a recent study, the detailed molecular arrangement of the *thin film phase* was resolved by a grazing incidence truncation rod scattering study [25]. The crystal structure was found to be triclinic [26] with the following unit cell parameters: $a = 5.958 \pm 0.005$ Å, $b = 7.596 \pm 0.008$ Å, $c = 15.61 \pm 0.01$ Å, $\alpha = 81.25 \pm 0.04°$, $\beta = 86.56 \pm 0.04°$ and $\gamma = 89.80 \pm 0.10°$ [27]. A detailed analysis of the Bragg peak intensities for pentacene on SiO_2 allowed to determine the tilt between the long molecular axis of the two unit cell molecules and the surface normal to $5.6 \pm 0.05°$ and $6.0 \pm 0.4°$ and the Herringbone angle to $54.3 \pm 1.3°$ [27]. The unit cell is shown in Figure 15.2b. Here, the unit cell vectors $\boldsymbol{a}$ (red), $\boldsymbol{b}$ (green) and $\boldsymbol{c}$ (blue) are colour coded. The Herringbone angle (green arc) is defined as the intersection angle between molecular planes (red planes). In one case, also a molecular axis (red dotted line) is indicated.

Surface energy calculations [28] reveal that the (001) cleaving plane is the surface with the lowest surface energy. In turn, the formation of (001) oriented films can be expected, if the interaction of the pentacene molecules with the surface is negligible to the pentacene–pentacene interaction. Experiments show that this condition apparently fulfilled for various inert substrates such as reduced and oxidised Si, as well as many polymeric films used as gate dielectric. Note that the surface energy of the *thin film phase* is rather isotropic

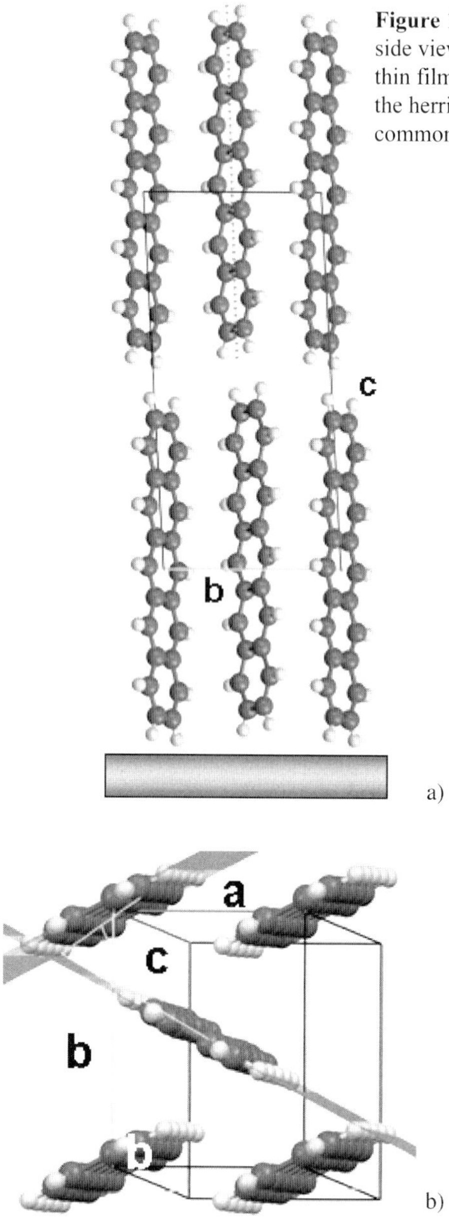

Figure 15.2 Thin film phase unit cell. (a) The side view illustrates the layered structure of the thin film phase. (b) The top view emphasises the herringbone ordering motive, which is a common feature of all pentacene polymorphs.

within the pentacene film plane [28], probably a direct result of the large herringbone angle. In turn, pentacene islands show on SiO_2 a rather isotropic shape, thus favouring the formation of closed films up to several monolayers [16, 29]. Beyond this thickness, strong roughening occurs [30] and competing phases start nucleating [31].

15.2.2 Film Formation on Metallic and Conductive Surfaces

Bottom contact TFTs involve either metallic or conducting polymer contacts. Therefore, the growth of pentacene on metallic surfaces has also been studied in detail [32–35]. The most commonly used contact material for pentacene TFTs is Au, whose work function value $\phi = 5.1$ eV [36] matches quite well with the ionisation energy of pentacene $\chi = 4.9$ eV [37]. Also, being a noble metal, Au contacts are sometimes considered insensitive to air exposure even if it has been observed that the Au work function is quite sensitive to ambient conditions. The observed growth behaviour for pentacene on Au depends strongly on the substrate, i.e. different growth is observed for single Au crystal surfaces [32] and for thin Au film composed out of polycrystalline grains [33]. Also the roughness and cleaning procedure influence the growth. For the technically most relevant polycrystalline Au film surfaces, pentacene shows a complex growth behaviour. If a Au(111) texture is present, a flat lying monolayer forms, which can be interpreted as a wetting layer. This layer then acts as a growth template for subsequently deposited pentacene forming microscopic grains. For these grains, the long molecular axis is oriented along the surface (*lying down phase*). Elevated temperatures promote a pronounced dewetting [34]. This behaviour is not unexpected, since the surface energy of a *lying down phase* is higher than the surface energy of the *thin film phase* [28] and thus dewetting is a direct result of the minimisation of the film surface area. Thus, surface diffusion is counter-productive for the formation of closed films and the formation of well ordered closed films seems quite impossible. To resolve this problem, two strategies have been pursued. One approach, which has been demonstrated for single crystal Ag(111) surfaces makes use of cryogenic substrate temperatures to suppress dewetting. In this case, the pentacene molecules are deposited making use of a hyperthermal He-beam [38] which provides the impinging pentacene molecules with energies up to 5 eV, well above thermal energies. Thus, a well ordered film can form at 200 K [35] (see Figure 15.3a).

Another way to avoid dewetting is to passivate the metallic surface by a self assembled monolayer (SAM), e.g. an alkane thiol monolayer (C-18). After passivation, the growth structure resembles the growth mode on inert surfaces [33], the same holds for growth of pentacene on conducting polymers such as PEDOT: PSS [poly(3,4-ethylenedioxythiophene)/poly(styrenesulfonate)] [39]. It is interesting to note that on bare Si, pentacene initially forms a flat lying monolayer, but on-top of this monolayer, the thin film phase readily forms [16] without need for passivation.

Other more exotic materials which have been studied include graphite and Bi. Graphite is an interesting substrate since it also promotes the formation of a *lying down phase* with Herringbone order textured in the (200) orientation [40] (see Figure 15.3b). This enabled an angular resolved photoelectron emission study (APES) for this pentacene polymorph along the (200) direction of the

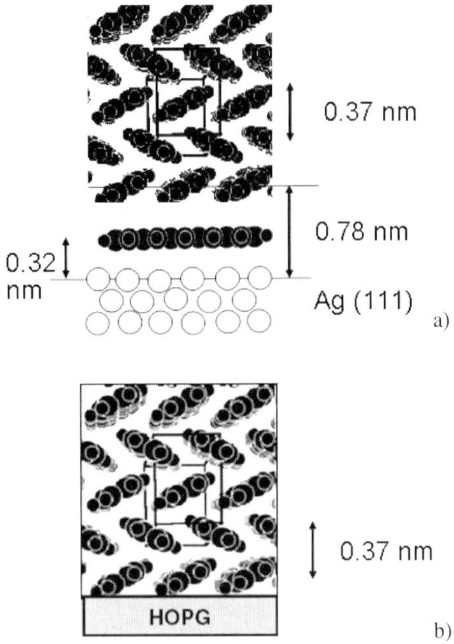

Figure 15.3 Pentacene growth on Ag and graphite. (a) Side view of the film structure of pentacene on Ag(111). A flat lying pentacene monolayer acts as a growth template. (b) Side view of the growth of pentacene on graphite. The structure resembles the growth on Ag, but it is not clear whether a pentacene wetting layer is present.

Brillouin zone. The experiments reveal indications of strong band dispersion [40]. Recently, a well defined epitaxial relationship between a pentacene film and the Bi substrate has been realised. This well defined epitaxial relationship allowed to explore the band dispersion of different crystallographic directions of the Brillouin zone of pentacene by APES, also revealing a pronounced band dispersion [41].

15.2.3
Mixed Films

For some applications such as organic photovoltaic devices or ambipolar transistors, a co-deposition of hole and electron conducting materials is needed, e.g. pentacene and C60. This gives rise to a whole zoo of possible growth scenarios, depending on how the different materials mix. For the co-deposition of pentacene and 6,13-pentacenequinone (an oxidised state of pentacene) a pronounced phase separation is observed, depending on deposition rates [42]. At the same time, the formation of the pentacene bulk phase is suppressed in mixed film.

15.3 Pentacene OTFT Properties

15.3.1 Mobility and Charge Carrier Density

The most simple pentacene OTFT test structure used in many labs is based on a Si wafer piece covered with a thermal oxide. Here, the heavily doped Si wafer takes the role of the back gate electrode, and the SiO_2 takes the role of the gate dielectric. A pentacene thin film is deposited as the semiconducting layer. Source and drain electrodes are deposited either on the silicon oxide (bottom contact) or on top of the pentacene film (top contact).

A pentacene TFT works in hole accumulation, i.e. a sufficiently negative voltage (V_G) is applied to the gate, accumulating holes at the pentacene/SiO_2 interface. If now a moderate voltage is applied between the source and drain electrode, a hole current flows. A characteristic output curve for a typical pentacene TFT from our lab is shown in Figure 15.4a. Here, as a substrate, heavily n-doped Si-wafers with a 150 nm thick thermal SiO_2 layer on top of it were used. Additionally, a very thin layer of polysterene (ca. 2–3 nm) was spin cast on top of the silicon oxide layer. A shadow mask was used to define 50 nm thick source and drain Au contacts that were evaporated in vacuum. Finally a 50 nm pentacene layer was evaporated at room temperature at a deposition rate of about 1 nm/min in a separate evaporation chamber, also using a shadow mask. This way, transistor channels with a length of $L = 25\ \mu m$ and a width of $W = 1000\ \mu m$ were obtained, cf. schematic in Figure 15.4. The output curve of this device shows a linear increase of the drain current for small drain voltages, and subsequently saturation occurs. The transfer characteristics of this device are shown in Figure 15.4b. The logarithmic scale reveals a sharp onset of the transistor activity and a subthreshold swing $S = V_G/\log(I_{SD}) = 1.1$ V per decade.

Pentacene OTFT curves are usually analysed within MOSFET theory [43]. For the geometry of the device used in Figure 15.4, which has a channel width $W = 1000\ \mu m$ and a channel length $L = 25\ \mu m$, the source-drain current (I_{SD}) in the linear regime ($V_{SD} \ll V_G - V_T$) can be written as:

$$I_{SD}^{lin} = W/L \cdot \mu^{lin} \cdot C_i \cdot (V_G - V_T) \cdot V_{SD} \,. \qquad (1)$$

Here, the graded channel approximation has been used. C_i is the capacitance per unit area of the dielectric and V_T is the threshold voltage. μ^{lin} is the mobility in the linear regime. The respective measurements are shown in Figure 15.5a.

For the saturation regime ($V_{SD} \gg V_G - V_T$) the accumulation within the channel is incomplete. This so called *pinch off* arises due to the superposition of the gate and drain potential. The source–drain current (I_{SD}) in the saturation regime reads:

$$I_{SD}^{sat} = W/(2L) \cdot \mu^{sat} \cdot C_i \cdot (V_G - V_T)^2 \,. \qquad (2)$$

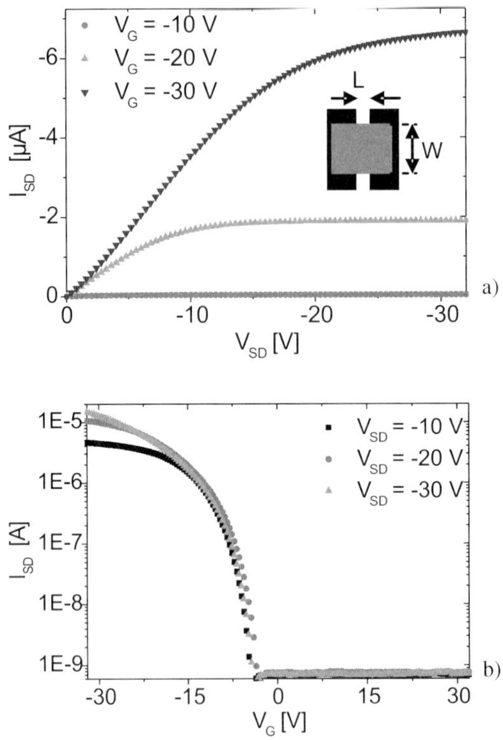

Figure 15.4 Characteristics of a bottom contact pentacene OTFT. Channel geometry: $L = 25\ \mu\text{m}$, $W = 1000\ \mu\text{m}$ (see text and inset). Sweep rate: 0.4 V/s, the measurement was performed in vacuum. (a) Output characteristics (I_{SD} vs. V_{SD}, V_G as indicated in the inset) and (b) transfer characteristics (I_{SD} vs. V_G in log. scale, V_{SD} as indicated in the inset).

The respective measurements are shown in Figure 15.5b. The mobility in the saturated regime (μ^{sat}) can be read off from the slope of the output curve (straight lines in Figure 15.5b, square root scale at y-axis) using Eq. (2). We obtain $\mu^{sat} = 0.06\,\text{cm}^2/(\text{V/s})$, which compares reasonably well with typical values for the bottom contact geometry ($\mu \approx 0.1 - 0.3\,\text{cm}^2/(\text{V/s})$). The mobility in the linear regime (μ^{lin}) can be read off from the slope of the output curve (Figure 15.5a, linear scale at y-axis) using Eq. (1). We obtain $\mu^{lin} = 0.04\,\text{cm}^2/(\text{V/s})$, which is slightly reduced compared with $\mu^{sat} = 0.06\,\text{cm}^2/(\text{V/s})$, indicating some non-ideal behaviour.

The most important physical parameter of an OTFT beyond the mobility μ is the number of charge carriers (n_h) induced by a given gate voltage (V_G). n_h can be estimated from the measured channel conductivity σ by

$$n_h \cdot q \cdot \mu = \sigma / d = L/(W \cdot R)\,. \tag{3}$$

Here, d is the active channel thickness (1–2 molecular layers) and q is the elementary charge of a hole. For the transistor used in Figure 15.4, we measure $R = 2.8 \times 10^6\ \Omega$ at $V_G = -30$ V and $V_{SD} = -10$ V. In turn, according to Eq. (3), the hole carrier density at this set point is $n_h = 0.9 \times 10^{12}\ \text{cm}^{-2}$.

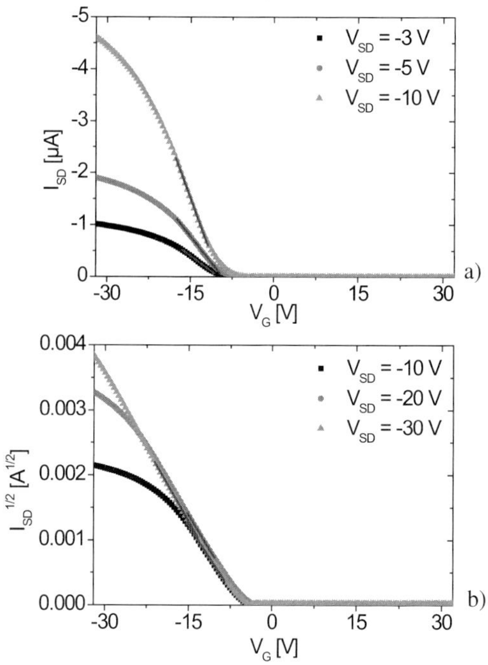

Figure 15.5 Transfer characteristics. (a) Linear regime (I_{SD} vs. V_G), the straight lines indicate the slope determining μ^{lin} according to Eq. (1). (b) Saturated regime ($\sqrt{I_{SD}}$ vs. V_G), the straight lines indicate the slope determining μ^{sat} according to Eq. (2). Same device as in Figure 15.4.

15.3.2
Influence of Trap States and Fixed Interface Charges

An example for the dynamic behaviour of a pentacene OTFT is shown in Figure 15.6a. Here, the gate voltage (V_G) is first swept from positive to negative voltages (sweep rate = 0.4 V/s, black data points in Figure 15.6a), and then immediately from negative to positive voltages (red data points in Figure 15.6a) for the same fixed value of the drain voltage (V_{SD} = −20 V). A pronounced hysteresis is observed. This hysteresis can be interpreted as an apparent shift of the threshold voltage (ΔV_T) induced by a charging Q_t of the pentacene/SiO_2 interface due to filling and emptying of trap states. The number of trap states can be estimated by

$$n_t = Q_t/q = \Delta V \cdot C_i/q \,. \quad (4)$$

The observed threshold shift of $\Delta V_T = 2.0$ V results in a trap density $n_t = 2.4 \times 10^{11}$ traps/cm². This number is not so small compared with the number of charge carriers n_h, indicating that traps largely influence the device performance. One may speculate whether structural defects within the pentacene film contribute to trap states. An X-ray analysis of the defect densities in pentacene films revealed defect densities in the order of $n_t = 2 \times 10^{11}$ defects/cm² [44], in good agreement with the trap density n_t inferred from the electronic characterisation. Thus, the observed structural defects apparently contribute to

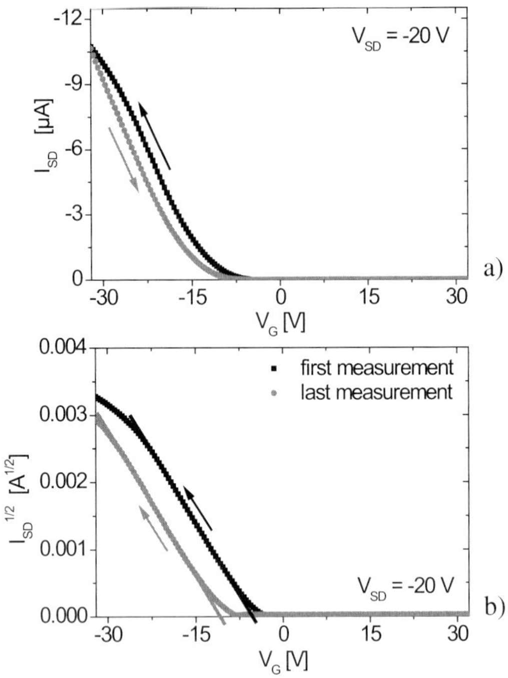

Figure 15.6 Hysteresis effects and voltage drifts in the transfer characteristics. Sweep rate is 0.4 V/s, same device as in Figure 15.4. (a) A gate sweep from positive to negative voltages (black data points), and back (red data points) for the same fixed value of the drain voltage ($V_{SD} = -20$ V) reveals a hysteresis effect. (b) After a couple of measurements (one hour later), a rather stable shift of the transfer curve due to fixed charges is established for the same device.

trap states. It is well known, that the hysteresis and thus the trap density depends on the nature of the gate dielectric surface. Especially hydroxyl groups are considered as potential trap states, therefore a passivation of the gate oxide by either silanisation [45] or polymer capping (here: polystyrene film) is needed to minimise trap densities.

A subsequent study of the pentacene TFT reveals also an irreversible drift of the transfer curve with operating time, cf. Figure 15.6b. This drift can be interpreted as an interface charging due to deep trap levels, also termed fixed charges. The observed voltage shift of $\Delta V = 6.5$ V implies a fixed charge density $n_{dt} = 7.7 \times 10^{11}$ cm^{-2}. Note that this density is already of the order of the charge carrier density n_h. In turn, voltage shifts in pentacene TFTs can completely redefine the working point of a transistor. For the application of such TFT devices in e.g. ring oscillators [46], one should find a way to control these drifts. Recent work suggests that controlled generation of fixed charges at the gate dielectric by UH-light exposure might be used to store electronically readable information in a pentacene TFT (one bit per transistor).

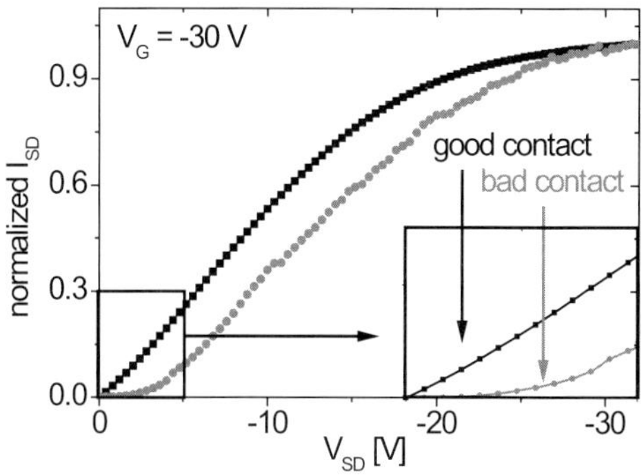

Figure 15.7 Injection efficiencies of pentacene TFTs. Comparison of ohmic and non-ohmic behaviour. The contacts of the so called *good device* were fabricated using shadow masks, and the contacts of the so called *bad device* were fabricated using optical lithography.

15.3.3 Injection

The whole TFT performance is largely influenced by injection efficiencies [see Chapter 20 by Scholz et al.). In particular surface contaminations such as remanent photo resist or organic adsorbates due to air or solvent exposure can modify the work function of the Au electrode up to a point, that charge injection is locally suppressed. The injection properties of two selected TFT devices from our lab are compared in Figure 15.7. Both devices are operated at $V_G = -30$ V. The output characteristics of the two devices show a pronounced difference for small drain voltage V_{SD}. While one device (so called *good device*, black data points in Figure 15.7) shows an ohmic increase of I_{SD} as a function of V_{SD}, the other device shows a damped response (so called *bad device*, red data points in Figure 15.7). Note that the amplitude of I_{SD} of the bad device has been rescaled for better comparison. Both devices are bottom contact TFTs of similar geometry, however, the preparation of the contacts was different. The contacts of the good device were fabricated using a shadow mask, while the contacts of the bad device were produced using optical lithography. Optical lithography involves resist and use of solvents such as acetone, which apparently can lead to less efficient contact properties. The physical mechanism which gives rise to the non-ohmic behaviour remains unclear at this level of analysis and thus using equivalent circuits [47] which take account of the observed deviations from ideal MOSFET behaviour seem to be the most practical way to deal with these effects.

15.4 Photo Response

Pentacene shows a strong absorption in the visible (see Figure 15.8), more optical properties of pentacene films can be found in [48]. If pentacene is combined with a proper n-conductor such as C60, it can be used as the active region in a solar cell [49]. Transient photoconductivity experiments using optical

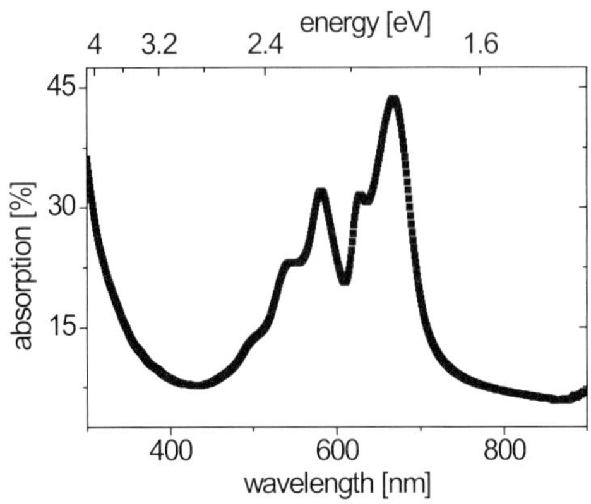

Figure 15.8 Absorption spectra. Wavelength dependency of the absorption of a 50 nm pentacene thin film on glass.

pump terahertz probe techniques suggest that charge photo generation occurs in pentacene on a subpicosecond timescale [50].

Here, we employ photo generated electron–hole pairs to analyse local properties of a TFT device. For this purpose, we have adapted a laser scanning confocal microscope experiment to spatially resolve the photo response of a TFT. The experimental set-up is shown schematically in Figure 15.9a. The laser spot size on the TFT is less than 1 μm. A x–y–z piezo-positioner stage [51] which is mounted on top of a x–y translation stage is used to translate the sample with respect to the laser. The reflected beam allows to identify the contact and the channel region. The laser beam is modulated by a chopper, and the difference signal is extracted using a lock-in amplifier.

We observe a strong photo response localised at the anode [52]. The response is inhomogenous along the contact indicating variations in the transport or injection properties of the device. These variations may be due to local variations of the contact work function, or due to bad physical contact of the pentacene grains adjacent to the electrode, or due to a local enhancement of defect densities in the pentacene film. Thus, a systematic study of the photo response in combination with the respective characteristic transistor curves allows visualising problematic regions of an OTFT, which is a key prerequisite for device optimisation.

15.5 Outlook

Pentacene has all the qualities needed for an OTFT. The main challenge is the control of the contact and dielectric interfaces to suppress the voltage drifts and hysteresis effects associated with trap states. Also the dependency of the injection efficiencies on contact preparation suggest that even if pentacene deposition technology can be applied under rather rough conditions, ultra high vacuum and clean environments may still be useful on the way of learning how to produce reliable devices.

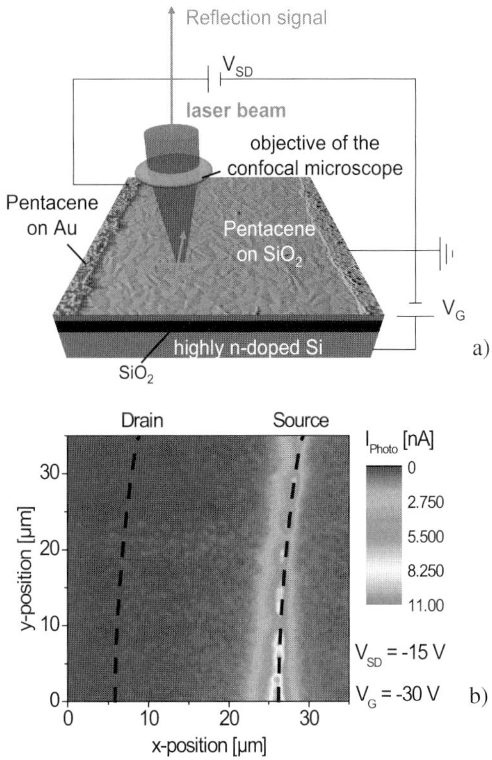

Figure 15.9 Spatially resolved photo response experiment. (a) Schematic of the experimental setup for spatially resolved photo response measurements. (b) Spatially resolved photo response data in the linear regime ($V_G = -30$ V and $V_{SD} = -15$ V). A Corbino device structure has been used (disk and ring geometry, $L = 20$ μm [channel length], $W = 1005$ μm [channel width, i.e. circumference]). The photo response measurement covers a small fraction of the whole device.

Acknowledgements

BN acknowledges financial support by Deutsche Forschungsgemeinschaft under contract no. Ni- 632/1-1 (research fellowship at Princeton U.) and Ni-632/2-1 (research proposal within SPP1121) and hospitality from G. Scoles and J. Rädler. Financial support was provided by Elite Netzwerk Bayern, CeNS and NIM. Purified pentacene was received from J. Pflaum. The X-ray experiments were performed at NSLS in Brookhaven and at HASYLAB in Hamburg. J. Kotthaus for provided access to clean room facilities at LMU. Finally, it is a pleasure to acknowledge the efforts of the students and collaborators involved in the experiments covered by this feature article, namely S. Schiefer, M. Fiebig, C. Erlen, P. Lugli, M. Göllner, M. Huth, F. Danisman, L. Casalis, and R. Ruiz. Fruitful discussions with the participants of SPP1112, and with N. Koch and U. Beierlein is acknowledged.

References

1. G. G. Malliaras and R. Friend, Phys. Today **58**, 53 (2005).
2. S. R. Forrest, Nature **428**(6986), 911–918 (2004).
3. C. D. Dimitrakopoulos, A. R. Brown, and A. Pomp, J. Appl. Phys. **80**(4), 2501–2508 (1996).
4. D. J. Gundlach, Y. Y. Lin, T. N. Jackson, S. F. Nelson, and D. G. Schlom, IEEE Electron Device Lett. **18**(3), 87–89 (1997).
5. Y. Y. Lin, D. J. Gundlach, S. F. Nelson, and T. N. Jackson, IEEE Electron Device Lett. **18**(12), 606–608 (1997).
6. D. Knipp, R. A. Street, A. Volkel, and J. Ho, J. Appl. Phys. **93**(1), 347–355 (2003).
7. C. D. Dimitrakopoulos and D. J. Mascaro, IBM J. Res. Dev. **45**(1), 11–27 (2001).
8. C. D. Dimitrakopoulos and P. R. L. Malenfant, Adv. Mater. **14**(2), 99–117 (2002).
9. G. R. Desiraju and A. Gavezzotti, Acta Crystallogr. B, Struct. Sci. **45**, 473–482 (1989).
10. R. Ruiz, D. Choudhary, B. Nickel, T. Toccoli, K.-C. Chang, A. C. Mayer, P. Clancy, J. M. Blakely, R. L. Headrick, S. Iannotta, G. G. Malliaras, B. Nickel, T. Toccoli, K.-C. Chang, A. C. Mayer, P. Clancy, J. M. Blakely, R. L. Headrick, S. Iannotta, and G. G. Malliaras, Chem. Mater. **16**(23), 4497 (2004).
11. G. Horowitz and P. Delannoy, J. Appl. Phys. **70**(1), 469 (1991).
12. G. Gu, M. G. Kane, J. E. Doty, and A. H. Firester, Appl. Phys. Lett. **87**(24), 243512 (2005).
13. S. R. Forrest, Chem. Rev. **97**(6), 1793–1896 (1997).
14. G. B. Blanchet, C. R. Fincher, and I. Malajovich, J. Appl. Phys. **94**(9), 6181–6184 (2003).
15. J. A. Venables, G. D. T. Spiller, and M. Hanbrücken, Rep. Prog. Phys. **47**, 339–459 (1984).
16. F.-J. Meyer zu Heringdorf, M. C. Reuter, and R. M. Tromp, Nature **412**, 517–520 (2001).
17. R. Ruiz, B. Nickel, N. Koch, L. C. Feldman, R. F. Haglund, A. Kahn, and G. Scoles, Phys. Rev. B **67**(12), 125406 (2003).
18. J. G. Amar, F. Family, and Pui-Man Lam, Phys. Rev. B **50**(12), 8781–8797 (1994).
19. J. G. Amar and F. Family, Phys. Rev. Lett. **74**(11), 2066–2069 (1995).
20. R. Ruiz, B. Nickel, N. Koch, L. C. Feldman, R. F. Haglund, A. Kahn, and G. Scoles, Phys. Rev. Lett. **91**, 136102 (2003).
21. S. E. Fritz, S. M. Martin, C. D. Frisbie, M. D. Ward, and M. F. Toney, J. Chem. Soc. **126**(13), 4084 (2004).
22. A. C. Malliaras, G. G. Nickel, B. Scoles, G. Kazimirov, A. Kim, H. Headrick, R. L. Islam, Z. Ruiz, and R. Mayer, Appl. Phys. Lett. **85**(21), 4926 (2004).
23. R. B. Campbell, J. Trotter, and J. M. Robertson, Acta Crystallogr. **14**(7), 705 (1961).
24. R. B. Campbell, J. Trotter, and J. Moneath, Acta Crystallogr. **15**(3), 289 (1962).
25. R. Feidenhansl, Surf. Sci. Rep. **10**(3), 105–188 (1989).
26. H. Yoshida, N. Inaba, and K. Sato, Appl. Phys. Lett. **90**(189), 181930 (2007).
27. S. Schiefer, M. Huth, A. Dobrinevski, and B. Nickel, J. Chem. Soc., in press (2007).
28. J. E. Northrup, M. L. Tiago, and S. G. Louie, Phys. Rev. B **66**(12), 121404 (2002).
29. R. Ruiz, B. Nickel, N. Koch, L. C. Feldman, R. F. Haglund, A. Kahn, and G. Scoles, Phys. Rev. B **67**(12), 125406 (2003).
30. S. Kowarik, A. Gerlach, W. Leitenberger, W. Hu, G. Witte, C. Wöll,

U. Pietsch, and F. Schreiber, Thin Solid Films **515**(14), 5606 (2007).

31. A. C. Mayer, A. Kazimirov, and G. G. Malliaras, Phys. Rev. Lett. **97**, 105503 (2006).

32. D. Käfer and G. Witte, Chem. Phys. Lett. **442**, 376 (2007).

33. D. Käfer, L. Ruppel, and G. Witte, Phys. Rev. B **75**, 085309 (2007).

34. G. Beernink, T. Strunskus, G. Witte, and Ch. Wöll, Appl. Phys. Lett. **85**, 398 (2004).

35. L. Casalis, M. F. Danisman, B. Nickel, G. Bracco, T. Toccoli, S. Iannotta, and G. Scoles, Phys. Rev. Lett. **90**(20), 206101 (2003).

36. D. R. Lide (ed.), Handbook of chemistry and physics (CRC Press, Cleveland, Ohio, 1981).

37. E. A. Silinsh and V. Čápek, Organic molecular crystals: Interaction, localization and transport phenomena (AIP, New York, 1994).

38. F. De Angelis, T. Toccoli, A. Pallaoro, N. Coppede, L. Mariucci, G. Fortunato, and S. Iannotta, Synth. Met. **146**(3), 291–295 (2004).

39. N. Koch, B. Nickel, J. Ghijsen, A. Elschner, J. Schwartz, J.-J. Pireaux, and A. Kahn, Mater. Res. Soc. Symp. Proc. **771**(20), L3.6.1 (2003).

40. N. Koch, A. Vollmer, I. Salzmann, B. Nickel, H. Weiss, and J. P. Rabe, Phys. Rev. Lett. **95**(15), 156803 (2006).

41. H. Kakuta, T. Hirahara, I. Matsuda, T. Nagao, S. Hasegawa, N. Ueno, and K. Sakamoto, Phys. Rev. Lett. **98**, 247601 (2007).

42. I. Salzmann, S. Opitz, R. Rogaschewski, J. P. Rabe, B. Nickel, and N. Koch, Phys. Rev. B **75**, 174108 (2007).

43. S. M. Sze, Physics of semiconductor devices (Wiley, Singapore, 1981).

44. B. Nickel, R. Barabash, R. Ruiz, N. Koch, A. Kahn, L. C. Feldman, R. F. Haglund, and G. Scoles, Phys. Rev. B **70**(12), 125401 (2004).

45. M. Shtein, J. Mapel, J. B. Benziger, and S. R. Forrest, Appl. Phys. Lett. **81**(2), 268 (2002).

46. C. Erlen, P. Lugli, M. Fiebig, and B. Nickel, J. Comput. Electron. **5**(4), 345 (2006).

47. P. V. Necliudov, M. S. Shur, D. J. Gundlach, and T. N. Jackson, J. Appl. Phys. **88**(11), 6594–6597 (2000).

48. D. Faltermeier, B. Gompf, M. Dressel, A. K. Tripathi, and J. Pflaum, Phys. Rev. B **74**(12), 125416 (2006).

49. S. Yoo, B. Domercq, and B. Kippelen, Appl. Phys. Lett. **85**(22), 5427 (2004).

50. O. Ostroverkhova, D. G. Cooke, F. A. Hegmann, J. E. Anthony, V. Podzorov, M. E. Gershenson, O. D. Jurchescu, and T. T. M. Palstra, Appl. Phys. Lett. **88**(12), 162101 (2006).

51. Atto Cube Systems AG in München.

52. M. Fiebig, C. Erlen, M. Göllner, P. Lugli, and B. Nickel, Spatially Resolved Photoresponse Measurements on Pentacene Thin-Film Transistors. Applied Physics A: Organic Materials for Electronic Applications **95**, 113 (2009).

16
Characteristics and Mechanisms of Hysteresis in Polymer Field-Effect Transistors

G. Paasch, S. Scheinert, A. Herasimovich, I. Hörselmann, and Th. Lindner

16.1
Introduction

Hysteresis effects usually occur in organic field-effect transistors (OFET) or MIS capacitors [1–4]. A detailed literature survey will be given in Section 16.2. In spite of the large number of publications, systematic investigations of the hysteresis effects are rare. Examples are discussed e.g. in Ref. [5].

It has been assumed that the main mechanisms leading to these hysteresis effects in OFETs might be either trap recharging or diffusion of mobile ions [3, 4]. Another proposed mechanism [6] is connected with the doubly charged state of polymer chains, formation and dissociation of bipolarons. Typical polymers used in devices are poly(3-alkylthiophene) (P3AT) or modified poly(phenylene-vinylene) (PPV). Charges on the polymer chains occur as polarons (P) since a chain distortion appears due to electron–phonon coupling. For a long time, bipolarons (BP) [7, 8] have been considered as doubly charged and spin-less states of the polymer chains. However, alternatives to BPs have been discussed, which are also ESR (electron spin resonance) silent, as are the BPs. One suggestion is the recent 'two polarons on a single chain' or 'polaron pair' (PP) model [9]. Giving a consistent explanation of optical data, the latter was proposed to be preferable compared with the BP model.

In the present chapter, after a literature survey (Section 16.2), we present in Section 16.3 experimental results on hysteresis and discuss typical trends. In OFETs, hysteresis occurs mainly during the gate-source voltage sweep, which is connected with depletion or accumulation of charge carriers in the region adjacent to the interface to the insulator. This effect can be observed already in the corresponding MIS capacitor by measuring the capacitance–voltage (*CV*) characteristics or the complex impedance.

In Section 16.4 we come back to hysteresis effects by trap recharging. We have shown in Ref. [10] that this mechanism can lead to hysteresis in quasi-static *CV*-curves of organic MIS capacitors. However, the form of the curves was qualitatively different from the observed ones. This analysis is extended

Organic Electronics. Structural and Electronic Properties of OFETs. Edited by Christof Wöll

ISBN: 978-3-527-40810-8

by a closer inspection of the necessary parameters for this mechanism. Of special interest are traps where the energetic distribution is exponential.

The BP mechanism for hysteresis in field-effect devices [6] proposed by us was strongly supported by the experimental determination of a rate constant for a second order reaction of Ps from voltage stress measurements [11, 12] which has been attributed to the BP formation. However, the above mentioned alternative of PPs requires a critical revisal of equilibrium for systems with Ps, and doubly charged polymer states, which is presented in Section 16.5. The kinetics and hysteresis in organic MIS devices due to formation and dissociation are then analysed in Section 16.6. Also a further mechanism is considered which is connected with the formation of different complexes between Ps, BPs and mobile ions. Conclusions are drawn in Section 16.7.

16.2 Literature Survey

Due to many unsolved problems with OFETs, connected e.g. with the required high mobility in solution processed materials, thin organic dielectrics, or contacts, questions connected with hysteresis effects, stress and ageing have been less at the centre of interest. Examples of reports on hysteresis can be found e.g. in Refs. [1–5] and [13–18]. Hysteresis in OFETs can occur by changing either the drain-source voltage or the gate-source voltage. In current devices the former is less pronounced or even negligible [14–18]. The latter is connected with depletion/accumulation of charge carriers near the interface to the gate insulator. This effect can be investigated directly by capacitance–voltage measurements in the corresponding MIS-capacitors [2, 13, 19]. The measured *CV*-curves are usually different for quasi static and dynamic measurements, alternatively one can in addition measure the complex impedance as a function of the frequency with the gate voltage as a parameter [13]. The hysteresis effect appears essentially as a difference between the flat-band voltages for the two sweep directions of the gate voltage. The hysteresis effects found in such measurements are of course related to processes in stress experiments [12, 20] and during ageing [13]. Up to now only a few detailed experimental characterisations of the field-effect hysteresis have been reported [2, 13] and correspondingly there is a lack of adequate understanding that is needed either to suppress such effects or to exploit them for specific purposes.

So far there exist basically verbal statements that either trap recharging (generally defects) or diffusion of mobile ions in the organic semiconductor (or even in the gate insulator) might be causal for the effects [3, 4, 15–17, 19, 21, 22]. Our investigations [2, 6, 13] indicate that at least two processes are needed to understand the observed hysteresis, namely a transport process and a reaction of the carriers taking part in the transport. Trap recharging would be an example, indeed a detailed simulation study [10] has shown that hysteresis effects are possible. However, the resulting forms of the hysteresis turn out to be qualitatively different from the observed ones. In Section 16.4 we extend this discussion. Another possibility would be transport of carriers (polarons) connected with formation/dissociation of bipolarons [6]. Experimental evidence for the formation of bipolarons (a second order reaction) was found in bias stress investigations [12, 20] of polymer based OFETs. The bipolaron mechanism has been extended in [23] by including possible reactions of polaron and bipolarons with counter ions. In Ref. [24], in addition the question was addressed whether the doubly charged state of the polymer chain formed by the polaron–polaron reaction is a bipolaron [7, 8] or a polaron pair [9]. In Section 16.5 we give a further analysis of this problem. Mobile ions alone can hardly explain the flat-band voltage difference between the two sweep directions. However, in connection with reactions e.g. at the interface they could be also causal for the hysteresis. Recently another model has been presented [25]. Here the assumption was made that the polarisation of the π-electron system of the organic materials depends on the sweep-direction. But this decisive assumption is not really reasoned. In order to achieve a real understanding, more and very detailed experimental investigations are needed with systematic variation of different parameters in connection with theoretical modelling and simulation of possible combinations of transport processes and reactions.

Due to the properties of the polymers, degradation and long time stability have already attracted interest in the first articles on OFETs. Thus, early measurements [26–28] had shown that the conductivity increases by storing the devices in air, surely due to oxygen. The effect could be reversed at least partly by tempering in vacuum. Later systematic investigations have been carried out with either repeated measurements after storage in air or other defined conditions [29–33], or due to bias-stress measurements for the investigation of the threshold voltage shift [34–39]. It has been shown in some cases that the transistor properties are stronger influenced by water than by air [30, 31], whereupon water reduces the conductivity due to the formation of traps [30, 32, 33]. The shift of the threshold voltage of a n-channel [6,6]-phenyl-C_{61}-butyric acid methyl ester (PCBM) transistor to smaller positive voltages after storing in air was also attributed [29] to traps. Not only the influence of air leads to changes in the transistor performance, but also defects in the organic semiconductor, at the interface to the gate insulator and also within the insulator can vary the current [40]. Thus, it has been found that traps at the silicon dioxide interface suppressed n-conduction for a long time [41]. Traps are also referred to as the origin for thc threshold voltage shift to negative values due to negative gate-voltage stress in p-channel transistors [34, 37, 38]. This shift can be reversed after a recovery time or by a positive gate-voltage stress. Bürgi et al. [38] claimed that traps causing the threshold voltage shift are located within the polymer. Salleo et al. [39] mention two origins, a fast, reversible trap-process and a very slow process enduring for a long time connected with traps in the polymer. Zilker et al. observed another reversible trend at negative gate-voltage stress [35]. Only for a short time (<10 … 100 s) is the threshold voltage slightly shifted to a negative gate voltage, presumably due to traps. But after a longer time a large threshold voltage shift in the positive direction appears, which has been attributed to mobile ions in the insulator.

Although all these effects influence the stability of the transistor performance, polymer FETs have been successfully prepared that were stable for a longer time [42–44]. Indeed, since the polymer poly(triarylamine) (PTAA) used in Ref. [42] is stable in air, and in addition a low-k-insulator has been used, one can expect for this transistor a large life time. A little surprising is that the poly(3-hexylthiophene) (P3HT) transistors presented in [43, 44] are stable for up to one year. The authors name the chosen design as a possible origin for the stability. They used a plastic substrate, deposited then source and drain, subsequently the polymer and finally the organic insulator. This way the active P3HT layer is protected from the surroundings by both the substrate and the insulator.

16.3 Experimental Results

In the following we discuss measured hysteresis effects occurring in OFETs and in the corresponding MIS capacitors. For this purpose measurements are presented which are representative of the different devices prepared by us.

16.3.1 Organic Field-Effect Transistors

In recent years we prepared rather different OFETs, either long channel devices with channel length between $L = 50\ \mu m$ and $25\ \mu m$ or short channel devices with channel length of a few μm or even with a special technology in the sub-micrometer range. As the active layer, mainly soluble polymers have been used, such as different modified PPV, different P3AT, but also pentacene using a precursor route. With these materials p-channel accumulation devices are prepared, whereas with PCBM n-channel accumulation is achieved. As source and drain contact materials, metals have been used, preferably Au, or Au with a Cr adhesion layer. The devices have been prepared usually on a silicon wafer with 30 … 50 nm thick silicon dioxide (SiO_2) as the gate insulator allowing for low operation voltage and suppression of short channel effects. Enhanced mobility is achieved by hexamethyldisilazane (HMDS) treatment of the silicon dioxide surface. In some cases also a plastic substrate and an organic insulator have been used. The device design for top (source and drain) contact and bottom contact OFETs is different for the two substrates. In any case both output and transfer characteristics were measured for both sweep directions of the gate-source voltage and/or the drain-source voltage.

16.3.1.1 Short Channel OFET Based on P3HT

A typical example of the observed hysteresis is shown in Figure 16.1 for a short channel transistor with channel length $L = 740$ nm and width $w = 1000\ \mu m$.

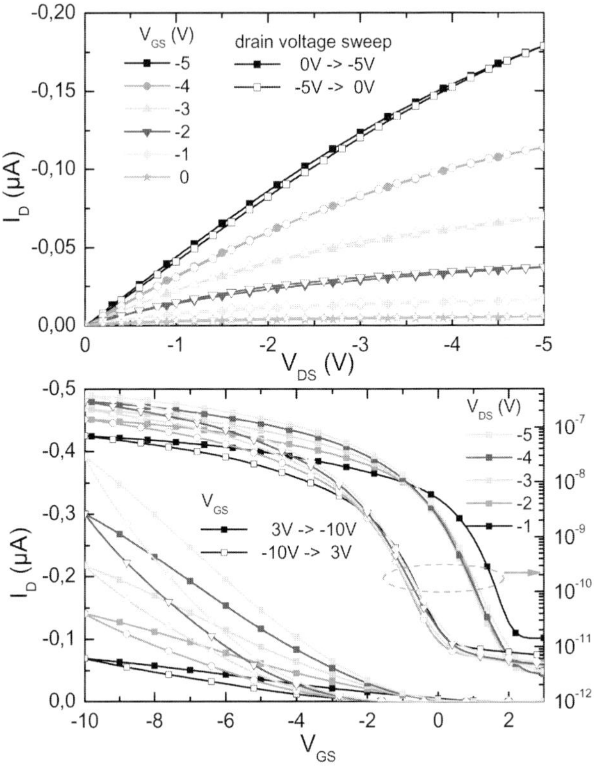

Figure 16.1 Output characteristics for different gate voltages and sweep directions of the drain voltage (top) and transfer characteristics on linear and logarithmic scales for different drain voltages and sweep directions of the gate voltage (bottom). OFET, 30 nm P3HT, 30 nm SiO_2, $L = 740$ nm and $w = 1000$ µm. Source/drain bottom contacts – one is from Au the other from Cr/Au.

The silicon dioxide gate insulator is 30 nm thick. This short channel has been prepared without high resolution lithography by using undercutting for the definition of the sub-micrometer channel length.

The output characteristics (Figure 16.1 (top)) exhibit for low gate voltages a good saturation behaviour without pronounced short channel effects, whereas at higher gate voltages, in spite of the thin gate oxide, saturation is not yet reached in the measured drain voltage region. The hysteresis connected with the different directions of the drain voltage sweep is almost negligible, which is characteristic of the present devices. The linear dependence of the current on small drain voltages indicates the absence of often observed non-linear contact resistances. In the transfer characteristics (Figure 16.1 (bottom)) there occurs a distinct hysteresis. It appears differently in the active and in the subthreshold regions. In the active region the characteristics show an almost linear dependency on the gate voltage only for the sweep from positive to negative gate voltage, whereas during the back sweep the current declines faster. On the logarithmic scale one finds for both sweep directions practically the same 'on/off' ratio of about 10^4 and a low inverse subthreshold slope of about $0.7 \ldots 0.8$ V/dec. Interestingly, in the subthreshold region the currents for both sweep directions are (with the exception of the first measurement at drain voltage –1 V) almost independent of the drain voltage indicating that the flat-band voltage is different for the two *sweep directions* of the gate voltage. From the parallel shift of the two groups of curves by ≈1.8 V one can estimate that interface charges for the two sweep directions differ by $\Delta Q'' = e\,\Delta N_{\mathrm{if}}$ with $\Delta N_{\mathrm{if}} \approx 1.3 \times 10^{12}\ \mathrm{cm}^{-2}$. This is of the order of magnitude of the total areal hole accumulation concentration.

Thus, one can phenomenologically describe the process as follows. During the waiting time before the back sweep, the fraction $\Delta Q''$ of the accumulated holes has been transformed from mobile to immobile at the interface. This gives no explanation of the microscopic processes. Different possibilities will be discussed in Sections 16.4 and 16.6.

16.3.1.2 **OFET Based on a Modified PPV and with Silanised Gate Oxide**

In our recent preparations with novel materials and advanced technological procedures the hysteresis is often less pronounced and partly also of a different form. An example is shown in Figure 16.2. Here the organic semiconductor is spin-coated from poly[1,4-phenylene-(4-methylphenyl)imino-4,48-diphenylene-(4-methylphenyl)imino-1,4-phenylene-vinylene-2-methoxy-5(2ethylhexyloxy)-1,4-phenylene-inylene] (TPD(4M)MEH-PPV). Source and drain (S/D) are prepared as top contacts (TOC) and bottom contacts (BOC) from gold. A highly doped silicon wafer with a thermal oxide was used for the gate and the insulator, respectively. The gate insulator was treated with HMDS. The measurements starts with switching the drain voltage to e.g. $V_{\mathrm{DS}} = -3$ V as in Figure 16.2. After a waiting time of 180 s the gate voltage is increased from $V_{\mathrm{GS}} = 0$ V in steps of 0.1 V to $V_{\mathrm{GS}} = -12$ V. At each voltage step the drain current is measured after a waiting time of 5 s. After a waiting time of 180 s at $V_{\mathrm{GS}} = -12$ V the reverse sweep starts in the same time regime.

The currents are considerably lower for the TOC OFET and just for this device the hysteresis is almost negligible in the transfer characteristics in the active and saturation regions visible on the linear scale. In this region the BOC transistor shows a clear hysteresis with lower currents for the sweep from negative to positive gate voltage. However, in contrast to the previous case, this is not connected with a significant difference between the flat-band voltages of the two sweep directions (see the logarithmic scale). On the other hand, the off-currents are rather different for the two sweep directions for both the TOC and BOC transistors with higher off-currents now for the sweep from negative to positive gate voltage. Such a difference in the off-currents has often been observed in our investigations.

A further peculiarity of these transistors has been observed and analysed in Ref. [45]. There is a striking difference between the TOC and the BOC transis-

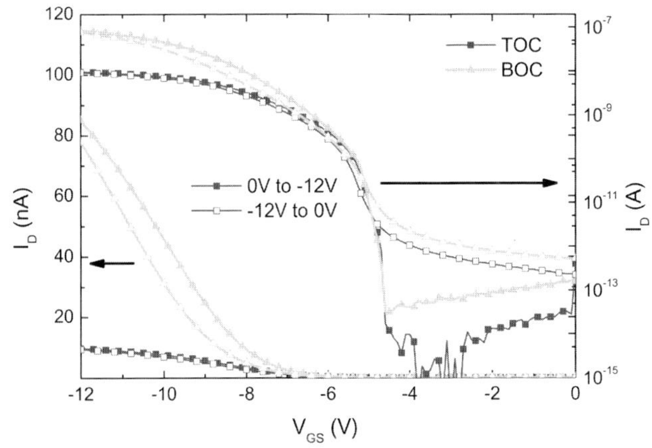

Figure 16.2 Transfer characteristics on linear and logarithmic scales for $V_{DS} = -3$ V and different sweep directions of the gate voltage. OFET with S/D gold contacts, 65 nm TPD(4M)-MEH-PPV, 30 nm SiO_2 gate insulator, $L = 25$ μm and $w = 1$ mm.

tors visible in the output characteristics. Whereas the drain current in the BOC transistor shows the normal linear dependency on the drain voltage for small drain voltages, one has in the TOC transistor a non-linear dependency with a positive curvature. It has been demonstrated by numerical two-dimensional simulations that such different behaviour can be caused by a high concentration of donor-like traps with a Gaussian or exponential distribution. The investigation of the influence of air confirms the presence of such traps. In Ref. [45] hysteresis has not been considered. Measured output characteristics for the transistor of Figure 16.2 show the just described peculiarity and in addition a hysteresis when the sequence of the gate voltages is from $V_{GS} = -12$ V in steps of 2 V to $V_{GS} = 0$ V, followed by the reverse sweep. After each voltage step the output characteristics is measured. Thus one has another time regime than in the case of the transfer characteristics. Now the output characteristics of both the BOC and TOC transistors show a similar hysteresis for large gate voltages.

Since the hysteresis is predominantly determined only by the gate-voltage sweep, i.e. accumulation and depletion at the interface to the gate oxide, it should be possible to study the effect mainly in the corresponding MIS capacitors. Results of such investigations will be discussed in the next section.

16.3.2 Organic MIS Capacitors

16.3.2.1 Quasi-Static *CV* Curves for a Capacitor with Arylamino-PPV

At first an example will be considered from an early systematic investigation [2] of hysteresis in organic field-effect devices. The quasi-static *CV* curves of an organic MIS capacitor are shown in Figure 16.3. The organic semiconductor is a 54 nm thick arylamino-PPV layer spin coated onto a 40 nm thick thermally grown oxide layer on a highly doped n-type silicon wafer used as the gate electrode. Afterwards gold contacts were evaporated (as bulk contacts) through a mask onto the organic layer with a diameter of 4 mm. The qualitative dependency of the *CV* curves shows that the polymer is (unintentionally) p-doped, since the oxide capacitance of 10.8 nF of the device is measured at negative gate-bulk voltages V_{GB}. For positive voltages one obtains the geometrical capacitance of 4.2 nF. This means that the whole organic layer is depleted. The *CV* characteristics in Figure 16.3 show a clear hysteresis. The transition between the geometrical and the oxide capacitance occurs for the gate-

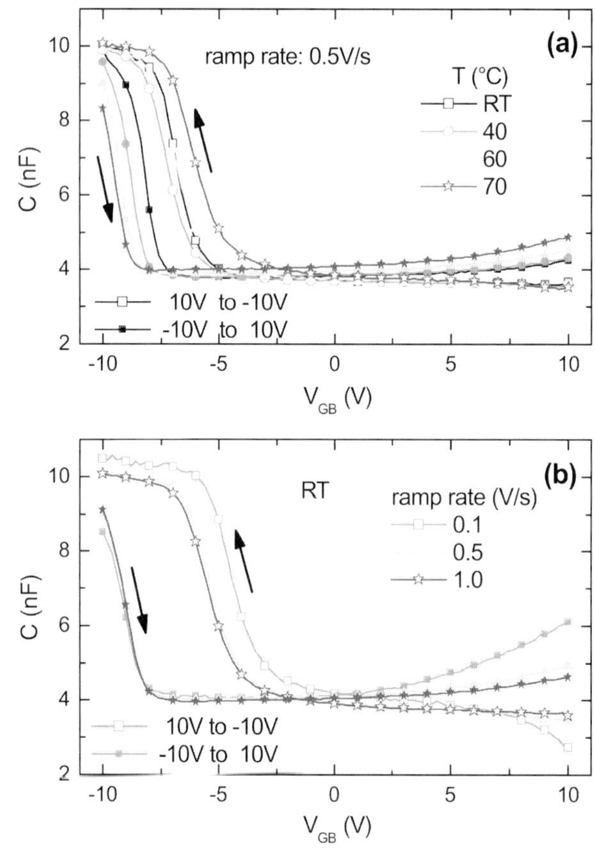

Figure 16.3 Measured quasi static *CV* curves of a MIS capacitor for different V_{GB} sweep directions at given ramp rate and different temperatures (a) and at room temperature for different ramp rates (b). The organic semiconductor is a 54 nm thick arylamino-PPV layer, the silicon dioxide insulator is 40 nm. Data from [2].

bulk voltage sweep direction from positive to negative voltages (i.e. from depletion to accumulation) at a more positive voltage than for the opposite sweep direction. Thus, as in the transistor discussed previously, the characteristics for the two sweep directions are essentially shifted against each other, which is equivalent to different flat band voltages for the two sweep directions and can be described formally by a charge at the interface between oxide and organic layer with a concentration depending on the voltage *sweep direction*.

In Figure 16.3(a) *CV* characteristics are depicted for a given voltage ramp rate *R* at different temperatures. With increasing temperature the curves for the sweep from negative to positive voltage are shifted to a more negative gate voltage, but for the sweep from positive to negative voltage they are first slightly shifted to a more negative voltage but from 60 °C to 70 °C to a more positive voltage. In Figure 16.3(b) the characteristics are depicted at room temperature for different voltage ramp rates. Here the curves for negative to positive voltage are essentially independent of the ramp rate, whereas for the opposite sweep direction the curve for the lowest ramp rate is shifted towards a positive voltage. The non-monotonous dependency on the temperature for the sweep from positive to negative voltages and the different dependency on the ramp rate for the two sweep directions indicate that the hysteresis can not be caused by one single process. From the parallel shift of the curves at room temperature by ≈4 V one can estimate that interface charge densities for the two sweep directions differ by $\Delta N_{\mathrm{if}} \approx 2.2 \times 10^{12}\ \mathrm{cm}^{-2}$. This is again comparable with the total hole accumulation charge.

16.3.2.2 Dynamic *CV* Curves

As another example we consider the hysteresis depicted in Figure 16.4. In this case the organic semiconductor is a 48 nm thick P3OT layer, the SiO_2 insulator is 50 nm thick. The dynamic *CV* curves were measured at 1 Hz for different gate-bulk voltage (V_{GB}) sweep directions at different temperatures [1]. In the preceding example the temperature range was from room temperature up to 70 °C, here the temperature is varied between 240 K and 300 K. But as in the former example, with increasing temperature here one has also a shift of the curve to more negative voltage for the sweep from negative to positive voltage and a shift to more positive voltage for the opposite sweep direction at the two higher temperatures.

In the third example, hysteresis of dynamic *CV* curves is compared for different active materials in Figure 16.5. The organic semiconductors are 50 nm thick P3OT layers, purified P3HT, in both cases with a silicon dioxide insulator of 50 nm, and pentacene prepared by a precursor route; here the layer thickness could not be determined accurately and the silicon dioxide insulator was 30 nm thick. The dynamic capacitance–voltage curves were measured at 10 Hz for different gate-bulk voltage (V_{GB}) sweep directions at room temperature. In the case of the P3OT capacitor one has again the typical distinct hysteresis with a parallel shift between the curves for the two sweep directions. On the other hand, the hysteresis is rather small for the capacitors with purified

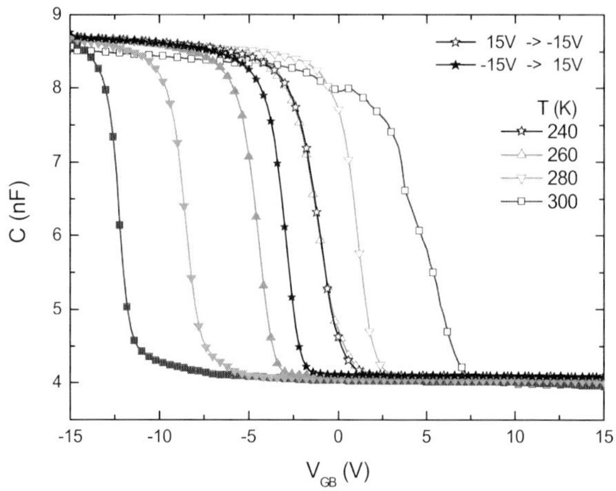

Figure 16.4 Dynamic capacitance–voltage curves of an organic MIS capacitor measured at 1 Hz for different gate-bulk voltage (V_{GB}) sweep directions at different temperatures. The organic semiconductor is a 48 nm thick poly(3-octylthiophene) (P3OT) layer, the silicon dioxide insulator is 50 nm thick. The curves for the two sweep directions are shifted against each other. Thus, one has different interface charges for the two sweep directions causing the different flat band voltages [1].

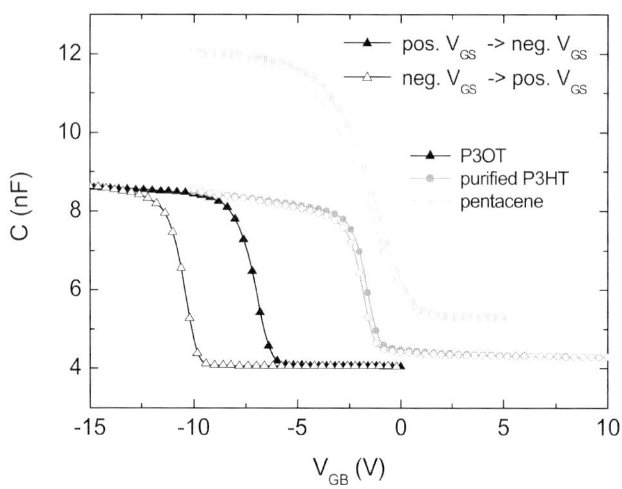

Figure 16.5 Dynamic *CV* curves of organic MIS capacitors measured at 10 Hz for different gate-bulk voltage (V_{GB}) sweep directions at room temperature. The organic semiconductors are P3OT purified P3HT, and pentacene prepared by a precursor route.

P3HT. But at the same time the mobility was reduced by about one order of magnitude by the purification. Reduction of the hysteresis does not necessarily mean that impurity induced traps must be causal for the hysteresis. The relative stability of polarons and bipolarons depend also on the presence of impurities and can be modified in this manner. Also for the capacitor with pentacene, prepared from a soluble precursor, the hysteresis is rather small. For such small molecules one can rule out a bipolaron mechanism, and apparently neither traps nor mobile ions lead to hysteresis.

As already mentioned, trap recharging alone can in principle lead to hysteresis [10]. But the shape of the simulated *CV* curves deviated significantly from the observed ones. The mechanism of trap recharging will be analysed further below in Section 16.4. In Section 16.6 processes will be considered in which doubly charged states on the polymer chains are involved. But first the equilibrium will by newly analysed.

16.4 Trap Recharging Mechanism

16.4.1 Simulations for the MIS Capacitor

In Ref. [10] we presented detailed numerical simulations carried out in order to check the presumption that the hysteresis is caused by trap recharging. These simulations were restricted to quasi-static *CV* characteristics for the MIS capacitor with one given thickness of the organic layer, 150 nm, which is larger than the depletion length for the assumed doping. Organic semiconductors with different types of traps, donor-like or acceptor-like, of different energetic positions, concentrations and spatial distributions are considered and their parameters were varied over a wide range. Trap recharging requires transport of the emitted (captured) carriers from (to) the traps during the voltage sweep. The combination of these processes leads to rather different types of hysteresis. For discrete bulk and interface traps the different forms of the hysteresis deviate clearly from the ones observed typically in experiments, or the hysteresis does occur for extremely small values $\sigma v_{\mathrm{th}} < 10^{-18}\ \mathrm{cm}^3\ \mathrm{s}^{-1}$ of the product from capture cross section and thermal velocity. (The cross section should be of the order $\sigma \approx 10^{-14}\ \mathrm{cm}^2$ and the thermal velocity for hopping of the order 1 . . . 10 cm/s for the chosen value of the mobilty [46].)

In principle the situation seems to be similar for energetically distributed traps with Gaussian or exponential density of states. But especially in the case of an exponential trap density of states in the gap with the maximum at a distance E_0 above the valence band edge E_{V}

$$D_{\mathrm{t}}(E) = D_{\mathrm{t0}}\, \mathrm{e}^{-\left|\frac{E-(E_{\mathrm{V}}+E_0)}{E_{\mathrm{s}}}\right|}, \qquad N_{\mathrm{t}} = E_{\mathrm{s}} D_{\mathrm{t0}}\,, \tag{1}$$

the form of the hysteresis is closer to the observed ones if one chooses extremely high total concentrations N_t of $10^{19} \ldots 10^{20}$ cm^{-3}, which is to be compared with the monomer concentration of e.g. 3×10^{21} cm^{-3} for P3HT (one trap per $3 \times 3 \times 3$ monomer units for the higher concentration!). Since the distribution increases towards the band edge, very strong accumulation is needed to recharge these states but recharging becomes faster nearer to the band edge. An example is shown in Figure 16.6. The traps are donor-like with the maximum of the density of states (DOS) at the valence band edge ($E_0 = 0$). The decay constant is $E_s = 0.1$ eV, and $\sigma v_{th} = 10^{-14}$ cm^3 s^{-1} has a reasonable value. Two concentrations are considered. Completely ionised acceptors with a concentration of 10^{17} cm^{-3} are present, the bulk contact work function is 5 eV. Further parameters: MIS capacitor with a 50 nm thick insulator and a thickness of the organic layer of 150 nm. The gate contact material has the work function $\Phi_G = 4.05$ eV corresponding to n$^+$-poly-silicon. The metal work function Φ_M of the bulk contact is with 5 eV near the valence band. The dielectric constant of silicon dioxide is $\varepsilon_{ox} = 3.9$, dielectric constant of the organic layer $\varepsilon = 3.24$, electron affinity $\chi = 3.0$ eV, band gap $E_g = 2.0$ eV, mobilities $\mu_n = \mu_p = 10^{-3}$ cm^2/Vs, effective density of states $N_C = N_V = 10^{21}$ cm^{-3} (monomer density). For a smaller trap concentration, the capacitance increases gradually from depletion to accumulation and near the flat band voltage a structure appears. For the larger trap concentration the structure disappears and the increase of the capacitance begins abruptly. This is a consequence of the reduced bulk hole concentration in the presence of the high concentration of the distributed traps. For the opposite sweep direction, at the beginning one has a steep decrease followed by a gradual decay until the geometrical capacitance is almost reached. The disagreement with experiment is directly seen as a crossing of the curves for the two sweep directions with the curve for the minus to plus sweep above the one from plus to minus, roughly in the region of positive voltage.

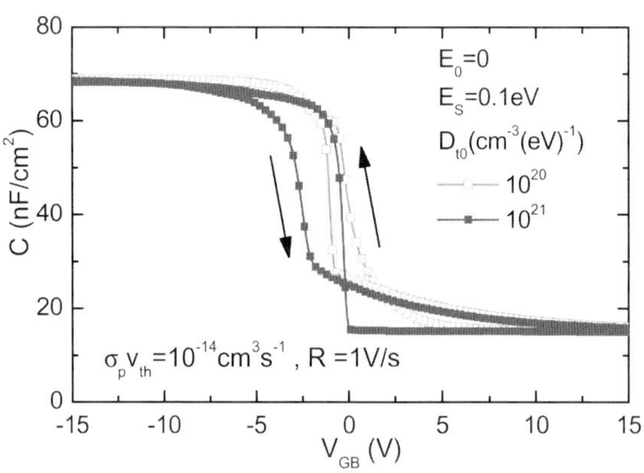

Figure 16.6 Simulated *CV* characteristics of a MIS capacitor with an exponential distribution of donor-like traps with different maximum concentration. Data taken from Ref. [10].

16.4.2
Simulations for Thin-Layer OFETs and the Corresponding Capacitor

In order to check further the influence of energetically distributed traps, additional simulations have been carried out, this time for both an OFET and the corresponding MIS capacitor, both with a thickness of the organic layer of 65 nm, allowing for full depletion as required for the operation of the OFET. The oxide is 30 nm thick. Results for the capacitor are depicted in Figures 16.7 and 16.8 for variations of different parameters. The remaining parameters are the same as before. The general form of the hysteresis is the same as in Figure 16.6 for the thicker organic layer. In particular, in all cases the curve for the sweep from positive to negative voltage lies in the region of positive voltages below the curve for the opposite sweep, which is usually not the case in the measured hysteresis. In Figure 16.7 the *CV* curves are depicted at first for different trap concentrations expressed by the maximum value D_{t0}. The corresponding total concentrations N_t of the trap states are 3×10^{18} cm^{-3} and 2×10^{19} cm^{-3}. Although even the first value is rather high, one has only a narrow hysteresis in this case and the shift between the two sweep directions increases of course with N_t. For the lower trap concentration also curves are compared for different sweep rates. A variation from $R = 0.1$ V/s (not shown in the figure) to $R = 0.02$ V/s leads to practically the same curves. Furthermore, the chosen product $\sigma v_{th} = 10^{-19}$ cm^3 s^{-1} is extremely small in order to obtain the hysteresis. Thus, in Figure 16.8 the larger concentration is chosen and comparison is made with the case $\sigma v_{th} = 10^{-14}$ cm^3 s^{-1}. In the latter case the hysteresis is narrow and one has a bump in the sweep to a negative voltage.

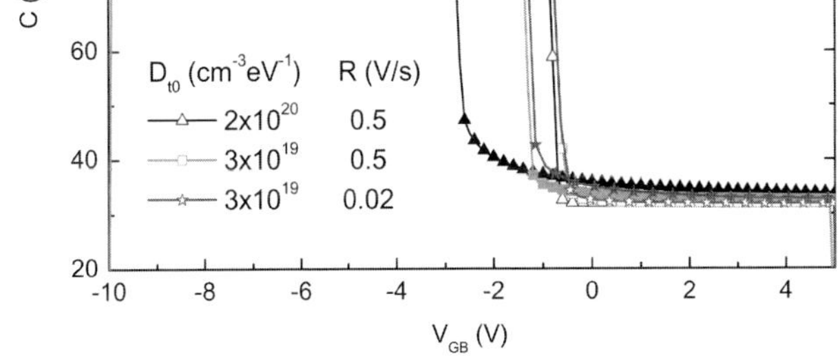

Figure 16.7 Simulated *CV* curves of a MIS capacitor with an exponential distribution of donor-like traps for different combinations of maximum DOS and ramp rates. $E_0 = 0$, $E_s = 0.1$ eV, $\sigma v_{th} = 10^{-19}$ cm^3 s^{-1}.

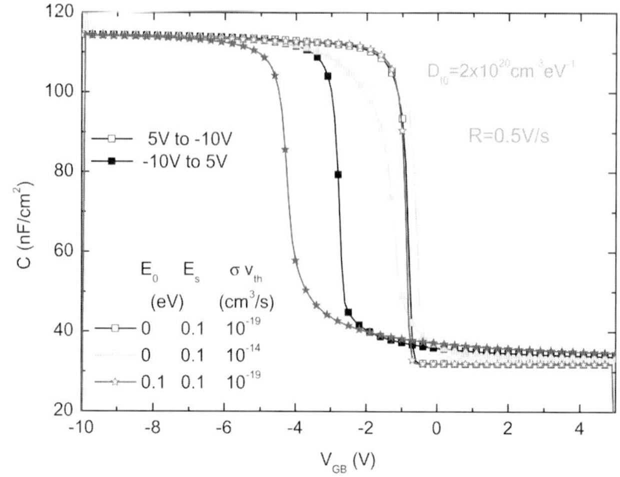

Figure 16.8 Simulated *CV* curves of a MIS capacitor with an exponential distribution of donor-like traps for different combinations of E_0, E_s, σv_{th}.

Finally, in Figure 16.8 it is also shown that the hysteresis becomes broader, when the maximum of the exponential DOS lies not just at the valence band edge but above it.

For the simulated hysteresis of a transistor an example is given in Figure 16.9. The transfer characteristics are depicted on the linear and logarith-

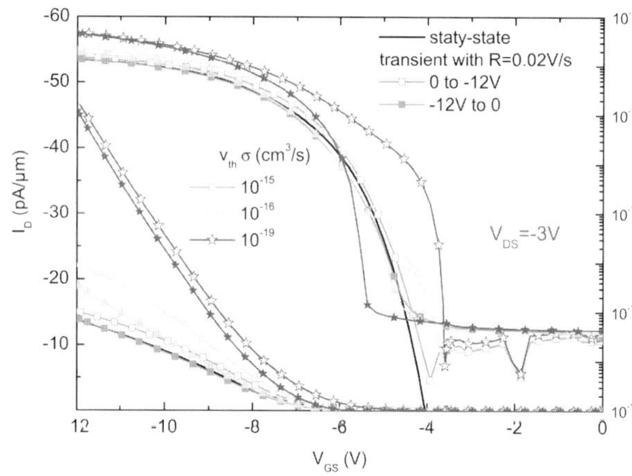

Figure 16.9 Simulated transfer characteristics of a TOC OFET for –3 V drain voltage. Organic semiconductor with an exponential distribution of donor-like traps for maximum DOS $D_{t0} = 2 \times 10^{20}\ \text{cm}^{-3}\ \text{eV}^{-1}$, $E_0 = 0$, $E_s = 0.1$ eV, and ramp rate $R = 0.02$ V/s for different values of σv_{th}. In addition an interface charge of areal density $2.33 \times 10^{12}\ \text{cm}^{-2}$ is taken into account.

mic scales for a TOC transistor. The curves are essentially the same for a BOC transistor, there is only a small difference in the off-current. In the simulation an additional interface charge has been included leading to a shift of the threshold voltage towards a negative gate voltage. For the trap density the larger value of the DOS maximum has already been chosen. Thus, a hysteresis does occur in the active and saturation regions (see linear scale), in this case with a non-monotonous dependency on the value of σv_{th}. It is seen on the logarithmic scale that a remarkable difference between the threshold voltages of the two sweep directions occurs only for the smallest chosen value. Due to the traps there remains for both sweep direction a higher off-current whereas the steady-state current decreases further. The off-current for the sweep from depletion to accumulation is lower than the current for the reverse sweep. This is a tendency also visible in the experimental result shown in Figure 16.2.

The extended simulation allows one to conclude that exponentially distributed traps can indeed lead to hysteresis in the field-effect. But the effects do occur only for extremely high trap concentrations and at the same time a very small value of σv_{th}, which is comparable to a value for the second order polaron reaction leading to bipolarons. This small value originates from the Coulomb repulsion which does not occur in the capture process at traps. In addition, the form of the resulting hysteresis shows partly clear deviations from the observed ones. Thus, traps might modify a hysteresis caused by other processes. But it does not seem plausible that the main effects are caused only by traps.

16.5 Equilibrium of Polarons With Doubly Charged States of the Polymer Chain

Charges on polymer chains connected with a chain distortion appear in the case of holes as (hole) polarons (Ps, charge +e and spin 1/2). Bipolarons have been considered for a long time as the doubly charged states [7, 8] (BPs, +2e and 0 -ESR silent). In order to investigate the possible occurrence of hysteresis due to the formation of doubly charged polymer chain states by polaron pair collision and dissociation of the doubly charged species, one needs the equilibrium concentrations as functions of the electrochemical potential and of temperature, that is the equations of state of both species, and also the kinetic equation for the reaction. The equations of state have been determined by Tang et al. [47] on the basis of Nernst equations in the low-density limit. The same result was later derived by Davids et al. [48] from statistics in the non-degenerate case. Extensions for the degenerate, high density limit due to the formation of bipolaron bands (that is a broadening of the bipolaron level) have been introduced in [8, 49, 50] and it has been shown that the low density limit extends to concentrations of the order of 10%. These modifications will not be considered here.

Alternatives to the bipolaron have been discussed earlier and especially the 'two polarons on a single chain' or 'polaron pair' (PP) model [9] (PPs, +2e and 0-ESR silent) was proposed to be preferable compared with the BP model, since the latter contradicts optical data whereas the former leads to a consistent explanation of optical data. The corresponding equation of state has been considered only recently [24] and led to a new contradiction. In Section 16.5.1 both models are summarised and compared with each other and in Sections 16.5.2 and 16.5.3 possible ways out of the contradiction are proposed.

16.5.1
Polarons and Bipolarons or Polaron Pairs

The chain distortion of the polymer chain due to excess charges extends over chain segments of a number of m monomer units ($m \approx 6$ to 10) for both Ps and BPs as proposed by [48]. Accordingly we suppose that the PP extends over $2m$ monomer units. We denote the concentrations of Ps, BPs, and PPs by c_{P}, c_{BP}, and c_{PP}, respectively. The concentration of neutral segments is c_0. The segment concentration follows, according to $mc_{\mathrm{s}} = c_{\mathrm{mon}}$, from the monomer concentration. For a concrete number P3HT is considered. From the assumed mass density of 1 g cm^{-3} and the molecular weight of about 166 one obtains $c_{\mathrm{mon}} \approx 3 \times 10^{21}$ cm^{-3} and $c_{\mathrm{s}} \approx 5 \times 10^{20}$ cm^{-3} for $m = 6$.

16.5.1.1 Polarons and Bipolarons

If one has only Ps and BPs, equilibrium is described by the Nernst equations [8] (in the following the symbol E is used for the potential with the unit volt, which is common in electrochemistry)

$$c_0 = c_{\mathrm{P}} Y_1 , \tag{2}$$

$$c_{\mathrm{P}} = c_{\mathrm{BP}} Y_2 , \tag{3}$$

$$Y_i \equiv \mathrm{e}^{-f(E-E_i)} , \tag{4}$$

where E_1 and E_2 are the standard potentials for the first and second oxidation steps, respectively. The electrochemical potential is connected with the Fermi energy by $-eE = \varepsilon_{\mathrm{F}}$, and $1/f = k_{\mathrm{B}}T/e$. The actual concentrations are obtained by solving (2), (3) together with the relation

$$c_{\mathrm{s}} = c_0 + c_{\mathrm{P}} + c_{\mathrm{BP}} , \tag{5}$$

between the concentrations. We notice here that (5) is equivalent to the assumption that Ps and BPs have the same extension m used explicitly in the statistical treatment [48]. Inclusion of the BPs is of interest if they are energetically favoured, i.e. for $E_2 < E_1$. It is convenient to introduce the equilibrium constant K_{BP} by considering the P–BP equilibrium as a result of polaron pair collisions

$$P^+ + P^+ \rightleftharpoons BP^{2+} + R^0 , \tag{6}$$

$$K_{BP} = e^{f(E_1 - E_2)} = \left(\frac{c_{BP} c_0}{c_P^2} \right)_{equ.} , \tag{7}$$

which is determined by the equilibrium concentrations. Notice that the neutral chain segment R^0 must be taken into account. The standard potentials are connected with the formation energies ε_P and ε_{BP} for the Ps and BPs (less than half of the band gap due to the energy gain by the chain distortion). Together with statistical degeneracy factors ($g_P = 2m,\ \ g_{BP} = m$) one has [8]

$$E_1 = \frac{\varepsilon_P}{e} - \frac{\ln g_P}{f} , \qquad E_2 = \frac{(\varepsilon_{BP} - \varepsilon_P)}{e} - \frac{\ln (g_{BP}/g_P)}{f} . \tag{8}$$

In (2), (3) the non-degenerate approximation (4) is valid up to more than a 10% oxidation degree [8]. Such high concentrations are not achieved in space charge layers in organic devices. The formation of bipolaron bands at larger oxidation degrees will not be considered here.

Explicit expressions for the P and BP concentrations are

$$c_P = c_s N^{-1} \, e^{f(E-E_1)} , \tag{9}$$

$$c_{BP} = c_s N^{-1} \, e^{2f(E-E_3)} , \tag{10}$$

$$E_3 = (E_1 + E_2)/2 , \tag{11}$$

$$N = 1 + e^{f(E-E_1)} + e^{2f(E-E_3)} . \tag{12}$$

A typical dependency of the P and BP concentrations on the potential is depicted in Figure 16.10 for energetically favoured BPs: the BP concentration saturates for high potential and the P concentration shows a maximum at the potential E_3 for the two-electron reaction. This dependency is in accordance with the maximum in the spin concentration observed in ESR spectro-electrochemical experiments. In spite of this agreement, the BP model is in disagreement with optical data and has been ruled out therefore by the PP model [9].

16.5.1.2 **Polarons and Polaron Pairs**

Now the thermodynamic consequences will be considered assuming that one has only Ps and PPs. The PP formation, according to

$$P^+ + P^+ \rightleftharpoons PP^{2+} , \tag{13}$$

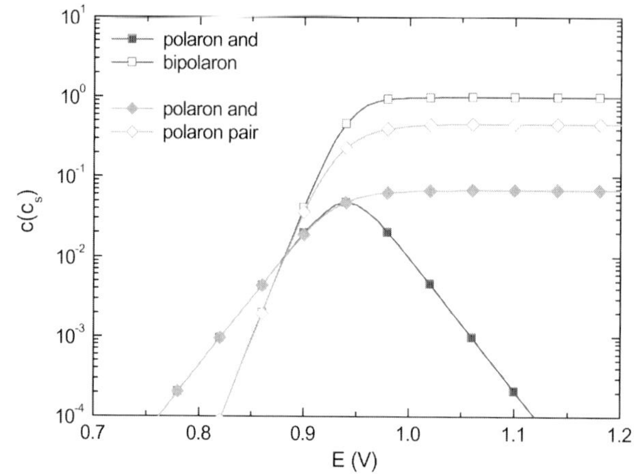

Figure 16.10 Equilibrium concentrations of either Ps and BPs, or Ps and PPs as functions of the electrochemical potential, zero at midgap position, oxidation potential $E_1 = 1$ V, $m = 6$. The equilibrium constant is 100, room temperature.

is not connected with an electron transfer – there is no Nernst equation for this process. Thus, equilibrium is determined by

$$c_0 = c_P Y_1 \,, \tag{14}$$

$$\bar{K}_{PP} = \left(\frac{c_{PP}}{c_P^2} \right)_{equ.} \,, \tag{15}$$

$$c_s = c_0 + c_P + 2c_{PP} \,. \tag{16}$$

Here, Eq. (16) reflects the use of $2m$ for the PP extension. To compare with the former case, the dimensionless form

$$K_{PP} \equiv \bar{K}_{PP} c_s = \left(\frac{c_{PP}/c_s}{(c_P/c_s)^2} \right)_{equ.} = \left(\frac{c_{PP} c_s}{(c_P)^2} \right)_{equ.} \tag{17}$$

is needed. In contrast to (7) the concentration of neutral segments does not occur here, a difference which becomes important for high oxidation degree. Moreover, (5) and (16) are different since the BP occupies only one chain segment and the PP just two. The resulting P concentration follows as

$$\frac{c_P}{c_s} = \frac{1}{4K_{PP}} \left(-(1+Y_1) + [(1+Y_1)^2 + 8K_{PP}]^{1/2} \right), \tag{18}$$

determining also the PP concentration via Eq. (17). The dependency on the potential is depicted also in Figure 16.10. For low potential/concentrations one cannot distinguish between both models. But both the PP and the P concentrations saturate at high potentials with $c_P + 2c_{PP} \to c_s$. The absence of a maximum in the P concentration for higher potentials contradicts the results of ESR spectro-electrochemical experiments.

One possible way out to resolve the discrepancy could be the assumption that PPs dominate at low and moderate concentrations, but an additional subsequent second oxidation step is possible at a standard potential near to the one of the first oxidation step.

16.5.2 Polarons, Bipolarons and Polaron Pairs

As an extension of the two former cases we consider now the possibility that besides the Ps both PPs and BPs can occur as doubly charged states of the polymer chains. Equilibrium is then described by (2), (3), (17) and, instead of (5) or (16) by

$$c_s = c_0 + c_P + c_{BP} + 2c_{PP} \,. \tag{19}$$

The resulting explicit expression for the P concentration is given by

$$\frac{c_P}{c_s} = \frac{1}{4K_{PP}}\left(-(1+Y_1+Y_2^{-1}) + [(1+Y_1+Y_2^{-1})^2 + 8K_{PP}]^{1/2}\right). \tag{20}$$

The BP and PP concentrations follow then from Eqs. (3) and (17). Typical dependencies of the three concentrations on the potential are depicted in Figure 16.11. Parameters are the oxidation potential $E_1 = 1\,\text{V}$, the segment concentration as estimated above (with $m = 6$), $c_s = 5\times10^{20}\ \text{cm}^{-3}$, and the two equilibrium constants (7), (17). Both equilibrium constants are chosen larger than unity since otherwise the influence of the PPs or BPs would be important only for potentials $E > E_1$. In all three cases shown in Figure 16.11, the P concentration dominates for lower potentials. As before, the BP and PP concentrations increase twice as fast with increasing potential. In all cases the BP concentration saturates for higher potentials ($c_{BP} \to c_s$) and both the P and PP concentrations decrease there exponentially with increasing potential whereby the decrease of the PP concentration is again twice as fast. In Figure 16.11(a) the same equilibrium constants have been chosen for BPs and PPs, $K_{BP} = K_{PP} = 300$. This leads to the same PP and BP concentrations for lower and medium potentials, which is in contrast to the optical evidence for PPs. Thus, this case must be ruled out. On the other hand, if the BP equilibrium constant is still larger than unity but much smaller than the PP equilibrium constant, in Figure 16.11(c) $K_{BP} = 0.01K_{PP}$, there is already an indication of a plateau in the P concentration, which contradicts the clear maximum of the spin concentration in ESR spectro-electrochemical measurements. Thus, the condition that on the one hand the PPs dominate at lower and medium concentrations as expected from the optical data, and that on the other hand the P concentration shows a clear maximum, is essentially fulfilled for $K_{PP} \gg 1$ and $K_{BP} \approx 0.1 K_{PP}$. This case is demonstrated in Figure 16.11(b): both PPs and BPs are possible doubly charged states of the polymer chains with the PPs favoured for lower potentials.

Nevertheless, one has two problems with this model. At first, the polaron peak is relatively broad whereas in experiments at least in some cases the peak of the spin concentration is indeed narrower than for a one-electron transition. Further, it seems to be hard to find a reason why the BP equilibrium constant should be just roughly one tenth of the PP equilibrium constant. A reasonable alternative will be presented in the next section.

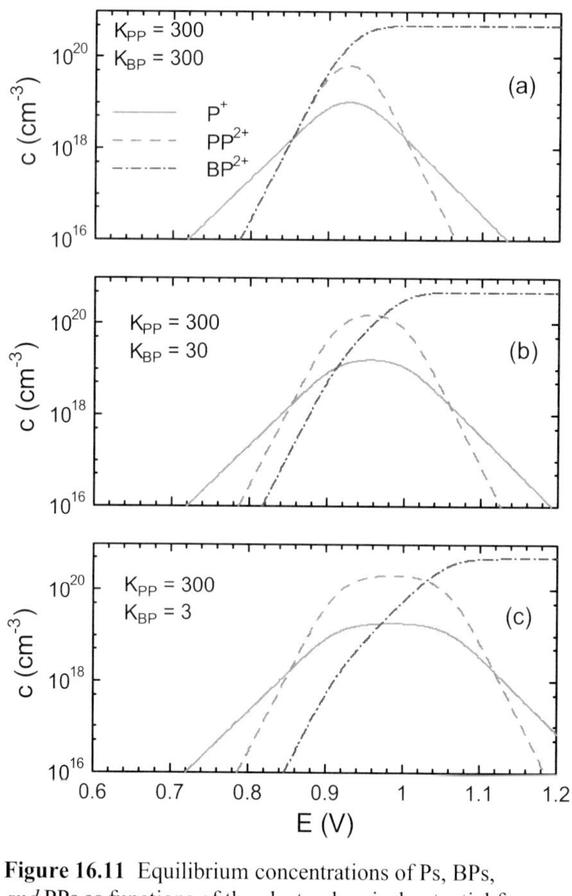

Figure 16.11 Equilibrium concentrations of Ps, BPs, *and* PPs as functions of the electrochemical potential for different equilibrium constants, zero at midgap position, oxidation potential $E_1 = 1\,\mathrm{V}$, $m = 6$.

16.5.3 Polarons and General Dipolarons

In the past, the relevance of the thermodynamic derivation of the P–BP equilibrium [47] or the equivalent statistical derivation [48] has not been questioned. However, there is one point which is crucial. This is the assumption of Eq. (5) for the equilibrium, equivalently Eq. (6) for the reaction, and equivalently in the statistical treatment the assumption that both Ps and BPs have the same extension over m monomer units. These assumptions are quite natural for molecules in solution, which can be singly or doubly charged. In the case of charges on the polymer chain connected with a chain distortion, there is no reason that the BP has the same extension as the P, nor that the PP has exactly twice the extension of the P.

Thus, as before we suppose that the P extends over m monomer units. Then a neutral chain segment of the same extension is defined. For the doubly charged species it is now supposed that it extends over m_2 monomer units. With $m < m_2 < 2m$ it can be larger than the BP and smaller than the PP. To distinguish this more general case we denote it as a 'dipolaron' (DP). It must be emphasised, that no information about the type of binding enters this supposition. Counting now the number of segments one finds

$$N_s = N_0 + N_P + \frac{m_2}{m} N_{DP} ,$$

$$c_s = c_0 + c_P + \alpha N_{DP} , \tag{21}$$

$$1 < \alpha \equiv \frac{m_2}{m} < 2 ,$$

which, in the second line, is expressed as a relation between the concentrations. Evidently, for $m_2 = m$ and $m_2 = 2m$ one has the Eqs. (5) and (16) for BPs and PPs, respectively. The reaction equation now takes the following form

$$\begin{aligned} &m\mathrm{P}^+ + m\mathrm{P}^+ \rightleftharpoons m\mathrm{DP}^{2+} + (2m - m_2)\,\mathrm{R}_0 , \\ &\mathrm{P}^+ + \mathrm{P}^+ \rightleftharpoons \mathrm{DP}^{2+} + (2 - \alpha)\,\mathrm{R}_0 , \end{aligned} \tag{22}$$

with the corresponding limits. According to this reaction equation the dimensionless equilibrium constant must be defined as

$$K = \left(\frac{(c_{DP}/c_s)\,(c_0/c_s)^{2-\alpha}}{(c_P/c_s)^2} \right)_{equ.} . \tag{23}$$

Finally, the Nernst equation for the first oxidation step remains unchanged as

$$c_0 = c_P Y_1 , \qquad Y_1 \equiv e^{-f(E-E_1)} . \tag{24}$$

In this manner the concentrations of the neutral chain segment, of Ps and of DPs are determined by the three Eqs. (21), (23), and (24). Analytical solutions for $m_2 = m$ (BPs) and $m_2 = 2m$ (PPs) have been given in Sections 16.5.1.1 and 16.5.1.2. In the general case a numerical solution is needed. An example is demonstrated in Figure 16.12. The DP concentration saturates at high potential and the P concentration lies in general between the two limits. Thus the P concentration has the symmetric maximum for $m_2 = m$, becomes asymmetric but still with a maximum for $m < m_2 < 2m$ and saturates for $m_2 = 2m$. A slower decrease of this concentration with increasing potential beyond the maximum is typical for the measured spin concentration, but is also influenced by the non-equilibrium and transition beyond the non-degenerate limit. For potentials below the maximum the different cases are almost indistinguishable. As will be shown below, this is just the region of interest for the hysteresis in field-effect devices.

As a result of these considerations, it seems likely that the doubly charged state of the polymer chain is something as the PP with optical properties in accordance with the observed spectra. A total extension of less than two Ps is

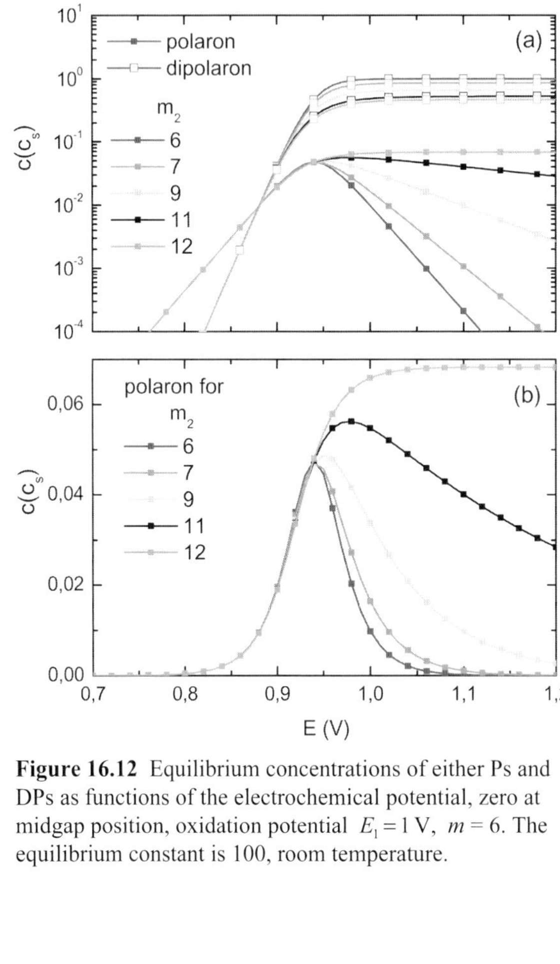

Figure 16.12 Equilibrium concentrations of either Ps and DPs as functions of the electrochemical potential, zero at midgap position, oxidation potential $E_1 = 1$ V, $m = 6$. The equilibrium constant is 100, room temperature.

needed to obtain the observed, asymmetric maximum in the spin concentration. Since this bound state is not simply the sum of two Ps, the notation dipolaron seems to be preferable.

16.6 Bipolaron Mechanism for Hysteresis

16.6.1 Formation and Dissociation of Bipolarons

16.6.1.1 Kinetics of Formation and Dissociation

To begin with we consider formation and dissociation of the generalised DPs, which is described by the reaction Eq. (22) with the equilibrium constant (23). The DP formation is a second order reaction due to pair collision of two Ps. The time dependency of the concentrations is given by the corresponding kinetic equation

$$2\frac{\mathrm{d}c_{\mathrm{DP}}}{\mathrm{d}t} = k\left[c_{\mathrm{P}}^2 - \frac{1}{K}c_{\mathrm{DP}}c_0^{2-\alpha}c_{\mathrm{s}}^{\alpha-1}\right] = -\left(\frac{\mathrm{d}c_{\mathrm{P}}}{\mathrm{d}t}\right)_{\mathrm{DP}}. \tag{25}$$

Here the rate constant is k and the c_i are of course non-equilibrium concentrations. The first term on the right hand side of the first line describes formation and the second one dissociation of the DPs. In the last expression the lower index DP denotes that it is only the polaron concentration that is changed by this process. The rate constant for the dissociation is $k' = k/K$. These processes can be connected with hysteresis only for energetically favoured DPs, this means for $K \gg 1$. In this case the rate constant for dissociation is much lower than the one for formation, $k' \ll k$. In (25) the dissociation is a first order reaction only for $\alpha = 2$ (the limiting case of PPs) since otherwise the concentration c_0 of the neutral chain segments does occur, which depends also on the P and DP concentrations. However, for analysing the hysteresis effects the depletion and accumulation regions in the MIS structure are of interest. The achievable maximum concentration in accumulation is about 10^{19} cm^{-3}. This limitation is caused by the maximum break-through field of the oxide. Considering a value of the segment concentration of $c_{\mathrm{s}} \approx 5\times10^{20}$ cm^{-3}, one has $c_0 \approx c_{\mathrm{s}}$ and the dissociation is practically a first order process. Since under this condition Eq. (25) does not depend on α, that is on the extension of the dipolaron, it is valid for the bipolaron as well the polaron pair. Similarly, Figure 16.12 shows that up to these maximum concentrations the equilibrium concentrations are also independent of α, i.e. one cannot distinguish between BP, PP or DP. From (25) one obtains the relaxation times τ for the initial process of formation and τ' for dissociation under the just mentioned condition

$$\tau = \frac{2}{kc_{\mathrm{P},0}}, \qquad \tau' = \frac{2K}{kc_{\mathrm{s}}} = \frac{2}{k'c_{\mathrm{s}}}. \tag{26}$$

Whereas τ for the DP formation depends on the initial P concentration $c_{P,0}$, τ' contains the smaller rate constant and the larger segment concentration c_S.

In bias stress experiments in polythiophene OFETs direct experimental evidence for the formation of a doubly charge state of the polymer chains has been found [11, 12], with $\mathrm{d}c_P/\mathrm{d}t = kc_P^2$ which was attributed by the authors to the formation of bipolarons. In [12] the rate constant has been determined as $k = 3 \times 10^{-19}\ \mathrm{cm}^3\ \mathrm{s}^{-1}$. This value will be used here for numerical estimates, and is many orders of magnitude lower than the product of cross section ($\approx 10^{-14}\ \mathrm{cm}^2$) and average velocity for hopping of the polarons (of the order of 10 cm/s). This deviation was attributed to a potential barrier for the pair collision due to Coulomb repulsion. As already explained, for the processes in the accumulation layer one cannot distinguish between doubly charged states with different extensions. Thus, in the following we will use the notation '*bipolaron mechanism*' introduced in Refs. [11, 12].

16.6.1.2 **The Bipolaron Mechanism**

In order to discuss this mechanism, in Figure 16.13(a) the P and BP concentrations are depicted for three values of the equilibrium constant $K \gg 1$ as functions of the Fermi energy, which is shifted in the gap near the interface downwards in the transition from depletion to accumulation. In the relevant region limited by the maximum concentration in accumulation as indicated in the figure, the P concentration is practically independent of the equilibrium constant whereas the BP concentration increases at a given position of the Fermi energy with increasing equilibrium constant. Using as the initial P concentration in the relaxation time for the BP formation (26) the equilibrium concentration from Figure 16.13(a), both relaxation times are depicted in Figure 16.13(b). Here the relaxation time for formation is essentially independent of the equilibrium constant whereas the relaxation time for the BP dissociation increases with increasing value of the equilibrium constant.

Now one can compare in detail the sweep from positive to negative voltage with the reverse sweep. During the sweep from positive to negative voltage, the increase of the capacitance (see e.g. Figure 16.3) is connected with the filling of the depleted layer, the polaron concentration is less than the doping, typically $c_P < 10^{17}\ \mathrm{cm}^{-3}$ and the BP formation is determined by τ (26), the time constant is in this region rather large (Figure 16.13(a) and (b)) $\tau > 70\ \mathrm{s}$ and therefore the BP formation is retarded (even the equilibrium concentration (Figure 16.13(a)) is negligible). With the initial accumulation, $c_P > 10^{17}\ \mathrm{cm}^{-3}$, the capacitance turns into saturation but accumulation proceeds and the relaxation time for BP formation is reduced to $\tau < 1\ \mathrm{s}$ as the accumulation concentration increases to its maximum value. Thus, during this sweep and the subsequent waiting time (typically up to 100 s) the BP concentration approaches the high equilibrium value, larger than the P concentration. For the subsequent sweep from negative to positive voltage with ramp rate of $0.1 \ldots 1\ \mathrm{V/s}$, the time constant τ' for the PP dissociation is given by the second equation in (26). For the large equilibrium constants of the order $K = 100, 300, 1000$ one

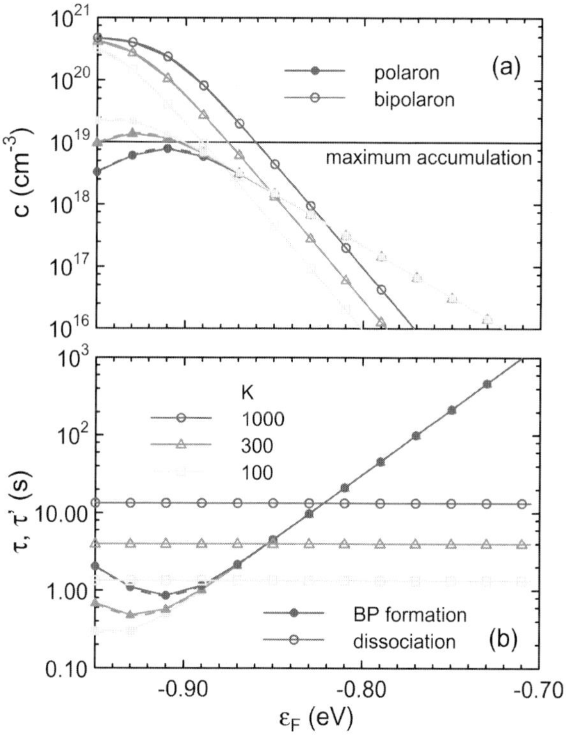

Figure 16.13 Polaron and bipolaron concentrations (a) and relaxation times for BP formation and dissociation (b) as functions of the Fermi energy (energy zero at midgap position) for different values of the equilibrium constant. In (a) the maximum concentration in an accumulation layer is indicated.

gets $\tau' \approx 1.3, 4, 13.3$ s. Due to the instantaneous depletion of mobile polarons at the beginning of this voltage sweep, the capacitance curve turns down into the descending depletion branch at the beginning of the sweep, the temporarily fixed immobile BP accumulation layer determines the shifted flat band voltage, and the following dissociation of the BPs occurs already on this branch.

16.6.2 Formation of Complexes With Counter Ions

16.6.2.1 The Kirova–Brazovskii Scenario of Complex Formation

When the total concentration of holes is changed by charge injection of single holes at a contact, Ps are immediately formed and the BP formation is needed in order to achieve equilibrium. Kirova and Brazovskii (KB) [7] assumed that this process is practically excluded since the BP formation by pair collision of Ps will be kinetically hindered due to the Coulomb repulsion. The estimates given in the preceding section have indeed shown that this process is rather slow for a low initial concentration and is still of the order of 100 s when the doping level (of the order 10^{17} cm^{-3}) is reached. Thus, the assumption of KB [7] that this process is practically excluded, is not really justified. Nevertheless, it is worth discussing the consequences [23] of a scheme proposed by them. They supposed that a bound state between a P^+ and a dopant or counter ion C^- can be formed, a neutral complex $(P^+ - C^-)^0 \equiv D_1^0$. For a second polaron the Coulomb repulsion is strongly reduced and a further P^+ can be captured leading to an overcharged $(BP^{2+} - C^-)^+ \equiv D_2^+$ complex. This process is also fast. Of course, these processes can take place only where the counter ions are present and up to their concentration, which means, in the bulk semiconductor up to the doping concentration. But in the accumulation layer only the direct process of bipolaron formation is possible as described in the preceding section. One further process was proposed by KB. Since in the polymers the dopants are mobile to some extent, in a long term process neutral $(BP^{2+} - 2C^-)^0 \equiv D_3^0$ complexes will be formed. The three reactions are described by

$$P^+ + C^- \rightleftharpoons D_1^0 \,, \tag{27}$$

$$P^+ + D_1^0 \rightleftharpoons \equiv D_2^+ + R^0 \,, \tag{28}$$

$$D_2^+ + C^- \rightleftharpoons D_3^0 \,. \tag{29}$$

The corresponding concentrations are denoted by $c_{D_1^0}$, $c_{D_2^+}$, and $c_{D_3^0}$ and the concentration of the mobile ions is c_{C^-}.

16.6.2.2 Slow Ion Capture by an Overcharged Complex

For analysing slow processes it is sufficient to consider the last reaction (29) describing the capture of a mobile ion by the overcharged complex. The equilibrium constant is given by

$$K_3 = \left(\frac{c_{D_3^0}}{c_{D_2^+} c_{C^-}} \right)_{\text{equ.}} \tag{30}$$

and the kinetic equation is

$$-\frac{\mathrm{d}c_{D_2^+}}{\mathrm{d}t} = \frac{\mathrm{d}c_{D_3^0}}{\mathrm{d}t} = k_3 \left[c_{C^-} c_{D_2^+} - \frac{1}{K_3} c_{D_3^0} \right]. \tag{31}$$

This process can be important only if the equilibrium lies on the side of the complex D_3^0, which means for $K_3 (c_{C^-})_{\text{equ.}} = (c_{D_3^0}/c_{D_2^+})_{\text{equ.}} \gg 1$. The time constant for the formation of the neutral complex D_3^0 is, according to (6), given by $\tau_3 = (k_3 c_{C^-})^{-1}$. On the other hand, the time constant for the decay of this complex into a mobile ion and the overcharged complex D_2^+ is given by $\tau_3' = (K_3/k_3)$. Thus the ratio $\tau_3'/\tau_3 = K_3 c_{C^-} = (c_{C^-}/(c_{C^-})_{equ.})\,(c_{D_3^0}/c_{D_2^+})_{equ.} \gg (c_{C^-}/(c_{C^-})_{equ.})$ depends on the non-equilibrium concentration of the ions. Assuming now that the capture of a mobile ion by the immobile overcharged complex is determined by the Langevin bimolecular reaction rate [51]

$$k_3 = \frac{e}{\varepsilon\varepsilon_0}\,\mu_{C^-}\,, \tag{32}$$

one can estimate the typical ranges for τ_3. With $\varepsilon = 3$ and a counter ion mobility of the order of magnitude of $\mu_{C^-} = 10^{-11}$ cm^2/Vs one has $k_3 = 6\times10^{-18}$ cm^3/s which is of the same order of magnitude as the corresponding constant for the direct BP formation. If the density of the dopants is about 10^{17} cm^{-3}, an initial time constant for the formation of the neutral complex D_3^0 is ≈ 2 s. However with the equilibrium on the side of this complex, the concentration of the mobile ions decreases and the process slows down. Thus it can really influence long term changes in the devices. On the other hand, a shift of the flat band voltage within the time regime of *CV* measurements can hardly be expected. But if another process leads to the main shift of the flat band voltage, the complex formation can modify the run of the capacitance curves in the two sweep directions especially in the depletion region.

16.7 Conclusion

As explicated in a literature survey, a lot of information on the hysteresis effects in polymer field-effect devices has been published. Recent observations are reported in [52–54] (included also in this book). However, systematic investigations with a variation of many parameters over a wide range are rare. Our previous and recent measurements show clearly, that the hysteresis effects are mainly characterised by a flat band voltage and hence a corresponding interface charge that depends on the *sweep direction* of the gate voltage. But details of the hysteresis in the different characteristics can be rather complex. This becomes clear especially by varying parameters as ramp rates in the *CV* measurements or the temperature. Moreover, in recent devices with improved technology and materials, the hysteresis is often reduced and other forms are possible. Thus, in some OFETs the off-current can depend on the sweep direction.

Several mechanisms for the hysteresis effects have been proposed in the literature, in most cases as verbal statements without mathematical formulation and numerical evaluation. Typical statements are that trap recharging or mobile ions are likely to cause the hysteresis. Additionally we have proposed that formation and dissociation of bipolarons in the accumulation layer can cause this effect.

Numerical simulations on the trap recharging mechanism in MIS capacitors and transistors indicate that energetically distributed traps in particular can lead to hysteresis in these devices. On the other hand, the form of the simulated hysteresis deviates from the observed one and extreme parameter values are needed either for the total trap concentration or for the product from capture cross section and thermal velocity. We conclude that it is more likely that trap recharging can modify a hysteresis caused by another mechanism than being the main origin of the hysteresis.

A revisal of the equilibrium between polarons and doubly charged states of the polymer chains shows that the previous assumptions on the extension of polaron, bipolarons and polaron pairs were not justified. A more general formulation is presented for the equilibrium concentrations and the kinetics. But the differences are almost negligible up to the maximum charge concentrations that can be achieved in accumulation layers. The resulting rate constants for formation and dissociation of (immobile) bipolarons can be estimated using a rate constant for the bipolaron formation determined recently by Salleo and Street, and indicate that these processes can cause the hysteresis on the time scale of the measurements.

Finally, it is shown that a process involving mobile ions can also contribute to hysteresis. After fast formation of complexes between a polaron and a counter ion, subsequent capture of a further polaron leads to complexes formed by a bipolaron and a counter ion. They can attract slowly moving ions. The Langevin bimolecular reaction rate for this process again yields time constants on the time scale of the measurements.

Although the last two processes occur on a time scale which is characteristic for the hysteresis, a full description of course requires complete numerical simulation of theses processes and also of the role of mobile ions and of their reactions. Such simulations can only be the subject of future investigations.

Acknowledgements

The authors would like to thank the Deutsche Forschungsgemeinschaft (DFG) for financial support within the Schwerpunktprogramm 1121 OFET.

References

1. S. Scheinert, Technologie und Eigenschaften organischer Halbleiterbauelemente, Habilitationsschrift, Technische Universität Ilmenau (2006).
2. S. Scheinert, G. Paasch, S. Pohlmann, H.-H. Hörhold, and R. Stockmann, Solid-State Electron. **44**, 845 (2000).
3. A. R. Brown, A. Pomp, D. M. de Leeuw, D. B. M. Klaassen, E. E. Havinga, P. Herwig, and K. Müllen, J. Appl. Phys. **79**, 2136 (1996).
4. G. Horowitz, R. Hajlaoui, D. Fichou, and A. E. Kassmi, J. Appl. Phys. **85**, 3202 (1999).
5. S. Scheinert and G. Paasch, phys. stat. sol. (a) **201**, 1263 (2004).
6. G. Paasch, Solid State Ion. **169**, 87 (2004).

7. N. Kirova and S. Brazovskii, Synth. Met. **76**, 229–232 (1996).
8. G. Paasch, P. H. Nguyen, and A. Fischer, Chem. Phys. **227**, 219 (1998).
9. J. A. E. H. van Haare, E. E. Havinga, J. L. J. van Dongen, R. A. J. J. Janssen, J. Cornil, and J.-L. Brédas, Chem. Eur. J. **4**, 1509 (1998).
10. Th. Lindner, G. Paasch, and S. Scheinert, J. Appl. Phys. **98**, 114505 (2005).
11. R. A. Street, A. Salleo, and M. L. Chabinyc, Phys. Rev. B **68**, 085316 (2003).
12. A. Salleo and R. A. Street, Phys. Rev. B **70**, 235324 (2004).
13. S. Scheinert and W. Schliefke, Synth. Met. **139**, 501 (2003).
14. G. Paasch and S. Scheinert, Mater. Sci. **22**, 423 (2004).
15. M. Matters, D. M. de Leeuw, M. J. C. M. Vissenberg, C. M. Hart, P. T. Herwig, T. Geuns, C. M. J. Mutsaers, and C. J. Drury, Opt. Mater. (Netherlands) **12**, 189 (1999).
16. A. R. Brown, C. P. Jarrett, D. M. de Leeuw, and M. Matters, Synth. Met. **88**, 37 (1997).
17. C. J. Drury, C. M. J. Mutsaers, C. M. Hart, M. Matters, and D. M. de Leeuw, Appl. Phys. Lett. **73**, 108 (1998).
18. S. Scheinert, A. Scherer, T. Doll, G. Paasch, and I. Hörselmann, Appl. Phys. Lett. **84**, 4427 (2004).
19. P. J. Brown, H. Sirringhaus, M. Harrison, M. Shkunov, and R. H. Friend, Phys. Rev. B **63**, 125204 (2001).
20. R. A. Street, A. Salleo, and M. L. Chabinyc, Phys. Rev. B **68**, 085316 (2003).
21. G. Horowitz and P. Delannoy, J. Appl. Phys. **70**, 469 (1991).
22. Y. S. Yang, S. H. Kim, J. L. Lee, H. Y. Chu, L. Do, H. Lee, J. Oh, T. Zyung, M. K. Ryu, and M. S. Jang, Appl. Phys. Lett. **80**, 1595 (2002).
23. G. Paasch, J. Electroanal. Chem. **600**, 131 (2007).
24. G. Paasch, S. Scheinert, A. Petr, and L. Dunsch, Russ. J. Electrochem. **42**, 1161 (2006).
25. D. Niemann, N. Gunther, Ch. Kwong, M. Barycza, and M. Rahman, Solid-State Electron. **50**, 1097 (2006).
26. G. Horowitz, D. Fichou, X. Peng, Z. Xu, and F. Garnier, Solid State Commun. **72**, 381 (1989).
27. A. Assadi, G. Gustafsson, M. Willander, C. Svensson, and O. Inganäs, Synth. Met. **37**, 123 (1990).
28. D. M. Taylor, H. L. Gomes, A. E. Underhill, S. Edge, and P. I. Clemenson, J. Phys. D, Appl. Phys. **24**, 2023 (1991).
29. Ch. Waldauf, P. Schilinsky, M. Perisutti, J. Hauch, and Ch. J. Brabec, Adv. Mater. **15**, 2084 (2003).
30. Y. Qiu, Y. Hu, G. Dong, L. Wang, J. Xie, and Y. Ma, Appl. Phys. Lett. **83**, 1644 (2003).
31. S. Hoshino, M. Yoshida, S. Uemura, T. Kodzasa, N. Takada, T. Kamata, and K. Yase, J. Appl. Phys. **95**, 5088 (2004).
32. D. Li, E.-J. Borkent, R. Nortrup, H. Moon, H. Katz, and Z. Bao, Appl. Phys. Lett. **86**, 042105 (2005).
33. O. D. Jurchescu, J. Baas, and Th. T. M. Palstra, Appl. Phys. Lett. **87**, 052102 (2005).
34. M. Matters, D. M. de Leeuw, P. T. Herwig, and A. R. Brown, Synth. Met. **102**, 998 (1999).
35. S. J. Zilker, C. Detcheverry, E. Cantatore, and D. M. de Leeuw, Appl. Phys. Lett. **79**, 1124 (2001).
36. D. B. A. Rep, A. F. Morpurgo, W. G. Sloof, and T. M. Klapwijk, J. Appl. Phys. **93**, 2082 (2003).
37. H. L. Gomes, P. Stallinga, F. Dinelli, M. Murgia, F. Biscarini, D. M. de Leeuw, T. Muck, J. Geurts, L. W. Molenkamp, and V. Wagner, Appl. Phys. Lett. **84**, 3184 (2004).
38. L. Bürgi, T. Richards, M. Chiesa, R. H. Friend, and H. Sirringhaus, Synth. Met. **146**, 297 (2004).
39. A. Salleo, F. Endicott, and R. A. Street, Appl. Phys. Lett. **86**, 263505 (2005).
40. H. Sirringhaus, Adv. Mater. **17**, 2411 (2005).

41. L.-L. Chua, J. Zaumseil, J.-F. Chang, E. C.-W. Ou, P. K.-H.Ho, H. Sirringhaus, and R. H. Friend, Nature **434**, 194 (2005).

42. L.-L. Chua, P. K. H. Ho, H. Sirringhaus, and R. H. Friend, Appl. Phys. Lett. **84**, 3400 (2004).

43. J. Ficker, A. Ullmann, W. Fix, H. Rost, and W. Clemens, J. Appl. Phys. **94**, 2638 (2003).

44. H. Rost, J. Ficker, J. S. Alonso, L. Leenders, and I. McCulloch, Synth. Met. **145**, 83 (2004).

45. A. Herasimovich, S. Scheinert, and I. Hörselmann, J. Appl. Phys., accepted.

46. G. Paasch, T. Lindner, S. Scheinert, Synt. Met. **132**, 97 (2002).

47. J. Tang, R. D. Allendoerfer, and R. A. Osteryoung, J. Phys. Chem. **96**, 3531 (1992).

48. P. S. Davids, A. Saxena, and D. L. Smith, J. Appl. Phys. **78**, 4244 (1995).

49. G. Paasch, P. H. Nguyen, and S.-L. Drechsler, Synth. Met. **97**, 255 (1998).

50. G. Paasch, P. H. Nguyen, S.-L. Drechsler, and J. Malek, Synth. Met. **104**, 197 (1999).

51. M. Pope and C. E. Swenberg, Electronic Processes in Organic Crystals (Oxford University Press, New York, 1982), p. 502.

52. K. Haubner, E. Jaehne, H.-J. P. Adler, D. Koehler, C. Loppacher, L. M. Eng, J. Grenzer, A. Herasimovich, and S. Scheinert, phys. stat. sol. (a) **205**, 430 (2008).

53. Niels Benson, Christian Melzer, Roland Schmechel, and Heinz von Seggern, phys. stat. sol. (a) **205**, 475 (2008).

54. Bert Nickel, Matthias Fiebig, Stefan Schiefer, Martin Göllner, Martin Huth, Christoph Erlen, and Paolo Lugli, phys. stat. sol. (a) **205**, 526 (2008).

17
Ambipolar Charge Carrier Transport in Organic Semiconductor Blends

Markus Bronner, Andreas Opitz, and Wolfgang Brütting

17.1
Introduction

Organic semiconductors have attracted considerable interest due to their growing potential as active materials in electronic and optoelectronic devices. A long-standing paradigm, however, has been their unipolar transport of electrical charges. This means that, apart from very few exceptions, until recently organic semiconductors have shown electrical conduction for one carrier species only, with positive carriers being preferentially transported in most materials. Nevertheless, both electron and hole transport were observed many years before in organic single crystals with photo generated charge carriers [1]. Thereby almost equal electron and hole mobilities with a temperature dependence indicative for band transport were obtained. This indicates that there is *a priori* no intrinsic asymmetry between the transport of electrons and holes in the bulk of high-purity organic semiconductors.

In organic field-effect transistors (OFETs) the transport of carriers is limited in addition to the intrinsic material properties by the injection from the electrodes and by traps at the organic/insulator interface. Thus for many years, OFETs were reported to show either p- or n-type behaviour (see, e.g., Ref. [2] for an overview of materials). Ambipolar transport in OFETs was observed for the first time by Meijer et al. using mixtures of a soluble poly(phenylene-vinylene) derivative (MDMO-PPV) and a fullerene derivative (PCBM) [3], which are known from bulk-heterojunction solar cells [4]. The phenomenon of ambipolar transport itself was reported already in the 1970s in the context of hydrogenated amorphous silicon thin film transistors [5]. Meanwhile ambipolar OFETs, where two different materials for electron and hole transport are used, have been realised with a variety of material combinations, including polymer/fullerene blends [3], mixtures of soluble oligomers [6] as well as evaporated molecular hetero-layer structures and mixed layers [7–11]. However, even in neat films of a single organic semiconductor ambipolar transport can be observed provided that the barrier for electron injection is reduced by choosing low-work function metals as source/drain electrodes. It has been

Organic Electronics. Structural and Electronic Properties of OFETs. Edited by Christof Wöll

ISBN: 978-3-527-40810-8

demonstrated, e.g., that electron transport can be obtained in pentacene when Ca is used as electrode material [12, 13]. Another possibility to get electron transport in traditionally p-conducting materials is to suppress electron trapping at the interface to the gate-dielectric by surface functionalisation or suitable polymeric dielectrics. A detailed report on polymeric dielectrics is given in Chapter 24 by Benson et al. For example, if silicon oxide is used as the insulator, the surface usually contains hydroxyl groups acting as electron traps. It has been shown that terminating these OH-groups by silanisation [14] or interface doping by calcium [15] are suitable tools to achieve ambipolar transport in a large variety of materials. Recently, also low-band gap materials have been suggested as further candidates for ambipolar OFETs [16, 17]. Due to their small energy gap (sometimes below 1 eV) and suitable molecular energy levels (ionisation energy and electron affinity) the injection of both carrier species is easily obtained with a single metal for source and drain electrodes.

In this work we have chosen the combination of hole-conducting copper-phthalocyanine (CuPc) and the n-conducting fullerene C_{60}, which are both known from organic photovoltaic cells either as heterolayer structures or bulk-heterojunctions [18–21]. They can be considered as model systems for ambipolar transport where the asymmetry of the electron and hole mobilities is adjustable by the concentration of both materials in the mixture.

Figure 17.1 shows a sketch of the ambipolar operation of an FET and the respective charge carrier distribution. The unipolar regime with $|V_D| < |V_G - V_T|$ is dominated by the transport of only one charge carrier species – either electrons or holes, depending on the material and the electrodes. In the ambipolar regime both electrons and holes are injected into the semiconductor. This occurs in the regime where in the transfer characteristics (constant drain voltage) the unipolar FET is normally in the OFF state. Furthermore, for the opposite sign of V_D and $V_G - V_T$ even reversed unipolar transport can be observed in this case. The charge carrier distribution shows the decrease of charge carrier concentration inside the channel [22] and indicates the location where electrons and holes recombine. The position of this recombination zone depends on the applied voltages and can be shifted inside the channel as observed in ambipolar light-emitting FETs [23–25].

17.2
Materials, Device Preparation and Experimental Methods

Copper-phthalocyanine (CuPc, purchased from Aldrich as sublimation grade) and buckminster fullerene (C_{60}, purchased from Hoechst as super gold grade) were used as hole and electron conducting materials, respectively. The structural formulae are given in Figure 17.2a. The materials were deposited by thermal evaporation from low-temperature effusion cells in a vacuum of better than 1×10^{-7} mbar to form neat and mixed layers with a thickness of about 25 nm on prestructured substrates to realise a bottom-gate and bottom-contact

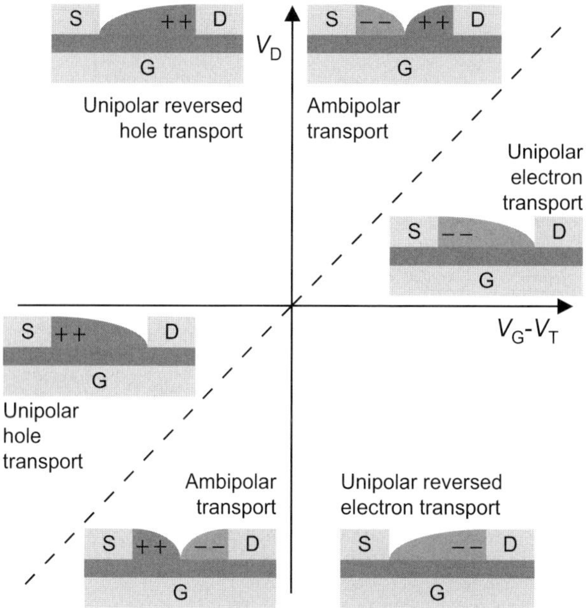

Figure 17.1 Sketch for the unipolar and ambipolar operation regimes of an organic field-effect transistor. V_D denotes the drain voltage, V_G and V_T are gate and threshold voltage, respectively. (see colour plates p. LXXX)

geometry (see Figure 17.2d). The deposition rates were between 0.35 Å/s for neat films and 1.4 Å/s for layers with 1:3 stoichiometry. The given mixing ratios in this chapter are volume percentages, as the evaporation process was controlled via two independent deposition monitors using quartz microbalances. They are always given in the form C_{60}:CuPc. Additionally, different substrate temperatures (300 K and 375 K) were used during evaporation of the materials.

Organic field-effect transistors were fabricated on highly conductive Si wafers (1–5 mΩ cm) with a 200 nm or 320 nm thick thermally grown oxide, which acts as the gate insulator. Photolithographically patterned Au (100 nm, using 1 nm Ti as adhesion layer) source and drain electrodes were made by electron-beam evaporation and by a lift-off process. These structures were cleaned in an ultrasonic bath with solvents (acetone, isopropyl) and ultra-pure water. The substrates were dried with pure nitrogen, treated with an O_2-plasma for 60 s at 200 W and 0.6 mbar, and heated in a fore-vacuum at 400 K for 2 hours. On some of the samples additionally a treatment of the silicon oxide surface by octadecyltrichlorosilane (OTS) was made before they were heated [26]. This OTS treatment was carried out at room temperature, in a solution of OTS in n-heptane (0.86 mM), in an exsiccator. Subsequently the samples were cleaned of residual OTS in chloroform in an ultrasonic bath.

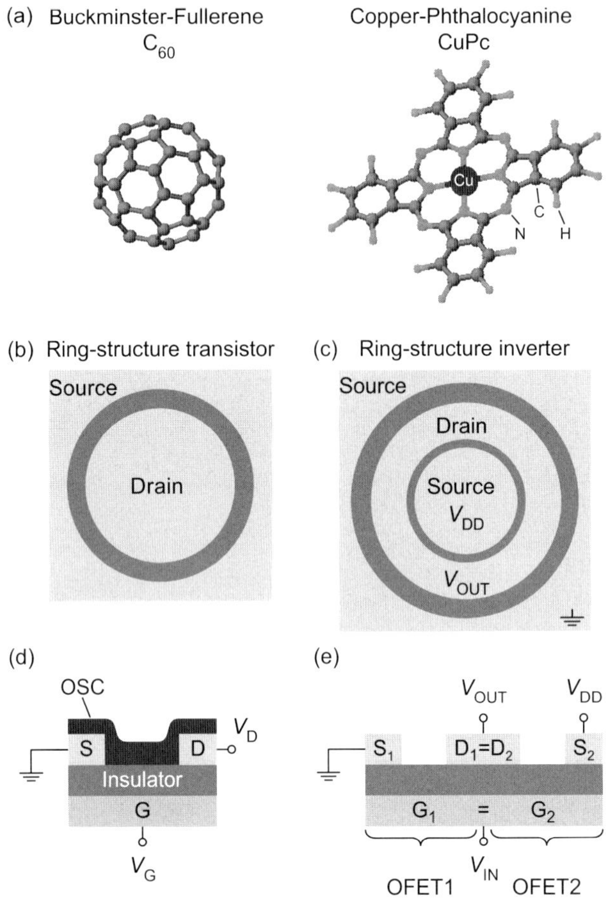

Figure 17.2 (a) Chemical structures of the used materials: fullerene (C_{60}) and copper-phthalocyanine (CuPc). Top view (b) and cross section (d) of the ring-type transistor in bottom-gate and bottom-contact geometry. Top view (c) and cross section (d) of the ring-type inverter. The silicon substrate acts as the gate electrode for transistors and inverters. (see colour plates p. LXXXI)

Transistors with a ring-type geometry (see Figure 17.2b) were used, whose source electrodes form a closed ring around the drain electrodes. This prevents parasitic currents from outside of the active transistor channel without the necessity of structuring the organic semiconductor [28]. The channel length and width were 5 μm and 2500 μm, respectively. Ambipolar inverters were made from two transistors stacked into each other (see Figure 17.2c) and have an additional ring channel around the first one. Thus both transistors share the common silicon substrate gate electrode as input of the inverter and the drain electrode as output (see Figure 17.2e). Length and width of the outer channel were 10 μm and 2500 μm, respectively, and of the inner channel 8 μm and 2000 μm, respectively. Consequently, the ratio of length to width was the same for both channels. Each substrate (20 mm × 20 mm) contained 24 individual transistors and 12 inverter structures to allow for a comparison of several devices. Complementary inverters were fabricated by evaporating neat CuPc and C_{60} each on one half of a prestructured substrate. For this purpose two evaporation steps were necessary with different shadow masks. In this case two separated transistors, one p-type and one n-type, from different areas on the substrate were connected together to form an inverter.

Without air-exposure the devices were transferred into a vacuum-chamber providing a pressure of less than 5×10^{-6} mbar for characterisation. The output and transfer characteristics of the transistors were measured using two independent source-meter units (Keithley 236). In order to measure inverter transfer curves an additional source-meter unit (Keithley 2400) was implemented.

The charge-carrier mobilities μ and the threshold voltages V_{T} were extracted from the slope of the transfer characteristics in the saturation region $|V_{\mathrm{D}}| > |V_{\mathrm{G}} - V_{\mathrm{T}}|$ using the standard relationship:

$$I_{\mathrm{D,sat}} = \frac{W}{2L} \cdot \mu \cdot C_{\mathrm{Ox}} \left(V_{\mathrm{G}} - V_{\mathrm{T}}\right)^2 \tag{1}$$

Here W is the channel width, L is the channel length, C_{Ox} is the gate-oxide capacitance per unit area, V_{G} is the gate voltage, and additionally V_{D} is the drain voltage. Mobility μ and threshold voltage V_{T} were determined from the linear regression of the measured data plotted as $\sqrt{I_{\mathrm{D,sat}}}$ vs. V_{G}. The inverter characteristics include the transfer curve V_{OUT} vs. V_{IN} as well as the current dissipation I_{DD} vs. V_{IN}.

X-ray diffraction (XRD) patterns were obtained in θ–2θ geometry using a Siemens D-500 diffractometer (Cu K_{α} radiation with a wavelength of 0.1542 nm) for analysing the crystallinity of the films. The morphology investigations were performed using a Thermo Microscopes Auto Probe scanning force microscope (SFM) operated in non-contact (tapping) mode.

For photoelectron spectroscopy organic layers (neat films and mixtures) with a nominal thickness of 25 nm were deposited on 100 nm thick gold films which were thermally evaporated onto oxidised Si wafers. The electronic properties of the films were characterised using X-ray and ultraviolet photoelectron spectroscopy (XPS, UPS) by employing monochromated Al K_{α} radiation ($h\nu = 1486.7$ eV) for measurement of the core levels as well as ultraviolet radiation [He I ($h\nu = 21.2$ eV) and He II ($h\nu = 40.8$ eV)] for an analysis of the occupied states near the Fermi level. For a measurement of the secondary electron cut-off to determine the sample work function exactly, an additional bias (–2 V and –5 V) was applied to the sample.

17.3 Unipolar Field-Effect Transistors

Output and transfer characteristics of neat C_{60} and CuPc FETs are shown in Figure 17.3. The output characteristics show the typical unipolar transistor behaviour with linear increase at low drain voltage and saturation at higher V_D. The transfer characteristics display an off-regime followed by an increase of drain current with increasing absolute value of the gate voltage exceeding the switch-on voltage.

The analysis of the shown transfer characteristics yielded saturation mobilities for these preparation conditions (O_2 plasma treatment, 375 K during evaporation) of 7×10^{-2} cm²/Vs for the C_{60} FET and 6×10^{-4} cm²/Vs for the CuPc FET. The threshold voltages were + 63 V and − 31 V, respectively.

The measured I–V characteristics show considerable hysteresis between increasing and decreasing voltage sweeps, which can be attributed to dynamic processes in the charging of the semiconductor-dielectric interface [29], originating e.g. from polaron–bipolaron reactions [30] or trap recharging [31]. For further details on this topic refer to Chapter 16 by Paasch et al. The largest hysteresis is present in the output characteristics of the C_{60} FET. The hysteresis in all other measurements, including the ambipolar FETs, and in particular for the transfer characteristics which were used for the analysis, are significantly smaller.

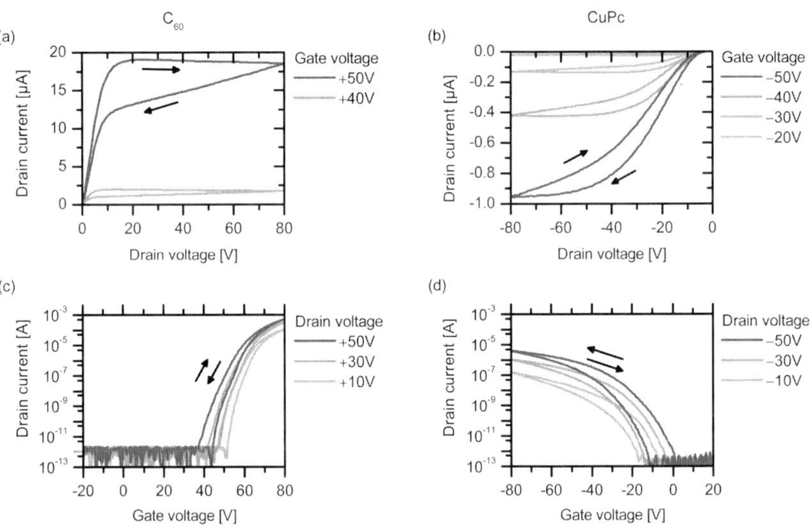

Figure 17.3 Output and transfer characteristics of unipolar field-effect transistors with neat C_{60} (a, c) and neat CuPc (b, d) films. The substrates were treated with O_2-plasma and the films evaporated at 375 K substrate temperature. The direction of the hysteresis is indicated by arrows. (Figure adopted from Ref. [27].) (see colour plates p. LXXXII)

The curvature in the output characteristics of CuPc at the origin of the $I-V$ diagram suggests a considerable injection barrier of the contact [32], whereas the linear increase in the C_{60} FET indicates a significantly smaller one. As we will show later on (see Section 17.8), this can be related to the electronic structure of the Au/CuPc and the Au/C_{60} interface, respectively [33–38]. However, it should be noted that other possibilities such as an influence of the adhesion layer [39], or from traps in the semiconductor [40] or from a field-dependent charge carrier mobility [41] have been put forward to account for nonlinearities in the $I-V$ characteristics.

17.4 Ambipolar Field-Effect Transistors

Ambipolar FETs with mixing ratios of 1:3, 1:1 and 3:1 between C_{60} and CuPc have been investigated. All of them showed ambipolar transport with the same qualitative features as the 1:1 mixture displayed in Figure 17.4. Here a strong increase of the current on saturation of the output characteristic (Figure 17.4a, b) is measured for both the n- and the p-channel regime. This is a clear signature of ambipolar behaviour. Electrons are injected at higher drain volt-

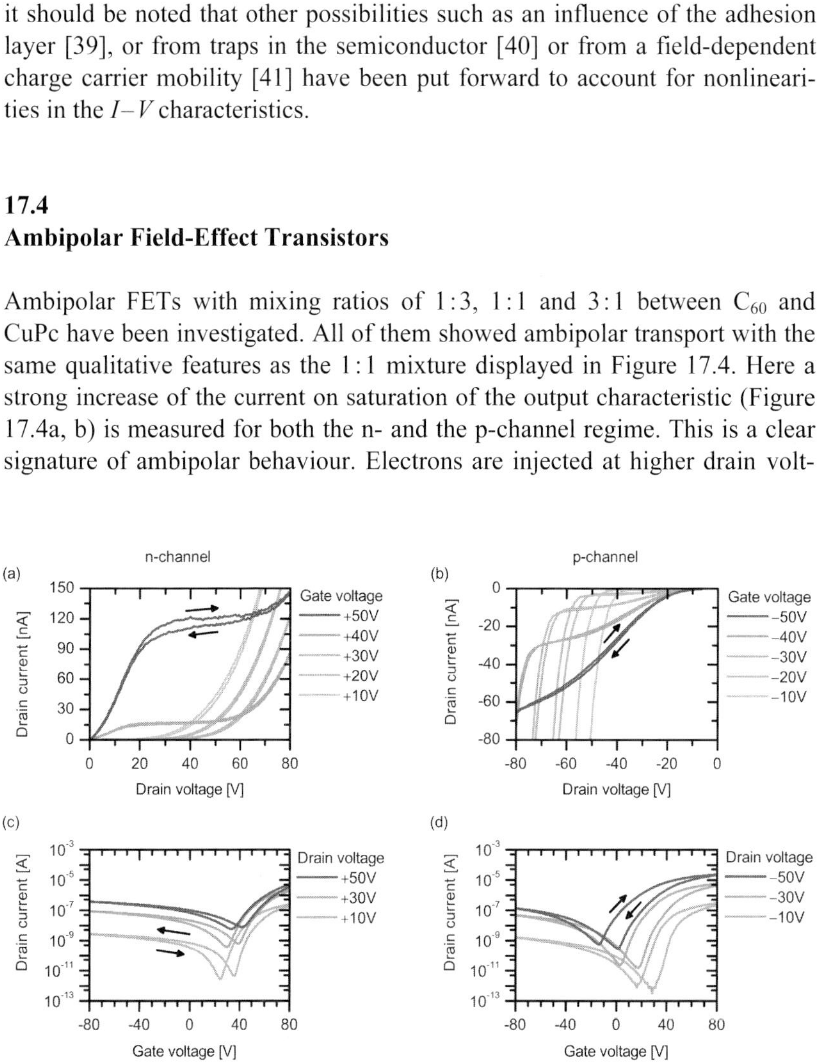

Figure 17.4 Output and transfer characteristics of ambipolar field-effect transistors for a mixing ratio between C_{60} and CuPc of 1:1 measured in the n-channel regime (a, c) as well as the p-channel regime (b, d). The substrate was O_2-plasma treated and the film evaporated at 375 K substrate temperature. The direction of the hysteresis is indicated by arrows. (Figure adopted from Ref. [27].) (see colour plates p. LXXXII)

ages into the hole-conducting channel, and vice versa. Consequently, the transfer characteristics (Figure 17.4c, d) do not show any off regime as the ambipolar increase takes place in the regime where the FET with the neat materials is switched off.

The magnitude of the drain currents in both regimes differs significantly for this mixing ratio. The p-channel (negative V_G) shows a current which is three orders of magnitude lower than the current in the n-channel (positive V_G). In both measurements the curvature in the output characteristics at the origin of the $I-V$ diagram suggests injection barriers with non-linear contact resistances. Here this contact resistance is also visible in the electron channel and even more pronounced in the hole channel because the conductivity of both channels is reduced due to mutual dilution of the conducting material by the other species.

17.5
Charge Carrier Mobility and Threshold Voltage

For the analysis of the mobility and the threshold voltage the saturation regime is used here. An analytical model based on the Shockley equation by Schmechel et al. [13] assumes that the electron and hole mobilities as determined from the linear or saturation regime are also valid in the ambipolar regime. As previously shown for the system under consideration, the mobilities determined from the saturation and ambipolar regime are in excellent agreement [27].

Figure 17.5 shows the mobility and the threshold voltage determined from the saturation regime as a function of concentration in the blend for different preparation conditions. We note that these data have been obtained by averaging over several identically prepared samples. The substrate temperature during evaporation of the blends was 25 °C and 100 °C for the O-plasma treated silicon oxide surface. The surface chemistry was modified additionally for the 100 °C substrate temperature by silanisation of SiO_2 with OTS.

In all cases, an exponential decrease of both electron and hole mobility is observed upon dilution of the corresponding conducting material with the other species. Remarkably, all blends with different mixing ratios show charge carrier transport for both charge carrier types. This means that there is always a percolation path for both electrons and holes. However, the hopping distance between molecules of the same type is increased upon mixing with the other species thus leading to a decreasing wave-function overlap. This is well-established for molecularly doped polymers [42], where a homogenous dilution of conducting molecules in an inert matrix is present. This scenario might even hold for nano-phase separated granular films, for which we have indications from SFM data (see below). In this case the mobility limiting step would be the hopping between grains where the average distance should also increase upon dilution.

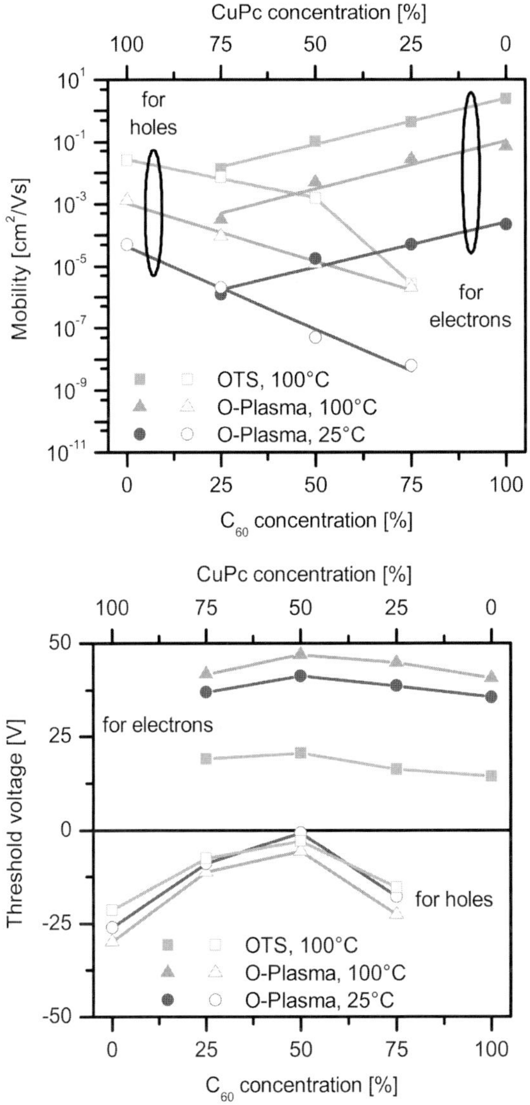

Figure 17.5 Mobility (above) and threshold voltage (below) as determined from transfer characteristics in the saturation regime of OFETs with different composition, substrate temperature and substrate treatment. The filled symbols are related to the electron transport, the open symbols to the hole transport. (see colour plates p. LXXXIII)

As compared with film growth at room temperature, an increased mobility is found for the higher substrate temperature and a further increase is realised by lowering the surface energy with OTS. This increase of mobility was reported for unipolar OFETs [43] and is also valid for these blends. Interestingly, for all treatments balanced mobilities are found at about 25% C_{60} content.

At this point it is noteworthy to compare our results on the variation of charge carrier mobilities with composition of the blends to other ambipolar mixed systems. There are on the one hand polymer-fullerene blends used for photovoltaic applications (in particular MDMO-PPV:PCBM) [44, 45] and on the other hand blends of molecular materials implemented again in photovoltaic devices (mostly phthalocyanine:C_{60}) [46] and ambipolar light-emitting transistors [47]. In both types of blend systems one observes as a common feature a strong (in many cases exponential) decrease of the mobility of one carrier type with increasing dilution by the other component. (A remarkable exception from this rule is the hole mobility in polymer-fullerene blends in which the addition of fullerene molecules even improves μ_h. In this case a stretching of the polymer chains occurs, leading to improved interchain hopping [45].)

It is remarkable, however, that despite a roughness which is larger than the film thickness in some of our films (see below), all of them show n-channel as well as p-channel transport. Obviously, there still exists a percolation path for conduction in both materials independent of phase separation and order formation. We also note that the bulk morphology does not necessarily provide precise information for organic field-effect transistors since in these devices the active channel is restricted to the first few molecular layers at the interface to the gate dielectric [48–50].

The threshold voltage shows a dependence on the mixing ratio for the hole channel, but not for the electron channel. By contrast, the threshold voltage for the electron channel changes with the preparation conditions, especially with the OTS passivation of the silicon oxide. The reason is that the O-plasma treated oxide surface contains OH-groups which are acting as electron traps and increase the threshold voltage for electrons [14]. By surface treatment with OTS these traps are at least partially passivated and the threshold voltage decreases. The change of threshold voltage for the hole transport with the concentration is related to the organic/organic interface in the CuPc/C_{60} blend [27] and can be explained by a hole accumulation at the CuPc side of the C_{60}/CuPc interface [36]. Because there is no charge transfer from CuPc to C_{60} in the ground state this shift should be related to a redistribution of charges within the CuPc molecules in the presence of C_{60} as demonstrated by calculations [51]. As expected, this charge displacement is independent of the preparation conditions. The threshold voltage shift is related to an interface charge [32] at the organic/organic interface in the proximity of the insulator by:

$$\Delta V_T = \frac{e \times N_{If}}{C_{Ox}} \tag{2}$$

By modelling the molecular packing [27] a charge accumulation of about 0.012 charges per molecule is estimated. The calculated charge transfer in an ideal complex of one C_{60} and one CuPc molecule was calculated to be 0.06 charge per molecule [51], 5 times higher than determined here.

17.6 Film Morphology and Structure

In order to get more insight into the concentration dependent electron and hole mobility, structure and morphology of neat and mixed films (evaporated at 375 K) were investigated by X-ray diffraction and scanning force microscopy. XRD measurements (Figure 17.6) show a strong (200) peak corresponding to the α-phase for the neat CuPc film and a weak (200) peak for the film with a 1:3 mixture, whereas for all other mixtures including neat C_{60} films no diffraction peaks are detectable, indicating a structure without coherent diffraction, e.g. amorphous or very small crystallites. These observations are in full agreement with measurements reported on CuPc:C_{60} mixtures for photovoltaic applications by Rand et al. [46]. The authors observed a disappearance of the CuPc diffraction peak for a fullerene content of more than about 15% and asserted no peak corresponding to C_{60}, although, as in our case, the electron mobility in neat C_{60} films was about two orders of magnitude higher than the hole mobility in CuPc. Other groups have obtained crystalline C_{60} films, however, only at elevated substrate temperatures of about 440 K during evaporation [52].

Figure 17.7 shows the surface morphologies for different mixing ratios as investigated by non-contact SFM for small areas. Neat CuPc films have a needle-like structure, corresponding to the α-phase structure observed in XRD. This film shows a low root mean square roughness (RMS) of only 2.2 nm. Also the Laue oscillations beside the diffraction peak for the neat CuPc film (see Figure 17.6) are an indication of a well defined film thickness. For neat C_{60} films a rough and granular structure is observed. Compared with the CuPc film, the roughness of the films is increasing with increasing C_{60} content.

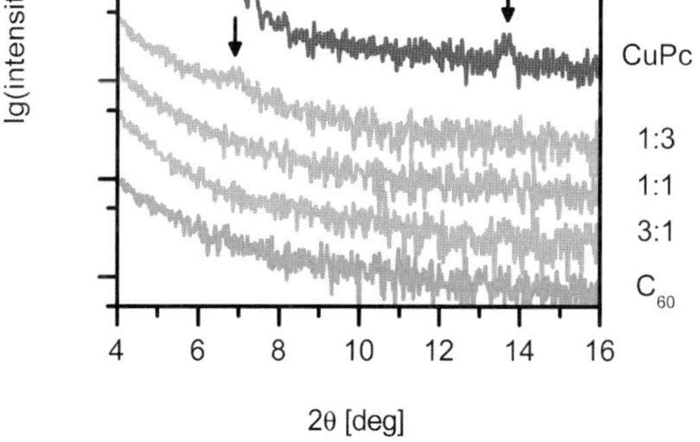

Figure 17.6 X-ray diffraction patterns for neat CuPc and C_{60} films as well as for three mixed films grown at 375 K. The arrows indicate the weak (200) peak in the sample with a mixing ratio of 1:3 and the second order diffraction peak of the neat CuPc sample. (see colour plates p. LXXXIV)

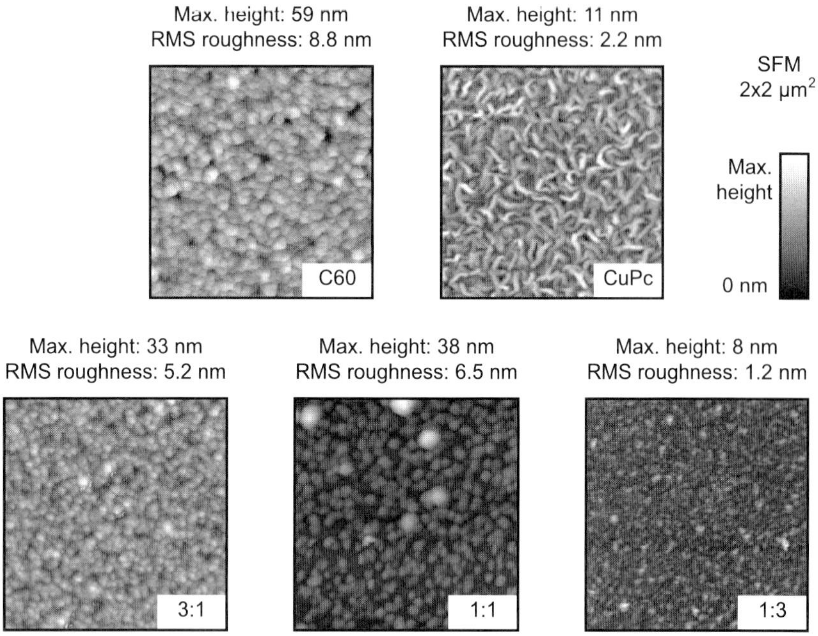

Figure 17.7 Scanning force microscopy images taken in non-contact mode for neat C_{60} and CuPc films as well as for three mixed films grown at 375 K. The total image size is $2 \times 2\ \mu m^2$. The max. height is given as the difference between the lowest value (dark blue) and the highest value (white) in each of the images. (see colour plates p. LXXXIV)

We note that we have also investigated films deposited without substrate heating by XRD and SFM (not shown here). These films are much smoother (RMS roughness of only 2.3 nm in all cases) and do not have such large grains. In particular, the neat CuPc film does not show a needle-like morphology in agreement with work by Schultes et al. [54]. However, in contrast, to this work there is no indication of an α to β conversion for our CuPc film grown at elevated temperature. For both preparation conditions the XRD data indicate the presence of the α-phase.

The film roughness has also been studied by XPS measurements of neat films and blends deposited on Au covered Si wafers (for details see next section). In this technique the intensity ratio I_{Au4f}/I_{C1s} of the Au4f and the C1s core level peaks can be used as a measure of the homogeneity in the film thickness. As shown in Figure 17.8 the Au substrate signal is zero for a closed and flat CuPc film and increases with increasing C_{60} content indicating film roughening. Here it is important to remember that measurements using SFM are a very local probe whereas XPS measurements sample a much larger area. Nevertheless both techniques show a similar trend for the roughness of the blended films.

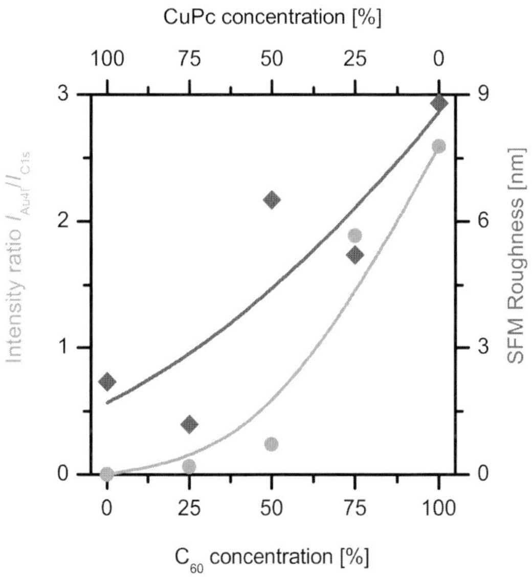

Figure 17.8 Comparison of the RMS roughness measured by SFM and the ratio of the substrate to film signal measured by XPS. The solid lines are to guide the eyes. (see colour plates p. LXXXV)

Furthermore phase separation of C_{60} and CuPc is described at elevated temperatures [55] which affects the molecular arrangement, but can not be detected by the techniques used here. Altogether, film morphology and structure of blends of flat CuPc and spherical C_{60} molecules are still not very well understood and need further investigation. This will become particularly important in photovoltaic cells, where this material combination is a potentially promising candidate for so-called bulk-heterojunction cells [21, 56].

17.7 Electronic Structure

For analysing the electronic structure the neat and mixed organic films were deposited on gold films that were pretreated in an oxygen plasma prior to the deposition of the organic materials. This treatment was chosen instead of *in situ* preparation of the gold layer under high-vacuum conditions to realise the same electrode properties as in OFETs. It should be noted that transporting the gold films through ambient air after the oxygen plasma treatment leads to the formation of a conductive hydrocarbon layer [57]. The work function of gold covered by this hydrocarbon layer is determined by photoelectron spectroscopy (PES) to be 4.2 eV, whereas a clean gold substrate (after removal of the adsorbed species by Ar^+ ion bombardment) exhibits a work function of 5.4 eV.

Figure 17.9 shows the important results of the XPS and UPS measurements. The CuPc C1s spectrum contains the C–C bond ($E_B = 284.2$ eV, indicated by small ticks) and the chemically shifted C–N ($E_B = 285.6$ eV) bond, as well as their π–π* satellites. For C_{60} only one pronounced feature is visible at $E_B = 284.7$ eV. Both spectra are in agreement with published results on pristine materials [33, 58]. The composition of the blends and the CuPc and C_{60} related C1s peak positions are determined by fitting the spectra of the mixed films with Gauss–Lorentz-functions using the shape of the neat C_{60} and CuPc core levels. As shown in Figure 17.9a the energetic positions of the C1s features are slightly shifting upon addition of C_{60}. The N1s spectra (not shown) were also analysed for the peak position and found to behave in a similar manner.

The electronic structure of the occupied states near the Fermi level (measured using He II radiation) is shown in Figure 17.9b. The maxima of the HOMO levels of the neat materials are indicated by small ticks. Whereas the CuPc HOMO level is clearly separated for all mixed films, the HOMO level of C_{60} is only visible in films with a concentration of 77% and 100%. For smaller amounts of C_{60} in the films the HOMO level is present only as a shoulder in the occupied molecular orbitals of CuPc. Again, the spectra can be described as a linear superposition of neat material spectra if a shift of the energetic position is included.

The work function of the films was determined from the secondary electron cut-off in the He I spectra with applied bias (see Figure 17.9c). As for the core and the HOMO levels an energy shift of the secondary electron cut-off with the film composition is present.

The graph in Figure 17.10 shows a comparison of the obtained vacuum levels of the organic films and the substrate (E^{Org}_{Vac} and E^{Sub}_{Vac}), of the HOMO levels (E^{Org}_{HOMO}) (defined as the HOMO edge near the Fermi level), as well as of the C1s and N1s core levels (E^{Org}_{C1s} and E^{Org}_{N1s}) depending on the mixing ratio. The ionisation energy of the neat films is 6.2 eV for C_{60} and 5.0 eV for CuPc in agreement with the literature [36] and the difference of the work function amounts to about ~0.5 eV.

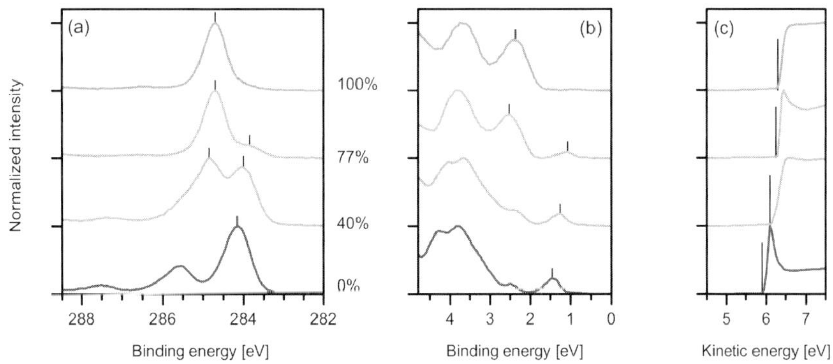

Figure 17.9 XPS and UPS measurements of neat and different mixed films of C_{60} and CuPc: (a) C1s spectra (excitation: monochromated Al K), (b) HOMO levels (excitation: He II) and (c) secondary electron cut-off (excitation: He I and $V_{Bias} = -2$ V). The numbers on the right side of the diagram (a) give the C_{60} concentration as determined from the measurements shown in part (a). (Figure adopted from Ref. [53].) (see colour plates p. LXXXV)

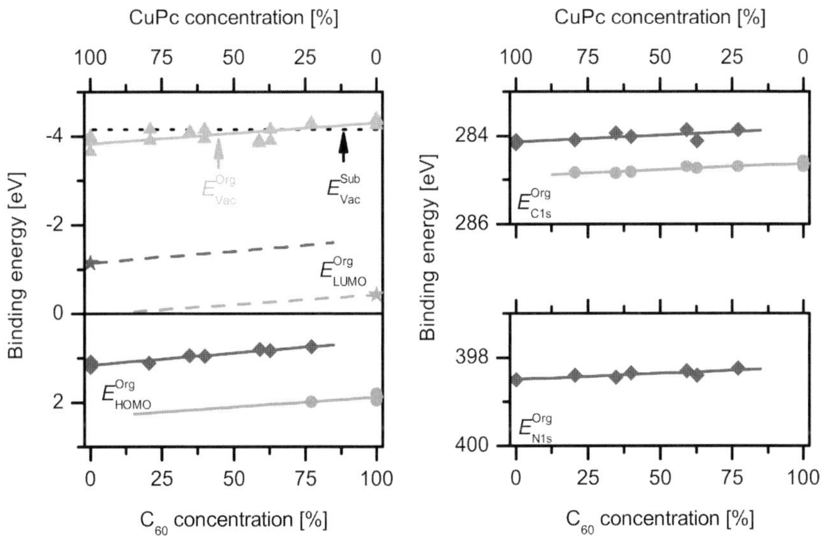

Figure 17.10 Position of the vacuum level, the high energy edge of the HOMO level, and the C1s and N1s core levels as a function of the C_{60}/CuPc mixing ratio. The ◆ and the ● symbols are representing the measured values for C_{60} and CuPc related levels, respectively, and the ▲ symbols the measured work functions. The solid lines are linear fits of the corresponding measured values. The constant dotted line is the vacuum energy of the gold substrate (covered by a conductive hydrocarbon layer). The LUMO levels of the neat materials are marked with the ★ symbols. The dashed lines are extrapolated LUMO levels for the blends. (Figure adopted from Ref. [53].) (see colour plates p. LXXXVI)

It is remarkable that the secondary electron cut-offs used to determine the work function of all mixtures were sharp and did not show a double step that would reflect two different local surface potentials [59], although the films have a nanogranular morphology as determined by atomic force microscopy [27]. Thus, the blends show a common vacuum level which shifts linearly with the concentration between the work functions of the neat materials given by $\Phi_{CuPc} = 3.8$ eV and $\Phi_{C_{60}} = 4.3$ eV. The linear change of the work function in our studies suggests that CuPc/C_{60} mixtures are electronically non-interacting (unless they are optically excited).

Taking all the results together, there is no evidence for a ground state charge transfer in blends of CuPc and C_{60}. The core level spectra can be described solely with the scaled features of the neat materials. Also the occupied molecular orbitals do not show additional structures unknown from the neat materials [36, 51]. The analysis of the difference between the occupied levels (core and HOMO levels) and the vacuum level shows for all compositions a constant ionisation energy. This is manifested in identical shifts of the core levels, the HOMO levels and the vacuum level with the concentration of the blends. Thus the molecular levels are simply following the change of the common vacuum energy.

Figure 17.10 allows some important conclusions for charge carrier injection in organic semiconductor blends to be drawn. For this purpose one also needs information on the lowest unoccupied molecular orbitals (LUMO) of the films. Unfortunately, up till now there are no direct measurements of the LUMO level for blends but only for the neat materials [37, 38]. However, one can take these data obtained from inverse photoemission and extrapolate them towards blends using the same shift of the LUMO levels that was observed for all other levels. In other words, we assume that the energy gaps of both materials of 2.3 eV [37, 38] stay constant in the mixture. This seems to be a reasonable hypothesis, as there is no indication for charge transfer between both materials in the ground state that would lead to new features in the electronic structure. Postulating that this model is correct, one can derive the injection barriers for both electrons and holes from the shown energy level diagram. The obtained values are given in Figure 17.11a. First, one notices that the injection barrier for holes into CuPc, i.e. the difference between the HOMO of CuPc and the Fermi level of Au (which is the reference level at binding energy zero) is larger than the respective injection barrier for electrons into C_{60}. This is in excellent agreement with the output characteristics discussed before (see Figure 17.3). Furthermore, both injection barriers are predicted to decrease from neat materials towards blends. This feature will be investigated separately in the next section when we analyse contact resistances in OFETs.

Before going to this, we would like to point out another interesting conclusion from the energy level diagram of these ambipolar blends which is related to photovoltaic cells based on CuPc and C_{60} as a donor–acceptor system. Upon photoexcitation, excitons generated in either of the two materials (within the reach of the respective exciton diffusion lengths) will be dissociated at the organic–organic interface leading to electrons in the LUMO of C_{60} and holes in the HOMO of CuPc. This charge separation leads to a gradient of the chemical potential at the interface that drives the photocurrent through the cell [60]. Thus the magnitude of the HOMO–LUMO offset between both materials will be related to the driving force, for which the open-circuit voltage can serve as an easily accessible experimental quantity. The obtained electronic structure therefore allows the conclusion that the HOMO-LUMO offset in blends is significantly smaller than at the heterojunction between neat layers of CuPc and C_{60} [53]. Preliminary measurements have indeed shown that the open-circuit voltage in a bulk-hetero junction organic photovoltaic cell (1 : 1 mixture) is about 0.2 V smaller than in a cell comprising a planar interface between CuPc and C_{60} [61, 67].

17.8
Charge Carrier Injection

Information on charge carrier injection was obtained from analysing contact resistances in OFETs. In contrast to Section 17.5, we now use transfer curves

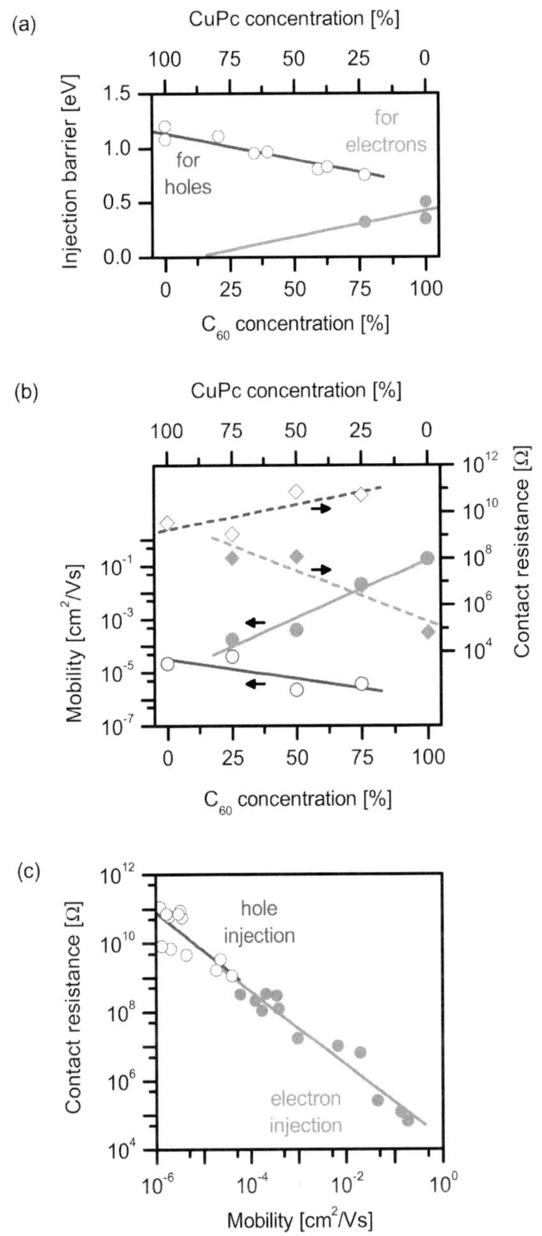

Figure 17.11 (a) Injection barrier determined from UPS (compare with Figure 17.10). (b) Mobility (solid line) and contact resistance (dashed line) as determined from transfer characteristics in the linear regime for $|V_G - V_T| = 33$ V. (c) Relation between mobility and contact resistance for different $|V_G - V_T| > |V_D|$ and different mixing ratios (0, 25, 50, 75 and 100% C_{60}). The straight lines are linear fits related to the vacuum level shift in part (a) and to guide the eyes in the parts (b) and (c). (see colour plates p. LXXXVII)

in the linear range. Following a procedure suggested by Horowitz [62] the drain voltage in the Shockley equation is replaced by the drain voltage corrected by the contact resistance ($V_D \rightarrow V_D - I_D \cdot R_C$). As a result the drain current is given by

$$I_D = \frac{(W/L) \cdot C_{Ox} \cdot \mu \cdot (V_G - V_T) \cdot V_D}{1 + (W/L) \cdot C_{Ox} \cdot \mu \cdot (V_G - V_T)} \tag{3}$$

Using the channel conductance $g_d = \partial I_D / \partial V_D$ and the transconductance $g_m = \partial I_D / \partial V_G$ the mobility μ and the contact resistance R_C were calculated from

$$\sqrt{\mu} \cdot (V_G - V_T) = \frac{g_d}{\sqrt{g_m}} \cdot \sqrt{\frac{LV_D}{WC_{Ox}}} \tag{4}$$

and

$$R_C = \frac{1}{g_d} - \frac{L}{W \cdot \mu \cdot C_{Ox} \cdot (V_G - V_T)} . \tag{5}$$

In this calculation the dependence of μ and R_C on V_G is neglected for the derivatives. To minimise the error of this simplification the mobilities and the contact resistances are plotted in Figure 17.11b for $V_D = 10$ V at a constant $|V_G - V_T| = 33$ V (using O-plasma treatment and 100 °C substrate temperature). Once again, one observes decreasing mobilities with decreasing concentration of the transport material, as in the saturation regime shown in Figure 17.5 and the contact resistance increases upon dilution. This behaviour is in contrast to the injection barrier shown in Figure 17.11a, which decreases with decreasing amount of the respective transport material, and can be taken as a clear indication that an additional mechanism has to be taken into account in the injection process.

First, there is a pure geometrical effect, meaning that the effective contact area will decrease if less molecules of one type are present in the mixture. However, this would only lead to a linear change of both mobility and contact resistance with the mixing ratio that can not account for the observed variation over several orders of magnitude. Thus as a second possibility, we consider the case that the injection process is influenced by the same increase of the intermolecular hopping distance as discussed before for the mobility. Therefore, we plot in Figure 17.11c the contact resistance vs. mobility for all presented mixing ratios and for different charge carrier densities (related to different values of $|V_G - V_T|$). A reciprocal relation $R_C \sim \mu^{-1}$ is observed indicating that in the presented system the mobility limits the injection of charge carriers. This behaviour can be explained by diffusion limited injection [32] following the equation

$$j_{inj} \sim \mu \cdot \exp\left[-\frac{\Phi_B}{kT}\right] . \tag{6}$$

This mechanism was already identified in organic photoconductors [63] varying the mobility by mixing semiconducting and insulating molecules and in unipolar OFETs [64] varying the mobility by the charge carrier concentration. Thus our investigations show that mixing of organic semiconductors has important consequences for both charge carrier injection and transport which will be relevant in various kinds of devices, including not only OFETs, but also OPV cells and OLEDs. (Actually, it was already demonstrated in OLEDs that grading interfaces can enhance device efficiency by reducing energy barriers at organic/organic interfaces and probably also enhancing recombination rates by slowing down carriers [65].)

17.9 Ambipolar and Complementary Inverter

As already mentioned, there was the suggestion of using ambipolar FETs to realise complementary-like organic integrated circuits [3, 10, 17, 66]. Here we investigate ambipolar inverters consisting of mixed-layer ambipolar FETs and compare their characteristics to a complementary inverter made of discrete p- and n-channel transistors from neat materials.

Ambipolar inverters or complementary-like inverters are based on two ambipolar transistors (Figure 17.12a). These inverters are working with a positive and a negative driving voltage whereby the sign of the driving voltage determines which of the two transistors works as n- and p-transistor (Figure 17.12b). Complementary inverters in contrast to this are based on unipolar n- and p-transistors with a defined power supply, as shown in either Figure 17.12b or Figure 17.12c. However, to realise such a circuit patterning of the organic semiconductor or locally different surface functionalisation are required to achieve spatially separated p- and n-conducting regions, which is not necessary in the case of ambipolar inverters. In order to evaluate the performance of these different concepts we have fabricated both ambipolar and complementary inverters and compared their output characteristics both by experimental studies und simulations.

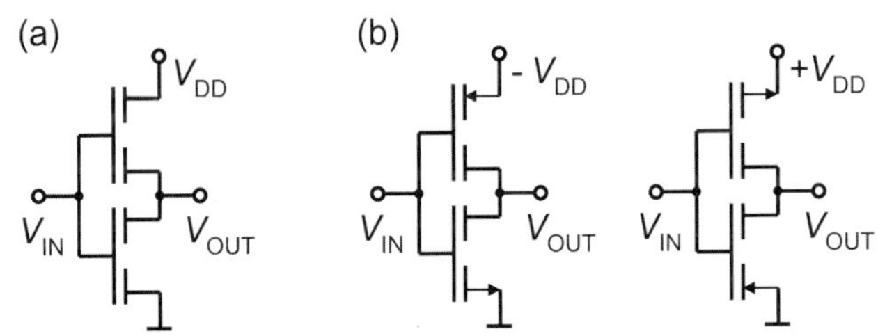

Figure 17.12 (a) Electrical circuit for an ambipolar inverter. (b) The two operation regimes (positive and negative supply voltage) for ambipolar and complementary inverters.

Figure 17.13a, b show the characteristics of ambipolar inverters using layers with mixing ratios of 3 : 1 and 1 : 3. The upper part presents the transfer characteristics of the inverters (output voltage vs. input voltage) and the lower part the corresponding current supplied by $V_{DD} = \pm 90$ V, which directly gives the dissipated current in the circuit.

A characteristic feature of ambipolar inverters is their operation in the first as well as in the third quadrant of the output-vs.-input diagram, depending on the sign of the supply voltage only. Ideally an inverter should have a sharp transition from the low to the high state at half of the driving voltage and the dissipated current should be negligibly small except for a narrow range around the transition voltage. Both inverters, based on two ambipolar transistors, show this transition at about half of the supply voltage ($V_{DD}/2 = \pm 45$ V) and reach high gain (about 13 for the 1 : 3 mixture and about 18 for the 3 : 1 mixture), which is defined as the steepness of the characteristics at the transitions between the high and the low states. However, they do not reach zero in the low state and the driving voltage in the high state. Also the voltages at the high and the low state are not constant.

The noise margin is an indicator of the tolerance of cascaded inverter stages. To demonstrate this, Figure 17.13 also contains the inverter characteristics mirrored at the bisecting line, which can be considered as the output of a following inverter. The noise margin is then defined as the largest possible square between these two curves. In this case the noise margin amounts to about 14 V for the 1 : 3 mixture and about 19 V for the 3 : 1 mixture.

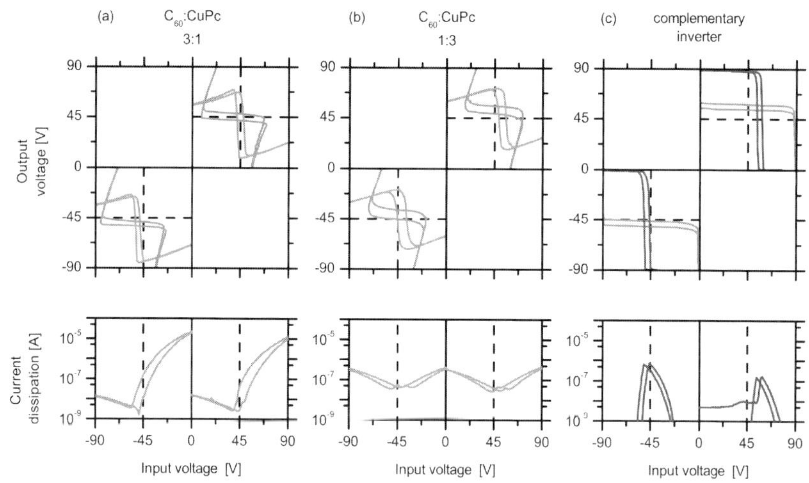

Figure 17.13 Characteristics of ambipolar inverters with mixing ratios of 3 : 1 (a) and 1 : 3 (b) and a complementary inverter (c) consisting of a neat C_{60} and a neat CuPc transistor. The driving voltage is $V_{DD} = \pm 90$ V. Transfer characteristics (top) and current dissipation (bottom) are shown. The substrates were O_2-plasma treated and the films evaporated at 375 K substrate temperature. The grey lines are shown to explain the noise margin (see text). (Figure adopted from Ref. [27].) (see colour plates p. LXXXVIII)

Significant differences between the two ambipolar inverters are observed in the dissipated current. Whereas for the inverter with a 1:3 mixing ratio the power dissipation is symmetric around ±45 V, the device with a 3:1 mixture shows an asymmetry of about three orders of magnitude. Moreover, the dissipated current is as high as 10^{-5} A, which is one-and-a-half orders of magnitude larger than in the previous case. Thus, it is remarkable that a huge asymmetry in electron and hole mobilities of more than three orders of magnitude, as observed for the 3:1 mixture (see Figure 17.5), has drastic consequences for the power dissipation of the inverter, although it does not lead to a significant asymmetry in the transition voltage.

For comparison we have also fabricated a complementary inverter by connecting a p-channel transistor (neat CuPc) and an n-channel transistor (neat C_{60}) together, its characteristics being shown in Figure 17.13c. In order to make it operate in the first quadrant it is necessary to connect the p-channel transistor to $+V_{DD}$ and the n-channel transistor to ground and vice versa, the n-channel transistor to $-V_{DD}$ and the p-channel transistor to ground in order make it operate in the third quadrant, as seen in Figure 17.12. These inverters also show slightly asymmetric transitions with respect to $\pm V_{DD}/2$ due to the unbalanced electron and hole mobilities in neat CuPc and C_{60}, but they reach the ground potential in the low state and the supply voltage in the high state. The gain is about 38 and the noise margin is about 29 V for a positive supply voltage, respectively, 34 V and 32 V for a negative supply voltage. So these values are about twice as high as those of the ambipolar inverters. A characteristic difference of the ambipolar inverter type is the current dissipation being high only in the vicinity of the transition. This is because at any time one of the two transistors is being switched off in each of the logic states whereas an ambipolar transistor always shows a non-negligible current.

For a better understanding of inverter characteristics we performed numerical simulations based on the analytical model of Schmechel et al. [13] for both ambipolar and complementary inverters. The output voltage as well as the current dissipation was calculated to demonstrate the differences between ambipolar and complementary inverters and the influence of device parameters like mobility and threshold voltage.

Figure 17.14 shows simulated transfer characteristics of ambipolar and complementary inverters. In the left picture balanced mobilities ($\mu_e = \mu_h$) and symmetric threshold voltages ($V_{T,e} = -V_{T,h} = 30$ V) were used for the simulation. The picture in the middle shows the characteristics for symmetric threshold voltages ($V_{T,e} = -V_{T,h} = 30$ V) but different mobilities ($\mu_e = 100\,\mu_h$). Whereas, in the right picture balanced mobilities ($\mu_e = \mu_h$) and different threshold voltages ($V_{T,e} = +30$ V, $V_{T,h} = -10$ V) were used. In all cases, as observed before in the measurements, the output voltage of the complementary inverter reaches the driving and the ground voltage in the high and the low state, respectively, whereas the output voltage of the ambipolar inverter does not [10]. Thus the noise margin of the ambipolar inverters are lower than for the complementary inverters. Nevertheless the gain for both types of inverters is comparably high. As expected, the current dissipation in the complementary inverter has a maximum at the transition between the logic levels, whereas the ambipolar inverter has its minimum at this point.

A closer look at the simulations in Figure 17.14a, which are based on symmetric mobilities and threshold voltages, shows that the transition between the logic states is at half of the driving voltage for both the ambipolar and the complementary inverter. An asymmetry in carrier mobilities (Figure 17.14b) and/or in threshold voltages (Figure 17.14c) leads to a shift of the transition voltage in both cases. Nevertheless, the transition voltages of the measured inverters (Figure 17.14a, c) are still more or less symmetric although the mobilities are largely different. The reason is that in both cases the shift of the transition voltage due to the higher mobility in the n-channel is compensated by a lower threshold voltage in the p-channel. One can further see that different mobilities have no influence on the gain, whereas different, especially lower, threshold voltages decrease both the gain and the noise margin. Experimentally, we have a decrease of the threshold voltage in the p-channel by diluting CuPc with C_{60} (Figure 17.5) and accordingly the gain in the measured ambipolar inverters is lower than in the complementary inverter. Finally, it is evident that the current dissipation of an ambipolar inverter is increasing drastically in the case of asymmetric mobilities. Altogether, complementary inverters are clearly superior to their ambipolar counterparts, but in both cases equal mobilities and symmetric threshold voltages are required to achieve optimal inverter characteristics.

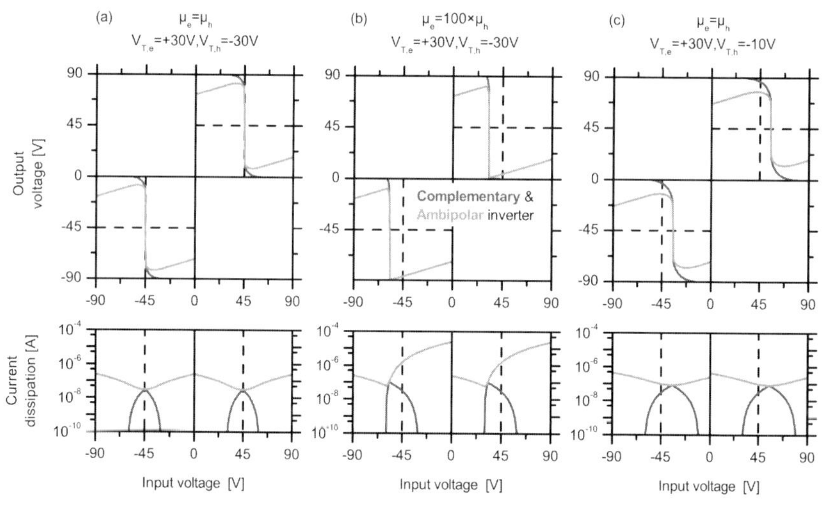

Figure 17.14 Simulations of ambipolar and complementary inverter transfer characteristics and current dissipation with (a) symmetric mobility and threshold voltage for p- and n-channel in comparison with (b) asymmetric mobilities and (c) asymmetric threshold voltages. (see colour plates p. LXXXVIII)

17.10 Summary

Mixtures of hole conducting CuPc and electron conducting C_{60} show ambipolar transport for all investigated mixing ratios. However, the mobilities of electrons and holes are strongly dependent on the composition indicating that percolation is a critical issue. Unlike in polymer/fullerene blends, there is no evidence for strong structural interactions, rather, the materials simply dilute each other. This is manifested also by the electronic structure of the blends, which does not show any significant evidence for charge transfer in the ground state.

Due to the higher electron mobility in neat C_{60} (as compared with the hole mobility in neat CuPc), an excess of CuPc (75%) is required to achieve balanced mobilities. But even with equal electron and hole mobilities an ambipolar inverter is inferior to a complementary one with discrete p- and n-channel transistors. Thus realising true complementary logic circuits will require devising efficient patterning methods, e.g. by locally different functionalisation of the gate insulator. In this context we note that it has recently been shown that neat films and single crystals of CuPc can exhibit both electron and hole transport as well as ambipolar transport, if the surface of the gate dielectric is passivated in order not to trap electrons and if contacts with suitable work function are used [68, 69].

Furthermore, our investigations of the electronic structure allow the conclusion that the HOMO–LUMO offset is significantly smaller in blends as compared with planar interfaces between both materials. Together with the lowering of the charge carrier mobility this reduction of the "built-in field" could have important consequences for photovoltaic cells based on these molecular materials. Bulk heterojunction devices with homogeneously mixed donor/acceptor layers will not necessarily be the optimal film morphology. The challenge will be to grow suitable blend morphologies with interdigitated donor/acceptor phases that simultaneously have large interfacial area and high exciton diffusion and charge carrier drift length for efficient charge carrier generation and extraction.

Acknowledgements

This work was supported by the Deutsche Forschungsgemeinschaft through Sonderforschungsbereich 484 and Schwerpunktprogramm 1121. The authors thank M. Himmerlich, J. A. Schaefer, and S. Krischok (TU Ilmenau, Germany) for the cooperation concerning photoelectron spectroscopy and J. Pflaum (Uni Stuttgart, Germany) for XRD measurements.

References

1. W. Warta and N. Karl, Phys. Rev. B **32**, 1172–1182 (1985).
2. C. D. Dimitrakopoulos and P. R. L. Malenfant, Adv. Mater. **14**, 99–117 (2002).
3. E. J. Meijer, D. M. De Leeuw, S. Setayesh, E. Van Veenendaal, B. H. Huisman, P. W. M. Blom, J. C. Hummelen, U. Scherf, and T. M. Klapwijk, Nature Mater. **2**, 678–682 (2003).
4. C. J. Brabec, V. Dyakonov, J. Parisi, and N. S. Sariciftci, Organic Photovoltaics (Springer Verlag, 2003).
5. G. W. Neudeck and A. K. Malhotra, J. Appl. Phys. **46**, 239– 247 (1975).
6. J. Locklin, K. Shinbo, K. Onishi, F. Kaneko, Z. N. Bao, and R. C. Advincula, Chem. Mater. **15**, 1404–1412 (2003).
7. J. G. Xue and S. R. Forrest, Phys. Rev. B **69**, 245322 (2004).
8. J. Wang, H. B. Wang, X. J. Yan, H. C. Huang, and D. H. Yan, Appl. Phys. Lett. **87**, 093507 (2005).
9. R. B. Ye, M. Baba, Y. Oishi, K. Mori, and K. Suzuki, Appl. Phys. Lett. **86**, 253505 (2005).
10. Y. Inoue, Y. Sakamoto, T. Suzuki, M. Kobayashi, Y. Gao, and S. Tokito, Jpn. J. Appl. Phys. **144**, 3663–3668 (2005).
11. M. A. Loi, C. Rost-Bietsch, M. Murgia, S. Karg, W. Rieß, and M. Muccini, Adv. Funct. Mater. **16**, 41–47 (2006).
12. T. Yasuda, T. Goto, K. Fujita, and T. Tsutsui, Appl. Phys. Lett. **85**(11), 2098–2100 (2004).
13. R. Schmechel, M. Ahles, and H. von Seggern, J. Appl. Phys. **98**, 084511 (2005).
14. L. L. Chua, J. Zaumseil, J. F. Chang, E. C. W. Ou, P. K. H. Ho, H. Sirringhaus, and R. H. Friend, Nature **434**, 194–199 (2005).
15. M. Ahles, R. Schmechel, and H. von Seggern, Appl. Phys. Lett. **85**, 4499–4501 (2004).
16. E. C. P. Smits, T. D. Anthopoulos, S. Setayesh, E. van Veenendaal, R. Coehoorn, P. W. M. Blom, B. de Boer, and D. M. de Leeuw, Phys. Rev. B **73**, 205316 (2006).
17. T. D. Anthopoulos, S. Setayesh, E. Smith, M. Cölle, E. Cantatore, B. de Boer, P. W. M. Blom, and D. M. de Leeuw, Adv. Mater. **18**, 1900–1904 (2006).
18. J. Rostalski and D. Meissner, Sol. Energy Mater. Sol. Cells **61**, 87–95 (2000).
19. T. Stübinger and W. Brütting, J. Appl. Phys. **90**(7), 3632– 3641 (2001).
20. P. Sullivan, S. Heutz, S. M. Schultes, and T. S. Jones, Appl. Phys. Lett. **84**(7), 1210–1212 (2004).
21. J. G. Xue, B. P. Rand, S. Uchida, and S. R. Forrest, J. Appl. Phys. **98**, 124903 (2005).
22. G. Paasch, T. Lindner, C. Rost-Bietsch, S. Karg, W. Rieß, and S. Scheinert, J. Appl. Phys. **98**, 084505 (2005).
23. A. Hepp, H. Heil, W. Weise, M. Ahles, R. Schmechel, and H. von Seggern, Phys. Rev. Lett. **91**, 157406 (2003).
24. C. Rost, S. Karg, W. Rieß, M. A. Loi, M. Murgia, and M. Muccini, Appl. Phys. Lett. **85**, 1613–1615 (2004).
25. J. Zaumseil, R. H. Friend, and H. Sirringhaus, Nature Mater. **5**, 69–74 (2006).
26. S. Grecu, M. Roggenbuck, A. Opitz, and W. Brütting, Org. Electron. **7**, 276–286 (2006).
27. A. Opitz, M. Bronner, and W. Brütting, J. Appl. Phys. **101**, 063709 (2007).
28. E. J. Meijer, C. Detcheverry, P. J. Baesjou, E. van Veenendaal, D. M. de Leeuw, and T. M. Klapwijk, J. Appl. Phys. **93**, 4831–4835 (2003).
29. N. Marjanovic, T. B. Singh, G. Dennler, S. Gunes, H. Neugebauer, N. S. Sariciftci, R. Schwödiauer, and S. Bauer, Org. Electron. **7**, 188–194 (2006).

30. R. Tecklenburg, G. Paasch, and S. Scheinert, Adv. Mater. Opt. Electron. **8**(6), 285–294 (1998).
31. T. Lindner, G. Paasch, and S. Scheinert, J. Appl. Phys. **98**, 114505 (2005).
32. S. Sze, Physics of Semiconductor Devices (Wiley, New York, 1982).
33. M. Knupfer and H. Peisert, phys. stat. sol. (a) **201**, 1055–1074 (2004).
34. O. V. Molodtsova, T. Schwieger, and M. Knupfer, Appl. Surf. Sci. **252**, 143–147 (2005).
35. S. C. Veenstra and H. T. Jonkman, J. Polym. Sci. B **41**, 2549–2560 (2003).
36. O. V. Molodtsova and M. Knupfer, J. Appl. Phys. **99**, 053704 (2006).
37. I. Hill, A. Kahn, Z. Soos, and R. Pascal, Chem. Phys. Lett. **327**, 181–188 (2000).
38. R. W. Lof, M. A. Vanveenendaal, B. Koopmans, H. T. Jonkman, and G. A. Sawatzky, Phys. Rev. Lett. **68**, 3924– 3927 (1992).
39. Y. Nobuhide, N. Makoto, H. Nobukazu, N. Kazumasa, W. Masaru, and K. Jiro, Appl. Phys. Lett. **85**(20), 4663– 4665 (2004).
40. T. Lindner, G. Paasch, and S. Scheinert, J. Mater. Res. **19**(7), 2014–2027 (2004).
41. S. Cherian, C. Donley, D. Mathine, L. LaRussa, W. Xia, and N. Armstrong, J. Appl. Phys. **96**, 5638–5643 (2004).
42. P. M. Borsenberger and D. S. Weiss, Organic Photoreceptors for Imaging Systems (Marcel Dekker Ltd, New York, 1993).
43. C. K. Song, B. W. Koo, S. B. Lee, and D. H. Kim, Jpn. J. Appl. Phys. **141**, 2730–2734 (2002).
44. S. M. Tuladhar, D. Poplavskyy, S. A. Choulis, J. R. Durrant, D. D. C. Bradley, and J. Nelson, Adv. Funct. Mater. **15**, 1171–1182 (2005).
45. V. D. Mihailetchi, L. J. A. Koster, P. W. M. Blom, C. Melzer, B. de Boer, J. K. J. van Duren, and R. A. J. Janssen, Adv. Funct. Mater. **15**, 795–801 (2005).
46. B. P. Rand, J. G. Xue, S. Uchida, and S. R. Forrest, J. Appl. Phys. **98**, 124902 (2005).
47. F. Dinelli, R. Capelli, M. A. Loi, M. Murgia, M. Muccini, A. Facchetti, and T. J. Marks, Adv. Mater. **18**, 1416 (2006).
48. G. Horowitz, R. Hajlaoui, and P. Delannoy, J. Phys. III (France) **5**, 355–371 (1995).
49. G. Paasch, P. H. Nguyen, and S. L. Drechsler, Synth. Met. **97**(3), 255–265 (1998).
50. F. Dinelli, M. Murgia, P. Levy, M. Cavallini, F. Biscarini, and D. M. de Leeuw, Phys. Rev. Lett. **92**, 116802 (2004).
51. L. Lozzi, V. Granato, S. Picozzi, M. Simeoni, S. La Rosa, B. Delly, and S. Santucci, J. Vac. Sci. Technol. A **24**, 1668–1675 (2006).
52. D. Faiman, S. Goren, E. A. Katz, M. Koltun, N. Melnik, A. Shames, and S. Shtutina, Thin Solid Films **295**, 283–286 (1997).
53. A. Opitz, M. Bronner, W. Brütting, M. Himmerlich, J. A. Schaefer, and S. Krischok, Appl. Phys. Lett. **90**, 212112 (2007).
54. S. M. Schultes, P. Sullivan, S. Heutz, B. M. Sanderson, and T. S. Jones, Mater. Sci. Eng. C, Biomim. **25**, 858–865 (2005).
55. M. Stöhr, T. Wagner, M. Gabriel, B. Weyers, and R. Möller, Adv. Funct. Mater. **11**, 175–178 (2001).
56. T. Stübinger and W. Brütting, SPIE Proc. **4465**, 102–112 (2002).
57. S. Rentenberger, A. Vollmer, E. Zojer, R. Schennach, and N. Koch, J. Appl. Phys. **100**, 053701 (2006).
58. T. R. Ohno, Y. Chen, S. E. Harvey, G. H. Kroll, J. H. Weaver, R. E. Haufler, and R. E. Smalley, Phys. Rev. B **44**, 13747–13755 (1991).
59. G. Koller, B. Winter, M. Oehzelt, J. Ivanco, F. P. Netzer, and M. G. Ramsey, Org. Electron. **8**, 63–68 (2007).

60. B. A. Gregg, MRS Bull. **30**, 20–22 (2005).
61. W. Brütting, M. Bronner, M. Götzenbrugger, and A. Opitz, Macromol. Symp. **268**, 38–42 (2008).
62. G. Horowitz, R. Hajlaoui, D. Fichou, and A. El Kassmi, J. Appl. Phys. **85**, 3202–3206 (1999).
63. Y. L. Shen, M. W. Klein, D. B. Jacobs, J. C. Scott, and G. G. Malliaras, Phys. Rev. Lett. **86**, 3867–3870 (2001).
64. B. H. Hamadani and D. Natelson, Appl. Phys. Lett. **84**, 443–445 (2004).
65. H. Riel, S. Barth, T. Beierlein, W. Brütting, S. Karg, P. Müller, and W. Rieß, SPIE Proc. **4105**, 167–174 (2001).
66. G. W. Neudeck, K. Y. Chung, and H. F. Bare, IEEE Trans. Electron Devices **34**, 866–871 (1987).
67. A. Opitz, M. Bronner, J. Wagner, M. Götzenbrugger, and W. Brütting, SPIE Proc. **7002**, 70020J (2008).
68. A. Opitz, M. Kraus, M. Bronner, J. Wagner, and W. Brütting, New J. Phys. **10**, 065006 (2008).
69. R. W. I. de Boer, A. F. Stassen, M. F. Craciun, C. L. Mulder, A. Molinari, S. Rogge, and A. F. Morpurgo, Appl. Phys. Lett., **86**, 262109 (2005).

18
Gate Dielectrics and Surface Passivation Layers for Organic Field Effect Transistors

T. Diekmann and U. Hilleringmann

18.1
Introduction

Field effect transistors based on organic semiconductors provide the opportunity to build cheap electrical devices and integrated circuits on mechanically flexible substrates [1–3]. In the area of displays, organic light emitting diodes (OLEDs) conquer more and more applications, for example in mobile telephones and in MP3 players. An important topic for organic field effect transistors (OFETs) could be the application in simple electrical circuits for single use, but also in radio-frequency tags as replacements for bar-code labels on consumer products [4, 5]. At present, however, organic semiconductors do not reach the electrical performance of inorganic crystalline semiconductors.

Besides the organic semiconductor, the choice of the gate dielectric layer is a crucial point in the processing of the devices, because it affects the electrical performance of the organic field effect transistors. In this chapter different inorganic gate dielectrics on a silicon substrate are presented in order to study their influence on the electrical parameters of the OFETs. These oxides and nitrides require high processing temperatures, and the mechanical flexibility is significantly limited. Hence, it is impossible to employ them as gate dielectrics on polymer foils.

This is the reason why a polymer dielectric film that can be deposited at low process temperatures is used. For the integration of OFETs on polymer dielectrics, such as polyimide films or commercial insulating varnishes, the initial studies still use silicon substrates. Subsequently, a new high-*k* resist based on hydrolysed and partially condensed ethyl silicates is used; it is filled with nanoparticles of an inorganic material. Applying this special resist, transistors on a polyester substrate could be built successfully.

The second part of this chapter discusses degradation experiments on OFETs. Unprotected organic devices suffer from degradation due to water vapour or oxygen incorporation. The influence of ambient air on the transistor parameters, such as charge carrier mobility, the on–off ratio and the threshold voltage, is discussed. In order to protect the organic semiconductors against

Organic Electronics. Structural and Electronic Properties of OFETs. Edited by Christof Wöll

ISBN: 978-3-527-40810-8

water vapour, OFETs are encapsulated with a hydrophobic polytetrafluoroethylene layer. In this way, effects of degradation mechanisms on the threshold voltage can be referred to the influence of water or oxygen.

18.2 Experimental

18.2.1 Transistor Device

The OFETs using inorganic gate dielectrics were prepared on a heavily doped silicon substrate, which also serves as the common gate electrode for all fabricated devices. The semiconducting film consists of evaporated pentacene, which is commercially available. In Figure 18.1 the transistor architecture with bottom gate and bottom drain and source contacts is shown.

The integration of the devices begins with an initial standard cleaning process (SC1 for the silicon substrate) and the gate dielectric deposition. A lithography step defines the resist mask for the metal lift-off. Subsequently, the metal for the bottom drain and source contacts was sputtered in a dc-sputter coater up to a thickness of 30 nm. The lift-off uses acetone and ultrasonic agitation. To complete the transistor device, a 30 nm thick pentacene layer is thermally evaporated in high vacuum at a pressure below 1×10^{-6} mbar at a very low deposition rate.

18.2.2 Inorganic Dielectrics

Different silicon dioxide layers are grown at temperatures between 850 °C and 1000 °C by wet thermal oxidation. During the tests, the best electrical parameters could be achieved at a process temperature of 960 °C and a maximum layer thickness of 150 nm.

Different layers deposited by low pressure chemical vapour deposition (LPCVD) or by plasma enhanced chemical vapour deposition (PECVD) were examined as further inorganic gate dielectrics. Insulating layers of tetraethylorthosilicate (TEOS) oxide, deposited by the thermal pyrolysis of the vapour

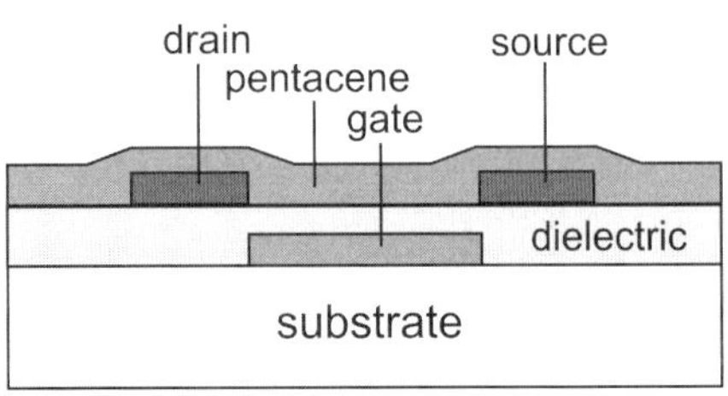

Figure 18.1 Organic field effect transistor with bottom gate and bottom drain and source contacts.

of the TEOS liquid at process temperatures of about 750 °C, combine the advantages of a high deposition rate at moderate process temperatures and electrical stability. However, on the negative side is the high trap density in comparison with thermally grown silicon dioxide.

Additionally, low-temperature oxides (LTO) were examined. The deposition process uses triethylsilane and oxygen at 550 °C. Table 18.1 gives an overview of the different inorganic gate dielectrics, their deposition temperatures, the deposited layer thicknesses and their permittivities.

The LTO oxide can be used to build transistors with insulated metal gates, which are necessary for simple logic and integrated circuits. For the integration of separated metal gates, a titanium layer is structured by wet chemical etching in ammonium hydroxide on top of a thermally oxidised silicon substrate. An additional temperature step in a nitrogen atmosphere at 550 °C converts the titanium into conductive titanium nitride. This film is thermally stable, so the LTO oxide can be deposited on top. For TEOS oxide and silicon nitride the application of a titanium nitride gate electrode is unsuitable, as it will be converted into insulating titanium oxide at process temperatures of more than 700 °C.

18.2.3 Polymer Dielectrics

Polymer electronics on foils require mechanically flexible gate dielectric layers. Unfortunately, inorganic insulating films suffer from high deposition temperatures and a lack of mechanical elasticity. In a first step the inorganic gate dielectric is substituted by a polymer film, still using a silicon substrate because of their smooth and well-known surface.

Initially, the experiments apply a commercially available polyimide coating, which is well known from many microelectronic applications. The polyimide (PI 2545, HD Microsystems GmbH) is a high-temperature coating that can be patterned by a positive photoresist. It is dissolved in the same process step as the exposed resist using an alkaline photoresist developer [6], but by different etching rates.

Table 18.1 Different inorganic gate dielectrics.

dielectric	process temperature	thickness	permittivity
SiO_2 (wet thermal)	960 °C	150 nm	3.9
TEOS oxide	750 °C	120 nm	3.9
LTO	550 °C	180 nm	3.9
PECVD/LTO oxide	300 °C	30/110 nm	3.9
Si_3N_4	800 °C	75 nm	8
Ta_2O_5	room temperature	165 nm	23

To improve the spreading of the polyimide precursor, the dehumidified wafer is flooded with a special adhesion promoter (VM 651, HD Microsystems GmbH), diluted in deionised water (content of VM 651 in mixture: 0.1%). After a residence time of 20 s the wafer is spin dried. For further improvement of adhesion, a post-bake step is performed at 120 °C for 60 s on a hot plate [7]. Following this, the diluted polyimide precursor (one part polyimide to two parts of a suitable thinner (T9039, HD Microsystems GmbH)) is spin coated on the wafer, resulting in a 190 nm thick film after curing. The curing process requires temperatures up to 350 °C in a nitrogen atmosphere; this is described elsewhere in detail [8].

Because of the high curing temperature of the introduced polyimide film, it is inapplicable to polymer substrates. Therefore, a commercially available coating varnish Bectron® (based on modified alkyd chemistry), favoured because of the low curing temperature of about 80 °C, is spin coated onto a silicon substrate. It is cured in a convection oven at 80 °C for 30 min, resulting in a 1 μm to 1.5 μm thick film. In contrast to the OFETs mentioned in this chapter so far, the transistor on the Bectron® varnish uses bottom gate and titanium top drain and source contacts, structured by a shadow mask. The latter is due to the lack of chemical resistance of the varnish towards solvents.

Moreover, a new high-epsilon resist was used as the gate dielectric layer. It is based on hydrolysed and partially condensed ethyl silicates, charged with an inorganic material to adapt the permittivity [16]. The resist contains different amounts of zirconium dioxide or titanium dioxide, resulting in dielectric constants between 9 and 12. It can be deposited by spin coating at 1000 rpm, followed by a soft-bake step in a convection oven at 80 °C for 10 min. The insulating film with a layer thickness of about 200 nm to 300 nm will be stable against developer solution (diluted NaOH) or acetone after an additional UV-curing step, so that drain and source contacts can be formed by UV lithography and lift-off in acetone.

Because of the low process temperatures the resist can be easily adapted on plastic films as the substrate material. Here OFETs are first integrated on a 23 μm thick polyester foil that is cut to a diameter of 100 mm, cleaned in acetone and deionised water, dried and put on a silicon wafer in order to be processed by the equipment of the silicon semiconductor technology. A 150 nm thick aluminium layer serves as the common gate electrode for the electrical analysis of the devices.

18.3 Results and Discussion

In this section the characteristics of transistors using different gate dielectrics will be shown. For the interpretation of the results, the way of determining the transistor characteristics is introduced first.

According to the IEEE standard test methods for the characterisation of OFETs [26], the Shockley equations for insulated gate field effect transistors (IGFETs) are used to approximate the field-induced drain current in the organic material between the drain and source contacts. It should be recommended that the following assumptions form the basis of the applicability of the equations:

- constant charge carrier mobility in the transistor channel,
- the channel length is much larger than the thickness of the gate dielectric layer,
- only one species of charge carriers is responsible for the current in the transistor channel,
- sufficiently insulating gate dielectric, low leakage current,
- charge injection is not restricted by the contacts (ohmic contact electrodes).

For OFETs it is often not possible to keep all of the conditions mentioned above. Therefore, the application of the equations is a more or less well fitting approximation to characterise the transistor behaviour. In the case of the linear regime ($V_{GS} < V_{th}$ and $V_{DS} > (V_{GS} - V_{th})$) of the hole-conducting transistor the drain current can be described by

$$I_D = \varepsilon_0 \varepsilon_r \mu_p / t_{ox} (V_{GS} - V_{th} - V_{DS}/2)\, V_{DS} \,. \quad (1)$$

If the OFET operates in saturation the drain current can be estimated by

$$I_D = \varepsilon_0 \varepsilon_r \mu_p / (2 t_{ox})\, (V_{GS} - V_{th})^2 \,, \quad (2)$$

where $\varepsilon_0 = 8.85 \times 10^{-12}$ A s/V m, ε_r is the relative permittivity of the gate dielectric, μ_p is the charge carrier mobility and t_{ox} is the thickness of the gate dielectric.

18.3.1 Inorganic Gate Dielectric Layers

There are some important demands for gate dielectrics in organic field effect transistors. For example:

- high electrical strength at thin layer thickness,
- smooth surface for improved pentacene growth [9–11],
- high dielectric constant to reduce the operating voltage,
- low deposition temperature with a view to deposition on polymer substrates,
- low trap density at the interface for a well-defined threshold voltage,
- non-polar surface for improved adsorption of pentacene molecules.

18.3.1.1 Thermally Grown Silicon Dioxide

The characteristics of pentacene OFETs are extracted using structures with thermally grown silicon dioxide as the gate dielectric layer for the reference data. The films offer a good electrical strength at thin film thicknesses, which can resist the high gate voltages of up to 40 V in OFETs. In the past, some groups even used supply voltages up to 100 V [12–14]. Moreover, through its smooth surface (0.1–0.2 nm rms) silicon dioxide offers ideal conditions for the growth of well-ordered pentacene films [9, 15, 17]. In contrast, rough surfaces will result in a lot of seed crystals for the pentacene molecules, giving an amorphous pentacene film with bad charge carrier mobility. Another preference for silicon dioxide is the low number of oxide charges that could influence the threshold voltage.

Disadvantages of thermally grown silicon dioxide are the high deposition temperature of about 960 °C and the exigency of a silicon substrate. Moreover, an adhesion layer of 4–8 nm of nickel is necessary to improve the adhesion of the 30 nm thick gold electrodes deposited on top of the oxide.

Figure 18.2 shows the output characteristic of an OFET using a thermally grown silicon oxide gate dielectric. The device depicts a drain current of −81 μA at $V_{DS} = -40$ V and $V_{GS} = -40$ V ($W/L = 1\,000$). The threshold voltage is about 1.5 V, and the on–off ratio reaches 10^3 at $V_{DS} = -40$ V. Even at a gate–source voltage of −40 V the silicon dioxide layer is still insulating and does not show any breakdown.

The surface of the pentacene film was examined by atomic force microscopy (AFM). It exhibits crystallites with a diameter of about 250 nm in the transistor channel. On the gold surface of the drain and source contacts, the dimensions are twice the size as on silicon dioxide. In comparison with growth studies in [18] where the surface was pretreated by an adhesion promoter, the presented pentacene film consists of very small crystallites.

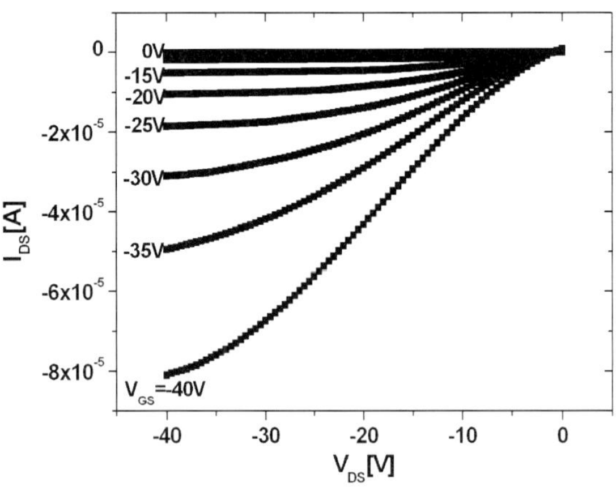

Figure 18.2 Output characteristic of an OFET with a channel length of 1 μm and a channel width of 1000 μm, using silicon dioxide as the gate dielectric.

In order to improve the transistor characteristics by increasing the pentacene grain size, the oxide surface was cleaned in an oxygen plasma to remove residues of the photoresist that remained on top after the lift-off step. The cleaning is performed directly before coating the structures with the organic material. The dielectric surface was exposed to the oxygen plasma for 30 s at 100 W in a parallel-plate reactor. A longer process time could result in etching of the silicon dioxide, which would be counterproductive.

After this treatment a 30 nm thick pentacene layer was deposited onto the samples at a process pressure of 8×10^{-7} mbar and a deposition rate of about 0.1 nm/s, resulting in a pentacene film with a grain size of about 1 μm, as confirmed by AFM [19].

Using such a pentacene film, a transistor with a channel length of 1 μm and a channel width of 1000 μm can drive drain currents of more than −600 μA at $V_{DS} = -80$ V and $V_{GS} = -25$ V ($W/L = 1000$) (see output characteristic in Figure 18.3). Due to the larger grain size of the pentacene there are fewer grain boundaries in the transistor channel, which is why the charge carrier mobility (calculated by Eq. (2) in the saturation regime) increases up to 2.4×10^{-2} cm²/V s. In comparison with values given in the literature of up to 2 cm²/V s [20], the mobility is still very low. This could be caused by a bad contact between the pentacene molecules and the drain and source contacts, resulting in a non-ohmic contact. Figure 18.3 proves this assumption by a nonlinear behaviour in the drain current in the linear region of the output characteristic, which indicates contact-limited charge carrier injection [44] and leads to an underestimation of the true mobility.

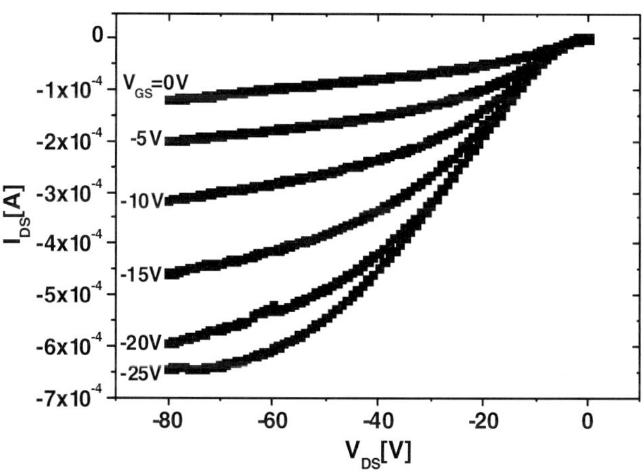

Figure 18.3 Output characteristic of an OFET with a channel length of 1 μm and a channel width of 1000 μm, using a 100 nm thick silicon dioxide layer as gate dielectric. Before evaporating the pentacene in high vacuum the sample was pretreated in an oxygen plasma for 30 s at a plasma power of 100 W.

The output characteristic in Figure 18.3 shows one more disadvantage: if no gate voltage is applied there is a big off-current through the transistor channel, caused by a high positive threshold voltage of 17.2 V. The reasons for such threshold voltages could be found in the generation of free bonds at the dielectric surface, which result in polarisation of the dielectric [21].

Because of the promising transistor characteristics, further investigations of silicon dioxide apply interdigitated drain and source electrodes, resulting in a channel width of 16.8 cm and a channel length of 10 μm. In this way the maximum drain current could be improved to drive electrical circuits that require higher currents. The surfaces of the 127 nm thick silicon dioxide layer and the drain and source electrodes are pretreated in an oxygen plasma before evaporating the pentacene film.

In Figure 18.4 the output characteristic of the transistor that can supply drain currents of up to −15 mA at $V_{GS} = -40$ V and $V_{DS} = -40$ V is shown ($W/L = 16800$). At a value of 11.7 V, the threshold voltage is still far away from 0 V for enhancement-mode transistors. The low charge carrier mobility in the saturation regime of 3.15×10^{-3} cm²/V s could be due to a bad crystalline order in the organic layer.

18.3.1.2 TEOS Oxide

TEOS oxide was tested first as a CVD dielectric in order to examine the compatibility of the pentacene and the TEOS oxide. This dielectric offers a conformal deposition on lateral and vertical surfaces at high deposition rate but

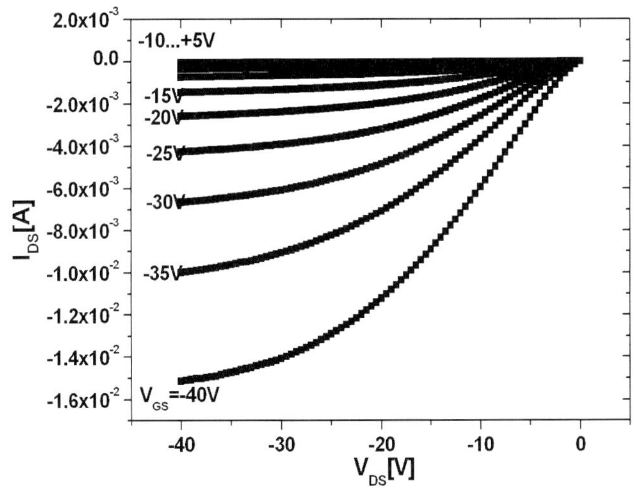

Figure 18.4 Output characteristic of an OFET with a channel length of 10 μm and a channel width of 16.8 cm, using a 127 nm thick silicon dioxide layer as gate dielectric. Before evaporating the pentacene in high-vacuum ambient, the sample was pretreated in an oxygen plasma for 30 s at 100 W plasma power.

still it requires high process temperatures of about 750 °C. Because of the chemical vapour deposition technique there are many traps due to unsaturated bonds in the dioxide layer and at its surface.

The investigations of TEOS oxide use a gate dielectric that is deposited at a process pressure of 0.2 mbar. By ellipsometry, the film thickness could be determined to be 120 nm. The surface of the dielectric was investigated by AFM and exhibited a smooth surface with a roughness of 1–3 nm rms.

Figure 18.5 shows the output characteristic of an OFET with TEOS oxide as the gate dielectric. At $V_{DS} = -20$ V and $V_{GS} = -10$ V the transistor with a channel length of 5 μm and a channel width of 1000 μm exhibits a drain current of about –500 nA ($W/L = 200$). Higher gate voltages caused an electrical breakdown; this was due to the large number of traps in the oxide layer. Although the off-state current is very low the on–off ratio achieves a value of only 10^2.

From the output characteristic in the saturation region the charge carrier mobility was calculated to be 1.3×10^{-3} cm²/V s. In comparison with the mobility of the transistor on thermally grown silicon dioxide and with regard to the larger channel length, this value is only slightly smaller. Although there are traps at the surface of the gate dielectric the threshold voltage of this device was ascertained to be 1.2 V, which is near to the target threshold voltage of close to zero volts. Due to the bad electrical strength of the deposited layers and the high deposition temperature, TEOS oxide cannot be recommended to be used as the gate dielectric in OFETs on foils.

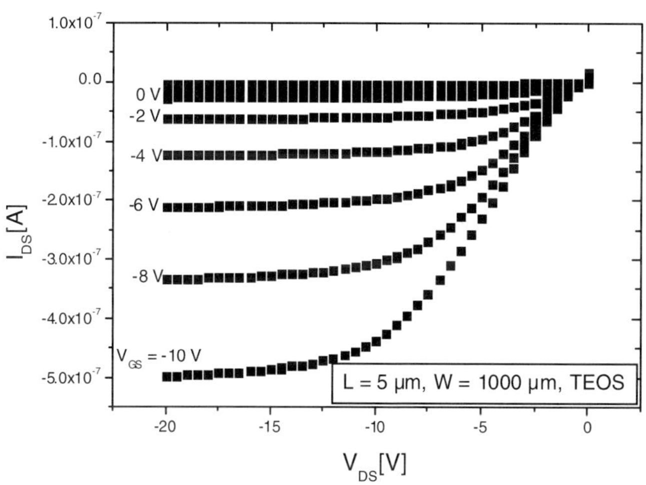

Figure 18.5 Output characteristic of an OFET with 1000 μm channel width and 5 μm channel length on a 120 nm thick TEOS oxide layer. The thickness of the pentacene film is about 60 nm, deposited at a substrate temperature of 60 °C and a deposition rate of 0.3 nm/s at 7.4×10^{-7} mbar chamber pressure.

18.3.1.3 Silicon Nitride

The Shockley equations describe a linear dependence between drain current and permittivity of the gate dielectric. In order to lower the supply voltage of OFETs, silicon nitride was examined because its permittivity $\varepsilon_{nitride} = 8$ is almost twice as large as that of silicon dioxide. Moreover, with a roughness of 1–2 nm rms it has the smoothest surface of the chemical vapour deposited dielectrics, which is one essential requirement for the growth of a well-ordered pentacene film.

In the following, a silicon nitride layer was deposited as the gate dielectric on a thermally oxidised silicon wafer. The nitride layer was re-oxidised to enhance the electrical stability. The silicon dioxide below the nitride film adopted the function of a buffer layer to reduce mechanical stress between the silicon and silicon nitride due to different thermal coefficients of expansion. To deposit the dielectric film, ammonia gas and triethylsilane were put into the process tube in a ratio of 1:5, at 800 °C and at a process pressure of 0.3 mbar. The thickness of the deposited dielectric film was about 75 nm in total.

In Figure 18.6 the output characteristic of a transistor with a channel width of 1000 μm and a channel length of 1 μm is shown. The drain current at $V_{DS} = -40$ V and $V_{GS} = -40$ V was about −19 μA ($W/L = 1000$), whereas the charge carrier mobility in saturation could be calculated to be 3.8×10^{-4} cm²/V s. The on–off ratio was smaller than 10^2 due to a large parallel conductivity that caused high drain currents in the off-state of the transistor, although the device had a threshold voltage of −2.1 V.

One reason for the high off-current of the transistor at $V_{GS} = 0$ V could be interface charges at unsaturated interface bonds of the gate dielectric. Another cause for the off-current could be found in carbon bonds, for example ethyl groups, that were separated from the chemical product triethylsilane during the

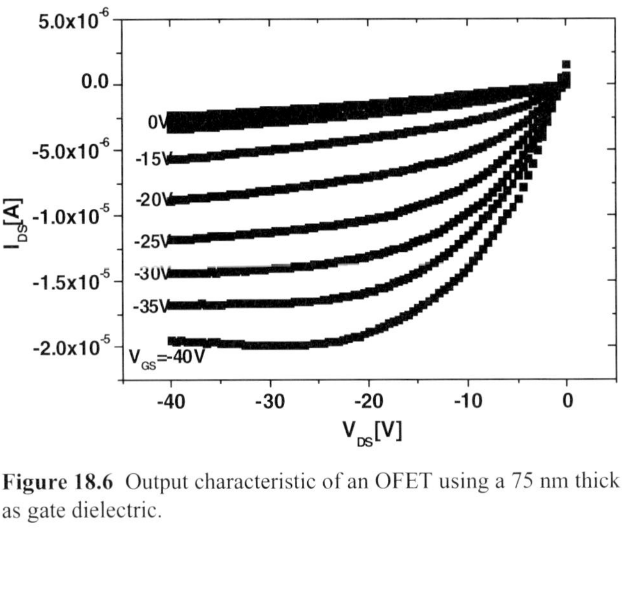

Figure 18.6 Output characteristic of an OFET using a 75 nm thick silicon nitride layer as gate dielectric.

processing. The assumption of impurities in the semiconducting pentacene film could be neglected, because reference measurements at samples that were deposited with pentacene purified several times did not show any difference. Therefore, it was concluded that the purity of the pentacene was not responsible for the observed off-current. One more possibility for the parasitic currents could be gate leakage currents through the gate dielectric, but then there would be positive drain currents, as can be observed in Figure 18.7.

Off-currents have also been measured at some organic field effect transistors using silicon dioxide as the gate dielectric. During the investigations it was not possible to detect the real reason for this sporadically occurring behaviour, which was only detectable after the pentacene deposition. One explanation could be the generation of interface traps as a consequence of the pentacene deposition.

The high deposition temperature excluded the manufacturing of simple circuits using silicon nitride as gate dielectric for transistors with a metal gate electrode. Therefore, another gate dielectric, deposited at lower process temperatures, is necessary.

18.3.1.4 **Low-Temperature Oxide: LTO**

To deposit oxide layers at lower process temperatures LTO oxide was introduced by a LPCVD process (low-pressure chemical vapour deposition) at a temperature of 550 °C using triethylsilane and oxygen as the gas sources. In a tube of a three-zone oven the ethyl groups were separated from the silane residues and reacted under a flow of oxygen at a pressure of 0.4 mbar to SiO_2, which was deposited on the silicon substrate. This deposition temperature was sufficient to use titanium nitride as the conducting material for transistors with single gates.

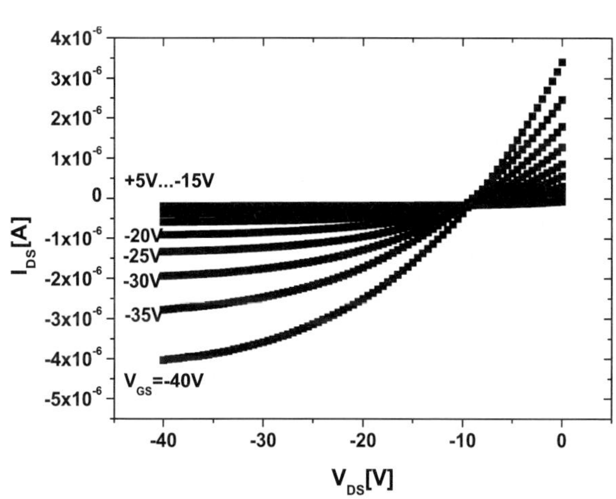

Figure 18.7 Output characteristic of an OFET with a 180 nm thick LTO oxide layer as gate dielectric.

The output characteristic of a transistor with a channel length of 5 μm and a channel width of 1000 μm is shown in Figure 18.7 using LTO oxide as the gate dielectric. Especially at high gate–source voltages and drain–source voltages near zero, a high gate leakage current through the gate dielectric could be observed by positive currents in the characteristic lines.

The origin of this parasitic current can be found in a large contingent of carbon in the oxide layer caused by the ethyl groups due to a bad ratio between oxygen and triethylsilane (1 : 3) during the deposition. A too large ratio of carbon in the oxide layer was detected by ellipsometry as a result of a shift in the refractive index from pristine silicon dioxide ($n = 1.46$) to larger values (in this case up to 1.49).

In summary, the electrical strength of the LTO oxide was not sufficient for use as the gate dielectric in organic circuits. Therefore, another inorganic insulator of a better electrical quality is necessary, which can be deposited at lower process temperatures.

18.3.1.5 **PECVD**

In the area of CVD oxide layers the PECVD (plasma-enhanced CVD) offers the deposition of dielectric films at low process temperatures of about 130–300 °C. During the investigations several oxide layers with different ratios of the process gases argon–2% silane and oxygen were tested as gate dielectrics for the OFETs.

Because of the low deposition temperature, PECVD oxide can be deposited on metal layers. An additional advantage is a high conformity that results in good edge coverage. As a disadvantage a high trap density occurs that is caused by the high deposition rate. This results in a shift of the threshold voltage and reduces the electrical strength. Therefore, dielectric breakdowns have been observed at several samples during the usage as gate dielectric with layer thicknesses of more than about 100 nm and at operating voltages up to −40 V. PECVD oxides were also reported in [22]; there, OFETs could be built on thick films with very small pentacene crystallites and low charge carrier mobilities.

Nevertheless, in this work PECVD oxide was used in combination with LTO oxide because of the possibility of depositing oxide layers on a substrate covered with titanium gate electrodes. Therefore, a 30 nm thick PECVD oxide layer was deposited at 275 °C at 12.5 W RF power, a process pressure of 500 mTorr and a gas flow of 350 sccm argon–2% silane and 20 sccm oxygen. In a subsequent process step a 110 nm thick LTO oxide film was applied.

Figure 18.8 shows the output characteristic of such an OFET with a titanium gate electrode and the combined oxide layers as the gate dielectric. At $V_{DS} = -40$ V and $V_{GS} = -40$ V the device with a channel width of 1000 μm and a channel length of 1 μm depicts a drain current of about −20.2 μA ($W/L = 1000$). In comparison with Figure 18.7, only a very small gate leakage current could be observed. The threshold voltage was determined to be 5.4 V, the subthreshold slope was 6 V/dec and the on–off ratio was only 30 at $V_{DS} = -40$ V.

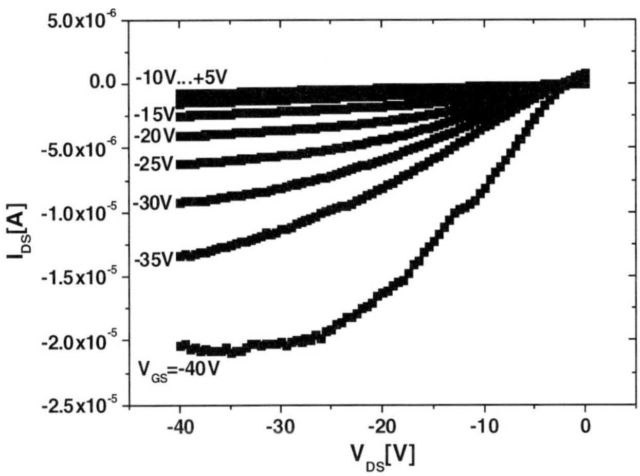

Figure 18.8 Output characteristic of an OFET with a titanium gate electrode, using 30 nm PECVD oxide and 110 nm LTO oxide as gate dielectric. The channel length of the device is 1 μm at a channel width of 1000 μm.

In summary, the PECVD oxide as a stand-alone insulating layer was not suitable as the gate dielectric because of its low electrical strength. On silicon substrates with titanium gate electrodes it simplified the sequence of processes. In combination with a LTO oxide film, gate leakage currents through the dielectric layer could be reduced as compared with LTO films as in Figure 18.7.

18.3.1.6 Ta_5O_2

In order to deposit dielectric layers on metallised silicon or polymer substrates, process temperatures have to be adapted. Here, reactive cathode sputtering was used to insulate the silicon gate by a tantalum pentoxide film at room temperature. Besides the deposition temperature the dielectric layer possesses a high permittivity of up to $\varepsilon_r = 23$, dependent on the quantitative compounds. This offers the possibility to accumulate more charge carriers in the transistor channel in comparison with silicon dioxide at the same layer thickness and the same gate–source voltage. In this way operating voltages should be reducible down to 5 V [23]. Although sputtered dielectric films have a rough surface, in [24] transistors were built with charge carrier mobilities of up to 0.36 cm²/V s.

As shown in Figure 18.9 a transistor with 1 μm channel length and 1000 μm channel width could drive drain currents of up to −9.1 μA at $V_{DS} = -40$ V and $V_{GS} = -40$ V ($W/L = 1000$). In spite of a 165 nm thick dielectric film a major gate leakage current was noticed. In [23, 24] the quantitative compounds of oxygen with tantalum were made responsible for a bad electrical isolation. If the fraction of the tantalum was too large, the metallic character dominated the dielectric properties of the deposited film resulting in a high conductance.

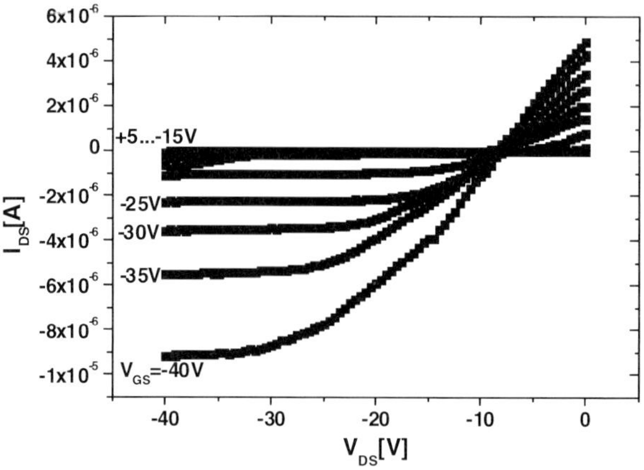

Figure 18.9 Output characteristic of an OFET with 1 μm channel length and 1000 μm channel width, using 165 nm Ta_2O_5 as gate dielectric.

During several experiments it was not possible to improve the electrical characteristics of the dielectric layer to avoid gate leakage currents. Setting up the film thickness did not solve the problem because mechanical stress led to cracks in the insulating film.

Contrary to the assumptions, the transfer characteristic (not shown in this chapter) indicated a low subthreshold slope of about 0.8 V/dec and a big on–off ratio of more than 10^4 even though there have been high parasitic currents. The threshold voltage could be determined by the transfer characteristic and reached a nice value of 2.5 V. Here, one must not neglect that the high gate leakage current presumably influences the determined value. This could be a reason for why impurities often observed at cathode-sputtered dielectrics did not affect the threshold voltage in this case.

To draw conclusions from the obtained results, sputtered Ta_2O_2 layers cannot be recommended for usage in OFETs as a gate dielectric. The disadvantages of the low electrical strength and the bad mechanical elasticity dominated the characteristics of this material. These effects cannot be eliminated by the larger electrical permittivity [25].

18.3.1.7 **Conclusion**

In order to investigate OFETs with pentacene, thermally grown silicon dioxide can be recommended as a gate dielectric. Because of its smooth surface it offers ideal conditions for the growth of pentacene films. The high electrical strength results in low gate leakage currents and the transistor characteristics obtain a high reproducibility. Therefore, it is well suited as a gate dielectric for degradation investigations on the organic semiconductors. If transistors with a metal gate are required, tantalum pentoxide has to be chosen, but it is very difficult to obtain insulating layers by setting the matching quantitative compounds of tantalum and oxygen. That is why this process has to be further optimised. As a result of this analysis, only the multilayer dielectric of 30 nm PECVD oxide and 110 nm LTO oxide can be recommended as a gate dielectric.

Especially with a view to the integration of OFETs on polymer substrates, none of the introduced deposition processes can deposit a suitable gate dielectric. Therefore, the examination of polymer dielectrics is of major interest.

18.3.2
Polymer Dielectrics

From the area of polymer dielectrics, a sample is presented here using a commercially available and (from many microelectronic applications) well-known high-temperature polyimide. The deposition and the subsequent curing process have been described in Section 18.2.3, resulting in a 190 nm thick insulating film that showed a good chemical resistance towards diluted developer solution and acetone, so that the cathode-sputtered Au drain and source contacts can be structured as described in Section 18.2.1. The 30 nm thick pentacene layer is thermally evaporated at a deposition rate of about 0.1 nm/s and a process pressure of 1×10^{-6} mbar.

In Figure 18.10 there is shown the output characteristic of an OFET with a channel length of 3 μm and a channel width of 1000 μm. At $V_{DS} = -40$ V and $V_{GS} = -40$ V it can push drain currents up to -15.8 μA ($W/L = 333$). The on–off ratio reaches a value of 10^2. Its magnitude is not sufficient for electrical circuits that require 10^4 or more. For example, in applications for matrix displays a value equal to or greater than 10^6 is necessary. In the shown transistor device it could be useful to reduce the off-current, which is in the order of 10^{-9} A. Its origin can be found in the hydrophilic characteristics of the polyimide layer, tending to incorporate water molecules.

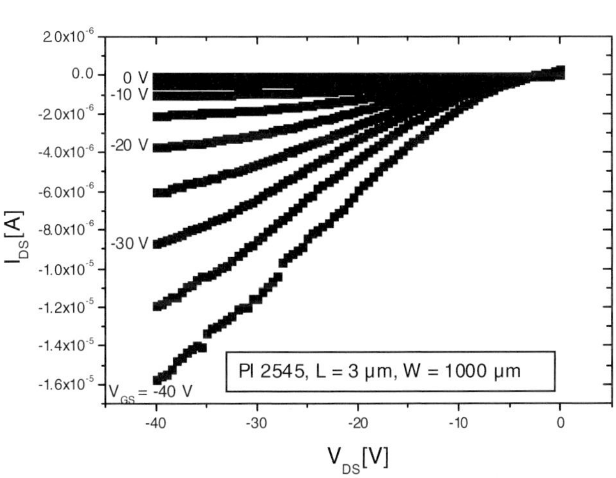

Figure 18.10 Output characteristic of an OFET with 3 μm channel length and 1000 μm channel width, using a 190 nm polyimide layer as gate dielectric.

In contrast to the IEEE conventions [26] the transfer characteristic is measured in the linear regime at $-5V_{DS}$ with a view to low operating voltages. That is why the values of the threshold voltage (–12.1 V) and the on–off ratio (10^2) differ from their values in the saturation regime. The subthreshold slope is about 5.5 V/dec and the charge carrier mobility in the linear regime is 1.7×10^{-3} cm^2/V s.

Besides differences in the method to determine the threshold voltage, its value deviates from the desired magnitude of zero volts because of residual impurities in the pentacene film that can act as uncontrollable doping. A further cause could be free charges at the dielectric surface, resulting in an applied electrical field that superposes with the field of the gate electrode.

In order to investigate the quality of the pentacene film, its surface was scanned by AFM in contact mode. In Figure 18.11 is shown a 10 × 10 μm sector of the pentacene film with dendritic crystallites up to 1 μm in diameter in the transistor channel. In comparison with [9], the size of the crystallites is equal to those grown on silicon dioxide. However, in [27] crystallites of several micrometers size with a charge carrier mobility of about 1 cm^2 on another polyimide film have been grown. That material has been deposited at 180 °C on a polyester substrate. The charge carrier mobility was much greater than the results achieved in this work. It is assumed that a bad contact between the pentacene molecules and the bottom drain and source contacts, observed in the nonlinear behaviour of the output characteristic in Figure 18.10, is responsible for the behaviour observed in this article. Another cause can be found in the pentacene layer itself. The AFM investigations of an evaporated 30 nm thick pentacene film did not allow statements about the first three monolayers on the substrate material, which are believed to be crucial for the charge carrier transport in the semiconducting layer [28].

Additionally, the lift-off structured drain and source electrodes have been examined by scanning electron microscopy (SEM) and AFM. Instead of the

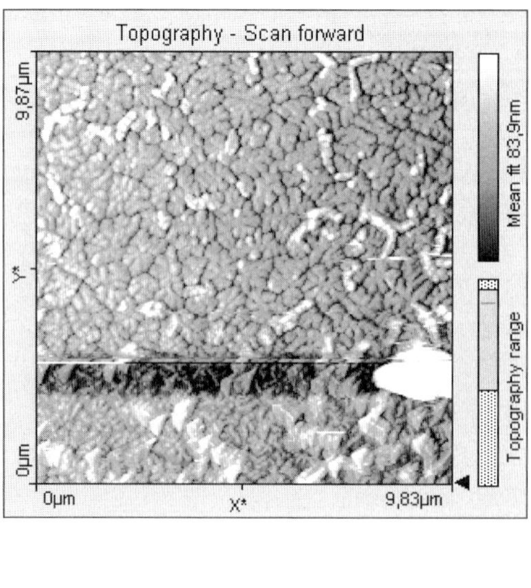

Figure 18.11 AFM scan of a pentacene film on a polyimide film as gate dielectric in the transistor channel of the introduced transistor. Below the black line, one electrode of the drain and source contacts can be observed.

required perpendicular edges, the pictures showed slightly outwardly inclined edges that shadow the lower areas of the contacts and inhibit a good adsorption of the pentacene molecules. This problem can mainly be observed at contacts deposited at high conformity. Because of this the perpendicular edges of the masking photoresist layer will be covered as well as the horizontal areas during the metal deposition. At the following lift-off step the edges at the upper side of the contacts did not pull down into the whole, remained at the top face of the contact (see Figure 18.12a) and jutted out over the bottom parts of the contacts, which are crucial for the charge carrier injection in the semiconducting film. For future contact electrodes structured by UV lithography and lift-off, a deposition process at low conformity can be recommended (Figure 18.12b), resulting in approximately ideal contacts with perpendicular edges.

In summary, a transistor on a polyimide film as gate dielectric has been shown that can drive drain currents of about −15.8 μA at $V_{DS} = -40$ V and $V_{GS} = -40$ V. The size of the pentacene crystallites reached a diameter of up to 1 μm. However, the large crystallite size transistors exhibit only small charge carrier mobilities and bad output characteristics. During the investigations there was shown a lack of electrical contact between the organic molecules and the drain and source electrodes due to the processing of the contacts. It is believed that the contacts have a crucial influence on the transistor characteristics, also shown in [44]. In spite of promising results in the growth of the pentacene molecules, another gate dielectric is required, coated at process temperatures below 100 °C, to be deposited on polymer substrates.

18.3.2.1 **Bectron® Varnish**

Differently to transistors introduced up to now, another transistor architecture was used for the investigations of the varnish Bectron® as gate dielectric. Instead of bottom drain and source contacts, electron beam evaporated contacts of titanium have been formed by a shadow mask, resulting in a channel length of about 125 μm and a channel width of 2800 μm. The application of a shadow mask demands a metal deposition at low conformity to avoid coating below masked areas. At the date of those tests there was no possibility in the laboratory available for thermal evaporation of gold. Therefore, another metal layer was required to be deposited onto the semiconducting film. In this case titanium was selected that was evaporated by an electron beam in high vacuum because the material was applied by several groups as an adhesion promoter for gold layers on the gate dielectric [28, 29].

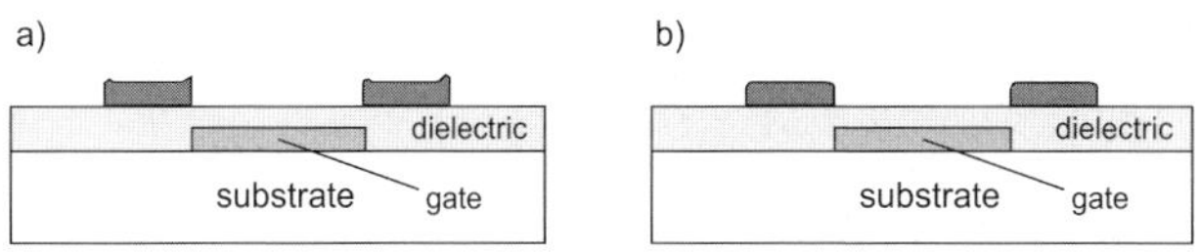

Figure 18.12 a) Simplified model of lift-off structured contacts where the metal layer is coated at high conformity. b) Same model for a metallisation deposited by low conformity.

Such a transistor device can drive a drain–source current of –17.1 nA at $V_{GS} = -40$ V and $V_{DS} = -80$ V ($W/L = 22.4$). The output characteristic did not show good behaviour, possibly due to a large charge carrier barrier, caused by the difference of the work function of the titanium (4.3 eV [30]) and the HOMO (highest occupied molecule orbital) of the pentacene film (4.8 eV [31]) that are not well adapted. Au, Pd or Pt would fit much better [32].

Furthermore, there could have been negative effects on the OFETs by the deposition of the metal layers. In [33] a diffusion of metal atoms into the organic semiconductor or a formation of a mixing layer on the rough surface of the pentacene film was reported. This can excite an additional shift of the Fermi level that need not shift in direction to improve the charge carrier injection.

Moreover, thermal effects could have an impact on the organic semiconductor due to a temperature of more than 80 °C during the deposition process of titanium as a consequence of thermal radiation. It could not be ruled out that this process step entails a premature degradation.

The dielectric resist only offers superior electrical insulation at sufficient layer thicknesses. Required thicknesses of 1 μm and more seem to be too large with a view to reducing the operating voltages. Because of this and the bad electrical transistor characteristics, the insulating resist is not well suited as the gate dielectric in OFETs with pentacene. This is why in the next paragraph a high-*k* resist will be examined.

18.3.2.2 **High-*k* Resist**

The high-*k* resist, based on hydrolysed and partially condensed ethyl silicates, offers the opportunity of deposition at process temperatures below 80 °C, resulting in a well-insulating film. Its permittivity can be adjusted by adding inorganic components. In this way advantages of mechanically flexible polymer coatings and the high permittivity of inorganic dielectric materials are interconnected (see also Table 18.2 for achieved dielectric constants).

In Figure 18.13 there is shown an output characteristic of an OFET integrated on a 200 nm thick high-*k* resist, using titanium dioxide particles to obtain a dielectric constant of 12. The channel width of the device was 1000 μm and the channel length was 10 μm. A pentacene film of 30 nm thickness was evaporated in high vacuum at a chamber pressure of about 8×10^{-7} mbar at low deposition rate. It could drive a drain current of –27 μA at $V_{DS} = -40$ V and $V_{GS} = -35$ V ($W/L = 100$). The charge carrier mobility in the saturation regime is about 0.022 cm²/V s, the on–off ratio is greater than 10^2, the threshold voltage is about 7.4 V and the threshold slope is 1.33 V/dec.

In the saturation regime in Figure 18.13 an overshoot in the drain current at high gate–source voltages can be observed, resulting in a decrease of the drain current at increasing values of the drain–source voltage. This decrease is believed to be a result of a further increase of the drain–source voltage, well known by conventional MOS technology at power devices in high-current operation [34]. At high power demand there is a thermal back-coupling that reduces the charge carrier mobility and the drain current.

Table 18.2 Different investigated polymer gate dielectrics.

dielectric	process temperature	thickness	permittivity
polyimide (PI254)	350 °C	190 nm	3.5
Bectron® varnish	100 °C	1–1.5 μm	3.5
high-*k* resist	80 °C	200–300 nm	9–12

To investigate the influence of the permittivity on the transistor characteristics a transistor was produced on a resist film with a smaller content of titanium dioxide in the resist. Its dielectric constant was 9.5 and the layer thickness was about 300 nm. This transistor could drive drain currents of up to –65 μA at $V_{DS} = V_{GS} = -40$ V ($W/L = 100$). Although the maximum drain current was larger than in Figure 18.13, the charge carrier mobility in the saturation regime was 0.02 cm²/V s also. In contrast to investigations in [35, 36], a dependency between the charge carrier mobility and the dielectric constant during these experiments could not be observed.

18.3.2.3 **OFET on Foil Substrates**

The initial promising results and low process temperatures demand an investigation of this high-*k* material on foil substrates. Therefore, a polyester film was cut to a diameter of 100 mm. After several cleaning steps in deionised water and acetone a 150 nm thick aluminium layer was deposited by electron-beam evaporation for the transistor's gate electrodes. Next, a 400 nm thick insulating resist film was spin coated and cross linked by an additional UV step, as described in Section 18.2.3. Drain and source electrodes were structured by UV lithography and lift-off in acetone. Before evaporating a 30 nm thick pentacene film in high-vacuum conditions, the surfaces of the gate dielectric and the

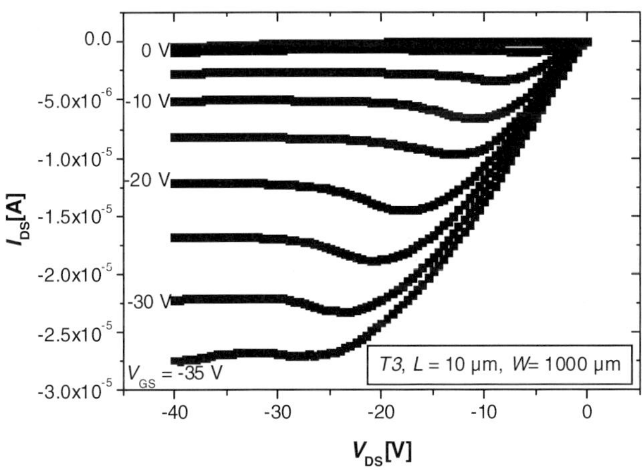

Figure 18.13 Output characteristic of an OFET with 10 μm channel length and 1000 μm channel width, using a 200 nm thick high-epsilon resist film as gate dielectric.

drain and source electrodes were pretreated in an oxygen plasma in order to improve the molecular order of the pentacene film.

Figure 18.14 shows the characteristics of a resulting transistor device. The channel dimensions were 1000 μm in width and 10 μm in length. The maximum drain current is about −36.7 μA at $V_{DS} = -40$ V and $V_{GS} = -40$ V ($W/L = 100$). From the transfer characteristic in saturation the on–off ratio was determined to be more than 10^3, the threshold voltage was −7.5 and the charge carrier mobility in the saturation regime could be calculated to be 0.35 cm^2/V s. The high charge carrier mobility could be due to a high crystalline order but also a lower contact resistance. At pretreated samples on this dielectric film, using a silicon substrate, a crystallite size of about 1.8 μm in diameter could be aimed at, but the charge carrier mobility of those devices was much smaller. The estimated mobility was possibly reduced by a charge carrier injection barrier, which was observed in the nonlinear behaviour in the linear regime of those output characteristics.

18.3.2.4 **Conclusion**

Transistor characteristics on the polymer dielectric films such as polyimide, Bectron® varnish and a high-epsilon resist have been demonstrated. On the polyimide film, pentacene crystallites reach diameters comparable to those on silicon dioxide. With a view to the application in polymer electronics, a major disadvantage of this material is the high cross-linking temperature that is not suitable for polymer films such as polyester. On the part of the cross-linking temperature the varnish Bectron® could be an alternative, but the not so good transistor characteristics that were achieved led to the conclusion of a bad molecular order in the organic semiconductor. Additionally, well-insulating films result only at thicknesses of more than 1 μm.

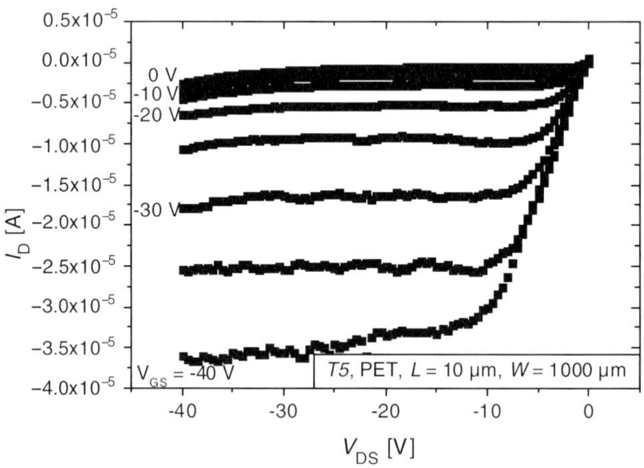

Figure 18.14 Output characteristic of a transistor with a channel length of 10 μm and a channel width of 1000 μm on a polyester substrate.

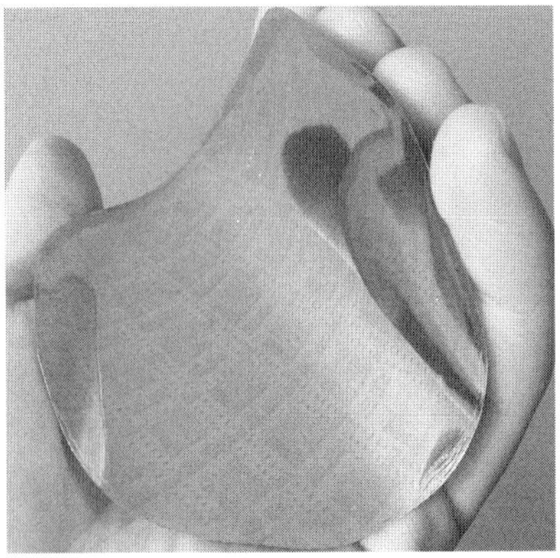

Figure 18.15 OFETs on a polyester substrate, using high-epsilon resist as gate dielectric.

The third investigated material, a high-*k* resist, showed the best results at high dielectric constant. Its coating temperature is ideal for the application on foil substrates. The material offers the advantage to set up the dielectric constant by the added amount of inorganic components. Besides transistors on silicon substrates, transistors were built on polyester substrates using this resist as gate dielectric (see Figure 18.15). On that material, pentacene crystallites achieved more than 1.8 μm in diameter and transistors reached a charge carrier mobility of about 0.35 cm^2/V s.

18.4 Degradation

Organic semiconductors suffer from lack of stability due to reactive gases such as oxygen [45] and water vapour [46] that can be found in ambient atmosphere. In order to investigate the degradation of OFETs the well-analysed architecture of bottom gate and bottom drain and source contacts on a silicon substrate with silicon dioxide as gate dielectric was used. Therefore, a 110 nm thick silicon dioxide layer was thermally grown on the silicon substrate. In the next step, bottom drain and source electrodes of gold were structured. Finally, a 58 nm thick pentacene film was evaporated at a deposition rate of 0.1 nm under high-vacuum conditions. The details of the preparation can be found elsewhere [37]. In order to evaluate the degree of degradation an unprotected OFET was measured directly after the deposition and then it was stored under ambient dark atmosphere in a shielded metal box.

Over periods of three months the device was measured again, up to nine months. During this time interval the drain current decreased from −61 μA directly after the deposition to a value of −187 nA (see also Table 18.3) ($W/L = 1000$). The channel length of this transistor was 1 μm and the channel width was 1000 μm. The mean charge carrier mobility in the linear regime declined from 2×10^{-3} cm^2/V s to 1.2×10^{-5} cm^2/V s and the threshold voltage shifted in the negative direction from 4.8 V to −8 V. Recently, a comparable

Table 18.3 Degradation of a non-encapsulated OFET, stored for a period of nine months under dark ambient atmosphere.

	I_{DS} (μA)	V_{th} (V)	μ (cm^2/V_s)
0 months	−60.9	4.8	2.0×10^{-3}
3 months	−6.8	2.3	2.4×10^{-4}
6 months	−0.457	−3.4	2.2×10^{-5}
9 months	−0.187	−8	1.2×10^{-5}

behaviour of the threshold voltage and the charge carrier mobility has been observed [48]. The cause of the shift in the threshold voltage will be discussed later in this text. At that moment it was not possible to distinguish whether the influence on the threshold voltage is dominated by water vapour or oxygen.

Therefore, further experiments have been done to differentiate between the particular gases. A chip containing an OFET of 1000 μm channel width and of 1 μm channel length was bonded, connected to the measurement setup and put into a vacuum chamber. Then, an approximately 30 nm thick layer was thermally evaporated at low deposition rate and a process pressure of 7×10^{-7} mbar. Next, the sample was characterised in high vacuum before the chamber was flooded by pure technical oxygen so that the OFET did not come into contact with the ambient atmosphere. The device was measured a second time before the chamber was evacuated again.

Regarding the output characteristic, the drain current in the saturation regime decreased from −15.9 μA to −6.4 μA. In Figure 18.16 there are presented the transfer characteristics of the OFET measured at high vacuum and in pure technical oxygen atmosphere. They confirm the decrease in the on-current by a diminished on–off ratio (1000–500), while the off-current remained at the same level. Moreover, a shift of the threshold voltage from −2.6 V to −0.7 V in the positive direction could be observed.

During these measurements the influence of water can be excluded. Throughout, the threshold shift was caused by the influence of oxygen, which is well known to connect at the middle ring of the pentacene molecule, the most reactive position [38]. Forming a new molecule, acceptor states to electrons were introduced and made available negative charges at the grain boundaries and the dielectric surface [39], resulting in a shift of the threshold voltage. So, the formation of scattering centres by the oxidised pentacene seems to be responsible for the reduction of the on-current.

The influence of the oxygen causes a shift of the threshold voltage in the opposite direction as observed by long term degradation experiments, where a detected shift in the negative direction was believed to be caused by water vapour [47] in ambient atmosphere. Therefore, further samples had to be investigated, exposed to atmospheres with defined air humidity.

In the literature there are several reports on the degradation of OFETs and the effects on their electrical characteristics due to water vapour (e.g. [49–51]). The reaction of pentacene single crystals with water molecules from am-

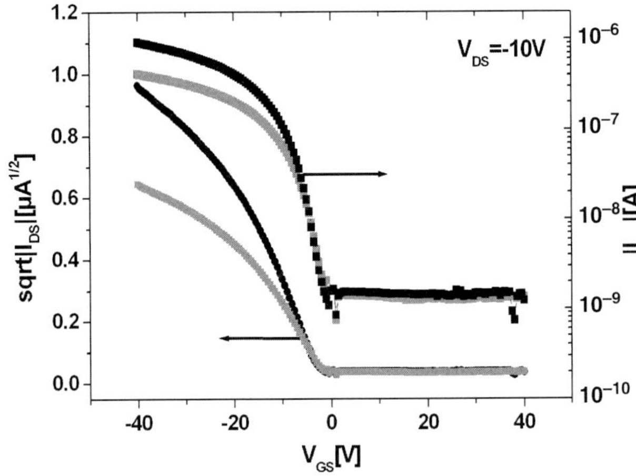

Figure 18.16 Transfer characteristic of an OFET of 1 μm channel length and 1000 μm channel width, measured at 7×10^{-7} mbar (black curves) and in oxygen atmosphere at 1000 mbar (red curves).

ambient air is sufficiently known [52, 54]. Water molecules diffuse into the organic semiconductor and reduce the on-current. The decrease in the current could be explained by chemical reaction due to an electrical field [40]. Because of the applied gate voltage, water molecules can dissociate and can attach themselves at the pentacene molecules. Due to covalent bonds, the H^+ and the OH^- ions form defects in the pentacene crystal and act as traps for the positive charge carriers in the pentacene. In this way the interface between the gate dielectric and the semiconductor is polarised and shifts the threshold voltage in the negative direction, as shown in Figure 18.17. This is why trapped positive charges act as mirror charges, decrease the electrical gate field and require a high gate voltage to accumulate charge carriers in the semiconducting transistor channel [40]. Through the sign of a permanent stress voltage the direction of the shift in the threshold voltage can be influenced by an effect on the H^+ and the OH^- ions [41].

For OFETs, stored at dark ambient atmosphere for a longer period of time without an applied bias field, the threshold shifted in the negative direction [42]. Regarding Table 18.3 and Figure 18.4, the influence of water seems to dominate the degradation of the OFETs in comparison with an oxygen atmosphere.

This is the reason for a suitable protection of the organic material to reduce the degradation of the organic semiconductor. While in [53] a capping layer was spin coated, in this report the possibility to encapsulate the organic material by a capping layer directly in high vacuum should be shown. In future this proceeding offers the opportunity to avoid contact to ambient air of the evaporated organic semiconductor.

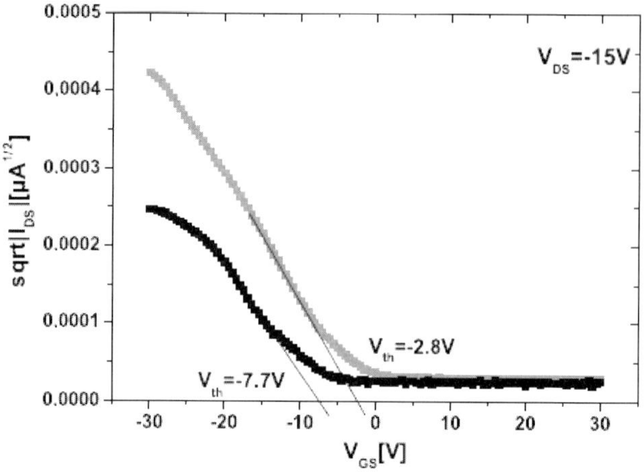

Figure 18.17 Transfer characteristic of an OFET of 1 μm channel length and 1000 μm channel width, measured at low (red curve) and at high air humidity (black curve).

Besides the deposition of the capping layer in vacuum, it should have the following characteristics:

- deposition at low process temperatures to minimise thermal stress to the organic film,
- low deposition rates in order to avoid a destruction of the sensitive layer,
- hydrophobic characteristics to reduce the inclusion of water molecules,
- deposition of a dense film to avoid penetration of oxygen.

In this report polytetrafluoroethylene (PTFE), well known as Teflon, was selected. This material can be deposited by cathode sputtering in high vacuum onto the OFET device (see also Figure 18.18). Moreover, it can be used as a charged electret layer, which may be used to set the threshold voltage to a defined value [43]. As there is a dependence on the launched power during the sputter process and the substrate temperature, the degree of the cross linking can be adjusted, resulting in different densities of the polymer capping layer. Additionally, by co-sputtering of for example titanium dioxide, the permittivity could be set to a defined value, which could be important for use as a gate dielectric for top gate transistors.

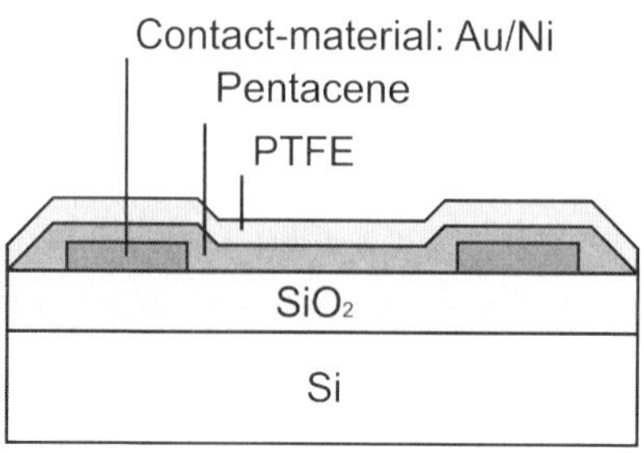

Figure 18.18 Bottom gate and bottom drain and source OFET on silicon substrate with a capping layer of PTFE.

An analysed transistor was produced on silicon and thermally grown silicon dioxide as the gate dielectric where the layer thickness was 127 nm. The pentacene film was evaporated in high vacuum at 6×10^{-7} mbar at a deposition rate of 0.1 nm/s. During the deposition the substrate temperature was 25 °C. The device with interdigitated drain and source contacts of 46.8 cm channel width and of 20 μm channel length, characterised at ambient atmosphere, could drive drain currents of about −6.8 mA directly after the pentacene deposition under ambient atmosphere ($W/L = 23400$). The on–off ratio reached a value of 10^4 and the threshold voltage was 12.3 V. After that the sample was encapsulated by a 1.5 μm thick Teflon layer. To avoid damage of the organic semiconductor, a low deposition rate of 21 nm/min was selected by a thickness gradient of less than 10% [19].

To analyse the degradation of the sample it was characterised directly after the Teflon deposition and after a gap of one month. Meanwhile, the device was stored in a dark ambient atmosphere in a shielded metal box like the OFET presented in Table 18.3. Just after the deposition of the capping layer, there could be a small decline in the on-current at −40 V gate–source voltage observed, as shown in Figure 18.19.

During the following two months the drain current decreased dramatically, while in the two following months a reduction in the decrease could be detected. In comparison with an unprotected OFET with a decrease in the on-current of up to 10%, during the first three months the encapsulation reduced the on-current to 25% of the start value (see Table 18.4).

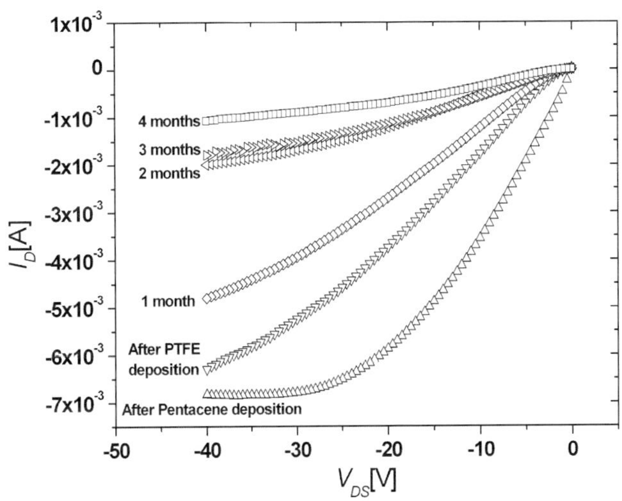

Figure 18.19 Output characteristic of an encapsulated OFET of 20 μm channel length and 46.8 cm channel width at −40 V gate–source voltage, measured at different points in time in order to analyse its degradation.

Table 18.4 Electrical parameters of an OFET with PTFE capping layer, characterised after storage for different periods of time at dark ambient atmosphere.

	I_{DS} (mA)	V_{th} (V)	μ (cm^2/V s)
before PTFE deposition	−6.8	12.3	8.3×10^{-3}
after PTFE deposition	−6.2	13.2	7.3×10^{-4}
1 month	−4.7	18.2	4.8×10^{-4}
2 months	−1.9	–	–
3 months	−1.7	18.9	1.7×10^{-4}
4 months	−1.1	23.2	1.0×10^{-4}

The observed effect was fed back to the influence of degradation mechanisms where the device was stored without doing anything. The exposure of the organic film to air humidity could be excluded because of the hydrophobic characteristics of the PTFE film. However, the non-polar oxygen molecules could still diffuse through the thick capping layer and affect the degradation. This assumption was improved due to the shift of the threshold voltage in the positive direction. If water vapour would have caused the degradation, the shift of the threshold voltage would have had to go in the negative direction, as explained in the context of Figure 18.17.

In conclusion, we have analysed degradation mechanisms due to oxygen and water vapour. An exposure of the unencapsulated OFET to water vapour caused a shift of the threshold voltage in the negative direction, whereas the threshold voltage was shifted to positive values because of the influence of oxygen. Through a capping layer of PTFE, the degradation due to water vapour could be clearly decreased, while some oxygen still diffuses through the encapsulation film.

18.5
Conclusion

The influence of the dielectric layer material on the characteristics of organic field effect transistors with the organic molecule pentacene has been shown. Inorganic insulators have been investigated using silicon substrates due to their high deposition temperature. Because of the deposition temperature and the well-known surface, silicon dioxide layers are predestined for growth studies of the organic material or degradation mechanism. For the application on polymer foils, they are not convenient. In this case, polymer insulators have been investigated as gate dielectrics. Promising OFETs on foil substrates have been built applying a high-k resist, reaching charge carrier mobilities of up to 0.35 cm^2/V s. Furthermore, a polymer capping layer has been shown to inhibit the degradation due to water vapour and to reduce the degradation owing to atmospheric oxygen.

Acknowledgements

The authors gratefully acknowledge Prof. Faupel and coworkers for sputtering PTFE layers, the IPMS Fraunhofer Dresden for supplying the purified pentacene and the DFG for financial support in the framework of the OFET-Schwerpunktprogramm for Contract No. Hi 551/7-3.

References

1. H. Klauk, M. Halik, U. Zschieschang, F. Eder, G. Schmid, and C. Dehm, Appl. Phys. Lett. **82**, 4175 (2003).
2. P. F. Baude, D. A. Ender, M. A. Haase, T. W. Kelley, D. V. Muyres, and S. D. Thiess, Appl. Phys. Lett. **82**, 3964 (2003)
3. F. Eder, H. Klauk, M. Halik, U. Zschieschang, G. Schmid, and C. Dehm, Appl. Phys. Lett. **84**, 2673 (2004).
4. C. Pannemann, T. Diekmann, and U. Hilleringmann, Kleinheubacher Berichte (2002), pp. 219–221, ISSN 1684–9965.
5. C. D. Sheraw, L. Zhou, J. R. Huang, D. J. Gundlach, T. N. Jackson, M. G. Kane, I. G. Hill, M. S. Hammond, J. Campi, B. K. Grenning, J. Francl, and J. West, Appl. Phys. Lett. **80**, 1088 (2002).
6. HD Microsystems, Product Bulletin PI-2545 (2005).
7. HD Microsystems, Product Information VM651 (2005).
8. Th. Diekmann, Ch. Pannemann, and U. Hilleringmann, Conf. Proc. Technologies for Polymer Electronics (TPE 06), Rudolstadt, Germany, 2006, pp. 148–152.
9. H. Klauk, D. J. Gundlach, J. A. Nichols, and T. N. Jackson, IEEE Trans. Electron Devices **46**, 1258 (1999).
10. R. Ruiz, B. Nickel, N. Koch, L. C. Feldman, R. F. Haglund, A. Kahn, and G. Scoles, Phys. Rev. B **67**, 125406 (2003).
11. S. Steudel, S. De Vusser, S. De Jonge, D. Janssen, S. Verlaak, J. Genoe, and P. Heremans, Appl. Phys. Lett. **85**, 4400 (2004).
12. Y.-Y. Lin, D. J. Gundlach, S. F. Nelson, and T. N. Jackson, IEEE Trans. Electron Devices **44**, 1325 (1997).
13. Y.-Y. Lin, D. J. Gundlach, S. F. Nelson, and T. N. Jackson, IEEE Electron Device Lett. **18**, 606 (1997).
14. C. D. Dimitrakopoulos, S. Purushithaman, J. Kymissis, A. Callegari, and J. M. Shaw, Science **283**, 822 (1999).
15. D. Knipp, R. A. Street, B. Krusor, R. Apte, and J. Ho, J. Non-Cryst. Solids **299–302**, 1042 (2002).
16. A. Berni, Inoflex Lacke, Technical data via mail, inomat GmbH, Bexbach (March 2007).
17. S. Steudel, S. De Vusser, S. de Jonge, D. Janssen, S. Verlaak, J. Genoe, and P. Heremans, Appl. Phys. Lett. **85**, 4400 (2004).
18. H. Klauk and T. N. Jackson, Solid State Technol. **43**, 63 (2000).
19. C. Pannemann, T. Diekmann, U. Hilleringmann, U. Schürmann, M. Scharnberg, V. Zaporojtchenko, R. Adelung, and F. Faupel, IEEE Polytronic Conference (2005), p. 63.
20. D. J. Gundlach, C.-C. Kuo, S. F. Nelson, and T. N. Jackson, 57th Annual Device Research Conference (1999), p. 164.
21. M. W. Lee and C. K. Song, Jpn. J. Appl. Phys. **42**, 4218 (2003).
22. D. Knipp, R. A. Street, B. Krusor, R. Apte, and J. Ho, J. Non-Cryst. Solids **299**, 1042 (2002).
23. Y. Liang, G. Dong, Y. Hu, L. Wang, and Y. Qiu, Appl. Phys. Lett. **86**, 132101 (2005).

24. Y. Iino, Y. Inoue, Y. Fujisaki, H. Sato, M. Kawakita, S. Tokito, and H. Kikuchi, Jpn. J. Appl. Phys. **42**, 299 (2003).
25. C. Pannemann, Prozesstechnik für organische Feldeffekt-Transistoren mit Pentacen: Kontakte, Dielektrika und Oberflächenpassivierung, Ph.D. thesis, University of Paderborn (2006).
26. D. Gamota, P. Brazis et al., IEEE Standard Test Methods for the Characterization of Organic Transistors and Materials (The Institute of Electrical and Electronic Engineers, New York, 2004).
27. Y. Kato, S. Iba, R. Teramoto, T. Sekitani, T. Someya, H. Kawaguchi, and T. Saruai, Appl. Phys. Lett. **84**, 3789 (2004).
28. N. Yoneya, M. Noda, N. Hirai, K. Nomoto, M. Wada, and J. Kasahara, Appl. Phys. Lett. **85**, 4663 (2004).
29. A. Schoonfeld, Transistors Based on Ordered Organic Semiconductors, Ph.D. thesis, University of Groningen (1999).
30. H. Ibach and H. Lüth, Festkörperphysik (Springer, Berlin, 1995)
31. T. Minakata, I. Nagoya, and M. Ozaki, J. Appl. Phys. **69**, 7354 (1993).
32. D. J. Gundlach, J. A. Nichols, C. C. Kuo, H. Klauk, C. D. Sheraw, T. N. Jackson, and D. G. Schlom, 41st Electronic Materials Conf. Dig. (1999), p. 16.
33. A. Kahn, N. Koch, and W. Gao, J. Polym. Sci. B, Polym. Phys. **41**, 2529 (2003).
34. R. Paul, MOS-Feldeffekttransistoren (Springer, Berlin, 1994).
35. J. Veres, S. Ogier, G. Lloyd, and D. de Leeuw, Chem. Mater. **16**, 4543 (2004).
36. D. Zielke, A. C. Hübler, U. Hahn, N. Brandt, M. Bartzsch, U. Fügmann, and T. Fischer, Appl. Phys. Lett. **87**, 123508 (2005).
37. Ch. Pannemann, T. Diekmann, and U. Hilleringmann, Proc. 16th Int. IEEE Conf. Microelectronics, ICM (2004), p. 79.
38. O. D. Jurchescu, J. Baas, and T. T. M. Palstra, Appl. Phys. Lett. **84**, 3061 (2004).
39. S. Verlaak, V. Arkhipov, and P. Heremans, Appl. Phys. Lett. **82**, 745 (2003).
40. J. E. Northrup and M. L. Chabinyc, Phys. Rev. B **68**, 041202 (2003).
41. D. Knipp, R. A. Street, A. Vökel, and J. Ho, J. Appl. Phys. **93**, 347 (2003).
42. S. J. Zilker, C. Detcheverry, E. Cantatore, and D. M. de Leeuw, Appl. Phys. Lett. **79**, 1124 (2001).
43. M. Scharnberg, V. Zaporojtchenko, R. Adelung, F. Faupel, C. Pannemann, T. Diekmann, and U. Hilleringmann, Appl. Phys. Lett. **90**, 013501 (2007).
44. B. H. Hamadani and D. Natelson, J. Appl. Phys. **97**, 064508 (2005).
45. O. D. Jurchescu, J. Baas, and T. T. M. Palstra, Appl. Phys. Lett. **85**, 3061 (2004).
46. T. N. Jackson, Y. Y. Lin, D. J. Gundlach, and H. Klauk, IEEE J. Sel. Top. Quantum Electron. **4**, 100 (1998).
47. Z.-T. Zhu, J. T. Mason, R. Dieckmann, and G. G. Malliaras, Appl. Phys. Lett. **81**, 4643 (2002).
48. S. Cipolloni, L. Mariucci, A. Valetta, D. Simeone, F. De Angelis, and G. Fortunato, Thin Solid Films **515**, 7546 (2007).
49. D. Li, E.-J. Borkent, R. Nortrup, H. Moon, H. Katz, and Z. Bao, Appl. Phys. Lett. **86**, 042105 (2005).
50. C. Goldmann, D. J. Gundlach, and B. Batlog, Appl. Phys. Lett. **88**, 063501 (2006).
51. M. L. Chabinyc, F. Endicott, B. D. Vogt, D. M. DeLongchamp, and E. K. Lin, Appl. Phys. Lett. **88**, 113514 (2006).
52. O. D. Jurchescu, J. Baas, and T. T. M. Palstra, Appl. Phys. Lett. **87**, 052102 (2005).
53. S. H. Han, J. H. Kim, J. Jang, S. M. Cho, M. H. Oh, S. H. Lee, and D. J. Choo, Appl. Phys. Lett. **88**, 073519 (2006).
54. R. Scholz, D. Lehmann, A.-D. Müller, F. Müller, and D. R. T. Zahn, Phys. Stat. Sol. A **205**, 591 (2008).

19
Influence of Metal Diffusion on the Electronic Properties of Pentacene and Diindenoperylene Thin Films

M. Scharnberg, R. Adelung, and F. Faupel

19.1
Introduction

In recent years there has been a major research effort in the field of organic electronics, i.e. organic light emitting diodes (OLEDs) [1], organic solar cells [2] and organic field effect transistors (OFETs) [3, 34]. But whereas OLEDs are already implemented in commercial products, for OFETs a number of important problems need to be solved before they can be implemented. Metallisation of OFETs can lead to metal diffusion into the organic semiconductor thin film in certain OFET structures, affecting the electronic properties. Understanding of the interface formation between the metal and the organic thin film as well as of the diffusion behaviour can help to improve the processing of OFETs. With better control of the interface formation the device performance of the FET can be improved [4].

First, noble metal diffusion in organic crystalline semiconductors is studied and a method to control the diffusion is tested. Therefore, two organic films, pentacene (Pc) and diindenoperylene (DIP) are used as substrates and Au and Ag are used as top contact materials. DIP was chosen for the diffusion studies because it not only possesses good charge carrier mobilities but its microstructure can be controlled precisely, ranging from amorphous films to crystalline films in which almost all molecules are aligned in parallel. This allows studying the influence of the film microstructure (e.g., grain boundaries) on the diffusion of noble metal atoms in organic crystalline semiconductors. Pc was selected for the diffusion studies because it shows mobilities that are in the range of amorphous silicon (a-Si:H). It is one of the most intensively studied organic semiconductors due to these high reported mobilities and active matrix liquid crystal displays (AMLCDs) using Pc field effect transistors have already been demonstrated [5], which makes it directly relevant for technical applications.

More importantly, the influence of the metal diffusion on the electronic properties of contacts in Pc field effect transistors is examined. Besides the influence of the metal layers, the influence of Teflon-based functional organic thin films was studied. They can improve the performance of organic field ef-

Organic Electronics. Structural and Electronic Properties of OFETs. Edited by Christof Wöll

ISBN: 978-3-527-40810-8

fect transistors by acting as one part of a protective multilayer against aging and as electret gate layer for control of the threshold voltage of the transistors.

19.2 Experimental

19.2.1 Organic Semiconductors

Two molecular crystalline organic thin films were used in this study, one is diindenoperylene (DIP) and the other is pentacene (Pc) (see Figure 19.1).

Unpatterned DIP films with a high degree of structural order [6] were grown on flat SiO_2 substrates under HV conditions at a pressure of 10^{-7} mbar by the group of Pflaum at the University of Stuttgart. Purified DIP was sublimed at a rate of 0.02 nm/s with the substrate temperature kept at 130 °C. The total thickness of the organic layer is about 90 nm as was estimated from post growth of X-ray diffraction (XRD) studies of these films.

The Pc samples were provided by two different groups. The group of Hilleringmann at the University of Paderborn, fabricated bottom contact Pc-FETs. The OFETs were prepared on 100 mm diameter p-type silicon wafers providing a thick thermally grown SiO_2 layer in the non-transistor areas and a 150 nm thin thermally grown SiO_2 dielectric layer in the active channel region. The pentacene that was obtained from Aldrich 99.5% was used as purchased, without further purification. The organic semiconductor was evaporated in high vacuum (HV) 6×10^{-7} mbar on a heated substrate (60 °C). The deposition rate was adjusted to 0.2 nm/s and the film reached a nominal total thickness of 58 nm. The electrical properties can be found in [7]. These transistors were used to study the influence of Teflon capping layers.

The other films were produced to study the influence of the interface between the metal top contact and the organic semiconductor. Here, silicon wafers with a native oxide were used as substrates. The samples were provided by the group of Pflaum at the University of Stuttgart. To minimise impurities, like in the case of DIP, Pc (purchased from Fluka) is purified twice by gradient sublimation before being used as starting material. The films were prepared in UHV at a base pressure of about 7×10^{-10} mbar from a graphite effusion cell. The evaporation rate was about 3 Å/s; it was controlled by a quartz microbalance located next to the sample [8].

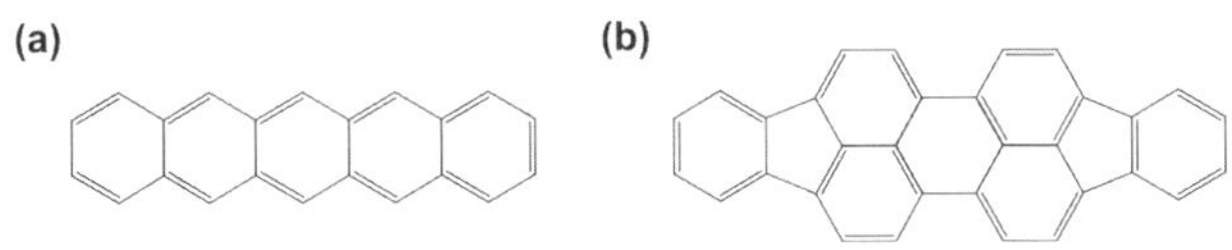

Figure 19.1 Chemical structure of (a) pentacene (Pc) and (b) diindenoperylene (DIP).

19.2.2
Thin Film Deposition

In order to study the diffusion of metal atoms in the organic films, Ag containing radioactive ^{110m}Ag tracer atoms was evaporated at a crucible temperature of 680 °C at a base chamber pressure of 1.8×10^{-8} mbar. During evaporation the pressure increased to 10^{-7} mbar. The Au tracers were evaporated at 860 °C at a base pressure of 1.8×10^{-8} mbar.

The source and drain contacts of the examined OFETs were deposited by thermal evaporation of Au as described above for deposition of the radiotracers. Deposition of the contacts was not performed in the same chamber as the radiotracer deposition in order to avoid contamination of the sample with radioactive isotopes. Patterning of the contact structures was obtained using a stainless steel shadow mask. By deposition of Au an array of nine contacts was formed. The contact area of the Au was 50 × 50 μm and the distance between the contacts varied from 300 μm to 3290 μm. Three Au contact arrays with a thickness of 50 nm were deposited onto a 40 nm Pc film at a substrate temperature of 75 °C. The first contact array (Array 1 in the following) was deposited at a rate of 0.8 nm/min. For the second set of contacts (Array 2 in the following) first a submonolayer of Au was deposited very slowly (<1 ML/h) on top of the Pc film in order to allow strong diffusion. Afterwards, the contacts were deposited at the same rate of 0.8 nm/min as the first set. The third array (Array 3 in the following) was deposited at 0.8 nm/min with the substrate at room temperature.

The Teflon AF films for the electrets were deposited by thermal evaporation. Other than the thermal evaporation of metals, in general, polymer evaporation is a more complicated process. A special property of the here used Teflon AF is the fact that upon annealing, cleaving of the bonds between two dioxole fragments in the backbone between the monomers is preferred and the created free radical is stabilised by interaction with the adjacent oxygen atom [9]. During thermal evaporation single monomers are therefore evaporated and these monomers polymerise on the substrate surface. While the chemical structure of Teflon AF is not changed upon deposition only the molecular weight decreases resulting in a lower glass transition temperature T_g.

The polymer films were charged using a corona discharge. Between a fine tungsten tip and a steel grid a 5 kV AC voltage is applied. Due to the high density of the electric field near the tip, molecules from the ambient air are ionised. The DC voltage that is applied between the grid and the sample allows the extraction of ions from the discharge region. By choosing the polarity of the field either positive or negative ions can be extracted. In air, the negative ions are mainly CO^{-3} of thermal energy and the positive charge carriers are $(H_2O)_nH^+$ [10]. The sample is charged by ions until the field created by the charges in the sample compensates the extraction voltage. For Teflon it was found that the charges penetrate just a little or not at all into the sample [10]. This is important for the desired application in electret-based dual-gate OTFTs as the charge should not penetrate into the TFT, where they might affect the device performance.

19.2.3 Radiotracer Measurements

The radiotracer technique has been used to study diffusion in various systems including metals, metallic glasses and also polymers [11, 12]. It uses radioactive isotopes of a given element in order to study its diffusion in the material of interest. In order to achieve that, the tracers must be deposited onto the substrate in which diffusion is to be studied. Depending on the tracer/substrate system, annealing might be necessary after the tracer deposition in order to facilitate and accelerate diffusion. Depth profiles, i.e., the tracer concentration with respect to the distance from the film surface, allow the examination of the diffusion of the metal atoms into the organic film during deposition. In order to obtain the depth profiles, serial sectioning of the sample is performed. During serial sectioning the sample is removed in layers by methods like grinding (the different layers are collected by changing the grinding paper) or, as in this work, by ion beam sputtering. In the case of ion beam sputtering, the material that is sputtered off the surface is caught on a polymer film. The film is installed in an apparatus that is similar in design to a film camera. A part of the polymer film is exposed to the material that is sputtered off from the substrate. After a layer of a certain thickness is sputtered off the sample, the polymer film is advanced exposing a new piece of the film to the sputtered material. In this way, sections of the substrate can be caught on different pieces of the film. The activity of the sections can be determined by detection of the radioactive decay of the tracer atoms in the film pieces. Through normalisation of the activity in a section with respect to the thickness of the section a depth profile can be constructed. The normalised activity is proportional to the metal concentration of the section. In this work, two types of radiotracers were used. The first, silver (Ag) containing a fraction of 1.5×10^{-4} ^{110m}Ag radio-isotopes, was used in order to compare the results with those obtained for polymers. This allows examining if the diffusion mechanisms for Ag atoms which are well understood for polymers also apply to organic crystalline materials [13]. ^{110m}Ag is a metastable state of the ^{110}Ag isotope and decays to ^{110}Cd by β^--decay with a half-life of 249.9 days. A neutron is converted into a proton and an electron under emission of a γ-ray with a probability >0.95. These γ-photons were detected in special detectors and used for analysis. The second tracer used was gold (Au) containing a fraction of 1.5×10^{-5} ^{198}Au isotopes. Au is used as contact material for Pc and DIP because of the good alignment of the Au work function and the HOMO of the organic semiconductors. A good alignment is important for a low charge carrier injection barrier. ^{198}Au also decays by β^--decay with a half-life of 2.69 days:

$$^{110m}\mathrm{Ag} \xrightarrow{\lambda=249.9d} {}^{110}\mathrm{Cd} + \beta^- + \gamma\,, \tag{1}$$

$$^{198}\mathrm{Au} \xrightarrow{\lambda=2.69d} {}^{198}\mathrm{Hg} + \beta^- + \gamma\,. \tag{2}$$

19.2.4
Serial Sectioning by Ion Beam Sputtering

In order to measure a diffusion profile, the organic thin film has to be cut layer by layer. Although the sputtering behaviour of polymers is well established in the Ref. [14] the sputtering of organic crystalline materials in this low-energy range is not well-studied. Since the sputtering behaviour of the organic films is determining the resolution in the analysis of the depth profiles, it was necessary to examine the sputtering behaviour of the two organic crystalline materials in detail. Only the knowledge of the sputtering behaviour allows the correct interpretation of the depth profiles. If the surface topography changes during sputtering, this would have to be taken into account when determining the depth resolution. Sputter experiments with Pc and DIP films show that the grain size and the surface topography are perfectly maintained within the resolution of the instrument. Simple constant-force contact-mode as well as non-contact mode AFM measurements (Figure 19.2) were performed for various sputter depths and rates, but no change in the topography was found.

In Figure 19.2(a) the structure of an unsputtered Pc film is shown. At the higher magnification the terrace structure of the film can be clearly seen. During sputtering (see e.g. Figure 19.2(b)–(d)) the roughness of the surface increases slightly, however the terrace structure can still be seen.

19.2.5
Electrical Measurements

All electrical measurements for the electret films were carried out in the group of Hilleringmann in Paderborn. A dark shielded metal box using a HP 4156A parameter semiconductor analyser identified the transistor characteristics. The influence of the interface between the Au top contacts and the Pc film was measured with a Keithley picoamperemeter in darkness as well as under illumination.

19.3
Results and Discussion

19.3.1
Radiotracer Measurements

First, noble metal diffusion in DIP and Pc were carried out. As Ag diffusion in polymers is well studied and understood [12, 13, 15–17, 32, 33], depth profiles for Ag diffusion in Pc and DIP were obtained. Comparison with the depth profiles of polymers will help to further the understanding of the nature of noble metals in these organic semiconductors.

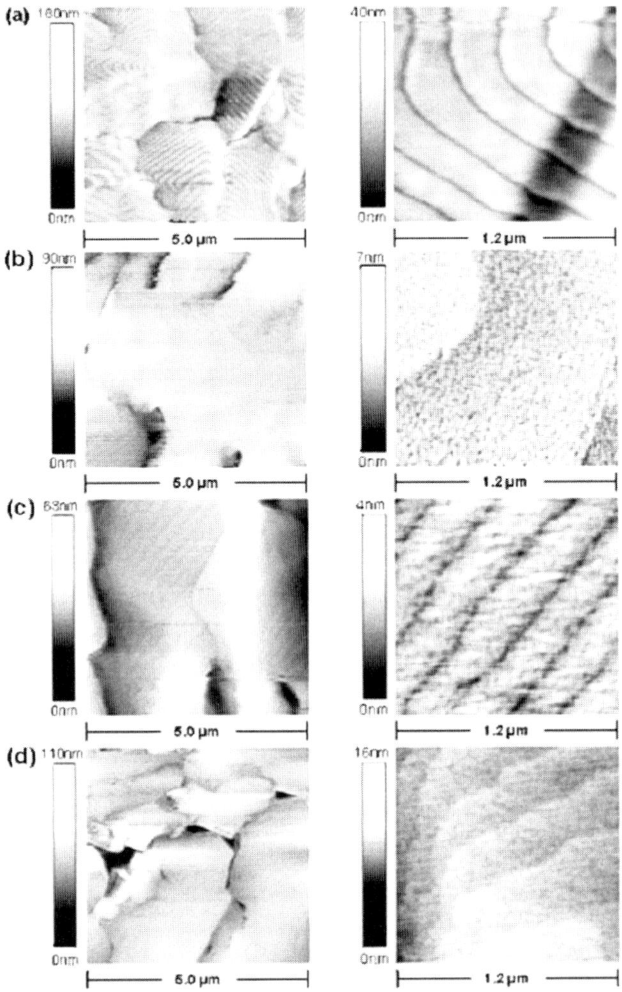

Figure 19.2 AFM images of (a) an unsputtered and (b)–(d) sputtered Pc film. Though the overall film topography is not influenced by sputtering a roughening of the surface on the molecular scale is observed.

About one monolayer of Ag + ^{110m}Ag was evaporated for 80 min at 680 °C onto Pc and DIP thin films with thicknesses of 60 nm and 80 nm respectively. The substrate temperature during evaporation was 75 °C. After evaporation the samples were annealed for 120 min at the same temperature. The depth profiles that were obtained are shown in Figure 19.3.

The examined functional organic thin films did not only differ in their molecular structure but also in their microstructure. The inset of Figure 19.2 shows AFM images of the films. The grain size is about 1 μm for the DIP film and 0.3 μm for the Pc thin film. Even though the films are very different the

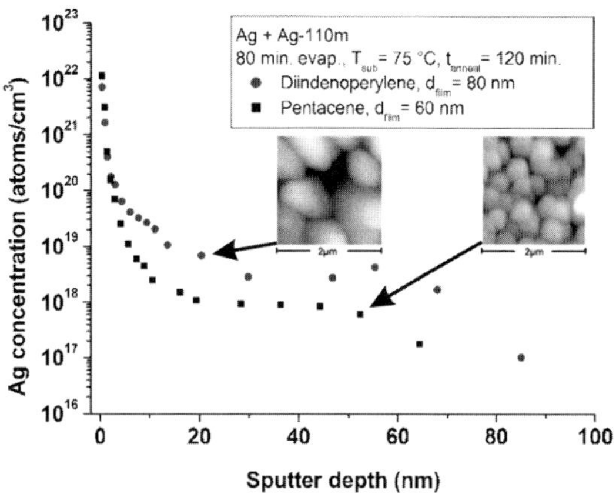

Figure 19.3 Depth profiles for Ag diffusion in thin films of Pc and DIP at a substrate temperature of 75 °C. For both materials, most metal atoms remain near the surface as the metal concentration drops by three orders of magnitude within the first 5 nm.

depth profiles share some common features. A drop in metal concentration of three orders of magnitude within the first 5 nm was observed for both films. This indicates that most metal atoms (>99%) remain on or near the surface, even though the total amount of material deposited was only about a monolayer. The metal concentration in the films drops to the level of the natural background at a depth of 60 nm for Pc and 80 nm for DIP which agrees very well with the corresponding film thicknesses. In addition to these common features some differences were observed. After the first initial drop, the metal concentration with respect to depth decreases more strongly in Pc than in DIP. At 20 nm the metal concentration in DIP is about one order of magnitude higher than in Pc. With increasing depth the difference in concentration decreases slightly but the Ag concentration in DIP is still much higher than in Pc.

As the same amount of Ag was deposited onto both films, this indicates that Ag diffuses faster and more easily in DIP. In the DIP depth profile, around 40 nm a slight increase in activity is observed, which drops towards the natural background after reaching a maximum. This concentration peak can be explained by an accumulation of Ag atoms at the DIP/Si interface (agglomeration was already observed for Au/DIP [18] and other systems [19]), the surface topography, and the sputtering behaviour of DIP. If Ag agglomeration at the interface takes place, this should ideally lead to a sharp peak in the detected metal concentration at the depth of the interface. However, as mentioned in the experimental section, the natural grain structure and the surface topography of the DIP film with a maximum height difference of around 40 nm are maintained. That is why some parts of the interface are sputtered prior to other parts (see Figure 19.4(a)).

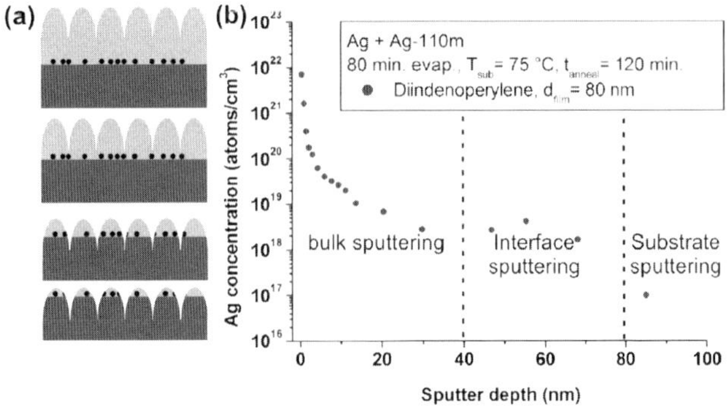

Figure 19.4 (a) Due to the surface topography of the film which is maintained during sputtering, some parts of the interface are sputtered prior to the rest of the interface. Material agglomerated at the interface will be sputtered likewise. (b) Due to non-uniform sputtering of the interface, three sputter regimes can be distinguished in the depth profile: after sputtering of the bulk of the DIP thin film, first parts of the interface are sputtered (~40 nm). Due to clusters agglomerated at the interface the Ag concentration increases while more of the interface is sputtered. After 80 nm the complete DIP film was sputtered and the Ag concentration decreases to the natural background.

The Ag atoms and clusters in these parts are also sputtered prior to the rest of the metal atoms. Instead of a sharp peak the metal concentration is smeared out over a depth range which correlates with the surface roughness of 40 nm due to the non-uniform sputtering of the interface.

A comparison of a typical polymer profile, in this case trimethylcyclohexane polycarbonate (TMC-PC) [13], and a DIP depth profile is shown in Figure 19.4. The diffusion of Ag in polymers is well-studied and the comparison of the depth profiles can provide a better understanding of the physical processes of metal diffusion in DIP. The profiles obtained for DIP are qualitatively very similar to the results obtained for TMC-PC, e.g., the first steep drop in concentration occurs over a very small depth range. In polymers, the cohesive energy of the organic matrix is smaller than that of the metal, and the interaction between metal and organic matrix is weak. Single atoms are very mobile in the polymer, but metal atoms encountering each other will form stable clusters which will impede further diffusion. The larger the clusters grow the stronger diffusion is impeded. This leads to effective immobilisation of the metal atoms by clustering [13, 20]. In these terms, the drop in intensity with increasing sputter depth can be interpreted as the superposition of different diffusion rates of clusters of different size (see Figure 19.5).

This similarity of the depth profiles for the DIP and the TMC-PC indicates that the interplay of diffusion and aggregation in organic crystalline films and polymers is basically the same. This was expected as the cohesive energy of DIP and TMC-Pc is of the same order. The region with decreased slope is as pronounced as in polymers, suggesting a similar interpretation. In contrast

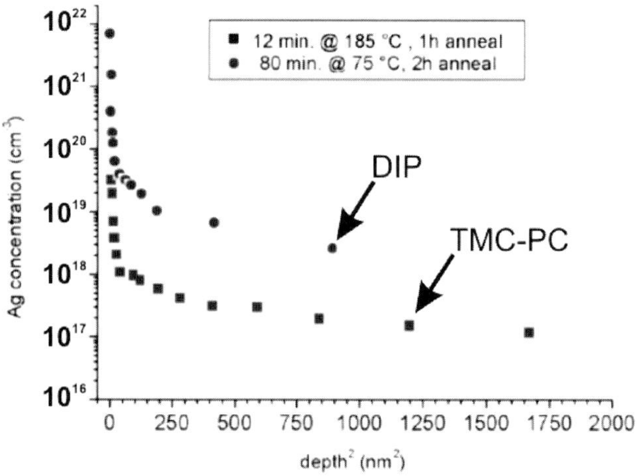

Figure 19.5 (a) Comparison of depth profiles for Ag diffusion in DIP and TMC-PC (data for TMC-PC from [46]). Both depth profiles show the same basic features like the steep drop in metal concentration within the first few nanometers. This indicates that the diffusion behaviour of noble metals is the same for both materials.

to the thin DIP film (the organic crystalline film has the typical thickness used in OFET transistors of about 80 nm) the polymer film is much thicker (>400 nm). Consequently, no agglomeration at the interface can be observed here. To further investigate the dynamics of noble metal diffusion in DIP, Ag + ^{110m}Ag was deposited onto a DIP thin film with a thickness of 80 nm at 680 °C for 8 min. The resulting depth profile was compared to the depth profile obtained for 80 min of tracer deposition (see Figure 19.6(a)). Increasing the evaporation time by a factor of ten leads to a metal concentration which is one order of magnitude higher, just as expected. Otherwise the general shape of the profiles is similar. With continuing film deposition the surface coverage with Ag increases as more than 99% of the Ag atoms is trapped on the surface. Due to the clusters growing on the surface, the probability of immobilisation for an Ag atom diffusing across the surface increases. If the atoms diffuse across the surface before penetrating into the bulk, the relative metal concentration in the DIP film with respect to the concentration on the surface should be smaller for the longer deposition time. In Figure 19.6(b) the relative concentration with respect to the surface concentration is shown. As can be seen, there is no difference in the depth profiles. That means either the metal coverage of half a monolayer is too low to impede the metal diffusion into the bulk, or the metal atoms do not diffuse across the surface but diffuse into the bulk almost directly.

In order to examine the temperature dependence of the metal diffusion, Ag + ^{110m}Ag was deposited onto 80 nm thick DIP films at substrate temperatures of 50 °C, 75 °C and 100 °C. For the deposition at 75 °C and 100 °C the

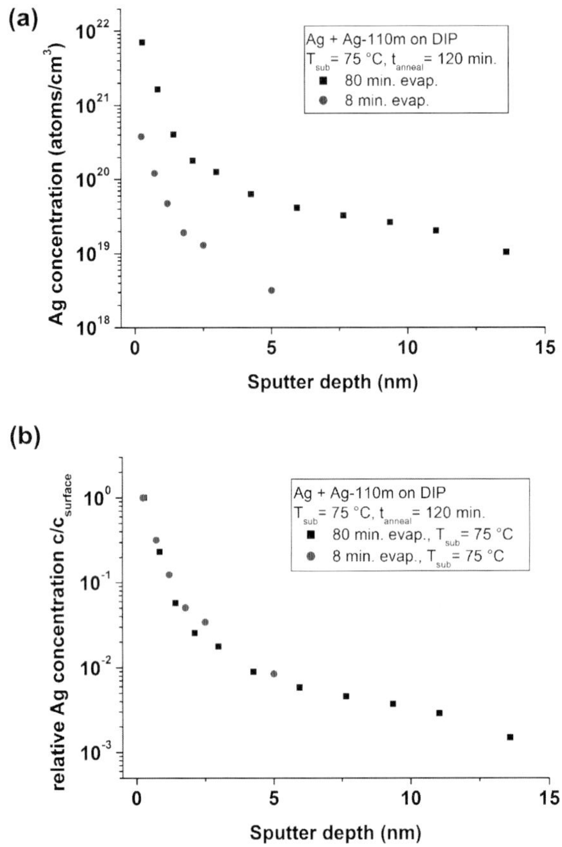

Figure 19.6 (a) Comparison of depth profiles obtained for 8 and 80 min of tracer deposition. The general shape of the profiles is similar. Increasing the evaporation time by a factor of ten leads to an increase in the metal concentration by one order of magnitude. (b) Relative Ag concentration with respect to the surface concentration. No difference in the depth profiles is observed indicating that the different surface coverage does not influence the diffusion of metal atoms into the bulk.

evaporation time was 80 min and the samples were annealed for another 120 min at the same temperature. For the sample kept at a temperature of 50 °C, the evaporation time was 176 min with no post-annealing treatment. Due to the longer evaporation time and the decreasing condensation coefficient with increasing temperature, more metal is deposited onto this sample. For the sample with a substrate temperature of 50 °C, about one monolayer was deposited onto the film. Because of the Volmer–Weber growth of noble metals on the organic films, it is not a closed layer, but the film consists of clusters distributed over the surface. The resulting depth profiles are shown in Figure 19.7. All three DIP profiles have some general features in common. Within the first 5 nm the metal concentration drops by three orders of magnitude, indicating that most metal atoms (>99%) remain on or near the surface, even though

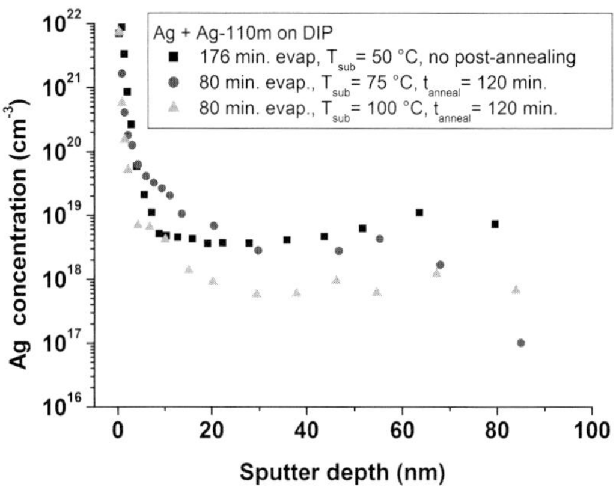

Figure 19.7 Temperature dependence of the Ag depth profiles obtained at substrate temperatures of 50 °C, 75 °C and 100 °C. After the initial drop in metal concentration, a region with a reduced negative slope is observed for all three temperatures. The negative slope decreases with increasing temperature, indicating increased metal diffusion with increasing temperature.

the total amount of material deposited was only about a monolayer. Following this first drop there is a region with a drastically reduced negative slope which is proof of a decreasing metal concentration. Considering the large difference in the sputter yields of Ag and the DIP films discussed earlier, Ag clusters on the surface should not contribute to the Ag signal in this region, meaning that it originates from metal atoms diffusing into the film. For substrate temperatures of 75 °C and 100 °C, the change in the slope is more pronounced as compared with the depth profile obtained at 50 °C. This indicates increased diffusion of metal into the organic layer at higher temperatures. The absolute concentrations cannot be compared directly because the condensation coefficient of Ag on the organic film is expected to drop drastically at elevated temperatures [21].

It was found that Ag diffuses faster into DIP than Pc. The reason could either be the different crystal structure or the different microstructure, i.e. the smaller grain size and resulting higher grain boundary density. To study the influence of the microstructure of the organic film on noble metal diffusion, DIP films deposited at substrate temperatures of –180 °C and 60 °C were examined. The different substrate temperatures resulted in different average grain sizes of the polycrystalline films. Details on the film structure are given in Table 19.1, schematic diagrams and AFM images are shown in the inset in Figure 19.8.

The film with the smaller grains has a higher grain boundary density. The radiotracer measurements can help to determine their influence on the diffusion. A higher grain boundary density might lead to more metal atoms diffus-

Table 19.1 Two DIP films were deposited at different substrate temperatures T_{sub}. The resulting films allowed the examination of the microstructures influence on the metal diffusion. The film deposited at a substrate temperature T_{sub} of 60 °C showed a crystallite size d_{cry} similar to the film thickness d_{film}. The crystallites of the film deposited at –180 °C are half as large as the film thickness. The rocking width of the XRD spectra is a measure for the structural order of the film. It is small for both films, indicating that both films are well ordered and most molecules are aligned in parallel.

sample	T_{sub}	d_{film}	d_{cry}	rocking width
DIP01	60 °C	183 nm	183 nm	0.09°
DIP02	–180 °C	147 nm	76 nm	0.04°

ing deeper into the organic film, if they act as fast diffusion paths. On the other hand, the metal diffusion would decrease, if the grain boundaries act as trapping sites for noble metal atoms, i.e., nucleation site for clusters. If the bulk diffusion is dominating the grain boundary diffusion, the depth profiles should show no dependence on the microstructure. In Figure 19.9 the depth profiles obtained for Ag diffusion in the two different DIP films are shown. For the DIP film with the smaller crystallites, a larger metal concentration with respect to the depth is observed, indicating that grain boundaries act as fast diffusion paths.

The comparison of the depth profiles for Ag diffusion in DIP and TMC-PC indicated a similar diffusion mechanism for polymers and organic crystalline materials. For TMCPC, it was shown that a submonolayer of Cr can effectively block the diffusion of Ag into the polymer [13]. Since Cr is a transition metal, it is very reactive. The Cr atoms immediately react with the polymer and do not diffuse into the bulk. On the surface they act as nucleation centres for Ag atoms impinging onto and diffusing across the organic surface. As more Ag atoms are trapped on the surface, diffusion into the material is reduced. If

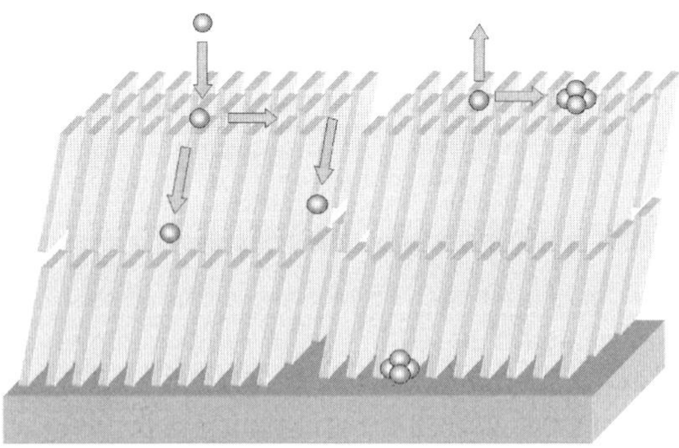

Figure 19.8 Schematic visualising the possible influence of the film structure on the metal diffusion. Grain boundaries might either act as fast diffusion paths or act as trapping and nucleation sites for clusters.

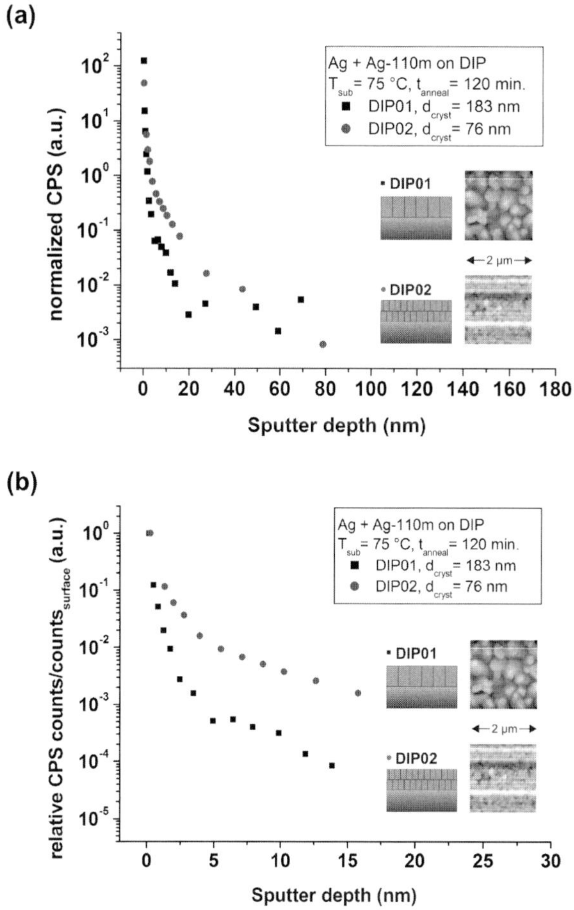

Figure 19.9 (a) Depth profiles for Ag diffusion in DIP films with differing microstructure. (b) The magnification of the first 15 nm shows the dependence of the metal diffusion on the film structure. In the film with the lower grain boundary density (DIP01 in inset), the relative Ag concentration with respect to the surface concentration is one order of magnitude lower as compared with the film with the higher grain boundary density (DIP02). This indicates that grain boundaries are acting as fast diffusion paths.

the diffusion mechanisms for DIP and TMC-PC are similar, a submonolayer of Cr should reduce Ag diffusion into the DIP film.

In order to test the feasibility of a Cr barrier layer, a submonolayer of Cr with a nominal thickness of 0.1 nm was deposited onto the DIP film prior to evaporation of the Ag radiotracers. In Figure 19.10(a) the depth profiles for DIP films with and without Cr barrier are shown. In Figure 19.10(b) the data are normalised with respect to the activity on the surface of each sample. As for TMC-PC, the deposition of Cr leads to a reduced Ag concentration in the film. However, the reduction of the Ag diffusion is not as effective as observed

(a)

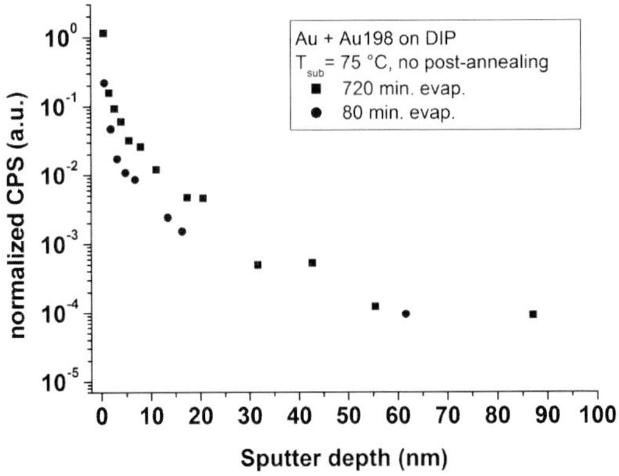

(b)

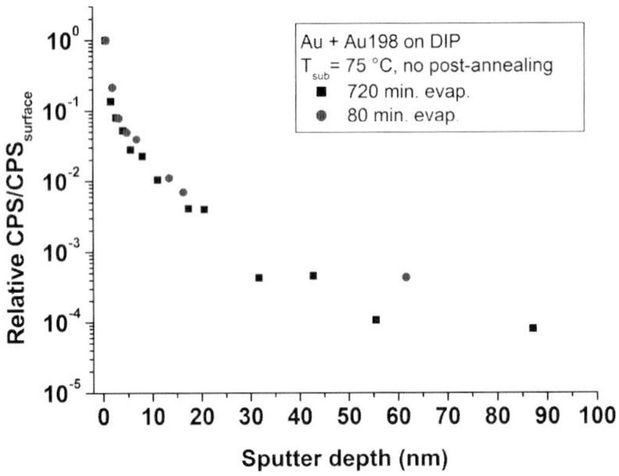

Figure 19.10 (a) Depth profiles for Au diffusion in DIP for 80 min and 720 min of tracer evaporation. The obtained depth profiles show the same characteristic features observed for the Ag depth profiles. (b) Relative CPS with respect to the CPS on the surface. As for the Ag diffusion no time dependence is observed.

for TMC-PC. Considering this together with the fact that an increasing surface coverage with Ag clusters had no effect on the metal diffusion indicates that the ratio of surface diffusivity $D_{surface}$ and bulk diffusivity D_{bulk} is smaller for DIP as compared with TMC-PC.

It was found that grain boundaries act as fast diffusion paths. While DIP and TMC-PC have the same cohesive energy they differ in their microstructure. DIP forms polycrystalline, well ordered films and TMC-PC films are amorphous. The existence of grain boundaries in DIP and their absence in TMC-PC can explain the difference in the ratios $D_{surface}/D_{bulk}$ and the varying effectiveness of Cr barrier layers because the grain boundaries act as fast diffusion paths in DIP. Submonolayers of Cr act as diffusion barriers on polymers because the Cr atoms dispersed across the surface act as nucleation centres for metals atoms which diffuse across the surface. If noble metal atoms diffuse right into the bulk along grain boundaries rather than to move across the surface first, dispersed Cr atoms will trap the metal atoms less effectively. While Cr can decrease metal diffusion, it still has to be examined whether the Cr barrier influences the electronic properties of the metal–organic interface.

As mentioned before, Au is used for metallisation of Pc- and DIP-TFTs due to the good alignment of the Au work function and the HOMO level of the organic semiconductors. Depth profiles for Au + ^{198}Au evaporated at 860 °C for 80 min and 720 min are shown in Figure 19.11(a). A comparison with the depth profiles obtained for Ag diffusion will show whether the diffusion pro-

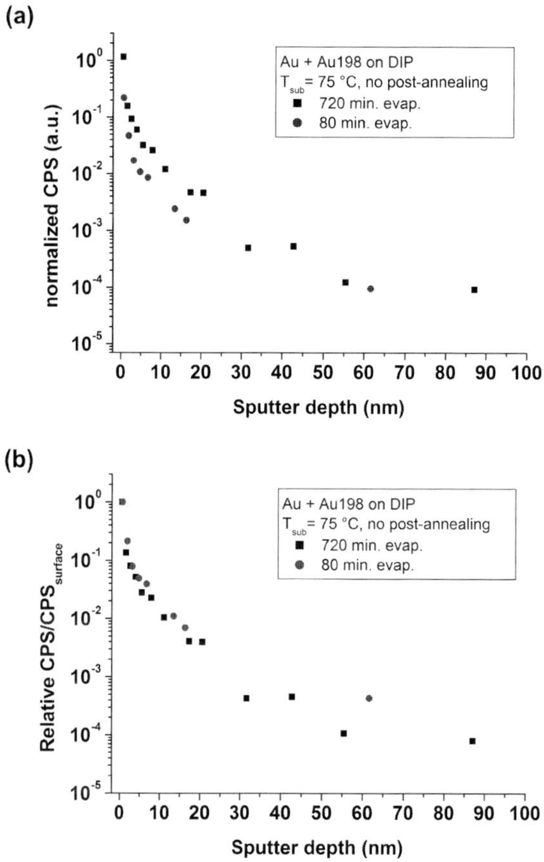

Figure 19.11 (a) Depth profiles for Au diffusion in DIP for 80 min and 720 min of tracer evaporation. The obtained depth profiles show the same characteristic features observed for the Ag depth profiles. (b) Relative CPS with respect to the CPS on the surface. As for the Ag diffusion no time dependence is observed.

cess is comparable for the two noble metals. As for Ag, a drop in the counts per second (CPS) is observed in the first 5 nm followed by a region of decreasing negative slope. In Figure 19.11(b) the relative CPS with respect to the surface are plotted against the sputter depth. No time dependence is observed. The low CPS imply that the surface coverage is too low to impede metal diffusion by trapping in clusters.

19.3.2 Correlation Between Metal Diffusion and Device Properties of OFETs

The radiotracer measurements provided details on metal diffusion into the organic semiconductor and the mechanism of the interface formation and the resulting structure. For fabrication of actual devices it is important to understand how these structures influence the electronic properties of the interface and thus the device. Charge carrier injection might be either enhanced or impeded depending on how the energy alignment and the injection barriers are influenced. Comparison of the current–voltage curves of contacts formed by slow and fast metal deposition combined with the information of the interface derived from the radiotracer depth profiles can further the understanding of the influence of the interfacial structure on the device performance. For the electronic characterisation of the Au contacts voltages between 0 V and 6 V were applied between the contacts. Different channel length could be measured by selecting different contact pads. The resulting currents were in the order of $0-10^4$ nA. Higher voltages were not used as the resulting currents could damage the contact between the metal and the organic semiconductor. For the $I-V$ curves of the contacts a switching process was observed for voltages mostly in the mV-range. This transition was observed for all three interface structures. In Figure 19.12 the transition in the onset of the $I-V$ curves for contacts of Array 2 are shown. Below the switching voltage the current is very small (<20 nA) and shows no dependence on the voltage. Above the switching voltage there is a sharp increase in the current and it continues to increase with increasing voltage. This switching process is reversible if only small voltages below 2 V are applied. If the voltage is increased up to 2 V and then ramped down to 0 V again, the switching is not as pronounced as before and the switching voltage changed from 700 mV to 300 mV (see Figure 19.13). A similar effect is observed in nanocomposites and nanowires [22, 23]. In plasma polymers that contain silver nanoparticles a switching process was observed if the voltage was varied from −2 V to +2 V.

With each cycle the switching was less pronounced and the switching voltage (the so-called threshold voltage) decreased [23a]. While still several mechanisms are discussed as the origin for the switching, it seems to be clearly linked to the microstructure as it is only observed for nanocomposites near percolation [23a]. Therefore, the switching observed for the Au contacts may be related to the morphology of the interface. Due to diffusion during contact deposition the interface is rough with dispersed metal clusters near the inter-

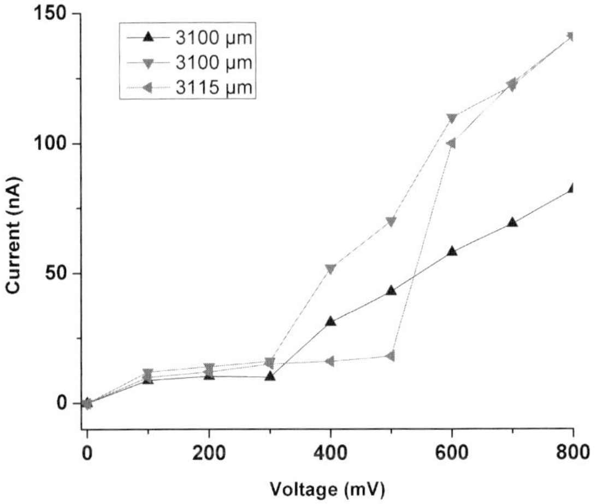

Figure 19.12 Transition in the $I-V$ curves of the contacts of Array 2 with a channel length varying between 3100 μm and 3115 μm. Below the switching voltage, the current is almost zero and independent of the applied voltage. Above a critical voltage the current increases with increasing voltage. The switching is reversible if the applied voltage is not increased above 2 V.

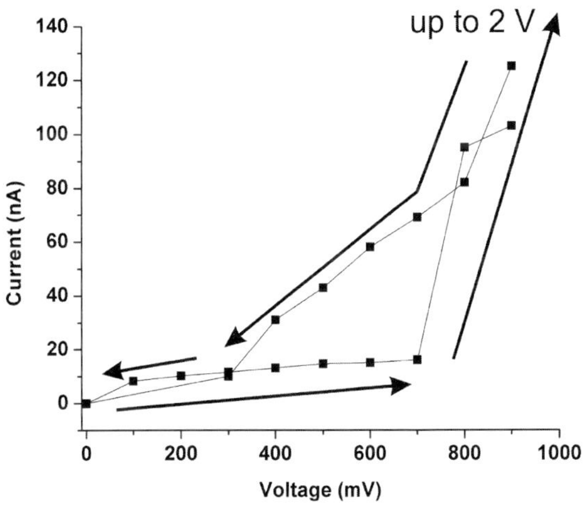

Figure 19.13 Graph showing the $I-V$ curve of a contact of Array 2 with a channel length of 3100 μm. A switching is observed as the voltage is increased above 700 mV. It is not observed anymore after the applied voltage was raised above 2 V. In that case, the current continuously decreases to zero when the voltage is decreased again.

face. If this interface region is comparable to a percolated network of clusters in a nanocomposite, the observed switching process could result. The large variation in the switching voltage and the slope of the I–V curve upon switching are attributed to differences in the individual contacts. As already small voltages seem to permanently change the electronic properties of the contact, the maximum voltage was limited to 6 V so that the influence of the metal diffusion on the contact properties could be examined and not the changes due to high voltages and currents. Furthermore, since the native oxide layer of the Si wafer only has a thickness of about 2 nm, no gate voltage was applied in order to avoid an electrical breakthrough of the insulating layer.

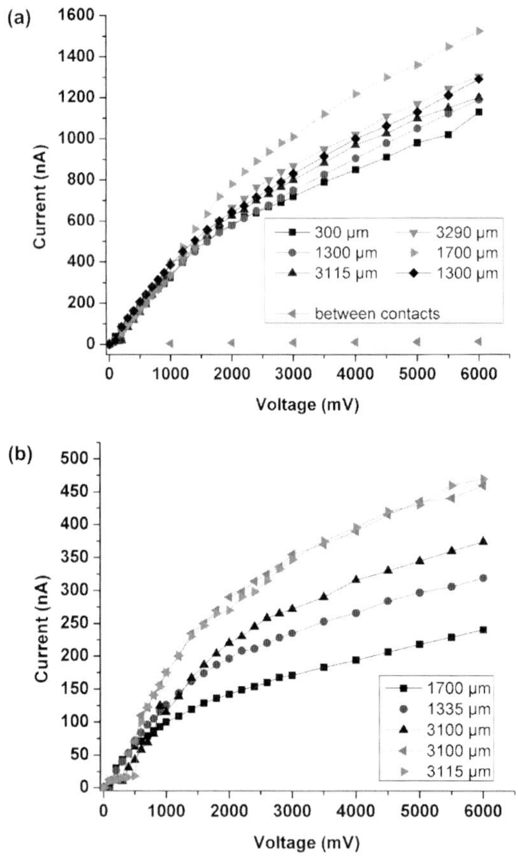

Figure 19.14 Current–voltages dependence of the Au contacts deposited with (a) high (Array 1) and (b) low (Array 2) deposition rates at a substrate temperature of 75 °C. The general shape of the I–V curves is similar for both contact arrays. The deviation within one array might be caused by differences in the individual top contacts. The channel length does not seem to have an influence on the I–V curves.

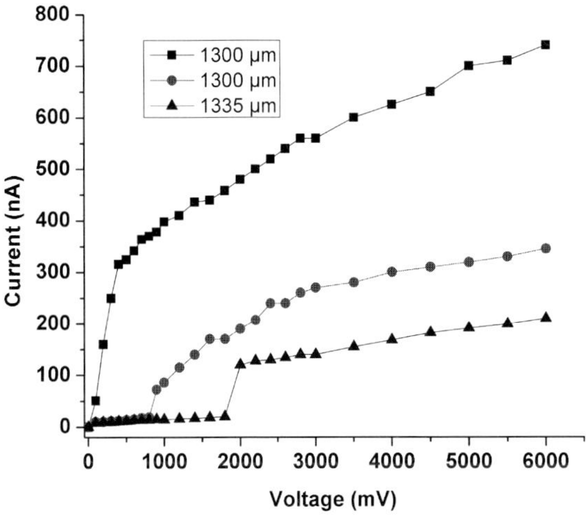

Figure 19.15 Current–voltages dependence of the Au contacts deposited with a high deposition rate with the substrate being at room temperature. The general shape of the *I–V* curves is similar to that of Arrays 1 and 2.

The general features that can be seen in Figure 19.14(a) and (b) show the *I–V* curves for Array 1 and Array 2 respectively. The *I–V* curves of Array 3 are shown in Figure 19.15. All the curves have a similar shape but the values of the *I–V* curves of different contacts within the same array can deviate from each other by about 50%. This proves that the formation of reproducible top contacts is extremely difficult even if the contacts are formed on the same film under the same conditions. Within each array, the channel length between the contacts does not seem to have an influence on the *I–V* curve. This indicates that the contact resistance between the Au pad and the Pc film is dominating the overall resistance so that the channel resistance can be neglected.

But even with this large deviation within the same array, the comparison of the *I–V* curves of the two different contact arrays in Figure 19.16 shows that the extent of the metal diffusion has a clear influence on the contact properties. The arrays were formed on the same Pc film, the only difference being the increased metal penetration for Array 2. For this array the contact resistance appears to be larger and charge carrier injection is impeded as the maximum current injected is smaller as compared with Array 1. Therefore, the rough interface formed by increased diffusion results in inferior contacts as the charge carrier injection is impeded. This agrees well with the results obtained by Scholz et al. [23b] (see also Chapter 20). They suppressed the diffusion of Au atoms into pentacene due to deposition at substrate temperatures of –150 °C. The resulting contacts showed almost no contact resistance.

However, the contacts of Array 3 feature the smoothest interface but its *I–V* curves are comparable to Array 2 and inferior to those of Array 1. This shows that a certain interface roughness could be favourable as the alignment of the

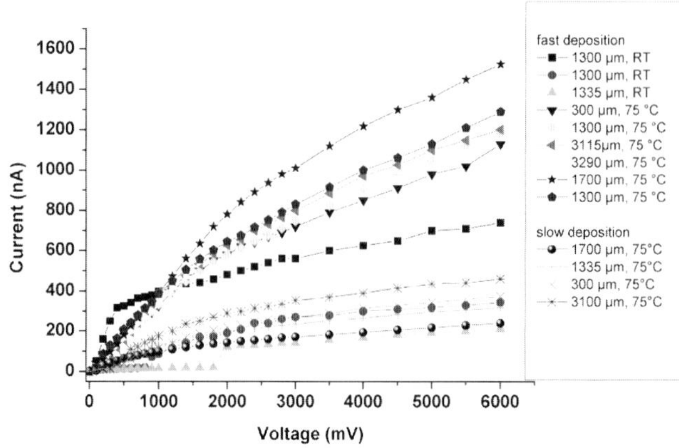

Figure 19.16 Comparison of *I–V* curves of different contacts. The contacts deposited at 75 °C at the higher deposition rate showed the best *I–V* characteristics of the three different contact arrays.

Au contacts and the HOMO of the organic semiconductor is improved and the injection barrier decreased. But if the roughness exceeds a certain limit the injection barrier is increased again. In Figure 19.14(a) it is also shown that almost no current is measured if the organic semiconductor is contacted directly. Under illumination from a bright light source a current of about 3 µA was measured for the contacts of Arrays 1 and 2 independent of the applied voltage or the deposition conditions of the contacts. This shows that the low intrinsic charge carrier concentration and the charge injection barrier are responsible for the much smaller currents measured without illumination. For Array 3, currents in the order of 25 µA were measured. This indicates that the alignment of metal work function and HOMO is optimised for extraction of the charge carriers from the organic semiconductor and that the current without illumination is limited by the injection.

Another interesting effect that was only observed for the contacts of Array 2 was a change of the *I–V* curve after 24 hours. The *I–V* curves of the contacts of Array 2 showed currents that were much smaller than the currents measured for the contacts of Array 1. After storage in ambient air for 24 hours, the *I–V* curves of Array 2 were comparable to the curves of Array 1 (see Figure 19.17). Since the *I–V* curves of Array 1 did not change upon exposure to air this improvement cannot be caused by simple oxidation of the interface as this would occur for both arrays. This change of the *I–V* characteristic may rather reflect room temperature diffusion in the organic material which was also observed in polymers [13, 24]. In this case, single atoms and small clusters might diffuse back to the interface where they are incorporated into the closed layer. This process would smooth the interface and its morphology would be similar to an interface that was formed by fast deposition.

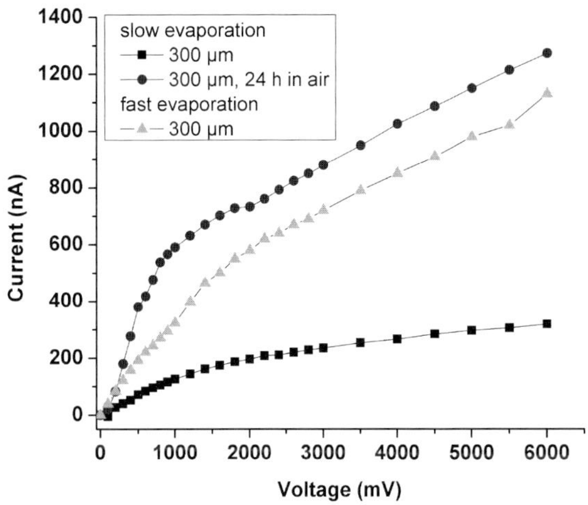

Figure 19.17 After storage in air for 24 hours the *I–V* characteristic of a contact from Array 2 changes considerably. Its *I–V* curve is now comparable to that of a contact of Array 1 with the same channel length. This might indicate the presence of room temperature diffusion as the clusters of the rough surface of Array 2 diffuse back to the interface. This process would result in a smoother surface comparable to that of Array 1.

The results presented above indicate that different interface morphologies can optimise the injection or extraction of charge carriers. Therefore, it might be possible to optimise the performance of an organic semiconductor by deposition of two contacts with different interfaces. This approach was tested by deposition of one contact similar to those in Array 1 (best injection behaviour) and one similar to Array 3 (best extraction). In Figure 19.18 the resulting *I–V* curves are shown. In one direction a very high current is measured. If the current direction is reversed the resulting current is much smaller. In the high-current direction the photo-induced current was 25 μA and in the low-current direction only 3 μA. These results prove that of the two contacts one is indeed optimised for injection and the other for extraction of charge carriers. This optimisation is not possible for bottom contact devices as the organic semiconductor is deposited on top of the metal and no diffusion occurs and the structure of the interface cannot be varied for individual contacts.

19.3.3
Teflon-Based Electret Layers for Threshold Voltage Tuning

An encapsulating layer is required if the organic semiconductor is susceptible to degradation by the ambient air or moisture. However, deposition of an additional film means that an additional processing step would be required in the fabrication, which results in increased production costs. If the capping layer could be further functionalised to enhance the device performance, the cost–

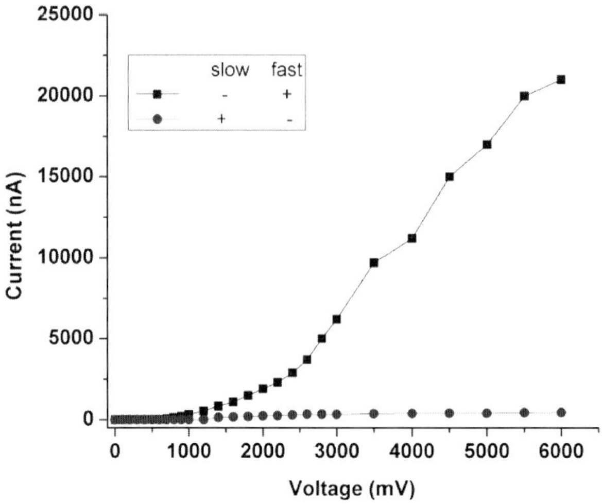

Figure 19.18 *I–V* curves of an asymmetric contact. Depending on the direction of the current show, the resulting *I–V* curves differ greatly from each other. This indicates that one contact is optimised for charge carrier injection while the other is optimised for extraction of charge carriers.

value ratio of the additional deposition step would improve. A capping layer containing a Teflon-based layer as introduced in [29, 30] might serve as such a multifunctional organic thin film. If a Teflon capping layer could also serve as second gate in a dual-gate OTFT, this bifunctional layer would address two problems faced in fabrication of OTFTs. In the following experiment, Teflon AF 1600 is deposited on top of a bottom gate Pc-TFT in order to examine whether the threshold voltage can be tuned by controlled charging of the electret film. In order to determine the shift in the threshold voltage due to charging of the electret the square root of the drain-source current as a dependence of the gate-source voltage VGS was measured at a mixed drain-source current IDS of −5 V. The influence of the electret gate on the field effect mobility and on-current of the TFT was determined by measuring IDS as a dependence of the drain-source voltage VDS for VGS ranging from +5 to −40 V directly before and after deposition of the electret gate. A 2 μm thick Teflon AF 1600 film was deposited onto the Pc-TFT by thermal evaporation and was charged to a charge density of -1.3×10^{8} C/cm^2 using the corona discharge method. This corresponds to a compensation voltage of −15 V. After fabrication of the BOC-Pc-TFTs the output and transfer characteristics were measured to determine their initial performance. The output current at VDS of −40 V and VGS of −40 V was as large as −800 μA (see e.g. 4.9(a)) and the calculated charge carrier field-effect mobility was 4.6×10^{-4} cm^2/Vs. The threshold voltage was extracted from the square-root of the on-current characteristics (see Figure 19.19(a)) and was found to be +13.1 V due to a modification of the surface potential of the dielectric induced by the oxygen plasma [25]. After charging of the electret, the OTFTs were characterised again showing a reduced on-current of −42 μA and a decreased charge carrier field effect mobility of 6×10^{-5} cm²/Vs (see e.g. Figure 19.19(b)). This reduction might be caused by the deposition of the additional layer onto the organic semiconductor as was already reported before [26] or due to aging of the film in between charging of the film and the *I*–*V* measurements. The threshold voltage can be extracted from the transfer characteristics in Figure 19.19(b) and was determined to be −2.3 V as compared with +13.1 V prior to electret deposition. The shift in the threshold voltage agrees very well with the charging of the electret to −15 V indicating that the shift is caused by the depletion of the channel region due to the positive charge carriers being pulled towards the electret top gate. The ex-

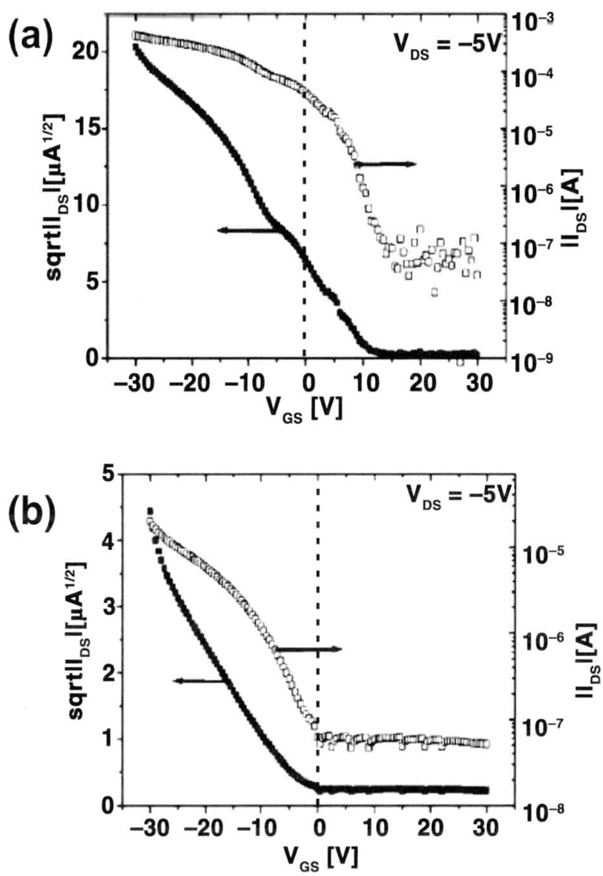

Figure 19.19 Drain–source current versus gate–source voltage of the Pc-TFT (a) prior to deposition of the Teflon AF 1600 thin film and (b) after deposition and charging of the electret layer. The threshold voltage shifted from (a) +13.1 V to (b) −2.3 V.

cess holes might then be trapped in the top of the pyramidal structure of the pentacene crystallites which are only connected at the base. The shift of the threshold voltage towards more negative values cannot be attributed to aging since a negative shift could only be caused by degradation due to moisture. This is prevented by Teflon films as was shown earlier. In addition to the shift in the threshold voltage a qualitative change in the shape of the *I–V* curves is observed, e.g., a minimum in the IDS−VGS curve. These changes are attributed to aging of the organic semiconductor due to exposure to oxygen as it was shown earlier, that Teflon-based capping layers cannot block oxygen. The shift in the threshold voltage cannot result from this aging as it would lead to a shift of the threshold voltage towards more positive values.

19.4 Conclusions

In this work, which reviews and extends the content of the publications [26–31], two types of top layer thin films were examined with respect to their influence on molecular organic crystalline layers for OFETs. One of the challenges arising from the implementation of organic semiconductors is the penetration of metal atoms into the organic thin film during metallisation. The diffusion behaviour of noble metals in organic molecular thin films of pentacene and diindenoperylene was examined with high resolution depth profiles obtained by radiotracer measurements. It was found that the basic diffusion processes are comparable to noble metal diffusion in polymers while differences are observed due to the different microstructures. The experiments showed that most metal atoms remain on or near the surface of the organic film, but a small fraction of metal atoms diffuses through the film and in some cases they accumulate at the interface between organic film and substrate. Grain boundaries in the crystalline small molecule semiconductors were identified as fast diffusion paths. These findings are of relevance to technical applications. Even at the high deposition rates used in industrial applications small amounts of metal can diffuse into the organic material in the early stages of contact deposition before a closed layer is formed. These small amounts can strongly alter the device performance and the charge carrier injection at the metal–organic semiconductor interface. In fact, the correlation of *I–V* measurements and radiotracer measurements of Au contacts on pentacene indicate that increased diffusion during deposition of the contacts leads to an increased contact resistance. This is a very important result since the charge carrier injection is of great importance for the device performance. In order to control diffusion, a submonolayer of Cr was deposited as a barrier layer and it was shown that just a submonolayer of Cr can even influence the noble metal diffusion but is less effective than for amorphous polymers because the grain boundaries of the polycrystalline films act as fast diffusion paths. The feasibility of radiotracer diffusion measurements for organic crystalline semiconductors was demonstrated.

It is also of interest for other organic functional thin films, i.e. organic semiconductors in organic solar cells and organic light emitting diodes. In these devices, metal deposition onto the organic thin films is often required, and metal diffusion might degrade the device performance. Examination and control of metal diffusion in these systems might be an interesting and important application of the radiotracer technique.

The second top layer system that was examined is Teflon-based functional thin films. The Teflon films can be further functionalised by charging them as electrets, allowing control of the threshold voltage of organic thin film transistors. This electret-based dual gate field-effect transistor structure allowed the threshold voltage of a transistor to shift to the low voltages that are required for technological applications like portable multimedia devices.

In general top layers can have a pronounced influence on the properties of organic thin films.

Acknowledgements

We gratefully acknowledge the DFG for financial support in the framework of the SPP 1121 for Contract Nos. AD 183/3-1 and AD 183/3-2 and for a grant for a Heisenberg professorship (R.A.).

References

1. J. H. Burroughes, D. D. C. Bradley, A. R. Brown, R. N. Marks, K. Mackay, R. H. Friend, P. L. Burns, and A. B. Holmes, Nature **347**, 539 (1990).

2. H. Hoppe and N. S. Sariciftci, J. Mater. Res. **19**(7), 1924 (2004).

3. C. Dimitrakopoulos and P. Malenfant, Adv. Mater. **14**, 99 (2002).

4. Y. Park, J. Lim, H. Lee, and K. Cho, Mater. Today **10**(3), 46 (2007).

5. S. Lee, B. Koo, J.-G. Park, H. Moon, J. Hahn, and J. Kim, MRS Bull. **31**, 455 (2006).

6. R. Farchioni and G. Grosso (eds.), Organic Electronic Materials, Part 2: Low Molecular Weight Organic Solids (Springer, Berlin, 2001), chap. 3.

7. Ch. Pannemann, T. Diekmann, and U. Hilleringmann, J. Mater. Res. **19**, 1999 (2004).

8. D. Faltermeier, B. Gompf, M. Dressel, A. K. Tripathi, and J. Pflaum, Phys. Rev. B. **74**, 125416 (2006).

9. T. Nason, J. Moore, and T. Lu, Appl. Phys. Lett. **60**, 1866 (1992).

10. G. Sessler (ed.), Electrets, 2nd ed. (Springer, Berlin, Heidelberg, 1987).

11. D. Gupta and R. Tsui, Appl. Phys. Lett. **17**(7), 294 (1970).

12. F. Faupel, A. Thran, V. Zaporojtchenko, T. Strunskus, and M. Kiene, in: Diffusion in Advanced Technological Materials, edited by D. Gupta (Noyes Publication, William Andrew Publishing, 2004).

13. A. Thran, T. Strunskus, V. Zaporojtchenko, and F. Faupel, Appl. Phys. Lett. **81**, 244 (2002).

14. H. Gokan, S. Esho, and Y. Ohnishi, J. Electrochem. Soc. **130**, 143 (1984).

15. F. Faupel, D. Gupta, B. Silverman, and P. Ho, Appl. Phys. Lett. **55**, 357 (1989).

16. F. Faupel, P. Hüppe, K. Rätzke, R. Willecke, and T. Hehenkamp, J. Vac. Sci. Technol. A **10**, 92 (1992).

17. A. Thran, M. Kiene, V. Zaporojtchenko, and F. Faupel, MRS Bull. 3 (1999).

18. A. Dürr, F. Schreiber, M. Kelsch, H. Carstanjen, and H. Dosch, Adv. Mater. **14**, 961 (2002).

19. F. LeGoues, B. Silverman, and P. Ho, J. Vac. Sci. Technol. A **6**, 2200 (1988).

20. F. Faupel, R. Willecke, and A. Thran, Mater. Sci. Eng. R **22**, 1 (1998).

21. A. Thran, M. Kiene, V. Zaporojtchenko, and F. Faupel, Phys. Rev. Lett. **82**, 1903 (1999).

22. S. K. Saha, Phys. Rev. B **69**, 125416 (2004).

23a. A. Kiesow, J. Morris, C. Radehaus, and A. Heilmann, J. Appl. Phys. **94**, 6988 (2003).

23b. R. Scholz, D. Lehmann, A.-D. Müller, F. Müller, and D. R. T. Zahn, Phys. Stat. Sol. (a) **205**, 591 (2008).

24. A. Thran, Diffusion von Gasen und Metallen in Polymeren: Datenanalysen, Experimente und Simulationen, Ph.D. thesis, Christian-Albrechts-Universität zu Kiel (2000).

25. M. Lee and C. Song, Jpn. J. Appl. Phys. **42**(7A), 4218 (2003).

26. M. Scharnberg, J. Hu, J. Kanzow, K. Rätzke, R. Adelung, F. Faupel, C. Pannemann, U. Hilleringmann, S. Meyer, and J. Pflaum, Appl. Phys. Lett. **86**, 024104 (2005).

27. M. Scharnberg, J. Hu, J. Kanzow, K. Rätzke, R. Adelung, F. Faupel, and J. Pflaum, Defect Diffus Forum **237–240**, 993 (2005).

28. M. Scharnberg, J. Kanzow, K. Rätzke, R. Adelung, S. Meyer, J. Pflaum, and F. Faupel, in: Organic Thin-Film Electronics, edited by A. C. Arias, N. Tessler, L. Burgi, and J. A. Emerson, Mater. Res. Soc. Symp. Proc., Vol. 871E (MRS, Warrendale, PA, 2005), p. I6.31.

29. C. Pannemann, T. Diekmann, U. Hilleringmann, U. Schürmann, M. Scharnberg, V. Zaporojtchenko, R. Adelung, and F. Faupel, IEEE Proc. of the IEEE Polytronic 2005 conference, IEEE Catalog No. 05TH8855C (2005), p. 63.

30. C. Pannemann, T. Diekmann, U. Hilleringmann, U. Schürmann, M. Scharnberg, V. Zaporojtchenko, R. Adelung, and F. Faupel, Mater. Sci., in press (2007).

31. M. Scharnberg, V. Zaporojtchenko, R. Adelung, F. Faupel, C. Pannemann, T. Diekmann, and U. Hilleringmann, Appl. Phys. Lett. **90**, 013501 (2007).

32. W. A. Lopes, Phys. Rev. E **65**, 031606 (2002).

33. S. B. Darling and A. Hoffmann, J. Vac. Sci. Technol. A **25**, 1048 (2007).

34. C. Kim, A. Facchetti, and T. J. Marks, Science **318**, 76 (2007).

20
Potentiometry on Pentacene OFETs: Charge Carrier Mobilities and Injection Barriers in Bottom and Top Contact Configurations

R. Scholz, D. Lehmann, A.-D. Müller, F. Müller, and D. R. T. Zahn

20.1
Introduction

The field of organic field effect transistors is now approaching a mature state, i.e. the main challenge is no longer to present working devices, but merely to understand the physical limitations of charge transport arising from intrinsic properties of ultrapure crystals, imperfections like grain boundaries, deep trap levels, and contact resistances.

For different classes of organic semiconductors, including polyacenes, thiophenes, fullerenes, and variants of these molecules with sidegroups like rubrene, it has been demonstrated that room temperature time-of-flight and field-effect mobilities can exceed 1 $cm^2 V^{-1} s^{-1}$ [1–6]. The temperature dependence of the lattice mobility in pure crystals can be understood from a combination of the narrowing of the polaron bands with rising temperature [7] together with the influence of electron–phonon coupling on the mobility [8]. However, even for field effect transistors employing single crystals, the observation of a rising mobility with decreasing temperature is more the exception than the rule [3, 6] because in most cases this dependence of the intrinsic mobility is strongly altered by trapping of mobile charge carriers at deep traps, resulting in an activated behaviour of the mobility. In addition, poly-crystalline samples suffer from a reduction of the intrinsic mobility due to grain boundaries, and it was demonstrated that annealing of the OFET structure may result in larger grains, reducing the influence of the grain boundaries [9].

Especially for OFETs with high mobility in the channel region, even small injection barriers may severely obscure the intrinsic properties of the organic semiconductor. Both the source and the drain contacts can be understood as Schottky barriers, so that the injection into the channel requires emission of electrons or holes from the metal over the barrier. In addition, the lower charge density in the depleted regions close to the source and drain contacts requires a larger electric field to carry the same current density as in the accumulation channel, resulting in a significant voltage drop. Three techniques have been proposed to quantify the corresponding contact resistances: a comparison of

Organic Electronics. Structural and Electronic Properties of OFETs. Edited by Christof Wöll

ISBN: 978-3-527-40810-8

similar devices with different channel lengths [10], four probe measurements [2, 4, 5, 11], and potentiometry [12–16]. The last potential profiling method is quite elegant because it allows to determine the local electrostatic potential on top of the channel with a resolution better than 100 nm, revealing asymmetries between the injection resistance and the drain resistance [12, 15], the increase of the electric field towards the drain contact in the saturation regime [13], different injection barriers depending on the contact metal [14, 15], and even potential drops over imperfections like cracks in the channel region [16].

In OFETs, the accumulation channel forms directly at the interface between the organic semiconductor and the gate insulator. Therefore, irrespective of the preparation technique, the density of traps in this region can be much larger than in the bulk of ultrapure single crystals. Both for poly-crystalline OFET channels and for devices based on single crystals of organic semiconductors, one observes typically a temperature-activated behaviour of the mobility and a drain current proportional to exp $(-E_{act}/k_B T)$ [2, 11, 17, 18], where the latter may have activation energies depending on gate voltage [18]. This so-called multiple trapping and release model assumes that the mobile carriers are trapped with a high probability, whereas the rate for de-trapping decreases when the temperature is lowered. An established technique to characterise trap states is charge transient spectroscopy (QTS) or deep level transient spectroscopy (DLTS) [19]. With these methods, the transient trapping or de-trapping dynamics of electrically active defect states can be observed. It was demonstrated that temperature-dependent QTS gives direct access to the density of states of charge traps in organic devices [20] and even to the energetics of several groups of trap states at once [21–23].

It has been shown previously that a treatment of the SiO_2/pentacene interface with an n-octadecyltrichlorosilane (OTS) self assembling monolayer (SAM) leads to improved device characteristics of OFETs and prolongs the lifetime under ambient conditions [24], increasing the mobility by a factor of up to 100 compared with an untreated sample [25, 26]. This enhancing effect has also been reported for gate insulators other than SiO_2, like Al_2O_3 [27]. The OTS treatment thereby creates a hydrophobic gate insulator surface, which is known to prevent trap formation related to molecular layers of water at the interface [28]. Potentiometry provides a sophisticated way to get deeper insight into the electric behaviour of the untreated and OTS treated channel region, respectively.

In the present work, we summarise our activities over the last few years within a project funded by the Deutsche Forschungsgemeinschaft within the scientific priority programme *Organic field effect transitors: Structural and dynamical properties*. The strategy to investigate the potential profile with potentiometry was inspired by the need to understand the influence of the injection barrier at the source contact on the performance of the entire device, and by previous investigations addressing this issue in terms of voltage-dependent contact resistances [12, 13, 16, 29, 30]. From the pronounced degradation of bottom-contacted pentacene OFETs stored under atmospheric conditions, it became clear that the amount of traps in the channel region increases with time, reducing the mobility in the channel region [31, 32]. A QTS characterisation of such non-ideal devices reveals a large amount of monoenergetic trap levels [33], presumably related to water molecules or to unintentional impurities like pentacenequinone.

Even though such bottom-contacted pentacene OFETs using unpurified source material were considered to represent a viable route towards commercial applications some years ago, they no longer represent the state of the art. Instead, it was demonstrated by various investigations that a pre-purification of the source material is mandatory. Moreover, there is clear evidence that top-contacted pentacene OFETs show a much better device performance, mainly because the deposition of gold on the organic material does not strongly deteriorate the morphology of the organic material, so that the source and drain contacts are improved significantly. Therefore, the subsequent step in the project consisted in the growth of the active layer from prepurified pentacene, with gold contacts deposited on top of the pentacene without breaking the vacuum. In potentiometry, these samples do not show a pronounced potential drop at the injecting contact, and QTS reveals a substantial reduction of the trap density, resulting in a much larger hole mobility.

Section 20.2 presents the sample geometries and preparation techniques, and Sections 20.3 and 20.4 summarise our findings for bottom and top contact geometries, respectively. In the latter case, we compare gate oxides with and without application of an OTS self-assembled monolayer. The main achievements of the present work are summarised in Section 20.5.

20.2
Device Geometries and Sample Preparation

The bottom-contacted pentacene OFETs were grown on an oxidised p-conducting silicon wafer with 5 nm Ni adhesion layers covered by about 50 nm thick Au contacts for source and drain [31, 33]. A similar sample geometry has been applied to a careful comparison of the impact of different gate dielectrics onto the overall device performance [32]. In our case, the gate oxide was grown by wet oxidation at 960 °C, resulting in a thickness of 125 nm, as determined by ellipsometry. The unpurified pentacene was thermally evaporated as received from the supplier (Aldrich), with a deposition rate kept at about 1 Å s^{-1} until the film had reached a thickness of about 30 nm. During the deposition, the sample was kept under high vacuum conditions of 2×10^{-6} mbar at room temperature. For the bottom-contacted OFETs, all measurements presented in the following were performed after two months storage under atmospheric conditions. The degradation of similar samples over an even longer time interval has been discussed elsewhere [32].

For the top-contacted OFETs, highly p-doped silicon substrates with 100 nm thermally grown SiO_2 were used. For contacting the structure *in situ*, it was necessary to apply macroscopic silver contacts with a thick (several μm) buffer layer of CaF_2 between silver and SiO_2 to protect the thin gate oxide against the force of the contact pins, compare Figure 20.1b. After this preparation step the substrates were cleaned by wiping with acetone, isopropanol and deionised water. One of the samples was treated with octadecyltrichlorosilane (OTS) while the other sample remained untreated. A 10 nm or 30 nm thick film of pentacene was produced by organic molecular beam deposition (OMBD) from a Knudsen cell under high vacuum conditions ($p < 4 \times 10^{-7}$ mbar) and at room temperature. The sublimation-purified pentacene (Sigma Aldrich) was carefully degassed prior to deposition through a mask. The layer thickness was controlled by a calibrated quartz micro-balance mounted close to the sample position. After deposition of pentacene, gold was thermally evaporated from a tungsten boat for the source and drain contacts.

From detailed radiotracer investigations of the diffusion of metal atoms into organic films, it became clear that gold atoms can diffuse substantially into the molecular layers [34]. In the case of pentacene OFETs with gold top contacts, it was shown that the formation of metal clusters below the surface of the organic material tends to increase the injection barrier [34]. Therefore, in order to minimise diffusion of gold into our pentacene films, during deposition of the top contacts the sample temperature was kept below $T = -150$ °C by cooling with liquid nitrogen. The channel structure for the OFETs was defined by thin wires used as a shadow mask, resulting in well-defined contact edges. By using different wire diameters, we could simultaneously obtain up to four OFETs from the same gold deposition process, but with different channel lengths ($L = 17$ μm, 38 μm, 86 μm, 187 μm). A channel width of $W = 3$ mm is also defined by this kind of mask. All electrical measurements were carried out *in situ* using a HP4140A pA meter/DC voltage source unit. The ground potential had to be applied to the gate back contact, so that formally the drain current is obtained as a function of U_{GD} and U_G, compare Figure 20.1b.

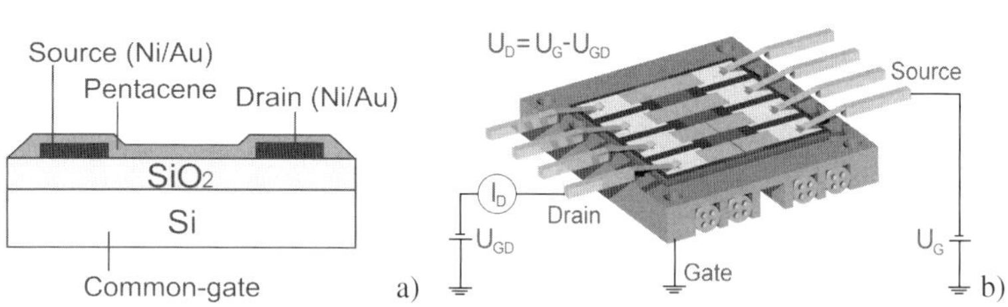

Figure 20.1 Sample geometries: (a) Schematic contact layout for bottom-contacted pentacene OFETs, and (b) for top-contacted pentacene OFETs, with sample holder and electrical contacts.

20.3 Pentacene OFETs With Bottom Contacts

20.3.1 Potentiometry and Electrical Probes

Assuming that the drain current I_D is reduced due to the voltage drop over the contact resistance R_C, an estimate for the contact resistances and the mobility in the channel region can be obtained from drain conductance $g_D = \partial I_D / \partial U_D$ and transconductance $g_m = \partial I_D / \partial U_G$ [35]. Based on a drain current according to

$$I_D = \mu C_i \frac{W}{L} U_G (U_D - I_D R_C), \tag{1}$$

where $C_i = \varepsilon_0 \varepsilon / d$ is the capacitance per area of the gate insulator of thickness d, the resulting hole mobility

$$\mu = \frac{g_D^2}{g_m} \frac{L}{W} \frac{U_D}{C_i U_G^2} \tag{2}$$

can be estimated to be about $\mu = 9 \times 10^{-3}$ cm^{-1} V^{-1} s^{-1} from the measured drain current, with a contact resistance $R_C = 0.6$ GΩ, indicating that the performance of our bottom-contacted pentacene OFET is contact-limited [33]. Therefore, a microscopic investigation of the voltage profile between source and gate contacts is highly desirable.

In Figure 20.2a, we present potentiometry traces obtained with an atomic force microscope (AFM) built by Anfatec [36] together with the output characteristics. The geometric positions of the contact edges are estimated to be at 0.2 μm from the potential kink close to the source and close to the position where the drain potential is reached. An earlier assignment based on the topography was somewhat misleading because the pentacene accumulates on top of the contact edges, inhibiting a precise geometric definition [33]. From the potentiometry traces at large gate voltages, there is strong evidence for a large voltage drop of up to 19 V related to the injection of holes from the source (S) contact.

In the AFM, the laser diode used to control the movement of the cantilever induces photogenerated charge carriers, so that the threshold voltage shifts from about $U_T = +6$ V to $U_T = +11$ V, compare Figure 20.2b for the output characteristics with and without the laser diode.

20.3.2 Mobility Estimates

The electric field $E_x(x)$ can be deduced from a differentiation of the smooth fitting function displayed in Figure 20.2a, revealing rather large electric fields

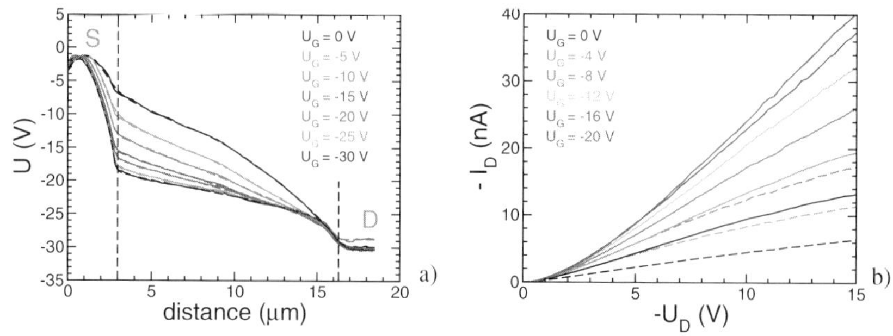

Figure 20.2 (a) Measured surface potential $U(x)$ (solid line) as obtained with an AFM Kelvin probe at $U_D = -30$ V, for different gate voltages U_G, as annotated, and fit to a smooth interpolation formula (dashed); (b) output characteristics with (solid line) and without (dashed line) light from the laser diode in the AFM.

close to the source and drain contacts together with a strongly reduced electric field in the main part of the channel. For the largest gate voltage of $U_G = -30$ V, an accumulation of majority charge carriers is expected to occur over the entire channel region, so that it can be estimated from the gate capacitance. From a self-consistent solution of the one-dimensional Poisson equation, it can be shown that a large part of the gate field is screened by the accumulated majority charge carriers in the first monolayer (ML) of the organic film, whereas the fraction of the gate field penetrating deeper into the active material accumulates smaller charge densities in the subsequent layers [37]. For simplicity, in the following analytical estimates, this situation will be replaced by a complete screening of the gate field within the first ML, so that the density of positive charges in the first ML of pentacene is defined by the gate capacitance:

$$n(x) = \frac{C_i}{qd_{\mathrm{ML}}}[U(x) - U_G], \tag{3}$$

where $U(x)$ is the voltage along the channel and $d_{\mathrm{ML}} \approx 1$ nm the monolayer thickness. Together with the continuity of the current density,

$$\frac{I_D}{Wd_{\mathrm{ML}}} = j(x) = en(x)\,\mu(x)\,E_x(x) = \text{const.} \tag{4}$$

and the electric field $E_x(x)$ along the channel, Eqs. (3), (4) allow the definition of a microscopic hole mobility, and according to Figure 20.3, the mobility would drop by two orders of magnitude close to the source contact. When exchanging the completely equivalent source and drain contacts, the voltage drop at the injecting contact occurs at the other end of the channel. As a consequence, the above analysis results in a reversed mobility profile, a rather counter-intuitive result. Between the middle of the channel and the drain contact, the capacitance formula is compatible with a nearly constant mobility. Therefore, we conclude that the quality of the pentacene is not severely deteriorated due to a different morphology induced by the metal contacts.

Assuming instead a constant mobility over the entire channel region as visualised by the second set of curves in Figure 20.3, we find that the charge density obtained from the continuity of the current density remains close to the capacitance formula between the middle of the channel and the drain contact. Close to the source contact, the assumption of constant mobility results in a strongly reduced density of majority carriers. Analysing this region with the model of space charge limited current from a region of low charge density at the injecting contact towards a region of high charge density at the beginning of the accumulation channel, we can deduce that the region of reduced charge density at the beginning of the channel extends over about 0.2 μm. In the two-dimensional device simulation discussed below, it will be demonstrated that this transport regime is closely related to hole injection over a rather large Schottky barrier.

20.3.3
Two-Dimensional Device Simulation

In order to assess the two key parameters of an OFET, i.e. the mobility of the charge carriers and the injection barrier, we have performed a two-dimensional

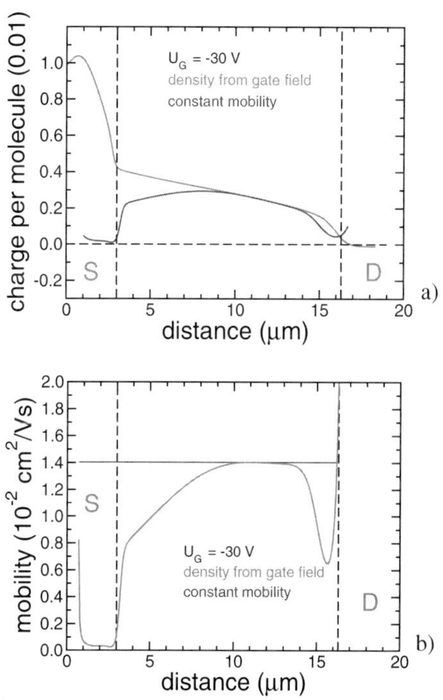

Figure 20.3 (a) Charge per molecule in the ML of pentacene closest to the gate oxide, as estimated with Eq. (3) from the gate capacitance (red), and hole density resulting for a constant mobility of $\mu = 1.4 \times 10^{-2}$ cm^2 V^{-1} s^{-1} (blue). (b) Mobility of holes, for a charge density in accordance with the gate capacitance (red), and a constant mobility throughout the entire channel region (blue).

simulation for an OFET with length $L = 16$ µm, width $W = 100$ µm, and a gate oxide with a thickness of $d = 150$ nm. The Poisson equation and the current continuity equation including drift and diffusion currents are solved self-consistently. Due to the restricted number of about 350 000 discretisation points we were able to use in our implementation of the FEMLAB simulation package, it was not possible to resolve the length scale of the barrier lowering related to the image charges in the Au contact. Instead, we use a lower *effective* barrier including the influence of the mirror charges. The hole mobility and the effective injection barrier were varied until we found the best agreement with the potentiometry trace observed at $U_G = -30$ V, compare Figure 20.4, resulting in a hole mobility of $\mu = 0.014$ cm^2 V^{-1} s^{-1} and an effective injection barrier of 0.42 eV. Taking into account the injection field, the dielectric constant of $\varepsilon = 3.1$ of pentacene, and the resulting mirror charge in the source contact, the effective injection barrier of 0.42 eV is equivalent to an injection barrier of 0.73 eV, in good agreement with a value of 0.85 eV obtained with spectroscopic studies of gold/pentacene interfaces [38]. For smaller gate voltages, the electric field close to the source contact remains smaller, resulting therefore in an effective barrier above 0.42 eV.

Close to the drain contact, the simulated potential profile for $U_G = -30$ V is in good agreement with the observed potentiometry trace shown in Figure 20.2. Based on previous potentiometry measurements on polymer OFETs revealing a substantial potential drop at the drain contact [13], two-dimensional device simulations have suggested that the mobility close to the contacts

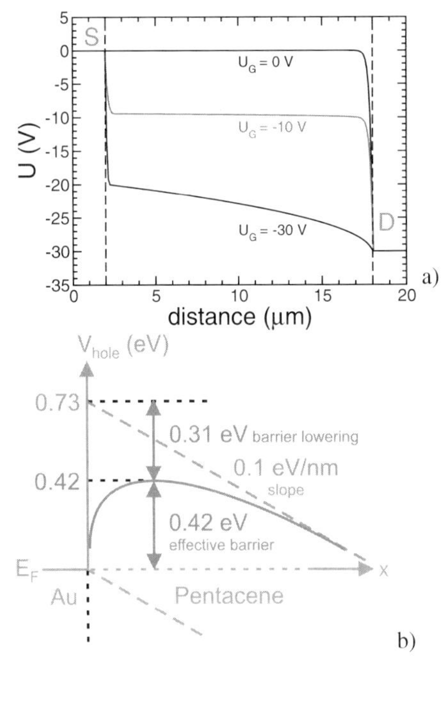

Figure 20.4 (a) Results of a two-dimensional simulation of a pentacene OFET with geometric parameters close to the bottom-contacted device investigated with potentiometry, for $\mu = 0.014$ cm^2 V^{-1} s^{-1} and an effective injection barrier of 0.42 eV, and (b) construction of an injection barrier of 0.73 eV out of the effective barrier of 0.42 eV and the electric field close to the source contact, as obtained in the simulation for a gate voltage of $U_G = -30$ V .

is strongly reduced [39]. Neither our potentiometry traces nor the simulation in the accumulation regime give evidence of a large potential step at the drain contact, so that we assume that the mobilities are similar in the channel region and close to the contacts, in agreement with the interpretation of Figure 20.3.

The close-up of the simulated electrostatic potential in Figure 20.5a reveals that the potential drop near the source contact is steeper at the bottom of the pentacene channel, whereas at the pentacene/air interface, it is smeared out over a somewhat longer distance. Moreover, in the experimental potentiometry traces, an uncontrolled amount of pentacene crystallites with a height exceeding 0.1 μm close to the gold contacts [33] results in a modified surface potential, so that in this region, the observed surface potential is no more a fair representation of the situation at the interface between the channel and the gate dielectric. Due to the injection over a large potential barrier, the hole density close to the source contact is very low, compare Figure 20.5b, so that the hole accumulation expected from Eq. (3) develops only at about 250 nm from the injecting contact. The existence of such a depletion layer was invoked already in earlier potentiometry studies of polymer-based OFETs with a poor alignment of the metal work function and the transport level in the organic material [15]. Only behind this intermediate region governed by space charge limited current, the gate field is entirely screened by the accumulated hole density, so that the electrostatic potentials at the bottom and at the top of the pentacene channel agree.

For a gate voltage $U_G = -30$ V, the entire channel is in the accumulation regime, so that the photogeneration of charge carriers in the AFM modifying the output characteristics in Figure 20.2b should be of minor importance, and as a consequence, our transport model without background doping and photoexci-

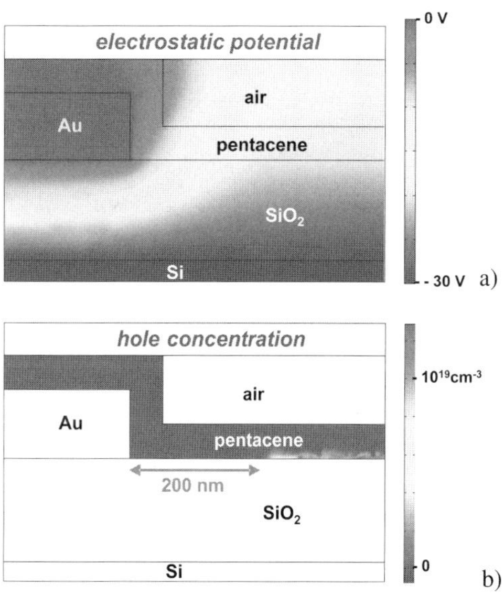

Figure 20.5 (a) Electrostatic potential and (b) hole density obtained in the two-dimensional device simulation for $U_G = -30$ V, within the first 0.4 μm from the source contact.

excitation is in reasonable agreement with the observed potentiometry traces. On the other hand, as evidenced in Figure 20.4a, the simulation for the same mobility and injection barrier fails for smaller gate voltages: The pinchoff point is pushed towards the drain contact, in sharp contrast to the potentiometry traces where it occurs close to the source. This failure of the simulation gives clear evidence that photogenerated carriers and unintentional doping of the organic material do not only modify the output characteristics in Figure 20.2b, but that they are also required for a microscopic interpretation of details of the electrostatic potential landscape as determined in an AFM Kelvin probe generating electron–hole pairs. From preliminary results based on a two-dimensional device simulation including a background density of charge carriers resulting from doping or illumination, it was found that potential profiles resembling the observed potentiometry traces can be obtained [40].

From the agreement of the simulated electrostatic potential for $U_G = -30$ V with the measured potentiometry profile, we conclude that our values of $\mu = 0.014$ cm^2 V^{-1} s^{-1} for the hole mobility and 0.73 eV for the injection barrier are suitable material parameters, *under the assumption* that the pentacene has *similar properties* in the channel region and close to the contacts. From high resolution AFM studies of pentacene layers grown on pre-patterned gold bottom contacts on SiO_2, it was found that the channel region consists of large pentacene crystallites forming a closed film, but the morphology on top of the gold contacts is totally different [41]. This disparate growth behaviour can be assigned to a first ML of pentacene where the molecules nucleate in a flatlying geometry, followed by a subsequent regime during which the organic layer grows in a herringbone geometry, with the long axes of the molecules parallel to the substrate plane. Even for large film thicknesses, a morphology with molecules standing upright like in the channel region is not recovered [41, 42].

Obviously, for such a complicated morphology of the organic layer on top of the contacts, it cannot be guaranteed that it may be described with similar microscopic parameters. From the continuity of the current density in Eq. (4) where the electric field can be estimated from the potentiometry traces, the assumption of a *reduced mobility* would require a *larger hole density*, necessitating in turn a *lower injection barrier*. In any case, a microscopic interpretation of the observed large potential drop at the injecting source contact will still require a rather large injection barrier paired with a rather low mobility of the charge carriers.

20.3.4
Charge Transient Spectroscopy

In the following, we present the results of charge transient spectroscopy performed on the bottom contacted pentacene OFETs, a variant of DLTS where the current transient is integrated, yielding a charge transient [43, 44]. In combination with capacitance DLTS, this technique can also provide information on the depth profile of the trap distribution [45].

The comparison between the signature of a single decay rate with the measured QTS signal allows an assignment of the density of states of the deep traps, as demonstrated for CuPc [46]. The DLTS traces have been obtained after switching a source–drain voltage of $U_D = 0$ to $U_D = -6$ V, with floating gate voltage. The charge transients are evaluated according to a filter function compiled from three consecutive measurements at the times t_1, $2t_1$ and $4t_1$ after switching the voltage:

$$\Delta Q = Q(t_1) - \tfrac{3}{2}Q(2t_1) + \tfrac{1}{2}Q(4t_1). \qquad (5)$$

If each entry $Q(t)$ in Eq. (5) results from the integration of a current transient with a single exponential decay time τ,

$$Q(t) = Q_0\left[1 - \exp\left(\frac{-t}{\tau}\right)\right], \qquad (6)$$

Eq. (5) provides a filter for $\tau \approx t_1$. Using the filter function in Eq. (5) together with a single decay constant τ according to Eq. (6), we have fitted the measured charge transients in Figure 20.6, excluding data points for the longest values of t_1. The good quality of the fit demonstrates that the charge transients are dominated by a trap level with a discrete decay time τ.

If the density of states (DOS) of the trap states were broadened, the charge transient $Q(t)$ would not follow a single exponential rise as in Eq. (6), resulting in turn in broadened QTS traces with respect to the fit based on Eqs. (5), (6). Such a behaviour was observed for polymer-based diodes [20] and for phthalocyanines [46], but for our bottom-contacted pentacene OFETs, we found no evidence of a broadened DOS of the trap states with a corresponding distribution of de-trapping rates.

According to Figure 20.6, the decay constant τ depends on temperature, a behaviour which can be rationalised with an Arrhenius dependence for the de-trapping rate,

$$\gamma(T) = \frac{1}{\tau(T)} = \exp\left(\frac{-E_{act}}{k_B T}\right), \qquad (7)$$

resulting in an activation energy of $E_{act} = 125 \pm 8$ meV. This shallow trap has a much lower activation energy with respect to earlier investigations of hole traps in pentacene, where the dominating trap states had activation energies of 0.24 eV, 0.31 eV, and 0.67 eV [22, 47]. As discussed elsewhere in more detail, larger pulses applied to the drain voltage result in more complex DLTS data, revealing several further trap energies [23]. We assign the dominating trap level at $E_{act} = 125 \pm 8$ meV to unintentional dopants in the unpurified pentacene source material like pentacenequinone [48] or dihydropentacene [49]. However, we cannot exclude that this trap might be related to adsorption of water at the SiO_2/pentacene interface [28], creating trap states after completing the growth of the sample. This second hypothesis would be in keeping with the

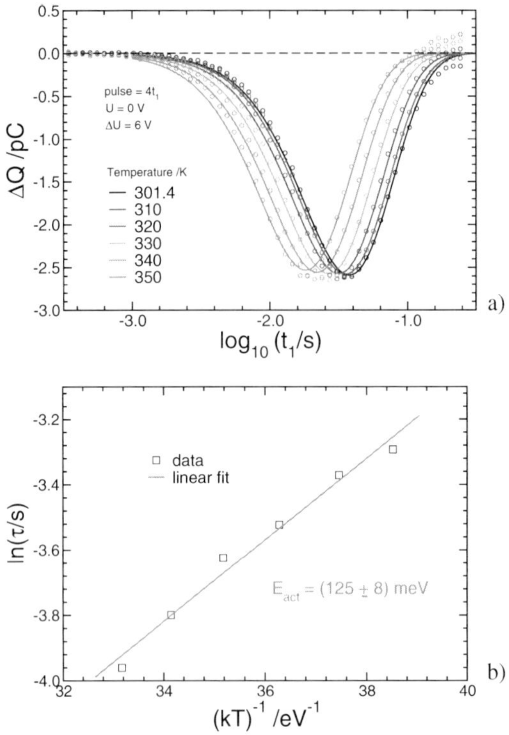

Figure 20.6 (a) Charge transients calculated according to Eqs. (5), (6) from three consecutive times t_1, $2t_1$ and $4t_1$ after switching the drain voltage, for floating gate voltage. At each temperature, the measured data points are fitted with a single decay constant τ. b) Arrhenius plot of the temperature-dependent decay times according to Eq. (7), revealing an activation energy of $E_{act} = 125 \pm 8$ meV for de-trapping.

observed degradation of similar devices over a time scale of several months [32].

An alternative method to quantify the amount of trap states is based on the pronounced hysteresis of the output characteristics, interpreted in terms of emptying and filling of traps at the pentacene/SiO_2 interface. From the observed shift of the threshold voltage, it was found that the amount of trap states resembles the density of structural defects determined independently [41]. Finally, from temperature-dependent measurements of the device characteristics, it could be checked if the activation energy for de-trapping obtained from DLTS coincides with the activation energy determined directly from the drain current [49]. Deviations between the so determined activation energies would reflect the temperature dependence of the charge injection.

20.4
Investigations of Top-Contacted Pentacene OFETs

20.4.1
Electrical Characterisation *In Situ*

The comparative *in situ* electrical measurements on the top-contacted pentacene OFETs with and without OTS treatment reveal an increased charge carrier mobility for the OTS-treated sample, leading in turn to higher drain currents, compare Figure 20.7. Even though the OTS treatment has only a tiny influence on the threshold voltage, the mobility rises substantially from $\mu_{lin} = 0.13$ cm^2 V^{-1} s^{-1} to $\mu_{lin,OTS} = 0.32$ cm^2 V^{-1} s^{-1} in the linear regime, and from $\mu_{sat} = 0.17$ cm^2 V^{-1} s^{-1} to $\mu_{sat,\,OTS} = 0.69$ cm^2 V^{-1} s^{-1} in saturation. This increase reveals that the OTS treatment improves the morphology of the pentacene film, reducing the number of grain boundaries and interface trap states between the SiO_2 gate dielectric and the pentacene channel.

From the comparison with OFETs grown on other oxidised Si wafers, we conclude that the large threshold voltages determined in Figure 20.7b can be related to a specific deficiency of the SiO_2 used in the present samples, because other substrates did not show a similar feature. However, the microscopic origin of the observed variation of the threshold voltages between different substrates is not understood yet.

20.4.2
Potentiometry Measurements *Ex Situ*

Comparing the potentiometry measurements in Figure 20.8 obtained on top-contacted pentacene OFETs with the data for the bottom-contacted sample in Figure 20.2, the most striking difference is the absence of a substantial potential drop close to the source contact. The reduction of the contact resistance

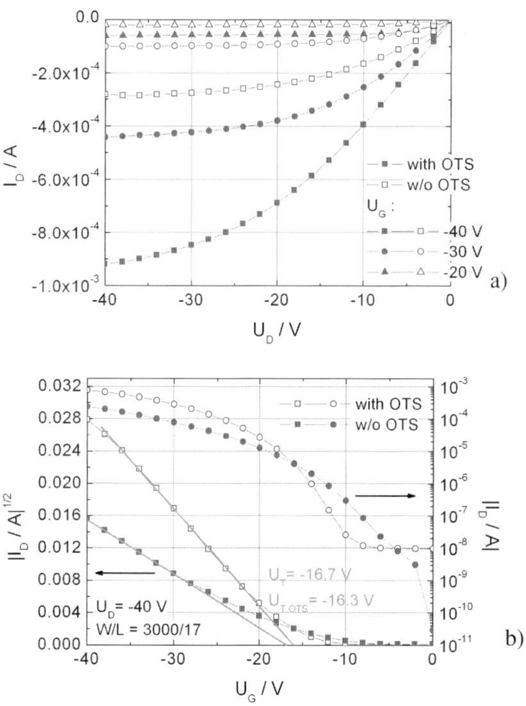

Figure 20.7 (a) Comparison between the output characteristics $I_D(U_D)$ obtained on an OTS-treated sample ($L = 17$ µm, $W = 300$ µm) with an untreated reference sample, for different gate voltages U_G, as annotated. (b) Transfer characteristics $|I_D(U_G)|$, and determination of threshold voltage U_T from $\sqrt{|I_D(U_G)|}$, for fixed drain voltage $U_D = -40$ V.

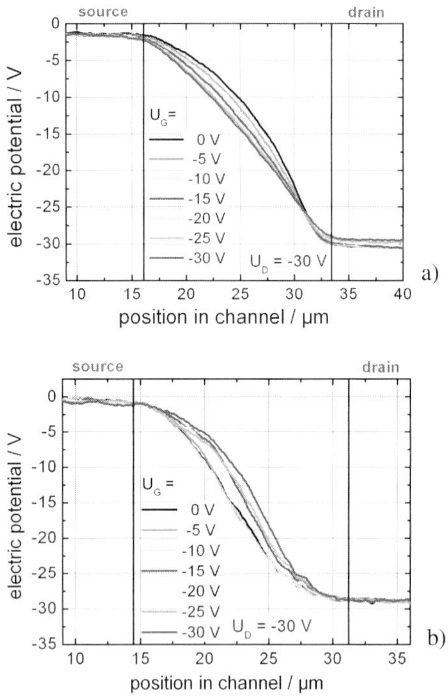

Figure 20.8 Comparison between the potentiometry traces obtained (a) without OTS treatment of the gate oxide, and (b) with OTS self-assembled monolayer, for different gate voltages, as annotated.

and the underlying height of the Schottky barrier reveals an improved contact interface when gold is deposited on top of pentacene at a very low substrate temperature of $T = -150$ °C. The AFM picture in Figure 20.9b shows the region of the 17 μm long channel, where the lighter stripe in the centre represents the channel and the darker borders on the left and right side the source and drain contacts. Obviously, gold adapts quite well to the rough surface morphology defined by the pentacene grains, so that the gold and pentacene surfaces become almost indistinguishable from each other. Even though diffusion of gold into the pentacene film cannot be excluded from the AFM picture, the low injection barrier gives clear evidence that the formation of metal clusters inside the organic film remains very small because such clusters would increase the injection barrier [34]. This result demonstrates that the very low substrate temperature of $T = -150$ °C during the deposition of the gold contacts inhibits the diffusion of metal atoms into the organic material, defining instead a rather sharp interface.

By differentiating the electric potential U obtained from potentiometry measurements, one obtains the electric field $E_x(x) = -\partial U/\partial x$ in the channel region, compare Figure 20.9a. In contrast to the asymmetric electric field distribution in the untreated pentacene OFET, in the OTS treated sample the field is rather symmetric between source and drain contacts.

Taking into account the large threshold voltages shown in Figure 20.7b, it is clear that the potentiometry traces in the range $U_G = 0$ V to $U_G = -15$ V do not

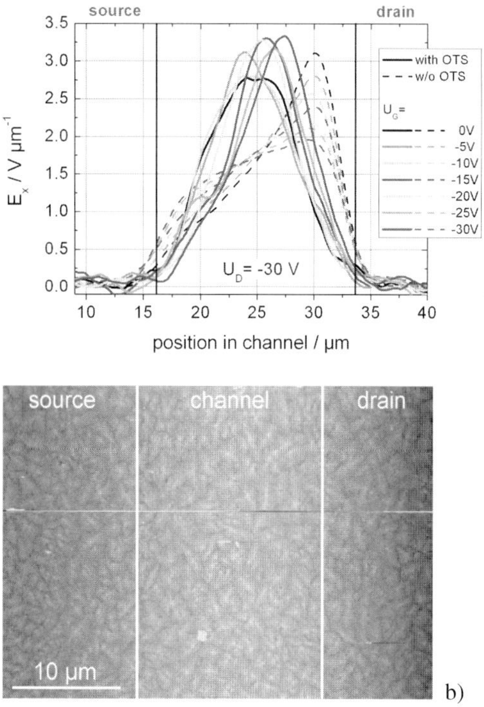

a)

b)

Figure 20.9 (a) Comparison of electric field distribution with (solid lines) and without (dashed) OTS treatment of the gate oxide, (b) AFM picture of a pentacene OFET with untreated gate oxide, revealing a channel length of about 17 μm.

correspond to charge accumulation in the channel region. However, due to the charge carriers generated by the laser diode in the AFM, even in this range of gate voltages, the transistor is not in the off-state, but merely in an ambipolar transport regime. Hence, the photogeneration in the AFM Kelvin probe results in a positive shift of the threshold voltage, as observed for the bottom-contacted samples, compare Figure 20.2b.

20.4.3 Charge Transient Spectroscopy

For the top-contacted OFET without OTS treatment of the gate oxide, the QTS spectra in Figure 20.10 do not show any impurities with a de-trapping time around 20 ms, compare Figure 20.6, where deep traps with a response time in this range were dominating. Instead, the QTS data show much smaller features at faster response times. From the comparison of the QTS data obtained on the bottom-contacted and the top-contacted OFETs, we relate the low hole mobility in the former with a kind of impurity not present in the latter. Presumably, this type of impurity is related to the low quality of the unpurified pentacene material used for growing the bottom-contacted pentacene film because a trap with similar properties can no longer be observed when growing the films from pentacene purified by sublimation.

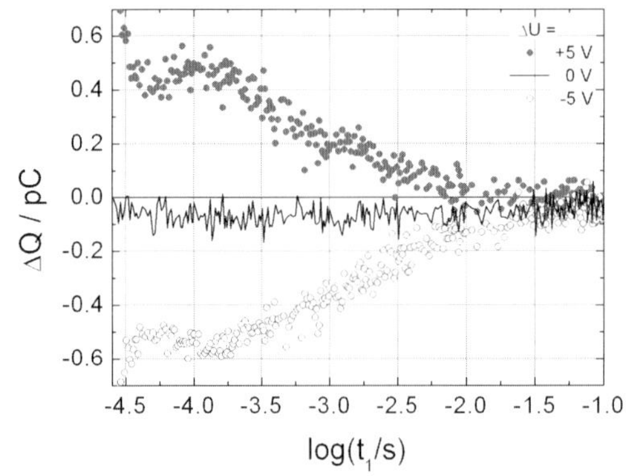

Figure 20.10 QTS data for a top-contacted pentacene OFET without OTS treatment, with $L = 17$ μm and $W = 3000$ μm.

20.5 Conclusion

In the present work, we have investigated pentacene-based OFETs with electrical probes, potentiometry, and DLTS. For bottom-contacted pentacene, the comparison between two-dimensional device simulations and the observed potentiometry traces reveals a substantial misalignment of about 0.73 eV between the work function in the Au bottom electrodes and the hole transport level in pentacene, in good agreement with a value of 0.85 eV obtained earlier [38]. The failure of the device simulation in the pinchoff regime gives clear evidence that the potential profile obtained with potentiometry is modified by photogenerated carriers. In principle, such an influence of the measuring device on the potential profile obtained could be avoided by using an AFM Kelvin probe with an infrared laser working in the transparent region of pentacene.

For top-contacted samples, the potentiometry measurements give clear evidence for a strongly reduced injection barrier. A treatment of the gate oxide with OTS prior to deposition of the pentacene increases the hole mobility to 0.32 $cm^2 V^{-1} s^{-1}$ in the linear regime, or 0.69 $cm^2 V^{-1} s^{-1}$ in saturation. In both variants of the top-contacted OFETs investigated, i.e. with and without passivation of the gate oxide with OTS, there is no evidence for a substantial Schottky barrier at the pentacene/Au interfaces. Charge transient spectroscopy reveals that the unintentional doping and the density of deep traps are strongly reduced. The substrates used for these top-contacted OFETs produce a substantial threshold voltage assigned to a deficiency of the oxide layers, not to traps at the various organic/inorganic interfaces. From a comparison of the QTS spectra obtained on different qualitites of pentacene, we conclude that the dominating trap level in non-purified pentacene can be eliminated by sublimation of the source material prior to the growth of the organic device.

Acknowledgements

We thank J. Pflaum for providing pentacene purified by sublimation, C. Pannemann and U. Hilleringmann for the bottom-contacted OFETs, C. Kaufmann for the Si/SiO_2 substrates, T. Baumgärtel and H. Graaf for preparing the self-assembled monolayers of OTS, and I. Thurzo and A. Fechner for expert advice concerning the DLTS/QTS method. This work was funded by the Scientific Priority Programme 1121 *Organic field effect transistors: Structural and dynamical properties* (projects Scho 521/5 and Za 146/14) financed by the Deutsche Forschungsgemeinschaft.

References

1. N. Karl, Organic Semiconductors, in: Landolt–Börnstein (New Series), Group III, Vol. 17i (Springer, Berlin, 1985), pp. 106–218.

2. V. Podzorov, V. M. Pudalov, and M. E. Gershenson, Appl. Phys. Lett. **82**, 1739 (2003).
V. Podzorov, S. E. Sysoev, E. Loginova, V. M. Pudalov, and M. E. Gershenson, Appl. Phys. Lett. **83**, 3504 (2003).

3. R. W. I. de Boer, M. Jochemsen, T. M. Klapwijk, A. F. Morpurgo, J. Niemax, A. K. Tripathi, and J. Pflaum, J. Appl. Phys. **95**, 1196 (2004).

4. V. C. Sundar, J. Zaumseil, V. Podzorov, E. Menard, R. L. Willett, T. Someya, M. E. Gershenson, and J. A. Rogers, Science **303**, 1644 (2004).

5. C. Goldmann, S. Haas, C. Krellner, K. P. Pernstich, D. J. Gundlach, and B. Batlogg, J. Appl. Phys. **96**, 2080 (2004).

6. V. Podzorov, E. Menard, A. Borissov, V. Kiryukhin, J. A. Rogers, and M. E. Gershenson, Phys. Rev. Lett. **93**, 086602 (2004).

7. K. Hannewald, V. M. Stojanovic, J. M. T. Schellekens, P. A. Bobbert, G. Kresse, and J. Hafner, Phys. Rev. B **69**, 075211 (2004).

8. K. Hannewald and P. A. Bobbert, Appl. Phys. Lett. **85**, 1535 (2004).

9. S. J. Kang, M. Noh, D. S. Park, H. J. Kim, C. N. Whang, and C.-H. Chang, J. Appl. Phys. **95**, 2293 (2004).

10. J. Zaumseil, K. W. Baldwin, and J. A. Rogers, J. Appl. Phys. **93**, 6117 (2003).

11. R. J. Chesterfield, J. C. McKeen, C. R. Newman, C. D. Frisbie, P. C. Ewbank, K. R. Mann, and L. L. Miller, J. Appl. Phys. **95**, 6396 (2004).

12. K. Seshadri and C. D. Frisbie, Appl. Phys. Lett. **78**, 993 (2001).

13. L. Bürgi, H. Sirringhaus, and R. H. Friend, Appl. Phys. Lett. **80**, 2913 (2002).

14. J. A. Nichols, D. J. Gundlach, and T. N. Jackson, Appl. Phys. Lett. **83**, 2366 (2003).

15. L. Bürgi, T. J. Richards, R. H. Friend, and H. Sirringhaus, J. Appl. Phys. **94**, 6129 (2003).

16. K. P. Puntambekar, P. V. Pesavento, and C. D. Frisbie, Appl. Phys. Lett. **83**, 5539 (2003).

17. C. R. Newman, R. J. Chesterfield, J. A. Merlo, and C. D. Frisbie, Appl. Phys. Lett. **85**, 422 (2004).

18. A. B. Chwang and C. D. Frisbie, J. Phys. Chem. B **104**, 12202 (2000).

19. D. K. Schroeder, Semiconductor material and device characterization (Wiley, New York, 1998), chap. 5.6.

20. A. J. Campbell, D. D. C. Bradley, E. Werner, and W. Brütting, Org. Electron. **1**, 21 (2000).

21. O. Gaudin, R. B. Jackman, T.-P. Nguyen, and P. Le Rendu, J. Appl. Phys. **90**, 4196 (2001).
22. Y. S. Yang, S. H. Kim, J.-I. Lee, H. Y. Chu, L.-M. Do, H. Lee, J. Oh, T. Zyung, M. K. Ryu, and M. S. Jang, Appl. Phys. Lett. **80**, 1595 (2002).
23. I. Thurzo, B. Paez, H. Méndez, R. Scholz, and D. R. T. Zahn, phys. stat. sol. (a) **203**, 2326 (2006).
24. Y.-Y. Lin, D. J. Gundlach, S. F. Nelson, and T. N. Jackson, IEEE Electron Device Lett. **18**, 606 (1997).
25. K. S. Pyo and C. K. Song, Thin Solid Films **485**, 230 (2005).
26. S. Grecu, M. Roggenbuck, A. Opitz, and W. Brütting, Org. Electron. **7**, 276 (2006).
27. L. A. Majewski, R. Schroeder, M. Voigt, and M. Grell, J. Phys. D, Appl. Phys. **37**, 3367 (2004).
28. C. Goldmann, D. J. Gundlach, and B. Batlogg, Appl. Phys. Lett. **88**, 063501 (2006).
29. P. V. Necliudov, M. S. Shur, D. J. Gundlach, and T. N. Jackson, Solid State Electron. **47**, 259 (2003).
30. H. Klauk, G. Schmid, W. Radlik, W. Weber, L. Zhou, C. D. Sheraw, J. A. Nichols, and T. N. Jackson, Solid State Electron. **47**, 297 (2003).
31. C. Pannemann, T. Diekmann, and U. Hilleringmann, J. Mater. Res. **19**, 1999 (2004).
32. T. Diekmann, C. Pannemann, and U. Hilleringmann, phys. stat. sol. (a) **205**, 564 (2008); (Chapter 18, this book).
33. R. Scholz, A.-D. Müller, F. Müller, I. Thurzo, B. A. Paez, L. Mancera, D. R. T. Zahn, C. Pannemann, and U. Hilleringmann, Proc. SPIE **5940**, 594001 (2005).
34. M. Scharnberg, R. Adelung, and F. Faupel, phys. stat. sol. (a) **205**, 578 (2008); (Chapter 19, this book).
35. G. Horowitz, R. Hajlaoui, D. Fichou, and A. El Kassmi, J. Appl. Phys. **85**, 3202 (1999).
36. More detailed specifications can be found under http://www.anfatec.de.
37. G. Horowitz, R. Hajlaoui, R. Bourguiga, and M. Hajlaoui, Synth. Met. **101**, 401 (1999).
38. N. Koch, J. Ghijsen, A. Elschner, R. L. Johnson, J.-J. Pireaux, J. Schwartz, and A. Kahn, Appl. Phys. Lett. **82**, 70 (2003).
39. T. Li, P. P. Ruden, I. H. Campbell, and D. L. Smith, J. Appl. Phys. **93**, 4017 (2003).
40. C. Erlen et al., unpublished results.
41. B. Nickel, M. Fiebig, S. Schiefer, M. Göllner, M. Huth, C. Erlen, and P. Lugli, phys. stat. sol. (a) **205**, 526 (2008); (Chapter 15, this book).
42. G. Witte and C. Wöll, phys. stat. sol. (a) **205**, 497 (2008); (Chapter 11, this book).
43. J. W. Farmer, C. D. Lamp, and J. M. Meese, Appl. Phys. Lett. **41**, 1063 (1982).
44. I. Thurzo, D. Barančok, and M. Haluška, Rev. Sci. Instrum. **66**, 5360 (1995).
45. I. Thurzo, R. Beyer, and D. R. T. Zahn, Semicond. Sci. Technol. **15**, 378 (2000).
46. I. Thurzo, G. Pham, and D. R. T. Zahn, Semicond. Sci. Technol. **19**, 1075 (2004).
47. D. V. Lang, X. Chi, T. Siegrist, A. M. Sergent, and A. P. Ramirez, Phys. Rev. Lett. **93**, 076601 (2004).
48. O. D. Jurchescu, J. Baas, and T. T. M. Palstra, Appl. Phys. Lett. **84**, 3061 (2004).
49. M. Voigt, J. Pflaum, and M. Sokolowski, phys. stat. sol. (a) **205**, 449 (2008); (Chapter 8, this book).

21
Microscopic and Spectroscopic Characterisation of Interfaces and Dielectric Layers for OFET Devices

K. Müller, Y. Burkov, D. Mandal, K. Henkel, I. Paloumpa, A. Goryachko, and D. Schmeißer

21.1
Introduction

For the characterisation of inorganic materials, photoemission electron microscopy (PEEM) is a well-known method. Photoelectrons are collected and accelerated in a high-voltage field (in our system 7 kV). Visualisation is realised by a lens system and a multi-channel plate. Sufficient resolution and intensity is possible by excitation with standard Hg lamps. The method is used for characterisation of adsorbates [1–3], the investigation of magnetic domains [4], the mapping of doping inhomogeneities [5, 6] or the characterisation of surface morphology or elemental distributions [7–9].

The bright areas of a PEEM image represent regions with high electron density in states with binding energies low enough for photo-excitation. Typically, contrast is a result of electron yield. The contrast is generally influenced by the work function (different materials or reduced work function at edges, doping level), chemical composition or topography (shadowing) [10]. Giesen et al. [11] show a wealth of information for PEEM, in the range of the space charge region in Si.

Especially for organic field effect transistors (OFETs), the investigation of the potential distribution and charge carrier density inside the channel region of the transistor is of interest. First of all, an interpretation of the PEEM contrast of an organic device is essential. Here, we present PEEM measurements of organic devices under applied voltages and give an interpretation of the contrast.

A further spatially resolved method, also based on work function contrast, is scanning Kelvin probe microscopy (SKPM). As an extended version of atomic force microscopy (AFM), additional information on the local surface potential is revealed by a second feedback circuit. The method delivers information depending on the value $\varphi = \varphi(x) + \Delta\phi(x)$. Here, $\Delta\phi(x)$ is the difference in work function between the sample and the AFM tip and $\varphi(x)$ is the local electric potential [12]. $\varphi(x)$ itself gives information on additional surface charges due to

Organic Electronics. Structural and Electronic Properties of OFETs. Edited by Christof Wöll

ISBN: 978-3-527-40810-8

applied fields, gradients or inhomogeneities of the local source–drain field or additional potential distributions due to an applied gate voltage.

The method was introduced by Bürgi et al. for the characterisation of OFETs in the working state with P3HT as the active semiconductor [12]. A continuous variation with an almost constant slope of potential inside the channel region is reported for transistors operating in the linear regime. This behaviour is confirmed by our own measurements, discussed below.

In addition to lateral information, depth profiling of interface chemistry is important for the understanding of organic functional structures. X-ray photoelectron spectroscopy (XPS) delivers chemical information for an interface region of around 2 nm in depth. Especially for extended applications, the knowledge of interface reactions, band offsets or charge transfer processes is important.

An actual example of an extended application with technological potential is the organic and non-volatile memory, based on an OFET. The main advantage of a non-volatile memory is that its information is maintained during the readout procedure (non-destructive readout, NDRO). A ferroelectric polymer is introduced as a dielectric layer in the OFET. Due to the ferroelectric alignment of dipoles the threshold voltage of the transistor is affected by their dipole moments and can be used as stored information.

For optimised operation voltages of the memory, a downscaling of the ferroelectric film thickness is necessary. During this scaling, the influence of interfaces and interface reactions becomes important. Here, we present an XPS study of interface reactions between the organic ferroelectric layer and the material of the electrode.

As ferroelectric material we use poly(vinylidene fluoride-trifluoroethylene) (P(VDF-TrFE)). This copolymer is soluble in non-toxic reagents, for example 2-butanone. The preparation of organic and ferroelectric thin films via spin coating from solution is possible [13]. The polarisation field of P(VDF-TrFE) is relatively high, about 50 MV/m [14]. Here, a downscaling of the P(VDF-TrFE) film thickness into a range below 100 nm is necessary in order to use small bias voltages for polarisation.

Just in this range of film thickness, the investigated properties are rather diffuse. Some authors report an increase of the coercive field strength with decreasing film thickness [15–18]. However, this effect seems to depend on the electrode materials. Normally, thermally evaporated aluminium is used, due to its good adhesive strength, its low leakage current and its high conductivity. An alternative for aluminium was introduced by Naber et al. [14]. In their work, poly(3,4-ethylenedioxythiophene):poly(styrenesulfonic acid) (PEDOT:PSS) was used as a polymeric electrode. For this organic electrode, no reduced remanent polarisation for films below 100 nm in thickness was found, in contrast to electrodes made of aluminium. Switching time measurements at commonly used field strengths of 80–140 MV/m show a reduction of this value by three orders of magnitude, compared with aluminium electrodes. Fixed dipoles at the Al/P(VDF-TrFE) interface were taken into account by the authors as a reason for this switching time difference. As a consequence, the interface interactions in such systems with low film thickness seem to be very important indeed. This was the motivation for a comparative XPS study of the two interfaces P(VDF-TrFE)/Al and P(VDF-TrFE)/PEDOT:PSS, described below.

In addition, we report on results regarding electrical measurements of metal ferroelectric insulator semiconductor (MFIS) and OFET structures. Here we demonstrate the ferroelectric switching resulting in a bistable operation in capacitor as well as in transistor devices (ferroelectric OFET). The bistable behaviour is a prerequisite for a memory; the ferroelectric material offers the non-volatile and NDRO attributes. Leakage currents may destroy the stored information in such devices. For that reason an insulating buffer layer is introduced between the semiconductor and the ferroelectric layer. Generally, a part of the programming voltage drops over the buffering insulator. For low-voltage operation this programming loss should be as small as possible. Furthermore, in the 'off state' of the memory (grounded gate) due to charge neutrality the electric charges caused by the polarisation of the ferroelectric layer will be neutralised by an established field in the insulator, leading to depolarisation fields, which can destroy the saved information [19]. Therefore, a high capacitance of the buffer layer and a ferroelectric material with a moderate permittivity value are needed to reduce this effect, but also a charge injection in the case of thin buffer layers should be avoided. Also the thickness of the ferroelectric layer can be increased, but this contradicts the low voltage operation. Investigations on the optimisation of buffer layer thickness and material will be published soon [20]. The reported permittivity values of P(VDF-TrFE) are in the region of 10 [21, 22]. This fact fulfils very well the above-mentioned condition. In this chapter we will focus on the principle of operation of these devices.

21.2 Experimental

21.2.1 Microscopic Methods

21.2.1.1 PEEM

As organic semiconductor for the PEEM investigation, we used regio-regular poly(3-hexylthiophene) P3HT (Aldrich). Thin films of a thickness smaller than 100 nm were prepared by spin-coating from solutions. Printed source–drain electrodes for PEEM at applied voltages and stainless steel for resonant photoelectron spectroscopy (RPES) were used as substrate. We used stainless steel to prevent charging of the sample during measurements. As solvent for P3HT, dried chloroform was used. The electrodes were patterned by a printing technique with carbon black as the conductor. With this technique, a resolution of 20 ± 10 μm is possible [23]. The carbon black was nano-dispersed in water with an average grain size of about 20 nm, glycol was used as dispergator (Derrusol An1, Degussa, Frankfurt, Germany). The substrate for printing was a commercial printing foil (poly(ethylene terephthalate), PET).

The preparation of the organic semiconductor by spin-coating was performed in an argon atmosphere; film thickness and handling conditions were comparable to those applied for OFETs [23]. The samples were contacted in ultrahigh vacuum (UHV) to allow imaging in the channel while voltages (max. ±10 V) at the source and drain electrode were applied. We operated the PEEM instrument (Focus) by illumination with a Hg lamp (4.9 eV) or a D_2 lamp (7.3 eV). By using synchrotron radiation at the BESSY U49/2 beam line, we recorded X-ray absorption spectra (μ-XAS) with PEEM and we collected RPES data using an HA125 (Omicron) analyser [10].

21.2.1.2 **SKPM**

The organic semiconductor was again regio-regular P3HT (Aldrich), spin-coated into thin films from chloroform, on printed source–drain structures, as described above.

The distribution of electrical potential on the surface was measured with the SKPM technique, using an Omicron VT AFM instrument in UHV environment and an NT-MDT Smena AFM instrument in ambient environment. Both instruments operated in the basic non-contact AFM mode, but used different schemes to extract the Kelvin signal. The UHV AFM instrument utilised a single-pass technique, meaning that surface topography was acquired simultaneously with surface potential distribution, the latter using the frequency modulation (FM) technique [24]. The ambient AFM instrument was operated in the dual-pass mode, meaning that each line in the raster was scanned twice. During the first pass, topography was obtained in the usual non-contact AFM mode, and after completing a line and digitally storing its topography $z(\boldsymbol{r})$, the tip was elevated 100 nm above the first pass trajectory and moved along the trajectory $z(\boldsymbol{r}) + 100$ nm. During this second pass the surface potential was acquired using the amplitude modulation (AM) technique [24, 25].

21.2.2 **Ferroelectric Devices**

21.2.2.1 **Interface Characterisation**

We present an XPS study of the interfaces P(VDF-TrFE)/Al and P(VDF-TrFE)/PEDOT:PSS. As copolymer, we used P(VDF-TrFE) in a molar ratio of 70 : 30. The material was delivered as film from Piezotech SA, France. As solvent for spin coating, we used 2-butanone. We prepared different concentrations (as wt% P(VDF-TrFE) in 2-butanone) to investigate the dependence on layer thickness. After spin-coating at 6000 rpm, the film was annealed at 135 °C for 2 h to improve the crystallinity of the all-*trans* conformation, the so called β-phase [21]. The thickness of the spin-coated film was measured using a Taylor Hobson (Talystep) profilometer.

To determine interface reactions of the ferroelectric copolymer and the electrodes, we prepared different sample geometries: (i) 'top electrode geometry' and (ii) 'bottom electrode geometry'. In the top electrode geometry, first the copolymer was spin-coated; in a second step, we evaporated a layer of aluminium of about 1 nm in thickness on top of the copolymer layer. As substrate, we used silicon wafers. In the bottom electrode geometry, first the electrode was prepared and then an ultrathin (< 10 nm) layer of the copolymer was spin-coated on top.

The spectroscopic characterisation of the P(VDF-TrFE)/Al interface was performed in top and bottom electrode geometries, the investigation of the P(VDF-TrFE)/PEDOT:PSS interface only in bottom electrode geometry: a silicon wafer, covered with a film of spin-coated PEDOT:PSS and then an ultrathin layer of P(VDF-TrFE).

As reference to monitor changes induced by interaction with the electrode material, we prepared bulk film samples (P(VDF-TrFE) 100 nm in thickness). Infrared spectra were obtained as Fourier transform infrared (FTIR) spectra, measured in transmission mode using a FTS 60B BioRad spectrometer. The FTIR spectra of these films show all typical features for the β-phase, but no indication for alternating *trans–gauche* conformation is observed. For this phase, a strong absorption feature at 1196 cm^{-1} occurs. The XPS spectra around the C1s edge show three main peaks, at binding energies of −287.0 eV, −289.4 eV and −291.5 eV. The first peak at −287.0 eV has to be attributed to the (H–C–H) group, the peak at −289.4 eV should be attributed to the (H–C–F) group and the peak with highest binding energy at −291.5 eV has to be assigned to the (F–C–F) group. The peak intensities reflect the copolymeric ratio of VDF:TrFE 70:30 [26].

With FTIR analysis, we determined the correct window in X-ray illumination time (for our structures, max. 20 min). Under X-ray irradiation, the copolymer is slowly degenerated from the β-phase or all-*trans* conformation into the *trans–gauche* conformation.

For XPS characterisation, we used an Omicron UHV system with hemispherical analyser [27]. Spectra were obtained with Mg K_α excitation at room temperature and were corrected for differences in contact potential. Elemental ratios were calculated from the XPS intensities, modified with the specific atomic sensitivity factors (ASFs) when necessary [28].

21.2.2.2 **Electrical Characterisation (CV, IV)**

For the measurements of the ferroelectric hysteresis of P(VDF-TrFE) via the flatband shift, we used capacitors with oxidised p-type (~10^{15} cm^{-3}) silicon substrate (100–235 nm SiO_2) to prevent large amounts of leakage current. The copolymer film was prepared as described above. We used films of thickness from 100 nm to 1 μm. The structures were prepared in 'top electrode geometry', with thermal evaporated aluminium, patterned via a shadow mask. The measurements of capacitance versus voltage (CV) were carried out with an Agilent 4284A LCR meter at a frequency of 1 MHz with sweep rates from

12.5 mV/s to 50 mV/s. Here, we present high-frequency investigations of these structures; in the low-frequency mode we find the same effect [16].

All measurements were started in accumulation and finished there too, after driving the voltage in the investigated range to inversion and back (e.g. – 10 V to 10 V and backwards to – 10 V; we call this a '±10 V loop'). Due to the polarisation of the ferroelectric layer, the CV curves show a hysteresis loop [30, 31], which depends on the maximal applied voltage in the CV mode.

Ferroelectric field effect transistors were prepared on a highly n-doped Si wafer with 50 nm thermally grown SiO_2. We used spin-coated copolymer films of a thickness of about 100 nm. On top of this copolymer/Si stack, the P3HT organic semiconductor was spin-coated from a solution with chloroform, as described above. The transistor was completed by the evaporation of two gold electrodes through a shadow mask. The gap between the source and drain electrode was 40 μm. P3HT was obtained from Aldrich, Germany, and Plextronics Inc., PA, USA (P3HT, average molecular weight 87000). The preparation and the measurement of the transistor characteristics were performed in a glove box under argon atmosphere with an oxygen content of 200 ppm. This leads to a low on/off ratio of the transistors, but the aim of this work is to show the bistable operation of the OFET. A combination of a HP power supply (E3631A) and two multimeters (HP 34301A, PREEMA4001) was used for measuring the current–voltage characteristics.

21.3 Results and Discussion

21.3.1 Microscopic Methods

21.3.1.1 PEEM

In this section, we give an interpretation of the PEEM intensities of organic source–drain structures in operation mode, with applied voltages.

Figure 21.1 shows a schematic of the samples investigated (side view), as well as a sequence of three PEEM images taken with D_2 excitation (7.3 eV, top view). The resolution is a few micrometres. In the first PEEM image the two (carbon) electrodes are grounded and the image is set to have two small stripes of the electrodes on the right and left side while the main field of view is used to image the channel in between. It appears with comparable brightness while the general appearance is much smoother than the electrode areas. In the next image we display the same area but now with voltage applied to the left electrode. The channel is visible with a much lower intensity as a broad stripe in between the left electrode (with voltage applied) and the right electrode (at grounded potential). Very similar, when we change the polarity of the applied voltage, the whole region of the channel becomes darker again and this seems to be independent of the polarity of the voltage applied. For both situations the

PEEM intensity within the channel appears to be homogeneous without any significant distributions or gradients.

As we see in Figure 21.1, the width of the imaged channel differs in the three images. This is due to the modification of the applied electric field at the sample, which changes the optical behaviour and the magnification parameters of the PEEM imaging system. We note that these modifications do not influence the contrast of the area in the channel at all.

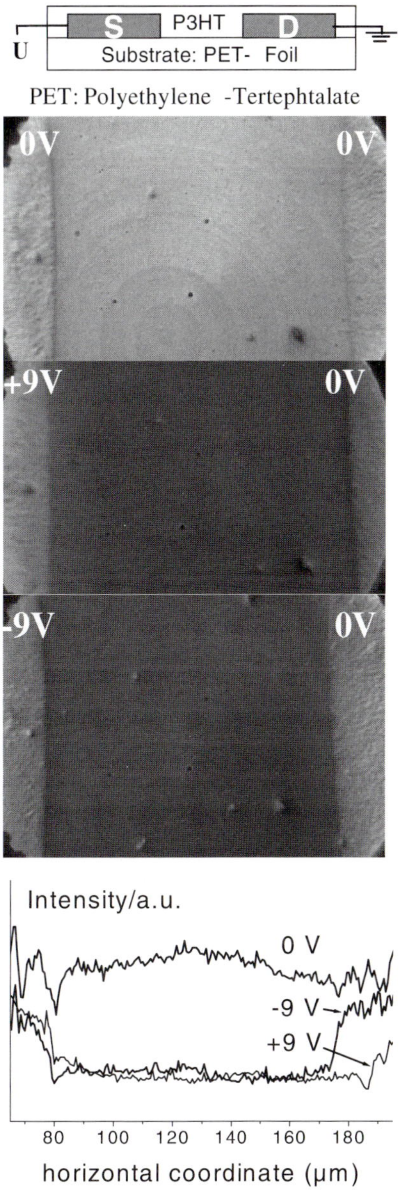

Figure 21.1 Schematic of the device structure used (top). PEEM images with D_2 excitation of the channel visible as a broad stripe, with 0 V, +9 V and −9 V applied at the left electrode. The lower diagram shows the intensity profiles across the channel for the three images shown above. Different channel lengths are due to strains.

With the same sample, we performed micro-photoelectron spectroscopy (μ-PES) at selected positions in the channel. With this technique a built-in spectrometer is used to record an electron distribution curve (EDC) with an iris aperture limiting the area of the detected photoelectrons to a range of about 30 μm. The spectra represent the integrated energy distribution of the photoelectrons.

In Figure 21.2, we show the shift of the main peak in the Hg μ-PES for different voltages applied (effect for D_2 excitation is equal). For a discussion of the EDC, see [33]. The shift is normalised to the main peak of the μ-PES spectrum at the grounded electrode, here at position 5, for zero voltage applied.

Position 1 corresponds to the drain level (for positive voltages, at position 1, left carbon electrode). Position 5 is equal to the source level, at zero (grounded) volt level. Positions 2, 3, 4 represent levels inside the channel, and we use the position of the main peak of the EDCs taken in the channel at these positions. We performed these measurements for different voltages applied at the two electrodes and plotted the corresponding shifts in the diagram shown above the images of the individual positions in the channel. We notice that the shift of the main peak is almost linear over the channel and the straight lines suggest that it is equal to the local potential at positions 1 and 5 for all positions.

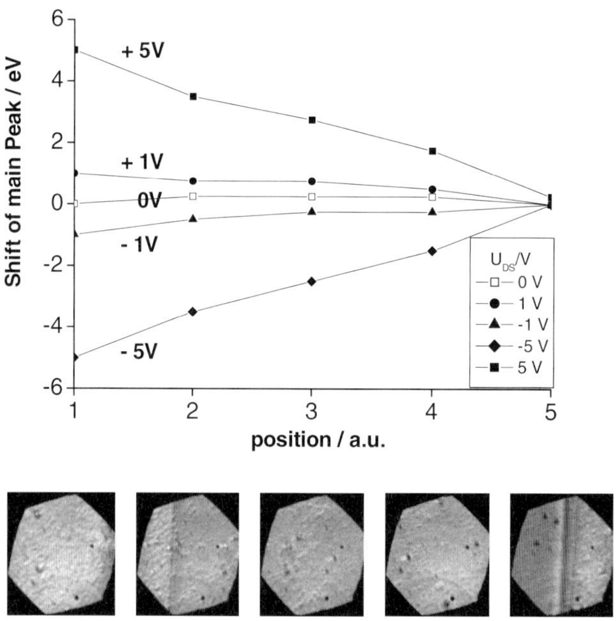

Figure 21.2 Shift of the main peak in the Hg (4.9 eV) μ-PES spectrum for different voltages applied. The shift is related to the main peak of the μ-PES spectrum at position 5, for zero voltage applied. Bottom: PEEM images of positions 1–5, diameter 30 μm.

In order to explain the difference between PEEM contrast and the surface potential of the organic structures, we focus on the electronic structure of P3HT and show in Figure 21.3a combined spectra of the occupied and the empty valence states, all energies being referred to the Fermi level.

The occupied states are derived from energy distribution curves of the photoelectrons, excited by synchrotron illumination. We present XPS spectra at resonant excitation energies at the C1s (285 eV), and at the S2p (165 eV) ionisation thresholds. For the C1s threshold, we in addition show the spectrum taken off resonance. These valence band spectra are dominated by two broad features at −7 eV and −11.5 eV which are due to carbon-derived (H_C) and sulfur-derived (H_S) HOMOs of the thiophene monomer. These levels are pronounced in all data, as marked by the dashed lines. The weaker emission of the highest band at −3.1 eV is attributed to emission from electrons out of the π-band, which is no longer assigned to individual monomers but is delocalised along the polymeric chain; see also [33].

The empty states are derived from corresponding μ-XAS spectra and the constant-initial-state (CIS) spectra again recorded in the range of the absorption edges of C1s and S2p. The CIS spectrum at the C1s edge is taken at a binding energy close to the valence band maximum. The CIS spectrum at the S2p edge is taken for the binding energy at the S3p HOMO (H_S). For both excitations (C1s or S2p) a pronounced peak occurs at energies just above the Fermi energy. It coincides with the excitonic feature in the μ-XAS curves.

These data indicate that within the π-band, both the C2p and the S3p derived states contribute. In the occupied part, π-bands appear in the resonant XPS spectra. From the CIS data we learn that also in the empty π*-band both the C1s and the S2p wavefunctions contribute. It should be pointed out that there

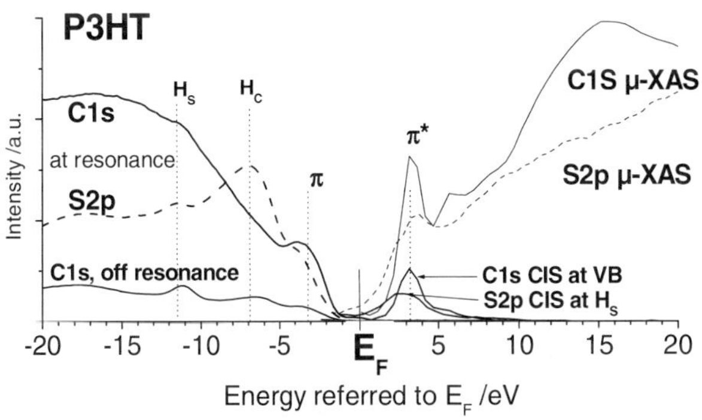

Figure 21.3 Electronic structure of P3HT. The valence band states are derived from UPS-EDC taken with resonant excitation (285 eV, C1s; 165 eV, S2p). The empty conduction band states show up in the μ-XAS and CIS spectra as recorded around the C1s and S2p threshold. All data are referred to the Fermi level.

is a significant difference between the two. While the C2p-derived states come out of the delocalised π-band, those from the hetero-atom arise out of the S3p HOMO states. Nevertheless, the data indicate that in the empty π*-band the contribution of the S3p states is even lower in excitation energy than the contribution from the C2p states. This is evident from both the μ-XAS and CIS data.

Coming back to the PEEM images, these were obtained with an excitation energy of 4.9 eV/7.2 eV. With these low excitation energies just electrons out of the π-band can be excited, evidently. The photon energy is not large enough to excite the electrons out of the monomeric HOMOs. As the electrons in the π-band are the only electrons that can be involved in the transport properties, it is quite evident that the number of these charge carriers depends on the applied voltage within the channel of our devices. Any voltage, independent of the polarity, will cause the charge carriers to move out of the channel towards one of the two electrodes. Therefore, we find a decrease in intensity independent of the polarity in our PEEM data.

In contrast, measurements with Kelvin AFM give information on the surface potential, as for our measurements with μ-XPS, but with much higher lateral resolution. Here, a constant slope of the surface potential depending on the applied voltage has to be expected, if we use a source–drain structure with a semiconductor working as an ohmic resistor, like in our case, respectively for organic transistors, operating in the linear regime [12–15].

21.3.1.2 **SKPM**

In this section we report the results of SKPM measurements on source–channel–drain structures with P3HT semiconductor with voltage applied between source and drain, without gate dielectric and gate electrode. Figure 21.4 shows potential profiles of two source–channel–drain structures with graphite and carbon black electrodes at different source–drain voltages, namely 5 V and 8 V, obtained in ambient air. The lateral resolution of the potential profile is in the region of 200 nm, as revealed by the smallest potential features on top of the electrodes.

Marked features of the profiles in Figure 21.4 are the abrupt potential drops ΔV_d at drain and ΔV_s at source at the electrode/P3HT interface, summarised in Table 21.1. Inside the channel between the two carbon black electrodes the potential drop is linear with a small inhomogeneity in slope; in the case of graphite electrodes the potential drop in the channel has a more pronounced nonlinearity. At smaller length scales, in the range of the above mentioned resolution of 200 nm, an additional superimposed heterogeneity was measured, here especially in the range of the carbon black source electrode. The value for this variation is of the order of 300 mV and not due to a signal noise of the SKPM.

The information revealed by our SKPM measurements in air are split into three main scales or ranges with information on (i) the voltage drop at the electrode/polymer interface, (ii) linearity and (iii) fluctuations.

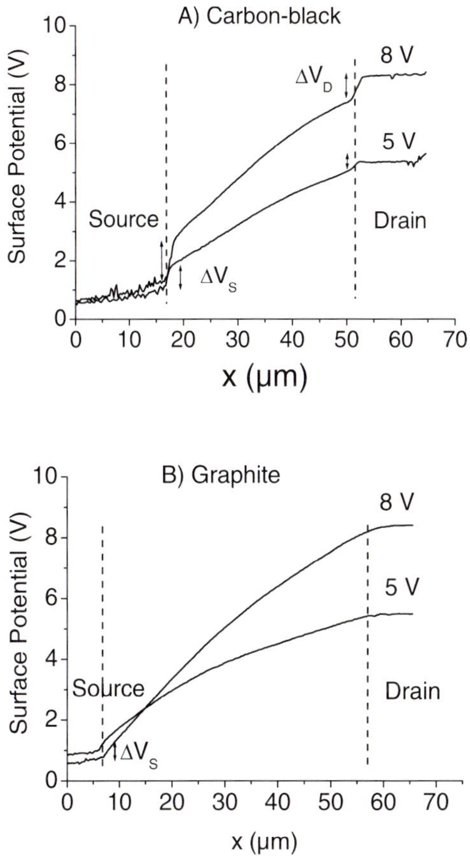

Figure 21.4 Potential profiles of source–drain structures at applied source–drain voltage; measurements performed in ambient air. Electrode material was a) carbon black and b) colloidal graphite.

The voltage drop between the electrode/semiconductor interface indicates a contact resistance due to different work functions of the two materials and makes it possible to characterise the quality of the measured contact.

At larger scales the measurement of potential gives information on the homogeneity of the transistor structure, especially the spin-coated P3HT layer and, in consequence, the related transistor performance. The curvature observed in Figure 21.4b could be caused by variations in thickness of the P3HT-layer: a thicker layer of the spin-coated semiconductor should generate a smal-

Table 21.1 Potential drops at the interfaces.

material	ΔV_d	ΔV_s	voltage
carbon black	1.0	0.3	5
carbon black	1.5	1.0	8
graphite	0	0	5
graphite	0.3	0	8

ler potential drop. For a space charge as a result of charge injection we expect a reversed curvature corresponding to the $x^{3/2}$ law of the related potential distribution. A calibration procedure was performed to exclude nonlinearities of the cantilever, but for further considerations we need better statistics, compared with simulations of the potential distribution. Short-range information like the above mentioned heterogeneities in the carbon black electrode region should be due to a lateral distribution of a surface potential related to single grains or agglomerations of the carbon black particles in this case.

The quality of the contact can be compared with the reported value for the contact resistance $R_c = L(\Delta V_d + \Delta V_s)/I_{ds}$, reported by Bürgi et al. [12] for a P3HT Cr/Au structure, which is in the region of 50 kΩ cm. For our carbon black we determined a value of about 10 kΩ cm for the two terminal devices used. The data measured in [12] are revealed for transistor structures with an applied gate voltage of −30 V and with drain–source voltages in the linear (ohmic) regime ($|U_{dc}|$ max. 8 V). The gate voltage causes an accumulation layer with approximately constant charge density in this range. In our measurements, the almost ohmic behaviour of the semiconductor was measured due to doping effects (oxidation).

In addition, we determined the work function directly by taking UPS measurements under UHV conditions for the three different pure materials on a Fe substrate. We obtain the following values: graphite, pure, 3.8 eV; carbon black, pure, 4.1 eV; P3HT, pure, on stainless steel (thickness <500 μm), 3.8 eV. This small difference of the work function is related to the better adaptation of systems with carbon or graphite as source–drain electrodes in terms of the contact resistance. Differences between carbon black and colloidal graphite should be due to different ink additives (graphite, NH_3; carbon black, glycol).

Compared to the PEEM technique, the SKPM signal is directly proportional to the local work function of the organic device. Both methods, the spectromicroscopic approach and the SKPM technique, offer lateral information in terms of homogeneity and interfaces. In the next section, we present a depth profiling of interfaces, performed using XPS.

21.3.2 Ferroelectric Devices

21.3.2.1 Interface Characterisation

As an example for depth profiling of interfaces, an XPS investigation of the ferroelectric copolymer P(VDF-TrFE) is presented. For a memory application, low operation voltages and a downscaling of the film thickness are necessary. For thin films of the copolymer, interface phenomena become important. We compare the two interfaces P(VDF-TrFE)/Al and P(VDF-TrFE)/PEDOT:PSS. For the organic electrode, a better functionality of P(VDF-TrFE) is reported, as described above.

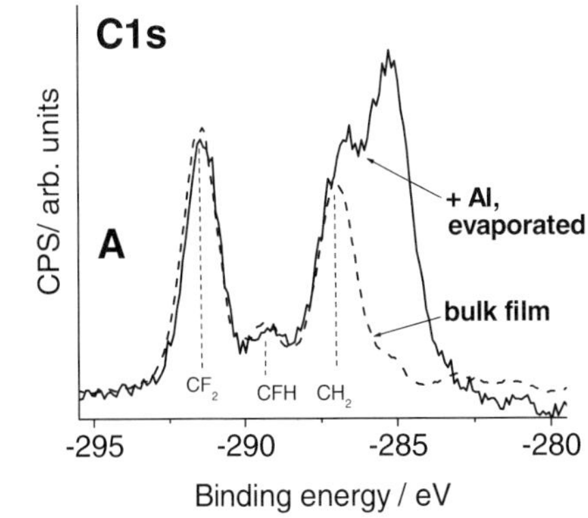

Figure 21.5 C1s XPS spectrum of a P(VDF-TrFE) film, after thermal evaporation of aluminium (solid line), compared with the spectrum of the bulk film (dashed line). The intensity after deposition of Al is normalised with reference to the CF_2 peak of the bulk film.

First, we show an XPS analysis of the P(VDF-TrFE)/Al interface, in top electrode geometry. Figure 21.5 shows an XPS spectrum of a P(VDF-TrFE) film after evaporation of a thin aluminium layer (around 1 nm). Compared to the spectrum of the bulk P(VDF-TrFE) film without aluminium, we find the following modifications: (i) the relative intensity between CH_2 and CF_2 peaks is modified towards lower fluorine content, approximately from $CH_2/CF_2 = 0.8$ before and $CH_2/CF_2 = 1.0$ after evaporation; (ii) a small shift of CH_2, CFH and CF_2; and (iii) a new peak at lower binding energies. This is a clear indication of a surface reaction.

The XPS spectrum of the F1s level with reference to the bulk copolymer film without aluminium reveals a slight asymmetry, indicating the formation of a thin AlF_3 layer. This is confirmed by the Al2p spectrum: here, we have a second small peak with a very high binding energy of −76.8 eV, due to the presence of fluorine (metallic Al2p: −72.65 eV). An oxidised surface of aluminium would have a binding energy of around −75.4 eV [28].

The next question we address is to whether there is any surface reaction in the case of a bottom electrode of aluminium. We measure twice: a set of samples with aluminium as bottom electrode, without any annealing, just after spin coating; then we anneal the samples at 135 °C for 2 h and measure again. The related XPS spectra for the C1s level are given in Figure 21.6.

An additional peak for the C1s level is obvious, even for the sample without any annealing procedure. If we compare the C1s XPS results of the two geometries, aluminium as top and bottom electrode, we find a similar behaviour: the relative intensity between the CH_2 and CF_2 peak is modified and a new peak occurs at binding energies near the CH_2 feature. This means that when we talk about a surface reaction for an evaporated electrode, in top electrode geometry, we also have to talk about a surface reaction in the bottom electrode geometry. An interface layer is obviously built up even at room temperature. For elevated temperatures, for example the common annealing procedure (135 °C, 2 h), we have to expect an increased amount of this additional C1s feature. This is exactly the case: the relative intensity of the additional peak is increased after annealing. The concentration for spin-coating of this sample is

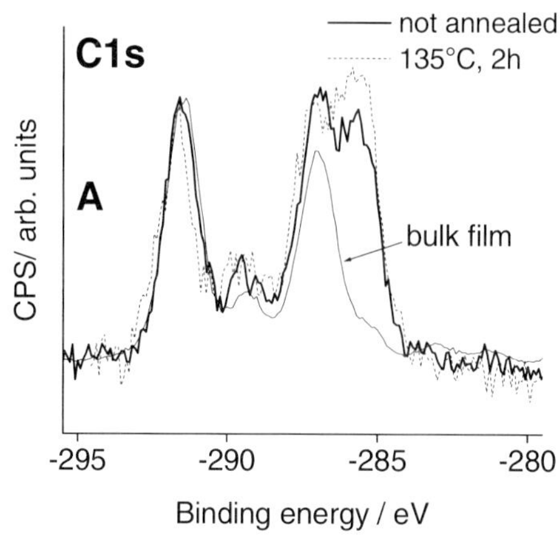

Figure 21.6 C1s XPS spectra of a P(VDF-TrFE) film, spin-coated onto an Al/Si substrate (bottom electrode structure). Thick solid curve: just after spin coating. Dashed curve: after thermal annealing at 135 °C for 2 h. The concentration for spin coating is 0.1 wt% P(VDF-TrFE) in 2-butanone. Spectra are corrected for 0.4 eV and normalised with reference to the bulk film, also shown (thin solid curve).

0.1 wt% P(VDF-TrFE) in 2-butanone. For a set of samples with 0.3 wt% and 0.5 wt% P(VDF-TrFE), we find the same behaviour, but the relative intensity of the additional C1s feature is reduced proportional to concentration, indicating a layered structuring. The F1s and the O1s levels are also analysed for the same set of samples. Here no relevant modification is visible, before and after annealing. The Al2p spectrum indicates the formation of an oxidised layer, as revealed by an additional peak at a binding energy of −75.4 eV. This oxidation is due to the experimental procedure: we have to transfer the samples into our glove box, and oxidation is not excluded.

For the bottom geometry we have to surmise that we have an obviously degenerated copolymer at the interface, not only at room temperature, even though we have an oxidised surface of aluminium.

For investigations of the P(VDF-TrFE)/PEDOT:PSS interface, samples in bottom electrode geometry were prepared. Thin films of P(VDF-TrFE) are spin-coated on PEDOT:PSS, as described before. Figure 21.7 summarises the C1s spectra of four samples with thin films of P(VDF-TrFE), spin-coated in different concentrations in 2-butanone, compared with a pure PEDOT:PSS spectrum (No. 1). As revealed from the figure, no additional structure near the CH_2 feature occurs.

Like aluminium, a possible interface reaction should have an influence on the relative intensities, also (for example CF_2 and CH_2 of P(VDF-TrFE)). Here an analysis of peak attenuation is helpful. As can be seen, also in Figure 21.7, the intensity of the PEDOT-related signal at −284.8 eV is attenuated propor-

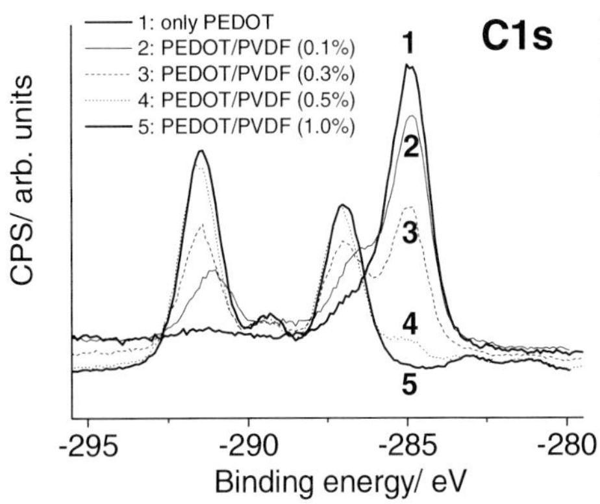

Figure 21.7 C1s spectra of samples with thin films of P(VDF-TrFE), spin-coated in different concentrations in 2-butanone (for the concentrations (wt%) indicated).

tional to P(VDF-TrFE) concentration in the 2-butanone solution. The relative intensity of this PEDOT signal is plotted in Figure 21.8. Here we extract a linear dependence, and from that we can conclude that no interface reaction with a modification of intensities occurs. For a concentration of 0.6 wt%, the straight line intersects with the *x*-axis. In this case, the thickness of the P(VDF-TrFE) film is larger than the information depth of photoelectrons.

If we use a value of 1 nm as mean free path of photoelectrons [28], an estimation of the P(VDF-TrFE) film thickness is possible. We obtain the following values: 0.4 nm for 0.1 wt%, 1.0 nm for 0.3 wt% and 2.5 nm for 0.5 wt%. Assuming a monolayer thickness of 0.5 nm [29], these data reflect a coverage from one to 5 monolayers. Also the film thickness corresponds linearly to the P(VDF-TrFE) concentration in 2-butanone.

For PEDOT:PSS as the electrode, we clearly have no interface modification and an ideal sandwich layer. For the spectra of S2p, F1s and O1s this is also confirmed.

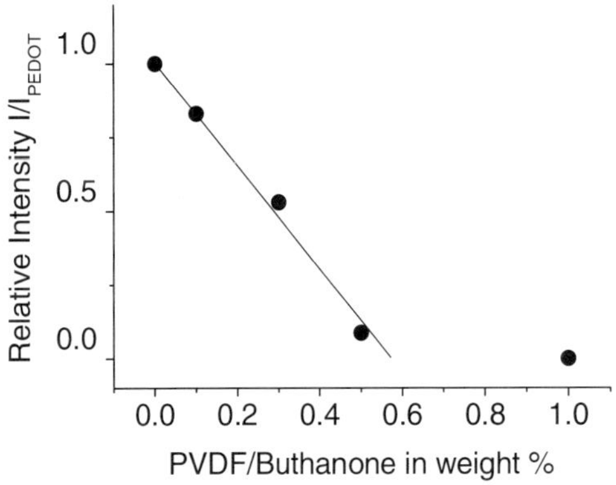

Figure 21.8 Relative intensities of the PEDOT:PSS-attributed C1s feature versus P(VDF-TrFE) concentration in 2-butanone solution, as I/I^0_{PEDOT}. The line is drawn as a guide for the eyes.

21.3.2.2 Electrical Characterisation of MFIS Capacitors (CV Measurements)

In Figure 21.9, a typical CV characteristic of a Si/100 nm SiO_2/110 nm P(VDF-TrFE) capacitor is shown. The CV plot shows sections of the ±10 V measurements at room temperature (RT) and at 100 °C. The inset shows the flatband shift ΔV_{FB} (memory window) versus temperature.

We note in the RT line a clear hysteresis for the measurement loop. The hysteresis of the CV line indicates the presence of polarisation charges and a ferroelectric behaviour of the P(VDF-TrFE) copolymer, resulting in a shift of the CV line along the voltage axis (flatband voltage shift), as is known for fixed charges in the insulator for MIS devices [34]. The shift of the flatband voltage is a result of ferroelectric polarisation of the copolymer. This is confirmed by CV measurements at elevated temperatures. The reported Curie temperature of P(VDF-TrFE) is in the region of 100 °C [21]. Figure 21.9 also summarises these investigations. We note the disappearance of the hysteresis at 100 °C. Furthermore, in the CV data we find an increased permittivity value at 100 °C by 80%, signified by a higher capacitance in accumulation. This is in agreement with observations in the literature, the permittivity is increasing up to the Curie point [34, 35]. We calculate a permittivity value of P(VDF-TrFE) of 6.7 ± 0.9 at room temperature and at a frequency of 1 MHz, which is in the same range as in other works [36, 37]. This value is used for a calculation of the ferroelectric charge density. The flatband voltage shift (ΔV_{FB}) is also needed to calculate the 'fixed' charge (N_{fix}) using

$$N_{fix} = \frac{\Delta V_{FB} \varepsilon_0 \varepsilon_{PVDF}}{q t_{PVDF}},$$

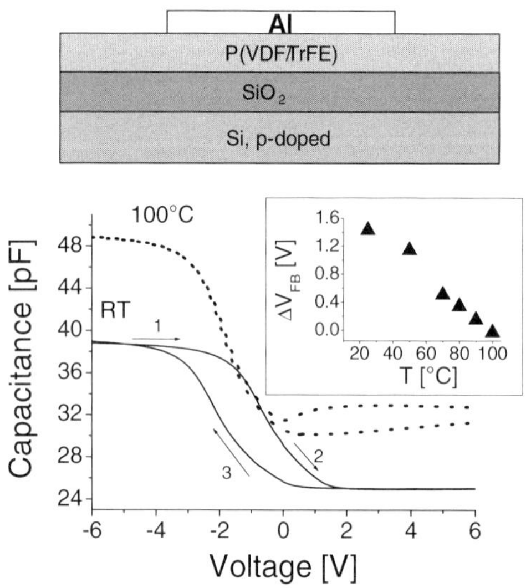

Figure 21.9 Top: schematic of our capacitors. Bottom: capacitance–voltage characteristic of a 100 nm SiO2/110 nm P(VDF-TrFE) sample at room temperature and at 100 °C. The inset shows the flatband voltage shifts dependent on temperature.

where ε_0 is the permittivity of vacuum, ε_{PVDF} is the relative permittivity value for P(VDF-TrFE), t_{PVDF} the thickness of P(VDF-TrFE) and q is the elementary charge.

In a next step, we study the flatband shift, the polarisation as a function of copolymer film thickness. For the measurements, a relatively thick SiO_2 buffer layer of 235 nm was used and we reveal a clear thickness dependence of the ferroelectric polarisation [32]. We calculated ΔE_{PVDF} as the voltage drop only over the P(VDF-TrFE) layer. For accumulation, negative voltages, the voltage is divided into two parts: $U_1 = U_{1PVDF} + U_{1SiO_2}$. For depletion, we have to calculate an additional voltage drop over the depletion layer, according to $U_2 = U_{2PVDF} + U_{2SiO_2} + U_{2D}$. The field, applied only at the P(VDF-TrFE) layer, is calculated as field amplitude $(U_{2PVDF} - U_{1PVDF})/t_{PVDF} = \Delta E_{PVDF}$, with a layer thickness t_{PVDF}.

Figure 21.10 shows the calculated N_{fix} values versus the total field amplitude ΔE_{PVDF}. We find an almost linear dependence of the 'fixed' charges versus the electric field strength applied. For the remanent polarisation, which is proportional to the amount of the 'fixed' charges, a saturation at relatively high field strength is to be expected. For our samples, even at high applied fields, no saturation occurs. However, we should refer to the fact that here the sum of the electric field in both branches of the polarisation loop is shown, so it might be plausible that the field is not yet high enough for saturation. Indeed, in our work related to the buffer optimisation we could find saturation in these kinds of plots by reducing the buffer layer thickness or using high-k material [36].

We further recognise a significant reduction of polarisation for the P(VDF-TrFE) thickness below 100 nm, while between 200 nm and 950 nm the dependence is very similar. Generally it has been postulated that the coercive field increases with decreasing of its thickness [38]. For P(VDF-TrFE) a range of 70–100 nm has been proposed as a critical thickness due to reduced crystal-

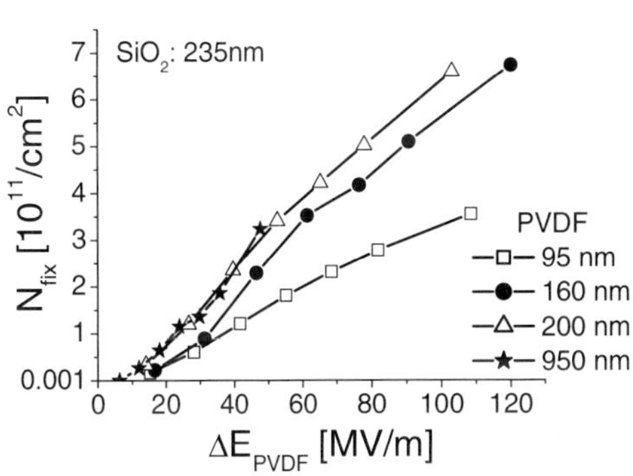

Figure 21.10 Calculated 'fixed' charges versus the electric field change of the ferroelectric layer inside one CV loop for SiO_2/P(VDF-TrFE) stacks with different thickness of the ferroelectric layer. Thickness of buffer layer is 235 nm. Thickness of P(VDF-TrFE) is as indicated.

linity [39] or due to interface interactions [14]. It must be pointed out here that a critical thickness is found for electrodes made of aluminium, not for PEDOT:PSS [14]. Our data are consistent in this context. In Section 21.3.3 we already showed reactive interactions between P(VDF-TrFE) and aluminium, not for the P(VDF-TrFE)/PEDOT:PSS interface. This becomes even more important when the thickness of P(VDF-TrFE) film is further downscaled.

Based on these results, we build up OFETs with a ferroelectric gate insulation, consisting of P(VDF-TrFE).

21.3.2.3 **Ferroelectric OFET**

In Figure 21.11, we show a schematic of the ferroelectric transistor structure. For the experiments, we use two field effect devices, one with the ferroelectric copolymer, and one without the ferroelectric layer. Both structures are prepared on a Si/SiO_2 substrate with 235 nm oxide thickness. We use P3HT from Aldrich for these structures. Due to an amount of 200 ppm oxygen inside our glove box, an ohmic current was measured without applied gate voltage. The residual current with ap-plied gate voltage is a superposition of both the amount of accumulation charge and the ohmic current [40]. The on/off ratio of these devices is in the range of 5 because of this fact. The influence of a ferroelectric hysteresis is obvious, nevertheless. In Figure 21.12 we show a measurement of the source–drain current of a ferroelectric OFET without applied gate voltages, but after an applied gate voltage pulse of ±73 V for 2 min. As reference, we measure the second device without the copolymer. A comparison of the relative drain–source current $(I^+ - I^-)/I^-$ after application of a positive voltage pulse (+73 V, I^+) and a negative voltage pulse (−73 V, I^-), both for 2 min, is shown in Figure 21.13. Here, a small shift for the structure without the copolymer is measured, but the switching of the device with the copolymer as additional gate isolation is approximately one order of magnitude higher. Here we confirm that a realisation of an 'on' and 'off' state seems to be possible. As the figure demonstrates, after application of negative voltage pulses, the channel resistance increases. After application of positive voltage pulses, the channel resistance decreases, when measured at zero voltage gate bias again. This behaviour is measured in several devices and is also found for devices based on Si/SiO_2 substrates with 50 nm oxide thickness and cleaned P3HT, delivered by Plextronics, USA.

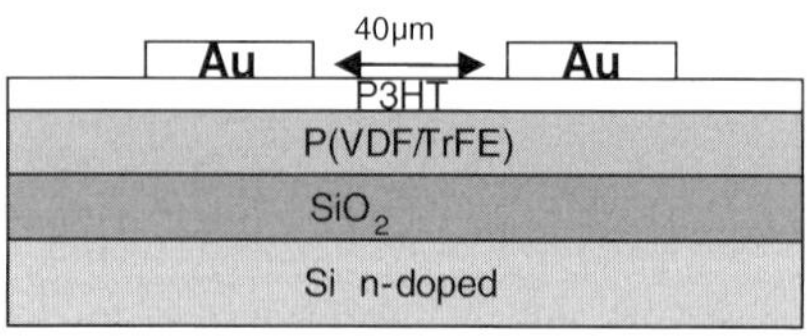

Figure 21.11 Schematic of the ferroelectric OFET. We use P3HT as active layer and thermal evaporated Au electrodes for source and drain. The channel length is 40 μm, the channel width is 1 cm.

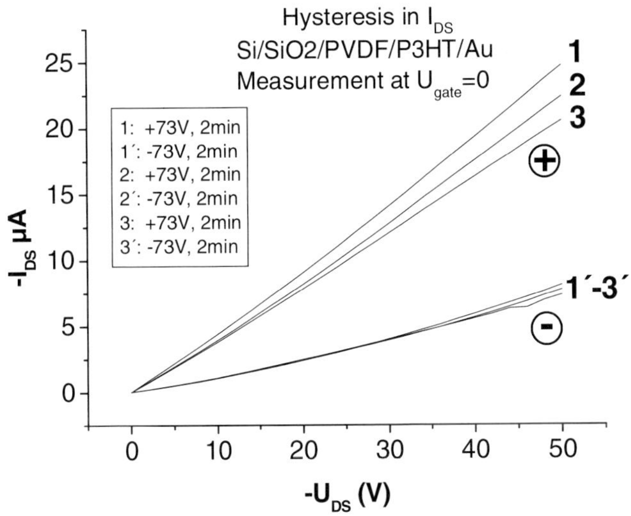

Figure 21.12 Source–drain current of a ferroelectric OFET without applied gate voltages, but after an applied gate voltage pulse of ±73 V for 2 min. After application of negative voltage pulses, the channel resistance increases. After application of positive voltage pulses, the channel resistance decreases.

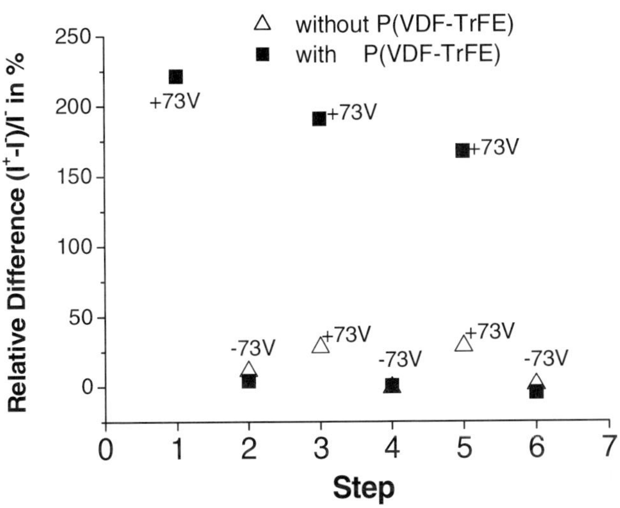

Figure 21.13 Relative drain–source current $(I^+ - I^-)/I^-$ after application of a positive voltage pulse (+73 V, I^+) and a negative voltage pulse (−73 V, I^-), both for 2 min. We compare two devices: the ferroelectric OFET of Figure 21.12 and a similar OFET, without the ferroelectric P(VDF-TrFe) copolymer, for $U_{DS} = -50$ V, without gate bias.

A transfer characteristic of such a device is shown in Figure 21.14. The drain–source voltage is constant at −5 V. As in our CV measurements, we apply different gate voltage sweeps. For example, a sweep of ±10 V starts at 0 V, runs to +10 V, to −10 V and back to 0 V. The ramp rate is 10 V/min. The characteristic is parabolic, demonstrating that the field effect induced accumulation charge is saturated (saturation regime of the transistor). For the transfer characteristics, plotted as $I_{DS}^{1/2}$ (U_{gate}), the intersection of the curves with the x-axis gives directly the threshold voltage of the transistor. The slope of the $I_{DS}^{1/2}$ (U_{gate}) plot is directly proportional to the square root of the mobility [40]:

$$\frac{d\sqrt{I_{DS}}}{dU_{gate}} = \left(\frac{\mu_0 C}{2l^2}\right)^{1/2} .$$

If we use a measured isolator capacitance per unit area of 28.7 nF/cm^2 at 20 Hz, the resulting value for mobility is $(3.9 \pm 0.28) \times 10^{-4}$ cm^2/(Vs). This is a typical value for P3HT [41].

In the transfer characteristic, we observe a shift of threshold voltage, which can be distinguished in two parts. (i) A shift of the transfer characteristic itself.

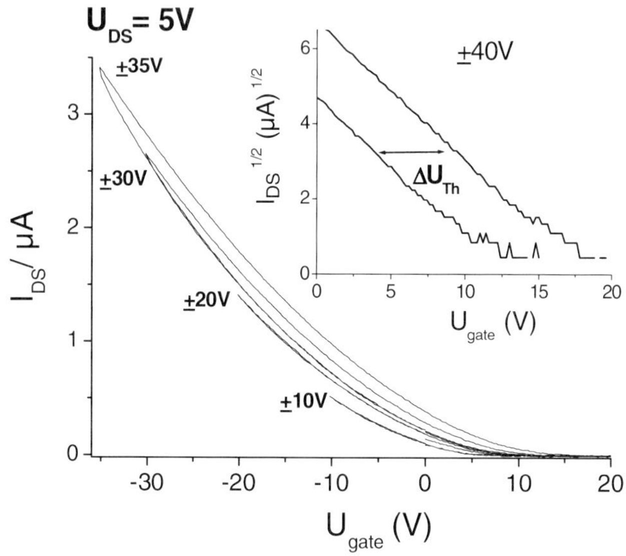

Figure 21.14 Transfer characteristic of an Au/P3HT/100 nm PVDF/50 nm SiO_2/Si device at constant drain–source voltage of −5 V. We show different voltage sweeps (voltage scan windows), which are defined as follows: for example, a sweep of ±10 V starts at 0 V, runs to +10 V, to −10 V and back to 0 V. The ramp rate is 10 V/min. Inset: transfer characteristics, plotted as $I_{DS}^{1/2}$ (U_{gate}), for a sweep of ±40 V.

This could be due to an accumulation of residual polarisation charges or due to a partial oxidation of the organic semiconductor (p_{O_2} is about 200 ppm). (ii) A shift of threshold voltage in the course of a voltage sweep (from positive to negative voltages). This shift, here called ΔU_{th}, could be due to the ferroelectric hysteresis. For voltage sweeps of about ± 10 V or smaller, ΔU_{th} is equal to zero. This is further evidence for ferroelectric behaviour: here, the field strength applied is not strong enough for polarisation. In the case of charge injection, a shift of threshold ΔU_{th} should be not equal zero at ± 10 V.

Finally, we compare the flatband voltage shift of the MFIS capacitor and the threshold voltage shift of the ferroelectric OFET versus the applied voltage amplitude in Figure 21.15. The P(VDF-TrFE) layer thickness is nearly equal for both structures, about 200 nm; this is also the case for the thickness of the SiO_2 buffer layer. The shift of flatband voltage and the threshold shift are of the same order of magnitude, but the voltage shift for the OFET is lower. This difference between both devices should be due to the different geometry of both devices. The programming field distribution in the capacitor structure is in principle perfectly perpendicular to the surface, while this is not the case for the OFET geometry. The ferroelectric copolymer in the OFET is polarised in the stray field between the gate and the source and drain electrode. Effective for the shift of threshold voltage is a polarisation only in the area of the channel. Programming losses for the ferroelectric OFET could be minimised by a buffer layer optimisation [20], or a reduction of channel length.

21.4 Summary and Conclusions

Using PEEM, we have performed a spectromicroscopic characterisation of organic devices under applied voltages. We show that the information gained from the PEEM intensity and from the related electron distribution curve is

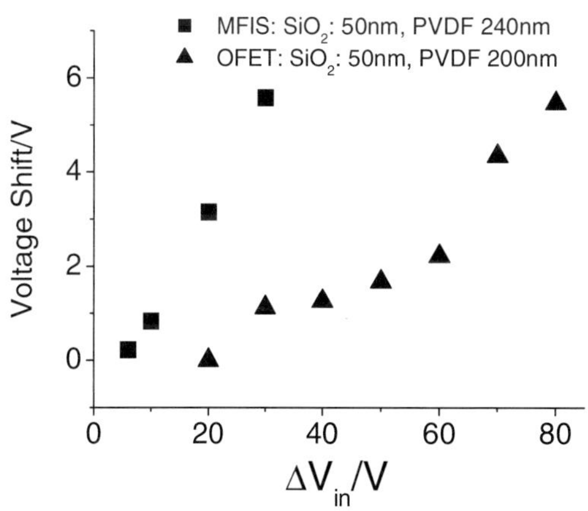

Figure 21.15 Comparison of ferroelectric MIS capacitors (MFIS) and ferroelectric OFET, as flatband shift and shift of threshold voltage, respectively, versus voltage amplitude $(V_{2PVDF} - V_{1PVDF}) = \Delta V_{in}$. For example, $\Delta V_{in} = 80$ V corresponds to a (bias or gate) voltage sweep from 0 V to +40 V, to −40 V and back to 0 V.

different. While the former represents the number of π-electrons in the organic P3HT semiconductor in the working state of the organic device, the shift in μ-PES spectra is directly related to the local surface potential.

We are able to distinguish between the charge carrier density, respectively, the density of π-electrons and the surface potential, on the other side. Thus the combination of μ-PES mode and the imaging mode of PEEM offers a new analytical method for the characterisation of active layers in operating devices.

As a complementary method for characterisation of organic devices, we apply SKPM. With SKPM, we characterise the following features of a completely organic source–drain structure: (i) the voltage drop at the electrode/polymer interface indicating a contact resistance; (ii) linearity; and (iii) local fluctuations of the surface potential due to inhomogeneities of the organic layers. Compared with the PEEM technique, the SKPM signal is directly proportional to the local work function of the organic device. Both methods, the spectromicroscopic approach and the SKPM technique, offer lateral information in terms of homogeneity and interfaces. We demonstrate that SKPM is a useful method for a failure mode analysis of organic devices.

As an example for depth profiling of interfaces, we present an XPS investigation of the ferroelectric copolymer P(VDF-TrFE). This material opens an opportunity for organic non-volatile memories. A prerequisite for low operation voltages is a downscaling of the film thickness, where interface phenomena between the electrode and the copolymer become important. We compare the two interfaces P(VDF-TrFE)/Al and P(VDF-TrFE)/PEDOT:PSS.

Under radiation damage-free conditions, we show a clear indication for a surface reaction of P(VDF-TrFE) with Al electrodes, not only for evaporated aluminium as top electrode, but also, at room temperature, for the aluminium as bottom electrode. In contrast, for PEDOT:PSS, the XPS measurements indicate a layer-by-layer structure of PEDOT:PSS/P(VDF-TrFE) without any interface modification. This could be the reason for lower relaxation times, higher switching frequencies and, in consequence, a better field dependence of the ferroelectric polarisation, if we choose PEDOT:PSS as material for the electrode, especially for thin films of the copolymer.

The ferroelectric hysteresis of P(VDF-TrFE) is directly investigated by MIS capacitors and OFETs. By using MIS capacitors, a systematic shift of flatband voltage is observed, after application of different voltage scan windows. The MIS structures are built up as Al/P(VDF-TrFE)/SiO_2/Si sandwich structure. The dependence of the remanent polarisation on thickness of the copolymer shows an elevated polarisation voltage for a copolymer film thickness below 100 nm, obviously due to the above mentioned interface reaction between the copolymer and aluminium.

Organic transistors with the copolymer P(VDF-TrFE) as ferroelectric gate insulation, P3HT as active and organic semiconductor are prepared. We show that by application of different gate voltages, a programmable on and off state of the drain–source current is possible. This bistable drain–source current is a result of the ferroelectric alignment of the dipoles inside the copolymer, which causes a shift of the threshold voltage in the same order of magnitude as the shift of flatband voltage for MIS structures.

Acknowledgements

We acknowledge the skilful experimental assistance of Giudo Beuckert, Patrick Hoffmann as well as the BESSY staff. This work was supported by the Deutsche Forschungsgemeinschaft (DFG) within priority program 1121.

References

1. H. Rotermund, W. Engel, S. Jakubith, A. von Oertzen, and G. Ertl, Ultramicroscopy **36**, 164 (1991).
2. P. Hoffmann, S. Wehner, D. Schmeißer, H. Brand, and J. Küppers, Phys. Rev. E **73**, 056123 (2006).
3. S. Wehner, P. Hoffmann, D. Schmeißer, H. Brand, and J. Küppers, Chem. Phys. Lett. **423**, 39 (2006).
4. R. Frömter, M. Seider, C. Schneider, Ch. Ziethen, W. Swiech, G. Schönhense, and J. Kirschner, BESSY Annu. Rep. 482–484 (1996).
5. R. Mikalo, P. Hoffmann, T. Heller, and D. Schmeißer, Solid State Phenom. **63**, 317 (1998).
6. P. Hoffmann, R. Mikalo, D. Schmeißer, and M. Kittler, phys. stat. sol. (b) **215**, 743 (1999).
7. K. Müller, Y. Burkov, and D. Schmeißer, Thin Solid Films **413–432**, 307 (2003).
8. K. Müller, S. Milko, and D. Schmeißer, Thin Solid Films **413–432**, 312 (2003).
9. K. Müller, R. Scheer, Y. Burkov, and D. Schmeißer, Thin Solid Films **451/452**, 120 (2004).
10. D. Schmeißer, P. Hoffmann, and G. Beuckert, in: Materials for Information Technology, Devices, Interconnects and Packaging, edited by E. Zschech, C. Whelan, and T. Mikolajick (Springer, 2005), pp. 449–460.
11. M. Giesen, R. J. Phaneuf, E. D. Williams, T. L. Einstein, and H. Ibach, Appl. Phys. A **64**, 423 (1997).
12. L. Bürgi, H. Sirringhaus, and R. H. Friend, Appl. Phys. Lett. **80**, 2913 (2002).
13. S. Zhang, Z. Liang, Q. Wang, and Q. Zhang, Mater. Res. Soc. Symp. Proc. **889**, W05-02.1 (2005).
14. R. Naber, P. Blom, A. Marsman, and D. de Leeuw, Appl. Phys. Lett. **85**, 2032 (2004).
15. J. Glatz-Reichenbach, F. Epple, and K. Dransfeld, Ferroelectrics **13**, 113 (1992).
16. K. Müller, K. Henkel, I. Paloumpa, and D. Schmeißer, Thin Solid Films **515**, 7683 (2007).
17. Q. Zhang, Xu, F. Fang, Z. Cheng, and F. Xia, J. Appl. Phys. **89**, 2613 (2001).
18. F. Xia, B. Razavi, H. Xu, Z. Cheng, and Q. Zhang, J. Appl. Phys. **92**, 3111 (2002).
19. F. P. Gnadinger, G. G. Huebner, G. F. Derbenwick, and A. D. Devilbiss, Ferroelectrics **268**, 729 (2002).
20. K. Henkel, B. Seime, I. Paloumpa, K. Müller, D. Mandal, and D. Schmeißer, Thin Solid Films (submitted).
21. T. Furukawa, Phase Transit. **18**, 143 (1989).
22. K. Urayama, M. Tsuji, and D. Neher, Macromolecules **33**, 8269 (2000).
23. K. Müller, I. Paloumpa, and D. Schmeißer, Synth. Met. **138**, 271 (2003).

24. S. Sadewasser, T. Glatzel, S. Schuler, R. Kaigawa, and M. C. Lux-Steiner, Thin Solid Films **431**, 257 (2003).

25. K. Müller, A. Goryachko, Y. Burkov, C. Schwiertz, M. Ratzke, J. Köble, J. Reif, and D. Schmeißer, Synth. Met. **146**, 377 (2004).

26. K. Müller, D. Mandal, and D. Schmeißer, Mater. Res. Soc. Symp. Proc. **997**, I6_2 (2007).

27. P. Hoffmann, R. P. Mikalo, and D. Schmeißer, Solid-State Electron. **44**, 837 (2000).

28. C. Wagner, W. Riggs, L. Davies, J. Moulder, and G. Mullenberg (eds.), Handbook of X-Ray Photoelectron Spectroscopy (Perkin Elmer Corporation, Physical Electronics Division, Minnesota, 1978).

29. S. Palto, L. Blinov, E. Dbovik, V. Fridkin, N. Petukhova, A. Sorokin, K. Verkhovskaya, S. Yudin, and A. Zlatkin, Eur. Lett. **34**, 465 (1996).

30. S. L. Miller and P. J. McWhorter, J. Appl. Phys. **72**, 5999 (1992).

31. K. Müller, I. Paloumpa, K. Henkel, and D. Schmeißer, Mater. Sci. Eng. C **26**, 1028 (2006).

32. K. Müller, Y. Burkov, and D. Schmeißer, Thin Solid Films **495**, 219 (2006).

33. D. Schmeißer, Synth. Met. **138**, 135 (2003).

34. E. H. Nicollian and J. R. Brews, MOS Physics and Technology (Wiley, New York, 1982).

35. Y. Tajitsu, Jpn. J. Appl. Phys. **34**, 5418 (1995).

36. K. Müller, I. Paloumpa, K. Henkel, and D. Schmeißer, J. Appl. Phys. **98**, 056104 (2005).

37. K. Kimura and H. Ohigashi, J. Appl. Phys. **61**, 4749 (1987).

38. M. Dawber, P. Chandra, P. B. Littlewood, and J. F. Scott, J. Phys.: Condens. Matter **15**, L393 (2003).

39. F. Xia, B. Razavi, H. Xu, Z. Cheng, and Q. Zhang, J. Appl. Phys. **92**, 3111 (2002).

40. G. Horowitz, X. Peng, D. Fichou, and F. Garnier, J. Appl. Phys. **67**, 528 (1990).

41. Z. Bao, A. Dodabalapur, and A. Lovinger, Appl. Phys. Lett. **69**, 4108 (1996).

22
Scaling Limits and MHz Operation in Thiophene-Based Field-Effect Transistors

A. Hoppe, T. Balster, T. Muck, and V. Wagner

22.1
Introduction

Within the last decade the field of organic thin film transistors (OTFTs) has developed rapidly. Devices based on organic semiconductors are of interest for a wide range of low-cost electronic applications. Such potential applications are electronic paper [1], active-matrix emissive displays [2], radio-frequency identity (RFID) tags [3], organic photovoltaic cells [4], chemical vapour sensors [5], or even low-cost integrated circuits [6]. Organic semiconductors have some specific advantages compared with their inorganic counterparts. They can be processed at low temperatures and therefore at lower cost with spin-coating [7], printing [8], or stamping techniques [9–11]. They also can be integrated on inexpensive plastic substrates, showing flexible and low-weight properties.

The electrical performance of the organic materials has increased constantly since the field started in 1986 with a device based on an electrochemically grown polythiophene film [12]. Since then a broad variety of different materials, small molecules, as well as polymers have been synthesized and tested for their electrical properties. Environmental stability and doping are additional important issues [13, 14]. An overview of different organic semiconductors can be found here for oligomers and for polymers [15–17]. One very promising type of organic molecules is the class of thiophenes. They show high charge carrier mobilities comparable to those in amorphous silicon, ordering phenomena, and the possibility of wet chemical processing. The properties of the oligothiophenes can be controlled by the number of thiophenes in the molecule, as shown later in this chapter, as well as by the type of the substitution. An introduction to the different types of thiophene-derivates will be given in Section 22.3.

Improving the OTFT performance by optimising the organic semiconductor properties and therefore the charge carrier mobility μ is one way to increase important device parameters. Another way is the reduction of the lateral size. The channel length L is hereby the critical device parameter. Downscaling of L

Organic Electronics. Structural and Electronic Properties of OFETs. Edited by Christof Wöll

ISBN: 978-3-527-40810-8

allows enhanced currents since the drain current I_d is proportional to μ/L. It also allows a higher density of integration which scales with $N \propto 1/L^2$ as well as higher operation frequencies f which are proportional to μ/L^2.

The extreme lateral downscaling to channel lengths in the sub-micrometer regime however increases the influence of the contact resistance between the organic layer and the metallic source and drain electrodes on the device characteristics. It also bears the risk of a deterioration of the electrical characteristics by disturbing the ratio between the lateral and the vertical geometry scale. With sufficient input parameters this effect can be modelled in detail by using a device simulator [18]. From scaling theories in inorganic field-effect transistors (FETs) it is known, that undesired short channel effects occur beneath a minimal channel length L_{min} [19]. For organic field-effect transistors (OFETs) a similar behaviour is expected. Figure 22.1 shows some important scaling parameters in the organic devices as the channel length L, the insulator thickness d_{ox}, and the thickness of the semiconductor layer d_{sc}. Furthermore, the region in the organic layer influenced by the contact is expected to affect the performance at small L and should be considered as important scaling parameter as well.

This chapter is structured as follows. The second part is an experimental section in which the typical sample preparation of the larger FETs as well as of the sub-micrometer structures is presented. The third section introduces the class of thiophene-based semiconductors divided into oligo- and polythiophenes to give an overview about recent and historical developments in the field. Section 22.4 focuses on organic transistors with downscaled L towards the sub-micrometer region [20]. General electric properties will be presented; short channel effects and the impact of different electrode materials on the performance will be discussed. Furthermore, the influence of the insulator thickness on the electrical properties will be elaborated. In Section 22.5 a contact resistance model is used to describe the resulting electrical data of the sub-micron devices to take the increasing influence of the contacts with decreasing L into account [21]. Different oligothiophenes are compared to find suitable materials with low contact resistance for sub-micron devices. Section 22.6 is about the influence of the semiconductor layer thickness on the charge carrier mobility. This section is divided into larger channels [22, 23] and downscaled OFETs which will be discussed separately. In the last experimental part, Section 22.7, optimisation strategies towards high frequencies are discussed and

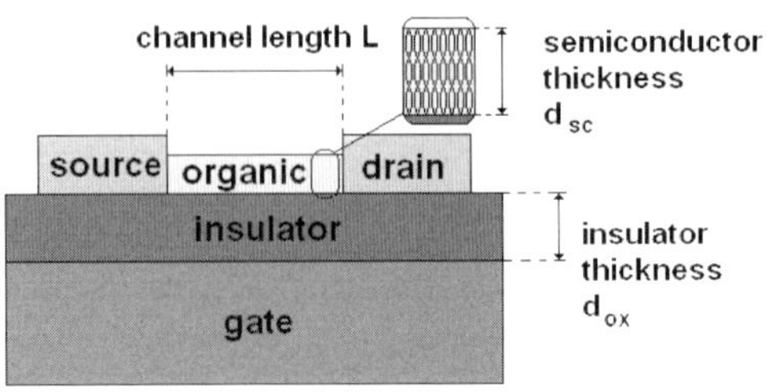

Figure 22.1 Important scaling parameters in OFETs.

optimised transistor structures with switching frequencies above one Megahertz are presented [24]. These devices are the fastest polymeric transistors reported in literature so far.

22.2 Device Preparation

22.2.1 Geometries

Organic field-effect transistors can be produced either in top or bottom contact architecture [16]. In top contact geometry the semiconductor is deposited first, followed by the electrodes. In bottom contact devices the electrodes are prepatterned and the organic material is layered on top. Both kind of structures have their own advantages. In top-contact devices the contact resistance is usually smaller due to the larger injection area from the electrodes to the semiconductor. Charge carrier mobilities in top contact devices tend to be higher than in their bottom-contact counterparts. It is also less difficult to prepare highly ordered films since substrates for top contact devices do not show the irregularities from the electrodes on the surface [16]. Due to the processing of the top electrodes however, which are mostly metallised through a shadow mask, shadowing effects usually circumvent channel lengths below 5 μm. Also, the process is not yet readily amenable for large scale production. For examinations on sub-micrometer channel length transistors, top contact architecture is therefore not suitable. Another advantage of bottom contact structures is the possibility to observe semiconductor thickness dependent electric properties in real-time during film growth. The setup used for these experiments is shown in Figure 22.2.

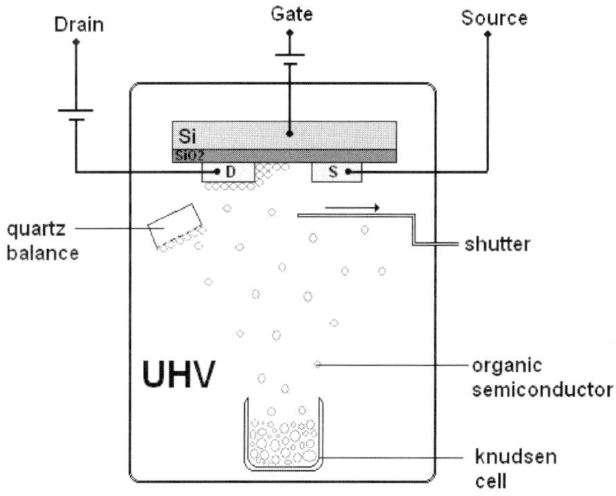

Figure 22.2 Setup for *in situ* measurements.

In general, bottom contact devices are more easily integrated into low-cost production processes since available patterning techniques such as photolithography and electron-beam lithography allow for smaller device feature sizes and therefore a higher density on the substrate. Since the investigations on small *L* devices as well as the dependence on the semiconductor layer thickness are two major topics within this paper, the focus will be on samples in bottom contact geometry.

22.2.2
Sample Preparation

The sample structures discussed in more detail were produced on highly n-doped silicon, which served as a common gate. Thermally grown silicon oxide ranging from 100 down to 30 nm thickness was used as insulator. Structures with feature sizes larger than one micrometer were typically produced by UV-lithography. Smaller feature sizes were produced by electron beam lithography (EBL). A two layer resist consisting of 50 K and 200 K poly(methylmethacrylate) (PMMA) was used for high resolution lithography. Electrode structures consisted typically of a 3 nm titanium adhesion layer and a 17 nm gold electrode. The surface of the sample was functionalised with –OH groups. To avoid leakage currents over the edges of the sample, the device sides' were covered with an insulating PMMA layer. The polymer deposition was performed in a four step dip-coating process. By dipping the device into a polymer solution from each side of the sample it was guaranteed that the later insulator/semiconductor interface was not in touch with the solution. In the last steps after cleaning, the transistor surface was functionalised either with hexamethyldisilazane (HMDS) from gas phase or with octadecyltrichlorosilane (OTS) from the liquid phase. The evaluation of the channel lengths of the sub-micrometer transistors was performed with an atomic-force microscope or (after electrical measurements) with SEM images. In Figure 22.3 an example of an electron beam written structure with interdigitated fingers can be seen.

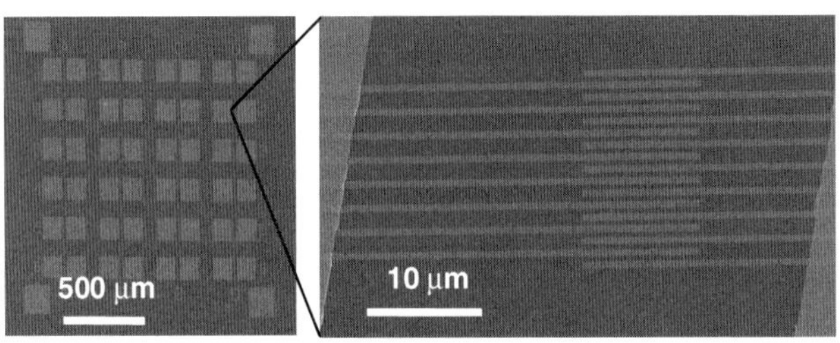

Figure 22.3 SEM images of contact pad structure (left) and electron beam written finger structure (right).

22.3
Thiophene-Based Semiconductors

22.3.1
Unsubstituted Oligothiophenes

Thiophene-based oligomers are one of the most studied π-conjugated oligomeric model systems [17]. Chemical tailoring of this class of molecules from groups worldwide has led to a huge variation of different types over the last twenty years [25]. The simplest molecule from this class is unsubstituted α-*n*T, where *n* stands for the number of thiophene units and α for the carbon-atom next to the sulfur in the thiophene, where the next ring is positioned. The chemical structure of α-*n*T is given in Figure 22.4. Different oligomers with *n* ranging from 1 to 16 are well known [26].

The first OFET produced with sexithiophene (α-6T) as active layer was already reported in 1989 by Horowitz and Garnier [27]. This was also the first report of an OFET made with a small conjugated molecule. Since then, the charge carrier mobility was constantly increased by different groups by optimising deposition parameters as well as surface and contact treatments from initially 10^{-4} cm^2/Vs to greater than 0.07 cm^2/Vs [28]. Even higher mobility values have been reported for octithiophene (α-8T). A mobility near 0.2 cm^2/Vs was measured when the active layer is deposited at a sample temperature of 150 °C and higher [29]. The influence of the chromophore length on the mobility is an intensively discussed topic in literature. While some claim, by further optimising the device parameters there is no influence on the mobility, systematically measured data usually show differences [30].

Due to different deposition details and other hidden parameters when producing OFETs and evaluating the field-effect mobility, those values are compared best when produced by the same groups and not in review papers with values from different groups and times as done in [31]. In the solid state, it is known from unsubstituted oligothiophenes to pack with their long axis parallel to each other. They form layers with the width equal to the length of the molecule, corrected for a small tilt angle [32].

Since this ordering can be valid over several monolayers, in thin films the layers can be seen as two-dimensional mediums. Charge transport in these mediums is favoured in the direction parallel to the organic layer, which was again shown by Horowitz and Garnier [33] with X-ray diffraction measure-

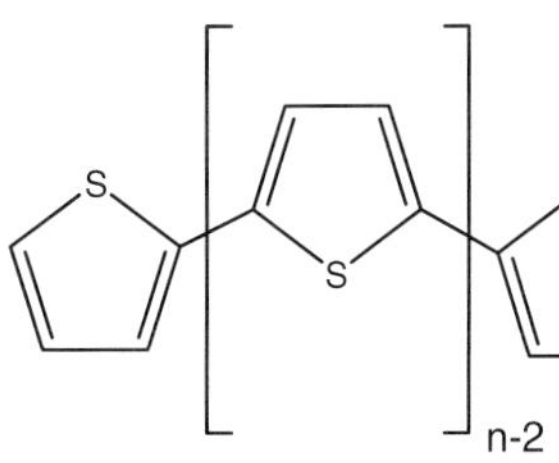

Figure 22.4 Chemical structure of α-*n*T, with *n* as the number of thiophene units in the oligomer.

ments on sexithiophene (α-6T) field-effect transistors. Highest mobility values are therefore obtained, when all molecules stand upright on the insulator. This behaviour was later also shown for other organic semiconductors, like pentacene.

Still, unsubstituted α-*n*T is a model system used in recent literature. One disadvantage towards application of the unsubstituted oligothiophenes however is the poor solubility of the rigid rods [34].

22.3.2
Substituted Oligothiophenes

The general chemical structure of αα′-alkylsubstituted oligothiophenes is shown in Figure 22.5. While *n* denotes again the number of thiophene units in the molecule, *m* gives the lengths of the alkyl spacers in the α and α′ positions. Due to the route of chemical synthesis to produce such molecules either by Grignard-coupling or by oxidative coupling the alkyl chains on both sides are usually of the same length.

By attaching alkyl spacers on both ends of the molecule the solubility of the compound is improved. While α-6T is nearly insoluble in most solvents, the dihexyl substituted derivate, DH6T, shows better properties and is soluble in acetone and in similar solvents [35]. Beside the solubility of the molecules, another improvement of the substitution became apparent, when αα′-substituted dimethyl ($-C_2H_5$) oligothiophenes ranging from DM3T to DM5T showed an increase of several orders of magnitude in the charge carrier mobility compared with their non-substituted counterparts α-3T to 5T [36]. This showed that the alkyl spacers help to improve ordering of the organic layer. Thiais and coworkers confirmed this by X-ray diffraction measurements on dihexyl ($-C_6H_{13}$) substituted DH6T [37] similar to those done on 6T [33]. It was shown, that the high mobility mainly derives from the very regular arrangement of the film. Thus, it should be possible to reach similar charge carrier mobilities with unsubstituted thiophenes when packing very carefully. Indeed, mobilities of α-6T meanwhile get very close to those measured in DH6T. Only the two-dimensional character of the layers is slightly decreased by the missing alkyl-layer in between.

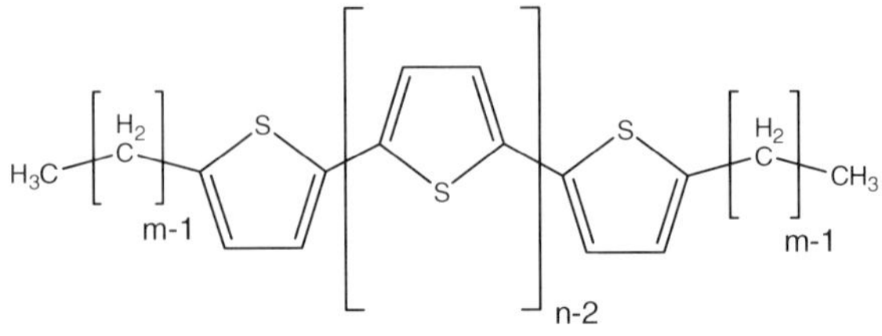

Figure 22.5 Chemical structure of αα′-alkylsubstituted *n*T, with *n* as the number of thiophene units in the oligomer, whereas *m* denotes the length of the alkyl chains.

In most of the recent literature on oligothiophenes however, substituted molecules become more and more favoured to those unsubstituted derivates since effects due to poor packing can be excluded much more easily and higher field-effect mobilities are usually easier to obtain.

After the first measurements on dimethyl-substituted *n*T in 1991 a large variety of different alkyl-lengths and substitution positions has been synthesised and tested for OFET performance. Some examples are didecyl ($-C_{10}H_{21}$), didodecyl ($-C_{12}H_{25}$), as well as γ,γ'-substituted oligothiophenes such as DP4T with dipentyl ($-C_5H_{11}$) groups in γ and γ' positions [38]. Due to the above mentioned data scattering caused by the large amount of parameters to be optimised as well as different setups, groups, surface treatments, insulator materials etc. it is difficult to compare the broad variety of such molecules. In conclusion it can be said, that the influence of the alkyl-group seems to be much less pronounced than thought initially. Different substitutes do change sublimation temperatures, position of the liquid crystal phase and other physical properties. However, huge differences in the charge carrier mobility are not reported, when first layers on the insulator are ordered properly. Besides all these different molecules the dihexyl-substituted derivates DH*n*T have become the most common ones and many of the recent publications focusing on fundamental properties of organic transistors use them as a semiconductor. Their properties have therefore been extensively optimised for different setups. Top values are reported to be in the range of 0.23 cm^2/Vs for DH4T and ~0.5 cm^2/Vs for DH5T and DH6T on polymer substrates with top-contacts [28].

The present chapter focuses on those α,α'-dihexyl-substituted oligothiophenes with n ranging from four to seven. DH4T serves in many of the experiments as a standard semiconductor model. The liquid crystal phase beyond 80 °C [39] allows for very large grain sizes (>10 μm). The influence of the chromophore length is also studied systematically in a comparative series of experiments with special focus on contact properties in sub-micrometer transistors. DH7T is a new material which was presented in literature first in 2007 [22, 40]. The materials used in this set of experiments discussed in greater detail were synthesised and purified either by the group of Prof. Bäuerle from the University of Ulm (DH5T, DH6T, DH7T) or bought commercially and purified by the group of Dr. Pflaum from the University of Stuttgart (DH6T). The DH4T compound was bought from SynCom B.V.

22.3.3
Polythiophenes

Besides the oligomers based on thiophene units, polymers with similar backbones have become very represented in literature since the late nineties. Due to the fact that pure thiophene polymer chains without any substitutes show very poor solubility, mainly alkyl substituted polythiophenes are used.

Figure 22.6 shows the chemical structure of P3AT, with n as the number of thiophene rings and m defining the length of the alkyl-group. Pioneering work

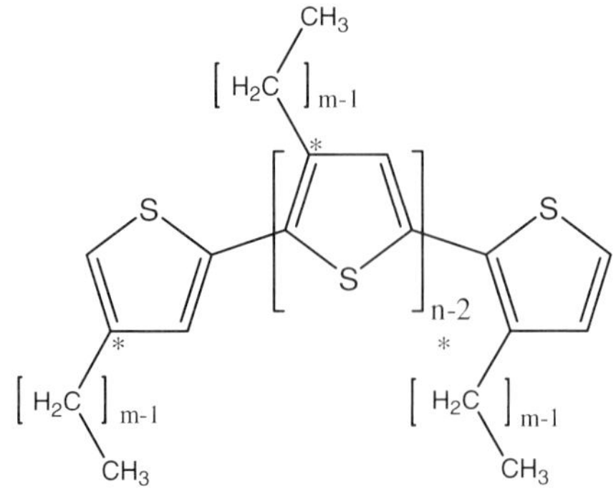

Figure 22.6 Chemical structure of poly(3-alkyl-thiophene) P3AT with *n* as the number of thiophene rings and *m* defining the length of the alkyl-group.

on this class of polymers was done by Sirringhaus and coworkers [41]. They used the meanwhile well established poly(3-hexyl-thiophene) and showed increased electrical performance related to structural ordering of the polymer. They demonstrated that the order intensively depends on the regiolarity of the polymer, which is the percentage of alkyl-groups in the beta-position of the thiophene rings (position marked as * in Figure 22.6). Highest performance with these polymers is reached when the chains stand etch on the surface, upright on their alkyl-spacers. The material shows ordering phenomena when processed from solution; lamella-like crystalline areas in different sizes can be produced with different solvents and deposition techniques like spin-coating or dip-coating [42]. P3HT does not usually show distinctive layer-by-layer growth, but thin films with only one lamella layer show thicknesses of about 2 nm, which was demonstrated with X-ray diffraction and AFM profiling [43]. Highest mobility presented so far is 0.2 cm^2/Vs, in thin, well ordered layers [42]. Another important parameter influencing the charge-carrier mobility is the molecular weight of the polymer [44, 45]. The trend usually shows that longer chains exhibit improved electrical properties.

22.4
L Dependence of OFETs

22.4.1
Influence of the Electrode Material

When shrinking the channel length of the transistor device, the channel resistance is reduced. Beneath a certain threshold, contact properties become visible in the present materials. To obtain optimal characteristics for the submicrometer transistors, the choice of the electrode material is crucial. Figure 22.7 shows mobility values $\mu(L)$ obtained from three series of OFETs with different electrode materials and DH4T as semiconductor. The thickness of the semiconductor was 10 nm (~4 monolayers), deposited at an evaporation speed of 2 pm/s. The electrode materials were Ti/Au (thickness 5/20 nm), Ti/Pt (5/20 nm), and Pd (25 nm). Samples were produced by optical lithography and channel lengths were ranging from 1 to 50 μm. Mobility values were extracted from transfer characteristics in the saturation regime ($V_{DS} = -10$ to -20 V) using the standard MOSFET theory [19]:

$$I_d = \frac{W}{2L} C_i \cdot \mu \cdot (V_{gs} - V_{th})^2 \ , \tag{1}$$

with W as the channel width, the insulator capacitance per area C_i (33 nF/cm^2 for 100 nm silicon-oxide), the threshold voltage V_{th}, and the gate voltage V_{gs}. Similar mobility values were obtained also in the linear regime. Obviously, the mobility μ in Figure 22.7 decreases with shrinking L for the Ti/Au and the PT/Ti electrodes, which is in accordance with the increasing role of the contacts. For the Pd electrode however, only a weak dependence on the channel length can be seen. The obtained mobility values include the serial resistance of the organic channel R_{Ch} and the total contact resistance R_C, the sum of contact resistances at drain and source contacts. Assuming a constant mobility, the channel resistance scales with L, $R_{Ch} = \rho_{Ch} \times L$ with ρ_{Ch} as the specific channel resistance. For the drain current and the mobility we obtain therefore:

$$I_d = \frac{V_{0,\,ds}}{R_C + \rho_{Ch} \cdot L} \propto \frac{\mu}{L} \tag{2}$$

and

$$\mu(L) \propto \left[\rho_{Ch} + \frac{R_C}{L} \right]^{-1} , \tag{3}$$

with $V_{0,ds}$ as the applied voltage between source and drain contact. The contacts affect the apparent mobility values especially at small L. Contact resistance values can be derived from $\mu(L)$ with decreasing L (transfer line method). The resulting values are 0.2 mm MΩ, 1.4 mm MΩ and 2 mm MΩ for the Pd, Ti/Pt and Ti/Au structures, respectively. These values are expected to depend on the work function of the contact metal ϕ_m. Since these values for ϕ_m are

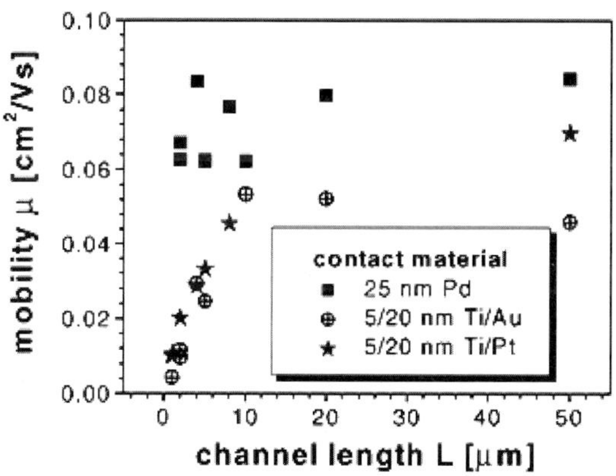

Figure 22.7 Scaling behaviour of DH4T transistors with different electrode materials (from [20]).

comparable (4.8 eV, 5.0 eV, and 5.3 eV for Au, Pd, and Pt, respectively [46]) they cannot explain the different behaviour of Pd contacts. The absence of the titanium adhesion layer in the Pd structures appears to play an important role. The work function of titanium (4.1 eV) is significantly lower. Furthermore, exposed titanium is expected to be oxidised in air.

In a second set of experiments the influence of the titanium adhesion layer on the mobility was tested. For this instance, the titanium layer thickness was varied from 1 nm to 25 nm within the 25 nm thick Ti/Au structures. DH4T was again used as organic material, deposited under the same conditions as stated above. The scaling behaviour of the mobility for these series is presented in Figure 22.8. As in Figure 22.7 the mobility drops with shrinking channel length due to increased influence of the contact resistance. However, the trend shows that this effect is strongly reduced with decreasing adhesion layer thickness. These results show the important influence of the titanium for the carrier injection at the organic to metal contact. Apparently, best transport properties are realised, when the first monolayers of the organic materials are in direct contact with the Au electrode layer [20].

22.4.2
Influence of the Insulator Thickness

A further lateral downscaling of OFETs towards channel lengths comparable to the gate insulator thickness d_{ox} causes deviations from the classical long-channel MOSFET behaviour [18]. The electric field E_y along the channel becomes comparable to the transverse field E_x induced by the gate voltage and the gradual channel approximation $E_x \gg E_y$ is not valid anymore. Figure 22.9 shows exemplarily output characteristics of DH4T transistors with different channel lengths to oxide thickness ratios. The device characteristics shown on the left side were produced on 100 nm thick oxide. At the 200 nm structure, on the upper left, the onset of short channel effects becomes visible; the current does not saturate completely. The 100 nm structure on the lower left side even

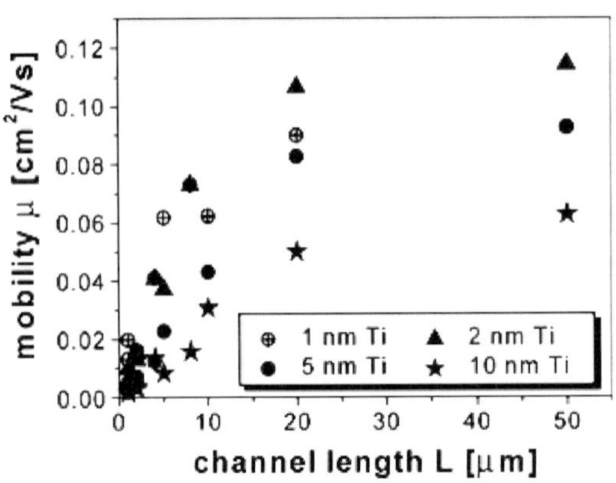

Figure 22.8 Scaling behaviour of DH4T transistors with different titanium thickness in the 25 nm thick Ti/Au electrodes (from [20]).

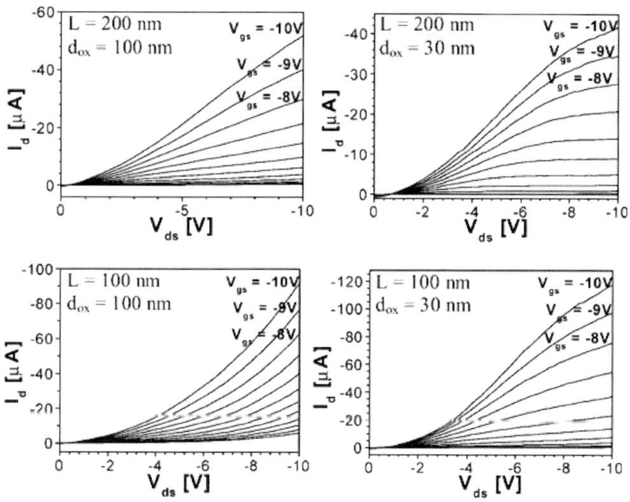

Figure 22.9 Output characteristics for various DH4T devices, showing the changeover to short-channel behaviour. The channel length varies from 200 nm (upper part) to 100 nm (lower part) and the oxide thickness varies from 100 nm (left side) to 30 nm (right side) (from [20]).

shows finite current flow in the saturation regime, also under "off" conditions. The behaviour is best described as diode-like. For the improved devices produced on 30 nm thick oxide, optimised behaviour is observed. The 200 nm transistor with sufficient thin oxide displayed on the upper right side shows saturation in contrast to the one produced on thicker oxide. The non-linear behaviour for small V_{ds} is contact related and will be discussed below. Even more significant is the change in the behaviour of the 100 nm transistor produced on 30 nm thick oxide shown in the lower right. The previous observed "off" current vanishes and reveals almost long-channel behaviour. The results confirm the requirement of an appropriate reduction of the insulator thickness d_{ox} when shrinking the channel length L. Empirically, the ratio between the channel length and the oxide thickness needs to be in the range of 3 : 1 to show reasonable output characteristics.

22.5 Optimised Sub-micron OFETs

22.5.1 Semiconductor Related Performance

As already shown in the previous section, short channel effects such as lack of saturation can be avoided by a sufficiently thin gate insulator and improved charge injection could be realised by a thinner titanium adhesion layer. Nevertheless, an undesired bending of the transfer curves is observed for sub-micron devices below a critical channel length, even with sufficiently thin gate insulators (see Figure 22.11). This effect was investigated systematically by analysing 32 transistors with channel lengths between 50 nm and 400 nm [21]. As the semiconductor a series of four different oligothiophenes was used. While all are α,α′-dihexyl-substituted, the number of thiophene rings in the molecule varies from four (DH4T) to seven (DH7T), which shifts the transport energy level [47]. To separate material- and contact-related effects the different materials were tested initially with a long channel length of 40 μm, where contacts effects are expected to be negligible. I–V characteristics were recorded in ambient air directly after material deposition in vacuum. Figure 22.10 displays the mobilities derived from transfer curves in the linear regime at $V_{ds} = -1\text{V}$ using standard MOSFET theory:

$$I_d = -\frac{W}{L} \cdot \mu \cdot C_i \cdot \left(V_{gs} - V_t - \frac{V_{ds}}{2} \right) \cdot V_{ds}\,, \tag{4}$$

with C_i the insulator capacitance of 67 nF/cm² for 50 nm oxide. All measured transfer curves showed ideal behaviour according to the standard theory.

Obviously, the charge-carrier mobility in air raises with an increasing number of thiophene units almost two orders of magnitude, ranging from 5×10^{-3} (DH4T) to 10^{-1} (DH7T) cm²/Vs.

A similar trend was reported for unsubstituted oligothiophenes [48]. The highest measured mobility was 0.12 cm²/Vs for DH7T in the linear regime [22, 40].

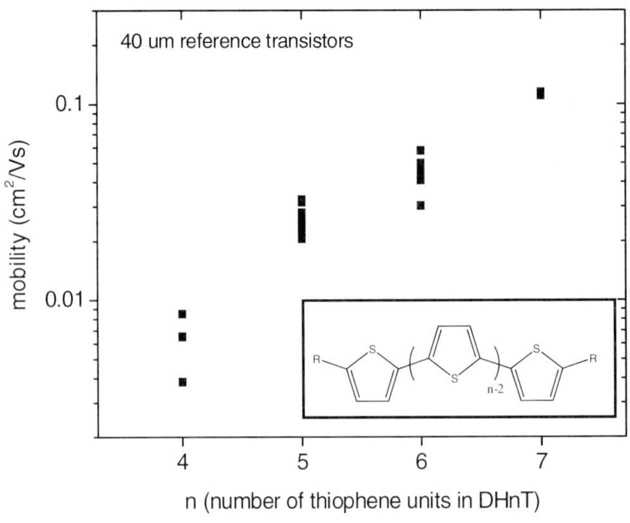

Figure 22.10 Charge carrier mobility as a dependence on n, the number of thiophene-rings in DHnT.

22.5.2 Tuning the Contact Resistance

A change in the shape of the transfer curves is observed for channel lengths below 500 nm. This effect is most pronounced for DH4T and gets smaller towards DH6T and DH7T. Figure 22.11 demonstrates the behaviour in a DH5T transistor with a channel length of 200 nm. This kink in the transfer-curve prevents the data from being described within the standard model, which predicts a simple linear dependence in the on state. It should be mentioned that the observed shape of the transfer-curves is neither a hysteresis effect nor is it dependent on the measurement speed. It is rather attributed to be caused by contact properties – visible only in short channels, since here the channel resistance gets sufficiently small.

A closer analysis of an individual transfer curve is used to extract information about the contacts for each device. To describe the electrical characteristics in a proper way, the gradual channel approximation model (Eqs. (1) and (4)) is extended by an ohmic contact resistances R_s and R_d in series at source and drain. For a p-type channel we obtain:

$$I_d = \begin{cases} -\frac{W}{L} C_i \mu \left([V_{gs} - R_s I_d] - V_{th} - \frac{[V_{ds} - (R_s + R_d) I_d]}{2} \right) \\ \times [V_{ds} - (R_s + R_d) I_d], \\ -\frac{W}{2L} C_i \mu ([V_{gs} - R_s I_d] - V_{th})^2 . \end{cases} \tag{5}$$

The upper and lower equations stand for the linear and saturation regimes, respectively. Since in the linear regime the small correction $R_s I_d$ to the large gate voltage term $(V_{gs} - V_{th})$ might be neglected, only the total contact resistance

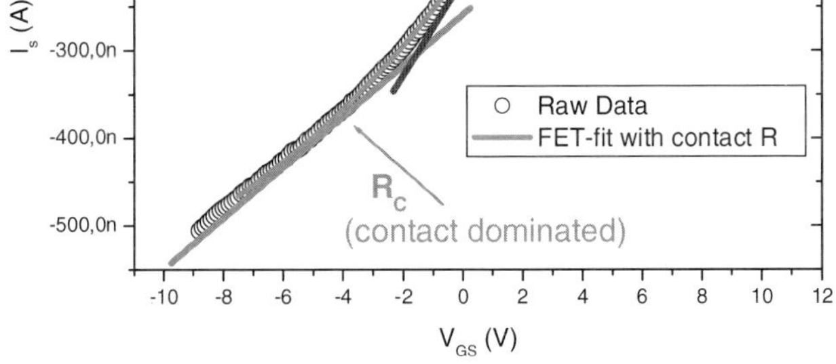

Figure 22.11 Transfer curve of a 200 nm DH5T transistor.

matters. Thus, the width normalised total contact resistance $r_c = W \times (R_s + R_d)$ is used as an additional fitting parameter besides μ and V_{th}. Solving Eq. (5) for larger current levels (linear case) results in:

$$I_d = -\frac{W}{L} C_i \mu \frac{[V_{gs} - V_{th} - V_{ds}/2]}{1 - r_c C_i \mu / L \cdot [V_{gs} - V_{th} - V_{ds}/2]} \tag{6}$$

(note that the voltage expression in brackets is negative). The expression allows a critical gate voltage to be defined:

$$V_{gsx} = V_{th} + V_{ds}/2 - L/(r_c C_i \mu) \,, \tag{7}$$

where the channel resistance, which is tuned by the gate voltage, is equal to the contact resistance. V_{gsx} defines the transition in the transfer curve from transistor dominated to contact resistance dominated regime. If the applied gate voltage is far beyond the critical gate voltage ($V_{gs} \ll V_{gsx}$) the transistor is fully switched on and the expression reduces to an ohmic current governed by the contact resistance $I_d = W/r_c V_{ds}$. For the reference transistors with large L the critical gate voltage V_{gsx} is far outside the measurement range. This is different for nanoscale transistors, where V_{gsx} is close to or even within the measurement range. Figure 22.11 exemplarily shows a measured transfer curve with raw data (hollow circles), and the corresponding fit with contact resistance (red line). The two lines demonstrate the different dominated regions of the device. The blue line is only driven by the channel resistance and therefore the mobility of the device, while at the magenta line the channel resistance has been sufficiently decreased by opening the channel, that the transistor is governed by the contacts. In this example, V_{gsx} is close to $V_{gs} = -2$ V.

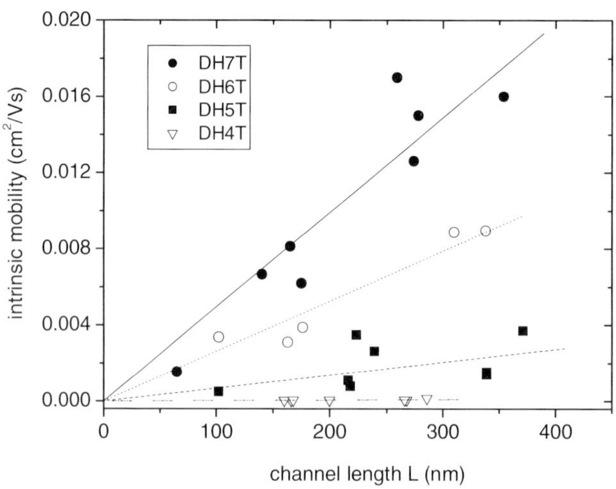

Figure 22.12 Intrinsic charge carrier mobilities of DHnT as a dependence on the channel length for sub-micrometer devices.

The mobility, as one of the fit parameters, is now a corrected value which displays the intrinsic charge carrier mobility. Figure 22.12 shows the intrinsic mobility values as a dependence on the channel length. Even with correction for the contact resistance in place the intrinsic mobility values obviously depend on the channel length. The effect however is much less pronounced than for example in Figure 22.7 but still, the mobility decreases linearly with shrinking channel length. This behaviour might indicate that the film quality is affected in the proximity of the contacts even over a length scale of several 100 nm.

As already seen in the 40 μm structures, DH7T shows the highest mobility values (10^{-2} cm²/Vs for $L > 200$ nm) while the mobility decreases systematically with lower number of thiophene units. In contrast to the charge carrier mobility, no dependence of the contact resistance on the channel length was found. Variation of the applied drain-source voltage showed no significant influence on the mobility and the contact resistance data. Figure 22.13 shows the averaged R_C values as a dependence on the different semiconductors.

Obviously contact resistance drops from DH4T to the larger molecules more than one order of magnitude and saturates between DH6T and DH7T at a value of 1 kΩ cm. In comparison with other organic semiconductors, this is a rather low value, especially for untreated gold contacts. Pentacene for example is reported to show a contact resistance of about 100 kΩ cm in comparable bottom contact devices [49].

22.6
Influence of the Semiconductor Thickness

As already mentioned in the introduction, oligothiophenes exhibit a distinctive layer-by-layer growth with a high ordering in the first layers. In addition, the

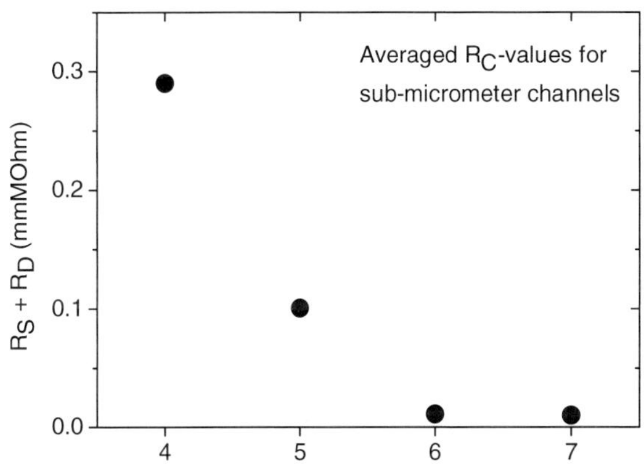

Figure 22.13 Averaged contact resistance values as function of *n*, the number of thiophene units in DH*n*T.

liquid crystal phase of DH4T above 80 °C might be utilised to grow films with grain sizes larger than ten micrometers [50]. In this chapter, the electrical properties and the morphology of the organic layer in the channel and close to the contacts as a function of the film thickness are investigated for the various oligothiophenes.

22.6.1 Large Channels

To examine the influence of the organic layer thickness, the experimental setup in Figure 22.2 is used. The electrical characteristics are measured *in situ* and in real-time during semiconductor evaporation. In parallel, a quartz balance is used to obtain the corresponding layer thickness. Figure 22.14 shows mobility values extracted from transfer curves measured in the linear regime at $V_{ds} = -1$ V according to Eq. (4) as a function of the film thickness. In the graph, data for three different semiconductors are shown, i.e. DH4T, DH5T, and DH6T. The measured transistor devices had a long channel length of 50 μm. The individual optimised evaporation parameters can be found in [21].

The onset of transistor behaviour is observed at 0.5–0.7 monolayers for the different semiconductors. In a previous experiment for DH4T the current onset was found to be at 0.57 monolayers, which reflects the percolation threshold of the growing DH4T islands [22]. For lower coverage, no significant current levels were observed. Beyond the current onset a quadratic increase of the mobility versus the coverage was found. Obviously all measured materials

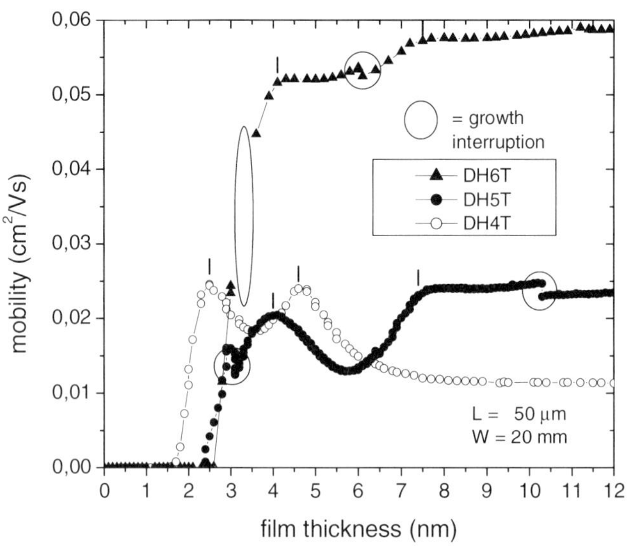

Figure 22.14 Field-effect mobilities deduced from transfer curves in the linear regime of devices with 50 μm channel length during DH*n*T layer growth.

show oscillations of the mobility with the film thickness. The peaks in the mobility reflect the lengths of the different molecules. All show maxima with the completion of the first monolayer. This behaviour is more pronounced for the smaller oligothiophenes and reduced towards DH6T. For DH4T and DH5T a decrease in the mobility was observed after the first monolayer while DH6T remains on the same level. A second peak is seen for all materials when closing the second monolayer. Then, DH5T and DH6T remain on the same level for further coverage while DH4T shows a drop in the mobility attributed to contact effects [22]. The transistors show best performance for closed layers. The unfinished second layer obviously disturbs the flow of the charges, thus decreasing the performance of the device. The data shown in Figure 22.14 suggest that charge transport happens in the first two monolayers only. Thicker films do not increase mobility values anymore. In the case of the DH4T the performance is even lower for thicker films.

22.6.2 Sub-micron Channels

While in the previous section the influence of the contacts on the growth of the organic material inside the 50 μm channel structures can be neglected, they are expected to become more important in smaller devices. The samples for this set of experiments were prepared as described in Sections 22.2 and 22.5. Figure 22.15 shows the intrinsic mobility and contact resistance data obtained during DH4T layer growth. The sub-micron transistor structure had a channel length of $L = 830$ nm. Like in Figure 22.14, each data point is obtained from a fit to a transfer curve recorded at $V_{ds} = -1$ V. In contrast to Figure 22.14 the

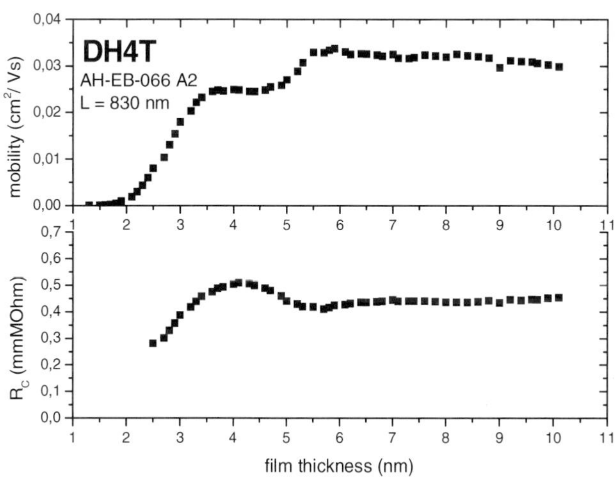

Figure 22.15 Growth curve of a DH4T transistor ($L = 830$ nm, $W = 300$ μm). Each data point is deduced from a transfer curve measured at $V_{ds} = -1$ V.

contact resistance model (Eq. (5)) was applied to obtain the data in Figure 22.15. Similar to the data presented in Figure 22.14, monolayer oscillations can be seen. Obviously, this behaviour is not only observed for μ, but for r_C as well. Here, a maximum in the contact resistance can be seen between the first two monolayers.

The same behaviour can be seen in DH5T devices with similar oscillations in the charge carrier mobility as well as in the contact resistance. DH6T transistors differ from the smaller oligothiophenes. As already displayed in Figure 22.14, the trend towards oscillations is less pronounced. In channels down to 600 nm, only a small shoulder in the mobility is visible for the closed first monolayer. Overall values of the contact resistance for the individual materials from this set of experiments agree very well with those presented in Figure 22.13. It appears that channels down to 500 nm exhibit the same oscillations as observed in the 50 μm channels. A further downscaling however indicates a change in the growth characteristics. Figure 22.16 shows the growth curves of two different sub-micrometer transistors with DH4T as active material. The channel lengths are 830 nm and 250 nm, respectively. Each data point is taken from a transfer curve at $V_{ds} = -1$ V according to Eq. (5).

It can clearly be seen, that the oscillations in the mobility are suppressed by a further shrinking of the channel length. While the larger structure shows clear oscillations, the 250 nm structure demonstrates a different growth mode. The same behaviour was observed for DH5T. Since DH6T even in large channels has a less pronounced trend towards oscillations, the step to the different growth mode is more difficult to see. AFM-studies at sub-micrometer finger structures with different DH4T coverages indicate an increased thickness of the DH4T layer next to the gold electrode. For a coverage of 0.7 monolayer the

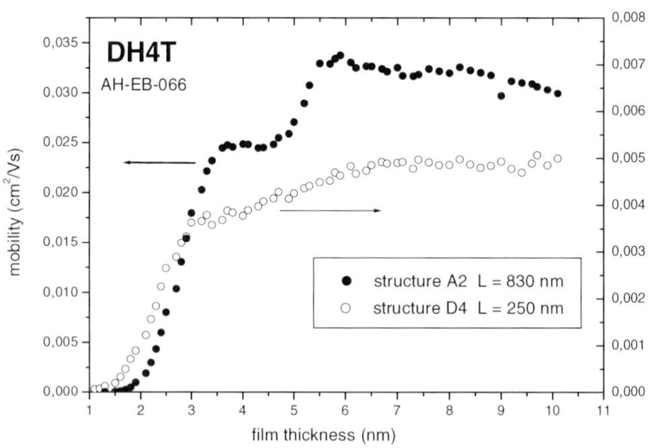

Figure 22.16 Growth curves of DH4T transistors with channel length of 830 nm and 250 nm, respectively. Each data point is deduced from a fit to a transfer curve measured at $V_{ds} = -1$ V according to Eq. (5).

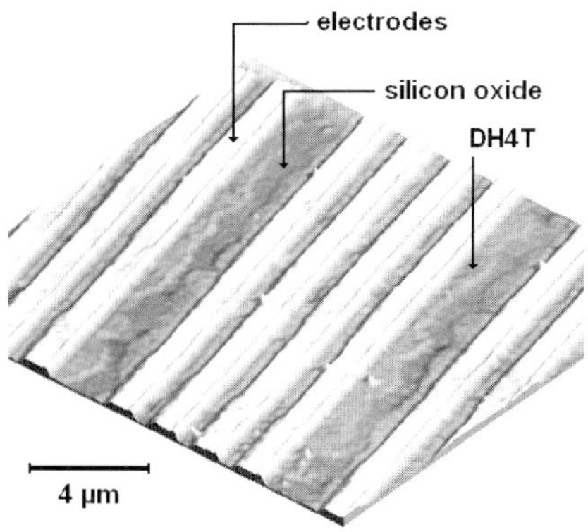

Figure 22.17 AFM image of 20 nm thick Ti/Au electrodes and coverage of 0.7 ML (2 nm) DH4T. Note that exclusively areas next to the fingers are covered.

organic material arranges exclusively next to the gold electrodes (Figure 22.17). With higher coverage, material gathers in the area around the contacts, e.g. an overall coverage of 2 ML results in 4 ML coverage next to the electrodes. The effect is attributed to the different surface diffusion of DH4T on the electrodes and in the channel area at the given growth temperature. With a sufficient close distance between two electrodes the whole channel area will be affected by the contact-modified growth mode.

Experimentally it is found, that for channels smaller than 500 nm the gap is filled with contact-related organic material and the 2D layer-by-layer growth of DH4T is therefore hindered. The growth of the organic material is then electrode dominated. This behaviour is clearly seen in Figure 22.18.

In conclusion, oscillations observed in the mobility as well as in the contact resistance are related to the layer-by-layer growth inside the free area of the channel which is not influenced by material diffusion from the contacts. Below 500 nm (for DH4T and DH5T) a different growth mode is observed by electrical measurements and by AFM imaging. A transition towards electrode dominated growth is found.

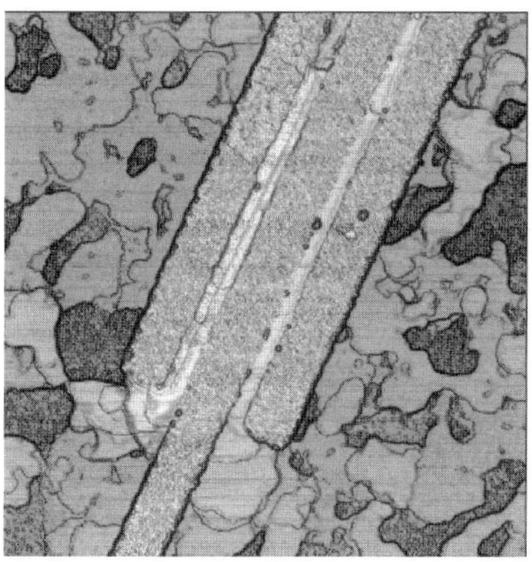

Figure 22.18 $5 \times 5\ \mu m^2$ AFM image of sub-micrometer finger structure. The 100 nm gap between the fingers is filled with DH4T.

22.7 Megahertz Operation

New applications of organic transistors demand increased switching speed which might be achieved mainly either by new materials or by reduced dimensions of the device. For cheap production, materials are usually limited to those processible from solution. In this section we will follow the second route of reduced dimensions and restrict the material basis to oligio- and polythiophenes, i.e. DH7T and rr-P3HT.

22.7.1 Theoretical Considerations

Transistors with bottom-contact architecture with a large gate electrode (Figure 22.1) require only one high resolution lithographic step to produce drain and source electrode structures with feature sizes down to the nanometer range. While EBL was applied for structuring the test samples under investigation, those structures can also be manufactured by cheap printing techniques like imprint lithography or microcontact printing [9–11].

A given transistor structure has an intrinsically frequency limit for its operation. While there is ideally no gate current in DC-mode, the FET input current (gate current) increases linearly with frequency for applied AC-voltages due to the finite gate capacitance. The output current (drain current) is mainly defined by the DC-characteristics at not too high frequencies. The unity gain bandwidth f_t is the upper frequency limit (transfer frequency), where gate and drain current exhibit the same amplitude (unity-gain condition).

$$\left.\frac{i_d}{i_g}\right|_{\omega=\omega_t} = 1\,. \tag{8}$$

The angular frequency ω denotes hereby the frequency $\omega = 2\pi f$. Beyond the frequency of transition f_t, the device is not able to efficiently drive a second transistor of the same kind. Generally, no circuitry of such transistors can operate beyond f_t.

To obtain high transfer-frequencies small gate voltage modulations should create huge drain current modulations, i.e. high transconductance conditions. As already presented in Eqs. (1) and (4), the DC response for a transistor within the gradual channel approximation can be precisely predicted. When applying a constant drain voltage V_{ds} and modulating the gate voltage V_{gs} around a DC offset V_{g0},

$$V_{gs}(t) = V_{g0} + v_g(\omega)\cdot\cos(\omega\cdot t)\,, \tag{9}$$

the AC drain current component for the linear regime becomes:

$$i_d(\omega) = -\frac{W}{L}\cdot\mu\cdot C_i\cdot V_{ds}\cdot v_g(\omega)\,, \tag{10}$$

and for the saturation regime and small modulations $v_g(\omega)$ holds:

$$i_d(\omega) = -\frac{W}{L} \cdot \mu \cdot C_i \cdot (V_{g0} - V_t) \cdot v_g(\omega). \quad (11)$$

The contribution to the drain current from the gate drain capacitance was found to be small for frequencies within the unity-gain bandwidth and is therefore ignored. The gate current becomes:

$$i_g(\omega) = i \cdot \omega \cdot (C_{gs} + C_{gd}) \cdot v_g(\omega), \quad (12)$$

with i the square root of -1. The above mentioned geometry with the gate electrode covering the whole area beneath the finger structure will affect the maximum switching frequency f_t due to parasitic capacities. Figure 22.19 sketches a capacity free transistor with parasitic capacitances. C_{gs} and C_{gd} are the capacities from the gate to the drain and source electrode, respectively. They include the channel capacity C_{ch} as well as the direct electrode-gate capacitances (right part in Figure 22.19). In the linear regime we obtain:

$$C_{gs} = C_{gs0} + \frac{C_{ch}}{2}, \quad (13)$$

$$C_{gd} = C_{gd0} + \frac{C_{ch}}{2}. \quad (14)$$

The drain–source capacitance C_{ds} can be neglected since only a constant drain voltage V_{ds} is applied. The total gate capacitance is given by $C_g = A_{tot}\, C_i$ with A_{tot} the total gate area. The unity-gain bandwidth can now be calculated from Eqs. (8), (10) and (12):

$$f_t = \left(\frac{a}{2 \cdot \pi}\right) \cdot \frac{\mu \cdot V_{ds}}{L^2}, \quad (15)$$

with the correction factor a defined by the contributing capacitances and the FET-layout:

$$a = L \cdot W \cdot \frac{C_i}{(C_{gs} + C_{gd})}. \quad (16)$$

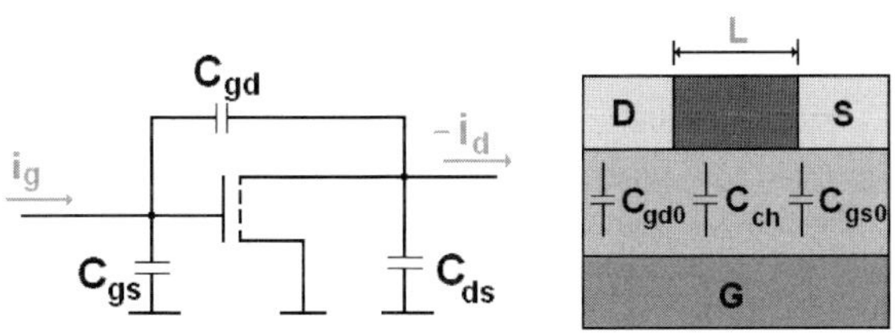

Figure 22.19 Left: sketch of a capacity free transistor with parasitic capacitances; right: side view of the transistor structure indicating the capacities between the channel and the electrodes to the gate electrode.

As one can immediately see from Eq. (15) the transfer frequency f_t depends linearly on the drain–source voltage V_{ds} and the charge carriere mobility μ. Regarding potential applications, the drain–source voltage should be reasonably low. The mobility is mainly a material-dependent parameter and high frequencies are tried to be realised with given semiconductors, e.g. DH7T or P3HT. The two remaining parameters to influence the switching speed of the device are therefore the correction factor a and the channel length L. Both parameters are design-based and can be changed by optimising the geometry of the device.

The total gate area in common gate architecture is given by the channel area and the related drain and source electrode area including additional areas where the gate is covering contact areas for outside connection. For a given structure layout with interdigitated electrode fingers (see Figure 22.20) the contributions from the contact pads and the fingers as well as from the channel can be easily identified and subsequently optimised to achieve higher transfer frequencies. The areas A_S and A_D in Figure 22.20 are the contact pads for the drain and source electrodes (or represent lithographically needed overlap areas in a multi-transistor circuitry layout). Together, as $A_0 = A_S + A_D$, they represent a contribution to the overall gate area, in addition to the interdigitated finger structure and the channel area. In Figure 22.20 B determines the width of the fingers, while L stands for the channel length between the fingers. The total gate capacitance $C_g = C_{gs} + C_{gd} = C_i A_{tot}$, which influences the transfer frequency as outlined above, is given by the equation:

$$C_g = C_i \cdot W \cdot (L + B) + C_i \cdot A_0 \,. \tag{17}$$

In combination with Eqs. (15) and (16), the transfer frequency can be expressed in terms of the structural layout parameters B, A_0 and W as:

$$f_t = \left(\frac{1}{2 \cdot \pi}\right) \cdot \mu \cdot \frac{V_{ds}}{L^2} \cdot \frac{1}{1 + \frac{B}{L} + \frac{A_0}{W \cdot L}} \,. \tag{18}$$

According to Eq. (18) it is obvious how the transfer frequency can be increased by the layout. Mainly the channel length L allows higher f_t to be achieved. The minimisation of this parameter is therefore the major goal, as

Figure 22.20 Schematic top view on the electrode geometry with interdigitated finger structure.

long as the mobility is not affected too much by the contacts. Further more, correction factor a should be maximised, i.e. become almost unity. This means the terms B/L and $A_0/(W \times L)$ should be as small as possible. Since the channel length has already tried to be minimised, for reasons of lithography, the finger width B will already have to be roughly of the same size or larger. This means the term B/L will be at least one and the correction factor a is at most 0.5. The contact pads A_S and A_D have typically a minimum size of $50 \times 50\ \mu m^2$ each. In real circuitry application there are no contact pads but similar areas are required for proper overlap of the electrode layers. To reduce the second term $A_0/(W \times L)$ (with a given channel length L), the only option is to maximise the channel width W. Figure 22.21 shows an optimised transistor structure with a channel length of 120 nm and a width of several cm. High resolution processing over a rather large area is necessary to obtain such defect-free devices.

22.7.2 Experimental Results

To achieve an optimal frequency response f_t for a given device the conditions for optimised transconductance values have to be determined first. These are found at the position of the steepest slope of transfer curve ($= V_{g0}$) for a given drain–source voltage V_{ds}. A small gate voltage modulation $v_g(\omega)$ around V_{gs0} should correspond now to the largest possible drain current modulation $i_d(\omega)$.

A systematic set of measurements were performed on samples with a series of channel lengths ranging from $L = 1\ \mu m$ to $50\ \mu m$. The devices were produced by optical lithography and the transistors were contacted by pads of normal size, i.e. side length of 200 μm or larger. The vacuum-deposited organic material for these devices was DH7T (Section 22.5 or in [40, 22]).

For the frequency dependent measurements, the applied drain DC voltage V_{ds} was 5 V. A typical measurement of such a device with a channel length of $L = 2\ \mu m$ is shown in double logarithmic scale in Figure 22.22. While the drain current (filled circles) remains fairly constant the gate current increases line-

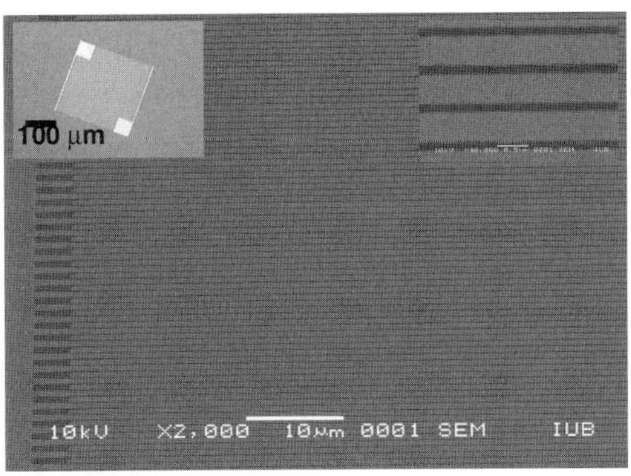

Figure 22.21 SEM image of optimised structure. Inset (left): overview image of the structure taken by an optical microscope. Inset (right): detail of the channel area to estimate channel length and finger width.

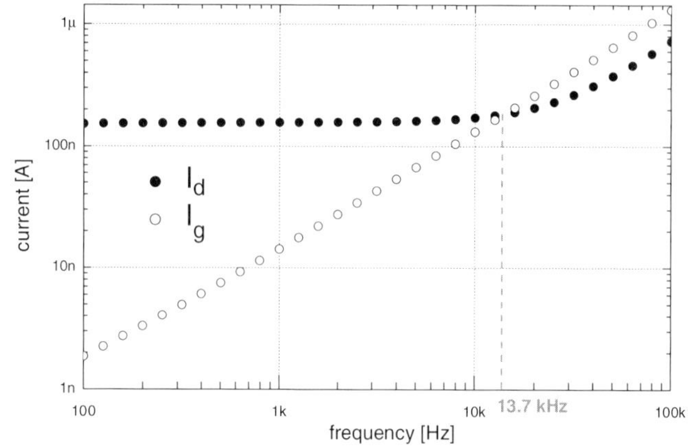

Figure 22.22 Input and output peak-to-peak currents vs. modulation frequency for an $L = 2$ µm DH7T transistor and non-optimised values.

arly with the frequency due to the capacitive coupling of the gate. In the presented measurement, the gate voltage was $V_{g0} = -1.5$ V and the peak-to-peak modulation was 60 mV, resulting in a 160 nA peak-to-peak drain current modulation. The drain current level remains constant within the unity-gain bandwidth while increased current levels are seen later, i.e. beyond 20 kHz, caused by the finite drain–gate capacitance. Both current amplitudes are equal for the given transistor and voltage settings at the transfer frequency of 13.7 kHz.

The highest measured frequency in this series structured by optical lithography was 46 kHz for a $L = 2$ µm structure. In agreement with Eq. (15) structures with larger channel length showed generally lower transfer frequencies. Since the different devices have slightly different corresponding correction factors a_j a direct comparison of the obtained frequencies versus channel length is misleading. Instead normalised frequencies, i.e. frequencies corrected for their differences in a_j, have to be compared. According to Eq. (18) the measured frequency values f are normalised to a typical correction factor, i.e. $a = 0.1$, which results in the normalised frequency $f_{j,\,\mathrm{norm}} = (0.1/a_j) \cdot f_j$ for a specific devices j.

Figure 22.23 shows in a double-logarithmic plot the normalised frequency versus channel length. The black circles are measured values. It is obvious that for larger channel length theory agrees very well with the measured data – a linear dependence of the transfer frequency on the channel length is found. These results show that DC measurements can be used in the presented experiment to predict the transfer frequency directly measured by AC measurements. From the data in Figure 22.23 one can also see, that for short channel lengths (one and two micrometer structures) the measured data start to differ from predicted values showing experimentally lower frequencies.

Figure 22.24 shows a measurement of a transfer frequency optimised structure. The device has a sub-micron channel length of 480 nm, a channel width

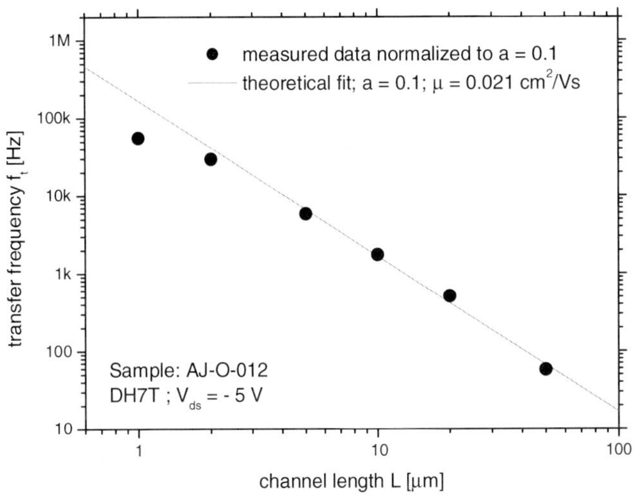

Figure 22.23 Comparison of normalised measured frequencies (black circles) with theory assuming a charge carrier mobility of 0.021 cm^2/Vs (red line).

of 46.2 mm and small contact pads. The semiconductor is rr-P3HT, which was deposited in air via spin-coating from solution. Since P3HT does not show good environmental stability in air, the electrical measurements were performed within 1 hour after preparation. The applied drain and gain voltage was −10 V and −4.5 V, respectively, with a gate AC modulation of 0.5 V peak-to-peak. These voltage settings resulted in a high peak-to-peak current modulation of 430 μA. Correspondingly the frequency of transition extrapolates into the Megahertz range, i.e. 2.0 MHz, which is to our knowledge the fastest reported polymer-based transistor in literature [24]. This frequency range of the experimental data in Figure 22.24 is limited to below 100 kHz due to the measurement setup. A 20 MHz oscilloscope was used to check the validity of the given extrapolation.

As expected, larger channel devices showed e.g. for $L = 740$ nm a lower transfer frequency of 1.2 MHz. However, reduction of the channel length below 0.48 μm did not show any further increase of the transfer frequency.

This finding is attributed to a finite contact resistance in the devices, which imposes an upper limit to the transfer frequency independent of the actual channel length [24]. Similar contact limited nano-devices were also reported for pentacene-based transistors [51]. In long channel devices and air ambient P3HT exhibit a mobility of 0.01 cm^2/Vs or higher. For the $L = 480$ nm structure this would calculate to a transfer frequency of 2.9 MHz. The mismatch to the actually measured 2.0 MHz is attributed to the onset of these contact effects. Experimentally we observe a scaling limit, where further channel length decrease does not show better performance with the given devices and semiconductors.

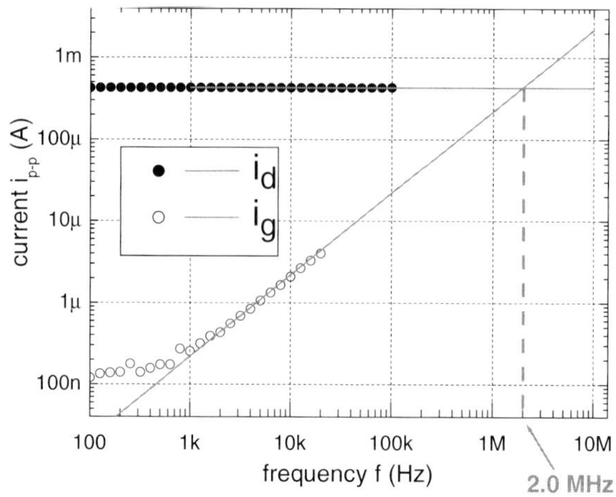

Figure 22.24 Input and out-put peak-to-peak currents vs. modulation frequency for an $L = 480$ nm P3HT transistor.

22.8 Summary

Different potentially limiting parameters for down-scaling of thiophene-based field-effect transistors, motivated by an empirical scaling law for silicon transistors, were systematically analysed. It was found, that the SiO_2 gate insulator thickness should stay smaller than a third of the channel length to avoid short channel effects, i.e. to avoid lack of saturation. Furthermore, it was shown that the organic semiconductor layer might be as thin as two monolayers without compromising mobility values. Such a small organic transport layer thickness imposes no limit for down-scaling the channel length far below 100 nm. However, in thin films maximal mobility values are only observed after deposition of an integer number of monolayers in a layer-by-layer growth mode. Especially for channel lengths in the sub-micron range the standard thickness calibration was found to be rendered invalid due to material diffusion from the electrodes into the channel. The corresponding critical channel length for DH4T and DH5T layers grown at 100 °C is as large as 500 nm. In addition, the layer growth mode is disturbed by the material diffusion from the electrodes.

A severe limitation for the down-scaling of the channel length was identified by the finite contact resistance to the organic semiconductor. Several approaches to minimise this contact resistance were discussed as thinning the titanium adhesion below 5 nm to allow for direct carrier injection into the first monolayer or to tune the transport level with respect to the contact metal by changing the chromophore length of oligiothiophenes. Furthermore, a method was presented to determine the contact resistance from a single transfer curve of a single nanoscale transistor. An extraordinary low contact resistance value of 1 kΩ cm could be demonstrated for DH6T and DH7T and gold electrodes.

By optimising the layout of a sub-micron transistor based on P3HT above Megahertz operation could be achieved. To the best of our knowledge no faster polymer-based transistor has been reported in literature so far. We found once again contact resistance to be the limiting factor for further improvement of the transfer frequency.

Acknowledgements

The authors would like to thank Prof. Bäuerle from the University of Ulm for synthesising and purifying the DH5T, DH6T, and DH7T compounds, Dr. Pflaum from the University of Stuttgart for purifying of the DH6T compound, Prof. Geurts from the University of Würzburg and Prof. Terfort from the University of Marburg for their close collaborations and the DFG for the financial support of project Wa1039/2–1,2 within SP1121 as well as the Nanomolecular Science Graduate Program at Jacobs University.

References

1. G. H. Gelinck, H. Edzer, A. Huitema, E. van Veenendaal, E. Cantatore, L. Schrijnemakers, J. B. van der Putten, T. C. Geuns, M. Beenhakkers, and J. B. Giesbers, Nature Mater. **3**, 106 (2004).

2. C. D. Sheraw, L. Zhou, J. R. Huang, D. J. Gundlach, and T. J. Jackson, Appl. Phys. Lett. **80**, 1088 (2002).

3. R. Rotzoll, S. Mohapatra, V. Olariu, R. Wenz, M. Grigas, K. Dimmler, O. Shchekin, and A. Dodabalapur, Appl. Phys. Lett. **88**, 123502 (2006).

4. S. Günes, H. Neugebauer, and N. S. Sariciftci, Chem. Rev. **107**, 1324 (2007).

5. L. Torsi, A. J. Lovinger, B. Crone, T. Someya, A. Dodabalapur, H. E. Katz, and A. Gelperin, J. Appl. Phys. **91**, 10140 (2002).

6. M. Matters, D. M. de Leeuw, M. J. C. M. Vissenberg, C. M. Hart, P. T. Herwig, T. Geuns, C. M. J. Mutsaers, and C. J. Drury, Opt. Mater. **12**, 189 (1999).

7. H. Sirringhaus, N. Tessler, and R. H. Friend, Science **280**, 1741 (1998).

8. H. Sirringhaus, T. Kawase, R. H. Friend, T. Shimoda, M. Inbasekaran, W. Wu, and E. P. Woo, Science **290**, 2123 (2000).

9. J. A. Rogers, Z. Bao, M. Meier, A. Dodabalapur, O. J. Schueller, and G. M. Whitesides, Synth. Met. **5**, 1154 (2000).

10. M. Leufgen, A. Lebib, T. Muck, U. Bass, V. Wagner, T. Borzenko, G. Schmidt, J. Geurts, and L. W. Molenkamp, Appl. Phys. Lett. **84**(9), 1582 (2004).

11. A. Benor, V. Wagner, and D. Knipp, J. Vac. Sci. Technol. B **25**, 1321 (2007).
A. Benor, A. Hoppe, V. Wagner, and D. Knipp, Thin Solid Films **515**, 7679 (2007).

12. H. Koezuka, A. Tsumura, and T. Ando, Synth. Met. **18**, 699 (1987).

13. D. Knipp, A. Benor, V. Wagner, and T. Muck, J. Appl. Phys. **101**, 044504 (2007).
D. Knipp, T. Muck, A. Benor, and V. Wagner, J. Non-Cryst. Solids **352**, 1774 (2006).

14. K. Walzer, B. Maennig, M. Pfeiffer, and K. Leo, Chem. Rev. **107**, 1233 (2007).
15. A. R. Murphy and J. M. J. Fréchet, Chem. Rev. **107**, 1066 (2007).
16. C. D. Dimitrakopoulos and P. R. L. Malenfant, Adv. Mater. **14**, 99 (2002).
17. D. Fichou, Handbook of Oligo- and Polythiophenes (Wiley-Interscience, New York, 1999).
18. S. Scheinert and G. Paasch, phys. stat. sol. (a) **201**, 1263 (2004).
19. S. M. Sze, Physics of semiconductor devices (Wiley-Interscience, New York, 1981).
20. M. Leufgen, U. Bass, T. Muck, T. Borzenko, G. Schmidt, J. Geurts, V. Wagner, and L. W. Molenkamp, Synth. Met. **146**, 341 (2004).
21. A. Hoppe, J. Seekamp, T. Balster, G. Götz, P. Bäuerle, and V. Wagner, Appl. Phys. Lett. **91**, 132115 (2007).
22. T. Muck, J. Fritz, and V. Wagner, Appl. Phys. Lett. **86**, 232101 (2005).
23. T. Muck, V. Wagner, U. Bass, M. Leufgen, J. Geurts, and L. W. Molenkamp, Synth. Met. **146**, 317 (2004).
24. V. Wagner, P. Woebkenberg, A. Hoppe, and J. Seekamp, Appl. Phys. Lett. **89**, 243515 (2006).
25. P. Bäuerle, Adv. Mater. **4**, 102 (1992).
26. P. Bäuerle, T. Fischer, B. Bindlingmeier, A. Stabel, and J. P. Rabe, Angew. Chem. **107**, 335 (1995).
27. G. Horowitz, D. Fichou, X. Z. Peng, Z. G. Xu, and F. Garnier, Solid State Commun. **72**, 381 (1989).
28. M. Halik, H. Klauk, U. Zschieschang, G. Schmid, S. Ponomarenko, S. Kirchmeyer, and W. Weber, Adv. Mater. **15**, 917 (2003).
29. G. Horowitz and M. E. Hajlaoui, Adv. Mater. **12**, 1046 (2000).
30. M. Halik, H. Klauk, U. Zschieschang, G. Schmid, S. A. Ponomarenko, and S. Kirchmeyer, Mater. Res. Soc. Symp. Proc. **771**, L3.2.1 (2003).
31. G. Horowitz, Adv. Mater. **10**, 371 (1998).
32. G. Horowitz, J. Mater. Res. **19**, 1946 (2004).
33. B. Servet, G. Horowitz, S. Ries, O. Lagorsse, P. Alnot, A. Yassar, F. Deloffre, P. Srivastava, R. Hajlaoui, P. Lang, and F. Garnier, Chem. Mater. **6**, 1809 (1994).
34. H. E. Katz, J. G. Laquindanum, and A. J. Lovinger, Chem. Mater. **10**, 633 (1998).
35. H. E. Katz, W. Li, A. J. Lovinger, and J. G. Laquindanum, Synth. Met. **102**, 897 (1999).
36. H. Akimichi, K. Waragai, S. Hotta, H. Kano, and H. Sakati, Appl. Phys. Lett. **58**, 1500 (1991).
37. H. E. Katz, A. Dodabalapur, L. Torsi, and D. Elder, Chem. Mater. **7**, 2238 (1995).
38. M. Rittner, P. Bäuerle, G. Goetz, H. Schweizer, F. J. Baltá-Calleja, and M. H. Pilkuhn, Synth. Met. **156**, 21 (2006).
39. B. Wegewijs, M. P. De Haas, D. M. De Leeuw, R. Wilson, and H. Sirringhaus, Synth. Met. **101**, 534 (1999).
40. G. Götz, T. Muck, V. Wagner, L. Brassat, S. Kirchmeyer, and P. Bäuerle, in preparation.
41. H. Sirringhaus, P. J. Brown, R. H. Friend, M. M. Nielsen, K. Bechgaard, B. M. W. Langeveldvoss, A. J. H. Spiering, R. A. J. Janssen, E. W. Meijer, P. Herwig, and D. M. de Leeuw, Nature **401**, 685 (1999).
42. G. Wang, J. Swenson, D. Moses, and A. J. Heeger, J. Appl. Phys. **93**, 6137 (2003).
43. H. G. O. Sandberg, G. L. Frey, M. N. Shkunov, H. Sirringhaus, R. H. Friend, M. M. Nielsen, and C. Kumpf, Langmuir **18**, 10176 (2002).
44. A. Zen, J. Pflaum, S. Hirschmann, W. Zhuang, F. Jaiser, U. Asawapirom, J. P. Rabe, U. Scherf, and D. Neher, Adv. Funct. Mater. **14**(8), 757 (2004).
45. R. J. Kline, M. D. McGehee, E. N. Kadnikova, J. Liu, and J. M. J. Fréchet, Adv. Mater. **15**, 1519 (2003).

46. Values from Goodfellow, URL: www.goodfellow.com.

47. A. Chandekar and J. E. Whitten, Synth. Met. **150**, 259 (2005).

48. S. Nagamatsu, K. Kaneto, R. Azumi, M. Matsumoto, Y. Yoshida, and K. Yase, J. Phys. Chem. B **109**(19), 9374 (2005).

49. N. Yoneya, M. Noda, H. Nobukazu, K. Nomoto, M. Wada, and J. Kasahara, Appl. Phys. Lett. **85**, 4664 (2004).

50. K. R. Amundson, H. E. Katz, and A. J. Lovinger, Thin Solid Films **426**, 140 (2003).

51. G. S. Tulevski, C. Nuckolls, A. Afzali, T. O. Graham, and C. R. Kagana, Appl. Phys. Lett. **89**, 183101 (2006).

23
Aluminium Oxide Film as Gate Dielectric for Organic FETs: Anodisation and Characterisation

X.-D. Dang, W. Plieth, S. Richter, M. Plötner, and W.-J. Fischer

23.1
Introduction

Anodic oxide films on aluminium have a large variety of industrial applications [1]. The oxide film has a sandwich structure consisting of an anodic barrier aluminium oxide film on the aluminium surface and a porous film on the electrolyte side [2]. Currently, the anodic barrier aluminium oxide film can be used as gate dielectric for organic transistors, especially for large-size organic field-effect transistors on flexible substrates using printing technologies instead of traditional lithography [3–4]. For the development of organic transistors, the gate dielectric is of same importance as the organic semiconductor material [5].

The anodisation of aluminium is a well-established process [6–10]. For the formation of barrier aluminium oxide films, commonly used electrolytes are citric acid, tartaric acid and ammonium adipate [11]. Barrier films with high capacitance values and breakdown field strength were obtained in tartaric acid of pH 7 [8, 12].

Concerning the two-layer model, the thickness and properties of each layer depend on the nature of the electrolyte and the anodisation conditions. For the application, a permanent control of thickness and electrical properties is necessary. In the present chapter, electrochemical impedance spectroscopy (EIS) was used to study the film properties. The EIS measurements can provide accurate information on the dielectric properties and the thickness of the barrier layer [13–14]. The porous layer cannot be studied by impedance measurements because of the high conductivity of the electrolyte in the pores [15]. The total thickness of the aluminium oxide films was determined by scanning electron microscopy. The thickness of the single layers was then calculated. The information on the film properties was confirmed by electrical characterisation performed on metal/insulator/metal (MIM) structures.

Organic Electronics. Structural and Electronic Properties of OFETs. Edited by Christof Wöll

ISBN: 978-3-527-40810-8

23.2 Experimental

23.2.1 Preparation

Pure aluminium (99.99%) films on glass substrates were prepared by magnetron sputtering or electron beam evaporation. Anodisation of aluminium was carried out in a stirred electrolyte of 0.01 M tartaric acid. The pH value of the solution was adjusted to pH = 7 by ammonium hydroxide. All chemicals (Aldrich) were used without additional purification.

The aluminium oxide film was prepared at constant current density (potentiostat in constant current mode) with current densities between 0.3 and 8.5 mA/cm^2. The electrode potential increased continuously until the desired pre-defined formation voltage (between 5 and 100 V) was reached. Then, the potentiostat was switched to constant voltage mode and the current was recorded as a function of time. A more detailed description of the procedure was given in Ref. [6]. In order to minimise contamination, the processes were carried out under clean-room conditions.

23.2.2 Characterisation

A three-electrode cell was used for the electrochemical impedance measurements, consisting of the working electrode, a platinum counter electrode, and a saturated calomel electrode (SCE). The electrolyte was 0.1 M sodium chloride. A Zahner-Electric IM6d impedance spectrometer was used for the impedance measurements. The impedance spectra were recorded at open circuit potential (OCP) in a frequency range from 0.1 Hz to 50 kHz with an AC amplitude of 10 mV. The thickness of the barrier film was calculated from the capacitance, a dielectric constant of 6.5 estimated from electrical measurement (see below) was used.

The electrical properties of the films were determined on aluminium/insulator/gold (MIM) structures. The capacitance was measured with a LCR Meter (Agilent 4284 A) at a frequency of 1 kHz. The specific resistance and the breakdown field strength were measured using a Source-Measure-Unit (SMU, Keithley 6430). The film thickness was determined by scanning electron microscopy (SEM).

23.3 Results and Discussion

23.3.1 Influence of Formation Current Density

SEM was used for morphological studies of anodic aluminium oxide films, formed at various formation current densities up to a formation voltage of 60 V. The total thickness of the films was determined by cross-section SEM micrographs as shown in Figure 23.1, but it should be emphasised that one can not clearly identify barrier and porous layers of the oxide film by using this technique.

The total thickness of the films increases linearly with the logarithm of the current density. This is in agreement with Stein et al. [2], who explained their observation assuming increased ion transport rates with increasing current densities across the initial natural aluminium oxide layer. However, it is in contradiction to the results in [15, 16], where the thickness of the film formed in a borate solution decreased with increasing current density, but increased with temperature [15], and decreased with increasing current density at formation voltages higher than 40 V [16]. The total film thickness can be controlled precisely by formation voltage. The ratio of the total thickness (d_T) to the formation voltage (V_A) is the anodisation factor (c_{af}) [3].

$$c_{af} = \frac{d_T}{V_A}\,. \tag{1}$$

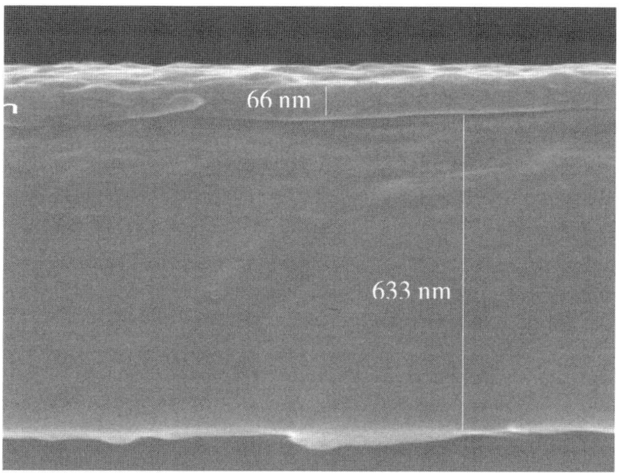

Figure 23.1 High-resolution cross-sectional SEM micrograph of an anodic aluminium oxide film of 66 nm formed at 0.3 mA/cm^2 and a formation voltage of 60 V (Al thickness remaining: 633 nm).

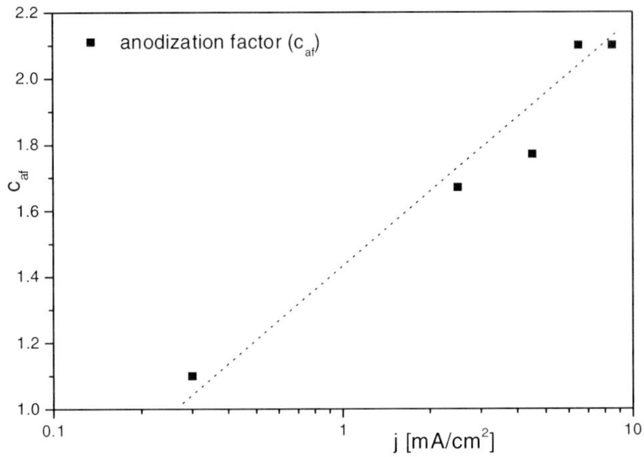

Figure 23.2 Dependence of the anodisation factor on the formation current density.

The anodisation factor can be considered as an indicator for the quality of the aluminium oxide films. In particular, the barrier aluminium oxide films formed with low anodisation factors exhibit high breakdown field strengths [3]:

$$E_{BD} \approx \frac{1}{c_{af}}. \tag{2}$$

Figure 23.2 shows the dependence of the anodisation factor on the formation current density. The anodisation factor increases from 1.1 to 2.1 nm/V when the current density is increased from 0.3 mA/cm^2 to 8.5 mA/cm^2. This was explained by the transformation of the amorphous oxide into crystalline oxide at high current densities [11].

Impedance measurements were used for the characterisation of the films. Figure 23.3 shows the Bode plots of the initial natural aluminium oxide and the aluminium oxide film formed at 2.5 mA/cm^2 (formation voltage 60 V). With the aluminium oxide film, a shift of the impedance and phase toward the high frequency range was observed, which was explained by increase in the thickness of the aluminium oxide film. The impedance spectra of the aluminium oxide films formed at other current densities showed a similar behaviour, excepting the film formed at 0.3 mA/cm^2. In this case, a new time constant (*R*, *CPE*) appeared in the low frequency range of the impedance spectrum. A similar second time constant was also observed in films formed at high formation voltages, which will be discussed later.

To interpret the impedance data, a model of the anodic aluminium oxide film must be established. The anodic aluminium oxide film is a sandwich film consisting of two layers, a barrier layer and a porous layer (Figure 23.4A [2]).

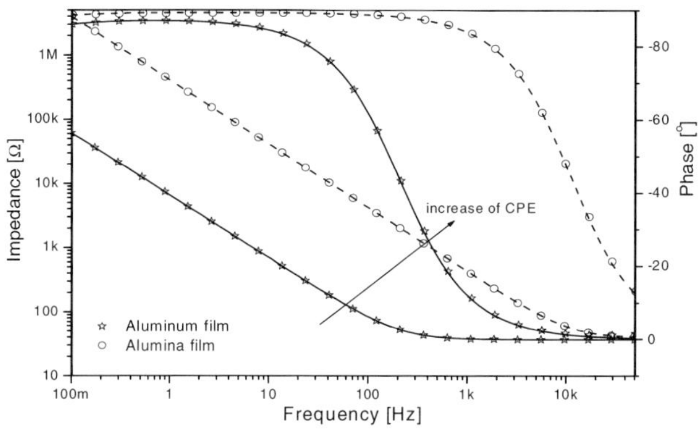

Figure 23.3 Bode plots of natural and anodic aluminium oxide films. The aluminium oxide film was formed at 2.5 mA/cm^2 and a formation voltage of 60 V, experimental (points) and fitting (lines).

Figure 23.4B shows the corresponding equivalent circuit. The equivalent circuit consists of a parallel combination of the resistance (R_1) and a constant phase element (CPE_1) in series with the electrolyte resistance (R_E). In general, the constant phase element (CPE_1) is used to replace the capacitance and is defined by:

$$CPE = A_{CPE}\left(j\omega\right)^{-n}. \tag{3}$$

This replacement was necessary to adapt the equivalent circuit to the non-ideal behaviour of the aluminium oxide film. The exponent n of the CPE element can be regarded as a measure of the inhomogeneity of the film structure [17]. For an ideal capacitor the exponent n is one. For the calculation of the CPE values, the fitting program in Ref. [18] was used.

The capacitance values of the aluminium oxide films formed at various formation current densities are shown in Figure 23.5. The capacitance (CPE_1) decreases from 0.19 μF/cm^2, for the film formed at 0.3 mA/cm^2, to 0.13 μF/cm^2, for the film formed at 2.5 mA/cm^2. At higher current densities, the capacitance of the films increases. The capacitance of the aluminium oxide films is much

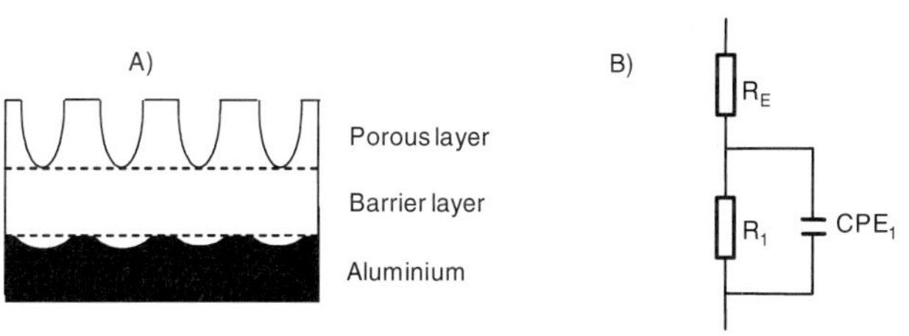

Figure 23.4 Two-layer model of the anodic aluminium oxide film (A) and associated equivalent circuit (B)

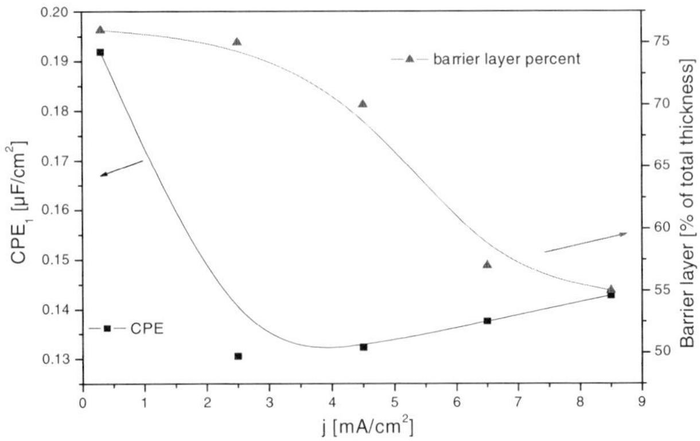

Figure 23.5 Effect of the current density on the capacitance and barrier layer thickness.

smaller than the capacitance of the natural aluminium oxide layer (7.1 μF/cm²). The capacitance is only reflecting the properties of the barrier layer. The variation in the capacitance shows the change in the barrier layer thickness d_b, determined by the equation:

$$d_b = \frac{\varepsilon_0 \cdot \varepsilon_r}{C_i} \tag{4}$$

with ε_0 the permittivity of free space (8.854×10^{-14} F/cm) and ε_r the dielectric constant.

The thickness of the initial aluminium oxide film is about 1.4 nm.

The variation of the thickness of the barrier layer with the current density is also shown in Figure 23.5. Increasing the current density, a thicker aluminium oxide film is formed. However, the thickness ratio of the barrier layer to the porous layer is almost unchanged for the films formed at low current densities. It reduces significantly for the films formed at current densities higher than 4.5 mA/cm².

For all aluminium oxide films the exponent n in the *CPE* was nearly constant, approximately 0.99 ± 0.04, determined for current densities between 2.5 and 8.5 mA/cm². This value is almost one and significantly higher than values found in the literature [15]. The n value is related to the layer inhomogeneity. The high value of n confirms the formation of homogeneous aluminium oxide films.

23.3.2 Influence of the Formation Voltage

A significant influence of the formation voltage on the film properties was found. The aluminium oxide films were formed at a constant current density of 0.5 mA/cm². The formation voltage was raised in steps to 5, 10, 15, 30, 60 and 100 V. When the formation voltage was reached, the potentiostat was switched

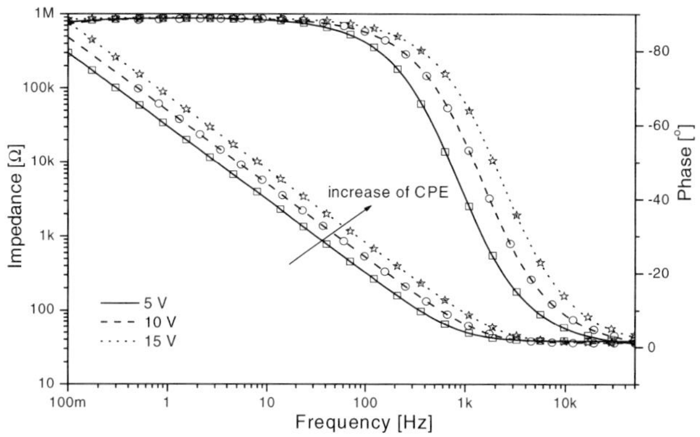

Figure 23.6 Bode plots of aluminium oxide films formed at 0.5 mA/cm², different formation voltages 5, 10 and 15 V, experimental points and fitting lines.

to constant voltage mode. The decay of the current density with time was recorded.

For low formation voltages, the current density approached a value of 0.01 mA/cm² after an anodisation time of 2400 s. However, above 30 V the current density reached values >0.01 mA/cm² and up to 0.02 mA/cm² at a voltage of 100 V.

The Bode plots confirm a change of the film properties around 30 V. The Bode plots of the films formed at 5, 10 and 15 V are shown in Figure 23.6. The impedance behaviour is similar to the initial natural aluminium oxide with a single time constant. The equivalent circuit in Figure 23.4B has been used to analyse the impedance data.

Figure 23.7 shows the Bode plots for 30, 60 and 100 V. A second time constant in the low frequency range appears, represented by an additional reso-

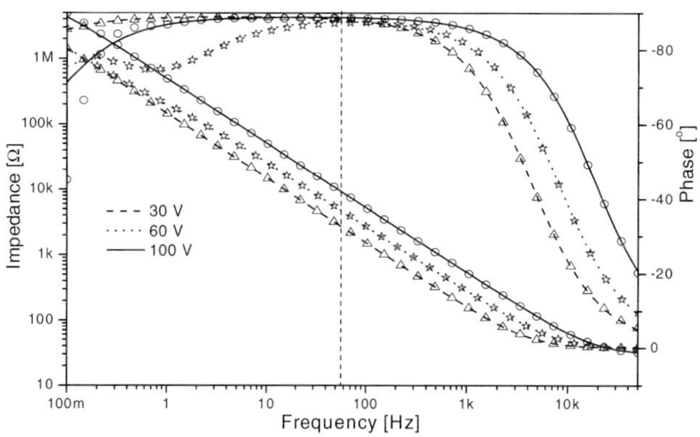

Figure 23.7 Bode plots of aluminium oxide films formed at 0.5 mA/cm², with different formation voltages 30, 60 and 100 V, experimental (points) and fitting (lines).

nance circuit of resistance (R_2) and constant phase element (CPE_2) shown in Figure 23.8. The origin of the change at 30 V has already been discussed in the literature [11, 15, 19]. In this chapter, we refer to this voltage as the *transition voltage* at which the formation of a crystalline phase within the aluminium oxide film starts. The appearance of the crystalline phase inside the amorphous matrix could create defects, micro-voids and other flaws.

The *transition voltage* depends on the formation current density. At formation current densities above 2.5 mA/cm^2, the *transition voltage* was greater than 60 V. Chang et al. [11] reported that the *transition voltage* was 150 V when the films were formed at 25 mA/cm^2 in 0.83 M ammonium adipate solution.

The transformation from an amorphous film to a crystalline film at 30 V was also observed for TiO_2 films [20]. The Raman spectrum of anatase was observed when the formation voltage was raised to 30 V. The Raman intensities increased with increasing formation voltages. A similar phenomenon can explain the change of the structural properties for the aluminium oxide films formed at voltages ≥ 30 V.

The significant reduction of the resistance R_2 of the aluminium oxide film from 166 MΩ/cm^2 at 30 V to 8.2 MΩ/cm^2 at 100 V could be explained by increasing crystallinity inside the amorphous phase. The resistance R_1 of the film decreases from 2.8 MΩ/cm^2 at 30 V to 1.6 MΩ/cm^2 at 100 V. Figure 23.9 shows the effect of the formation voltage on the capacitance and the barrier layer thickness. The capacitance decreases with increasing formation voltage, while the barrier layer thickness approaches a constant value around 40% of total thickness.

23.3.3 Influence of Anodisation Time

The dependence of the aluminium oxide film properties on the anodisation time was studied. Figure 23.10 shows the current–time transients during anodisation at the formation current density of 2.5 mA/cm^2. In the constant current

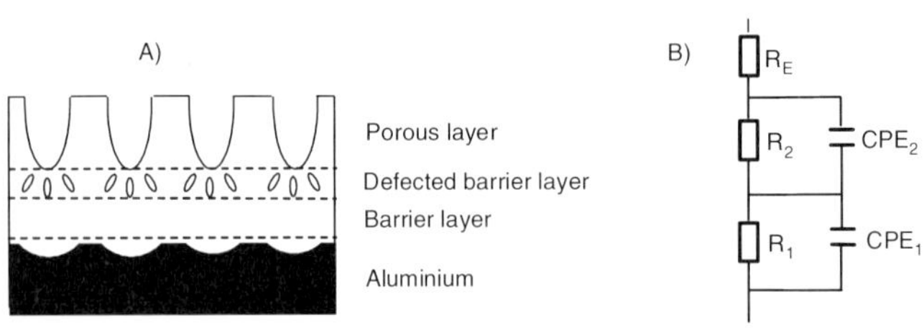

Figure 23.8 Model of anodic aluminium oxide film formed at formation voltage of 30–100 V (A) and equivalent circuit (B). R_E is electrolyte resistance, and R_1 and R_2 are resistance of barrier aluminium oxide layer and the barrier layer with defect, respectively. CPE_1 and CPE_2 are constant phase elements of the barrier layer and the defected barrier layer, respectively.

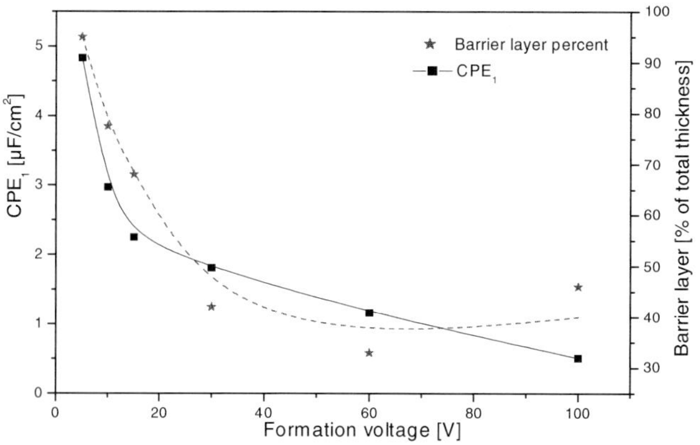

Figure 23.9 Effect of the formation voltage on capacitance and barrier layer thickness.

mode, the film thickness increases until the formation voltage reaches 60 V which takes about 100 s. During the constant voltage mode a rapid decrease of the current density was observed. This typical behaviour of the anodisation is discussed in the literature [16, 21].

The aluminium oxide films formed for different times (marked in Figure 23.10) were characterised by electrochemical impedance spectroscopy. For the aluminium oxide film anodised for 160 s, the open circuit potential (OCP) is not stable. This can be explained by instability of the film structure. The processes of the film formation were not yet completed. The OCP is more stable and positive for films anodised for more than 700 s. This can be explained by the formation of the compact barrier aluminium oxide layer.

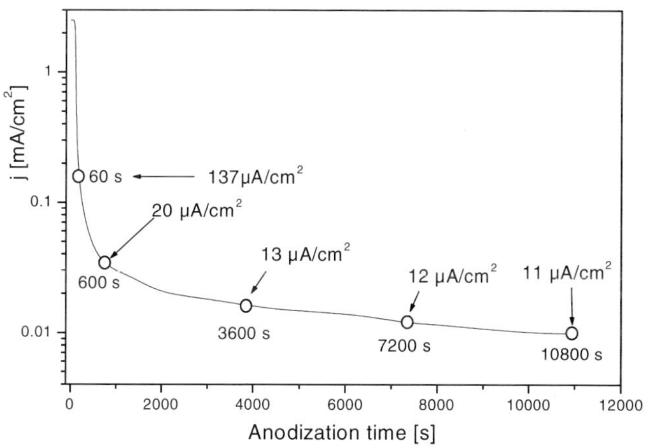

Figure 23.10 Current–time transient during anodisation at the constant current density of 2.5 mA/cm^2 and a formation voltage of 60 V.

The impedance response of the films was observed with just one time constant (R_1, CPE_1), except for the film formed for 10900 s. This film showed two time constants similar to the films formed at voltages ≥ 30 V. In this case, the aluminium oxide film might be contaminated by anions migrating from the surface into the film.

The equivalent circuit in Figure 23.4B was used to analyse the impedance data. The dependence of the capacitance on the anodisation time is shown in Figure 23.11. A large capacitance is formed after 160 s decreasing gradually to a minimum value around 3700 s, corresponding to the highest barrier layer thickness. The dependence of the barrier layer thickness on the remaining current density when the anodisation is stopped is not linear. At the remaining current density of 13 μA/cm² (after 3700 s), the barrier layer thickness has a maximum. The percentage of the barrier layer within the films is lowest when the anodisation was stopped at the current density of 11 μA/cm².

During the period of the current decay, there are two competitive processes, densification to form the barrier layer and dissolution of barrier layer to form the porous layer. Under the high electrical field strength (constant voltage mode), the densification of the aluminium oxide films is favoured for process durations shorter than 3700 s. When the constant voltage was kept for a long anodisation time (beyond 10900 s), the dissolution of the aluminium oxide film becomes more dominant. Thus, the film could be contaminated by inward migration of the electrolyte into the film or by formation of micro-voids.

23.3.4 Influence of Surface Roughness

The morphology of the sputtered aluminium film on a glass substrate (1000 nm thickness) before and after anodisation was studied by AFM microscopy. The surface roughness (R_a) of both films is approximately 35 nm. The

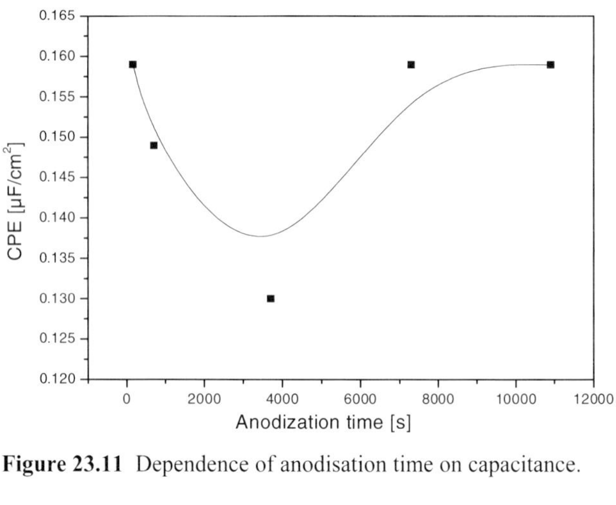

Figure 23.11 Dependence of anodisation time on capacitance.

Table 23.1 Relationship of the surface roughness and the film properties.

sample*	R_a (nm)	CPE_1 (nF/cm²)	barrier layer (% of total thickness)	appearance
S-200-Al	5.2	520	100	colourless, mirror-like
S-1000-Al	35.0	818	73	white, defected
E-200-Al	2.0	530	100	colourless, mirror-like

* S-200-Al and S-1000-Al: aluminium films sputtered on glass substrates with a thickness of 200 and 1000 nm, respectively. E-200-Al: aluminium film (200 nm thickness) evaporated on glass.

anodisation of aluminium films was performed at the current density of 0.5 mA/cm² up to a formation voltage of 10 V, which was kept constant for 2400 s. The dependence of electrical properties of the aluminium oxide films on the surface roughness is summarised in Table 23.1. Morphological properties of the aluminium oxide films are similar to those of aluminium substrates. The anodisation of 1000 nm thick sputtered aluminium films led to white-coloured and defected aluminium oxide films with the barrier layer occupying 73% of the total thickness. In contrast to that, the aluminium oxide films formed on the 200 nm thick sputtered or evaporated aluminium-on-glass-substrates looked colourless and mirror-like.

23.3.5 Barrier Aluminium Oxide Films as Gate Dielectrics for Organic Transistors

The above results showed that an aluminium oxide film with the best dielectric properties was prepared in neutral electrolyte of 0.01 M tartaric acid at low current densities and formation voltages < 30 V.

The film is formed on E-200-Al/glass at a current density of 0.5 mA/cm² for the anodisation time of 1800 s. The film thickness was controlled by the formation voltage of 10 V. The electrical measurements were carried out on different electrode areas ranging from 0.0025 to 0.04 cm² (see Figure 23.12). Below breakdown, low leakage currents of less than 13 nA/cm² at 1.67 MV/cm could be observed. The breakdown of the dielectric film occurs when the electrical field strength approaches 8 MV/cm.

The average capacitance and specific resistivity of the barrier aluminium oxide films are determined to be 430 … 470 nF/cm² and 1.3 … 2.4 · 10^{14} Ωcm, respectively. By using the anodisation factor of 1.2 nm/V for the films formed at low formation voltage, dielectric constants of 5.8 … 6.4 are calculated from the measured capacitance values. The comparatively low dielectric constant is in agreement with the formation of an amorphous anodic aluminium oxide film as discussed above rather than a crystalline structure for which a higher dielec-

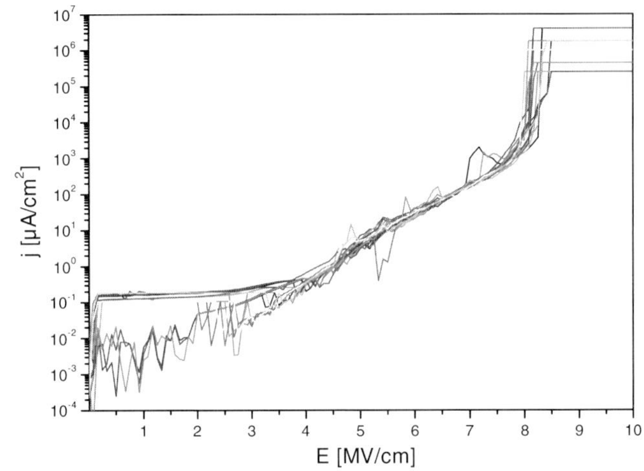

Figure 23.12 Current density (j) vs. field strength (E) of the barrier aluminium oxide film/Al(200 nm)/glass formed at a current density of 0.5 mA/cm^2, a formation voltage of 10 V and an anodisation time of 1800 s.

tric constant would be expected. The electrical properties of the film determined on MIM structures are comparable to those obtained from the EIS measurement as seen in Table 23.2.

23.4 Conclusion

The thickness and properties of the barrier aluminium oxide layer were investigated by electrochemical impedance spectroscopy. The total thickness of the films was determined by scanning electron microscopy of cross-sections. Then, the thickness of each layer within the aluminium oxide films was calculated. Formation current density, formation voltage, anodization time, and surface roughness of the substrate influenced the electrical and structural properties of the barrier aluminium oxide layer.

The breakdown field strength was approximately 8 MV/cm. The high capacitance (430 … 470 nF/cm^2) and resistivity (1.3 … 2.4 · 10^{14} Ω cm) of the barrier aluminium oxide film were determined. The leakage current density

Table 23.2 Comparison of data of capacitance and specific resistance determined by EIS and MIM.

capacitance (nF/cm^2)		specific resistivity (Ω cm)	
EIS	MIM	EIS	MIM
530	450	1.7 · 10^{14}	1.85 · 10^{14}

(7 … 12.7 nA/cm^2) through the dielectric film is very small at 1.67 MV/cm. The electrical properties of the film were comparable to that of silicon dioxide films. Apparently, the obtained barrier aluminium oxide film can be seen as an excellent gate dielectric for organic field-effect transistors on flexible substrates.

Acknowledgements

This work was supported by the Deutsche Forschungsgemeinshaft (DFG), Priority Programme 1121 "Organische Feldeffekt-Transistoren: strukturelle und dynamische Eigenschaften".

References

1. G. E. Thomson, and G. C. Wood, in: Treatise on Material Science & Technology. Vol. 23, edited by J. C. Scully (Academic Press, New York, 1983).
2. N. Stein, M. Rommelfangen, V. Hody, L. Johann, and J. M. Lecuire, Electrochim. Acta **47**, 1811 (2002).
3. L. A. Majewski, M. Grell, S. D. Ogier, and J. Veres, Organic Electronics **4**, 27 (2003)
4. L. A. Majewski, R. Schroeder and M. Grell, J. Phys. D: Appl. Phys. **37**, 21 (2004).
5. J. Veres, S. D. Ogier, S. W. Leeming, D. C. Cupertino, and S. M. Khaffaf, Adv. Funct. Mater. **13**, 199 (2003).
6. K. Shimizu, G. M. Brown, H. Habazaki, K. Kobayashi, P. Skeldon, G. E. Thompson, and G. C. Wood, Electrochim. Acta **44**, 2297 (1999).
7. J. R. Morlidge, K. Shimizu, P. Skeldon, G. E. Thompson, and G. C. Wood, Thin Solid Films **258**, 341(1995).
8. R.-L. Chiu, P.-H. Chang, and C.-H. Tung, Thin Solid Films **260**, 47 (1995).
9. Y. Xu, G. E. Thompson, G. C. Wood, and B. Bethune, Corros. Sci. **27**, 83 (1987).
10. V. Surganov, P. Morgen, J. G. Nielsen, G. Gorokh, and A. Mozalev, Electrochim. Acta **32**, 1125 (1987).
11. J.-K. Chang, C.-M. Liao, C.-H. Chen, and W.-T. Tsai, J. Electrochem. Soc. **150**, B266 (2003).
12. P. T. Nguyen, U. Rammelt, W. Plieth, S. Richter, M. Plötner, W.-J. Fischer, N. Kiriy, K. Potje-Kamloth, and H.-J. Adler, Electrochim. Acta **50**, 1757 (2005).
13. J. De Laet, J. Scheers, H. Terryn, and J. Vereecken, Electrochim. Acta **38**, 2103 (1993).
14. J. Bessone, C. Mayer, K. Jüttner, and W. J. Lorenz, Electrochim. Acta **28**, 171 (1983).
15. D. A. Brevnov, G. V. Rama Rao, G. P. López, and P. B. Atanassov, Electrochim. Acta **49**, 2487 (2004).
16. S. Ono, C. Wada, and H. Asoh, Electrochim. Acta **50**, 5103 (2005).
17. U. Rammelt, and C.-A. Schiller, ACH-Models Chem. **137**, 199 (2000).
18. X. D. Dang, C. M. Intelmann, U. Rammelt, and W. Plieth, J. Solid State Electrochem. **8**, 727 (2004) and **9**, 706 (2005).
19. N. Xu, G. E. Thompson, J. L. Dawson, and G. C. Wood, Corros. Sci. **34**,, 479 (1993).
20. L. D. Arsov, C. Kormann, and W. Plieth, J. Electrochem. Soc. **138**, 2964 (1991).
21. C. W. Liang, T. C. Luo, M. S. Feng, H. C. Cheng, and D. Su, Mater. Chem. Phys. **43**, 166 (1996).

24
Electronic States at the Dielectric/Semiconductor Interface in Organic Field-Effect Transistors

Niels Benson, Christian Melzer, Roland Schmechel, and Heinz von Seggern

24.1
Introduction

The control of charge flow by an electric quantity is a key issue of today's electronics. The concept to electrically specify the conductivity of a resistor by pure solid state effects was already proposed in 1928 by Julius Edgar Lilienfeld in Germany [1]. The basic idea was to control the charge carrier density in a solid by an electric field, applied over a third electrode. However, there is no evidence for a practical realisation by Lilienfeld. The first report about a pure electrically controllable solid state device was the well know Germanium transistor from William Shockley, John Bardeen and Walter Brattain [2]. The new term "transistor" was later explained as a combination of the words "transconductance" and "varistor". Meanwhile a broad variety of different transistor concepts exists, which, however, can be mainly subdivided in two basic operational principles:

1) Field-effect transistors (FETs) are based on the field effect, where charge carriers are accumulated or depleted by an electric field. This field is applied by a control electrode, usually called a gate. The change in charge carrier concentration in the solid results in a change in conductivity between two working electrodes, commonly called source and drain. Hence, a voltage applied at the gate electrode, controls the transistor resistance between source and drain, while ideally only a capacitive current flows through the gate.

2) Bipolar junction transistors are based on an n–p–n or p–n–p junction sequence. One of thc p–n junctions is reverse biased, which suppresses the current between the outer electrodes, commonly called emitter and collector. The middle layer, which is connected to a third electrode (base), is spatially very thin. Initiated by a small control current over the base electrode, charge carriers are injected over the forward biased p–n junction and minority carriers can flood directly over the reverse biased p–n junction. This causes an increased current between emitter and collector. In contrast to FETs, bipolar junction transistors are current controlled.

Organic Electronics. Structural and Electronic Properties of OFETs. Edited by Christof Wöll

ISBN: 978-3-527-40810-8

Since stable charge transfer (organic/inorganic) doping of organic semiconductors is not well established, mainly due to a high diffusion of the respective dopants, bipolar junction transistors cannot be built from organic materials. Therefore, organic transistors are mainly field effect devices (OFETs), typically thin film transistors (TFTs).

Cross sections of schematic top and bottom gate TFTs are illustrated in Figure 24.1. They consist of a thin semiconducting film, ideally forming ohmic contacts to the source and drain electrodes. In addition, the semiconductor is separated by an insulating layer from the gate electrode.

These transistor types are widespread in today's electronic applications. This is due to the fact that TFTs can be easily realised by common thin film techniques like physical or chemical vapour deposition, solution-based processes like spin coating, dip coating or by printing techniques. The main advantage of these transistors is that they are less restricted to a specific substrate and that they can be produced on large-area substrates. Therefore, TFTs are mainly used in active matrix flat panel displays and in simple logic circuits on flexible substrates. Most TFTs are currently made of amorphous or polycrystalline silicon, however, organic TFTs are an interesting alternative due to their simple processability.

The basic electrical characteristics of a field effect transistor can be described using a simple capacitor-resistor-equivalent circuit, as presented in Figure 24.2. Here the channel resistance R_c is determined by the accumulated amount of charge. As long as Ohmic source and drain contacts are assumed for a respective injection of electrons and holes, one obtains an extended Schockley equation [3] describing the FET current–voltage characteristic. In the following V_D is the drain–source voltage, V_G stands for the gate–source voltage and $V_{th,n}$, $V_{th,p}$ represent the threshold voltage for electrons (n) and holes (p) respectively. The extended Shockley equation is divided into three ranges, here formulated for the electron accumulation ($V_G > 0$):

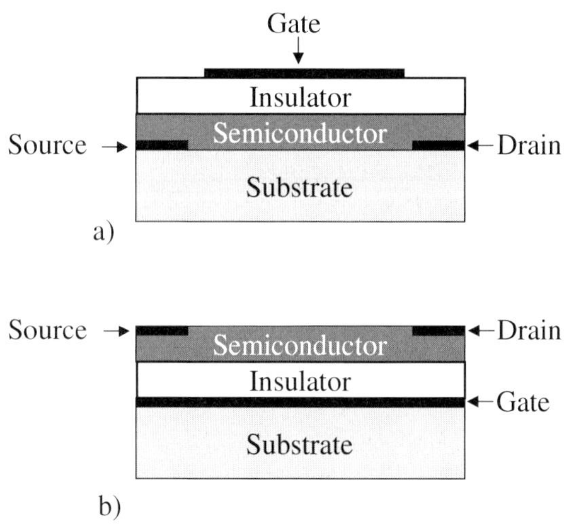

Figure 24.1 Basic configuration of top-gate (a) and bottom gate (b) TFT.

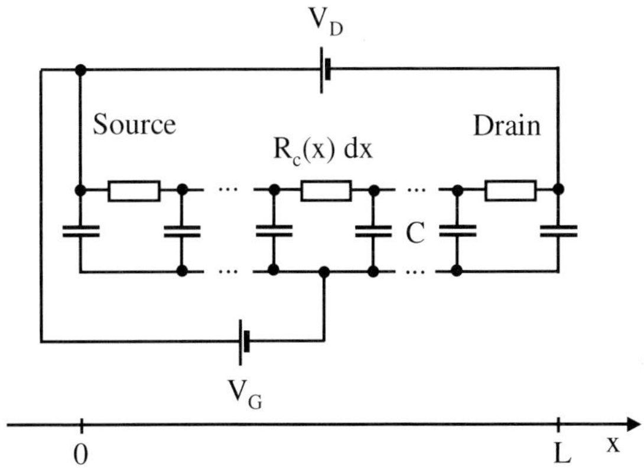

Figure 24.2 TFT channel equivalent circuit.

For $V_D \leq V_G - V_{th,n}$ (unipolar range)

$$|I_D| = \frac{WC}{L} \mu_n \left[(V_G - V_{th,n}) - \frac{1}{2} V_D \right] V_D , \tag{1}$$

for $V_D \geq V_G - V_{th,n}$ but $V_D \leq V_G - V_{th,p}$ (saturation range)

$$|I_D| = \frac{WC}{2L} \mu_n (V_G - V_{th,n})^2 , \tag{2}$$

for $V_D \geq V_G - V_{th,p}$ (ambipolar range)

$$|I_D| = \frac{WC}{2L} \{\mu_n (V_G - V_{th,n})^2 + \mu_p [V_D - (V_G - V_{th,p})]^2\} . \tag{3}$$

Here I_D represents the drain current and μ_n, μ_p the respective electron and hole mobility. C defines the area capacitance of the insulator. The channel geometry is defined by the channel width W and length L. The ambipolar range, described by Eq. (3), is only valid as long as both electrons and holes can be injected and further transported in the active layer of the transistor. However, in most cases the injection and/or the transport in the transistor channel are suppressed for one charge carrier type. In that case, the FET operates only in the unipolar and saturation range as described by Eqs. (1) and (2).

Of course these equations do not account for a field- and charge carrier density dependent mobility, which will result into slight deviations in the transistor characteristics. A similar model which is extended by the field dependent mobility was presented by Smits et al. [4]. Furthermore, the transistor subthreshold behaviour is not represented. Nevertheless, the extended Shockley equations are quite useful for a qualitative interpretation of obtained experimental results.

To understand the current–voltage behaviour of organic TFTs in more detail, it is essential to know about the region where the charge carriers are primarily conducted in the device. The following estimate will give an indication for the spatial distribution of the charge carrier transport in the organic semiconductor. For $V_G = 10$ V, a relative dielectric permittivity of 4 and an insulator thickness of 100 nm, a charge carrier area density of approx. 2×10^{12} cm^{-2} is accumulated in the active layer. The density of states can be estimated by assuming that each molecule in the organic solid provides an electronic transport state. This is reasonable since the wavefunction overlap between molecules is weak. The density of states is then comparable to the density of molecules, which can be approximated to the order of 10^{14} cm^{-2} [5]. Hence, the first monolayer of the organic semiconductor provides enough states to accommodate the accumulated charge as a result of the field effect. For a carrier to hop from the first monolayer to the second monolayer at an approximate distance of 0.8 nm, depending on size and orientation of the molecules, an additional energy of ~80 meV would be required to overcome the electric field from the gate electrode. Compared with the mean thermal energy (25 meV) of a charge carrier at room temperature, the hop to the second layer would be much less preferred. This rough estimate indicates that the main charge transport will take place within the first few monolayers of the organic semiconductor. Indeed, it is well known from thickness dependent measurements [6–10], that an increase of the film thicknesses above 3 nm does not essentially affect the transistor characteristics. On one side this demonstrates the crucial role of the first few semiconductor layers, while the volume properties of the semiconductor seem less important. On the other side it indicates that the transistor characteristics can be tuned and engineered by modifications of the dielectric interface.

The dielectric interface affects the electrical characteristics of OFETs mainly by (1) determining the morphology of the organic semiconductor, (2) the dielectric properties of the insulator layer and (3) the electronic states at the dielectric interface.

Aspect (1). The influence of the dielectric interface on the morphology of the semiconductor is probably the most investigated issue [11, 12]. Thereby the first goal is to obtain a smooth interface without morphological defects, in order to lower the interface roughness. A second goal is to gain control over the orientation of the molecules. This is important, because the charge transport in organic semiconductors is highly non-isotropic and requires a strong π–π overlap between the molecules along the transport direction. Self-assembled monolayers (SAMs) are often used on oxide dielectric interfaces to control the grain growth and the molecular orientation [13–22]. Recent studies by Halik [23] as well as Klauk et al. [24] have demonstrated that SAMs can also be utilised as a standalone gate dielectric. Such ultra thin ($d = 2.5$ nm), tailored dielectrics, allow for the realisation of low power OFET and circuitry applications.

Aspect (2). The effect of dielectric properties on the OFET performance is less investigated. However, there is clear experimental evidence for a lowering of the mobility with an increasing dielectric constant of the insulating layer

[25–27]. Most probably a higher dielectric constant of the insulating layer causes a broader distribution of the electronic states in the organic semiconductor, resulting in the reduced charge carrier mobility.

Aspect (3). The effect of electronic states at the dielectric interface has been underestimated for OFETs, but will be the focus of this chapter. The reason for this was the belief that organic semiconductors are not able to form dangling bonds at the interface, which is the main cause for interface states in inorganic semiconductors. However, there are other impurities or organic side groups that form interface states. A weak electron transfer between the organic semiconductor and fluorine- or amino-groups of SAMs at the dielectric interface has already been used to control the charge carrier density in the transistor channel [28]. Furthermore, it could be demonstrated that pentacene, well known for its hole transporting properties, also conducts electrons with a mobility comparable to that of holes [29]. The p-type behaviour of pentacene turned out to be the direct result of interfacial electron traps existing in particular insulators, such as SiO_2. In the past SiO_2 has been commonly used to study pentacene OFETs. However, if such electron traps are compensated or passivated, pentacene n-type OFETs with electron mobilities of ~0.19 cm^2/Vs are possible. Later, Chua et al. [30] identified the origin of these electron trap states as hydroxyl groups at the SiO_2 interface.

For OFET applications in electronic logic circuits unipolar n-type, p-type or ambipolar transistors can be considered. These transistor types allow for the realization of inverter stages, which represent one of the basic building blocks for modern day logic circuitry. Since inverter stages realised using ambipolar OFETs [31], or only one type of unipolar transistor, exhibit large power losses during operation, complementary inverter stages comprising unipolar n- and p-type OFETs are favoured. However, these types of inverters require the integration of complementary OFETs on a single substrate, which is a real technological challenge. With respect to the crucial role of interface states the question can be posed: how far can interface engineering be used to alter an ambipolar semiconductor to unipolar n-type or p-type? By insertion of selective interface states at the gate dielectric, the transport properties of electrons and holes in an ambipolar OFET should be controllable. Thus, without having to deposit spatially separated semiconductors unipolar n- and p-type transistors should be feasible. In this work we will summarise our current results on interface trap states and dielectric interface engineering.

24.2 Experimental

24.2.1 Device Structure

All of the devices described in the following were realised on highly p-doped $1.7 \times 1.7\ cm^2$ silicon substrates with a one sided 200 nm dry oxide SiO_2 layer.

Before being processed, the substrates were sonicated in a 5 vol% deconex solution and deionised water. The substrates were dried using a nitrogen gun.

The standard process to fabricate metal insulator semiconductor (MIS) diodes and organic field effect transistors (OFET) is illustrated in the following. Specific device modifications are described in greater detail in Section 24.3. In a first optional process step Ca traces or different types of insulating polymers are deposited onto the substrate, either by physical vapour deposition (PVD) or spin-coating. The different polymers, as well as their respective solvents and dilutions are listed in Table 24.1. The chemical structures of the used polymers are depicted in Figure 24.3c. Subsequently, a 50 nm or 200 nm thick layer of pentacene was deposited at a rate of 2 Å/s by PVD. The OFET structures were finalised by PVD of either 100 nm Au or Ca drain–source electrodes. For the MIS-diode devices Ca electrodes with a thickness of 60 nm were deposited. All electrodes were deposited at a rate of 2 Å/s and structured by the use of shadow masks. The OFET channel length and width was 0.8 cm and 100 μm, respectively. The MIS diode electrodes covered an area of 25 mm^2. All PVD steps in the production process were conducted at a chamber base pressure below 1×10^{-6} mbar. The respective profiles of the completed OFET and MIS-diode devices are shown in Figure 24.3a and b.

24.2.2
Device Measurement

All device measurements have been conducted in inert N_2 atmosphere. The OFET characterisation was performed using a HP4155A parameter analyser in the short-integration-time mode. For the capacitance–voltage measurements, on MIS diodes, a Solatron SI 1260 impedance analyser was used at a frequency of 100 Hz, an AC voltage amplitude of 500 mV and a 3 s integration time per applied DC voltage step. Static MIS diode charging experiments were conducted using a Keithley 6517A electrometer.

To determine the chemical composition of the SiO_2/Ca interface, X-ray photo electron spectroscopy (XPS) measurements were conducted at the Darmstadt Integrated System for Material Science (DAISY-MAT). For these experiments traces of Ca were deposited by PVD onto SiO_2 substrates, at a chamber base pressure of 5×10^{-10} mbar and a rate of 0.5 Å/s. The measure-

Table 24.1 Utilised polymers.

polymer	abbreviation	solution
polymethylmethacrylat	PMMA	2 wt% in tetrahydrofuran
polycarbonat	PC	2 wt% in tetrahydrofuran
polystyrene	PS	1.7 wt% in toluol
polyimid	PI	in JSR-AL-1054
poly(4-vinylphenol)	P4VP	2 wt% in tetrahydrofuran

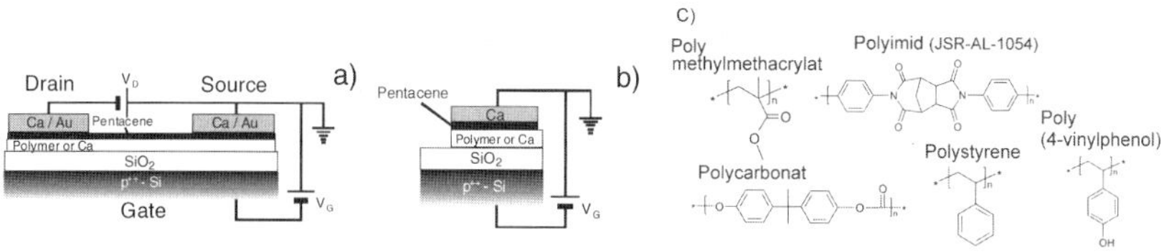

Figure 24.3 (a) Profile of the realised OFET structures, (b) realised MIS-diode profile, (c) polymer insulators used for the OFET realisation.

ments were carried out using monochrome Al K_α (1486.6 eV) or Mg K_α (1253.6 eV) radiation. For both types of excitation, a sample analyser angle of 45° was chosen.

24.3 Results and Discussion

Illustrated in Figure 24.4 is the output characteristic of a pentacene OFET with Au drain–source electrodes and a 200 nm SiO_2 dielectric [32]. The OFET exhibits unipolar p-type behaviour with a hole mobility $\mu_h = 0.165$ cm²/Vs, a threshold of $V_{th} = -4.5$ V as well as an On/Off ratio of $>10^5$. These parameters have been derived from the respective transfer characteristics. The absence of an s-shaped feature in the linear range of the characteristic indicates ohmic contacts between the Au electrodes and the pentacene active layer. This is attributed to the good matching of the ionisation potential of the organic semiconductor and the Au work function. However, employing a Ca drain–source metallisation, with an otherwise identical OFET device structure, the transistor did not exhibit any current in the electron accumulation mode. This is unexpected, since the metal work function is well matched to the electron affinity of pentacene.

Considering the results of MIS-diode impedance measurements [29, 33] this observation can be explained. In Figure 24.5a a differential MIS-diode capacitance measurement is presented. The diode consists of a Ca/pentacene/SiO_2/p^{++}

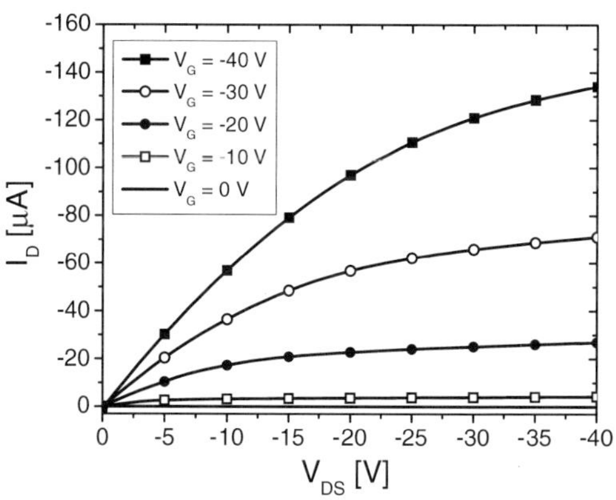

Figure 24.4 Output characteristic for a pentacene OFET, realised using Au drain–source contacts and a pristine SiO_2 dielectric [32].

Si layer stack, an analogue to the OFET device structure. C_{Diel} is the capacitance of the 200 nm SiO_2 dielectric, while C_{Tot} stands for the total device capacitance including the organic semiconductor. The MIS diode capacitance shows no dependence on the applied voltage V_G and is close to C_{Tot}. For reverse biases this is due to a mismatch in metal work function and the semiconductor ionisation potential of ~2.2 eV, suppressing the hole injection. For forward biases electrons are either not transported through the pentacene layer at a measurement frequency of 100 Hz, or electrons are not injected from the Ca electrode. Taking the inset of Figure 24.5a into account, illustrating the DC charging of a pristine MIS-diode at a forward bias of $V_G = 40$ V, an accumulated charge of 172 nC is determined. This amount of charge accounts for a device capacitance of 4.3 nF, which is close to the capacitance value of the SiO_2 dielectric. This proves that electrons can be injected into the pentacene active layer from Ca electrodes, and are accumulated at the SiO_2/pentacene interface. However, once accumulated, the negative charge carriers cannot follow the applied AC electric field of the impedance measurement. This is expressed by the lack of change in the recorded differential device capacitance. The result strongly suggests that charge carrier traps are present at the dielectric/semiconductor interface, impeding the electron transport in the transistor channel of pentacene OFETs.

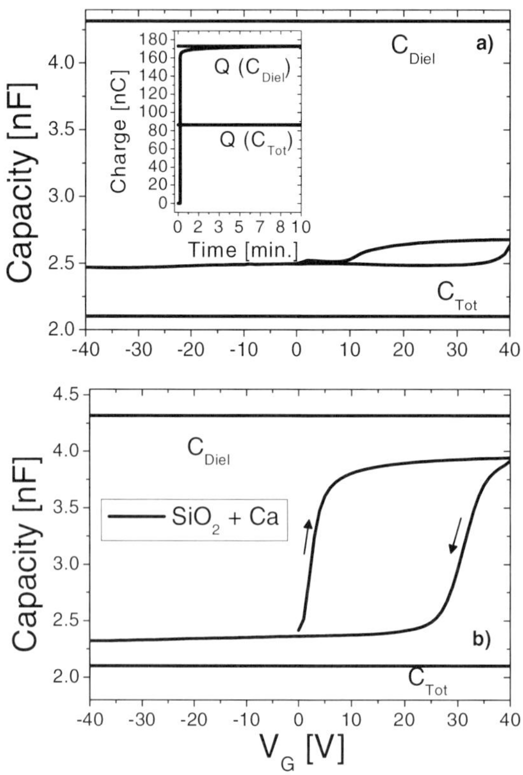

Figure 24.5 Impedance measurements for a Ca/pentacene/insulator/p^{++} Si layer stack. The insulators consist of (a) SiO_2 or (b) SiO_2 + 6 Å Ca. Inset a): MIS-diode electron charging at a sample DC bias of 40 V.

In order to compensate these interfacial electron traps, Ahles et al. [29] introduced traces of Ca, as a low work function metal, to the SiO_2 dielectric interface. The Ca was initially believed to act as an electron donor. For these MIS-diode experiments 0.6 nm of Ca were deposited prior to the pentacene deposition, at a rate of 0.4 Å/s. The result of the impedance measurements is shown in Figure 24.5b. Again, due to the high hole injection barrier the injection of holes is hindered, and a capacitance close to C_{Tot} was found for reverse biases ($V_G < 0$ V). However, a differential device capacitance close to C_{Diel} was measured for positive biases. This suggests that traces of Ca on the SiO_2 dielectric significantly reduce the electron trap density at the dielectric/semiconductor interface. Nevertheless, in spite of the deposited Ca traces, the strong hysteresis indicates that electron trap states are still present and active. We propose that negative charge carriers become localised in the remaining electron traps at the dielectric interface during the forward bias sweep. Hence, as soon as V_G is reduced, the measured device capacitance drops due to trapped electrons compensating the applied electric field and the hysteresis like behaviour is obtained. For negative gate voltages these traps are emptied, as is evident from the reproducibility of the hysteresis.

Considering the presented experimental results it can be concluded that neither the electron injection, nor the electron transport in the bulk of pentacene prevents the n-type device performance of OFETs with Ca source–drain electrodes. In fact, the trapping of accumulated electrons at the dielectric interface leads to an inhibited electron transport in the transistor channel. By depositing Ca traces on trap afflicted SiO_2 insulators, the interfacial trap density can be significantly reduced, allowing for the accumulation of mobile negative charge carriers.

In order to further investigate details of the Ca deposition on top of an SiO_2 insulator, an XPS experiment was conducted to examine the chemical nature of the Ca/SiO_2 bilayer [34]. Depicted in Figure 24.6 is the respective Ca2p core level emission spectrum, illustrating measurement results for several Ca layer thicknesses in ascending order. Surprisingly, for low adsorbate thicknesses only oxidized instead of metallic Ca was found at respective binding energies of 346.5–348.5 eV ($Ca2p_{3/2}$) and 349.6–352.0 eV ($Ca2p_{1/2}$) [35, 36]. As will be substantiated, using the O1s emission spectrum depicted in Figure 24.7, the oxidized Ca layer contains –CaOH as well as –CaO components. With an increase in film thickness, the adsorbate started to develop a metallic fraction in the oxide between layer thicknesses of 5–12 Å. This was initially identified for an adsorbate thickness of 12 Å, by the elevated ground level at high binding energies and a first indication of the characteristic metallic $Ca2p_{3/2}$ (345 eV) and $Ca2p_{1/2}$ (349 eV) emission lines. However, the fully developed metallic phase did not occur until the adsorbate thickness exceeded 115 Å. This interpretation was deducted from the appearance of the asymmetry in the Ca2p emission lines, the elevated ground level for high binding energies and the development of a distinct plasmon emission at a binding energy of 355 eV. The occurrence of a closed metallic layer could be confirmed for an

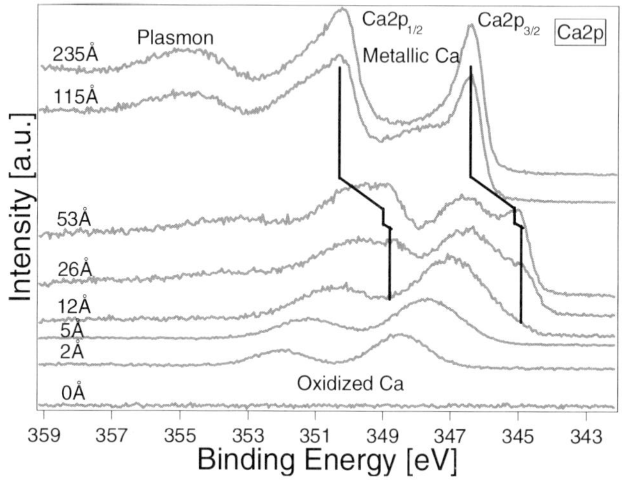

Figure 24.6 Ca2p emission spectra for different Ca layer thicknesses in ascending order. Metallic Ca is only detected for higher nominal thicknesses, while oxidised Ca, containing –CaOH and CaO, is obtained for low film thicknesses.

adsorbate thickness of 235 Å, by the complete attenuation of the Si2p and O1s substrate emission lines (not shown). By considering the intensity normalised O1s emission spectrum, as depicted in Figure 24.7, the respective oxide components contained in the adsorbate can be quantified. Here the intensity normalised O1s emission spectrum is depicted for adsorbate thicknesses between 0 Å and 53 Å. This spectrum is corrected in binding energy with respect to the oxygen substrate component at 533.3 eV for better comparability of the spectra. With increasing Ca coverage two emission lines at binding energies of ~529 eV and ~531 eV appear in addition to the oxygen substrate component. These lines were identified as CaO (529 eV) and –CaOH (531 eV) [36].

In Figure 24.6 a rigid shift in the emission spectrum to higher binding energies can be observed between Ca film thicknesses of 53 Å and 115 Å. In addition, by considering the respective valence band spectra (not shown) for adsorbate thicknesses <53 Å, the Fermi level is found at negative binding energies. However, with the appearance of the distinct metallic phase in the oxidized Ca film (between 53 Å and 115 Å), the Fermi level shifts back to a binding energy of 0 eV. While the final cause leading to the observed shift in energy is unclarified, we suggest this to be related to charge transfer effects between the substrate and the adsorbate.

In view of the discussed XPS experiments, the deposition of Ca on SiO_2 results in an oxidized Ca layer containing CaO and –CaOH. This holds true as long as the Ca layer is thin. Apparently, a chemical interface reaction between atomic Ca and the oxygen substrate component must occur, since atmospheric oxygen is absent in the UHV. This conclusion is supported by the Si2p emis-

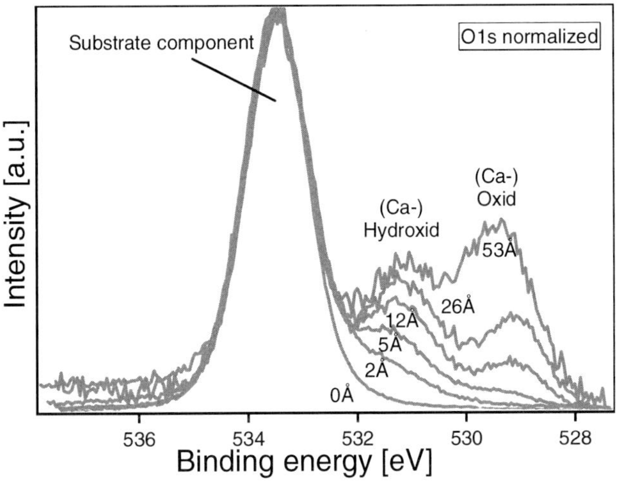

Figure 24.7 Intensity normalised O1s emission spectrum, corrected with respect to the substrate component, for increasing Ca film thicknesses. In addition to the substrate component, Ca-hydroxide and Ca-oxide components are identified in the adsorbate.

sion spectrum (not shown). Here, in addition to the Si^{4+} (103.7 eV) substrate component, an emission line at binding energies of ~102 eV was detected. Since the Si2p emission line is subject to a chemical shift of ~1 eV [36] per oxidation state, the emission is ascribed to either Si^{3+} or Si^{2+} as a Ca reduced substrate component. Furthermore, the appearance of –CaOH supports the suggested interface reaction of atomic Ca with inherent –OH groups, terminating the SiO_2 dielectric. Since –CaOH is known for its ionic binding, where Ca donates an electron to the –OH group, it seems likely that the original electron traps at the SiO_2 interface, as discussed above, are caused by SiOH. Therefore, the appearance of –CaOH in the adsorbate indicates a passivation of the orginal trapping nature of the hydroxyl groups by the ionic bond to the Ca atom. This is in agreement with a recent publication of Chua et al. [30], who appointed hydroxyl groups at the surface of SiO_2 as strong electron traps. As a result of the proposed reaction, Ca eliminates the surface traps. Hence, the accumulation of mobile negative charge carriers at the dielectric interface is possible. In the following, the reaction of Ca with –OH groups will be referred to as Ca passivation.

Empolying such a Ca passivated SiO_2 insulator in combination with Ca drain–source electrodes, n-type pentacene OFETs can be realised as has been demonstrated by Ahles et al. [29]. This, however, holds only for thin Ca layers as will be shown in the following, where the influence of the Ca passivation thickness on the electron transport in pentacene OFETs is discussed. Illustrated in Figure 24.8 is the electron field effect mobility in dependence of the Ca thickness. By considering the mobility of pristine devices, which have not been subject to thermal or electrical stress prior to the measurement, we find an increase in μ_e up to 0.028 cm²/Vs for an increase in Ca layer thickness between 0 Å and ~12 Å. For film thicknesses exceeding 12 Å, however, the mobility begins to decline until it almost completely vanishes at a thickness of 26 Å. The extracted On/Off ratios perform accordingly, exhibiting a maximum value of ~10^4. The behaviour of V_{th} differs only in the position of its minimum (42.2 V), which is obtained for a Ca passivation thickness of 8 Å. A short circuit between source and drain electrodes is obtained for a Ca layer thickness of ~250 Å, which correlates well with the value for a closed metallic Ca layer as derived from the XPS measurements (235 Å).

As a matter of fact, the observed degradation in the device parameters occurs with the first indication of a metallic fraction in the Ca adsorbate. Even more, the device performance is significantly degraded as soon as the Ca characteristic $Ca2p_{3/2}$ and $Ca2p_{1/2}$ emission lines are clearly developed. This implies that the availability of metallic Ca at the dielectric interface, even in small quantities, has a negative effect on the electron transport along the dielectric–semiconductor interface. It is likely that the metallic fraction in the oxidized Ca layer disturbs or even fully screens the electric field in the transistor channel.

As already indicated by the electron mobility depicted in Figure 24.8, the device performance improves once the OFET has been subject to an electrical cyclic stress. The stress is applied in the form of a constant drain voltage of $V_D = 80$ V for the duration of 1 h, while the gate voltage is pulsed in 5 s intervals between $V_G = 0$ V and 80 V. For such cyclic conditioned transistors, the electron mobility has increased irreversibly and the Ca passivation thickness

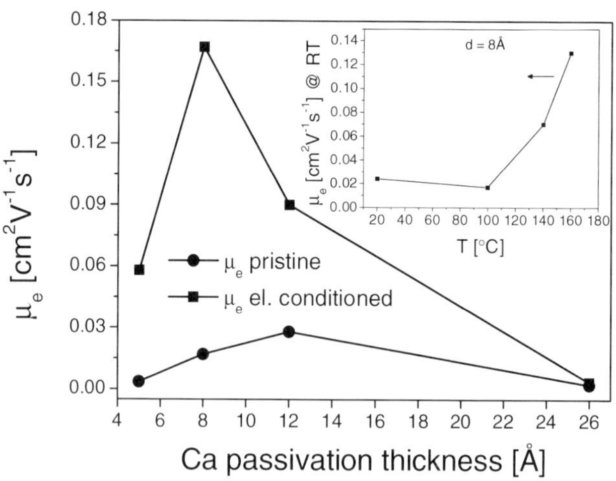

Figure 24.8 μ_e as a dependence on the Ca layer thickness, for OFETs in their pristine and cyclic electrically conditioned state. Inset: μ_e obtained at RT, after the substrate has been subject to a temper step prior to the source–drain metallization deposition.

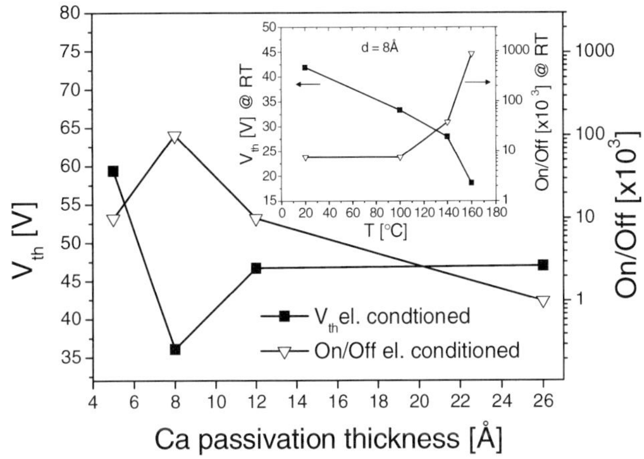

Figure 24.9 V_{th} and the On/Off ratio for different Ca passivation thicknesses of el. conditioned OFETs. Inset: V_{th} and On/Off ratio obtained at RT, after the substrate has been subject to a temper step prior to the source–drain metallisation deposition, at a Ca passivation thickness of $d = 8$ Å.

for a maximum in electron mobility shifted from ~12 Å to ~8 Å (see Figure 24.2). The improvement in the device performance is not limited to the mobility, but also applies to V_{th} and the On/Off ratio, as depicted in Figure 24.9.As for μ_e, a Ca passivation thickness of 8 Å yielded the best V_{th} and On/Off ratio.

Analogous to the electrical conditioning, the OFET characteristics can be improved by a thermal treatment prior to the deposition of the source–drain metallisation. The improvement in the device performance can be seen in the insets of Figures 24.8 and 24.9, showing the OFET parameters, measured at RT, for devices exposed to different annealing temperatures for $t = 1$ h. The examined devices comprised a 8 Å Ca passivation layer. It can be noted that an annealing temperature of 160 °C resulted in a device performance, comparable to the one obtained by the electrical cyclic conditioning.

In Table 24.2 the respective OFET parameters for pristine, electrically conditioned and annealed ($T = 160$ °C) devices are summarised for a 8 Å thick Ca passivation layer. We find that thermal and electrical stress both increase the electron mobility, as well as the On/Off ratio by approximately one order of magnitude. The parameters differ, however, in the reduction of the threshold voltage by $\Delta V_{th} = 19$ V.

In Figure 24.10a and b the O1s emission spectrum of a 53 Å Ca film on SiO_2 in its pristine and its annealed state is illustrated for comparison. The Ca deposition as well as the thermal treatment ($t = 1$ h @ $T = 180$ °C) were conducted without breaking the UHV prior to and in between the respective XPS measurements. We find that an annealing of the Ca adsorbate results in a significant

Table 24.2 Comparison of device parameters for 8 Å Ca passivated OFETs in pristine as well as in thermally or electrically stressed devices.

	μ_e (cm² V⁻¹ s⁻¹)	On/Off	$V_{th,e}$ (V)
pristine	0.017	~10^4	42.2
el. conditioned	0.167	~10^5	36
tempered @ 160 °C	0.14	~10^5	17

increase in the Ca oxide and hydroxide components, with respect to the oxygen substrate component. This implies that the Ca oxidation is promoted by heat, as a result of the discussed interface reaction. Taking the thermally promoted oxidation reaction of the SiO_2/Ca bilayer into account, an explanation for the enhanced device performance as a result of the substrate annealing can be given. Due to the heat treatment, the reaction of Ca with the –OH groups of the interface is promoted and the metallic fraction in the Ca layer is reduced. The former results in a reduction of the electron trap density and the latter in a reduction of screening effects by the metallic Ca.

In order to rule out an improved device performance due to heat induced morphological changes, the following experiment was conducted. A p-type pentacene OFET without Ca passivation and Au source–drain contacts was annealed at $T = 160$ °C for $t = 1$ h, in an inert nitrogen atmosphere. Even though

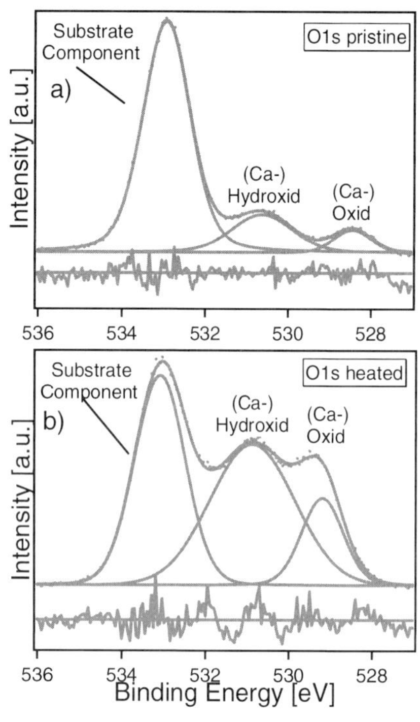

Figure 24.10 O1s emission spectrum for a 53 Å Ca layer on SiO_2 in a) its pristine and b) its annealed state. The annealing was conducted for $t = 1$ h and $T = 180$ °C.

this experiment yields straightforward information only on OFET hole transport properties, a morphological change should affect the hole as well as the electron charge carrier transport in a similar fashion. The experiment indicates a degradation in the device performance as summarized in Table 24.3. Upon annealing, μ_h is reduced by a factor of ~2 and the On/Off ratio by one order of magnitude. We therefore conjecture that the improvement in charge carrier transport for n-type pentacene devices is indeed the result of a thermally promoted Ca oxidation reaction and not the consequence of a thermally induced change in the pentacene morphology.

Even though the origin of the improvement in OFET device characteristic upon a thermal treatment seems unveiled; it is still unclear what causes the enhancement due to electrical cyclic stress. However, additional experiments have shown that a electrical cyclic conditioning of annealed transistors does not yield a significant additional improvement in the device performance. We therefore suggest that the observed OFET improvement for both device treatments, have the same consequence. This implies that the electrical cyclic conditioning also leads to a promoted oxidation of the metallic fraction in the Ca passivation, as discussed for the substrate annealing.

From Figure 24.8 it becomes evident that the improvement in the electron field effect mobility, due to a heat treatment prior to the measurement, is discriminative for different Ca layer thicknesses. For Ca passivation thicknesses < 12 Å the metallic fraction in the oxidized Ca layer is sufficiently small to be fully oxidized during the heat treatment. The increase in Ca layer thickness results on the one hand in an increased coverage of the dielectric interface, but on the other hand in an increase of the metallic fraction, impeding the device performance. Hence, a promoted oxidation reaction will lead to different device improvements for different Ca layer thicknesses. For Ca passivation thicknesses exceeding ~12 Å, the metallic fraction can no longer be sufficiently reduced by the heat treatment. Therefore, the device improvement due to an increasing coverage of the SiO_2 dielectric is counterbalanced by an increase in metallic content in the oxidized Ca layer. This effect is responsible for the deteriorating device performance, when compared with the Ca layer at thicknesses of ~12 Å.

By correlating XPS measurements with the OFET performance, we explained why Ca traces on SiO_2 dielectric surfaces allow for the accumulation of

Table 24.3 Comparison of device parameters for a standard p-type pentacene OFET, utilising a pristine SiO_2 dielectric with Au source–drain contacts and the respective thermally treated device.

	μ_h ($cm^2V^{-1}s^{-1}$)	On/Off	$V_{th,h}$ (V)
pristine	0.18	~10^4	−0.3
tempered @ 160 °C	0.1	~10^3	7.2

mobile negative charges in respective OFET structures. Ca deposited in a thin film (<12 Å) on silicon dioxide results in the formation of an oxidized Ca layer, as the result of an interface reaction between atomic Ca and the oxygen substrate components. This layer isolates and compensates available electron traps, in the form of hydroxyl groups [30], at the dielectric–semiconductor interface, allowing for the realisation of n-type OFET devices. A technological implication of this technique would be the selective deposition of Ca traces on a SiO_2 dielectric. By this technique we were able to realise unipolar p- and n-type pentacene OFETs on a single substrate, without the need to employ different organic semiconductors [37]. The obtained balanced unipolar p- and n-type characteristics will be discussed below.

As an alternative to the Ca passivation of a SiO_2 dielectric, an additional polymer insulator without hydroxyl groups in its chemical structure can be used to enable electron transport in a pentacene OFET, by simply covering the –OH groups available at the SiO_2 interface. Such a polymer is for instance PMMA. On the basis of a MIS-diode containing a SiO_2/PMMA bilayer dielectric, this can be confirmed experimentally. The investigated MIS-diode structure is depicted in Figure 24.3b. The thickness of the PMMA insulator is 90 nm. The result of the impedance measurement is shown in Figure 24.11. Here the capacitance of the SiO_2/PMMA double layer structure is given by C_{Diel}. The total device capacitance is defined as C_{Tot}. As for the case of the MIS-diode with a Ca passivated SiO_2 dielectric, a differential capacitance close to C_{Diel} is obtained for positive device biases. Thus the accumulation of mobile negative charge carriers at the polymer dielectric interface is possible. This result supports the above conclusions that one of the limiting factors in the performance of n-type OFETs and MIS-diodes is an electron trap afflicted semiconductor/dielectric interface.

In the following, the influence of several polymer dielectrics on the charge carrier transport properties in OFETs is discussed [38]. The chemical structures of the polymers are displayed Figure 24.3c and the respective polymer dilutions are given in Table 24.1. Except for P4VP none of the polymers contain hydroxyl groups, however, they differ in their surface polarity as has been de-

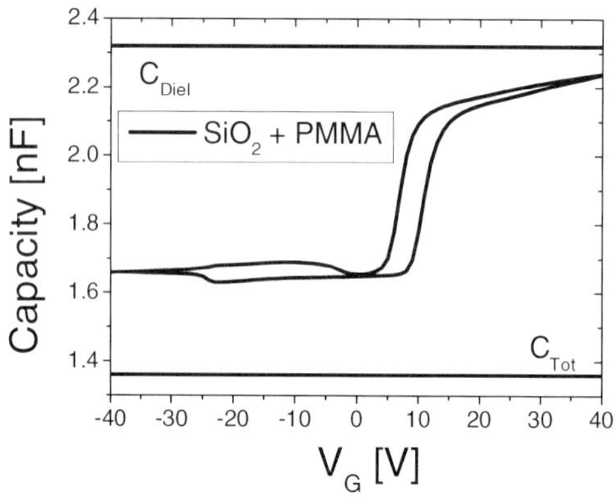

Figure 24.11 Impedance measurement for a Ca/pentacene/insulator/p^{++} Si layer stack. The insulator consists of 200 nm SiO_2/90 nm PMMA bilayer.

rived using water contact angle measurements. The difference in contact angle is a result of varying polar groups in the repeating chain of the polymers, yielding values of 97°, 92°, 81°, 74° and 67° for PS, PC, PMMA, P4VP and PI, respectively.

The implemented transistor structures are illustrated in Figure 24.3a. The charge polarity of the individual OFETs was defined as unipolar p- and n-type by the use of a respective Au and Ca drain–source metallization. Illustrated in Figure 24.12a is the determined OFET threshold charge carrier density $n_{th} = C_{tot}|V_{th}|/|q|$ as a function of the polymer dielectric contact angle. Here q is the elemental charge and C_{tot} is the total device capacitance of the SiO_2/polymer double layer structure. We have chosen n_{th} instead of V_{th}, in order to account for differences in the total device capacitance. While we obtain a decrease in $n_{th,e}$ for electrons with an increase in contact angle, no clear correlation between a change in contact angle and $n_{th,h}$ for positive charge carriers was obtained. Depicted in Figure 24.12b is the field effect mobility μ as a dependence on the contact angle, as derived from the respective transfer characteristics. While no clear correlation of the hole mobility μ_h and the contact angle was found, the electron mobility μ_e improved with increasing contact angle and therefore decreasing dielectric interface polarity. The obtained degradation in the electron transport properties with a decrease in contact angle could originate from the polar oxygen groups of the used polymers. For the investigated materials, as depicted in Figure 24.3c, such groups are available in the form of keto and hydroxyl groups. Indeed, the observed increase in μ_e with an increase in contact angle for PI, PMMA, PC and PS, seems to be related to a decrease in keto groups [39]. We also observe that in the case of P4VP the presence of one hydroxyl group per monomer unit completely inhibits the electron charge carrier transport. This indicates that the presence of a hydroxyl group influences the electron transport much stronger than the keto groups. This is supported by the fact that PI, which exhibits a higher surface polarity (lower contact angle) as well as a higher concentration of keto groups than available hydoxyl groups in P4VP, still allows for electron transport in an OFET, however, with a low electron mobility of $\mu_e \sim 3 \times 10^{-4}$ cm^2 V^{-1} s^{-1}. Hence, the pronounced electron trapping nature of hydroxyl groups is confirmed, in agreement with the findings discussed above and by others [30].

With the given delineation we have not considered the influence of varying pentacene morphologies on the different substrates with respect to the charge carrier transport. While the semiconductor morphology may influence the OFET transport properties, as has been observed by Karl [40] and Voigt et al. [12] for the case of holes, our experiment unveils no clear dependence of the hole mobility on the water contact angle of the different dielectrics and therefore possible changing morphologies of the pentacene active layer. Since we would expect the influence of the morphology to be similar for the transport properties of electrons and holes, we suggest the observed dependence of μ_e to be essentially influenced by the discussed electronic states at the dielectric interface.

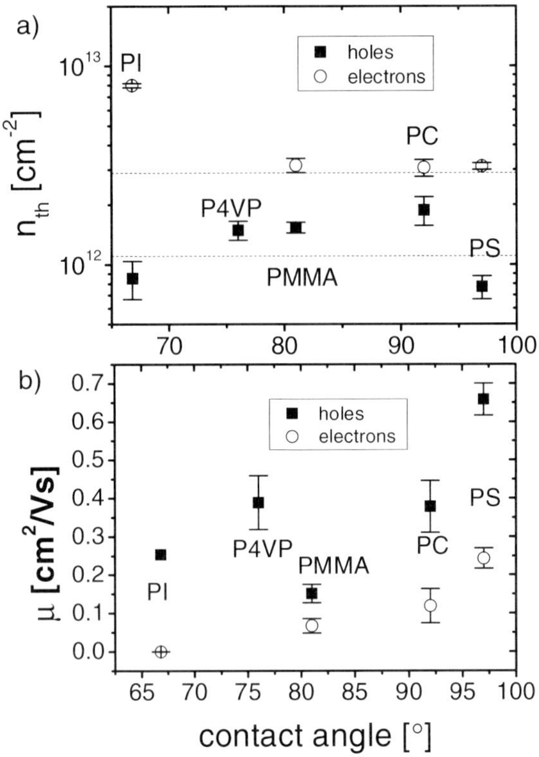

Figure 24.12 (a) n_{th} and (b) μ for electrons and holes as a dependence on the water contact angle. For the error bar statistic up to 4 samples out of different batches were utilised, with a maximum of 3 OFETs per sample. The dotted lines in (a) represent orientation lines.

At this point of our understanding, the question arises whether one can modify interfacial traps of a polymer insulator surface in such a way that one gains control over the electron transport in a transistor, e.g. by the introduction of –OH groups to an otherwise –OH free PMMA surface. Following earlier publications on modification of polymer surfaces, one can introduce keto and hydroxyl groups by exposing a PMMA dielectric to UV radiation in ambient atmosphere [42, 43]. Therefore, for the following experiments, the PMMA dielectric layer is exposed to 254 nm radiation for a time frame of 10 minutes and an optical power of 15 mW/cm^2 prior to the pentacene deposition.

However, before considering the impact of the UV irradiation on the electronic properties of the resulting OFETs, the surface properties of the pristine and UV irradiated PMMA dielectrics, such as surface roughness and contact angle are determined. We observe that the RMS roughness of the PMMA remains unchanged during the UV treatment, as determined by AFM tapping mode measurements. However a significant change in water contact angle from 80° to 30° was found, which indicates a change in chemical composition of the PMMA near surface layer [41]. To get a deeper insight into the chemical changes, XPS measurements using Mg K_α radiation were conducted. In Figure 24.13 the measurement result is illustrated in the form of an area normalised O1s emission spectrum, exhibiting pristine (dotted line) and UV modified PMMA (straight line). The emission line for the pristine PMMA surface is composed of two components, which can be ascribed to O=C and O–C groups at respective binding energies of 533.4–533.5 eV and 534.8–535.0 eV [42]. During UV exposure we clearly obtain a relative increase in keto groups (O=C) at the expense of O–C groups, as can be derived from the pronounced maximum of the O1s emission spectrum at lower binding energies for the UV-treated PMMA, when compared to the two maxima observed for the pristine polymer. By taking the work of Wei et al. [43] into consideration, who have conducted an XPS study of UV exposed PMMA under identical experimental conditions, the formation of hydroxyl groups has to be considered in addition to the obtained relative increase in keto groups.

Illustrated in Figure 24.14a and b is the output characteristic of a pentacene OFET with Ca source–drain electrodes, in the electron and hole accumulation mode. The PMMA dielectric has been modified using UV radiation. As expected, only a negligible hole current and no electron current was obtained (Figure 24.14b) for the first measurement. With respect to the hole current this can be explained by the implemented Ca electrodes. Ca contacts prevent an efficient hole injection due to an insufficient matching of the metal work function and the HOMO level of the organic semiconductor. The inhibited electron current, however, is the result of electron traps introduced to the dielectric/semiconductor interface by the UV radiation. Surprisingly, a significant increase in drain current was obtained for the second measurement in the electron accumulation. Depicted in Figure 24.14a are the output characteristics for the first eight measurements in electron accumulation mode. The plots are illustrated for $V_G = 0$ V, however, between measurements the device has been op-

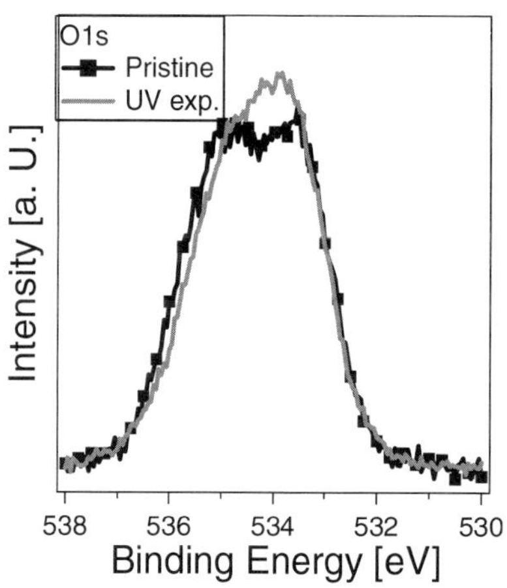

Figure 24.13 O1s emission spectrum for pristine and UV modified PMMA.

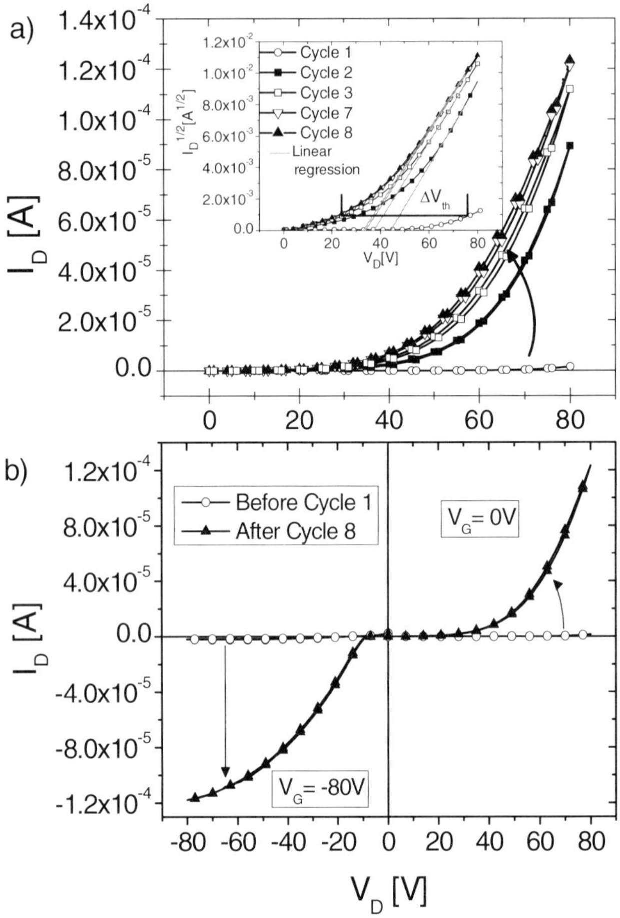

Figure 24.14 (a) Drain current as a function of V_D for a UV modified pentacene OFET in the electron accumulation mode. Inset: threshold voltage shift between the 1st and 8th successive measurement cycles at $V_G = 0$ V. (b) Electron and hole accumulation mode for a UV modified OFET during the 1st and after the 8th measurement cycle. The OFETs were realised using Ca electrodes.

erated in gate voltage steps of 20 V between $V_G = 20$ V and 80 V. At each step the drain voltage has been cycled between 0 V and 80 V. Apparently, after each operation cycle I_D is increased, until saturation is achieved between the 7th or 8th cycle of operation. The increase in drain current is not the result of an unsaturated electron or leakage currents, but due to holes injected from the drain electrode [44] for |VDS| ≥ |VG – Vth,p|. Such a characteristic is predicted by Eq. (3) for a vanishing electron mobility μn. The unipolar p-type behaviour of the UV modified OFET can be confirmed by considering the 1st and 3rd quadrant of Figure 24.14b. Here the output characteristic in the electron and hole accumulation mode is shown prior to the 1st and after the 8th operation cycle. The s-shaped drain current in the linear range of the transistor operation indicates a pronounced injection barrier for holes, as can be seen in the third quadrant of the device characteristic. It needs to be stressed that the hole injection occurs despite an energy difference between the work function of Ca (2.82 eV) and the pentacene HOMO level (5.02 eV) of 2.2 eV.

By considering the inset of Figure 24.14a, the origin of the increase in unipolar hole current injected from the drain electrode can be understood. In this graph the square root of the ambipolar drain current (Eq. (3)) [3],

$$\sqrt{I_{\mathrm{D}}} = \sqrt{\frac{wC_{\mathrm{tot}}\,\mu_{\mathrm{p}}}{2l}}\,(V_{\mathrm{DS}} + V_{\mathrm{th,p}} - \Delta V_{\mathrm{th}})\,, \qquad (4)$$

for $\mu_n = 0\ \mathrm{cm^2\,V^{-1}\,s^{-1}}$ and $V_G = 0$ V, as well as $|V_{DS}| \geq |V_G - V_{th,p}|$ is compared with the experimental data. Here $C_{tot} = 10.4\ \mathrm{nF\,cm^{-2}}$ represents the resulting total area capacitance of the SiO_2/PMMA bilayer dielectric. The comparison suggests that the improvement in hole current is the result of a positive threshold voltage shift of ~60 V, during the first eight operation cycles in the electron accumulation. The observed shift in ΔV_{th} has been derived from the respective output characteristics in the hole accumulation, and occurs with an almost constant hole mobility. We propose that trapping of electrons (n_t) in a near surface layer of the UV modified PMMA dielectric is the origin for the large ΔV_{th}. During the electron accumulation negative charge carriers are accumulated at the dielectric/semiconductor interface, and localized in the UV generated electron traps. This quenches the electron current and leads to a field enhancement below the Ca drain-source electrodes. As a result a field enhanced injection for holes is obtained. The density of trapped charge carriers n_t is estimated to be:

$$n_{\mathrm{t}} = \frac{\Delta V_{\mathrm{th}} C_{\mathrm{tot}}}{q} = 3.9 \times 10^{12}\ \mathrm{cm}^{-2}\,, \qquad (5)$$

This value is only valid if negative charges are trapped directly at the dielectric interface, representing the lowest possible estimate. If charges are also be trapped in the volume of the PMMA, the trapped charge density increases. The negligible hysteresis observed for the p-type characteristic, as shown in Figure 24.14a and b, indicates the absence of hole traps at the dielectric interface, as well as a missing recombination of mobile positive charge carriers with trapped electrons. This missing recombination suggests that the negative charges are localised in the volume of the PMMA dielectric, which then isolates trapped electrons from mobile positive charge in the transistor channel.

The demonstrated change in OFET polarity can be utilized to integrate unipolar p- and n-type pentacene transistors on a single substrate, without altering the organic semiconductor, or even the device cross section.

In the following the discussed dielectric interface engineering approaches used to fabricate organic complementary metal oxide semiconductor (O-CMOS) devices, namely an inverter stages. In a first approach we selectively passivated spatially separated areas of a SiO_2 dielectric using Ca traces [37]. This defined a p-type transistor in the region of the pristine SiO_2 and an n-type OFET in the region of the Ca passivated SiO_2, as long as Au source–drain contacts were used for the former and Ca contacts were used for the latter OFET. Illustrated in Figure 24.15a are the transfer characteristics of the individual p- and n-type devices in the electron and hole accumulation mode. In analogy to an inverter structure, V_G is termed as V_{In} and the source potential of the p-type transistor is kept at a constant potential value of $V_{DD} = 60$ V. As discussed above the n-type OFET has been electrically cyclic conditioned prior to the measurement, in order to improve the electron transport properties. From the transfer characteristics an electron and hole mobility of $\mu_e = 0.11$ cm^2 V^{-1} s^{-1} and $\mu_h = 0.10$ cm^2 V^{-1} s^{-1}, as well as an On/Off ratio $>10^5$ has been determined. The respective threshold voltages are $V_{th,e} = 34$ V and $V_{th,p} = -20$ V. The observed hysteresis for the n-type pentacene OFET most probably stems

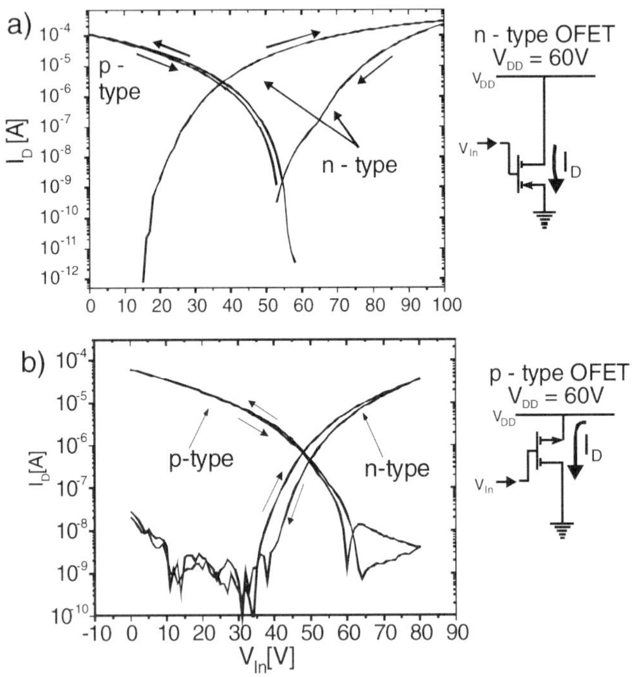

Figure 24.15 (a) Transfer characteristics for p- and n-type pentacene OFETs realized on SiO_2. The insulator area for the n-type transistor is Ca passivated. (b) Complementary OFET transfer characteristics, realised on a PMMA dielectric. The PMMA dielectric of the p-type transistor is UV modified, otherwise the transistor is identical to the n-type device. In analogy to an inverter structure, the source contact of the p-type OFET is held at a potential of 60 V, for both plots.

from residual electron traps being filled during the input voltage sweep. The obtained balanced charge carrier transport properties for electrons and holes have enabled us to realize an O-CMOS inverter stage. The inverter stage is implemented by connecting the two complementary pentacene OFETs, as shown in the inset of Figure 24.16a. The inverter demonstrated stable operation below its supply voltage of $V_{DD} = 60$ V, as well as a maximum gain of 24. The obtained hysteresis for the inverter characteristic reflects the drain current hysteresis of the n-type OFET, as can be seen in Figure 24.15a.

In a second approach to generate a CMOS inverter, OFETs with a PMMA insulator were utilized. For one of the OFETs the otherwise electron trap free PMMA dielectric was UV irradiated prior to the pentacene deposition [44]. By charging the introduced electron traps, the polarity of the UV irradiated OFET changes from unipolar n-type to unipolar p-type, as demonstrated in Figure 24.15b. The measurements were conducted in the same manner as described for the Ca modification. For both devices only a very low current hysteresis was registered. Electron and hole mobilities were determined to be $\mu_e = 0.078$ cm^2 V^{-1} s^{-1} and $\mu_h = 0.11$ cm^2 V^{-1} s^{-1}, respectively. Threshold voltages of $V_{th,e} = 46$ V and $V_{th,p} = -27.5$ V have been obtained, and the On/Off ratio exceeded 10^4 for both transistor types. The p- and n-type OFETs exhibited balanced charge carrier transport properties, enabling the realization of an O-CMOS inverter stage. The inverter transfer characteristic is displayed in Figure 24.16b. The inverter stage exhibits stable operation below its supply voltage of $V_{DD} = 60$ V and a maximum gain of 17 [44]. The observed hysteresis in the inverter transfer characteristic is ascribed to the current hysteresis of the utilised p- and n-type pentacene OFETs.

24.4 Conclusion

In the current chapter we investigated the importance of charge carrier traps at the interface between the gate dielectric and the organic semiconductor for the device performance of an organic field-effect transistor. Until recently pentacene OFETs were believed to operate exclusively as p-type devices due to the organic semiconductor transport properties. However, this p-type behaviour has to be assigned to electron traps at the dielectric interface and not to the pentacene transport properties. Using traces of calcium we were able to establish a trap free SiO_2/semiconductor interface and therefore n-type pentacene OFETs. XPS measurements have revealed the formation of an oxidized Ca layer for small Ca film thicknesses (< 12 Å) composed of calcium oxide and calcium hydroxide on the SiO_2 dielectric. This oxidized Ca layer represents an additional dielectric layer, which isolates and passivates electron traps in the form of hydroxyl groups. The resulting new dielectric/semiconductor interface enables the n-type transport in pentacene. Alternatively, we introduced electron traps to an otherwise trap free PMMA dielectric by the use of UV radia-

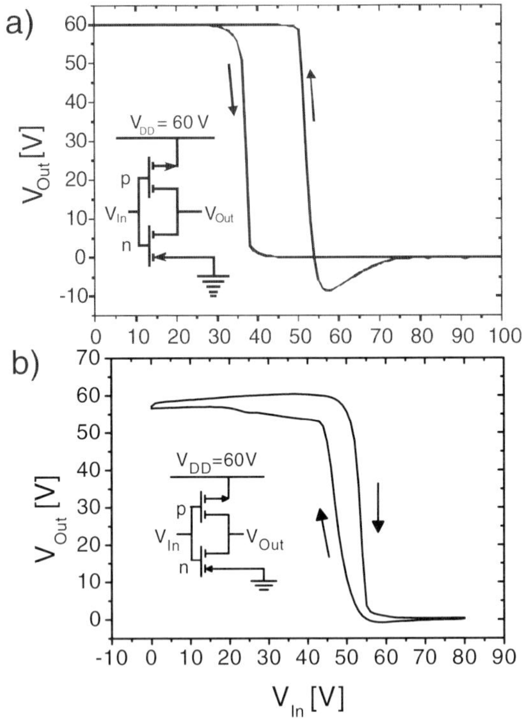

Figure 24.16 O-CMOS inverter stage transfer characteristic. The inverters have been realised using (a) the SiO_2 Ca passivation technique and (b) the UV modification approach of a PMMA dielectric. For plot (a) a max. gain of 24, and for plot (b) a max. gain of 17 was obtained.

tion in ambient atmosphere. The created traps are hydroxyl- and keto-groups. Once the electron traps are charged the polarity of the pentacene OFET changes from unipolar n-type to unipolar p-type. The polarity inversion is the result of trapped negative charge carriers in the bulk of the PMMA dielectric, leading to a field enhanced injection of positive charge carriers. This injection occurs despite an energy barrier of ~2.2 eV between the metal work function of Ca and the pentacene HOMO level. Both techniques were used to realise O-CMOS inverter stages, as the result of balanced charge carrier transport properties. The obtained charge carrier mobilities were in the order of $\mu_{e,h} \sim 0.1\ cm^2\ V^{-1}\ s^{-1}$, and the OFETs exhibited On/Off ratios of $>10^4$. Our work demonstrates that the engineering of the dielectric/semiconductor interface is an essential tool to control the performance of organic field effect transistors.

Acknowledgements

The authors gratefully acknowledge Dr. Thomas Mayer and Eric Mankel for the useful discussions about the XPS measurements, as well as the financial support of the German Research Foundation (DFG) in the framework of the OFET Schwerpunkt Programm (Schm 1523/3).

References

1. U.S. Patent 1.745.175, U.S. Patent 1.900.018, U.S. Patent 1.877.140.
2. W. Schockley, M. Sparks and G. K. Teal, Phys. Rev. **83**, 151 (1951).
3. R. Schmechel, M. Ahles, and H. von Seggern, J. Appl. Phys. **98**, 084511 (2005).
4. E. C. P. Smits, T. D. Anthopoulos, S. Setayesh, E. van Veenendaal, R. Coehoorn, P. W. M. Blom, B. de Boer, and D. M. de Leeuw, Phys. Rev. B **73**, 205316 (2006).
5. E. A. Silinsh and V. Capek, Organic Molecular Crystals (AIP Press, 1994).
6. A. Dodabalapur, L. Torsi, and H. E. Katz, Science **268**, 270 (1995).
7. H. G. O. Sandberg, Langmuir **18**, 10176 (2002).
8. R. Ruiz, A. Papadimitratos, A. C. Mayer, and G. G. Malliaras, Adv. Mater. **17**, 1795 (2005).
9. R. J. Kline, M. D. McGehee, and M. F. Toney, Nature Mater. **5**, 222 (2006).
10. A. Hoppe, T. Balster, T. Muck, and V. Wagner, Phys. Stat. Sol. (a) **205**, 612 (2008); Chapter 22 in this book.
11. A. Gerlach, S. Sellner, S. Kowarik, and F. Schreiber, Phys. Stat. Sol. (a) **205**, 461 (2008); Chapter 9 in this book.
12. M. Voigt, J. Pflaum, and M. Sokolowski, Phys. Stat. Sol. (a) **205** 449 (2008); Chapter 8 in this book.
13. K. Shankar and T. N. Jackson, J. Mater. Res. **19**, 2003 (2004).
14. Y. Y. Lin, D. J. Gundlach, S. F. Nelson, and T. N. Jackson, IEEE Electron Device Lett. **18**, 606 (1997).
15. H. Sirringhaus, P. J. Brown, R. H. Friend, M. M. Nielsen, K. Bechgaard, B. M. W. Langeveld-Voss, A. J. H. Spiering, R. A. J. Janssen, E. W. Meijer, P. Herwig, and D. M. de Leeuw, Nature **401**, 685 (1999).
16. A. Salleo, M. L. Chabinyc, M. S. Yang, and R. A. Street, Appl. Phys. Lett. **81**, 4383 (2002).
17. D. Knipp, R. A. Street, A. Völkel, and J. J. Ho, J. Appl. Phys. **93**, 347 (2003).
18. D. H. Kim, Y. D. Park, Y. Jang, H. Yang, Y. H. Kim, J. I. Han, D. G. Moon, S. Park, T. Chang, C. Chang, M. Joo, Y. R. Chang, and K. Cho, Adv. Funct. Mater. **15**, 77 (2005).
19. J. Veres, S. Ogier, and G. Lloyd, Chem. Mater. **16**, 4543 (2004).
20. S. Lukas, G. Witte, and Ch. Wöll, Phys. Rev. Lett. **88**, 028301 (2001).
21. F.-J. Meyer zu Heringdorf, M. C. Reuter, and R. M. Tromp, Nature **412**, 517 (2001).
22. K. Ihm, K. Bongsoo, T.-H. Kang, K.-J. Kim, M. H. Joo, T. H. Kim, S. S. Joon, and S. Chung, J. Appl. Phys. Lett. **89**, 033504 (2006).
23. M. Halik, H. Klauk, U. Zschieschang, G. Schmid, C. Dehm, M. Schütz, S. Malsch, F. Effenberger, M. Brunnbauer, and F. Stellacci, Nature **431**, 963 (2004).
24. H. Klauk, U. Zschieschang, J. Pflaum, and M. Halik, Nature **445**, 745 (2007).
25. J. Veres, S. D. Ogier, G. Lloyd, and D. M. de Leeuw, Chem. Mater. **16**, 4543 (2004).
26. J. Veres, S. D. Ogier, S. W. Leeming, D. C. Cupertino, and S. M. Khaffaf, Adv. Funct. Mater. **13**, 199 (2003).

27. I. N. Hulea, S. Fratini, H. Xie, C. L. Mulder, N. N. Iossad, G. Rastelli, S. Ciuchi, and A. F. Morpurgo, Nature **5**, 982 (2006).
28. S. Kobayashi, T. Nishikawa, T. Takenobu, S. Mori, T. Shimoda, T. Mitani, H. Shimotani, N. Yoshimoto, S. Ogawa, and Y. Iwasa, Nature Mater. **3**, 317 (2004).
29. M. Ahles, R. Schmechel, and H. v. Seggern, Appl. Phys Lett. **85**, 4499 (2004).
30. L.-L. Chua, J. Zaumseil, J.-F. Chang, E. C.-W. Ou, P. K.-H. Ho, H. Sirringhaus, and R. H. Friend, Nature **434**, 194 (2005).
31. M. Bronner, A. Opitz, and W. Brütting, Phys. Stat. Sol. (a) **205**, 549 (2008); Chapter 17 in this book.
32. M. Ahles, Tech. Univ. Darmstadt, Diss. (2006).
33. N. Benson, M. Ahles, M. Schidleja, A. Gassmann, E. Mankel, T. Mayer, C. Melzer, R. Schmechel, and H. v. Seggern, Proc. SPIE **6336**, 63360S (2006).
34. N. Benson, A. Gassmann, E. Mankel, T. Mayer, C. Melzer, R. Schmechel, and H. v. Seggern, J. Appl. Phys. **104**, 054505 (2008).
35. H. Van Doveren and J. A. T. Verhoeven, J. Electron Spectrosc. Relat. Phenom. **21**, 265 (1980).
36. J. F. Moulder, W. F. Stickle, P. E. Sobol, and K. D. Bomben, Handbook of X-ray Photoelectron Spectroscopy (Physical Electronics Inc. 1995).
37. M. Ahles, R. Schmechel, and H. v. Seggern, Appl. Phys. Lett. **87**, 113505 (2005).
38. N. Benson, M. Schidleja, C. Siol, C. Melzer, and H. v. Seggern, Proc. SPIE **6658**, 66580W (2007).
39. A. Kadashchuk, R. Schmechel, U. Scherf, A. Vakhnin, and H. v. Seggern, J. Appl. Phys. **98**, 024101 (2005).
40. N. Karl, Synth. Met. **133/134**, 649 (2003).
41. M. Morra, E. Occhiello, and F. Garbassi, Adv. Colloid Interface Sci. **32**, 79 (1990).
42. A. Hozumi, T. Masuda, K. Hayashi, H. Sugimura, O. Takai, and T. Kameyama, Langumir **18**, 9022 (2002).
43. S. Wei, B. Vaidaya, A. Patel, S. Soper, and R. McCarley, J. Phys. Chem. B **109**, 16988 (2005).
44. N. Benson, M. Schidleja, C. Melzer, R. Schmechel, and H. v. Seggern, Appl. Phys. Lett. **89**, 182105 (2006).

25
Aspects of the Charge Carrier Transport in Highly-Ordered Crystals of Polyaromatic Molecules

J. Pflaum, J. Niemax, S. Meyer, and A.K. Tripathi

25.1
Introduction

The successful application of low-weight organic molecules in thin film devices has raised questions about their ultimate performance, which is tightly correlated with the intrinsic transport characteristics of each individual material.

The amount of transport data collected to-date on various molecular compounds and under various experimental conditions [1], including pressure and temperature, provides reliable benchmark data that need to be described by appropriate theoretical concepts of the charge carrier conductivity in polyaromatic hydrocarbons (PAHs). This marks an important development, as until now there was either a lack of high-quality samples to prove theoretical predictions on selected molecular species, or simplified transport models had been applied that were not capable of accounting for the observed transport characteristics. As a concomitant difficulty it became obvious that models that successfully predict the transport parameters in covalently bound semiconductors, such as Si, are not suited for organic solids due the similar energy scales of the contributing processes such as polaron formation, bandwidth, reorganisation etc. [2–4].

A major constraint to theoretical modelling is the weak interaction between the van-der Waals bound molecular entities. This weak energy contribution causes principal problems in applying up-to-date density functional theory (DFT) concepts that require a sufficiently high wavefunction overlap between the adjacent entities, such as in the case of covalently bound inorganic semiconductors. As a consequence, the lattice spacing of the relaxed geometry estimated by DFT for molecular crystals in general is too small and hence the electron–phonon coupling becomes too high [5, 6].

Secondly, contribution energies are of similar magnitude making it difficult to discriminate the relevant microscopic mechanisms determining the electronic transport. For example, the energy of ~10 meV for the lowest-energy phonon modes in organic crystals turns out not to be negligible compared with

Organic Electronics. Structural and Electronic Properties of OFETs. Edited by Christof Wöll

ISBN: 978-3-527-40810-8

the inter-molecular exchange energies of ~100 meV, and the resulting anisotropy of charge carrier mobility is strongly affected by both contributions at temperatures relevant for applications [5, 7]. In the case of naphthalene – consisting of two linearly conjugated benzene-rings (Figure 25.1a) – the direction of preferred transport at room temperature does not coincide with the (ab)-plane, although it provides the largest molecular orbital overlap, but is aligned almost along the c′-direction. The exchange interaction between π-orbitals of adjacent molecules is much smaller along this axis, but the electron–phonon coupling proves to be diminished even more and results in a reduced scattering of charge carriers [8]. Lowering the temperature and, thereby, progressively freezing-out the thermally activated phonon modes, the direction of preferred transport continuously rotates into the (001)-plane, which yields the highest mobilities for electrons and holes at cryogenic temperatures [9].

Addressing the conductivity in PAHs as an intrinsic material dependent property, one has to bear in mind that for this class of wide-band gap semiconductors the effective transport is strongly affected by impurities (dopants, local oxidation), structural defects (vacancies, dislocations, grain boundaries) or electronic states at organic/metal and organic/oxide interfaces, both providing major contributions to the electronic performance of organic thin film devices [10]. To enable investigations on the material inherent transport parameters, viz. mobility, bandwidth, dispersion, the use of organics that can be sufficiently purified and that are stable with respect to degradation is inevitable. Moreover, the necessary electronic key parameters have to be extracted by means of experimental methods based on charge carrier injection as well as by contact-free techniques.

Accordingly, in this contribution we will address the transport properties of selected materials, tetracene, diindeno[1,2,3-cd:1′,2′,3′-lm]perylene (DIP), etc. (see Figure 25.1), which show different chemical stabilities and susceptibilities on (photo-)oxidation due to their respective aromaticity [11]. The temperature dependent charge carrier mobility proves to be an indicator of high sensitivity for chemical and structural irregularities and can be accessed by injection-free Time-Of-Flight (TOF) and injection-based Field-Effect Transistor (FET) and Space-Charge-Limited-Current (SCLC) studies.

We will begin with an introductory overview on purification and growth techniques, also considering the specific advantages and drawbacks, followed by a description of the applied methods for electronic characterisation (Section 25.2). In Section 25.3 the transport data on selected crystalline samples will be discussed. In Section 25.3.1, chemical effects on the surface vs. the bulk transport will be analysed for tetracene single crystals. The second part, Section 25.3.2, addresses effects on the surface and the bulk mobility upon structural changes, in this case the enantiotropic phase transition occurring in diindenoperylene single crystals. The conclusions derived from these results are representative for a broad range of molecular compounds as well as for their combinations, the latter being of particular interest for e.g. organic photovoltaics [12, 13].

25.2
Experimental

25.2.1
Material Selection

On account of the variety of organic molecules available, it is necessary to restrict the discussion of the transport phenomena to selected compounds, which comprise the characteristic properties of the related material class. In addition, the chosen PAHs should allow one to distinguish between the intrinsic behaviour and extrinsic effects caused by chemical or structural inhomogeneities. Experimentally, the molecules should feature sufficiently low vapour pressure at room-temperature and sublimation or melting points far below degradation. If possible, the solubility should be sufficiently good for chemical trace analysis. Molecules that fulfil most of these conditions and which will be referred to in the following are, for example, naphthalene or tetracene (see Figure 25.1a and 25.1d). The influence of the temperature dependent crystal structure will be demonstrated by transport studies on the organic semiconductor diindenoperylene (Figure 25.1e) in combination with crystallographic analyses. The molecular shape and the long-range order induced by this motif make DIP a prototypical candidate for thin film applications.

25.2.2
Purification

Dealing with organic semiconductors in most cases means dealing with wide band-gap materials of $E_{gap} \approx 2\text{–}3$ eV [1]. Therefore, any defect might cause intergap states that could effectively hamper the carrier transport [14]. Consequently, material purification is an essential requirement to attain high quality

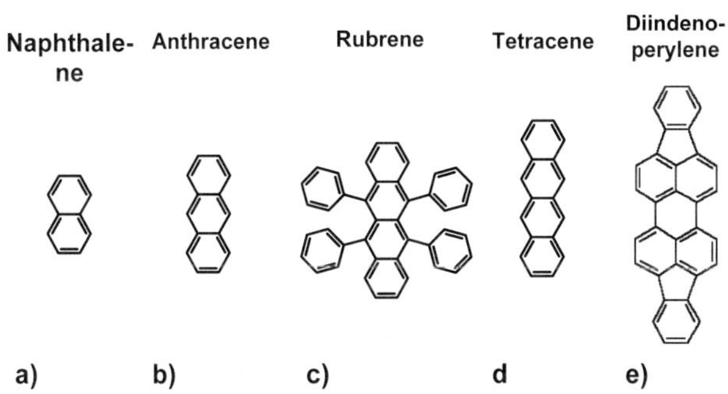

Figure 25.1 Selected organic molecules discussed in the course of this chapter. Special attention is drawn to tetracene and diindenoperylene both representing different classes of PAHs, viz. oligoacenes and perylene-derivatives, which are of great interest for organic electronic applications.

samples with high charge carrier mobilities for electrons and holes. It has been demonstrated on organic single crystals that chemical contaminations at relative concentrations of 1 ppm can already significantly degrade the electronic performance [15]. The influence can occur through the energy levels of the contaminants localised within the gap of the crystalline host and leading to traps of discrete energy, or by the steric constraints, the local distortion of the translational symmetry in the surrounding lattice results in a quasi-continuous (dispersive) distribution of trap states [16–19]. As a result, an impurity might provoke a different effect on moving electrons and holes in the polyaromatic host.

Purification proves to be a tedious task due to possible oxidation and thermal instability of the respective compounds, e.g. upon melting. For larger molecules like pentacene, rubrene or phthalocyanines such constraints make it difficult to achieve purity grades sufficient to extract the intrinsic electronic properties by macroscopic measurements. Appropriate purification methods depend on solubility, chemical and thermal properties of a given organic species. Therefore, we will briefly highlight two general approaches based on melting and on sublimation. Alternative techniques like re-crystallisation or column-chromatography are not considered here, as they cannot be applied to the molecules under discussion and as in many cases the resulting degree of purity cannot cope with that achieved by the techniques described in the following.

25.2.2.1 **Purification by Zone Refinement**

Zone refining was originally introduced by Pfann as a possible route to significantly reduce the impurity concentration in silicon [20]. The lower melting points of van-der Waals bound semiconductors (e.g. 352 K for tetra-methylene benzene, 489 K for anthracene and up to 551 K for perylene) renders this method also appropriate for organic polyaromatics [21]. An inevitable constraint imposed by zone refinement is the long-term stability of the batch over days or months at temperatures slightly above the melting point.

According to the Clausius–Clapeyron equation the transition temperature at the solid → liquid phase boundary of a binary mixture shifts by (∂T) upon a relative concentration change (∂c) between the two constituents:

$$\left(\frac{\partial c}{\partial T}\right)_{\text{coex}} = \frac{\Delta H}{\Delta V}\frac{1}{T}. \tag{1}$$

Here, ΔH and ΔV describe the change in latent heat and volume, respectively. *Vice versa*, Eq. (1) can be interpreted such that upon changing the temperature the relative concentration of the two components can be varied. An organic mixture (host and contaminants) travelling across a heating zone will experience an increasing stoichiometrical gradient across its volume. The efficiency of the zone-refinement is determined by the distribution coefficient $k = c_s/c_l$, which relates the susceptibility of the host material for incorporating an 'impurity-component' in the solid phase c_s to that in the liquid phase c_l. The higher the saturation concentration in the liquid phase versus that in the solid phase, the better the degree of purification.

25.2.2.2 Purification by Sublimation

Although under the best conditions zone-refinement provides the lowest density of chemical contaminants down to the sub-ppm range, many compounds do not fulfil the stability requirements and chemically degrade in the melt over time. In particular, lager conjugated molecules such as pentacene or rubrene, which turned out to be of interest for technical applications, were found to degrade even prior to reaching their melting point at viable pressures. Therefore, an alternative purification approach is established by sublimation, which for many compounds proceeds without significant chemical degradation.

Segregation via vaporisation and re-sublimation relies on the fact that, in general, impurities have sublimation temperatures different to that of the purified compound [22]. Depending on the sublimation temperature of the host, impurities are either more volatile or remain as residue. Purification by sublimation can be further distinguished by two modifications:

Step sublimation: In this arrangement the pristine material is placed inside a quartz tube and heated slightly (less than ~10 K) above its sublimation point under HV conditions at $\sim 10^{-6}$ Torr. Low-weight residues from the synthesis or pre-purification evaporate upon heating and re-sublime far apart from the pristine and the purified material. If their molecular weight is higher than that of the host (here we neglect effects by permanent dipoles on the sublimation temperature), the impurities will remain in the pristine material.

If re-contamination of the already purified fraction can be avoided, step sublimation offers the advantages of purifying several grams of material on the time scale of hours at defined thermal condition. The latter is of particular importance for newly synthesised compounds with unknown sublimation temperatures.

Gradient sublimation: To ensure a high material yield in combination with a high selectivity, sublimation over an extended temperature gradient (~ 500 K/m) is the preferred method if the sublimation temperature of the host material is known. Inside a glass tube separation across the temperature gradient takes place and, in most cases, the purified fraction occurs spatially well-separated from the contaminants. The efficiency can be improved in terms of yield and stability in the presence of an additional inert carrier gas, e.g. Ar or N_2, which at low pressure (10^{-2} Torr) reduces the molecular mean free path and equilibrates temperature fluctuations across the glass tube.

25.2.2.3 Control of Chemical Purity

A major challenge in the research on ultra-pure organic semiconductors is the quantification of their chemical composition by suitable tracer techniques. An interesting approach providing a resolution of 0.1% has been demonstrated by laser-assisted desorption in combination with TOF mass spectrometry on rubrene single crystals [23].

To gain higher resolution, charge carrier mobility or single molecule fluorescence are the most sensitive 'probes' for contaminants which are still present in the host [24, 25]. However, as a drawback, they either do not provide the necessary chemical selectivity or they can be applied only at concentrations far below the level relevant for macroscopic charge transport.

Optical studies of the triplet lifetime offer an alternative approach to analyse the chemical homogeneity of purified materials at a resolution far below 1 ppm [26, 27]. Due to the spin selectivity the lifetime of the triplet excitons depends on their diffusion and possible immobilisation at chemical or physical defects. Thus the lifetime of the triplet–triplet annihilation provides an indirect measure for the concentration of trapping sites.

A correlation between the triplet lifetime and the charge carrier mobilities has been successfully established on various ultra-purified compounds as indicated in Table 25.1. However, this technique can be interpreted only qualitatively because of the insensitivity to the chemical nature of the defects and the unknown probabilities of exciton localisation and recombination.

As a further chemical tracer method enabling ppm-resolution we have employed gas chromatography (GC). In combination with mass spectrometry this allows for material selective quantification of chemical impurities. The GC technique is based on the retention time of a given species on a functionalised column and varies with the respective molecular affinity to the column surfactants. By subsequent cycles of solvation and rinsing it is possible to obtain a depth profile of the chemical impurities contained in the crystalline host [10]. The main requirements imposed by gas chromatography are a sufficient solubility and separability of the analysed molecular species. For example, for tetracene to be solvable to an adequate amount, saturation in toluene is reached at 150 ng/1 μL [31].

25.2.3 Crystal Growth

As a representative example of the class of oligoacenes, tetracene crystals were grown by sublimation under streaming H_2 gas at a flow rate of 50 sccm. The

Table 25.1 Material dependent lifetime of triplet excitons and the highest TOF mobility obtained at low temperature. In parentheses are the respective temperature and crystallographic axis. Except for perylene, all mobilities refer to hole mobilities.

material	triplet lifetime (ms)	highest mobility (cm²/Vs)	ref.
naphthalene	320	400 (10 K, a)	[28]
anthracene	25	54 (30 K, b)	[9]
diphenylanthracene	20	10 (100 K, c′)	[29]
perylene	0.4	2 (150 K, c′)	[30]

growth of the crystalline platelets, displayed in Figure 25.2a, preferentially proceeds along the (001) plane perpendicular to the surface normal. Different carrier media such as H_2 or forming gas yield to similar structural and electronic properties of the crystals. Interference contrast images as shown in Figure 25.2b for a single terrace step at the surface, indicate optically smooth terraces of several hundred micrometer extension.

As discussed in previous work tetracene, although decomposing upon melting, can be grown from the saturated vapour phase in a modified Bridgman setup [31, 32]. Through this approach, crystals of cm^3 volume suite e.g. for neutron scattering can be prepared, as demonstrated in Figure 25.2c. However, the long-term thermal treatment in the presence of remaining chemical impurities as well as the constraints imposed by the glass wall upon cooling the crystals to room temperature produces low quality samples with respect to the mechanical and electronic properties. We therefore will restrict our discussion to sublimation grown samples of high crystalline quality.

For investigations of the structural influence on the electronic behaviour, crystals of the perylene-derivative diindenoperylene were prepared by physical-vapour-transport in a temperature gradient under streaming H_2 carrier gas

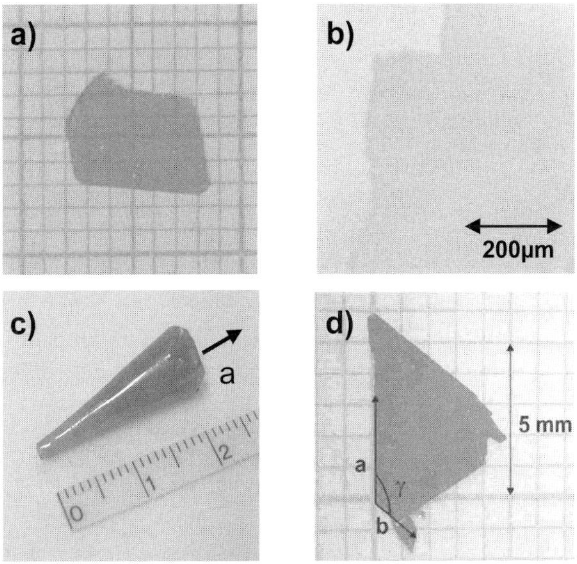

Figure 25.2 Photographs of a sublimation grown tetracene crystal with millimetre extensions along the a- and b-directions (a). Interference contrast imaging of the corresponding (001) surface reveals flat terraces of several hundred micrometers in width (b). Tetracene bulk crystals can be prepared from saturated vapour in a standard Bridgman-setup (c). In this case, the conical volume of 1 cm^3 is aligned with the cone axis along the a-direction. Similar to the oligoacenes, diindenoperylene crystals prepared by sublimation under streaming hydrogen show platelet-like morphologies with an [001] surface normal (d). The smooth terraces of 100 µm are a precondition for reliable FET measurements. (see colour plates p. LXXXIX)

at a flow rate of 100 sccm. Like tetracene, the crystalline DIP platelets are terminated by the (ab)-plane (see Figure 25.2d). Again, the preferred growth is driven by the lower surface energy of the (001) facets compared with that of the (h00) or (0k0) planes of better π-orbital overlap.

After growth, DIP crystals have to pass the structural phase transition upon cooling to room temperature and the occurring stress causes shattering of many platelets. Cooling at very slow rates of ~ 1 K/h results in fragments with lateral dimension up to 5×5 mm^2, as indicated in Figure 25.2d, and thicknesses between 20 and 50 μm. The lateral extension of the terraces amounts to 100 μm, which is sufficient to carry out reliable TOF, FET and SCLC investigations.

We will close this subsection by a short comment on the important aspect of polymorphism in crystalline organic samples. As a result of the competing contributions by the inter-molecular packing, the environment (polarisability of the solvent, molecule–wall interaction, etc.) and defects in the material, many compounds exhibit different crystallographic phases at the same temperature. In the case of the prominent organic semiconductor pentacene, up to four structural morphologies with different lattice spacings were reported [33, 34]. However, as it was demonstrated that many of these phases are either thermodynamically unstable upon annealing, caused e.g. by solvent molecules in the lattice, or result from incomplete phase transitions, e.g. due to strong temperature gradients during processing [35]. Therefore, careful sample preparation in terms of suitable growth conditions and process parameters is inevitable to obtain reliably the thermodynamically stable polymorphs and the associated properties at a given temperature.

25.2.4
Field-Effect-Transistor Fabrication

Handling of the purified organic compounds as well as Field-Effect-Transistor (FET) preparation has to proceed under yellow light to minimise photo-oxidation of the crystal surfaces. The source and drain electrodes are of silver paste painted on top of the DIP and tetracene crystals. The gate insulator of about 0.6–2 μm thickness is made of paracyclophane, which after pyrolysis at about 700 °C polymerises as poly-paraxylylene (PPX, or alternatively, Parylene N) on the crystal surface at room temperature. Beside its optical transparency in the visible, PPX forms electrically closed layers at thickness above 100 nm and enables electrical breakdown fields up to 3–10 MV/cm [36–38]. As an aside, the starting material paracyclophane has also indicated interesting electronic properties with respect to electron transport, presumably due to its molecular motif similar to that of fullerenes [39]. Finally, the PPX polymer gate was contacted by means of a 20 nm thick gold layer thermally evaporated under HV.

Figure 25.3 schematically indicates the individual steps of FET preparation. The described FET architecture enables electrical breakdown fields up to 5 MV/cm corresponding to gate voltages up to 500 V. Advantageously, in the

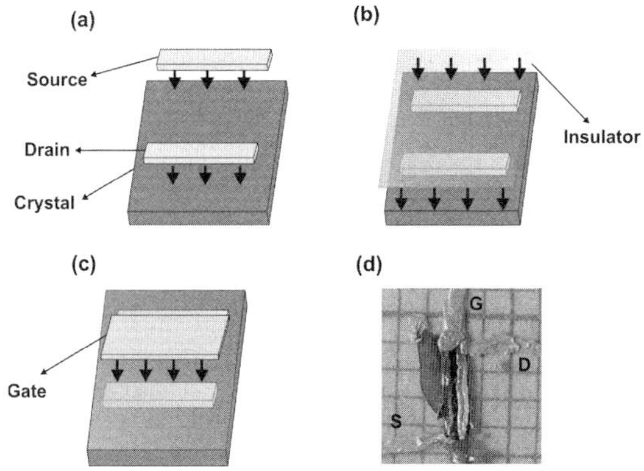

Figure 25.3 Scheme of the three major steps of FET fabrication. After preparing source drain contacts by Ag paste (a) the gate insulator is fabricated by a two-step temperature process of para-cyclophane (b). Finally, by thermal evaporation of Au the top-gate contact of 20 nm thickness is deposited (c). An image of the resulting structure is shown in (d) for a DIP crystal FET. (see colour plates p. LXXXIX)

case of organic compounds with substantial vapour pressure at elevated temperatures, the polymeric PPX gate enhances the thermal stability by its capping properties [40]. However, care has to be taken upon gate fabrication as monomeric units might remain after incomplete polymerisation and might generate electrically active trap states at the crystal surface, i.e. at the transport region. As a final comment, it was evidenced by Morpurgo and coworkers that the dielectric constant of the gate insulator material used might strongly affect the effective mass and the transport of the charge carriers at the interface in analogy to the formation of Froehlich-polarons [41, 42].

25.2.4.1 **Gate Insulator Thickness**

For a quantitative analysis of the accumulated interfacial charge carriers at a given gate voltage, knowledge of the insulator thickness is necessary. One possibility is the analysis by stylus measurements, which often lead to smaller thickness values due to the elastic response of polymer layer on mechanical pressure. An alternative approach of comparable precision, but being non-invasive, is given by optical absorption studies in the mid infra-red (MIR) regime. The detected Fabry–Perot oscillations are directly related to the polymer film thickness, d.

Figure 25.4 shows the MIR absorption spectra at normal incidence of a free standing 7 μm thick PPX film. The peaks occurring in the spectrum are superimposed by an oscillatory modulation of defined period. Whereas the individual peaks are related to vibronic excitations of the polymer, the periodic modulation of the baseline can be attributed to Fabry–Perot oscillations caused by refractive index changes at the PPX interfaces.

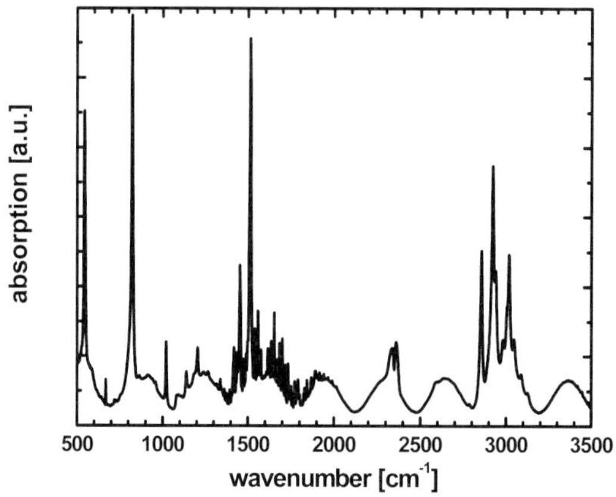

Figure 25.4 MIR absorption spectrum measured on a free standing ~7 μm thick poly-paraxylylene (PPX) film. The sharp peaks originate from vibronic excitations of the polymer and the periodic modulation attributed to Fabry–Perot oscillations.

According to the interference condition for the optical path difference the film thickness, *d*, can be estimated by

$$d = \frac{1}{2n}\left[\frac{1}{\lambda_1} - \frac{1}{\lambda_2}\right], \tag{2}$$

where $n = (\varepsilon_r)^{1/2} = 1.4$ is the refractive index of PPX [36] and λ_i refers to the wavenumber of the *i*-th interference maximum. As a condition for the validity of Eq. (2) we assumed a non-dispersive behaviour of the refractive index, i.e. $n(\lambda_1) \approx n(\lambda_2)$, which holds true as long as the analysis is carried out only for adjacent interference maxima. The adjusted gate thickness covers a range between 3 and 10 μm and can be reliably adjusted by the amount of PPX starting material and the deposition time.

We conclude this section by stressing the fact that due to the contact-gate geometry the electronic performance of single crystal FETs is governed by the carrier transport in the surface near regions. In contrast, injection-free Time-Of-Flight or injection based Space-Charge-Limited-Current measurements are sensitive to the bulk transport or, depending on the contact geometry, sensitive to the bulk and the surface mobility as in the case of SCLC. We do not intend to describe these two experimental techniques in further detail but refer to the literature [9, 43–45].

25.3
Results and Discussion

25.3.1
Tetracene Crystals: Surface *Versus* Bulk Transport

This section will begin with a particular focus on the surface versus the bulk transport in the case of tetracene single crystals. Single crystal FETs with PPX gate insulators were prepared according to the procedure described above and electrically characterised in a temperature range from room temperature (RT) up to 400 K. All samples discussed were electrically characterised at 10^{-7} Torr in the dark to minimise unintended effects through contaminations or by modifications of the Schottky barriers and the transport processes upon exposure to light.

In combination with contact metals such as Ag or Au, tetracene exhibits hole transport only, because of the matching between the metal workfunction and the tetracene HOCO (*highest occupied crystal orbital*). The hole mobility can be independently estimated from the saturation of the output curve

$$\mu_{\text{sat}} \sim \frac{1}{U_G - U_{\text{th}}} \frac{\partial I_{D,\text{sat}}}{\partial U_G}\bigg|_{U_D=\text{const}} \tag{3}$$

and from the linear regime of the transfer curve

$$\mu_{\text{lin}} \sim \frac{1}{U_D} \frac{\partial I_D}{\partial U_G}\bigg|_{U_D=\text{const}} . \tag{4}$$

The indices *D*, *G* and *th* indicate the currents and voltages at the drain, gate and the threshold, respectively. Figure 25.5a and 25.5b shows FET characteristics along the in-plane $[\bar{1}10]$ direction. The high reliability of the measurements is indicated by similar hole mobilities of about 0.78 cm^2/Vs calculated by Eqs. (4) and (5) from the respective curves.

To elucidate the underlying transport mechanisms it is essential to perform mobility studies over a broad temperature range. Figure 25.6 depicts the variation of the hole mobility as a function of temperature for measurements along the $[\bar{1}10]$ direction at the surface (FET) and the [001] axis in the bulk (TOF).

A major similarity is that both μ(T)-curves exhibit a maximum mobility at around 330 K, framed by a plateau-like depenence over a broad temperature range. This behaviour has been previously observed for inorganic semiconductors and has been successfully described by the Hoesterey–Letson (HL) model assuming multiple trapping and release events caused by Boltzman distributed trap levels [46]. According to this model, carriers frequently trapped and thermally released can contribute to the electronic transport and cause an effective mobility smaller than the material dependent intrinsic mobility. For a discrete trap level at depth E_T, which is assumed to be larger than $k_B T$, and with a relative concentration n_T the relative change between the effective and the intrinsic mobility can be approximated by:

$$\ln\left[\frac{\mu_{\text{int}}(T) - \mu_{\text{eff}}(T)}{\mu_{\text{eff}}(T)}\right] = \ln n_T + \frac{E_T}{k_B T} . \tag{4}$$

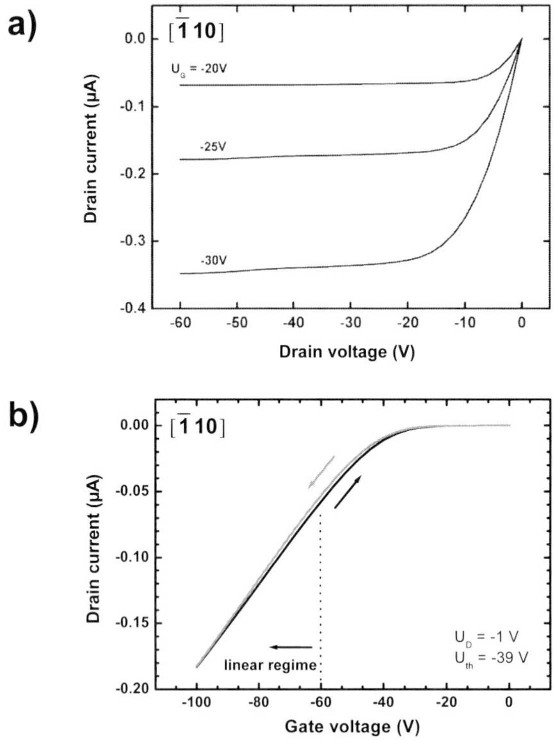

Figure 25.5 Input curve (a) and transfer-curve (b) measured on a tetracene single crystal FET along the $[\bar{1}10]$. According to Eqs. (4) and (5) the hole mobility can be deduced from different parts of the curve and is consistently estimated to $\mu = 0.78\ cm^2/Vs$. The small hysteresis of the transfer curve indicates a minor influence of trapping and charging effects above the threshold voltage. (see colour plates p. XC)

Furthermore, if we expect the electrically active trap states affecting the bulk and the surface transport to be of the same origin, i.e. of the same E_T, the higher mobilities observed in TOF can be generated by two possible effects: a higher intrinsic mobility and/or a lower density of these trap states in the bulk compared with that being active in FET measurements at the surface. Though both scenarios are qualitatively confirmed by the parameters adjusted from the HL-modelling of the experimental data (see caption to Figure 25.6), the exactness of these values should be considered with caution due to the number of degrees of freedom in the model.

For numerous samples investigated by TOF and FET, the hole mobilities in the bulk along the [001] direction, i.e. along the direction of weakest electronic overlap, tend to be slightly higher than that observed along the surface e.g. in the $[\bar{1}10]$ direction (see Figure 25.6 and Table 25.2). In contrast, for other oligoacene materials of superior quality, such as naphthalene or anthracene, the

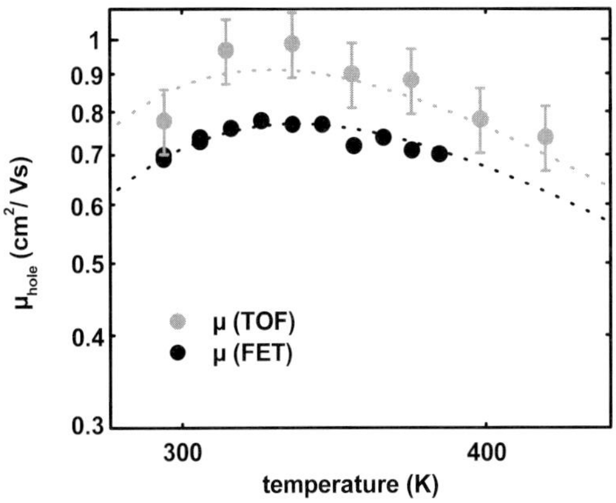

Figure 25.6 Hole mobility of tetracene single crystals measured at the surface (FET) and in the bulk (TOF). Both curves resemble a temperature behaviour that can be ascribed to the multiple-shallow-trapping and release of charge carriers. Parameter of the HL fits for TOF (FET) are $E_T = 200$ meV (190 meV), $n_T = 3 \times 10^{-4}$ (7×10^{-4}) and $\mu_{\text{int}} = 1.5$ cm²/Vs (1.4 cm²/Vs). Remarkably, the surface mobility in the plane of highest π-overlap, is lower than that along the [001] direction of smallest π-orbital overlap. (see colour plates p. XCI)

carrier mobility obeys a band-like temperature behaviour together with a high spatial anisotropy of the principal mobility values [28, 47]. According to literature, the latter can amount to up one order of magnitude with the highest values along directions of the largest π-overlap [3, 5, 6].

In addition, the experimental findings on the enhanced trap density at the tetracene e surface versus that in the bulk might stem from two possible sources of inhomogeneities, structural or chemical. Structural imperfections in van-der Waals bound crystals with a herring-bone packing motif are mainly related to dislocation lines with Burgers vector along the [110](001) or $[1\bar{1}0](001)$ in-plane directions. These dislocations lead to local compression and expansion of the crystal lattice and, thereby to local variations of the polarisation field. Analysing the dislocation line density, e.g. by selective surface etching, reveals a relative concentration of 10^{-7} trap states per molecule [10]. In contrast, for the deduced trap energy of about 200 meV, which agrees reasonably well with that reported for thermally activated transport in molecular stacks [15, 18, 48], the correlated trap density of 10^{-4} is significantly higher and therefore, cannot solely be explained by structural defects.

Alternatively, chemical inhomogeneities and degradation have to be considered. It is well established, that local chemical impurities might act as electrically active trap states in a wide-band gap organic semiconducting host and reduce the effective mobility in the crystal [15, 49]. To judge on possible variation in the impurity concentration at the crystal surface and in the bulk it is necessary to carry out depth dependent chemical analyses, which we accomplished by repeated GC (see Section 25.2.2).

In summary, Figure 25.7 shows the composition of the tetracene sublimation crystal determined at the top-most layers and in the bulk. It immediately becomes obvious that the crystal surface exhibits a much higher impurity concentration than the bulk. This tendency has also been confirmed for other polyaromatic materials, e.g. rubrene (see Figure 25.1c) [23]. Specifically, the stable oxidation product tetracene-quinone, indicated by the arrow in Figure 25.7, which can mainly be generated via three different intermediate products (5-hydroxy-tetracene, 5,12-tetracene-diol, 5,12-dihydro-tetracene) occurs at the surface at a concentration about one order of magnitude higher than in the bulk of the same crystal. This reaction pathway is common to many oligoacenes with more than two rings, according to the aromatic sextet rule proposed by Clar [11, 50].

It is important to stress certain aspects at this point. Firstly, although oxidation products are expected to behave preferentially as electron traps due to their higher electron affinity, the distortion of the surrounding crystal lattice due to steric constraints might also influence the transport of holes [18, 19].

Secondly, normalising and integrating all peak areas attributed to chemical impurities result in a concentration of similar order to the trap density obtained by fitting the measured $\mu(T)$-curves by multiple-shallow-trapping and release processes (dotted lines in Figure 25.6). Care has to be taken as reliable mobility values can only be achieved over a small temperature range limited by desorption of tetracene from the crystal surface and by the blurred TOF and FET signals occurring at lower temperatures.

Thirdly, signatures for the oxidation of polyaromatic molecules can be observed for many materials. However, the oxidised species might not necessarily result in electrically active trap states, but might cause in some cases effective p-doping of the host material. This yields an excess of free charge carriers which can fill the existing trap states in the material. The reduced trapping in combination with the effectively higher density of free charge in the transport channel will result in higher surface (FET) mobilities. Though these effects are still under intensive research one might expect a correlation between extrinsic surface states causing an effective p-doping and the high mobilities of up to 15–20 cm²/Vs observed at the (001) surface of laminated rubrene single crys-

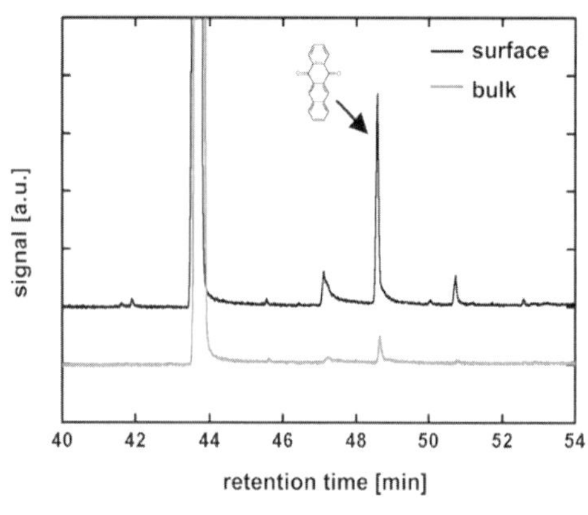

Figure 25.7 Depth dependent profile of the chemical composition of tetracene sublimation crystal. Specifically, the stable oxidation product 5,12-tetracene-quinone, indicated by the arrow, is concentrated to a much larger amount at the surface than in the bulk. (see colour plates p. XCI)

tals [51, 52]. In contrast, the mobility in the bulk though along the [001] axis of high orbital overlap turns out to be smaller by about one order of magnitude [53].

Finally, the uncontrolled degradation of tetracene and many other materials raises the question of strategies to conserve the chemical integrity of the deposited species. The above mentioned sextet rule by Clar provides a criterion on the chemical stability of a given species [50]. Briefly, this rule claims molecules to be less stable the higher the aromaticity, i.e. as more individual aromatic sextets can be formed, e.g. upon oxidation of the molecular skeleton. According to this rule tetracene is expected to show a higher stability against oxidation than pentacene. Furthermore, perylene-derivates or coronene will gain chemical stability through their planar delocalised conjugated system.

The higher stability of π-electron systems delocalised within the molecular plane versus linearly conjugated species can be verified by UV-Vis absorption measurements. For tetracene and diindenoperylene, the time-dependent spectral absorption curves and therewith the dynamics of degradation are shown in Figure 25.8.

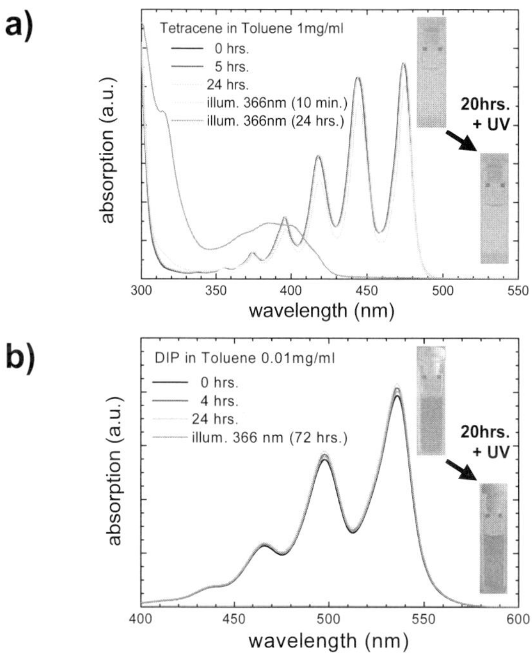

Figure 25.8 UV-Vis absorption spectra of tetracene (a), and diindenoperylene (b), both solved in toluene at saturation concentration. As indicated by the curves as well as by the cuvettes in the insets, there is a strong chemical degradation of tetracene whereas DIP remains intact, even after UV illumination at 366 nm for 3 days. Furthermore, the oxidation product generated upon UV illumination of tetracene can be identified as 5,12-tetracenequinone by its spectral signature (absorption maximum at 390 nm). (see colour plates p. XCII)

Obviously, diindenoperylene shows almost no degradation in solution even if exposed to UV light at 366 nm. In contrast, the intensity of the tetracene absorption peaks decreases with time and the process can be drastically accelerated by UV illumination. The photographs of the cuvettes in the inset show that the yellowish colour of the solution at the beginning almost vanishes after 20 h. The final spectral signature with an absorption maximum around 390 nm (red curve) can be identified with the stable oxidation product 5,12-tetracenequinone.

Summarising this paragraph we can deduce the following statements, applicable to a broad variety of polyaromatic materials:

- for many sublimation purified materials the influence of chemical inhomogeneities on the transport dominates that of structural imperfections
- molecules with planar extension of the conjugated electron systems provide a higher stability than linearly conjugated systems
- the surface mobility of organic single crystals can be strongly affected by chemical inhomogeneities at concentrations different to those in the volume
- the surface (FET) mobility along a given crystallographic direction does not necessarily resemble the intrinsic mobility, though it might show similar characteristics, e.g. an in-plane anisotropy

Again, with respect to the last two points, surface oxidation might result in effective charge-transfer doping even at room temperature and thereby enhancing the density of free carriers and the effective mobility.

25.3.2 Diindenoperylene Crystals: Structural Impact on Transport

In the second part of this section we will discuss the effects of structural ordering on the transport for the case of diindenoperylene (DIP). As we have demonstrated in the former section, DIP has proven to be less susceptible to chemical degradation and, therefore, represents a prototypical organic semiconductor to study the relations between structure and electronic transport.

Through its specific molecular shape – two indeno-groups are attached at opposite sides to a central perylene core – long-range order along the its long-molecular axis is sustained [54]. In the case of thermally evaporated DIP films on SiO_2 or AlOx., this results in single crystalline domains with vertical extensions of more than 150 nm. For example, photovoltaic applications have shown that DIP enables exciton diffusion lengths at least similar to the spatial extension of the crystallites [55].

A further interesting aspect of diindenoperylene is the solid-to-solid phase transition which occurs at around 403 K [56, 57]. Figure 25.9 shows the characteristics of this enantiotropic transition measured by X-ray diffraction in Bragg–Brentano geometry along the [001] direction. The transition comprises a change of the unit cell from a triclinic low-temperature structure with $P\bar{1}$ symmetry (α-phase) to a high-temperature monoclinic $P2_1/a$ symmetry (β-phase) [56]. This epitactic solid-to-solid transition proceeds via formation of crystalline nuclei, which continuously grow during the structural transition. However, the distortion of the local environment generates nucleation sites of different, e.g. strained (intermediate) phases. Upon further heating these phases merge into the thermodynamically stable structural phases, as can be seen in Figure 25.9 at 370 K for the peak at $2\theta = 5.17°$ (grey curve). In addition, these thermodynamically unstable intermediates are expected to also be responsible for the large hysteresis in the phase transition temperature detected for instance by differential scanning calorimetry.

Comparison with the crystallographic structure of DIP thin films deposited on oxide surfaces reveals an agreement with the high temperature bulk phase (β-phase), which indicates that the weakly interacting substrate is able to stabilise the DIP phase of higher energy already at room temperature. This is important for thin film applications, for which the device performance should not be affected by structural changes under the operating conditions.

As the charge carrier transport in organic polycrystalline layers is, to a large extent, dominated by the texture of the films [43], the temperature dependent spatial extension of the crystalline grains is an essential parameter. In the case of the diindenperylene crystals this relation is displayed in Figure 25.10, where the vertical size of the crystallites, estimated by the Scherrer formula, is plotted versus temperature. Completing the phase transition at 400 K, the average domain size of 220 nm almost resembles that of the low-temperature phase. Fur-

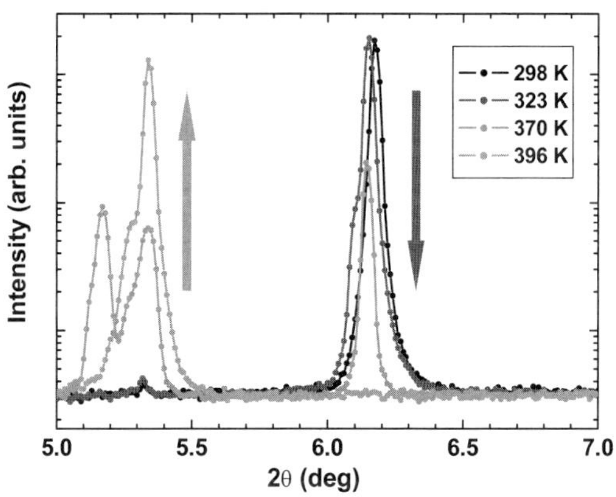

Figure 25.9 X-ray structural characterisation of DIP single crystal along the c′-direction at different temperatures. As indicated by the measurements performed in specular θ-2θ geometry, the structural phase transition is completed at around 400 K. The [001] lattice spacing of 1.68 nm ($2\theta = 5.31°$) in the high temperature phase coincides with that of DIP thin films grown on weakly interacting substrates such as SiO_2 at room temperature. (see colour plates p. XCIII)

thermore, the strained phase occurring at an intermediate temperature regime starts to gradually decrease in grain size towards higher temperatures. For the respective crystal analysed in Figure 25.10, the transition has been defined at 370 K due to the vanishing of the low-temperature phase and the rising of the high-temperature and intermediate phases at this point. Again, lattice distortions accompanied by stress and strain as well as steric hindrances are the main origins of the distribution of transition temperatures and for the co-existence of the low- and the high-T phase in many samples at elevated temperatures.

To analyse the effects of those static and dynamic structural modifications on the electronic properties at the phase transition, the mobility measured by TOF, SCLC and FET was chosen as an indicator of high sensitivity. By this concerted experimental approach the carrier transport can be studied independently for electrons and holes and with special focus on the structure–transport correlation along different crystallographic directions.

At first, by injection-free TOF spectroscopy transients for electrons and holes were observed between 300 K and 420 K, i.e. above the structural phase transition. The two charge carrier types obey a thermally activated transport behaviour with mobility maxima of 0.03 cm²/Vs for holes and 0.2 cm²/Vs for electrons both at 400 K. The detection of well-defined electron transients in a sublimation purified material already indicates its aforementioned chemical stability. For many polyaromatic materials the formation of the respective quinones, e.g. upon (photo-)oxidation, shifts the molecular levels to lower energies because of the higher electronegativity and causes stronger binding of electrons on the molecular skeleton [2]. The LUMO (lowest unoccupied molecular orbital) of those oxidation-products is often positioned below the

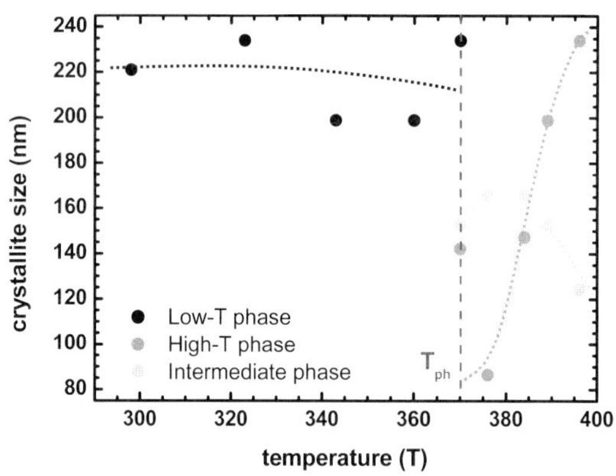

Figure 25.10 Temperature dependent evolution of the crystallite sizes in the respective structural phases of DIP. Above the transition point the previous crystallite size of about 220 nm is almost restored. In the vicinity of the phase transition, strong effects on the carrier transport might be expected from structural inhomogeneities that appear. (see colour plates p. XCIII)

LUCO of the host and acts as an effective electron trap [49]. In combination with the comparable mobility of electrons and holes, the chemical inertness evidences the potential of diindenoperylene for application in e.g. photovoltaics where a balanced charge transport is required.

In the following, we will focus our discussion on the hole transport, as the use of standard contact metals like gold or silver in FET and SCLC studies promotes a p-type semiconducting behaviour due to the matching of the workfunctions ($\Phi_{Au} = 5.4$ eV [58] and $\Phi_{Ag} = 4.5$ eV [59]) with the participating DIP HOCO at 5.8 eV [58].

Similar to FETs, space-charge limited current investigations rely on the injection of charge carriers and on the voltage dependent current. Care has to be taken of the fact that for many organic materials, even in single crystal form, the *so-called* trap-free region of transport where all trap states are filled and no longer hamper the conductivity cannot be reached [44]. In this case, the mobility estimated by the Mott–Gurney law defines a lower limit of the intrinsic mobility:

$$\mu_{int} \geq \mu_{min} = \frac{8}{9\varepsilon_r \varepsilon_0} \frac{I}{A} \frac{d^3}{U^2}. \tag{5}$$

Here, I is the current, flowing at a given voltage U through the contact area A and the sample thickness d. The relative permittivity ε_r is material dependent and for many non-polar aromatics is of the order of 3–4; ε_0 describes the vacuum permittivity.

Furthermore, the formula listed above holds true only for one-dimensional transport in plate capacitor geometry. Across the bulk Eq. (5) cannot be applied to the case of surface transport, where besides the edge effects through the finite contact dimensions, the electronic transport occurs in an infinitesimal slab and, therefore, is of 2D character. In this special geometry not only the is prefactor different with respect to the 1D bulk case but also the current dependence on the contact spacing is quadratic instead of cubic [39, 60].

Figure 25.11 shows a representative SCLC study performed on a DIP single crystal along the c′-direction between room-temperature and the phase transition. Voltage regimes with different current slopes can be easily identified.

Furthermore, sudden jumps occur in the $I(V)$-characteristics and can be attributed to the situation when the metal Fermi level passes a discrete trap level at increasing voltage and fills-up the respective trap states.

Two main effects can be established as functions of temperature. Firstly, the current jumps shift towards lower voltages upon temperature raising (s. horizontal arrow). As the voltage where the current jump occurs is directly proportional to the density of the trap states, this means that by the additional thermal energy at higher temperatures, certain traps become less relevant for the carrier transport. Secondly, referring to an applied voltage of 100 V (dotted line in Figure 25.11), the slope of the $I(V)$-curves at room temperature almost resembles that of a trap-free regime ($n \approx 2$) whereas the slope decreases at temperatures above 378 K (s. oblique arrow).

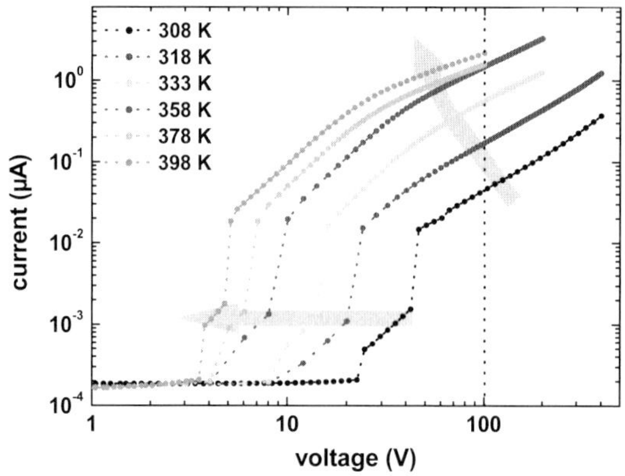

Figure 25.11 SCLC measurements along the c′-direction of a diindenoperylene single crystal at various temperatures. The mobility of the holes has been adjusted according to Eq. (5) at around 100 V where slope of the $I(V)$-curve approximates 2 at room temperature. The estimated values present a lower limit with respect to the intrinsic hole mobility in DIP. A shift of the trap-related current jump towards lower voltages and a change in the $I(V)$-slopes can be detected at elevated temperature (indicated by the arrows). (see colour plates p. XCIV)

According to Eq. (5) this means that the hole mobility approaches an almost plateau-like behaviour at this temperature which roughly coincides with the phase transition previously obtained by X-ray diffraction (see Figure 25.10).

Summarising the temperature dependent transport data obtained by TOF, SCLC and FET reveals the mobility behaviour displayed in Figure 25.12. In addition the temperatures $T_{ph,trans.}$ and $T_{ph,X\text{-}ray}$ which have been assigned to the phase transition from the electronic and the structural measurements, are indicated.

Initially, it becomes obvious from Figure 25.12 that starting from room-temperature all mobility curves can be referred to a thermally activated hole transport. From an intrinsic, material dependent point of view this can be understood by the increasing thermal motion of molecules, which leads to an enhanced contribution of the phonon-assisted transport across the lattice. With respect to extrinsic properties, the additional thermal energy also renders static inhomogeneities, such as impurities or grain boundaries, to become less relevant to the charge carrier movement. Over the entire temperature range the mobilities estimated by injection-free TOF are higher than those deduced by SCLC or FET, the latter even probing the transport within the (ab)-plane with highest orbital overlap. We explain this striking difference in mobility by the fact that contact preparation, which is usually performed via thermal evaporation of metal on top of the organic semiconductor (see Section 25.2.4), causes formation of inter-gap trap states in proximity to the injecting hetero-interface. These trap states reduce the average density of free charge carriers at a given voltage. Moreover, this effect is accompanied by Coulomb-scattering of charges moving in the transport level, e.g. the HOCO in the case of hole transport, at charges immobilised in the trap states.

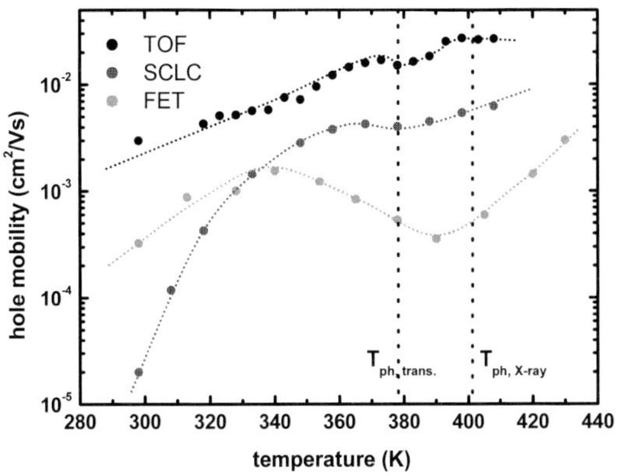

Figure 25.12 TOF, SCLC and FET measurements on DIP single crystals as a function of temperature. The dotted curves present a guide-to-the-eyes. The electronic characterisation by TOF and SCLC studies were performed along the c′-direction whereas hole transport in FETs occurs along the (ab)-plane. In addition, the phase transition temperatures deduced from transport data, $T_{\text{ph, trans}}$, and from X-ray structural analysis, $T_{\text{ph, X-ray}}$, are indicated by the dotted vertical lines. (see colour plates p. XCIV)

The existence of trap states that hamper the charge injection at the contact boundaries is further evidenced by the rapid increase of current with slight temperature changes. As shown in Figure 25.12, for a temperature increase of 40 K the SCLC current rises by about two orders of magnitude. This increase correlates directly with the decrease of electrically active trap states at higher temperature as discussed in the context of the current jump in Figure 25.11.

Comparing the results from different techniques, such as SCLC or FET, each affected similarly by the contact effects, the temperature dependent mobility renders the expected spatial anisotropy of the transport in and normal to the (ab)-plane. The SCLC mobility at room temperature along the c′-direction is found to be one order of magnitude smaller than the FET mobility along the (ab)-plane.

Finally, all three electronic characterisation methods indicate deviations from the expected Hoestery–Letson behaviour at temperatures between 360 K and 400 K where the structural phase transition was identified by X-ray diffraction. The deviation of the $\mu(T)$-characteristic from the expected trapping-and-release behaviour starts at a lower temperature for the in-plane FET transport than for the SCLC or TOF transport along the surface normal. Moreover, the temperature interval where the in-plane transport is affected by structural changes amounts to at least 60 K and appears to be much broader than that for the out-of-plane transport.

For a detailed analysis of the investigated hetero-system metal/organic/gate-dielectric the complexity of the temperature dependent electronic processes has to be considered, i.e. the injection across the Schottky-barriers, trapping, release and conduction of charges. The dynamics and the temperature of the phase transition might be strongly affected by pinning of molecules at the metal/contact interfaces, in contrast to the X-ray experiments on free standing DIP crystals. None the less, the higher susceptibility of the FET mobility to structural changes is evidenced by the lowest on-set temperature of the $\mu(T)$-deviation, which can be qualitatively correlated with the thermal distortion of the electronic orbital overlap in the (ab)-plane.

To clarify this in more detail, the fundamental vibronic excitations of the DIP molecules were calculated from X-ray diffraction analyses as a function of temperature. Affecting the in-plane distortion of the orbital overlap, Figure 25.13 shows the temperature evolution of the librations around the principal molecular axes.

All six fundamental librations of the DIP α-phase increase monotonously in amplitude at elevated temperatures. The molecules constituting the basis of the α-phase are labelled by (1) and (2) and merging into the one-molecule basis of the β-phase above the phase transition. Comparing the individual changes, the largest variation in temperature occurs for the libration L_1 around the long molecular axis. In single crystals this molecular axis is slightly inclined from the c′-direction and the L_1 libration should yield large distortion of the π-orbital overlap in the (ab)-plane already far below the structural transition [3, 61]. In agreement, the deviation of the TOF and SCLC mobilities from the expected HL-behaviour along the [001] direction occur at similar temperatures of 370 K over a similar range of 20 K.

To enable a microscopic analysis of the spatial anisotropy of the hole transport in diindenoperylen and to correlate the macroscopic observations quantitatively with the molecular orbital overlap, the purity of the samples as well as

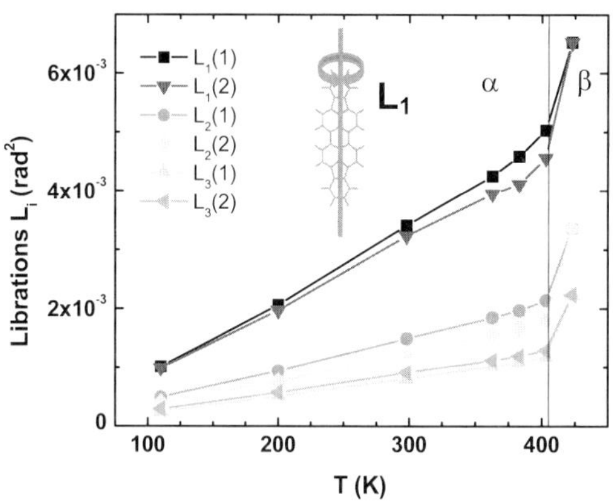

Figure 25.13 Temperature dependence of the fundamental libration modes determined by X-ray diffraction on DIP single crystals. The libration mode L_1 around the long molecular axis (inset) experiences the largest variation in amplitude with increasing temperature and, therefore, might significantly distort the charge carrier movement in the DIP (001) plane. The two molecules in the unit cell of the α-phase are indicated by (1) and (2); the basis of the β-phase is generated by just one molecule [56]. (see colour plates p. XCV)

models allowing for theoretical predictions on such large molecules have to be further improved.

25.4 Conclusion

We have demonstrated the impact of chemical and structural effects on the charge carrier transport for selected examples of polyaromatic molecules. If purification can only be achieved by sublimation, as is the case for several organic compounds, the electronic transport is governed by chemical inhomogeneities rather than by structural defects. As discussed in the first part, for tetracene as a representative of oligoacenes, the spatial anisotropy of the mobility originates to a large extent from the substantially higher density of contaminants at the surface, e.g. caused by (photo-)oxidation. This does not necessarily result in degradation of the electronic performance but can also be beneficial if charge transfer between the contamination and the host occurs.

In the second part we have illustrated, for the prototypical organic semiconductor diindenoperylene, how structural changes, in this case the enantiotropic phase transition, can influence the temperature depend transport. In this case, the mobility measured by different techniques acts as a sensitive probe for the electronic properties at the surface and in the bulk. This approach enables detailed insights into the underlying mechanisms relevant for the conduction at various temperature regimes.

As an overview Table 25.2 summarises the mobility data measured at the surface and in the bulk for selected, technically relevant organic crystalline materials. It is important to note that the surface mobility does not necessarily coincide with the mobility along the same crystallographic direction in the crystal bulk.

The independent estimation of the surface and the volume transport characteristics enables a thorough picture of the contributing processes relevant for modelling the macroscopic transport. On this basis, organic single crystals al-

Table 25.2 Representative hole mobilities reported for the bulk (TOF) and the surface (FET) of sublimation grown crystals. The FET data were obtained on samples with on-top deposited polymeric gate insulators. Bulk transport refers to the [001] direction and surface transport to the (001) plane. Except for rubrene, all materials offer comparable or lower FET mobilities at the surface.

material	bulk mobility (cm^2/Vs)	surface mobility (cm^2/Vs)
tetracene	~1 [62, 63]	0.78 [63]
pentacene	~1 [64]	0.3 [48]
rubrene	0.2 [53]	10 [41]
perylene	0.2 [30]	0.1 [30]
diindenoperylene	3×10^{-3} [57]	3×10^{-4} [65]

though a subject of research in solid state physics for more than a hundred years [66–68] remain the benchmark system to access the ultimate intrinsic performance of a given polyaromatic material and to understand and optimise the electronic transport also in organic thin films.

Acknowledgements

The authors acknowledge the financial support by the Deutsche Forschungsgemeinschaft (Project No. PF385/2 and KA427/8). We further appreciated N. Karl, S. Hirschmann, C. Ender-Vögele and C. Herb for intense discussion on the subject as well as for their assistance on purification, crystal growth and chemical trace analysis. T. Siegrist is gratefully acknowledged for his support on the X-ray diffraction of DIP single crystals.

References

1. N. Karl, in Group III, Vol. 17 Semiconductors (Springer-Verlag, Berlin, 1985), pp. 106.
2. E. A. Silinsh, Organic Molecular Crystals: Their Electronic States (Springer, Berlin, 1980).
3. M. Pope and C. E. Swenberg, Electronic Processes in Organic Crystals and Polymers (Oxford University Press, New York, 1999).
4. V. Coropceanu, J. Cornil, D. A. da Silva, Y. Olivier, R. Silbey, and J. L. Bredas, Chem. Rev. **107**, 926 (2007).
5. K. Hannewald and P. A. Bobbert, Phys. Rev. B **69**, 075212 (2004).
6. K. Hannewald and P. A. Bobbert, Appl. Phys. Lett. **85**, 1535 (2004).
7. J. L. Bredas, D. Beljonne, V. Coropceanu, and J. Cornil, Chem. Rev. **104**, 4971 (2004).
8. K. Hannewald, V. M. Stojanovic, and P. A. Bobbert, J. Phys.: Condens. Matter **16**, 2023 (2004).
9. N. Karl, in Organic Electronic Materials (Springer Verlag, Berlin, 2001), pp. 283.
10. J. Pflaum, J. Niemax, and A. K. Tripathi, Chem. Phys. **325**, 152 (2006).
11. E. Clar, Polycyclic Hydrocarbons, Organic Molecular Crystals: Their Electronic States (Academic Press, New York, 1964).
12. P. Peumans, S. Uchida, and S. R. Forrest, Nature **425**, 158 (2003).
13. S. R. Forrest, Chem Rev **97**, 1793 (1997).
14. N. Karl, in Festkörperprobleme/ Adv. Solid State Phys. XIV (Pergamon/Vieweg, Braunschweig, 1974), pp. 261.
15. K. H. Probst and N. Karl, Phys. Status Solidi A **27**, 499 (1975).
16. W. Helfrich and F. R. Lipsett, J. Chem. Phys. **43**, 4368 (1965).
17. W. Helfrich, Physics and Chemistry of the Organic Solid State (Interscience Publishers, New York, 1967).
18. E. A. Silinsh, Phys. Status Solidi A **3**, 817 (1970).
19. H. Baessler, G. Herrmann, N. Riehl, and G. Vaubel, J. Chem. Phys. **30**, 1579 (1967).
20. W. G. Pfann, Zone Melting (J. Wiley, New York, 1958).
21. R. C. Weast, CRC Handbook of Chemisty and Physics (CRC Press, 1981).
22. H. Inokuchi and H. Akamatu, Solid State Phys. **12**, 93 (1961).
23. D. Kafer and G. Witte, Phys. Chem. Chem. Phys. **7**, 2850 (2005).

24. W. E. Moerner and T. Basche, Angewandte Chemie-International Edition in English **32**, 457 (1993).
25. W. E. Moerner, Prog. Cryst. Growth Ch **33**, 11 (1996).
26. N. Karl, J Cryst Growth **51**, 509 (1981).
27. W. Warta, R. Stehle, and N. Karl, Appl. Phys a-Mater **36**, 163 (1985).
28. W. Warta, N. Karl, Phys. Rev. B **32**, 1172 (1985).
29. A. K. Tripathi, M. Heinrich, T. Siegrist, and J. Pflaum, Adv. Mater. **19**, 2097 (2007).
30. M. Kotani, K. Kakinuma, M. Yoshimura, K. Ishii, S. Yamazaki, T. Kobori, H. Okuyama, H. Kobayashi, and H. Tada, Chem. Phys. **325**, 160 (2006).
31. J. Niemax, A. K. Tripathi, and J. Pflaum, Appl. Phys. Lett. **86**, 122105 (2005).
32. J. Niemax and J. Pflaum, Appl. Phys. Lett. **87**, 241921 (2005).
33. C. C. Mattheus, G. A. de Wijs, R. A. de Groot, and T. T. M. Palstra, J. Am. Chem. Soc. **125**, 6323 (2003).
34. C. C. Mattheus, A. B. Dros, J. Baas, G. T. Oostergetel, A. Meetsma, J. L. de Boer, and T. T. M. Palstra, Synthetic Met. **138**, 475 (2003).
35. T. Siegrist, C. Besnard, S. Haas, M. Schiltz, P. Pattison, D. Chernyshov, B. Batlogg, and C. Kloc, Adv. Mater. **19**, 2079 (2007).
36. W. F. Beach, C. Lee, D. R. Basset, T. M. Austin, and O. Olson, Xylylene Polymers, Encyclopedia of Poymer Science and Engineering (John Wiley and Sons Inc., 1989).
37. F. E. Cariou, D. J. Valley, and W. E. Loeb, Ieee Tr Parts Mater **1**, S54 (1965).
38. R. W. I. de Boer, M. E. Gershenson, A. F. Morpurgo, and V. Podzorov, Phys. Status Solidi A **201**, 1302 (2004).
39. W. P. Hu, B. Gompf, J. Pflaum, D. Schweitzer, and M. Dressel, Appl. Phys. Lett. **84**, 4720 (2004).
40. S. Meyer, S. Sellner, F. Schreiber, H. Dosch, G. Ulbricht, M. Fischer, B. Gompf, and J. Pflaum, Mat. Res. Soc. Symp. Proc. **965E**, S06 (2007).
41. A. F. Stassen, R. W. I. de Boer, N. N. Iosad, and A. F. Morpurgo, Appl. Phys. Lett. **85**, 3899 (2004).
42. I. N. Hulea, S. Fratini, H. Xie, C. L. Mulder, N. N. Iossad, G. Rastelli, S. Ciuchi, and A. F. Morpurgo, Nat. Mater. **5**, 982 (2006).
43. N. Karl, Synthetic Met. **133**, 649 (2003).
44. K. C. Kao and W. Hwang, Electrical Transport in Solids (Pergamon Press, Oxford, 1981).
45. R. W. I. de Boer, M. Jochemsen, T. M. Klapwijk, A. F. Morpurgo, J. Niemax, A. K. Tripathi, and J. Pflaum, J. Appl. Phys. **95**, 1196 (2004).
46. D. C. Hoesterey and G. M. Letson, Journal of the Physical Chemistry of Solids **24**, 1609 (1963).
47. N. Karl and J. Marktanner, Mol. Cryst. Liq. Cryst. **355**, 149 (2001).
48. V. Y. Butko, X. Chi, D. V. Lang, and A. P. Ramirez, Appl. Phys. Lett. **83**, 4773 (2003).
49. N. Karl, Advances in Solid State Physics **14**, 261 (1974).
50. E. Clar, The Aromatic Sextet (Academic Press, London, 1972).
51. V. C. Sundar, J. Zaumseil, V. Podzorov, E. Menard, R. L. Willett, T. Someya, M. E. Gershenson, and J. A. Rogers, Science **303**, 1644 (2004).
52. V. Podzorov, E. Menard, A. Borissov, V. Kiryukhin, J. A. Rogers, and M. E. Gershenson, Phys. Rev. Lett. **93**, 086602 (2004).
53. W. G. Williams, Discuss Faraday Soc. **51**, 61 (1971).
54. A. C. Dürr, F. Schreiber, M. Münch, N. Karl, B. Krause, V. Kruppa, and H. Dosch, Appl. Phys. Lett. **81**, 2276 (2002).
55. D. Kurrle and J. Pflaum, Appl. Phys. Lett. **92**, 133306 (2008).

56. M. A. Heinrich, J. Pflaum, A. K. Tripathi, W. Frey, M. L. Steigerwald, and T. Siegrist, J. Phys. Chem. C **111**, 18878 (2007).

57. A. K. Tripathi and J. Pflaum, Appl. Phys. Lett. **89**, 125416 (2006).

58. A. C. Dürr, N. Koch, M. Kelsch, A. Ruhm, J. Ghijsen, R. L. Johnson, J. J. Pireaux, J. Schwartz, F. Schreiber, H. Dosch, and A. Kahn, Phys. Rev. B **68**, 115428 (2003).

59. N. Koch, I. Salzmann, R. L. Johnson, J. Pflaum, R. Friedlein, and J. P. Rabe, Org. Electron **7**, 537 (2006).

60. J. A. Geust, Phys. Status Solidi **15**, 107 (1966).

61. A. Troisi, G. Orlandi, and J. E. Anthony, Chem. Mater. **17**, 5024 (2005).

62. P. Berrehar, P. Delannoy, and M. Schott, Phys. Status Solidi B **77**, K119 (1976).

63. J. Pflaum, J. Niemax, and A. K. Tripathi, Mat. Res. Soc. Symp. Proc. **871E**, I7.2 (2005).

64. M. Muench, unpublished data (2001).

65. A. K. Tripathi, unpublished data (2008).

66. A. Pochettino, Acad. Lincei Rend. **XV**, 355 (1906).

67. A. Pochettino, Acad. Lincei Rend. **XV**, 171 (1906).

68. J. Koenigsberger and K. Schilling, Ann. Physik **32**, 179 (1910).

Section V
Novel Devices

Organic Electronics. Structural and Electronic Properties of OFETs. Edited by Christof Wöll

ISBN: 978-3-527-40810-8

26
Carbon Nanotube Transistors – Chemical Functionalisation and Device Characterisation

Kannan Balasubramanian, Eduardo J. H. Lee, Ralf Thomas Weitz, Marko Burghard, and Klaus Kern

26.1
Introduction

Carbon nanotubes (CNTs) are emerging as promising components of molecular-scale electronic devices [1], due to their excellent electrical properties including large current densities of $\geq 10^9$ A/cm^2 [2] and field-effect mobilities that considerably exceed the intrinsic mobility of silicon [3]. Basic components of electronic circuits such as field-effect transistors (FETs) [4], logic gates [5] and memory elements [6] have been demonstrated based on CNTs. Despite these impressive advancements, a number of factors have been hampering the widespread application of CNTs in the semiconductor industry. The major task of the present chapter is to demonstrate that chemical functionalisation represents a highly versatile strategy to overcome several of these hurdles. When applied to appropriate device architectures, the functionalisation approach yields CNT-FETs whose performance successfully rivals that of state-of-the-art metal oxide semiconductor FETs (MOSFETs).

The chapter is organised as follows. After the introduction, a brief overview of the fundamentals of CNTs and CNT-based FETs will be given. The next section describes a range of chemical functionalisation schemes that have been devised for the performance enhancement of CNT-FETs. The subsequent section is devoted to the characterisation of as-prepared and functionalised CNT-FETs through electrical transport measurements and scanning photocurrent microscopy. In this context, the relevant device parameters of the FETs such as saturation behaviour, field-effect mobility, transconductance and sub-threshold slope will be analysed and compared. The chapter concludes with future perspectives for the fabrication of CNT-based FETs.

Organic Electronics. Structural and Electronic Properties of OFETs. Edited by Christof Wöll

ISBN: 978-3-527-40810-8

26.2 Carbon Nanotubes – Fundamentals

26.2.1 Physical and Electronic Structure

Carbon nanotubes can be visualised as concentric cylinders of graphene sheets, which are single atomic layers of graphite. They are classified as single-wall (SWCNTs) comprised of just a single shell, and multi-wall (MWCNTs) comprised of at least two shells of the graphene sheet. The scope of this chapter is restricted to SWCNTs. Based on its orientation with respect to the hexagonal graphene lattice, the structure of a nanotube can be completely specified through the chiral vector $\boldsymbol{C}_{\mathrm{h}}$ given by $\boldsymbol{C}_{\mathrm{h}} = n\boldsymbol{a}_1 + m\boldsymbol{a}_2$ (n, m are integers), denoted as (n, m), where $\boldsymbol{a}_1$ and $\boldsymbol{a}_2$ are the real space unit vectors of the hexagonal lattice of graphene (see Figure 26.1).

SWCNTs can be produced by a variety of methods, namely arc-discharge [7], laser ablation [8], high pressure pyrolysis of carbon monoxide (HiPco [9]) and chemical vapour deposition [10, 11]. All of these methods combine a carbon source such as graphite, carbon monoxide, ethanol or methane with a catalyst in the form of metal particles [12]. The commonly investigated HiPco SWCNTs exhibit a relatively broad diameter distribution of 0.7–1.1 nm [9]. By controlling the size distribution of the catalyst particles, it is possible to restrict the diameter range of the synthesised nanotubes [13]. It is an intriguing property of CNTs that their electronic structure varies drastically even within a small diameter range. In a first approximation, it can be derived from that of graphene by applying the zone-folding approach [14]. Figure 26.2 displays the calculated electronic structure of a (9,0) and a (10,0) tube. Although the diameters of the two tubes differ by just one Ångstrom, the (9,0) tube is metallic

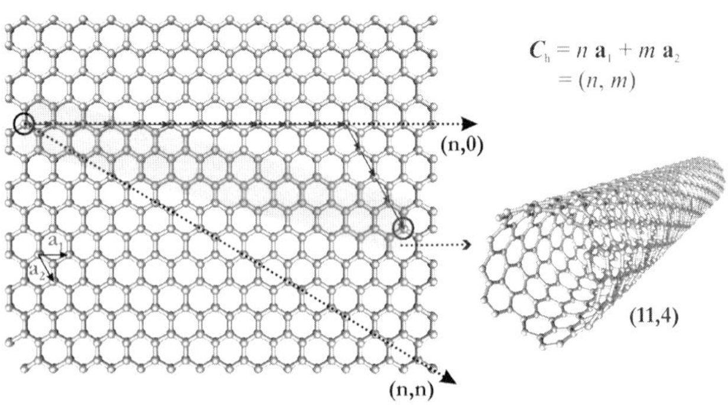

Figure 26.1 Physical structure of a carbon nanotube starting from a graphene sheet, whose unit vectors $\boldsymbol{a}_1$ and $\boldsymbol{a}_2$ are shown. Every nanotube is uniquely determined by its chiral vector $\boldsymbol{C}_{\mathrm{h}} = (n,m)$. The (11,4) nanotube is obtained by rolling the graphene sheet along the red shaded region.

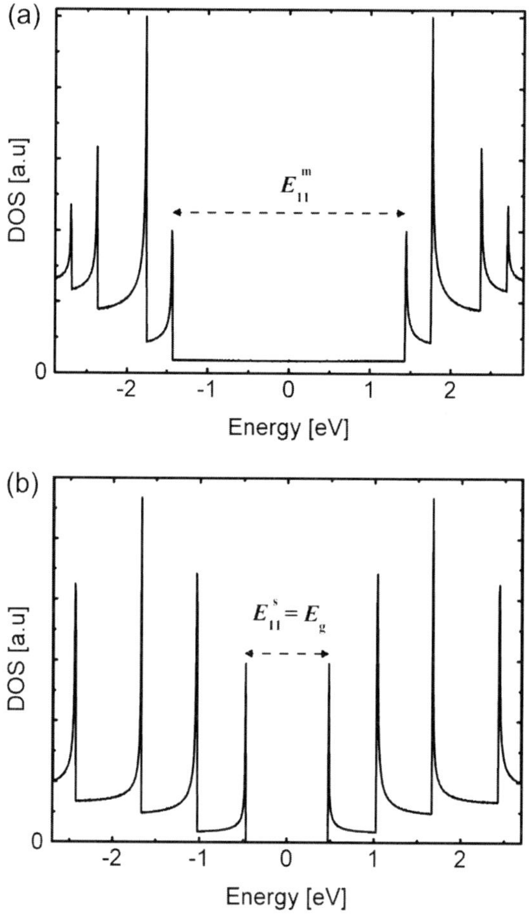

Figure 26.2 Calculated electronic density of states (DOS) of (a) the (9,0) and (b) the (10,0) tube as obtained from tight-binding calculations with the overlap energy $\gamma_0 = -2.7$ eV. While the (9,0) having a finite density of states at the Fermi level is metallic, the (10,0) tube is semiconducting.

while the (10,0) tube is semiconducting. As a general rule, (n, m) tubes with $(n - m)$ being an integer multiple of 3 are metallic (m-SWCNTs), while the remaining tubes are semiconducting (s-SWCNTs). The band-gap of the s-SWCNTs can be approximated by the relation $E_g = 0.8$ eV/d, where d is the diameter of the nanotube in nm [15].

26.2.2 Field-Effect Transistors Based on Single SWCNTs

The prediction that SWCNTs can be semiconducting motivated the fabrication of FETs containing individual SWCNTs as conducting channel. The first devices [4, 16] as well as most of the SWCNT-FETs realised so far are fabricated in a back-gate configuration. In such devices, a single semiconducting SWCNT is contacted by metal electrodes defined by e-beam lithography (see Figure 26.3), with the standard substrate being silicon covered by a thermally grown oxide of thickness in the range of 100 nm to 1 μm that serves as the gate insulator. Unlike a conventional MOSFET, the SWCNT-FET functions like a Schottky-barrier FET (SBFET) [17]. Moreover, compared with the former type of device, it is more difficult to implement ohmic contacts in a SWCNT-FET, since most electrode metals lead to Schottky-barrier formation at the contacts to the nanotube, due to mismatch in the work functions of the two materials and the formation of metal–carbon bonds [18]. As depicted in Figure 26.4(a), this barrier is pinned such that a variation in the gate potential merely shifts the nanotube sub-bands [19, 20]. Carrier injection thus takes place through thermionic emission and tunnelling across the Schottky barriers, whose width depends on the applied gate voltage [21]. Figure 26.4(b) compares the dependence of conductance for a contacted m- and s-SWCNT as a function of back-gate voltage. While the conductance of the m-SWCNT remains fairly constant over the entire gate voltage range, the s-SWCNT conductance can be modulated over more than three orders of magnitude.

The first SWCNT-FETs exhibited high resistances and poor switching characteristics [4, 16]. The height of the Schottky barriers, which limits the ON conductance of these devices, depends on the type of metal used for contacting the tubes. Improved device performance has been achieved with two different contact metals. In the first case, Ti is used as the metal and the device is annealed under argon at a temperature of 800 °C [22], whereupon TiC is formed at the contacts which facilitates carrier injection (both electrons and holes) leading to high ON currents. Alternatively, the use of Pd as contact metal directly yields devices displaying near-to-ballistic ON conductance [23]. The reason why especially Pd makes particularly good contacts with s-SWCNTs is not yet fully understood, although it appears that the extraordinary wetting properties of Pd play an important role [23]. However, it is important to keep in mind that even with Pd contacts, due to the large variation in the band-gap

Figure 26.3 An atomic force microscope (AFM) amplitude image of a field-effect transistor comprising of a single semiconducting carbon nanotube contacted with electrodes separated by ~ 1.5 μm.

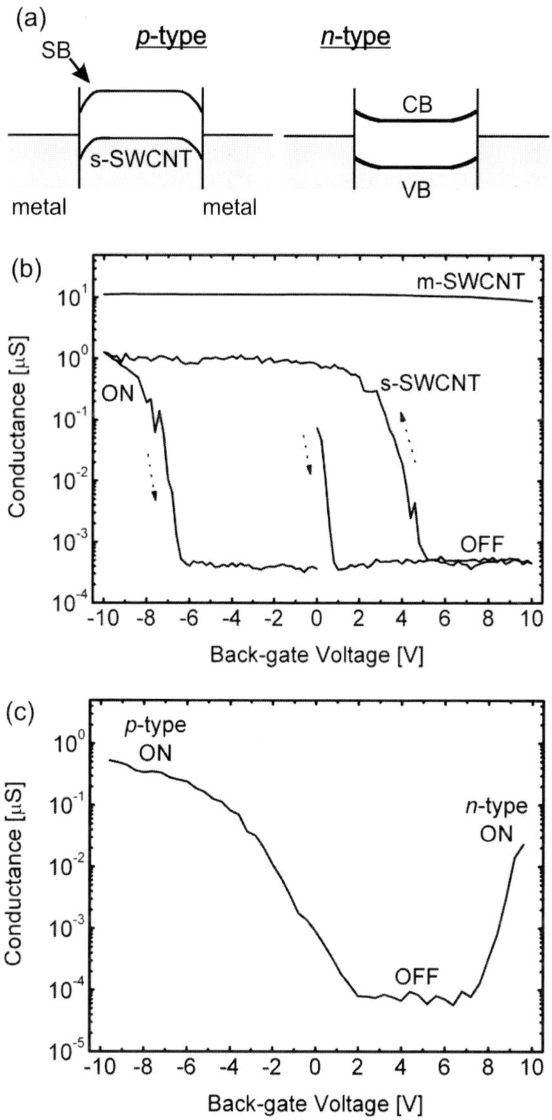

Figure 26.4 (a) Simplified view of the electronic band structure of a field-effect transistor made of a single semiconducting nanotube (s-SWCNT) as the channel. At negative back-gate voltages, the bands are bent upwards and injection of holes is favoured, while at positive back-gate voltages, electron injection is favoured. (b) Variation of conductance of an m-SWCNT and an s-SWCNT as a function of the gate voltage. While the m-SWCNT conductance remains unaltered over the entire gate voltage range, the conductance of the s-SWCNT can be modulated by at least three orders of magnitude. (c) Back-gate dependence of conductance for a large diameter (2 nm) s-SWCNT showing ambipolar behaviour. Drain–source bias (V_{ds}) for the m-SWCNT and s-SWCNTs is 1 mV and 100 mV, respectively.

of the semiconducting tubes, the fabricated devices exhibit a broad distribution of ON conductances (up to an order of magnitude).

Unless subjected to further treatment, SWCNT-FETs in general show p-type characteristics (Figure 26.4(b)), which has been attributed to oxygen adsorption causing p-type doping of the nanotube and/or a change in the work function of the metal contact [24, 25]. It is worth noting that for larger diameter s-SWCNTs, as a consequence of their small band-gap, the barrier heights are greatly reduced for both electrons and holes [26]. The corresponding FETs exhibit ambipolar characteristics, as exemplified in Figure 26.4(c) for an FET comprising a nanotube with a diameter of ~2 nm. Ambipolar SWCNT-FETs are promising for the implementation of novel device architectures such as XOR gates with a minimal device count of just one [27]. Figure 4(b) furthermore reveals that the transfer characteristics measured under ambient conditions are usually imprinted with a pronounced hysteresis. The origin of this hysteresis has been ascribed to charge traps in the vicinity of the nanotube, which may be related to surface-bound water molecules on the SiO_2 or water molecules trapped inside the oxide layer [28].

26.2.3 CNT-FETs Based on Electrochemical Field-Effect

An alternative design for CNT-FETs involves the use of a liquid electrolyte as gating medium. While this approach does not appear promising for semiconductor electronics, it offers a number of other advantages including a highly efficient gate coupling (see Section 26.2.4), as well as the possibility of fabricating FET-based nanoscale sensors in liquids [29]. An electric field-effect inside an aqueous electrolyte was first observed using a MWCNT [30], and later also for SWCNTs [31]. In the latter experiments, devices with individual s-SWCNTs were immersed in an aqueous electrolyte solution, and a standard reference electrode (e.g., Ag/AgCl) used as the gate electrode. Due to the ionic conductivity of the medium, application of a gate voltage leads to the formation of an electrochemical double layer (Helmholtz layer) at the nanotube/electrolyte interface [32]. Variation of the gate voltage leads to charging and discharging of this layer, which shifts the Fermi level of the contacted nanotube and thereby modulates its conductance. We have observed that even in the absence of an electrolyte gate switching can be achieved, as demonstrated by Figure 26.5 for a single s-SWCNT immersed in distilled water. It furthermore turned out that the key factor determining whether a solvent works as an effective gating medium is a certain polarisability, as is the case for water, ethanol, *N*-methyl formamide and *N,N*-dimethylformamide. Alternative gating media for the fabrication of electrochemical SWCNT-FETs are solid polymer electrolytes (SPE) [33, 34] or ionic liquids [35].

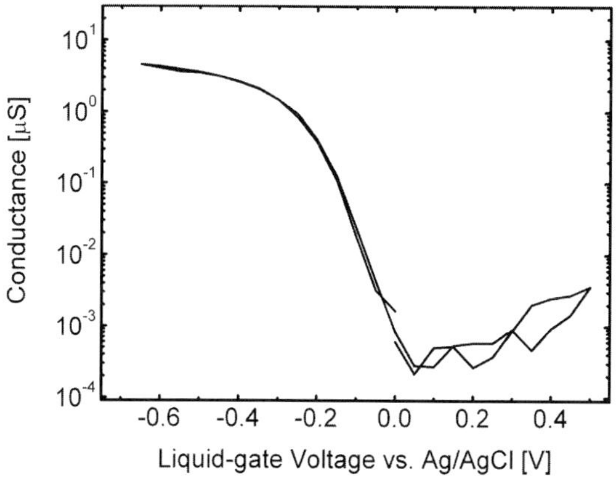

Figure 26.5 Variation of conductance of an individual s-SWCNT as a function of the liquid gate voltage ($V_{ds} = \pm 10$ mV). A water droplet placed on the contacted tube serves as the gating medium with a thin Ag/AgCl wire functioning as the gate electrode. Hysteresis-free field-effect behaviour is clearly observed in a very small gate voltage range. The water droplet (resistivity: 18.2 MΩ cm) was protected with a thin layer of non-volatile solvent (squalane) to avoid evaporation and to maintain its resistivity.

26.2.4
Role of Capacitances

The strength of the gating effect in CNT-FETs is governed by the capacitance occurring between the gate electrode and the nanotube channel [31]. As a basis for the discussions presented in the next sections, we first describe capacitance models for the two gating configurations described above. In the case of Schottky barrier CNT-FETs based on dielectric gate insulators, the gate electrode has to be positioned above or below the source and drain contacts, in order to be able to tune the transmission through the barriers. The gating effect here is thus determined by the dielectric capacitance of the insulating layer between the nanotube and the gate electrode, as shown in Figure 26.6(a). This capacitance can be best modelled by assuming cylindrical capacitor geometry [36]. The capacitance per unit tube length is then given by $c_{ox} = 2\pi\varepsilon_r\varepsilon_0/\ln(2(t_{ox} + r)/r)$, where ε_r is the dielectric constant of the oxide layer (~4 for SiO_2), ε_0 is the vacuum permittivity, t_{ox} is the thickness of the oxide layer, and r is the radius of the nanotube. Assuming an oxide thickness of 100 nm and a nanotube radius of 1 nm, a value of ~37 pF/m is obtained for c_{ox}. Importantly, due to the lower density of electronic states in the quasi-1D nanotubes in comparison with bulk semiconductors, the so-called quantum capacitance has to be taken into account as a second contribution in series with the geometric oxide capacitance [31]. The quantum capacitance per unit tube length can be calculated as $c_q = e^2 g(E)$, where $g(E)$ is the electronic density of states which can be approximated to 400 pF/m [31]. The lower of the two capacitances dominates

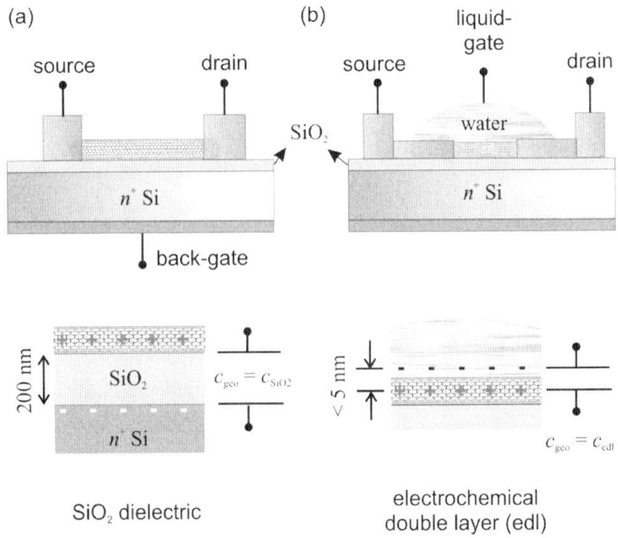

Figure 26.6 Model of geometric gate capacitances for (a) back-gated and (b) electrochemically gated SWCNT-FETs. In the case of back-gating, the geometric capacitance is determined by the capacitance of the SiO_2 dielectric. During electrochemical gating the electrochemical double layer (edl) formed at the interface between the tube and the liquid determines the geometric capacitance. The thickness of the edl is determined by the Debye length, which is in turn a function of the electrolyte concentration.

the total capacitance. Accordingly, the net capacitance for the above back-gated s-SWCNT-FET is given by $c_{\text{bg-single}} = (c_{\text{ox}}^{-1} + c_{\text{q}}^{-1})^{-1} = 34$ pF/m, very close to the value of the geometric capacitance.

Enhancement of the coupling efficiency requires increasing the gate capacitance, which can be attained either by increasing ε_r (high-κ dielectrics) or by reducing the thickness of the gate oxide. Both strategies have been realised in back- [37] and top- [38] gate geometry. For instance, an 8 nm thick ZrO_2 ($\varepsilon_r = 25$) layer has been used as top gate insulator, which yields a geometric capacitance of 550 pF/m [38]. In this case, the net capacitance is 230 pF/m, which is one order of magnitude larger than for the corresponding back-gated FET with a SiO_2 insulator of 100 nm thickness. However, the net capacitance at the same time determines the field-effect mobility of the device [36]. Since a decreased capacitance leads to increased mobility, there is a trade-off between improved gate coupling and higher field-effect mobility. It should furthermore be noted that although the gate voltage range is appreciably reduced in local top-gate devices, their operation still requires permanent application of a constant back-gate voltage. An alternative approach utilises aluminium electrodes as back-gate, whose insulating native oxide layer (~few nm thickness) ensures low operating voltages [5].

The scenario for a CNT-FET immersed in a liquid or coated with an SPE is illustrated in Figure 26.6(b), which reveals that the geometric capacitance is determined by the Debye length (λ_D) of the Helmholtz layer. A simple estimate for this double-layer capacitance is given by $c_{edl} = 2\pi\varepsilon_r\varepsilon_0/\ln(1 + \lambda_D/r)$ [31]. For water ($\varepsilon_r = 80$) and a low concentration of ions ($\lambda_D \sim 4$ nm to 5 nm) [39], c_{edl} is of the order of 2000 pF/m. Like for the back gate configuration, this capacitance occurs in series with the quantum capacitance, yielding a net capacitance of around 330 pF/m, very close to the theoretical limit of 400 pF/m. It thus becomes evident that the performance of liquid-gated transistors surpasses transistors comprising solid dielectric gate insulators. By increasing the electrolyte concentration and thereby reducing the Debye length, the net capacitance for the liquid-gated configuration can be increased further. Moreover, in contrast to back-gated CNT-FETs, liquid-gated CNT-FETs are independent of device geometry, since in the latter devices the quantum capacitance is dominating. However, in both gate configurations, the maximum capacitance limit is set by the quantum limit of 400 pF/m.

26.3 Chemical Functionalisation

26.3.1 Motivation and Strategies

In the last section, different fabrication approaches to CNT-FETs have been presented, along with device engineering strategies to improve the performance of the devices. Chemical functionalisation methodologies such as doping or the realisation of electrochemical FETs, offer an effective means for enhancing the device performance further. As a spin-off, (bio-)chemical sensors based upon FET operation become accessible, wherein functional groups are attached to the nanotubes in order to obtain specificity toward analytes [40, 41]. In this section, we demonstrate how simple chemical modification methods allow mitigating a number of problems inherent to back-gated CNT-FETs, as well as obtaining all-semiconducting nanotube ensembles suitable as FET channels.

Chemical functionalisation approaches for CNT-based devices can be classified into two categories, namely modification involving the whole device, and functionalisation of only the nanotube channel. For whole device modification, the FET is either immersed in a solvent containing the desired reagent, or the reagent is spin-coated or spotted from the solvent. Nanotube functionalisation, on the other hand, can be performed in three different ways, i.e., thermally-activated chemistry, photochemistry, and electrochemistry [42]. Among those, electrochemistry is emerging to be the most controllable and versatile method for obtaining functionalised nanotube devices. In the following subsections, first whole device modification will be discussed, followed by controlled electrochemical functionalisation of nanotubes and chemical doping. Finally, a brief description of chemical sensors based on functionalised CNT-FETs is given.

26.3.2
Chemically Modified Devices

An important requirement for efficient FET device operation is the absence of hysteresis. As mentioned above, hysteresis arises mainly due to the presence of traps in the vicinity of the nanotube. It can be eliminated by reducing the trap/surface defect density, as has been demonstrated through incorporation of a high-quality hydrophobic organic layer between nanotube and the SiO_2. For this purpose, the Si/SiO_2 ($t_{SiO_2} = 4$ nm) substrates have been covered by a self-assembled monolayer (SAM) prepared from 18-phenoxyoctadecyltrichlorosilane [43], as shown in Figure 26.7(a). The resulting SWCNT-FETs display an unprecedented combination of device parameters, most notably a very low operating voltage (~1 V), an excellent I_{ON}/I_{OFF} ratio (>10^5), and a very low subthreshold swing of 60 mV/dec at room temperature as is apparent from Figure 26.7(b). An alternative strategy is to coat the FETs with appropriate polymers such as polyethyleneoxide (PEO) that can neutralise the trap states due to their donor capability. One example of hysteresis reduction realised in this manner is provided by Figure 26.8.

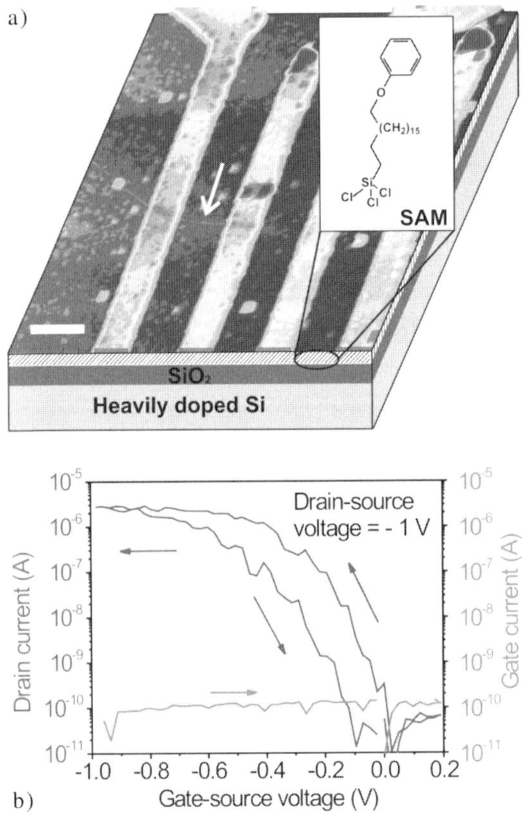

Figure 26.7 (a) Schematic depiction of an s-SWCNT-FET, with the gate insulator comprising of an organic self-assembled monolayer (SAM) made of (18-phenoxyoctadecyl)trichlorosilane. (b) Gate-dependence of the fabricated device as a function of the back-gate voltage showing low-hysteresis, high sub-threshold swing and low operating voltages. (Reprinted with permission from [43].)

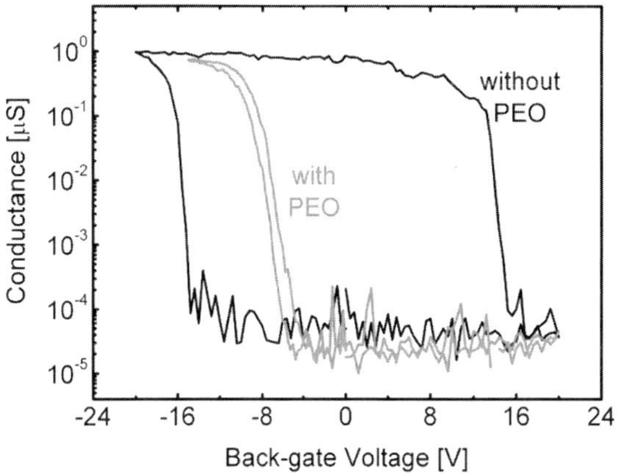

Figure 26.8 Removal of hysteresis through chemical modification demonstrated by the back-gate dependence of conductance of a single s-SWCNT FET (V_{ds} = 100 mV). A drop of 1 mg/mL of polyethyleneoxide (PEO) in acetonitrile was spotted on the substrate and left to dry.

Upon addition of lithium perchlorate ($LiClO_4$) to the PEO matrix, SWCNT-FETs with a solid polymer electrolyte (SPE) coating are obtained [33, 34]. In addition to hysteresis elimination, this imparts the possibility to electrochemically gate the FET. To this end, a micro-fabricated gold electrode on the surface of the substrate or an Ag/AgCl wire in contact with the polymer coating is used as the gate electrode. The resulting SWCNT-FETs can be operated within a gate voltage range that is approximately ten times smaller than for corresponding devices in back-gate configuration (100 nm SiO_2 thickness). By variation of the SPE composition, it is possible to tune the threshold voltage of the obtained transistors, and thus to fabricate either depletion- or enhancement-mode transistors. An example is the replacement of PEO by polyvinylchloride (PVC) [44, 45], whereby the threshold voltage is shifted to more positive voltages (see Figure 26.9). The SPE-coated FETs were found to be stable under ambient conditions for up to four weeks.

26.3.3 Electrochemical Functionalisation

As mentioned before, electrochemical functionalisation (ECM) is a suitable means to alter mainly the contacted tube regions of the FET [46]. ECM is performed either potentiostatically (at constant potentials) or galvanostatically (at constant currents) in a solution containing a reagent as precursor for a reactive chemical species (e.g., a radical) that will be formed via electron transfer with the nanotube [47]. The precursor is usually an organic molecule comprising a reactive end-group and a persistent functional group that the nanotube surface would possess after the modification. Since many organic radical species have

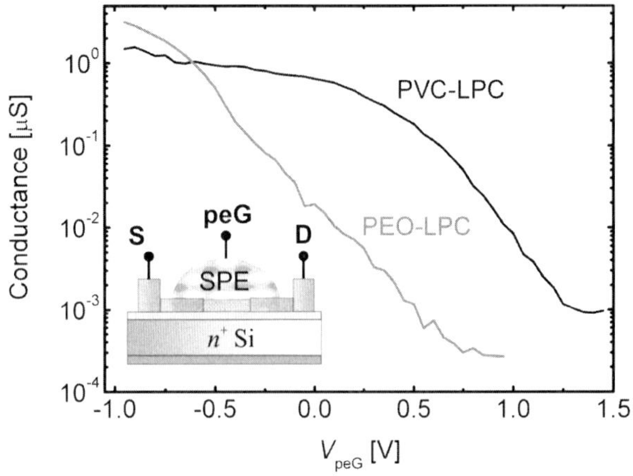

Figure 26.9 Electrochemical gating of SWCNT-FETs covered with two different solid polymer electrolytes (SPEs). The SPEs were spotted from solution on top of the FET surface and heated at 60 °C for an hour. Conductance of the same individual s-SWCNT ($V_{ds} = \pm 0.1$ V) with two different SPEs is plotted as a function of the voltage at the polymer-electrolyte-gate (V_{peG}). An adjacent microfabricated AuPd electrode (~200 μm distance) on the substrate was used as the gate electrode. PEO-LPC: 20 mg/mL polyethyleneoxide, 10 mg/mL lithium perchlorate in a mixture of methanol/water (v/v 4 : 1). PVC-LPC: 20 mg/mL polyvinylchloride, 2 mg/mL lithium perchlorate in tetrahydrofuran. It is apparent that by changing the composition of the polymer matrix, it is possible to shift the threshold voltage of the transistors. The inset shows the schematic of the fabricated FET (S – source, D – drain).

a tendency to react with the precursor or to self-polymerise, a polymer coating is often formed on the tubes.

Toward the coupling of organic moieties onto nanotubes, we have devised two schemes that enable grafting a variety of functional groups either covalently or non-covalently [48]. While the former scheme is based upon the electrochemical generation of substituted phenyl radicals from aromatic diazonium salts, the latter involves the electrochemical oxidation of a substituted aromatic amine. Electrical transport and confocal Raman spectroscopic measurements on the same individual SWCNTs have been used to determine the respective bonding interaction between the formed polymer coating and nanotube. This is illustrated in Figures 26.10 and 26.11, where the electrical resistance and Raman spectra of individual metallic nanotubes before and after the two kinds of modification are displayed. It is evident that the amine coupling brings in very few changes to the resistance and Raman spectra (Figure 26.10), whereas the nanotube subjected to the diazonium coupling displays a strong increase in resistance (at least two orders of magnitude) and a D-line of significantly increased intensity (Figure 26.11). Both features in the latter case signify the introduction of sp^3-bonded carbons into the nanotube framework, and hence confirm that the diazonium coupling results in the formation of covalent bonds.

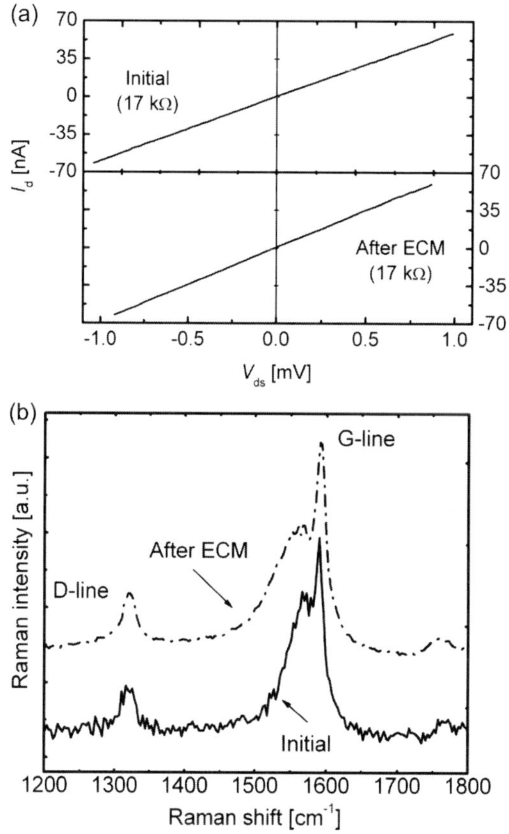

Figure 26.10 Oxidative electrochemical modification (oECM) of individual contacted m-SWCNTs using a substituted aromatic amine. (a) Electrical transport measurements and (b) Raman spectra before and after oECM on the same contacted tube. The oECM was performed at +0.75 V versus Pt for 120 s in a solution of 10 mM 4-aminobenzylamine in ethanol. 0.1 M lithium perchlorate was used as a background electrolyte. (Reprinted with permission from [48].)

26.3.4 Selective Electrochemical Functionalisation

The major bottle-neck in the fabrication of FETs incorporating a SWCNT network as conducting channel is the fact that the synthetically available SWCNT material is composed of a mixture of both m- and s-SWCNTs. This hurdle has stimulated the development of methods for selective elimination of the m-SWCNTs. One intuitive method for this purpose involves the destruction of m-SWCNTs with high electrical currents. This has been realised through application of a large back-gate voltage in order to switch off the s-SWCNTs while a high drain bias is applied [49]. When performed under ambient conditions, high electric currents flow only through the m-SWCNTs which are hence burnt away through oxidation. Despite its increasing use, this method faces several difficulties, in particular when large, dense nanotube networks are to be modified, as the Joule heat generated in the m-SWCNTs increasingly affects adjacent s-SWCNTs. In addition, it requires the presence of a back gate that cannot be easily implemented on plastic substrates. One way around this problem is the transfer of the SWCNT network onto a different substrate [50],

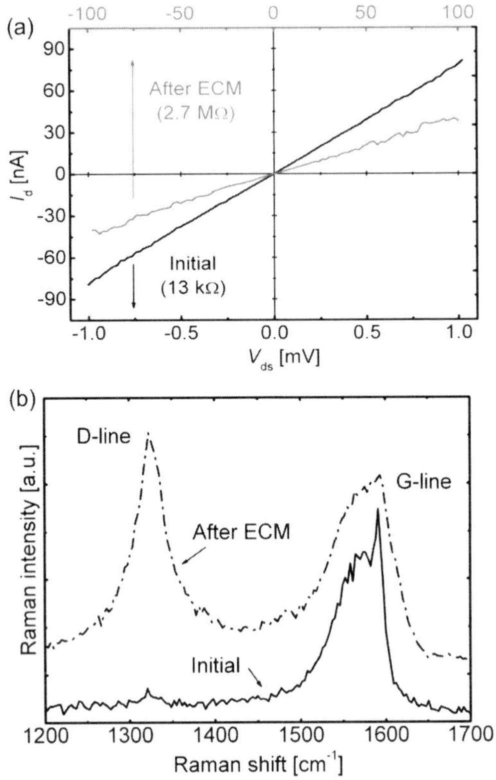

Figure 26.11 Reductive electrochemical modification (rECM) of individual contacted m-SWCNTs utilising an aromatic diazonium salt. (a) Electrical transport measurements and (b) Raman spectra before and after rECM on the same contacted tube. The rECM was performed at − 1.3 V versus Pt for 120 s in a solution of 10 mM 4-nitrobenzene diazonium tetrafluoroborate in *N*,*N*-dimethylformamide. 0.1 M lithium perchlorate was used as a background electrolyte. (Reprinted with permission from [48].)

which is however a tedious procedure. Furthermore, for higher tube densities substantial gate leakage can only be avoided if gate insulators of sufficiently large thickness are used, which in turn limits the electrostatic coupling of the back gate.

We have developed a selective electrochemical functionalisation route to eliminate the m-SWCNTs in nanotube ensembles [51, 52]. The functionalisation consists of the above described diazonium coupling. The covalent attachment of a high density of phenyl groups to the m-SWCNTs increases their electrical resistance by more than two orders of magnitude, whereby they are effectively eliminated from the electrical transport path. Selectivity of the electrochemical coupling to the m-SWCNTs has been achieved by two different methods. In the first approach [51], which has been originally developed for the back-gate configuration, advantage is taken of the hysteresis of the devices in order to switch the s-SWCNTs to the OFF state. After the completion of a gate cycle, these tubes remain switched off (at $V_G = 0$) for at least 15 minutes. During this time, the ECM is performed and mainly the m-SWCNTs are modified. Figure 26.12 compares typical transfer characteristics of devices containing between 10 and 20 SWCNTs before and after such electrochemical modification. With denser nanotube networks, however, problems arise due to a lower stability of the OFF state adjusted via the hysteresis.

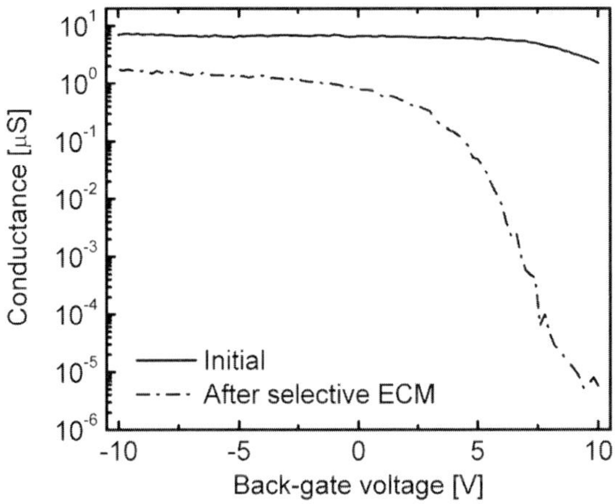

Figure 26.12 A nanotube FET obtained through back-gate assisted selective ECM of SWCNT bundles containing both m- and s-SWCNTs. The gate dependence of conductance initially shows metallic behaviour (solid line). After selective ECM, the conductance shows a variation of around five orders of magnitude (dash-dotted line) signifying that the m-SWCNTs have been eliminated. V_{ds} was 10 mV before ECM and 50 mV after ECM.

In order to avoid this drawback, we have recently developed an alternative ECM scheme which does not require the presence of a back gate and is applicable also to denser networks [52]. This method relies upon the fact that a voltage V_{lG} applied to an Ag/AgCl liquid gate electrode has the same effect as applying a potential of $-V_{lG}$ to the working electrode (the contacted nanotube) versus Ag/AgCl [53]. Selective electrochemical coupling is then achieved by choosing a diazonium salt whose reduction peak potential lies in a voltage region where the s-SWCNT is in its OFF state. In order to verify this concept, we have calculated the reduction rates by using the Gerischer–Marcus model [32] analogous to Ref. [53] at a (9,0) m- and a (10,0) s-SWCNT for a range of diazonium salts with different reduction potentials [52]. Figure 26.13(a) presents the magnitudes of reduction rates at the two tubes for a diazonium salt whose reduction potential is chosen such that the s-SWCNT is gated to its OFF state. Correspondingly, Figure 26.13(b) plots the relative reaction rate at the (9,0)- with respect to the (10,0)-SWCNT as a function of the liquid-gate voltage. It is apparent from the figure that by performing the ECM at the denoted potential, the coupling to the m-SWCNT is expected to be more than five orders of magnitude faster than to the s-SWCNT. Thus, when this diazonium salt is combined with the suitable potential applied vs. Ag/AgCl, selective functionalisation of dense networks of CVD-grown SWCNTs (see Figure 26.14(a)) becomes possible. The transfer characteristics before and after selective ECM of this tube ensemble are depicted in Figure 26.14(b), which reveal that the

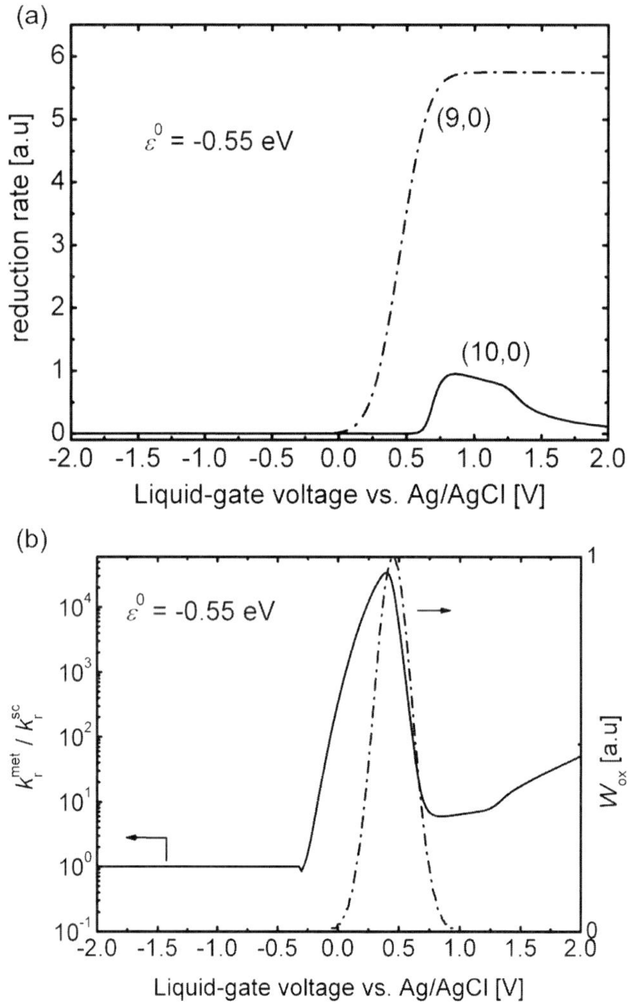

Figure 26.13 (a) Calculated magnitudes of electrochemical reduction rate (k_r) at a (9,0) m- and a (10,0) s-SWCNT with a coupling agent having a redox potential of $\varepsilon^0 = -0.55$ V. (b) Calculated relative reduction rate based on the k_r curves computed in (a). The dash-dotted line shows the calculated distribution of oxidised states (W_{ox}) for the redox couple with an ε^0 of -0.55 V. It is apparent that for a range of potentials around 0.3 V vs. Ag/AgCl, the reduction of the molecule at a (9,0) tube is around four orders of magnitude faster than at the (10,0) tube.

metallic tubes have been successfully eliminated and the device exhibits a good switching ratio of more than three orders of magnitude. It should be noted that the reproducibility of this method is low in case of SWCNT networks with a sizeable diameter distribution. The major complication arises from the presence of large-diameter metallic tubes that due to their small curvature have only a low chemical reactivity. This limitation may be overcome

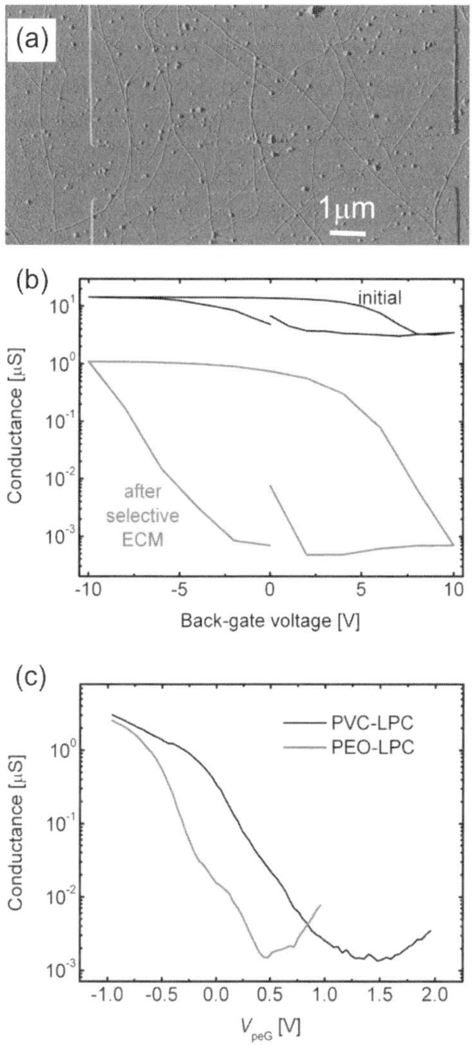

Figure 26.14 FETs based on SWCNT networks: (a) Atomic Force Microscope (AFM) amplitude image of a network of tubes grown by chemical vapour deposition. (b) Gate dependence of conductance of such a network before and after selective ECM. V_{ds} is 10 mV before and 100 mV after ECM. (c) Variation of conductance of the same device after spotting two different solid polymer electrolytes and using an electrode in contact with the SPE as the gate ($V_{ds} = \pm 100$ mV). The composition of the SPE is identical to that in Figure 26.9. V_{peG} is the voltage applied to the polymer electrolyte–gate.

by a stricter control over the catalyst particle size and hence the nanotube diameter [13]. The nanotube networks modified in this manner can be subsequently coated with SPEs to obtain FETs operating at low voltages as shown in Figure 26.14(c).

26.3.5 Chemical Doping

The implementation of complementary logic requires not only p-type but also n-type transistors. Chemical doping has proven to be a viable approach for converting p- into n-type SWCNT-FETs [54, 55], albeit the efficacy lacks significantly behind what would be needed for reliable technological application. Such conversion can be achieved with the aid of strong electron donors, among which the polymer polyethylenimine (PEI) has turned out to yield most reproducible results [56, 57]. Figure 26.15(a) presents the transfer characteristics of a SWCNT-FET before and after PEI doping. Due to its noticeable ionic conductivity, PEI serves at the same time as an SPE, and thus enables an adjacent electrode that is in touch with the PEI layer to act as a gate (see Figure 26.15(b)). However, the lifetime of PEI-coated SWCNT-FETs, in both the back gated and electrochemically gated configurations is very limited when the devices are stored under ambient, which is a consequence of the low air stability of PEI. In most cases, after more than 7 days the transistors show very high resistances (in the GΩ range) and loose their switching capability. Other donor compounds that have been studied for p- to n-type conversion of CNT-FETs are tetrathiafulvalene (TTF) [58], hydrazine [59], aminosilanes [60] and polyaniline [59]. However, like in the case of PEI, both the reproducibility of doping and the stability of the obtained n-type behaviour are still posing a serious problem.

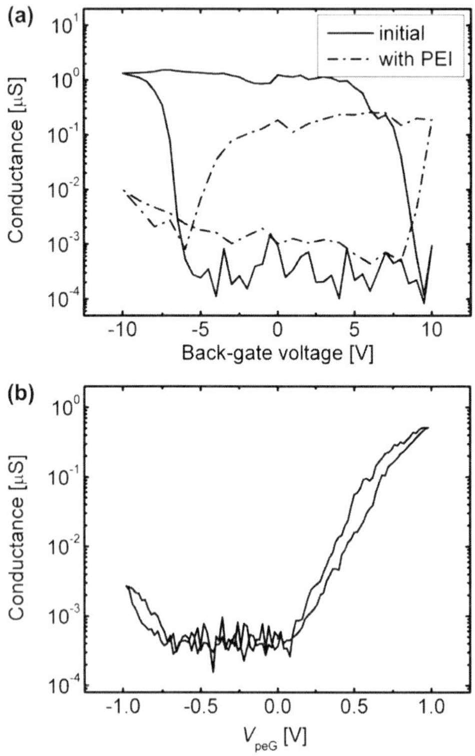

Figure 26.15 n-type doping of SWCNT-FETs. (a) Back-gate dependence of conductance ($V_{ds} = \pm 100$ mV) for an s-SWCNT-FET before and after chemical modification with polyethyleneimine (PEI). A 33 wt% solution of PEI in methanol was spotted on the device and left to dry. The FET shows n-type characteristics after the modification, albeit still with a large hysteresis. (b) PEI can be used as a solid polymer electrolyte. The panel shows the variation of channel conductance as a function of the polymer electrolyte gate voltage (V_{peG}). An adjacent AuPd electrode on the substrate was used as the gate electrode ($V_{ds} = \pm 100$ mV).

26.3.6
Sensors Based on Functionalised SWCNT-FETs

The electric field effect in SWCNT-FETs offers the possibility to detect chemical species by exploiting charge transfer to the nanotube or changes in the tube's surface potential occurring upon molecular adsorption onto the nanotube channel. In this manner, a wide range of different (bio-)chemical sensors with impressively high sensitivities have been realised during the past few years [40, 41]. One example is the use of a liquid-gated s-SWCNT-FET to detect ammonia in solution at concentrations as low as 1 mM [29]. Adsorption of ammonia molecules onto the s-SWCNT leads to charge transfer and thereby shifts the threshold voltage of the liquid-gated nanotube transistor.

Along these lines, we have extended the electrochemical functionalisation method to fabricate gas and chemical sensors based upon SWCNTs. For the aim of hydrogen sensing, SWCNT networks have been decorated with palladium nanoparticles via electroreduction of a Pd-containing salt [61]. Hydrogen is known to dissolve in Pd forming atomic hydrogen, with concomitant alteration of the work function of the Pd particles [62]. The resulting electron transfer onto the semiconducting tubes contained in the network triggers an increase of resistance of the Pd-modified nanotube network. The key advantage of the ECM method is the control over the Pd particle size and density, as well as the intimate coupling between particles and nanotube originating from their direct formation on the tube sidewalls. Toward the application of SWCNTs in liquid chemical sensors, we have developed a novel ECM strategy that can be applied to m-SWCNTs [63]. It comprises the covalent attachment of organic moieties bearing amino groups in a controlled density to individual metallic tubes. The resistance increase observed for the functionalised tubes upon increasing protonation of the appended basic groups has been assigned to an enhanced scattering strength for carriers in the nanotube. That this mechanism is dominant for the functionalised m-SWCNTs is confirmed by the low threshold shifts recently observed on pristine semiconducting nanotubes [39]. While the realisation of such pH sensors served as a proof-of-principle, a similar mechanism may be exploited for the detection of other chemical species such as metal ions.

26.4
Device Characterisation of CNT-FETs

Due to the significant distribution of diameter and chirality of the nanotubes in the raw material, it is important to characterise SWCNTs incorporated into an FET configuration individually. In this section, we discuss the important device parameters of CNT-FETs fabricated in the two gating configurations. Following this, we introduce a novel device characterisation technique based upon local photocurrent detection, which enables the estimation of electronic band profiles of back-gated CNT-FETs.

26.4.1 Back-Gated Devices

26.4.1.1 Saturation

In conventional MOSFETs, saturation is reached when the channel is pinched-off. It is an important device characteristic forming a key element of digital circuits [64]. In long-channel MOSFETs, saturation arises due to carrier–carrier interaction via Coulomb repulsion [65]. By contrast, the fact that SWCNT-FETs are unconventional SB-FETs brings in different mechanisms like velocity saturation [66]. In order for this mechanism to be effective, the carriers must experience negligible scattering by impurities or defects [67]. While this appears to be fulfilled in high-performance SWCNT-FETs [38, 43, 50], complete saturation is usually not observed in FETs fabricated on SiO_2 substrates [4, 16].

26.4.1.2 Transconductance

Transconductance is a measure of how fast the transistor is able to switch while sweeping the gate voltage, and is related to the gain of the transistor. It is given by the ratio of the change in drain current to the change in gate voltage at a constant drain-source bias. In order to aid the comparison between different devices, a normalised transconductance g_m is computed per unit channel width. FETs based on single s-SWCNTs in the back-gated configuration with varying dielectrics generally exhibit g_m values in the range of 10 S/cm to 30 S/cm [37, 43]. However, with top-gated s-SWCNT-FETs higher g_m values have been reported. For example, a transconductance of 60 S/cm was observed for a top-gated single s-SWCNT-FETs with ZrO_2 as the gate dielectric [38]. Interestingly, the g_m values of transistors based on nanotube networks also fall in the same range of 10 S/cm to 30 S/cm. FETs obtained through selective elimination of m-SWCNTs in a low density network displayed g_m values close to 10 S/cm [51]. For highly ordered networks of high tube density, where the m-SWCNTs are eliminated via electrical breakdown, a transconductance of 30 S/cm has been reported [50].

26.4.1.3 Sub-Threshold Swing

The sub-threshold swing is given by $S = \ln(10)\,[dV_G/d(\ln(I_d))] = (k_B T/e) \times \ln(10)\,(1/\alpha)$, where $k_B T$ is the thermal energy at a temperature T (~24 meV for room temperature), e is the elementary charge, α is the gate coupling factor and V_G and I_d correspond to the gate voltage and drain current, respectively. The theoretical limit for S is 60 mV/decade at room temperature [31], corresponding to a maximum coupling efficiency of $\alpha = 1$. For most back-gated SWCNT-FETs (with oxide thickness ≥ 100 nm), a high value of S in the range of 100–2000 mV/decade is observed, corresponding to a low coupling efficiency α of between 0.6 and 0.03. The value of 60 mV/decade has been closely approached [38] or reached [43, 68] for single s-SWCNT-based FETs with optimised net device capacitance.

26.4.1.4 **Mobility**

The mesoscopic size of CNTs renders it difficult to measure the intrinsic carrier mobility. Moreover, evaluation of the field-effect mobility from the transfer characteristics is hampered by the fact that the SWCNT-FETs normally behave as Schottky-barrier transistors. Nonetheless, for the purpose of device comparison, it is reasonable to extract an effective field-effect mobility using $\mu = (dI_d/dV_G) \cdot L/(c_{net}V_{ds})$, where L is the channel length and c_{net} is the net capacitance per unit length. On this basis, impressive values as high as 100 000 cm^2/Vs have been reported for holes in a 325 µm long s-SWCNT at room temperature [3]. However, it should be kept in mind that, since the transconductance of SWCNT-FETs is often contact-limited, high-quality nanotubes of sufficient length can easily display extremely large effective field-effect mobilities. In comparison, SWCNT-FETs with channel lengths in the range of 100 nm to a few microns display lower values of a few 1000 cm^2/Vs. Similar values have been found for our FETs obtained via selective ECM. Compared with the mobility of holes, the electron mobility derived from the transfer characteristics of ambipolar-SWCNT-FETs is approximately one order of magnitude smaller. In the case of back-gated SWCNT networks, the ordering and alignment of the nanotubes across the source/drain electrodes play a crucial role in determining the carrier mobility. Disordered networks exhibit mobilities less than 50 cm^2/Vs, while with highly ordered tube arrays mobilities higher than 3000 cm^2/Vs have been reached [50].

26.4.2 Electrochemically Gated Devices

Saturation in liquid-gated and polymer-electrolyte-gated devices is hardly observed, since the gate leakage increases exponentially for drain-source bias higher than 1 V. However, as the capacitances for the liquid-gated and polymer electrolyte-gated devices are very close to the quantum limit, the minimum possible sub-threshold swing can be reached. Typical values of subthreshold swing obtained for liquid-gated and polymer electrolyte-gated devices are ~80 mV/decade, signifying a coupling efficiency close to 0.75 [31]. For single SWCNTs within PEO-based electrolytes, a transconductance of 5 S/cm is observed, corresponding to a field-effect mobility of ~2500 cm^2/Vs [33]. In the case of SWCNT networks, the degree of ordering is a decisive factor, similar to the back-gated transistors. Highly-ordered networks [50] exhibit field-effect mobilities of the order of 3000 cm^2/Vs. However, a major disadvantage of the polymer electrolyte-gated transistors is that they cannot be operated at frequencies higher than 1 kHz [34].

26.4.3 Scanning Photocurrent Microscopy

We have developed photoelectronic transport imaging [69] or scanning photocurrent microscopy (SPCM) [70] as an easy and fast method for determining

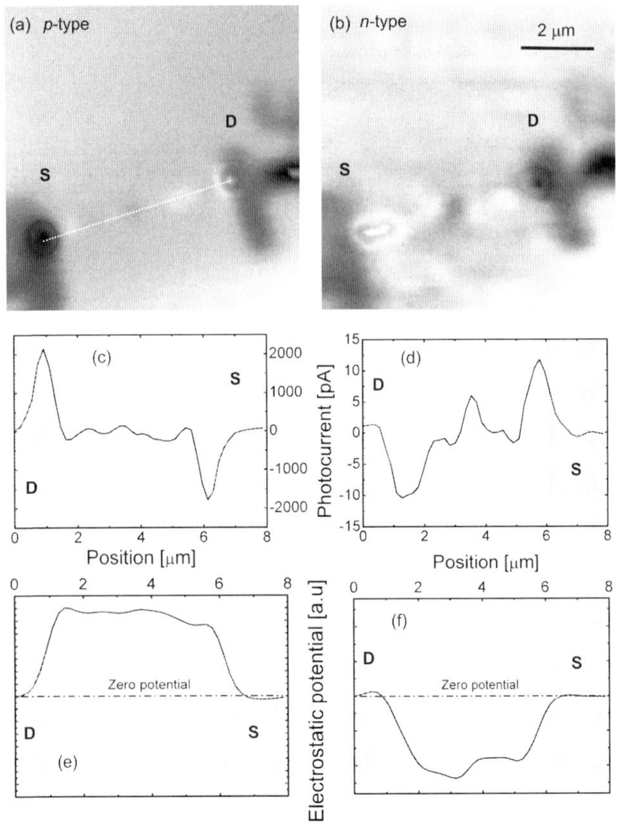

Figure 26.16 Scanning photocurrent microscopy (SPCM) of an ambipolar s-SWCNT-FET with data recorded in the p-type (a, c, e) and n-type (b, d, f) regimes. (a, b) SPCM images showing the spatial variation of the photocurrent at zero drain-source bias. An optical image of the device is overlaid in order to identify the position of the contacts (black regions). (c, d) Line profiles of the photocurrent along the nanotube marked as a white dotted line in (a). The shaded regions indicate the position of the contacts. (e, f) Electrostatic potential maps obtained by integrating the corresponding line profiles in (c) and (d). The position of zero potential is marked by the dash-dotted line.

band profiles of CNT-FETs with the aid of a confocal microscope. In this method, a diffraction-limited laser spot is used to illuminate the channel of CNT-FETs while simultaneously recording the local photoresponse. By choosing a channel length (>1.5 μm) much larger than the diameter of the diffraction-limited spot (~300 nm for $\lambda_{exc} = 514.5$ nm), it is possible to resolve the photocurrents originating from the middle of the channel from those at the contacts. In this manner, the Schottky barriers at the contacts of a p-type SWCNT-FET at zero drain-source bias can be directly visualised, as shown in Figure 26.16(a, c, e) [71]. When applied to ambipolar s-SWCNT-FETs, zero-bias photocurrent images acquired as a function of gate voltage reveal a reversal of the bending of the Schottky barriers upon transition from the p- to the n-type regime as shown in Figure 26.16(b, d, f) [70]. On this basis, determination of the Schottky barrier height at the contacts can be accomplished in a straightforward manner. The obtained values for holes of the order of ~200 meV are in close correspondence with results of low-temperature conductivity measurements. Furthermore, qualitative electrostatic potential profiles are accessible through spatial integration of the photocurrents measured along the nanotube channel (Figure 26.16(e, f)). Finally, photocurrent imaging has proven to be a versatile tool to identify tube defects [72] such as intratube junctions. The local electrical fields associated with such defects create additional signals along the nanotube (Figure 26.16(b)).

26.5 Future Perspectives

One important advantage of the selective electrochemical functionalisation approach is its direct applicability to nanotubes deposited on flexible (plastic) substrates. This would avoid the tedious transfer of the nanotubes between plastic and silicon substrates just for the sake of elimination of metallic tubes [50]. Another promising future perspective for ECM is the fabrication of gate insulators around the nanotubes, which could lead to the development of wrap-around gates offering optimal coupling of dielectric gates [73]. Moreover, the polymer electrolyte-based FETs could prove technologically relevant due to the ability to tune the threshold voltage just by varying the composition of the SPE. However, this task still requires a substantial improvement of the stability of the electrolyte material. One possible solution to this problem could be a hermetic sealing of the devices to avoid exposure to oxygen and ultraviolet (UV) light.

Further to this, the increasing activities in the field of biosensors based upon CNT-FETs could strongly benefit from the development of more elaborate (electro-)chemical functionalisation methods, in particular with regard to achieving ultimate selectivity and biocompatibility [74]. Another stimulating impact may be on the elucidation of the exact mechanism by which SWCNT-FETs detect analytes in solution, specifically whether charge transfer or enhanced carrier scattering is the predominant effect of molecule adsorption [39, 75].

A major hurdle for the widespread application of SWCNTs in FETs is the current lack of synthesis or separation methods yielding exclusively semiconducting nanotubes, though significant progress has already been achieved in this direction. For instance, methods based on ion-exchange chromatography have shown promising results [76]. With the future advent of large-scale production methods of SWCNT with specified chirality and further advancements in the controlled growth of oriented SWCNT ensembles, combined with the help of the available chemical functionalisation schemes, large-scale fabrication of high-performance thin film SWCNT-FETs should become feasible. The first steps in this direction have very recently been reported [50, 76]. Important clues toward the minimisation of defects in SWCNTs could be gained from microscopic techniques like scanning gate [77, 78] or scanning photocurrent microscopy [72, 79, 80]. The latter method offers the advantage of being applicable also to extended nanotube networks, and holds promise to provide valuable insights into the mechanism of inter-tube electrical transport.

26.6 Conclusion

In summary, it is now well-established that semiconducting SWCNTs are close-to-ideal components of high-performance nanoscale FETs, albeit some of their intrinsic features posing serious challenges to the device fabrication. Foremost, the high sensitivity of s-SWCNTs against electric charges in their neighborhood renders them susceptible to minute changes in their environment. Chemical functionalisation represents a valuable approach to improve the performance of SWCNT-FETs, with electrochemical methods being particularly useful due to their inherent capability to control the functionalisation extent of the tubes, as well as to restrict the modification to the nanotube channel. One of the most exciting developments in CNT electrochemistry is the selective functionalisation of metallic nanotubes contained within an ensemble, which provides for an effective route to FETs whose channel is comprised of a SWCNT network. Moreover, the emergence of novel powerful characterisation techniques such as scanning photocurrent microscopy promises further elucidation of the operation mechanism of SWCNT-FETs.

Acknowledgements

This work was supported by the Deutsche Forschungsgemeinschaft within the framework of the priority program SPP1121. We are grateful to F. Nan, M. Scolari, and A. Mews from the University of Siegen (Germany) for their continuous supply with CVD-grown SWCNTs and their close collaboration on confocal microscopy of individual nanotubes. We thank H. Klauk and U. Zschieschang for the help with the fabrication of nanotube transistors incorporating a SAM-based dielectric.

References

1. G. Cuniberti, G. Fagas, and K. Richter (eds.), Introducing Molecular Electronics – Lecture Notes in Physics, Vol. 680 (Springer Verlag, Berlin–Heidelberg, 2005).
2. Z. Yao, C. L. Kane, and C. Dekker, Phys. Rev. Lett. **84**, 2941 (2000).

3. T. Durkop, S. A. Getty, E. Cobas and M. S. Fuhrer, Nano Lett. **4**, 35 (2004).
4. S. J. Tans, M. H. Devoret, H. Dai, A. Thess, R. E. Smalley, L. J. Geerligs, and C. Dekker, Nature **386**, 474 (1998).
5. A. Bachtold, P. Hadley, T. Nakanishi, and C. Dekker, Science **294**, 1317 (2001).
6. J. B. Cui, R. Sordan, M. Burghard, and K. Kern, Appl. Phys. Lett. **81**, 3260 (2002).
7. T. W. Ebbesen and P. M. Ajayan, Nature **358**, 220 (1992).
8. T. Guo, P. Nikolaev, A. Thess, D. T. Colbert, and R. E. Smalley, Chem. Phys. Lett. **243**, 49 (1995).
9. M. J. Bronikowski, P. A. Willis, D. T. Colbert, K. A. Smith, and R. E. Smalley, J. Vac. Sci. Technol. A **19**, 1800 (2001).
10. J. Kong, A. M. Cassell, and H. J. Dai, Chem. Phys. Lett. **292**, 567 (1998).
11. Y. Murakami, Y. Miyauchi, S. Chiashi, and S. Maruyama, Chem. Phys. Lett. **374**, 53 (2003).
12. D. S. Bethune, C. H. Kiang, M. S. DeVries, G. Gorman, R. Savoy, and R. Beyers, Nature **363**, 605 (1993).
13. Y. Li, W. Kim, Y. Zhang, M. Rolandi, D. Wang, and H. Dai, J. Phys. Chem. B **105**, 11424 (2001).
14. R. Saito, G. Dresselhaus, and M. S. Dresselhaus, Physical Properties of Carbon Nanotubes (Imperial College Press, London, 1998).
15. J. W. G. Wilder, L. C. Venema, A. G. Rinzler, R. E. Smalley, and C. Dekker, Nature **391**, 59 (1998).
16. R. Martel, T. Schmidt, H. R. Shea, T. Hertel, and Ph. Avouris, Appl. Phys. Lett. **73**, 2447 (1998).
17. S. Heinze, J. Tersoff, R. Martel, V. Derycke, J. Appenzeller, and Ph. Avouris, Phys. Rev. Lett. **89**, 106801 (2002).
18. W. Kim, A. Javey, R. Tu, J. Cao, Q. Wang, and H. Dai, Appl. Phys. Lett. **87**, 173101 (2005).
19. V. Derycke, R. Martel, J. Appenzeller, and P. Avouris, Appl. Phys. Lett. **80**, 2773 (2002).
20. J. Appenzeller, J. Knoch, V. Derycke, R. Martel, S. Wind, and Ph. Avouris, Phys. Rev. Lett. **89**, 126801 (2002).
21. Y. F. Chen and M. S. Fuhrer, Nano Lett. **6**, 2158 (2006).
22. R. Martel, V. Derycke, C. Lavoie, J. Appenzeller, K. K. Chan, J. Tersoff, and Ph. Avouris, Phys. Rev. Lett. **87**, 256805 (2001).
23. A. Javey, J. Guo, Q. Wang, M. Lundstrom, and H. Dai, Nature **424**, 654 (2003).
24. P. G. Collins, K. Bradley, M. Ishigami, and A. Zettl, Science **287**, 1801 (2000).
25. F. Leonard and J. Tersoff, Phys. Rev. Lett. **84**, 4693 (2000).
26. Z. Chen, J. Appenzeller, J. Knoch, Y. Lin, and P. Avouris, Nano Lett. **5**, 1497 (2005).
27. R. Sordan, K. Balasubramanian, M. Burghard, and K. Kern, Appl. Phys. Lett. **88**, 053119 (2006).
28. W. Kim, A. Javey, O. Vermesh, Q. Wang, Y. Li and H. Dai, Nano Lett. **3**, 193 (2003).
29. K. Bradley, J. C. P. Gabriel, M. Briman, A. Star, and G. Gruner, Phys. Rev. Lett. **91**, 218301 (2003).
30. M. Krueger, M. R. Buitelaar, T. Nussbaumer, C. Schoenenberger, and L. Forro, Appl. Phys. Lett. **78**, 1291 (2001).
31. S. Rosenblatt, Y. Yaish, J. Park, J. Gore, V. Sazonova, and P. L. McEuen, Nano Lett. **2**, 869 (2002).
32. A. J. Bard and L. R. Faulkner, Electrochemical Methods: Fundamentals and Applications (John Wiley and Sons, New York, 2001).
33. C. Lu, Q. Fu, S. Huang, and J. Liu, Nano Lett. **4**, 623 (2004).
34. T. Ozel, A. Gaur, J. A. Rogers, and M. Shim, Nano Lett. **5**, 905 (2005).
35. J. Lee, M. J. Panzer, Y. He, T. P. Lodge, and C. D. Frisbie, J. Am. Chem. Soc. **129**, 4532 (2007).
36. P. Avouris, Acc. Chem. Res. **35**, 1026 (2002).

37. S. J. Wind, J. Appenzeller, R. Martel, V. Derycke, and P. Avouris, Appl. Phys. Lett. **80**, 3817 (2002).

38. A. Javey, H. Kim, M. Brink, Q. Wang, A. Ural, J. Gu, P. McIntyre, P. L. McEuen, M. Lundstrom, and H. Dai, Nature Mater. **1**, 241 (2002).

39. J. H. Back and M. Shim, J. Phys. Chem. B **110**, 23736 (2006).

40. G. Gruner, Anal. Bioanal. Chem. **384**, 322 (2006).

41. K. Balasubramanian and M. Burghard, Anal. Bioanal. Chem. **385**, 452 (2006).

42. M. Burghard, Surf. Sci. Rep. **58**, 1 (2005).

43. R. T. Weitz, U. Zschieschang, F. Effenberger, H. Klauk, M. Burghard, and K. Kern, Nano Lett. **7**, 22 (2007).

44. S. Rajendran and T. Uma, J. Power Sources **88**, 282 (2000).

45. V. S. Kolosnitsyn, G. P. Dukhanin, S. A. Dumler, and I. A. Novakov, Russ. J. Appl. Chem. **78**, 1 (2005).

46. S. E. Kooi, U. Schlecht, M. Burghard, and K. Kern, Angew. Chem. Int. Ed. **41**, 1353 (2002).

47. K. Balasubramanian and M. Burghard, Small **2**, 180 (2005).

48. K. Balasubramanian, M. Friedrich, C. Jiang, Y. Fan, A. Mews, M. Burghard, and K. Kern, Adv. Mater. **15**, 1515 (2003).

49. P. G. Collins, M. S. Arnold and P. Avouris, Science **292**, 706 (2001).

50. S. J. Kang, C. Kocabas, T. Ozel, M. Shim, N. Pimparkar, M. A. Alam, S. V. Rotkin and J. A. Rogers, Nature Nanotechnol. **2**, 230 (2007).

51. K. Balasubramanian, R. Sordan, M. Burghard, and K. Kern, Nano Lett. **4**, 827 (2004).

52. K. Balasubramanian, M. Burghard, and K. Kern, Phys. Chem. Chem. Phys. **10**, 2256 (2008).

53. I. Heller, J. Kong, K. A. Williams, C. Dekker and S. G. Lemay, J. Am. Chem. Soc. **128**, 7353 (2006).

54. C. W. Zhou, J. Kong, E. Yenilmez, and H. Dai, Science **290**, 1552 (2000).

55. P. Avouris and J. Chen, Mater. Today **9**, 26 (2006).

56. M. Shim, A. Javey, N. W. S. Kam, and H. Dai, J. Am. Chem. Soc. **123**, 11512 (2001)

57. G. P. Siddons, D. Merchin, J. H. Back, J. K. Jeong, and M. Shim, Nano Lett. **4**, 927 (2004).

58. T. Takenobu, T. Takano, M. Shiraishi, Y. Murakami, M. Ata, H. Kataura, Y. Achiba, and Y. Iwasa, Nature Mater. **2**, 683 (2003).

59. C. Klinke, J. Chen, A. Afzali, and P. Avouris, Nano Lett. **5**, 555 (2005).

60. J. Kong and H. Dai, J. Phys. Chem. B **105**, 2890 (2001).

61. U. Schlecht, K. Balasubramanian, M. Burghard, and K. Kern, Appl. Surf. Sci. **253**(20), 8394 (2007).

62. I. Lundstroem, M. Armgarth, and L. G. Petersson, CRC Crit. Rev. Solid State Mater. Sci. **15**, 201 (1989).

63. A. Maroto, K. Balasubramanian, M. Burghard, and K. Kern, Chem. Phys. Chem. **8**, 220 (2007).

64. S. M. Sze, Physics of Semiconductor Devices (John Wiley and Sons, New York, 1981).

65. W. Schockley, Proc. IRE **40**, 1365 (1952).

66. Y. F. Chen and M. S. Fuhrer, phys. stat. sol. (b) **243**, 3403 (2006).

67. T. Yamada, Appl. Phys. Lett. **76**, 628 (2000).

68. Y. R. Lu, S. Bangsaruntip, X. Wang, L. Zhang, Y. Nishi and H. Dai, J. Am. Chem. Soc. **128**, 3518 (2006).

69. K. Balasubramanian, M. Friedrich, Y. Fan, U. Wannek, A. Mews, M. Burghard, and K. Kern, Appl. Phys. Lett. **84**, 2400 (2004).

70. E. J. H. Lee, K. Balasubramanian, J. Dorfmueller, Ralf Vogelgesang, Nan Fu, A. Mews, M. Burghard, and K. Kern, Small **3**, 2038 (2007).

71. K. Balasubramanian and M. Burghard, Semicond. Sci. Technol. **21**, S22 (2006).

72. K. Balasubramanian, M. Burghard, K. Kern, M. Scolari, and A. Mews, Nano Lett. **5**, 507 (2005).

73. W. B. Choi, E. Bae, D. Kang, S. Chae, B. Cheong, J. Ko, E. Lee, and W. Park, Nanotechnology **15**, S512 (2004).

74. B. L. Allen, P. D. Kichambare, and A. Star, Adv. Mater. **19**, 1439 (2007).
75. D. S. Hecht, R. J. A. Ramirez, M. Briman, E. Artukovic, K. S. Chichak, J. F. Stoddart, and G. Gruener, Nano Lett. **6**, 2031 (2006).
76. M. Zheng and E. D. Semke, J. Am. Chem. Soc. **129**, 6084 (2007).
77. M. Bockrath, W. Liang, D. Bozovic, J. H. Hafner, C. M. Lieber, M. Tinkham, and H. Park, Science **291**, 283 (2001).
78. M. Freitag, A. T. Johnson, S. V. Kalinin, and D. A. Bonnell, Phys. Rev. Lett. **89**, 216801 (2002).
79. M. Freitag, J. C. Tsang, A. Bol, D. Yuan, J. Liu, and P. Avouris, Appl. Phys. Lett. **91**, 031101 (2007).
80. Y. H. Ahn, W. Tsen, B. Kim, Y. W. Park, and J. Park, cond-mat arXiv:0707.3176 (2007).

27
Contact Effects in Cu(TCNQ) Memory Devices

Artur Hefczyc, Lars Beckmann, Eike Becker, Hans-Hermann Johannes, and Wolfgang Kowalsky

27.1
Introduction

The increasing integration of different electronic systems requires new communication components which should not only suffice the technical requirements but should also take the economic aspect into consideration. Many communication systems are not inevitably dependent on very fast silicon semiconductor technology. A lot of tasks and applications can be managed with low complexity electronics.

One possible application for low complexity electronics is radio frequency identification-tags (RFID) for example in the food or drug packaging industry. In addition, for economic reasons the low cost electronics system should be able to operate without a permanent power supply. Many organic materials not only show conducting or semi-conducting but also bistable switching behaviour.

Switching has been observed in various organic materials, including semiconducting polymers, small molecules, and materials based on nanoparticles [1–6]. A nonvolatile organic memory device can be switched into states of either low or high resistance. The device retains this state even when no bias is applied. Simple fabrication and the properties of organic devices make them suitable for low cost non-volatile memory applications. Furthermore, organic electronics are printable and can be deposited on various substrate materials. A promising candidate for non-volatile organic memory applications is the charge-transfer salt coppcr-tctracyano-quinodimethane (Cu(TCNQ)). The Cu(TCNQ) units are planarly stacked and form columns. Within these columns the π-orbitals overlap and form a one dimensional metal. Switching in Cu(TCNQ) devices was first reported by Potember in a Cu/Cu(TCNQ)/Al sandwich structure [1].

Upon application of a suitable bias voltage, the resistance of the stacked Cu/Cu(TCNQ)/Al device decreased abruptly and remained in a highly conducting state. The device could be switched back into its high resistance state when an opposite bias was applied.

Organic Electronics. Structural and Electronic Properties of OFETs. Edited by Christof Wöll

ISBN: 978-3-527-40810-8

In a memory application, the information is stored in the resistance of the device. In the read-out operation, a voltage that is too small to initiate switching is applied to measure the resistance. Figure 27.1 shows a typical $I-V$ characteristic of an organic bistable Cu/Cu(TCNQ)/Al device. The graph contains one branch with a low and one with a high resistivity. At a threshold voltage, switching from the low to the high conducting state occurs. Switching back to the low conducting state happens either step-like (shown in Figure 27.1) or through a transition region of negative differential resistance (NDR).

The mechanism of conductivity change in the TCNQ-metal salts is usually described by the following equation:

$$[\mathrm{M}^{+}(\mathrm{TCNQ})^{-}]_{n} \leftrightarrow \mathrm{M}_{x}^{0} + (\mathrm{TCNQ}^{0})_{x} + [\mathrm{M}^{+}(\mathrm{TCNQ}^{-})]_{n-x}\,.$$

In the fully ionised form (all metal atoms M and all TCNQ-molecules are ionised), the systems does not possess free carriers for charge transport and it is in its high-resistance state. Switching transfers the system into the partially ionised form with neutral metal atoms and TCNQ-molecules. Neutral TCNQ-molecules correspond to free holes. These free carriers reduce the resistance of the material.

This model of the switching process was confirmed by Kamitsos et al. Using Raman spectroscopy, they showed that a fraction of 10–15 mol% neutral TCNQ exists in the high conducting state immediately after switching [7]. A field-induced redox reaction was postulated by Potember [8] to describe the phase transition described above. Many publications deal with the switching mechanism but the details are still not yet fully understood.

The following discussion is concerned with the influence of various contacts to Cu(TCNQ) and different device structures on the switching properties. As the switching process is activated electrically, an influence of the contact is

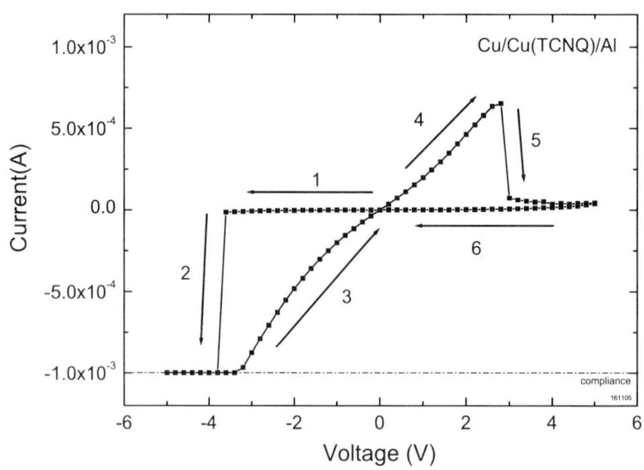

Figure 27.1 Typical $I-V$ characteristics of a Cu/Cu(TCNQ)/Al device with two stable resistance states. Numbers indicate the voltage sweep sequence.

to be expected. Beginning with Pt, we chose a set of metals with decreasing work function and investigated the influence on the *I*–*V* characteristics. We also noted an influence of the contact geometry and varied the contact size from a few mm^2 down to a metal-tip.

Further, we studied and traced the localisation of the switching in the device. Possible regions for switching could be the interfaces of the metal contact to the Cu(TCNQ) or in the Cu(TCNQ) bulk. In order to localise the switching we changed the stacked structure into a planar structure and extended it with an additional electrode.

There are also reports that switching is assisted by oxide layers formed at the metal–organic interface. It has been noted recently that typically aluminium is used for at least one of the metal contacts in many organic bistable devices. Aluminium is well known to readily form an oxide layer on its surface even at low oxygen concentrations. Additionally, it has been shown that aluminium oxide layers can improve the switching properties of Cu(TCNQ)-based devices [9, 10]. To further investigate the role of the contacts, devices with different contact configurations had been prepared, using either pure Au or Au with an oxide interlayer.

27.2 Experimental and Results

27.2.1 Device Preparation

Cu(TCNQ) devices used in our experiments were prepared as follows. Borofloat glass substrates were obtained from Schott and cleaned with boiling acetone, ultrasonically treated and then boiled in isopropanol. In addition, the substrates were treated with an oxygen plasma to remove organic contaminants. Next, a 300 nm Cu layer was deposited by e-beam evaporation onto the substrates and patterned by standard photolithography. After etching in $FeCl_3$ the Cu substrates were cleaned with acetone and isopropanol. According to Potember, the Cu(TCNQ) layer was formed by spontaneous electrolysis reaction. The Cu substrates were dipped into saturated TCNQ/acetonitrile solution kept at 40 °C. The immersion time was varied between 30 seconds and 120 seconds. While the Cu substrates were immersed a dark blue Cu(TCNQ)-layer was grown onto the Cu surface. The Cu(TCNQ) layer had a microcrystalline structure and a thickness of 3–4 μm. After the immersion into TCNQ/acetonitrile the Cu/Cu(TCNQ) device was rinsed with acetone and finally blown dry with nitrogen. Finally, a metal top contact was deposited by e-beam evaporation onto the Cu(TCNQ) surface through a shadow mask.

I–*V* characteristics of the devices were measured using a Keithley 238 Source-Measure Unit (SMU). A positive voltage corresponds to a positive potential on the top contact with respect to the bottom contact. The bottom Cu contact was grounded in all measurements. The current compliance of the SMU was set to avoid destruction of the devices in their low-resistivity state. No series resistor was used.

27.2.2
Contact Size

First, we analysed the influence of the top metal contact size on the switching behaviour. We used Au as the contact metal to exclude the possible influence of an unintentionally formed oxide interlayer between the metal and the Cu(TCNQ).

Devices with an Au top contact of ≥ 1 mm^2 did not show any kind of switching but only a slightly non-linear characteristic, as can be seen in Figure 27.2. With decreasing top contact size of the device the *I–V* characteristic changed and the difference between the two conductive states increased. As we reduce the size of the top contact down to $0.2 \times 0.2\ mm^2$ a weakly pronounced switching process could be observed (Figure 27.3). However, switching did not occur reproducibly at every sweep. Next, Figure 27.4 shows the behaviour of a 0.02 mm^2 Au pad size device. The branch of the low resistivity has been formed and the threshold and NDR region can be seen clearly.

Finally, the Au top contact was substituted by a sharp Au tip and placed onto the Cu(TCNQ) layer (see Figure 27.5 for *I–V* characteristics). In this case, pronounced switching occurred reproducibly continuously for many hundred sweep cycles. These results show that Cu(TCNQ) can be switched between two conductivity states with electrodes that do not form an oxide on their surface when the contact area is small enough.

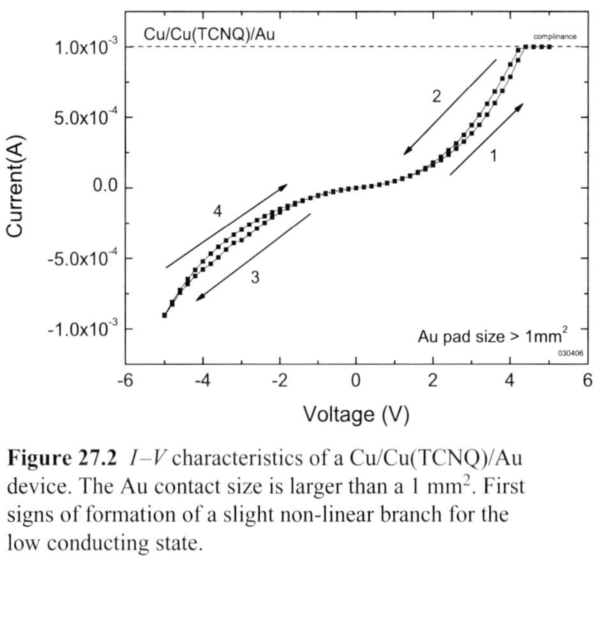

Figure 27.2 *I–V* characteristics of a Cu/Cu(TCNQ)/Au device. The Au contact size is larger than a 1 mm^2. First signs of formation of a slight non-linear branch for the low conducting state.

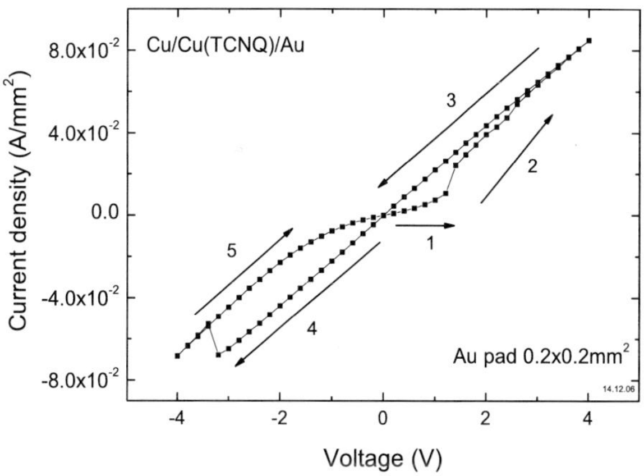

Figure 27.3 *I–V* characteristics of a Cu/Cu(TCNQ)/Au device with an Au contact size of 0.04 mm^2. The low conducting branch becomes more pronounced. Formation of threshold voltage at 1.2 V.

27.2.3 Oxide Interlayer Between Top Contact and Cu(TCNQ)

In the next experiment, an oxide interlayer was deposited between the Cu(TCNQ) layer and the Au top contact (0.5 × 2 mm^2). The Al_2O_3 layer had a thickness of 200 nm and was e-beam evaporated at a rate of 2–3 A/s before deposition of the Au pad. In spite of a 200 nm Al_2O_3 dielectric layer no expected insulating behaviour was observed. The rough surface profile (up to 1 μm) of a Cu(TCNQ) layer seemed not to be completely covered with insulating Al_2O_3.

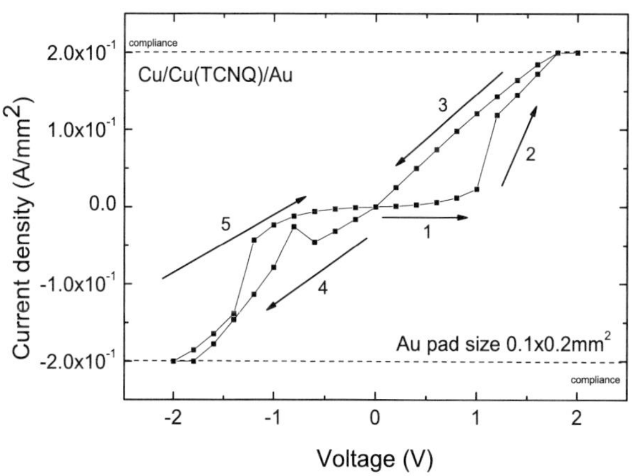

Figure 27.4 *I–V* characteristics of a Cu/Cu(TCNQ)/Au device with an Au contact size of 0.02 mm^2. The low conducting state has been developed. Threshold voltage and the NDR region are clearly visible.

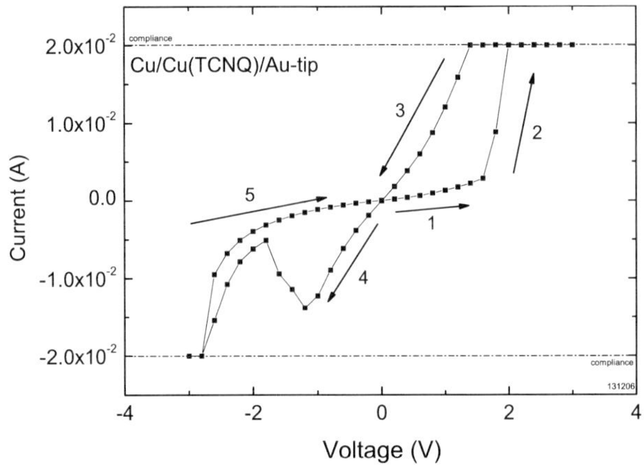

Figure 27.5 *I–V* characteristics of a Cu/Cu(TCNQ)/Au-tip device with two distinct stable conducting states.

These Cu/Cu(TCNQ)/Al_2O_3/Au devices showed distinct switching in ambient atmosphere even for large pad sizes. The average threshold voltage for switching was 2–3 V higher than in the Au device without an oxide interlayer. Similarly to the Au-tip devices without the oxide interlayer, the Al_2O_3 oxide layer device showed distinct and stable switching for many hundreds of sweep cycles. The return to the high resistive state was not always accompanied by a region of NDR but in some cases the switching occurred in a single step. Then the oxide interlayer was exchanged by e-beam evaporated ZrO_2 (stabilised with TiO_2) and Y_2O_3 layers. In fact, the ZrO_2 and Y_2O_3 interlayer showed similar and stable switching behaviour compared with the devices with Al_2O_3 interlayer.

The comparison between organic devices with an Al_2O_3 or ZrO_2 oxide interlayer and devices with small-area Au top contacts on the Cu(TCNQ) layer showed an inverse *I–V* characteristic. That is, the switching from low to high conducting state occurred at a positive threshold voltage applied at the top-contact in contrast to a negative bias voltage for devices without oxide interlayer.

27.2.4
Reversible Loss of Bistability in Oxygen-Free Ambience

A change of the atmosphere has a remarkable effect on the switching characteristic. For devices with an oxide interlayer the removal of oxygen leads to a reversible loss of bistability. After testing them in ambient atmosphere to confirm their bistability, devices were placed into a vacuum recipient, which was evacuated to a pressure of 10^{-7} mbar. Subsequently, the *I–V* characteristic was taken in vacuo. No switching was observed. The oxide interlayer devices showed a nearly ohmic characteristic with currents comparable to the high conducting state.

When dry nitrogen was filled into the recipient the behaviour of the Cu/Cu(TCNQ)/Al_2O_3/Au device did not change and the ohmic characteristic remained as seen in vacuo. Filling the recipient with pure oxygen or air restored the bistable behaviour. The effect could be repeated many times, with ohmic behaviour in oxygen-free ambience and bistability when oxygen was present. Figure 27.6 demonstrates the behaviour under different gas conditions. The $I-V$ characteristic in vacuum shows a slight ohmic behaviour without any threshold voltage or NDR region. Contrary to these curves, the same device behaves very similarly in air and dry O_2. In general, the threshold voltage and the NDR region varied from sweep to sweep and no relationship could be noticed between the condition in air or pure O_2.

Cu/Cu(TCNQ)/Al_2O_3/Au devices restored their bistability in a few seconds after contact with oxygen in contrast to ZrO_2. Devices with a ZrO_2 interlayer required more time to restore the bistability. During the $I-V$ characteristics sweep the high-resistivity branch slowly developed. The low conducting state was formed after several minutes. However, the Cu/Cu(TCNQ)Au-tip device (without oxide interlayer) did not change its behaviour in different atmospheres. The switching occurred in air, oxygen and nitrogen.

27.2.5 Tip Contacts of Various Metals to Cu(TCNQ)

The influence of the work function of several metals as top contact to Cu(TCNQ) was investigated. Pt was the metal with the highest work function in this series.

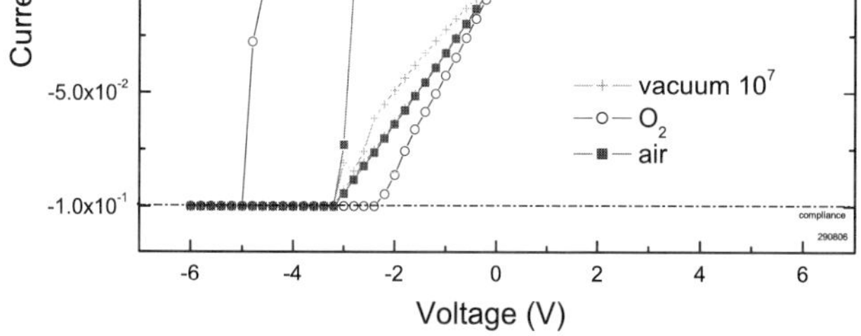

Figure 27.6 I–V characteristics of a Cu/Cu(TCNQ)/Al_2O_3/Au device in different ambient atmospheres. The removal of oxygen leads to the loss of bistability in vacuum. Without oxygen the device remains in the high conducting state. The bistability of the device can be restored by oxygen.

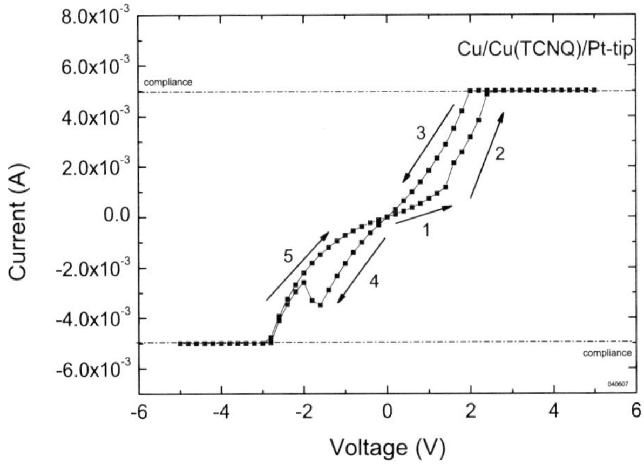

Figure 27.7 Typical $I-V$ characteristics of a Cu/Cu(TCNQ)/Pt-tip device.

The Cu/Cu(TCNQ)/Pt device showed similar $I-V$ characteristics as the Cu/Cu(TCNQ)/Au-tip device (Figure 27.7). In contrast to aCu/Cu(TCNQ)/Au-tip device the ratio between the low and high resistive state is smaller and the threshold voltage is not so sharp. The Pt tip was replaced by W, Mo, Cu, Al, and In tips and the $I-V$ characteristics were recorded. Cu, Al, In, and Zr showed similar behaviour as Cu/Cu(TCNQ)/Al_2O_3/Au devices. These devices showed an inverse $I-V$ characteristic as devices with Pt or Au tips. In Figure 27.8 representative $I-V$ characteristics of a Cu/Cu(TCNQ)/Cu-tip device are shown. Figure 27.9 shows $I-V$ characteristics of a Cu/Cu(TCNQ)/In-tip device. Compared with a device with an Au or Pt tip, the inverse characteristic can be well spotted. The $I-V$ characteristics of the W-tip device showed an unreproducible behaviour. The NDR regions of the $I-V$ characteristics were observed when positive or negative bias was applied. Occasionally, the device switched on immediately after passing through a NDR region.

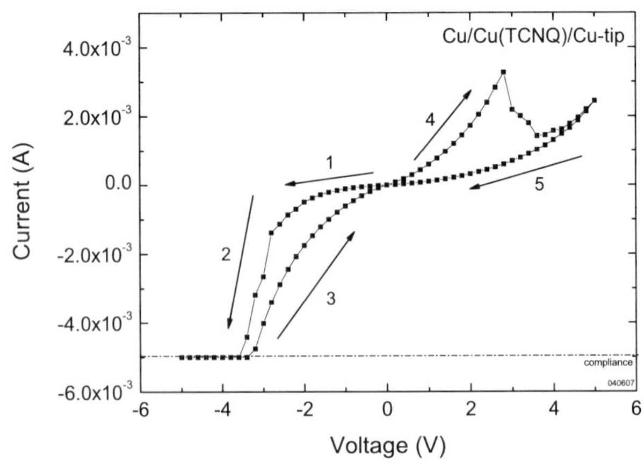

Figure 27.8 Typical $I-V$ characteristics of a Cu/Cu(TCNQ)/Cu-tip device.

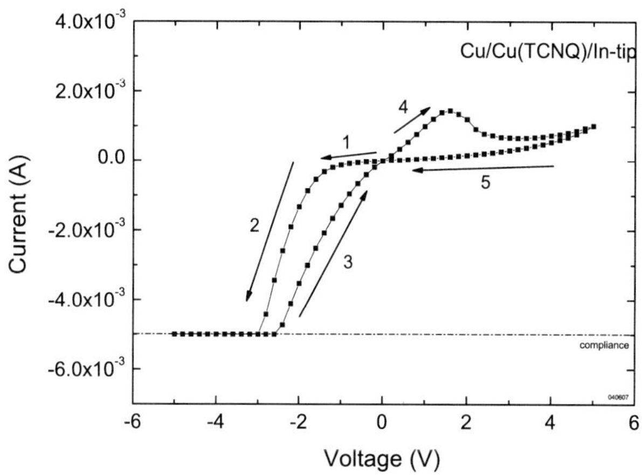

Figure 27.9 Typical *I–V* characteristics of a Cu/Cu(TCNQ)/In-tip device.

The threshold voltages for switching between the conducting states varied from metal to metal. A correlation between the work function and the value of the threshold voltage could not be determined because of significant fluctuations from measurement to measurement.

Table 27.1 shows the different metals for the tip contacts, their work functions and the type of characteristic that has been observed. The transition from regular to inverse characteristics occurs around 4.7 eV. We note that the work functions are literature values and may be slightly different for the tip-contacts used. No effort was made to remove any oxide or adsorbent layer from the tips.

Furthermore, Li- and Ca/Cu(TCNQ)Cu devices were fabricated and measured in vacuum to avoid oxidation because of their high reactivity. A Cu/Cu(TCNQ) substrate was placed into a substrate holder with two contact pins. The substrate was covered with a shadow mask that exposed only a part of the Cu stripe

Table 27.1 Work function [11] and switching behaviour for metals that have been used as tip contacts for Cu/Cu(TCNQ) devices.

top contact	work function (eV)	characteristic
Pt	5.7	regular
Au	5.1	regular
Mo	4.8	regular
Cu	4.7	inverse
W	4.6	undefined
Zn	4.4	inverse
Al	4.3	inverse
In	4.2	inverse
Zr	4.1	inverse

and a part of Cu(TCNQ). The first of the contact pins was positioned onto the Cu stripe and the second one beyond the Cu(TCNQ) stripe on a small spot of Ag conductive lacquer. The Ag spot and the Cu(TCNQ) stripe had no electrical connection. The evaporated metal covered the Cu/Cu(TCNQ) stripe and the Ag spot and the electrical connection was established. The *I–V* characteristics were recorded while a top layer of Li/Ca was being deposited by thermal evaporation at 10^{-6} mbar. However, no reproducible switching was observed in Li- and Ca/Cu(TCNQ)/Cu devices.

27.2.6 Planar Device Structure

A drawback of the stacked device structure is that the devices are not symmetrical. The contact area on top of the Cu(TCNQ) layer is much larger than underneath the layer. The surface of the bottom contact Cu layer is very smooth compared with the rough surface of the grown polycrystalline Cu(TCNQ) layer. Additionally, one can expect that the properties of the top contact are changed by adsorbates on the Cu(TCNQ) layer. In order to equalise the contacts a planar device structure was developed as shown in Figure 27.10. The planar device comprises two metal contacts on the substrate covered by a Cu(TCNQ) stripe. Through this planar symmetrical structure the devices do not have any structural preferences. If an irreversible 'forming' process [16] would take place in the device during the first measurement (known from bistable devices based on oxide layers sandwiched between metal contacts), it should have an unsymmetrical *I–V* characteristic with a polarity that would depend on the polarity of the first voltage sweep.

The planar structure was fabricated as follows: the glass substrate was carefully cleaned as described above. A 300 nm metal layer was deposited onto the substrate by e-beam evaporation. Next, the metal layer was patterned by standard photolithography. After metal etching, the device structure contained two parallel metal stripes that served as contacts. The distance between the two stripes was 10 μm. Then, a Cu-stripe with a thickness of 70 nm was e-beam evaporated onto the contacts. Finally, the substrate was immersed into the TCNQ/acetonitrile solution. The device was kept in the TCNQ/acetonitrile solution until the Cu layer was completely converted into Cu(TCNQ). Finally, the substrate was rinsed with acetone and dried with nitrogen. Figure 27.10 shows a schematic drawing of the device.

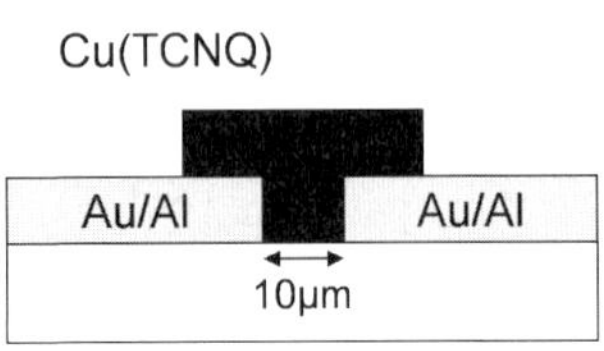

Figure 27.10 Schematic illustration of the planar symmetrical device.

First, an Au/Cu(TCNQ)/Au device was fabricated and the *I–V* characteristics were taken. No switching was ob-served and only a slightly non-linear behaviour was detectable. In contrast, Al/Cu(TCNQ)/Al devices showed switching, but the behaviour differed significantly from the previously measured *I–V* characteristics of stacked devices (Figure 27.11). It appears to be a superposition of one of the regular and one of the inverse characteristics, which were composed into one symmetrical curve. Applying positive voltage, the run of the *I–V* characteristics passes through a NDR region and the current increases significantly at 6–7 V. A similar behaviour was observed when a negative voltage was applied. This suggest that the switching occurs at both interfaces of Cu(TCNQ) and the metal contacts or in their vicinity.

27.2.7 Localisation of Switching Region

The prior experiments indicated switching near the interface of Cu(TCNQ) and the contacts. However, from the results with the symmetric device we can not conclude at which of the contacts (positively or negatively biased) the switching occurs. Hence, the previous planar structure was extended by an additional contact to facilitate the localisation of the switching. This electrode was placed into the gap between the two metal contacts and also covered with Cu(TCNQ) as described above.

The extended device was fabricated as follows. After substrate cleaning a 100 nm Au-layer was deposited by e-beam evaporation. After photolithographic patterning an Au middle contact stripe of 5 μm was patterned. Through the next deposition, photolithography and etching steps two Al contact stripes were formed. The two Al stripes were positioned parallel to the previous Au

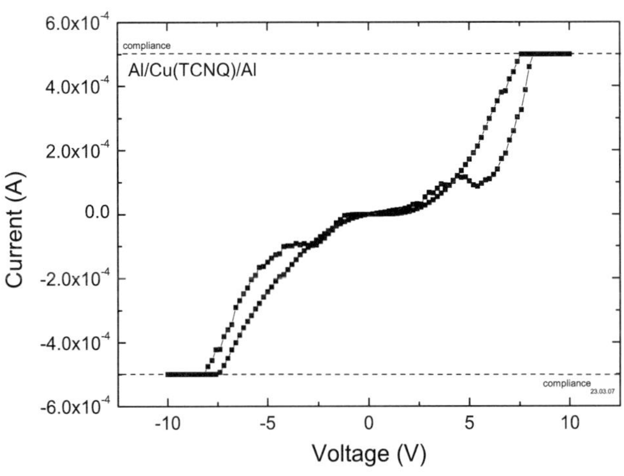

Figure 27.11 *I–V* characteristics of a planer Al/Cu(TCNQ)/Al device, corresponding to the superposition of the characteristics of two anti-parallel connected devices.

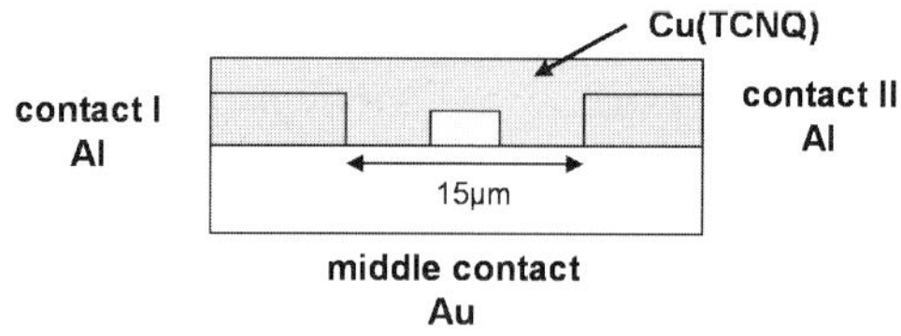

Figure 27.12 Schematic illustration of the extended planar structure. Through this extension, the device contains two device parts with one common centre electrode.

stripe contact and had a 15 µm gap between them. Finally, a 70 nm Cu stripe was deposited by e-beam evaporation through a shadow mask. This Cu stripe covered the gap and partly the Au and Al contacts. After the evaporation the device was immersed into the saturated TCNQ/acetonitrile solution. Again, the device was kept in the solution until the Cu was completely converted into Cu(TCNQ). Finally, the device was rinsed with acetone and dried by nitrogen. During the $I-V$ characterisation, the middle electrode was grounded.

Effectively, the extended device contained two symmetrical device parts with one common electrode. Figure 27.12 shows the design of the extended planar structure. Two outer metal contacts I and II with a gap of 15 µm can be recognised. The common electrode is positioned in the middle of the gap.

The additional centre contact allows us to use one common contact in the switching process of two different devices. The device with contacts I and II should operate as two separate device parts with one centre electrode with a common Cu(TCNQ)-layer. Applying a suitable voltage to one of the two device parts should switch this part from one conducting state into the other. If the switching would occur at the interface between the common centre electrode and the Cu(TCNQ), it should be possible to influence the properties of

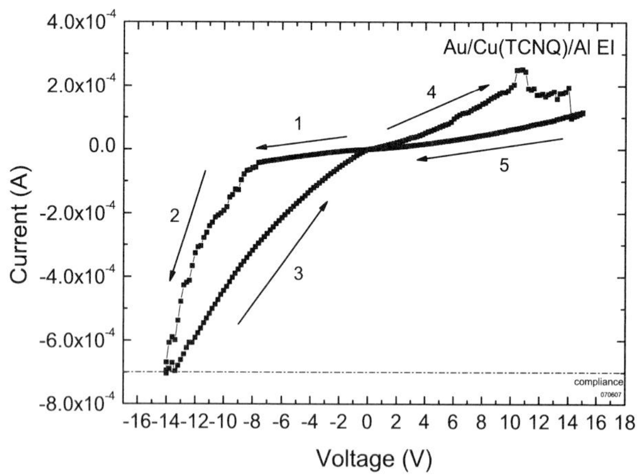

Figure 27.13 Typical $I-V$ characteristics of an Al/Cu(TCNQ)/Au device (Au-contact grounded). This device is a part of a symmetrical Al/Cu(TCNQ)/Au/Cu(TCNQ)/Al device.

the second device part. If switching in device part one would not evoke any change in device part two, then one would conclude that the switching occurs at the opposite interface or the nature of the switching is rather a bulk effect of the Cu(TCNQ). In order to investigate and localise the switching we fabricated devices with outer Au contacts and a centre Al contact as well as structures with outer Al contacts and a centre Au contact.

First, the proper switching function of the device was verified. Both device parts were tested and each of them showed bistable behaviour as it can be seen for device part Al I in Figure 27.13. The NDR region and the threshold bias were shifted to higher voltages and appeared at 10–14 V as compared with a few volts in the stacked devices. Additionally, the current did not rise as fast and abruptly at the threshold voltage.

Subsequently, a suitable voltage was applied to the device with the Al II contact and the high resistivity state was set. This high resistivity state has been checked and the *I–V* characteristics have been recorded (Figure 27.14). Applying a positive voltage the run of the curve did not pass the NDR region because the device was already in the low conducting state. After this check the Al II device was transferred into the high resistivity state, again.

Next, a negative voltage was applied to the Al I device part, switching this part into the high conducting state. A voltage sweep was applied to the Al II device part and the run of the curve was recorded to check its conductivity state. The measured current increased slowly and no NDR region was observed.

Indeed, the Al II device remained in the low conducting state and no switching was noticed. This behaviour was observed many times and no influence between the two devices with an Au common electrode was noted. Therefore, the *I–V* characteristics of the two symmetrical device parts indicate that the switching does not appear at the interface of the Au contact and the Cu(TCNQ)

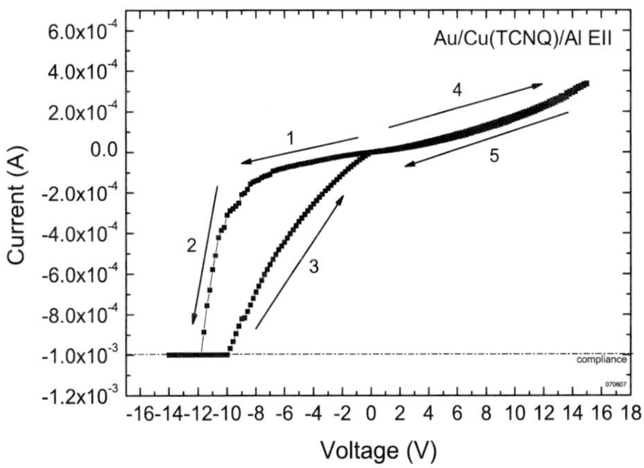

Figure 27.14 *I–V* characteristics of an Al/Cu(TCNQ)/Au device set to low conducting state. A voltage sweep was applied to verify the conductivity state.

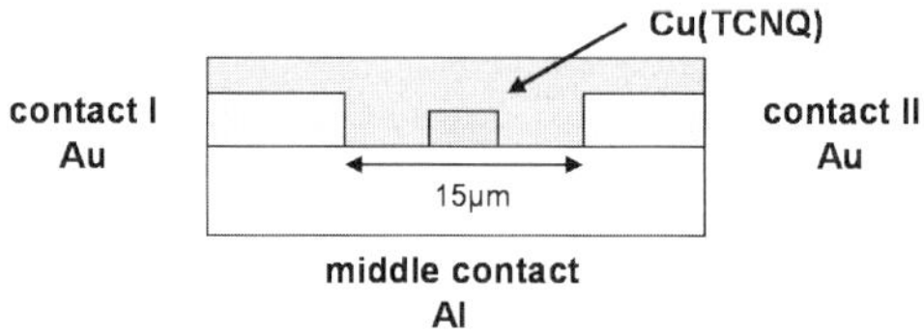

Figure 27.15 Schematic illustration of an extended planar Au/Cu(TCNQ)/Al/Cu(TCNQ)/Au device.

layer. A possible localisation for the switching could be the interface Al and Cu(TCNQ).

The following experiment has been done to verify the switching at the interface of the Al contact and Cu(TCNQ). We fabricated a planar Au/Cu(TCNQ)/Al/Cu(TCNQ)/Au device, as can be seen in Figure 27.15. The fabrication process was similar to the Al/Cu(TCNQ)/Au/Cu(TCNQ)/Al device described above. The Au common centre electrode was substituted by Al and the outer contacts Al I and Al II by Au electrodes. The device was connected as previously with the centre contact grounded.

Again, first of all proper switching behaviour of the device parts was checked. The Au I/Cu(TCNQ)/Al and Au II/Cu(TCNQ)/Al devices showed switching that was comparable to the behaviour of an Al/Cu(TCNQ)/Au device part but with inverse characteristic as a result of the exchange of the contacts. After the operation test the Au II device was transferred into the low conductivity state. The Au I device part was switched from low conducting state into the high conducting state, as can be seen in Figure 27.16.

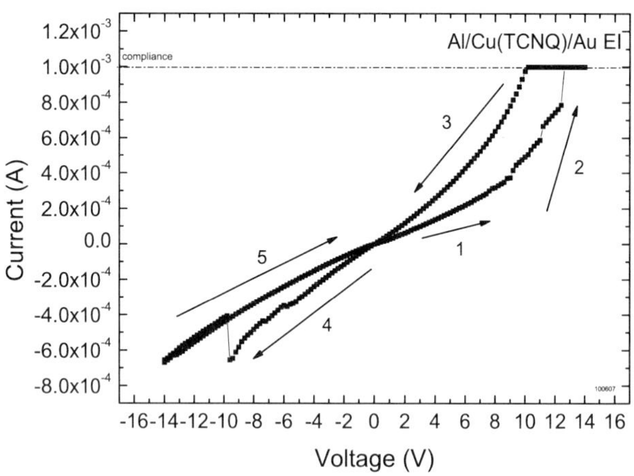

Figure 27.16 $I-V$ characteristics of an Au/Cu(TCNQ)/Al device part (Al contact grounded). A positive voltage sweep was applied and the device switched from low to high conducting state. After a negative voltage sweep the device was reversed into the low conducting state.

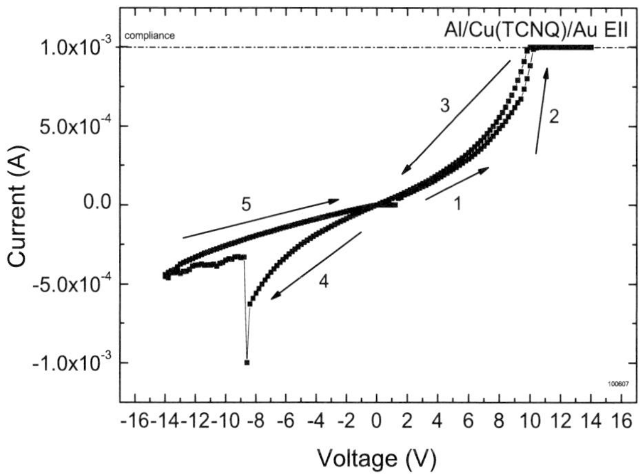

Figure 27.17 *I–V* characteristics of an Au/Cu(TCNQ)/Al device part (Al contact grounded) that has been in the high conducting state. Applying a positive voltage did not change the conductivity. The low conducting state was reached as a reverse voltage was applied.

Afterwards, we applied a positive voltage to the Au II device part and recorded the *I–V* characteristics. The run of this curve increased rapidly without a sudden rise of current or a NDR region, as can be seen in Figure 27.17. Through a negative voltage sweep the Au I device part was switched back into the low conducting state. This behaviour shows that the device part was already switched on before the measurement.

Indeed, the Au II device part had been switched into the high conducting state through switching of the Au I device part. Hence, these results showed that it is possible to initiate switching in one part of the device by switching the other part.

27.3 Discussion and Conclusion

From the results presented above, a number of conclusions for the nature of the switching mechanism can be drawn:

The majority of reports on switching in Cu(TCNQ) films relate to configurations with relatively large-area contacts on Cu(TCNQ) films in connections with unintentionally or deliberately formed oxide films between a metal contact and the organic layer.

Kever et al. [10] have proposed that the main reason for the switching is the field-assisted diffusion of Cu-ions into the oxide layer that is commonly found at the Cu(TCNQ)/metal interface in devices with aluminium top-contacts, leading to the formation of conductive filaments through the oxide. We have demonstrated switching of Cu(TCNQ) devices without an oxide interlayer. We could show that for Au contacts, reduction of the size of the contact can facilitate switching. When using an even smaller tip-contact, distinct switching was observed.

These results for contacts, which do not have an interfacial oxide layer, but do show switching properties, indicate that the oxide layer may assist the switching process, but can not be the main reason for the phenomenon. This is supported by other reports on switching in configurations without oxide interlayers [12, 13]. Remarkably, these reports also relate to structures with very small cross sections, i.e. an STM-tip and nanowires. Therefore, the oxide layer seems to be essential to achieve stable switching in devices with relatively large-area contacts.

In connection with the results obtained by Cölle et al. [14], who found a filamentary nature of current through devices comprising various organic materials (although not Cu(TCNQ)) and an interfacial oxide layer underneath the metal contact, we come to the conclusion that in the case of Cu(TCNQ) devices, the switching process takes place in the Cu(TCNQ) layer, while the main role of the oxide is to effectively reduce the contact area by providing highly localised current paths or to cause local inhomogenities in the electric field distribution.

We fabricated devices with various metal-tip contacts and investigated the switching behaviour dependence on the metal work function. In our experiments, the top contact with the highest work function was Pt. In addition, devices with Au, Cu, Mo, W, Zn, Al, In, and Zr-tips have been measured.

The experiments have shown that devices with tip-contacts from this set of metals can be separated into two groups, depending on the metal work function: The first group with metals above 4.8 eV (Au, Pt, and Mo) shows a switching characteristic with regular polarity. The second group (Zn, Al, In, Zr) with 4.7 eV and below, switches at reverse polarity, with unreproducible results for W (4.6 eV).

Interestingly, devices with a large-area Au contact and an additional oxide interlayer of Al_2O_3 and ZrO_2 show reverse switching behaviour as compared with the small-sized (or tip) Au contacts. If the oxide would only confine the current injection to localised paths, one would expect identical switching polarity as with an Au-tip contact. Obviously, the oxide also plays a role in modifying the effective work function of the contact.

This is supported by the fact that devices with a large Au top contact and an oxide interlayer show a dependence on atmosphere. The influence of oxygen on the conduction properties of oxide layers (in stacked devices comprising an Al_2O_3 layer sandwiched between an Al and Au contact) has already been reported by Emmer et al. [15]. He proposed that adsorbed oxygen could immobilise conducting electrons in the oxide layer that originate from structural defects and respective localised energy states. The energy states can form conductive filaments through the oxide layer. O_2 can easily penetrate the dielectric layer and can be chemisorbed at the defects, immobilising the free electrons.

Removal of oxygen leads to the formation of highly conductive paths through the oxide layer. An increasing conductivity of the oxide would definitely change the electrical field distribution at the interface and accordingly change the injection properties of the contact.

The planar device structure allowed us to investigate the switching behaviour of a completely symmetric device. Stacked devices have the disadvantage that the properties of the top and bottom contact are never exactly equal, e.g. due to different interface roughness or adsorbates on top of the Cu(TCNQ)-layer. In our planar device with Al-contacts, we were able to observe a switching behaviour that can be attributed to two independent, antiparallel devices: when an increasing bias was applied, the run of the curve passed a NDR region and increased slowly up to a certain voltage. At this voltage, the device left the high resistance state and jumped into a high conducting state. An identical behaviour was observed for the voltage sweep with reverse polarity.

This indicates that the switching process is not a bulk effect but located in the vicinity of the contacts. However, from this experiment we could not determine if switching happens at the positively or negatively biased contact. Planar devices with two Au contacts did not show switching but only a slightly non-linear behaviour.

Furthermore, we have extended the planar device structure to localise the interface where the switching appears. An additional contact stripe was located in the gap between the existing two metal contact stripes. In this case, the distance between the two parallel metal contacts was 15 μm. The middle electrode was positioned between them and had a width of 5 μm. The contacts were covered with Cu(TCNQ) by complete conversion of a previously deposited Cu layer.

We fabricated devices with Al as outer electrodes and Au as centre contact and device with the reverse contact configuration (Al centre contact/Au outer electrodes). For the measurements, the centre electrode was grounded and the outer electrodes were used alternately to connect the device. The unused contact was left floating.

For the device with Al outer contacts, the two parts of the device (one Al contact and the centre Au contact) each showed the expected switching behaviour. Each of the device parts switched independently from the other. When Au was used for the outer electrodes and Al for the centre contact, a different behaviour was observed: applying a voltage to one of the Au outer electrodes that switched the respective device into the On-state influenced the state of the other device part. When this device was in the Off-state before, it also changed into the On-state. This behaviour was reproducible many times.

From this effect we can conclude that the switching process occurs at the interface or in the near vicinity of the Al-contact. The device switches on when the Al-contact is negatively biased with respect to the Au-electrode.

In conclusion, our results show that the switching process occurs in the Cu(TCNQ). It is not a bulk phenomenon but localised at the interface to the metal contact. Initiation of the switching process is favoured when the interface comprises an oxide layer. Presumably, this oxide layer is required to achieve special carrier injection properties of the contact. Low work function metals readily form a thin oxide layer at their surface and therefore facilitate switching. The low-resistivity state is achieved when the contact with the lower work function is negatively biased. In the case of stacked devices with noble metals top-contacts, which show a reverse switching polarity as compared with those with an Al-top electrode, we suppose that switching occurs at the Cu-bottom contact.

Acknowledgements

We gratefully acknowledge the Deutsche Forschungsgemeinschaft for financial support within the priority program 1121.

References

1. R. S. Potember, T. O. Poehler, and D. O. Cowan, Appl. Phys. Lett. **34**, 405 (1979).
2. K. Sakai, H. Matsuda, H. Kawada, K. Eguchi, and T. Nakagiri, Appl. Phys. Lett. **53**, 1274 (1988).
3. N. Watanabe, Y. Iwasa, and T. Koda, Phys. Rev. B **44**, 11111 (1991).
4. H. J. Gao, Z. Q. Xue, K. Z. Wang, Q. D. Wu, and S. Pang, Appl. Phys. Lett. **68**, 2192 (1996).
5. D. Tondelier, K. Lmimouni, D. Vuillaume, C. Fery, and G. Haas, Appl. Phys. Lett. **85**, 5763 (2004).
6. M. Lauters, B. McCarthy, D. Sarid, and G. E. Jabbour, Appl. Phys. Lett. **89**, 013507 (2006).
7. E. I. Kamitsos, C. H. Tzinis, and W. M. Risen, Solid State Commun. **42**, 561 (1982).
8. R. S. Potember, T. O. Poehler, and R. C. Benson, Appl. Phys. Lett. **41**, 548 (1982).
9. T. Oyamada, H. Tanaka, K. Matsushige, H. Sasabe, and C. Adachi, Appl. Phys. Lett. **83**, 1252 (2003).
10. T. Kever, U. Böttger, C. Schindler, and R. Waser, Appl. Phys. Lett. **91**, 083506 (2007).
11. S. M. Sze, Physics of Semiconductor Devices (John Wiley & Sons, New York, 1981), p. 251.
12. K. Xiao, I. N. Ivanov, A. A. Puretzky, Z. Liu, and D. B. Geohegan, Adv. Mater. **18**, 2184 (2006).
13. S. Yamaguchi, C. A. Viands, and R. S. Potember, J. Vac. Sci. Technol. **9**, 1129 (1991).
14. M. Cölle, M. Büchel, and D. M. de Leeuw, Org. Electron. **7**, 305 (2006).
15. I. Emmer, Thin Solid Films **20**, 43 (1974).
16. G. Dearnaley, A. M. Stoneham, and D. V. Morgan, Rep. Prog. Phys. **33**, 1129 (1970).

28
Organic Field-Effect Transistors for Spin-Polarised Transport

M. Michelfeit, G. Schmidt, J. Geurts, and L. W. Molenkamp

28.1
Introduction

Storage and manipulation of information is one of most important issues in modern technology. Traditionally information is manipulated and controlled using conventional integrated circuits which are operated by the transfer of charge. In the past, e.g. CMOS devices kept becoming faster and smaller due to great progress in traditional semiconductor technology. In particular, improving the techniques of processing elementary semiconductors, initially namely silicon, has been one of the largest fields of research in past decades. However this technology will face fundamental physical limitations that are predicted to halt Moore's law of doubling computational speed every 18 months in approximately 10–15 years.

Thus it is clear that in order to preserve progress beyond this period, new ideas are required and new technologies have to be established. One approach to this task is based on the concept of manipulation of the spin of electrons, which brings a new degree of freedom to the conventional charge based electronics. This technology of spin-electronics, or spintronics, has been brought more and more into the focus of research in recent years [1].

Major tasks in this field are the injection, transport and detection of spin polarised currents. Thus new multifunctional devices including spin valves (magnetoresistive devices), quantum bits for quantum computing, spin-polarised light-emitting diodes (spinLEDs) and spin-polarised field-effect transistors (spinFETs) can be and have already been partially realised. The most important fundamental mechanisms that form the basis of these applications are namely the giant magnetoresistance (GMR) and the tunnelling magnetoresistance (TMR). It should be noted that the character of spintronic effects strongly depends on the choice of the materials. A variety of experiments have been made using different combinations of materials. While ferromagnetic metals and semiconductors are mostly used as spin injectors and detectors, insulators or non-magnetic metals and semiconductors are often employed as tunnel junctions or for spin transport, respectively.

Organic Electronics. Structural and Electronic Properties of OFETs. Edited by Christof Wöll

ISBN: 978-3-527-40810-8

One of the major ingredients for spintronics is the transfer of spin-polarised electrons into a non-magnetic semiconductor, the so called spin-injection. Highly efficient spin-injection into a semiconductor was realised for the first time in 1999 [2], using a dilute magnetic (II,Mn)VI semiconductor as spin aligner, n-doped AlGaAs for spin transportation and a spinLED geometry for spin detection by analysing the polarisation of the light emitted by polarised electrons recombining with unpolarised holes. Since then spin injection into semiconductors has been successfully detected in several experiments.

In the last few years, an increasing number of research groups have been focussing on the incorporation of organic materials in spintronics concepts. The application of this material class for spintronics purposes is motivated by the extremely weak spin scattering in most of these materials. At the focus of this chapter are organic field-effect transistors (OFETs) for spintronics applications. In the first part, some concepts and progress of spintronics activities and recent developments in the application of organic materials in this field are summarised. After the description of these aspects of the present status of spintronics research, our concept of OFETs for spin-polarised transport is explained. Subsequently, results are presented of our OFET research activities: (i) optimisation of bottom-contact OFETs with dihexylquaterthiophene (DH4T) as active layer, and (ii) the realisation of DH4T-based OFETs for achieving transport by spin-polarised electrons, injected from ferromagnetic source and drain electrodes, whose magnetisation directions can be tuned independently by an external magnetic field.

28.2
Concepts and Progress of Spintronics

The application of spin polarised transport in devices, which constitutes the basic idea of spintronics, was triggered in the late 1980s by the discovery of giant magnetoresistance (GMR) in metallic layer systems [3, 4]. This effect occurs in sandwich systems of ferromagnetic metal, non-magnetic metal, ferromagnetic metal. For parallel orientation of the magnetisation of both ferromagnetic layers, the electronic conductivity is strongly enhanced with respect to the antiparallel case. Conductivity switching is achieved by a reorientation of the magnetisation of one of the ferromagnetic layers by an external B-field. After only a few years, GMR devices were implemented as miniaturised harddisk reading heads.

Subsequently, the interest of spintronics research activities was increasingly focused on layer systems with a semiconductor material instead of the nonmagnetic metal. Spin-polarised electrons should be injected from a ferromagnetic metal into a semiconductor layer. The large spin-flip length in the semiconductor materials should be exploited for spin-polarised transport through the semiconductor layer. One of the first proposals for a spin-polarised transistor originates from Datta and Das [5]. In their concept, the externally driven switching of the transistor is performed by manipulating the spin orientation of the electrons on their way from source to drain by a gate voltage, which affects their spins through spin–orbit coupling.

Many attempts were made to realise this concept and related ones, most of them based on III–V semiconductor layers like InAs or (In,Ga)As with Fe or Co contacts. However, they brought no convincing demonstration of spin-polarised transport. Instead, their most relevant impact was a more profound knowledge of the specific challenges of spin-injection and detection, which finally resulted in a successful demonstration of spin-polarised transport through semiconductor layers, based on a distinctly modified injection concept.

The essential reason for the failure of spin injection from a ferromagnetic metal into a semiconductor turned out to be the conductivity difference between these materials, which amounts to several orders of magnitude. It was shown that for the usual parameters of a ferromagnetic metal (i.e. (i) non-complete spin-polarisation at the Fermi-level, and (ii) spin-flip length ~10 nm) the injection of spin-polarised electrons into a semiconductor is prohibited by fundamental laws of electrodynamics [6]. This impossibility of spin injection from a ferromagnetic metal into a semiconductor because of its conductance mismatch was derived consistently from models, based on electrochemical potential alignment or on a circuit of parallel conductivity channels for spin-up and spin-down polarisation, respectively [7].

A breakthrough in spin-injection into a non-magnetic semiconductor was achieved by replacing the ferromagnetic metal contact by another semiconductor layer, which should warrant conductance-matching, and, most importantly, supply electrons with 100% spin-polarisation. These requirements are fulfilled by dilute magnetic semiconductors (DMS), e.g. II–VI compounds with a few percent of manganese. Through (s–d) exchange interaction the Mn incorporated in the II–VI host material induces a giant Zeeman splitting of the conduction band states when an external magnetic field is applied. For (Zn,Mn)Se the splitting amounts up to 20 meV. Thus, at low temperature, the electron spin-polarisation can be tuned by an external B-field and a complete spin-polarisation can be achieved for fields in the order of one Tesla.

In 1999, successful spin injection from (Zn,Mn)Se into GaAs was demonstrated [2]. The degree of polarisation of the injected electrons in GaAs was optically determined from polarised luminescence radiation in the GaAs (spin-LED). Following symmetry selection rules of the optical recombination of electrons and holes in GaAs, the degree of circular polarisation corresponds to the spin-polarisation of the recombining electrons.

Voltage controlled spin-filtering was demonstrated in spin polarised resonant tunnelling diodes (RTD). The heart of RTDs is a semiconductor quantum well (QW), sandwiched between two tunnelling barriers. By an external voltage the energy levels of the well are tuned with respect to the electron emitter. Maximum current occurs for resonance, i.e. when the QW energy level coincides with the Fermi-level of the emitter. In the case of a spin-polarised RTD the QW consists of a diluted magnetic semiconductor. Thus, the giant Zeeman splitting between the QW energy states of different spin orientation can be tuned by an external B-field. This should induce a B-dependent splitting of the resonance maximum into one peak for spin-up electrons and one for spin-down. This concept was experimentally confirmed in RTDs with a (Zn,Mn)Se well, in which a splitting of the resonance peak up to 20 meV was observed [7].

An appealing alternative to DMS spin aligners are the above mentioned ferromagnetic metal contacts, applied in conjunction with a very thin metal–oxide tunnelling layer for conductance matching with the semiconductor material, e.g. $Fe/Al_2O_3/GaAs$. This ferromagnet-based approach is highly relevant for the concept of defining a tuneable, but non-volatile conductance state of a FET, which is in the scope of this chapter. The electron transmission probability of the oxide layer is spin-dependent, because it is proportional to the product of the densities of states at both sides of the tunnelling barrier. Beside the traditional application of Al_2O_3 barriers in metal–oxide tunnelling magnetoresistance devices [8], they were implemented into spin LEDs [9]. They have enabled a considerable spin injection efficiency even at elevated temperatures. An even more pronounced polarisation of injected spins in GaAs was achieved by applying MgO [10]. The main challenge in the concept of spin injection through oxide tunnelling contacts is the exact control of the oxide layer thickness. The exact thickness control is required in order to avoid that the tunnelling contact resistance distinctly exceeds that of the semiconductor material. An exaggerated oxide thickness would imply a conductance mismatch opposite to the one which commonly occurs between ferromagnetic metals and semiconductors and leads to increasing spin-flip in the latter one [11]. Besides, it would drastically reduce the current through the device.

28.3
Organic Semiconductors in Spintronics Applications

The application of organic semiconductors in spintronic devices combines two major innovative research fields. Apart from the general advantages of organic electronics like low cost materials, processing, tunability of electronic and optical properties, etc. the interest for its use in spintronics applications is mainly driven by the low probability of spin scattering in most organic semiconductor materials. The reason is that spin–orbit scattering and hyperfine interaction are both very weak in conjugated organic semiconductors because of the low atomic numbers of the elements involved. This leads to long spin relaxation times of about 1 microsecond or even higher at room temperature.

A number of different research groups have reported strong magnetoresistance in FM-OS-FM devices, where FM is a ferromagnetic electrode, and OS an organic semiconductor. A significant spin valve effect of about 40% associated with spin-polarised injection has been demonstrated at 11 K [12]. Perhaps the most intriguing fact is that in some cases the magnetoresistance is observed even at room temperature [13], although the nature of this magnetoresistance is still under discussion. In addition, among other results, an intriguing negative magnetoresistance has been reported in hybrid organic-inorganic devices with no magnetic electrodes [14].

Also organic light emitting diodes with spin-polarised electrodes have attracted a considerable interest of different groups [Naval Research Laboratory, IBM Zurich Research Laboratory, Cambridge University, ISMN-CNR Bologna]. One of the most interesting results was the observation of unusual spectral properties in OLEDs (i.e. a strong red shift) that has not yet been understood [15].

The physical processes in spintronic devices are strongly governed by interface properties. One example for this is given by [16] in which strong magnetic field effects in organic light emitting diodes were tentatively attributed to peculiar interface properties between the ferromagnetic manganite electrode and the organic emitter.

In the last few years the first theoretical models describing spin injection at hybrid organic-inorganic interfaces and spin transport in organic materials were proposed. Most models take into account the polaronic nature of carriers in organic semiconductors [17, 18]. The interface role was strongly underlined by, among others, Smith and coworkers [19].

In addition to all this, an intriguing possibility to use hybrid organic-ferromagnetic spin devices as very sensitive magnetic sensors (10^{-18} T) was also pointed out in [20].

28.4 OFET Concept for Spin-Polarised Transport

The concept of the organic field-effect transistor for spin-polarised transport (spinOFET) is mainly based on that of a conventional OFET. The essential modification is the employment of ferromagnetic materials for the source and drain electrode instead of the usually applied noble metals. Due to these ferromagnetic electrodes, an external magnetic field can be applied as an additional switching parameter. So even with an applied gate voltage above the threshold voltage, the FET can still be (de-)activated by changing the magnetisation of the source and drain electrodes. Parallel magnetisation of source and drain electrodes causes a small channel resistance (activated-state) whereas anti-parallel magnetisation leads to a high channel resistance (deactivated-state). The electrical output characteristics curves for both states are shown schematically in Figure 28.1.

A requirement for achieving both states is the ability to switch the magnetisation of source and drain electrodes separately by an external magnetic field. For this purpose source and drain contacts with different coercive fields are needed. This requirement is fulfilled through exploiting the size dependent domain wall nucleation energy. Narrow stripes tend to be magnetised parallel

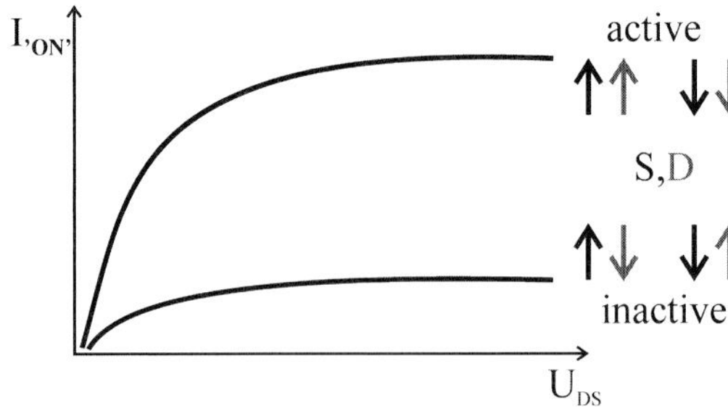

Figure 28.1 Schematic plot of two output curves of a spin OFET for parallel and antiparallel magnetisation of the source (S) and drain (D) contacts. Both curves correspond to the same value of the gate/source voltage above threshold. Thus electrically the OFET is in the 'ON'-state. However, substantial current flow only occurs for parallel magnetisation of the S and D contacts (active state).

to the stripe direction. The narrower the stripes are the more energy is needed to nucleate a domain wall which leads to a magnetisation reversal of the stripe. For this reason we apply contact stripes of two different widths for source and drain electrodes. The broad stripes should then have a smaller coercive field than the narrow ones.

The top panel of Figure 28.2 schematically shows the hysteresis curves of ferromagnetic stripes with two different widths. Without loss of generality we assign the magnetically softer, i.e. the wider stripe, to the drain (D) and the magnetically harder one, i.e. the narrower stripe, to the source electrode (S).

The effect of the S and D magnetisation direction on the channel resistance R of the FET is given in the lower panel of Figure 28.2. Obviously, the application of a strongly negative magnetic field $H < -H_{C,S}$ results in a parallel magnetisation state of source and drain. The FET is activated. If we subsequently increase the magnetic field into the positive regime barely over the coercive field of the drain contact $H_{C,D}$, it will change the drain magnetisation direction, whereas the source contact remains in its original state. Thus we have achieved antiparallel magnetisation of the electrodes resulting in an increased channel resistance, which means deactivation of the FET. Any further raise of the magnetic field above the coercive field of the source contact $H_{C,S}$ would cause parallel magnetisation (i.e. FET activation) again. However, a much more convenient route towards reactivation is the application of a negative magnetic field strength which exceeds the negative coercive field of the drain electrode $-H_{C,D}$ because here a lower field amplitude than $H_{C,S}$ is required. So the different coercive fields allow switching of the magnetisation of only the magnetically softer electrode (e.g. drain) while the magnetisation of the harder one (source) can be kept constant all the time.

Most relevant for applications is the memory effect of the spinFET: whenever the external magnetic field is turned off, the electrodes remain in their

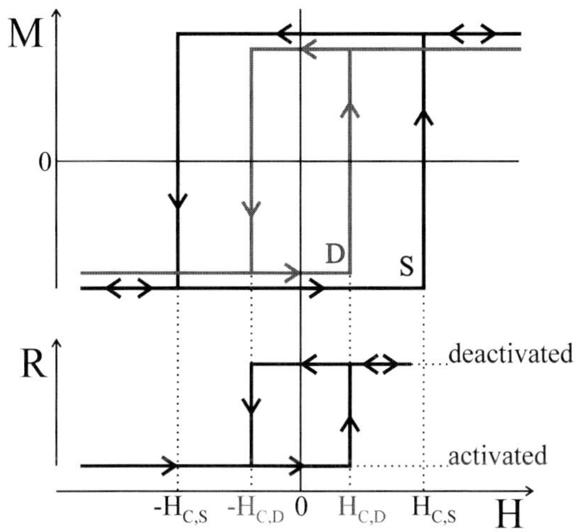

Figure 28.2 Spin FET switching concept. The top panel shows the hysteresis of two contact stripes with different widths. The magnetically softer stripe is assigned to the drain D and the other one to the source contact S. The bottom panel gives a scheme of the channel resistance dependence on the magnetisation of the contacts. Parallel magnetisation results in a low resistance *R*, which is attributed to an active FET state, whereas anti-parallel magnetisation leads to a high resistance, which represents the inactive state.

state of magnetisation and thus the FET keeps its actual state of activation or deactivation, respectively. This means in contrast to the electrical field, which has to be present all the time while driving a current through the channel (self-closing FET assumed), the magnetic field is only needed for the switching process. Therefore a spinFET can be used as a non-volatile memory device, allowing programmable logics.

In principle, only external magnetic field pulses of two amplitudes have to be generated: (i) initially, one pulse of amplitude $< -H_{C,S}$ (or $> H_{C,S}$) for defining the initial magnetisation state of the source electrode, which then will be kept constant and (ii) during operation, positive and negative pulses with an amplitude between $H_{C,D}$ and $H_{C,S}$ for switching the drain electrode, thus inducing deactivation and re-activation, respectively. Therefore it has to be guaranteed that the coercive fields $H_{C,S}$ and $H_{C,D}$ are sufficiently separated, so the magnetisation of the drain electrode can easily be switched without changing the magnetisation of the source electrode.

Besides, another demand on the magnetic properties of the contacts is related to switching speeds. As the time needed for a switching process strongly depends on the size of the switching fields, it is desirable to keep them small. This is another argument for switching only the electrode with the lower coercive field.

28.5 Experimental Realisation

The spinOFET can be considered as a further development of a conventional OFET. Thus, the first aim was the fabrication, investigation and optimisation of conventional OFETs. They consist of interdigitated gold electrodes with a titanium adhesion layer on thermally grown SiO_2 (50 nm) on highly n-type Si (cf. Figure 28.3). The Si wafer serves as the common gate electrode (G). The lithographical techniques employed to define the structures are electron beam lithography (EBL) for channel lengths L down to 100 nm and optical UV lithography using a photomask and a maskaligner for contact pad fabrication. The metallisation with Ti and Au was realised by electron beam evaporation at a base pressure of 10^{-8} mbar. In all cases the total thickness of the interdigitated contacts was kept between 10 nm and 25 nm (± 1 nm). For enhancement of the performance of OFETs a substrate surface preparation procedure was performed for optimising the organic layer growth. This procedure essentially consists of an OH-termination of the SiO_2-surface by a so called piranha solution of hydrogen peroxide and sulfuric acid (1 : 3–4) and the subsequent application of a silane monolayer which is chemically bonded on the surface by the exposure of the sample to octadecyltrichlorosilane (OTS) solved in chloroform. Since the preparation of OTS layers requires an ambient free of humidity [21], the latter step was performed in a glove box with a dry N_2 atmosphere. For removing possible residual OTS from the electrodes, the sample was afterwards rinsed with acetone and isopropanol in an ultrasonic bath, leaving only the chemisorbed OTS on the SiO_2 surface. The exposure of the sample to piranha solution after metallisation and lift-off also has the beneficial side effect of removing possible residual resist. For all transistors dihexylquaterthio-

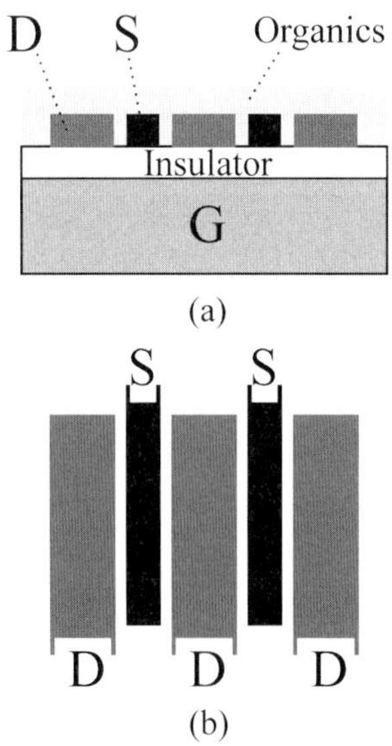

Figure 28.3 SpinOFET layout. (a) The sample cross section demonstrates the OFET bottom contact configuration: drain (D) and source (S) contacts are directly deposited on the gate insulator layer. Subsequently the organics are deposited on top. (b) The top view visualises the interdigitated arrangement of the contact stripes with different widths for source and drain.

phene (DH4T) was used as the organic semiconductor because of its potentially high mobility values [22]. The DH4T was deposited in an ultra high vacuum (UHV) organic molecular beam deposition (OMBD) chamber at a base pressure of 10^{-8} mbar. The substrate temperature was set to 90 °C during growth, which results in a good ordering of the organic film. On all samples a thin film of 11 nm DH4T (4 monolayers) was deposited with a rate of 2×10^{-3} nm/s. The electrical characterisation was done with an HP4145B semiconductor parameter analyser either *ex situ* in a probe station or *in situ* using a special sample holder.

For spinOFET fabrication the interdigitated source and drain stripes were structured in alternating widths (cf. Figure 28.3) of 0.3 μm and 2 μm, respectively. As ferromagnetic metal $Co_{60}Fe_{20}B_{20}$ was deposited by magnetron sputtering.

28.6 Results and Discussion

For initial examination and in order to optimise contact preparation and growth of DH4T thin films, conventional OFETs with gold contacts were fabricated and investigated in a first stage.

Using undercut lithography with a two-layer polymethylmethacrylat (PMMA) resist contact shapes were optimised. Further improvement of the OFETs was achieved by lowering the base pressure to 10^{-8} mbar during metallisation. Thus we were able to build high performing OFETs showing nearly ideal output characteristics and high charge carrier mobilities.

Figure 28.4 shows an example of output characteristics of an OFET with gold contacts. It can be observed that the I_D vs. U_{DS} curves start at the origin with a straight line slope before they show the typical saturation behaviour. Furthermore, the saturation currents exhibit very well the theoretically predicted quadratic dependence on the gate/source voltage. The charge carrier mobility was derived from the transfer characteristics of several of these transistors with different channel lengths and rather high values from 0.09 cm²/Vs up to 0.12 cm²/Vs were determined. All these results indicate that the desired properties, such as highly ordered organic thin film growth, low contact resistance and nearly vanishing leakage currents, are well fulfilled.

Also the stability of the transistors turned out to be remarkably good. In long term *in situ* studies they didn't show any degradation effects over 80 days. Even after prolonged voltage stressing the charge carrier mobility was not affected. Only the threshold voltage was shifted during the application of gate/source and drain/source voltages, but this effect was reversible within a short time of recovery.

The excellent performance of the OFETs was also confirmed by results of downscaling in the sub-μm range. Characteristic transistor behaviour was observed for channel lengths down to 180 nm [23].

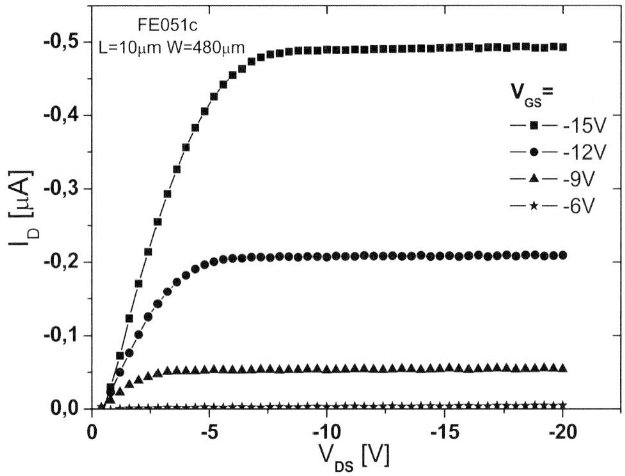

Figure 28.4 Output characteristics of a conventional OFET with gold electrodes and DH4T as organic semiconductor.

Having succeeded in the fabrication of high performance conventional OFETs, the next step was to realise ferromagnetic stripes of different widths for source and drain contacts. For this purpose, $Co_{60}Fe_{20}B_{20}$ was deposited by magnetron sputtering, after patterning by e-beam lithography. The resulting pattern of interdigitated source and drain contact stripes is shown in Figure 28.5. The active channel of the OFET is formed by the overlapping range of the source and drain fingers in the middle range of the picture. Well defined channel lengths L were achieved down to the sub-μm range. E.g. for a channel length $L = 90$ nm, the accuracy was ±5 nm.

In order to determine the coercive fields of ferromagnetic stripes with different widths, SQUID measurements were performed, which yield the magnetic momentum vs. external magnetic field. For achieving a sufficient sensitivity a very large number of magnetic stripes are required. For this purpose special

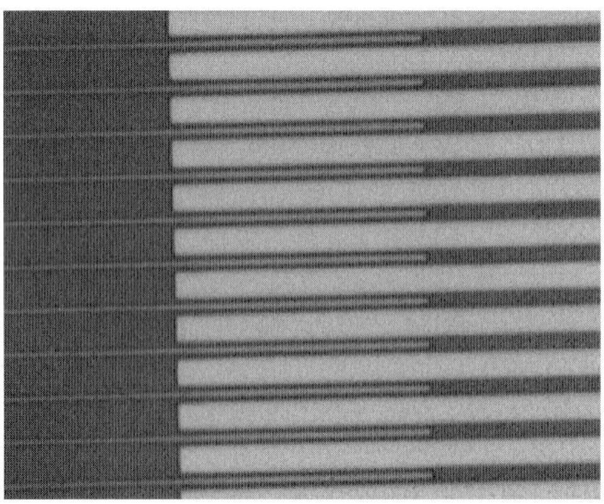

Figure 28.5 Optical microscopy picture of the interdigitated CoFeB source and drain electrodes stripes for a spin OFET. Stripe widths are 0.3 μm and 2 μm, respectively. External contacts are realised by contact pads from the sides (not shown).

samples were prepared by e-beam lithography. Each sample contains an array of 13600 $Co_{60}Fe_{20}B_{20}$ stripes of the same height (10 nm) and length (80 μm) but alternating widths (300 nm and 2 μm). Thus they simulate the source and drain electrodes of the spinOFETs. The number of narrow stripes chosen was twice the number of broad ones, in order to level the mass fractions of material deposited for each width. In this way, comparable contributions of both stripe widths to the total magnetisation should be obtained. Figure 28.6 shows the results of two SQUID measurements on two nominally identical samples at a temperature of 4 K. The magnetic field was first varied from −300 Oe to +300 Oe (upsweep) and subsequently from +300 Oe to −300 Oe (downsweep).

The almost perfect matching of the magnetisation curves for the two different samples shows the high reproducibility of the experiment and sample preparation. Changes of the magnetic momentum with the external field in two pronounced steps for the upsweep as well as for the downsweep are observed. When increasing the field from a highly negative value into the positive regime, a first clear step in the magnetic momentum is observed between 35 Oe and 48 Oe. At further increasing of the field a plateau is passed until a second step is found between 160 Oe and 198 Oe and maximum magnetisation is reached. For the downsweep the two steps are situated between −37 Oe and −51 Oe and between −162 Oe and −201 Oe. The obtained average step height values amount to 4.7×10^{-6} emu for the small step and 15.2×10^{-6} emu for the big step, which gives a step-height ratio of 0.31. As already mentioned before, the array contains 2 stripes of 0.3 μm for each stripe of 2 μm width. So if the ratio of the masses of CoFeB deposited in narrow and broad stripes is calcu-

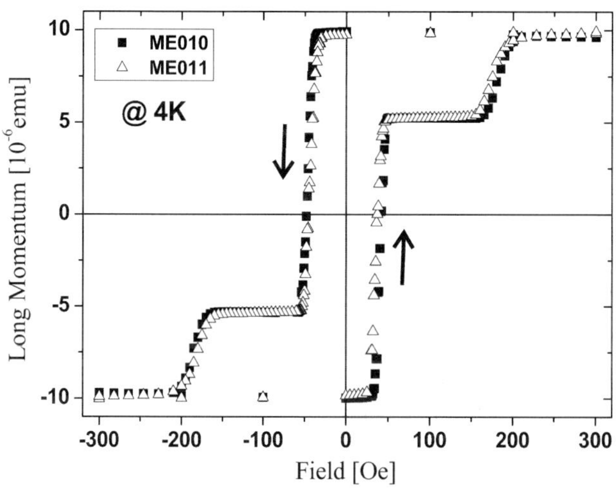

Figure 28.6 Magnetic momentum vs. the magnetic field for two samples, consisting of an array of FeCoB-stripes with alternating stripe widths of 0.3 μm and 2 μm. The momentum was determined by SQUID measurements at 4 K.

lated, considering that all stripes have the same length and height, it gives a value of 0.3. This is very consistent with the ratio of the step heights. Thus the big step in magnetic moment can clearly be attributed to the magnetic switching of the broad stripes while the small step shows the switching of the narrow stripes. This means that the plateaus between the steps represent the state of antiparallel magnetisation of the stripe sets. It was already mentioned before that it is crucial for experimental reasons to have a considerably large range in the external field for which antiparallel magnetisation is achieved, so this state can easily be generated. Extracting the widths of the plateaus for 4 K from Figure 28.5 gives an average value of 111 Oe which is a comfortable wide range in which antiparallel magnetisation can be induced. Thus, this architecture of source and drain stripes fulfils the required conditions for magnetical switching between the active and inactive state of spin-FETs.

It was already mentioned that one of the major advantages for the application of organic semiconductors in spintronics applications is the large spin diffusion length even at room temperature. While most of the spintronics experiments published up to now are performed at very low temperature, it is assumed that applying organics makes this effort non-essential. For this reason SQUID measurements were also carried out at room temperature in order to check if the magnetic behaviour of the contacts still allows spinFET operation.

Figure 28.7 compares two SQUID measurements on one sample, taken at 4 K and room temperature. The results show that at room temperatures the absolute values of the switching fields are significantly reduced, whereas the remnant magnetisation is hardly affected. The average width of the plateaus at room temperature is determined as 73 Oe. This is still considered sufficient for spinOFET application.

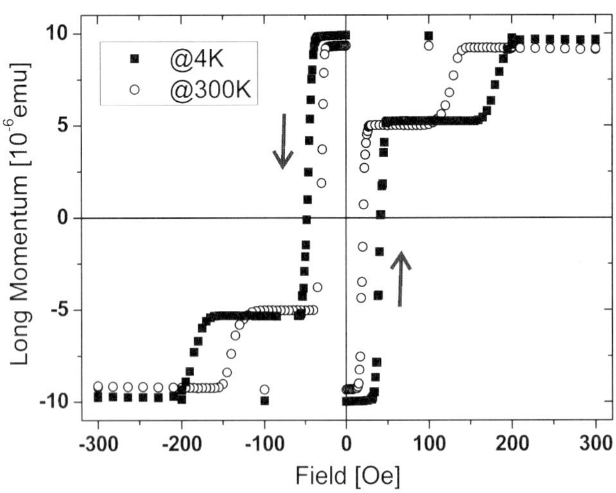

Figure 28.7 Comparison of the magnetic hysteresis at 4 K and room temperature for an array of FeCoB-stripes with alternating stripe widths of 0.3 μm and 2 μm.

In a further step the processibility of CoFeB with respect to OFET fabrication was examined. For this purpose samples with CoFeB contact stripes were subject to the commonly used surface treatment procedures for high-quality organic layer growth. They were exposed to piranha solution and to octadecyltrichlorosilane (OTS), alternatively in the gas phase or solved in chloroform. However, none of these processes turned out to be suitable for the CoFeB contacts. Both piranha and OTS strongly affect the processed electrodes. In the case of the piranha the contacts were totally dissolved even after an exposure of only some seconds. And also the treatment with OTS/chloroform solution caused an extreme roughening of the CoFeB surface. This can be clearly attributed to the OTS because tests with pure chloroform did not show any effect, whereas pure OTS applied by drop casting under N_2 atmosphere as well as by exposure to the gas phase also results in a macroscopic roughening.

Therefore, DH4T layers were deposited without the established surface treatment procedures. Special precautions were taken to guarantee the thorough removal of residual resist after the lift-off process, which is commonly ensured by the piranha treatment. In order to compensate the lack of this treatment, we used a second solvent. After the conventional lift-off process with acetone the sample was exposed to methylisobutylketone (MIBK) for one hour at 100 °C. Secondly, the time between lift-off and introduction of the sample in the UHV chamber for OMBD was kept as short as possible in order to protect the contacts from oxidising as much as possible. Making these arrangements allowed the fabrication of the first working OFETs with ferromagnetic contacts.

Figure 28.8 shows the output characteristics of an OFET with ferromagnetic CoFeB contacts and DH4T. Clear transistor behaviour is observed. However, the curves still show non-ideal features: (i) the current values are rather low,

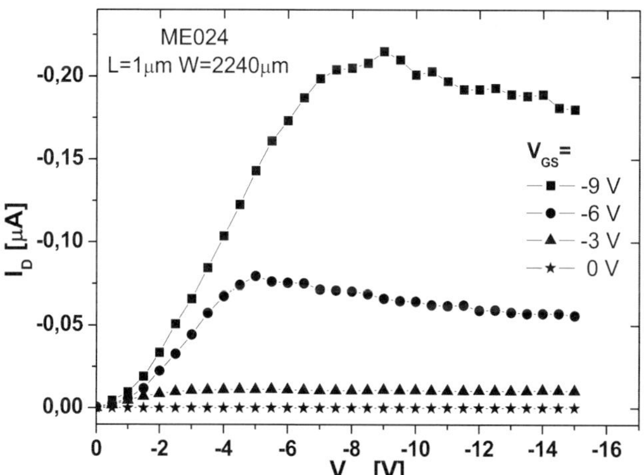

Figure 28.8 Output characteristics of a spinOFET with CoFeB as contact material and DH4T as organic semiconductor.

considering the channel length of only one 1 μm and the very large channel width, and (ii) the current decreases in the saturation regime. This non-ideal behaviour is attributed to the lack of OTS treatment, in accordance with published results comparing gold contact OFETs, whose organic layers were deposited either with or without OTS pre-treatment [24]. The OTS pre-treatment yielded a mobility increase by a factor of 500, and an improvement of the on/off ratio by a factor of 100.

Positive aspects are the pronounced straight line slopes, which are close to crossing the origin. Also the saturation current values of the different curves show a rather good quadratic dependence on the gate/source voltage.

This prototype of an OFET with ferromagnetic contacts is a major step towards spin-polarised transport and magnetic switching of OFETs. The demonstration of these effects requires a further improved transistor performance. In this respect, probably the main challenge is the improvement of the DH4T layer growth.

28.7 Conclusion

In this chapter, we presented a concept of an OFET for spin-polarised transport, which can be brought into a non-volatile active or inactive status by applying magnetic field pulses as a new steering parameter in addition to the steering by the electrical gate voltage. For this purpose, ferromagnetic source and drain contacts are employed instead of the usual noble-metal ones. Our strategy for the experimental realisation of this device is based on conventional bottom-contact OFETs with a dihexylquaterthiophene (DH4T) channel. Therefore, the performance and stability of conventional DH4T-OFETs were initially optimised. A mobility value $\mu \sim 0.12\ cm^2/Vs$ was achieved and a long-term stability over 80 days was demonstrated in *in situ* experiments.

Subsequently, working OFETs with ferromagnetic CoFeB stripes as source and drain contacts were fabricated. To achieve different coercive fields, the specific magnetic shape anisotropy for different stripe widths was exploited. In this way, the ability of a separate switching of the magnetisation directions of source and drain stripes with stripe widths of 0.3 μm and 2 μm, respectively, was proven by SQUID experiments.

For a successful demonstration of magnetic activation and deactivation, a further improvement of the electrical performance of the existing spinOFET prototype devices is required. In this respect, the main challenge is to improve the DH4T layer growth by applying a suitable substrate treatment, alternative to the common procedure, which has turned out to be incompatible with the ferromagnetic contact material.

Acknowledgements

The authors would like to thank the Deutsche Forschungsgemeinschaft (DFG) for financial support within the Schwerpunktprogramm 1121.

References

1. S. A. Wolf, D. D. Awschalom, R. A. Buhrman, J. M. Daughton, S. von Molnár, M. L. Roukes, A. Y. Chtchelkanova, and D. M. Treger, Science **294**, 1488 (2001).
2. R. Fiederling, M. Keim, G. Reuscher, W. Ossau, G. Schmidt, A. Waag, and L. W. Molenkamp, Nature **402**, 787 (1999).
3. G. Binasch, P. Grünberg, F. Saurenbach, and W. Zinn, Phys. Rev. B **39**, 4828 (1989).
4. M. N. Baibich, J. M. Broto, A. Fert, F. N. Vandau, F. Petroff, P. Etienne, G. Creuzet, A. Friedrich, and J. Chazelas, Phys. Rev. Lett. **61**, 2472 (1988).
5. S. Datta and B. Das, Appl. Phys. Lett. **56**, 665 (1990).
6. G. Schmidt, G. D. Ferrand, L. W. Molenkamp, A. T. Filip, and B. J. van Wees, Phys. Rev. B **62**, R4790 (2000).
7. A. Slobodskyy, C. Gould, T. Slobodskyy, C. R. Becker, G. Schmidt, and L. W. Molenkamp, Phys. Rev. Lett. **90**, 246601 (2003).
8. J. S. Moodera, L. R. Kinder, T. M. Wong, and R. Meservey, Phys. Rev. Lett. **74**, 3273 (1995).
9. V. F. Motsnyi, J. De Boeck, J. Das, W. van Roy, G. Borghs, E. Govaerts, and V. I. Safarov, Appl. Phys. Lett. **81**, 265 (2002).
10. S. S. P. Parkin, C. Kaiser, A. Panchula, P. M. Rice, B. Hughes, M. Samant, and See-Hun Yang, Nature Mater. **3**, 862 (2004).
11. A. Fert and H. Jaffres, Phys. Rev. B **64**, 184420 (2001).
12. Z. H. Xiong, Di Wu, Z. Valy Vardeny, and Jing Shi, Nature **427**, 821 (2004).
13. V. Dediu, M. Murgia, F. C. Matacotta, C. Taliani, and S. Barbanera, Solid State Commun. **122**, 181 (2002).
14. Ö. Mermer, G. Veeraraghavan, T. L. Francis, and M. Wohlgenannt, Solid State Commun. **134**, 631 (2005).
15. E. Arisi, I. Bergenti, V. Dediu, M. A. Loi, M. Muccini, M. Murgia, G. Ruani, C. Taliani, and R. Zamboni, J. Appl. Phys. **93**, 7682 (2003).
16. D. Wu, Z. H. Xiong, X. G. Li, Z. V. Verdeny, and Jing Shi, Phys. Rev. Lett. **95**, 016802 (2005).
17. S. J. Xie, K. H. Ahn, D. L. Smith, A. R. Bishop, and A. Saxena, Phys. Rev. B **67**, 125202 (2003).
18. J. F. Ren, J. Y. Fu, D. S. Liu, L. M. Mei, and S. J. Xie, J. Phys.: Condens. Matter **17**, 2341 (2005).
19. P. P. Ruden, D. L. Smith, J. Appl. Phys. **95**, 4898 (2004).
20. Z. G. Yu, M. A. Berding, and S. Krishnamurth, Phys. Rev. B **71**, 060408 (2005).
21. Y. Wang, M. Lieberman, Langmuir **19**, 1159 (2003).
22. H. E. Katz, A. J. Lovinger, and J. G. Laquindanum, Chem. Mater. **10**, 457 (1998).
23. M. Leufgen, U. Bass, T. Muck, T. Borzenko, G. Schmidt, J. Geurts, and V. Wagner, Synth. Met. **146**, 341 (2004).
24. C.-K. Song, B.-W. Koo, S.-B. Lee, D.-H. Kim, Jpn. J. Appl. Phys. **41**, 2730 (2002).

Index

Organic Electronics. Structural and Electronic Properties of OFETs. Edited by Christof Wöll

ISBN: 978-3-527-40810-8